Complete Solutions Guide for

THIRD EDITION

by
Steven S. Zumdahl

Kenneth C. Brooks
Thomas J. Hummel
Steven S. Zumdahl

University of Illinois at Urbana - Champaign

D. C. Heath and Company
Lexington, Massachusetts Toronto

Published simultaneously in Canada.

Printed in the United States of America.

International Standard Book Number: 0-669-32871-5

10 9 8 7

TO THE STUDENT: HOW TO USE THIS GUIDE

Chemistry is an applied science. Chemistry is valuable because a collection of facts about chemical behavior can be dealt with in a systematic manner and applied to solving the new problems that chemists encounter daily. In your study of chemistry you should give a priority to solving problems over other activities.

Solutions to all of the end of chapter exercises are in this manual. This "Solutions Guide" can be very valuable if you use it properly. The way NOT to use it is to look at an exercise in the book and then check the solution, often saying to yourself, "That's easy, I can do it." Chemistry is easy once you get the hang of it, but it takes work. Don't look up a solution to a problem until you have tried to work it on your own. If you are completely stuck, see if you can find a similar problem in the Sample Exercises in the chapter. Then look up the solution. After you do this, look for a similar problem in the end of chapter exercises and try working it. The more problems you do, the easier chemistry becomes. It is also in your self interest to try to work as many problems as possible. Most exams that you will take in chemistry will involve a lot of problem solving. If you have worked several problems similar to the ones on an exam you will do much better than if the exam is the first time you try to solve a particular type problem. No matter how much you read and study the text, or how well you think you understand the material, you don't really understand it until you have taken the information in the text and used it to solve a problem.

In this manual we have worked problems as in the textbook. We have shown intermediate answers to the correct number of significant figures and used that rounded answer in later calculations. Thus, some of your answers may differ slightly from ours. When we have not followed this convention, we have noted this in the solution.

We are grateful to Delores Wyatt for her outstanding effort in preparing the manuscript of this manual.

TABLE OF CONTENTS

CHAPTER ONE: CHEMICAL FOUNDATIONS

QUESTIONS

1. a. law: A law is a concise statement or equation that summarizes a great variety of observations.

   theory: A theory is an hypothesis that has been tested over a length of time.

   A law is less likely to be challenged or modified than a theory.

   b. theory versus experiment: A theory is our explanation of why things behave the way they do, while experiment is the process of observing that behavior. Theories attempt to explain the results of experiments and are, in turn, tested by further experiments.

   c. qualitative versus quantitative: A qualitative measurement only measures a quality while a quantitative measurement attaches a number to the observation. Examples:

   qualitative observations: The water was hot to the touch. Mercury was found in the drinking water. quantitative observations: The temperature of the water was 62 °C. The concentration of mercury in the drinking water was 1.5 ppm.

   d. hypothesis versus theory: Both are explanations of experimental observation. A theory is an hypothesis that has been tested over time and found to still be valid, with, perhaps, some modifications.

2. No, it is useful whenever a systematic approach of observation and hypothesis testing can be used.

3. a. No b. Yes c. Yes

4. Yes, Paracelsus is admonishing his students to learn by observation. Observation of facts is the foundation for all of the other steps in the scientific method.

5. accuracy: How close a measurement or series of measurements is to an accepted or true value.

   precision: How close a series of measurements of the same thing are to each other.

6. The results, avg. = 14.91% ± 0.03% are precise but not accurate.

7. Experimental results are the facts that we deal with. Theories are our attempt to rationalize those facts. If the experiment is done properly and the theory can't account for the facts, then the theory is wrong.

8. If the measurement is not precise, then it may be close to a particular "true" value only by chance. The next measurement may be way off or even the next several may be far from the true value.

9. Chemical changes involve the making and breaking of chemical forces (bonds). Physical changes do not. The identity of a substance changes after a chemical change, but not after a physical change.

10. Many techniques of chemical analysis need to be performed on relatively pure materials. Thus, many times a separation step is necessary to remove materials that will interfere with the analytical measurement.

EXERCISES

Uncertainty, Precision, Accuracy and Significant Figures

11. a. inexact b. exact c. exact

For c, $\frac{36\text{ in}}{\text{yd}} \times \frac{2.54\text{ cm}}{\text{in}} \times \frac{1\text{ m}}{100\text{ cm}} = \frac{0.9144\text{ m}}{\text{yd}}$ (All conversion factors used are exact.)

d. inexact - Although this number appears to be exact, it probably isn't. The announced attendance may be tickets sold but not the number who were actually in the stadium. Some people who paid may not have gone, some may leave early, or arrive late, some may sneak in without paying, etc.

e. exact f. inexact

12. a. exact b. inexact, 0.9144 m/yd is exact (see 1-11c). Thus, there are 1/0.9144 = 1.093613 ... yd/m.

c. exact d. inexact

13. a. 0.0012; 2 S.F., 1.2×10^{-3}

b. 437,000; 3 S.F., 4.37×10^5

c. 900.0; 4 S.F., 9.000×10^2

d. 106, 3 S.F.; 1.06×10^2

e. 125, 904,000; 6 S.F., 1.25904×10^8

f. 1.0012; 5 S.F., 1.0012×10^0

g. 2006; 4 S. F., 2.006×10^3

h. 3050; 3 S.F., 3.05×10^3

i. 0.001060; 4 S.F., 1.060×10^{-3}

14. a. 100; 1 S.F.

b. 1.0×10^2; 2 S.F.

c. 1.00×10^3; 3 S.F.

d. 100.; 3 S.F.

e. 0.0048; 2 S.F.

f. 0.00480; 3 S.F.

g. 4.80×10^{-3}; 3 S.F.

h. 4.800×10^{-3}; 4 S.F.

15. a. 6×10^8 b. 5.8×10^8 c. 5.82×10^8

d. 5.8200×10^8 e. 5.820000×10^8

16. a. 5×10^2

b. 4.8×10^2

c. 4.80×10^2

d. 4.800×10^2

17. a. 467; (25.27 - 24.16 = 1.11, only 3 significant figures)

b. 0.24; (8.925 - 8.904 = 0.021, 2 significant figures)

c. $(9.04 - 8.23 + 21.954 + 81.0) \div 3.1416 = 103.8 \div 3.1416 = 33.04$

d. $\dfrac{9.2 \times 100.65}{8.321 + 4.026} = \dfrac{9.2 \times 100.65}{12.347} = 7.5 \times 10^1$

e. $0.1654 + 2.07 - 2.114 = 0.12$

Uncertainty begins to appear in the second decimal place. Numbers were added as written and the answer was rounded off to 2 decimal places at the end. If you round to 2 decimal places and add you get 0.13.

f. $8.27(4.987 - 4.962) = 8.27(0.025) = 0.21$

g. $\dfrac{9.5 + 4.1 + 2.8 + 3.175}{4} = 4.9$ Uncertainty appears in the first decimal place.

h. $\dfrac{9.025 - 9.024}{9.025} \times 100 = \dfrac{0.001}{9.025} \times 100 = 0.01$

18. a. $6.022 \times 10^{23} \times 1.05 \times 10^2 = 6.32 \times 10^{25}$

b. $\dfrac{6.6262 \times 10^{-34} \times 2.998 \times 10^8}{2.54 \times 10^{-9}} = 7.82 \times 10^{-17}$

c. $1.285 \times 10^{-2} + 1.24 \times 10^{-3} + 1.879 \times 10^{-1}$

$= 0.1285 \times 10^{-1} + 0.0124 \times 10^{-1} + 1.879 \times 10^{-1} = 2.020 \times 10^{-1}$

d. $1.285 \times 10^{-2} - 1.24 \times 10^{-3} = 1.285 \times 10^{-2} - 0.124 \times 10^{-2} = 1.161 \times 10^{-2}$

e. $\dfrac{(1.00866 - 1.00728)}{6.02205 \times 10^{23}} = \dfrac{0.00138}{6.02205 \times 10^{23}} = 2.29 \times 10^{-27}$

f. $\dfrac{9.875 \times 10^2 - 9.795 \times 10^2}{9.875 \times 10^2} \times 100 = \dfrac{0.080 \times 10^2}{9.875 \times 10^2} \times 100 = 8.1 \times 10^{-1}$

g. $\dfrac{9.42 \times 10^2 + 8.234 \times 10^2 + 1.625 \times 10^3}{3}$

$= \dfrac{0.942 \times 10^3 + 0.8234 \times 10^3 + 1.625 \times 10^3}{3} = 1.130 \times 10^3$

Units and Unit Conversions

19. a. $1\text{ km} = 10^3\text{ m} = 10^6\text{ mm} = 10^{15}\text{ pm}$

b. $1\text{ g} = 10^{-3}\text{ kg} = 10^3\text{ mg} = 10^9\text{ ng}$

c. $1\text{ mL} = 10^{-3}\text{ L} = 10^{-3}\text{ dm}^3 = 1\text{ cm}^3$

d. $1\text{ mg} = 10^{-6}\text{ kg} = 10^{-3}\text{ g} = 10^3\ \mu\text{g} = 10^6\text{ ng} = 10^9\text{ pg} = 10^{12}\text{ fg}$

e. $1\text{ s} = 10^3\text{ ms} = 10^9\text{ ns}$

20. a. $1\ \text{Tg} \times \dfrac{10^{12}\ \text{g}}{\text{Tg}} \times \dfrac{1\ \text{kg}}{1000\ \text{g}} = 10^{9}\ \text{kg}$

b. $6.50 \times 10^{2}\ \text{Tm} \times \dfrac{10^{12}\ \text{m}}{\text{Tm}} \times \dfrac{10^{9}\ \text{nm}}{\text{m}} = 6.50 \times 10^{23}\ \text{nm}$

c. $25\ \text{fg} \times \dfrac{1\ \text{g}}{10^{15}\ \text{fg}} \times \dfrac{1\ \text{kg}}{1000\ \text{g}} = 25 \times 10^{-18}\ \text{kg} = 2.5 \times 10^{-17}\ \text{kg}$

d. $8.0\ \text{dm}^{3} \times \dfrac{1\ \text{L}}{\text{dm}^{3}} = 8.0\ \text{L}$

e. $1\ \text{mL} \times \dfrac{1\ \text{L}}{1000\ \text{mL}} \times \dfrac{10^{6}\ \mu\text{L}}{\text{L}} = 10^{3}\ \mu\text{L}$

f. $1\ \mu\text{g} \times \dfrac{1\ \text{g}}{10^{6}\ \mu\text{g}} \times \dfrac{10^{12}\ \text{pg}}{\text{g}} = 10^{6}\ \text{pg}$

21. $1\ \text{A} \times \dfrac{10^{-8}\ \text{cm}}{\text{Å}} \times \dfrac{1\ \text{m}}{100\ \text{cm}} \times \dfrac{10^{9}\ \text{nm}}{\text{m}} = 1 \times 10^{-1}\ \text{nm}$

$1 \times 10^{-1}\ \text{nm} \times \dfrac{10^{-9}\ \text{m}}{\text{nm}} \times \dfrac{1\ \text{pm}}{10^{-12}\ \text{m}} = 1 \times 10^{2}\ \text{pm}$

22. $134\ \text{pm} \times \dfrac{1\ \text{m}}{10^{12}\ \text{pm}} \times \dfrac{100\ \text{cm}}{\text{m}} = 1.34 \times 10^{-8}\ \text{cm} = 1.34\ \text{Å}$

or $134\ \text{pm} \times \dfrac{1\ \text{m}}{10^{12}\ \text{pm}} \times \dfrac{100\ \text{cm}}{\text{m}} \times \dfrac{10^{8}\ \text{Å}}{\text{cm}} = 1.34\ \text{Å}$

$134\ \text{pm} \times \dfrac{1\ \text{m}}{10^{12}\ \text{pm}} \times \dfrac{10^{9}\ \text{nm}}{\text{m}} = 0.134\ \text{nm}$

23. a. Since 1 in = 2.54 cm, then an uncertainty of $\pm \frac{1}{2}$ in is about $\pm$ 1 cm after we convert from in to cm. Thus, we should express the heights to the nearest centimeter.

7'6" = 7(12) + 6 in = 90 in

$90\ \text{in} \times \dfrac{2.54\ \text{cm}}{\text{in}} = 229\ \text{cm} = 229 \times 10^{-2}\ \text{m} = 2.29\ \text{m}$

5'2" = 5(12) + 2 in = 62 in

$62\ \text{in} \times \dfrac{2.54\ \text{cm}}{\text{in}} = 157\ \text{cm} = 157 \times 10^{-2}\ \text{m} = 1.57\ \text{m}$

b. $25{,}000\ \text{mi} \times \dfrac{1.609\ \text{km}}{\text{mi}} = 4.0 \times 10^{4}\ \text{km}$

We can derive this conversion factor from information in the chapter.

$1\ \text{mi} \times \dfrac{5280\ \text{ft}}{\text{mi}} \times \dfrac{1\ \text{yd}}{3\ \text{ft}} = 1760\ \text{yd}$ and

$$1760 \text{ yd} \times \frac{1 \text{ m}}{1.094 \text{ yd}} \times \frac{1 \text{ km}}{1000 \text{ m}} = 1.609 \text{ km}$$

or 1 mi = 5280 ft = 1760 yd = 1.609 km

$$4.0 \times 10^4 \text{ km} \times \frac{1000 \text{ m}}{\text{km}} = 4.0 \times 10^7 \text{ m}$$

c. $V = l \times w \times h$

$$V = 1.0 \text{ m} \times \left(5.6 \text{ cm} \times \frac{1 \text{ m}}{100 \text{ cm}}\right) \times \left(2.1 \text{ dm} \times \frac{1 \text{ m}}{10 \text{ dm}}\right) = 1.2 \times 10^{-2} \text{ m}^3$$

$$1.2 \times 10^{-2} \text{ m}^3 \times \left(\frac{10 \text{ dm}}{\text{m}}\right)^3 \left(\frac{1 \text{ L}}{\text{dm}^3}\right) = 12 \text{ L}$$

$$12 \text{ L} \times \frac{1000 \text{ cm}^3}{\text{L}} \times \left(\frac{1 \text{ in}}{2.54 \text{ cm}}\right)^3 = 730 \text{ in}^3$$

$$730 \text{ in}^3 \times \left(\frac{1 \text{ ft}}{12 \text{ in}}\right)^3 = 0.42 \text{ ft}^3$$

24. a. $18 \text{ gal} \times \frac{4 \text{ qt}}{\text{gal}} \times \frac{0.946 \text{ L}}{\text{qt}} = 68 \text{ L} = 68 \text{ dm}^3$

b. $4.00 \times 10^2 \text{ in}^3 \times \left(\frac{2.54 \text{ cm}}{\text{in}}\right)^3 = 6.55 \times 10^3 \text{ cm}^3 = 6.55 \text{ L}$

c. $0.25 \text{ in} \times \frac{2.54 \text{ cm}}{\text{in}} = 0.64 \text{ cm}$

d. $14{,}110 \text{ ft.} \times \frac{1 \text{ yd}}{3 \text{ ft}} \times \frac{1 \text{ m}}{1.0936 \text{ yd}} = 4301 \text{ m}$

e. $198 \text{ lb} \times \frac{1 \text{ kg}}{2.205 \text{ lb}} = 89.8 \text{ kg}$

f. $\frac{45 \text{ mi}}{\text{hr}} \times \frac{1.6093 \text{ km}}{\text{mi}} = \frac{72 \text{ km}}{\text{hr}}$

25. a. $928 \text{ mi} \times \frac{5280 \text{ ft}}{\text{mi}} \times \frac{1 \text{ fathom}}{6 \text{ ft}} \times \frac{1 \text{ cable length}}{100 \text{ fathoms}} \times \frac{1 \text{ nautical mi}}{10 \text{ cable lengths}}$

$$\times \frac{1 \text{ league}}{3 \text{ nautical miles}} = 272 \text{ leagues}$$

$$928 \text{ mi} \times \frac{5280 \text{ ft}}{\text{mi}} \times \frac{1 \text{ yd}}{3 \text{ ft}} \times \frac{1 \text{ m}}{1.094 \text{ yd}} \times \frac{1 \text{ km}}{1000 \text{ m}} = 1.49 \times 10^3 \text{ km}$$

b. $1.0 \text{ cable length} \times \frac{100 \text{ fathom}}{\text{cable length}} \times \frac{6 \text{ ft}}{\text{fathom}} \times \frac{1 \text{ yd}}{3 \text{ ft}} \times \frac{1 \text{ m}}{1.094 \text{ yd}} \times \frac{1 \text{ km}}{1000 \text{ m}}$

$$= 0.18 \text{ km}$$

$$1.0 \text{ cable length} = 0.18 \text{ km} \times \frac{1000 \text{ m}}{\text{km}} \times \frac{100 \text{ cm}}{\text{m}} = 1.8 \times 10^4 \text{ cm}$$

c. $315 \text{ ft} \times \frac{12 \text{ in}}{\text{ft}} \times \frac{2.54 \text{ cm}}{\text{in}} \times \frac{1 \text{ m}}{100 \text{ cm}} = 96.0 \text{ m}$

$37 \text{ ft} \times \frac{12 \text{ in}}{\text{ft}} \times \frac{2.54 \text{ cm}}{\text{in}} \times \frac{1 \text{ m}}{100 \text{ cm}} = 11 \text{ m}$

$315 \text{ ft} \times \frac{1 \text{ fathom}}{6 \text{ ft}} \times \frac{1 \text{ cable length}}{100 \text{ fathoms}} = 5.25 \times 10^{-1} \text{ cable length}$

$37 \text{ ft} \times \frac{1 \text{ fathom}}{6 \text{ ft}} = 6.2 \text{ fathoms}$

26. a. $1.25 \text{ mi} \times \frac{8 \text{ furlongs}}{\text{mi}} = 10.0 \text{ furlongs}$

$10.0 \text{ furlongs} \times \frac{40 \text{ rods}}{\text{furlong}} = 4.00 \times 10^2 \text{ rods}$

$4.00 \times 10^2 \text{ rods} \times \frac{5.5 \text{ yd}}{\text{rod}} \times \frac{36 \text{ in}}{\text{yd}} \times \frac{2.54 \text{ cm}}{\text{in}} \times \frac{1 \text{ m}}{100 \text{ cm}} = 2.01 \times 10^3 \text{ m}$

$2.01 \times 10^3 \text{ m} \times \frac{1 \text{ km}}{1000 \text{ m}} = 2.01 \text{ km}$

b. Let's assume we know this distance to ± 1 yard. First convert 26 miles to yards.

$26 \text{ mi} \times \frac{5280 \text{ ft}}{\text{mi}} \times \frac{1 \text{ yd}}{3 \text{ ft}} = 45{,}760 \text{ yd}$

26 mi 385 yd = 45,760 + 385 = 46,145 yards

$46{,}145 \text{ yard} \times \frac{1 \text{ rod}}{5.5 \text{ yd}} = 8390.0 \text{ rods}$

$8390.0 \text{ rods} \times \frac{1 \text{ furlong}}{40 \text{ rods}} = 209.75 \text{ furlongs}$

$46{,}145 \text{ yard} \times \frac{36 \text{ in}}{\text{yd}} \times \frac{2.54 \text{ cm}}{\text{in}} \times \frac{1 \text{ m}}{100 \text{ cm}} = 42{,}195 \text{ m}$

$42{,}195 \text{ m} \times \frac{1 \text{ km}}{1000 \text{ m}} = 42.195 \text{ km}$

27. a. $\frac{100.8 \text{ mi}}{\text{hr}} \times \frac{1760 \text{ yd}}{\text{mi}} \times \frac{1 \text{ m}}{1.0936 \text{ yd}} \times \frac{\text{hr}}{60 \text{ min}} \times \frac{1 \text{ min}}{60 \text{ s}} = \frac{45.06 \text{ m}}{\text{s}}$

b. 60 ft 6 in = 60(12) + 6 in = 726 in

$726 \text{ in} \times \frac{2.54 \text{ cm}}{\text{in}} \times \frac{1 \text{ m}}{100 \text{ cm}} \times \frac{1 \text{ s}}{45.06 \text{ m}} = 0.409 \text{ s}$

28. $1.71 \text{ warp factor} = 5.00 \left(\frac{3.00 \times 10^8 \text{ m}}{\text{s}}\right) \times \frac{1.094 \text{ yd}}{\text{m}} \times \frac{60 \text{ s}}{\text{min}} \times \frac{60 \text{ min}}{\text{hr}}$

$\times \frac{1 \text{ knot}}{2000 \text{ yds/hr}} = 2.95 \times 10^9 \text{ knots}$

29. a. $1 \text{ lb. troy} \times \frac{12 \text{ oz tr}}{1 \text{ lb troy}} \times \frac{20 \text{ pw}}{1 \text{ troy oz}} \times \frac{24 \text{ gr}}{\text{pw}} \times \frac{0.0648 \text{ g}}{\text{gr}} \times \frac{1 \text{ kg}}{1000 \text{ g}} = 0.373 \text{ kg}$

$1 \text{ troy lb} = 0.373 \text{ kg} \times \frac{2.205 \text{ lb}}{\text{kg}} = 0.822 \text{ lb}$

b. $1 \text{ oz tr} \times \frac{20 \text{ pw}}{\text{oz tr}} \times \frac{24 \text{ gr}}{\text{pw}} \times \frac{0.0648 \text{ g}}{\text{gr}} = 31.1 \text{ g}$

$1 \text{ oz tr} = 31.1 \text{ g} \times \frac{1 \text{ carat}}{0.200 \text{ g}} = 156 \text{ carats}$

c. 1 lb tr = 0.373 kg

$0.373 \text{ kg} \times \frac{1000 \text{ g}}{\text{kg}} \times \frac{1 \text{ cm}^3}{19.3 \text{ g}} = 19.3 \text{ cm}^3$

30. a. $1 \text{ gr ap} \times \frac{1 \text{ scruple}}{20 \text{ gr ap}} \times \frac{1 \text{ dram ap}}{3 \text{ scruple}} \times \frac{3.888 \text{ g}}{\text{dram ap}} = 0.06480 \text{ g}$

From the previous question we are given that 1 gr tr = 0.0648 g. So, the two are the same.

1 gr ap = 1 gr tr

b. $1 \text{ oz ap} \times \frac{8 \text{ dram ap}}{\text{oz ap}} \times \frac{3.888 \text{ g}}{\text{dram ap}} \times \frac{1 \text{ oz tr*}}{31.1 \text{ g}} = 1.00 \text{ oz tr}$ * See 29b

c. $5.00 \times 10^2 \text{ mg} \times \frac{1 \text{ g}}{1000 \text{ mg}} \times \frac{1 \text{ dram ap}}{3.888 \text{ g}} \times \frac{3 \text{ scruple}}{\text{dram ap}} = 0.386 \text{ scruple}$

$0.386 \text{ scruple} \times \frac{20 \text{ gr ap}}{\text{scruple}} = 7.72 \text{ gr ap}$

d. $1 \text{ scruple} \times \frac{1 \text{ dram ap}}{3 \text{ scruple}} \times \frac{3.888 \text{ g}}{\text{dram ap}} = 1.296 \text{ g}$

31. $2240 \text{ lb} \times \frac{1 \text{ kg}}{2.205 \text{ lb}} = 1016 \text{ kg} = 1 \text{ long ton}$

$1016 \text{ kg} \times \frac{1 \text{ tonne}}{1000 \text{ kg}} = 1.016 \text{ tonne}$

1 long ton = 1.016 metric tonne

32. $1.00 \text{ ns} \times \frac{1 \text{ s}}{10^9 \text{ ns}} \times \frac{3.00 \times 10^8 \text{ m}}{\text{s}} = 0.300 \text{ m}$

Temperature

33. $°C = \frac{5}{9}(°F - 32) = \frac{5}{9}(102.5 - 32) = 39.2°C$, $K = °C + 273.2 = 312.4 \text{ K}$ (Note: 32 is exact)

34. $°C = \frac{5}{9}(74 - 32) = 23°C$, $K = 23 + 273 = 296 \text{ K}$

35. $°F = \frac{9}{5}°C + 32 = \frac{9}{5}(25) + 32 = 77°F$, $K = 25 + 273 = 298 \text{ K}$

36. K = °C + 273.2; °C = 4 - 273.2 = -269

$$°F = \frac{9}{5}°C + 32 = \frac{9}{5}(-269) + 32 = -452\ °F$$

37. We can do this two ways.

First, we calculate the high and low temperature and get the uncertainty from the range.

20.6 °C ± 0.1 °C means the temperature can range from 20.5 °C to 20.7 °C.

$$T_F = \frac{9}{5}T_c + 32 \leftarrow \text{(exact)}$$

$$T_F\ (\text{min}) = \frac{9}{5}(20.5) + 32 = 68.9\ °F; \quad T_F\ (\text{max}) = \frac{9}{5}(20.7) + 32 = 69.3\ °F$$

So, the temperature ranges from 68.9 °F to 69.3 °F which we can express as 69.1 °F ± 0.2 °F.

An alternative way is to treat the uncertainty and the temperature in °C separately.

$$T_F = \frac{9}{5}T_c + 32 = \frac{9}{5}(20.6) + 32 = 69.1\ °F$$

$$\text{and} \pm 0.1\ °C \times \frac{9\ °F}{5\ °C} = \pm 0.18\ °F \approx \pm 0.2\ °F$$

Combining the two calculations: T_F = 69.1 °F ± 0.2 °F

38. 96.1 °F ± 0.2 °F; First convert 96.1 °F to °C

$$°C = \frac{5}{9}(°F - 32) = \frac{5}{9}(96.1 - 32) = 35.6\ °C$$

A change in temperature of 9 °F is equal to a change in temperature of 5 °C. So the uncertainty will be:

$$\pm 0.2\ °F \times \frac{5\ °C}{9\ °F} = \pm 0.1\ °C.$$ Thus, 96.1 ± 0.2 °F = 35.6 °C ± 0.1 °C

Density

39. $$\frac{2.70\ g}{cm^3} \times \frac{1\ kg}{1000\ g} \times \left(\frac{100\ cm}{m}\right)^3 = \frac{2.70 \times 10^3\ kg}{m^3}$$

$$\frac{2.70\ g}{cm^3} \times \frac{1\ lb}{453.6\ g} \times \left(\frac{2.54\ cm}{in}\right)^3 \times \left(\frac{12\ in}{ft}\right)^3 = \frac{169\ lb}{ft^3}$$

40. $$\frac{1.0\ g}{cm^3} \times \frac{1\ kg}{1000\ g} \times \left(\frac{100\ cm}{m}\right)^3 = 1.0 \times 10^3\ kg/m^3$$

$$\frac{1.0\ g}{cm^3} \times \frac{1\ lb}{454\ g} \times \left(\frac{2.54\ cm}{in} \times \frac{12\ in}{ft}\right)^3 = 62\ lb/ft^3$$

41. $D = \frac{\text{mass}}{\text{volume}}$; mass = 1.67×10^{-24} g; r = d/2 = 5.0×10^{-4} pm

$$V = \frac{4}{3}\pi r^3 = \frac{4}{3}(3.14) \times \left(5.0 \times 10^{-4}\ \text{pm} \times \frac{10^{-12}\ \text{m}}{\text{pm}} \times \frac{100\ \text{cm}}{\text{m}}\right)^3 = 5.2 \times 10^{-40}\ \text{cm}^3$$

$$D = \frac{1.67 \times 10^{-24}\ \text{g}}{5.2 \times 10^{-40}\ \text{cm}^3} = \frac{3.2 \times 10^{15}\ \text{g}}{\text{cm}^3}$$

42. $$\frac{0.045\ \text{g}}{\text{L}} \times \frac{1\ \text{L}}{1000\ \text{cm}^3} = \frac{4.5 \times 10^{-5}\ \text{g}}{\text{cm}^3}$$

$$\frac{0.045\ \text{g}}{\text{L}} \times \frac{1\ \text{L}}{1000\ \text{cm}^3} \times \left(\frac{100\ \text{cm}}{\text{m}}\right)^3 \times \frac{1\ \text{kg}}{1000\ \text{g}} = \frac{0.045\ \text{kg}}{\text{cm}^3}$$

43. $$5.0\ \text{carat} \times \frac{0.200\ \text{g}}{\text{carat}} \times \frac{1\ \text{cm}^3}{3.51\ \text{g}} = 0.28\ \text{cm}^3$$

44. $$2.8\ \text{mL} \times \frac{3.51\ \text{g}}{\text{mL}} \times \frac{1\ \text{carat}}{0.200\ \text{g}} = 49\ \text{carats}$$

45. a. both are the same mass

b. 1.0 mL of mercury. Mercury has a greater density than water.

Mass of mercury = $1.0\ \text{mL} \times \frac{13.6\ \text{g}}{\text{mL}} = 13.6$ g of mercury

mass of water = $1.0\ \text{mL} \times \frac{1.0\ \text{g}}{\text{mL}} = 1.0$ g of water.

c. same, 19.32 g

d. 1.0 L of benzene

46. a. 1.0 kg feathers, feathers are less dense than lead

b. 100 g water

c. same, both volumes are 1.0 L

Classification and Separation of Matter

47. Solid: own volume, own shape, does not flow

Liquid: own volume, takes shape of container, flows

Gas: takes volume and shape of container, flows

48. Homogeneous: only one phase present

Heterogeneous: more than one phase present

a. heterogeneous

b. heterogeneous: There is usually a fair amount of particulate matter present in the atmosphere (dirt) in addition to condensed water (rain, clouds).

c. heterogeneous d. homogeneous

e. homogeneous f. homogeneous

49. a. pure b. mixture c. mixture d. pure e. mixture

f. pure g. mixture h. mixture i. pure

50. Iron and uranium are elements. Water and table salt are compounds.

ADDITIONAL EXERCISES

51. a. 8.41 (2.16 has only three significant figures)

b. 16.1 (uncertainty appears in the first decimal place, 8.1)

c. 52.5 d. 5 (2 contains one significant figure)

e. 0.009 f. 429.59 (uncertainty appears in 2nd decimal, 2.17, 4.32)

52. a. $\frac{0.30\text{ g}}{\text{mL}} \times \frac{1000\text{ mg}}{\text{g}} = \frac{3.0 \times 10^2\text{ mg}}{\text{mL}}$

b. $\frac{0.30\text{ g}}{\text{mL}} \times \frac{1\text{ kg}}{10^3\text{ g}} \times \frac{1\text{ ml}}{\text{cm}^3} \times \left(\frac{100\text{ cm}}{\text{m}}\right)^3 = \frac{3.0 \times 10^2\text{ kg}}{\text{m}^3}$

c. $\frac{0.30\text{ g}}{\text{mL}} \times \frac{10^6\ \mu\text{g}}{\text{g}} = \frac{3.0 \times 10^5\ \mu\text{g}}{\text{mL}}$

d. $\frac{0.30\text{ g}}{\text{mL}} \times \frac{10^9\text{ ng}}{\text{g}} = \frac{3.0 \times 10^8\text{ ng}}{\text{mL}}$

e. $\frac{0.30\text{ g}}{\text{mL}} \times \frac{10^6\ \mu\text{g}}{\text{g}} \times \frac{1000\text{ mL}}{\text{L}} \times \frac{1\text{ L}}{10^6\ \mu\text{L}} = \frac{3.0 \times 10^2\ \mu\text{g}}{\mu\text{L}}$

53. a. Density = $\frac{\text{mass}}{\text{volume}}$; Volume of a sphere = $\frac{4}{3}\pi r^3$

$$D = \frac{2 \times 10^{36}\text{ kg}}{\frac{4}{3} \times 3.14 \times (6.96 \times 10^5\text{ km})^3} = \frac{2 \times 10^{36}\text{ kg}}{1.41 \times 10^{18}\text{ km}^3}$$

$$= \frac{1.4 \times 10^{18}\text{ kg}}{\text{km}^3} \approx \frac{1 \times 10^{18}\text{ kg}}{\text{km}^3}$$

b. $\frac{1 \times 10^{18}\text{ kg}}{\text{km}^3} \times \left(\frac{1\text{ km}}{1000\text{ m}}\right)^3 = \frac{1 \times 10^9\text{ kg}}{\text{m}^3}$

c. $\frac{1 \times 10^{18}\text{ kg}}{\text{km}^3} \times \frac{1000\text{ g}}{\text{kg}} \times \left(\frac{1\text{ km}}{1000\text{ m}}\right)^3 \times \left(\frac{1\text{ m}}{100\text{ cm}}\right)^3 = \frac{1 \times 10^6\text{ g}}{\text{cm}^3}$

54. a. No, because if the volumes were the same the gold idol would have a much greater mass.

b. Mass = $1.0\text{ L} \times \frac{1000\text{ cm}^3}{\text{L}} \times \frac{19.32\text{ g}}{\text{cm}^3} \times \frac{1\text{ kg}}{1000\text{ g}} = 19.32\text{ kg}$

It wouldn't be easy to play catch with the idol.

55. $\frac{36.1 \text{ miles}}{\text{gallon}} \times \frac{1760 \text{ yd}}{\text{mi}} \times \frac{1 \text{ m}}{1.094 \text{ yd}} \times \frac{1 \text{ km}}{1000 \text{ m}} \times \frac{1 \text{ gal}}{4 \text{ qt}} \times \frac{1.057 \text{ qt}}{\text{L}} = 15.3 \text{ km/L} = 15 \text{ km/L}$

56. Circumference = $2\pi r$

$$V = \frac{4\pi r^3}{3} = \frac{4\pi}{3}\left(\frac{c}{2\pi}\right)^3 = \frac{c^3}{6\pi^2}$$

$$\text{Largest density} = \frac{5.25 \text{ oz}}{\frac{(9.00 \text{ in})^3}{6\pi^2}} = \frac{5.25 \text{ oz}}{12.3 \text{ in}^3} = \frac{0.427 \text{ oz}}{\text{in}^3}$$

$$\text{Smallest density} = \frac{5.00 \text{ oz}}{\frac{(9.25 \text{ in})^3}{6\pi^2}} = \frac{5.00 \text{ oz}}{13.4 \text{ in}^3} = \frac{0.373 \text{ oz}}{\text{in}^3}$$

Maximum range is $\frac{(0.37 - 0.43) \text{ oz}}{\text{in}^3}$; or $\frac{0.40 \pm 0.03 \text{ oz}}{\text{in}^3}$

57. a. For $\frac{103 \pm 1}{101 \pm 1}$: Maximum = $\frac{104}{100}$ = 1.04; Minimum = $\frac{102}{102}$ = 1.00

So $\frac{103 \pm 1}{101 \pm 1} = 1.02 \pm 0.02$

b. For $\frac{101 \pm 1}{99 \pm 1}$: Maximum = $\frac{102}{98}$ = 1.04; Minimum = $\frac{100}{100}$ = 1.00

So $\frac{101 \pm 1}{99 \pm 1} = 1.02 \pm 0.02$

c. For $\frac{99 \pm 1}{101 \pm 1}$: Maximum = $\frac{100}{100}$ = 1.00; minimum = $\frac{98}{102}$ = 0.96

So $\frac{99 \pm 1}{101 \pm 1} = 0.98 \pm 0.02$

Considering the error limits, (a) and (b) should be expressed to three significant figures and (c) to two significant figures. The division rule differs in (b). The rule says (b) should be expressed to two significant figures. If we do this for (b) we imply that the answer is 1.0 or between 0.95 and 1.05. The actual range is less than this, so we should use the more precise way of expressing uncertainty. The significant figure rules only give us guidelines for estimating uncertainty. When we have a better handle on the uncertainty we should use it in precedence of the significant figure guidelines.

58. We need to calculate the maximum and minimum values of the density, given the uncertainty in each measurement. The maximum value is:

$$D_{max} = \frac{19.625 \text{ g} + 0.002 \text{ g}}{25.00 \text{ cm}^3 - 0.03 \text{ cm}^3} = \frac{19.627 \text{ g}}{24.97 \text{ cm}^3} = 0.7860 \text{ g/cm}^3$$

The minimum value of the density is:

$$D_{min} = \frac{19.625 \text{ g} - 0.002 \text{ g}}{25.00 \text{ cm}^3 + 0.03 \text{ cm}^3} = \frac{19.623 \text{ g}}{25.03 \text{ cm}^3} = 0.7840 \text{ g/cm}^3$$

The density of the liquid is between 0.7840 g/cm^3 and 0.7860 g/cm^3. These measurements are sufficiently precise to distinguish between ethanol and isopropyl alcohol.

59. $126 \text{ gal} \times \frac{4 \text{ qt}}{\text{gal}} \times \frac{1 \text{ L}}{1.057 \text{ qt}} = 477 \text{ L}$

60. a. $1 \text{ ha} \times \frac{10{,}000 \text{ m}^2}{\text{ha}} \times \left(\frac{1 \text{ km}}{1000 \text{ m}}\right)^2 = 1 \times 10^{-2} \text{ km}^2$

b. $5.5 \text{ acre} \times \frac{160 \text{ rod}^2}{\text{acre}} \times \left(\frac{5.5 \text{ yd}}{\text{rod}} \times \frac{36 \text{ in}}{\text{yd}} \times \frac{2.54 \text{ cm}}{\text{in}} \times \frac{1 \text{ m}}{100 \text{ cm}}\right)^2 = 2.2 \times 10^4 \text{ m}^2$

$2.2 \times 10^4 \text{ m}^2 \times \frac{1 \text{ ha}}{10^4 \text{ m}^2} = 2.2 \text{ ha}$

$2.2 \times 10^4 \text{ m}^2 \times \left(\frac{1 \text{ km}}{1000 \text{ m}}\right)^2 = 0.022 \text{ km}^2$

c. Area of lot = $120 \text{ ft} \times 75 \text{ ft} = 9.0 \times 10^3 \text{ ft}^2$

$9.0 \times 10^3 \text{ ft}^2 \times \left(\frac{1 \text{ yd}}{3 \text{ ft}} \times \frac{1 \text{ rod}}{5.5 \text{ yd}}\right)^2 \times \frac{1 \text{ acre}}{160 \text{ rod}^2} = 0.21 \text{ acre}, \quad \frac{\$6{,}500}{0.21 \text{ acre}} = \frac{\$31{,}000}{\text{acre}}$

We can use our result from (b) to get the conversion factor between acres and ha (5.5 acre = 2.2 ha.). Thus, 1 ha = 2.5 acre.

$0.21 \text{ acre} \times \frac{1 \text{ ha}}{2.5 \text{ acre}} = 0.084 \text{ ha}.$ The price is: $\frac{\$6{,}500}{0.084 \text{ ha}} = \frac{\$77{,}000}{\text{ha}}$

d. $1.4 \text{ furlongs} \left(\frac{40 \text{ rod}}{\text{furlong}}\right)\left(\frac{5.5 \text{ yd}}{\text{rod}}\right)\left(\frac{3 \text{ ft}}{\text{yd}}\right) = 924 \text{ ft} = 920 \text{ ft}$

Area = $27 \text{ ft} \times 920 \text{ ft} = 25{,}000 \text{ ft}^2$

$25{,}000 \text{ ft}^2 \left(\frac{1 \text{ yd}}{3 \text{ ft}} \times \frac{1 \text{ rod}}{5.5 \text{ yd}}\right)^2 = 92 \text{ rod}^2$

$92 \text{ rod}^2 \left(\frac{1 \text{ acre}}{160 \text{ rod}^2}\right) = 0.58 \text{ acre}$

$0.58 \text{ acre} \left(\frac{1 \text{ ha}}{2.5 \text{ acre}}\right) = 0.23 \text{ ha}$

$0.23 \text{ ha} \left(\frac{10{,}000 \text{ m}^2}{\text{ha}}\right) = 2.3 \times 10^3 \text{ m}^2$

61. a. $85 \text{ crowns} \times \frac{1 \text{ royal}}{20 \text{ crowns}} = 4.25 \text{ royals}$

$85 \text{ crowns} \times \frac{100 \text{ weights}}{\text{crown}} = 8.5 \times 10^3 \text{ weights}$

$$50 \text{ royals} \times \frac{1 \text{ horse}}{4.25 \text{ royals}} = 11.8 \text{ horses.}$$ So, 11 horses can be bought.

$$11 \text{ horses} \times \frac{4.25 \text{ royals}}{\text{horse}} = 46.75 \text{ royals}, \; 50 - 46.75 = 3.25 \text{ royals will be left}$$

b. $$\frac{13 \text{ crowns}}{\text{bundle}} \times \frac{100 \text{ weights}}{\text{crown}} \times \frac{1 \text{ bundle}}{25 \text{ haigus hides}} = \frac{52 \text{ weights}}{\text{haigus hide}}$$

$$288 \text{ bundles} \times \frac{13 \text{ crowns}}{\text{bundle}} \times \frac{1 \text{ royal}}{20 \text{ crowns}} = 187.2 \text{ royals}$$

62. We will calculate the largest and smallest values that the density can be.

Since, $D = \frac{\text{Mass}}{\text{Volume}}$; $D_{max} = \frac{M_{max}}{V_{min}} = \frac{16.52 \text{ g}}{15.3 \text{ cm}^3} = \frac{1.08 \text{ g}}{\text{cm}^3}$

$D_{min} = \frac{M_{min}}{V_{max}} = \frac{16.48 \text{ g}}{15.7 \text{ cm}^3} = \frac{1.05 \text{ g}}{\text{cm}^3}$; $D = \frac{M}{V} = \frac{16.50 \text{ g}}{15.5 \text{ cm}^3} = \frac{1.06 \text{ g}}{\text{cm}^3}$

Combining these results, we can express the density as

$$\frac{1.06 \text{ g}}{\text{cm}^3} \pm \frac{0.02 \text{ g}}{\text{cm}^3}$$

63. $D_{max} = \frac{M_{max}}{V_{min}}$; $V = V(\text{final}) - V(\text{initial})$

We get V_{min} from $9.7 \text{ cm}^3 - 6.5 \text{ cm}^3 = 3.2 \text{ cm}^3$

$$D_{max} = \frac{28.93 \text{ g}}{3.2 \text{ cm}^3} = \frac{9.0 \text{ g}}{\text{cm}^3}$$

$$D_{min} = \frac{M_{min}}{V_{max}} = \frac{28.87 \text{ g}}{9.9 \text{ cm}^3 - 6.3 \text{ cm}^3} = \frac{8.0 \text{ g}}{\text{cm}^3}$$

The density is $\frac{8.5 \text{ g}}{\text{cm}^3} \pm \frac{0.5 \text{ g}}{\text{cm}^3}$

64. With no rounding off:

$V = 29 \text{ mm} \times 35 \text{ mm} \times 100. \text{ mm} = 1.015 \times 10^5 \text{ mm}^3$

$V_{max} = 30. \times 36 \times 101 = 109{,}080 \text{ mm}^3$

$V_{min} = 28 \times 34 \times 99 = 94{,}248 \text{ mm}^3$

$V = (1.02 \pm 0.08) \times 10^5 \text{ mm}^3$

or the volume is $1.02 \times 10^5 \text{ mm}^3$ with an uncertainty of $8 \times 10^3 \text{ mm}^3$.

In cm^3: $1 \text{ mm}^3 \times \left(\frac{1 \text{ cm}}{10 \text{ mm}}\right)^3 = 10^{-3} \text{ cm}^3$ and $V = 102 \pm 8 \text{ cm}^3$

Note: If we round the volume to 2 significant figures ($V = 1.0 \times 10^2 \text{ cm}^3$) we imply the range is between $0.95 \times 10^2 \text{ cm}^3$ and $1.05 \times 10^2 \text{ cm}^3$. The actual range is larger and we should show that

$$D = \frac{615.0 \text{ g}}{102 \text{ cm}^3} = \frac{6.03 \text{ g}}{\text{cm}^3}$$

$$D_{max} = \frac{615.2 \text{ g}}{94 \text{ cm}^3} = \frac{6.54 \text{ g}}{\text{cm}^3} \ (6.03 + 0.51)$$

$$D_{min} = \frac{614.8 \text{ g}}{109 \text{ cm}^3} = \frac{5.64 \text{ g}}{\text{cm}^3} \ (6.03 - 0.39)$$

The best way to express the density is: $\frac{6.0 \text{ g}}{\text{cm}^3} \pm \frac{0.5 \text{ g}}{\text{cm}^3}$

65. $\frac{100. \text{ m}}{9.86 \text{ s}} = 10.1 \text{ m/s}$

$$\frac{10.1 \text{ m}}{\text{s}} \times \frac{1 \text{ km}}{1000 \text{ m}} \times \frac{60 \text{ s}}{\text{min}} \times \frac{60 \text{ min}}{\text{hr}} = 36.4 \text{ km/hr}$$

$$\frac{100. \text{ m}}{9.86 \text{ s}} \times \frac{1.094 \text{ yd}}{\text{m}} \times \frac{3 \text{ ft}}{\text{yd}} = 33.3 \text{ ft/s}$$

$$\frac{33.3 \text{ ft}}{\text{s}} \times \frac{1 \text{ mi}}{5280 \text{ ft}} \times \frac{60 \text{ s}}{\text{min}} \times \frac{60 \text{ min}}{\text{hr}} = 22.7 \text{ mi/hr}$$

$$1.00 \times 10^2 \text{ yds} \times \frac{1 \text{ m}}{1.094 \text{ yd}} \times \frac{9.86 \text{ s}}{100. \text{ m}} = 9.01 \text{ s}$$

66. Volume of lake $= 100 \text{ mi}^2 \times \left(\frac{5280 \text{ ft}}{\text{mi}}\right)^2 \times 20 \text{ ft} = 6 \times 10^{10} \text{ ft}^3$

$$6 \times 10^{10} \text{ ft}^3 \times \left(\frac{12 \text{ in}}{\text{ft}} \times \frac{2.54 \text{ cm}}{\text{in}}\right)^3 \times \frac{1 \text{ mL}}{\text{cm}^3} \times \frac{0.4 \text{ µg}}{\text{mL}} = 7 \times 10^{14} \text{ µg}$$

$$7 \times 10^{14} \text{ µg} \times \frac{10^{-6} \text{ g}}{\text{µg}} \times \frac{1 \text{ kg}}{10^3 \text{ g}} = 7 \times 10^5 \text{ kg of mercury}$$

67. $K = °C + 273.15; \ 0 = °C + 273.15; \ °C = -273.15$

$°F = \frac{9}{5}°C + 32; \ °F = \frac{9}{5}(-273.15) + 32 = -459.67 \ °F$ (32 exact)

68. $16 \text{ oz} \times \frac{1 \text{ lb}}{16 \text{ oz}} \times \frac{1 \text{ kg}}{2.205 \text{ lb}} = 0.45 \text{ kg}$

$$\frac{65 \text{ ¢}}{0.45 \text{ kg}} = \frac{1.4 \times 10^2 \text{ ¢}}{\text{kg}} \text{ or } \frac{\$1.4}{\text{kg}}$$

$$\frac{1.4 \times 10^2 \text{ ¢}}{\text{kg}} \times \frac{1 \text{ kg}}{1000 \text{ g}} = \frac{0.14 \text{ ¢}}{\text{g}}$$

69. The object that sinks has the greater density. Since it sinks its density is greater than that of water, 1.0 g/cm^3. The second object must have a density less than water since it floats. Both objects have the same mass. Thus, the sphere that sinks has the smaller volume. The sphere that floats has the larger diameter.

70. $V = 5.00\ cm \times 4.00\ cm \times 2.50\ cm = 50.0\ cm^3$

$$50.0\ cm^3 \times \frac{22.57\ g}{cm^3} = 1130\ g$$

$$1.00\ kg \times \frac{1000\ g}{kg} \times \frac{1\ cm^3}{22.57\ g} = 44.3\ cm^3$$

71. $D_{cube} = \frac{140.4\ g}{(3.00\ cm)^3} = \frac{5.20\ g}{cm^3}$

If this is correct to 1.00% then the density is: $\frac{5.20\ g}{cm^3} \pm \frac{0.05\ g}{cm^3}$

$$V_{sphere} = \frac{4}{3}\pi r^3 = \frac{4}{3}\pi(1.42\ cm)^3 = 12.0\ cm^3$$

$$D_{sphere} = \frac{61.6\ g}{12.0\ cm^3} = \frac{5.13\ g}{cm^3};\ D_{sphere} = \frac{5.13\ g}{cm^3} \pm \frac{0.05\ g}{cm^3}$$

D_{cube} is between 5.15 g/cm^3 and 5.25 g/cm^3

D_{sphere} is between 5.08 g/cm^3 and 5.18 g/cm^3

We can't decisively say if they are the same or different. The data are not precise enough to say.

CHALLENGE PROBLEMS

72. a. $\frac{2.70 - 2.64}{2.70} \times 100 = 2\%$ b. $\frac{16.48 - 16.12}{16.12} \times 100 = 2.2\%$

c. $\frac{1.000 - 0.9981}{1.000} \times 100 = \frac{0.002}{1.000} \times 100 = 0.2\%$

73. In a subtraction, the results get smaller, but the uncertainties add. If the two numbers are very close together, the uncertainty may be larger than the result. For example, let us assume we want to take the difference of the following two measured quantities, 999,999 ± 2 and 999,996 ± 2. The difference is 3 ± 4.

74. a. At some point in 1982, the composition of the metal used in minting pennies was changed.

b. It should be expressed as 3.08 ± 0.05 g. The uncertainty in the second decimal place will swamp any effect of the next two decimal places.

75. Heavy pennies (old): mean mass = 3.08 ± 0.05 g

Light pennies (new): mean mass $= \frac{(2.467 + 2.545 + 2.518)}{3}$

$= 2.51 \pm 0.04$ g

Average Density of old pennies:

$$D_{old} = \frac{\frac{95 \times 8.96\text{ g}}{\text{cm}^3} + \frac{5 \times 7.14\text{ g}}{\text{cm}^3}}{100} = \frac{8.9\text{ g}}{\text{cm}^3}$$

Average Density of new pennies:

$$D_{new} = \frac{\frac{2.4 \times 8.96\text{ g}}{\text{cm}^3} + \frac{97.6 \times 7.14\text{ g}}{\text{cm}^3}}{100} = \frac{7.18\text{ g}}{\text{cm}^3}$$

Since $D = \frac{\text{mass}}{\text{volume}}$ and the volume of old and new pennies are the same,

then $\frac{D_{new}}{D_{old}} = \frac{\text{Mass}_{new}}{\text{Mass}_{old}}$; $\frac{D_{new}}{D_{old}} = \frac{7.18}{8.9} = 0.807 = 0.81$

$$\frac{\text{Mass}_{new}}{\text{Mass}_{old}} = \frac{2.51}{3.08} = 0.81$$

To two decimal places the ratios are the same. We can reasonably conclude that yes, the difference in mass is accounted for by the difference in the alloy used.

76. a.

100°A	115°C
100°A	160°C
0°A	-45°C

A change in temperature of 160°C equals a change in temperature of 100°A.

So, $\frac{160\text{ °C}}{100\text{ °A}}$ is our conversion for a change in temperature.

0 °A = -45 °C or °C = °A -45 At the freezing point

Combining the two pieces of information:

$$°C = \frac{160\text{ °C}}{100\text{ °A}}\ °A - 45;\quad °C = \frac{8}{5}\ °A - 45$$

b. $°F = \frac{9}{5}\ °C + 32;\quad °C = \frac{5}{9}(°F - 32)$ (Assume 32 is exact)

$$\frac{5}{9}(°F - 32) = \frac{8}{5}\ °A - 45$$

$$°F - 32 = \frac{72}{25}\ °A - 81;\ °F = \frac{72}{25}\ °A - 49$$

c. $°C = \frac{8}{5}\ °A - 45$ and $°C = °A$

So, $°C = \frac{8}{5}\ °C - 45;\ \frac{3}{5}\ °C = 45;\ °C = 75 = °A$

d. $°C = \frac{8}{5}°A - 45$; $°C = \frac{8}{5}(86) - 45 = 93\ °C$

$°F = \frac{72}{25}°A - 49$; $°F = \frac{72}{25}(86) - 49 = 199\ °F$

e. $°C = \frac{8}{5}°A - 45$; $\frac{8}{5}°A = °C + 45$

$°A = \frac{5}{8}(°C + 45)$; $°A = \frac{5}{8}(45 + 45) = 56\ °A$

77. a. One possibility is that rope B is not attached to anything and rope A and rope C are connected via a pair of pulleys and/or gears.

b. Try to pull rope B out of the box. Measure the distance moved by C for a given movement of A. Hold either A or C firmly while pulling on the other.

CHAPTER TWO: ATOMS, MOLECULES, AND IONS

QUESTIONS

1. There should be no difference. It does not matter how a substance is produced, it is still that substance.

2. a. Atoms have mass and are neither destroyed nor created by chemical reactions. Therefore, mass is neither created nor destroyed by chemical reactions. Mass is conserved.

   b. The composition of a substance depends on the number and kinds of atoms that form it.

   c. Compounds of the same elements differ only in numbers of atoms of the elements forming them, i.e., NO, N_2O, NO_2.

3. Some elements exist as molecular substances. That is, hydrogen normally exists as H_2 molecules, not single hydrogen atoms.

4. Yes, many questions can be raised from Dalton's theory. For example: What are the masses of atoms? Are the atoms really structureless? What forces hold atoms together in compounds?, etc.

5. We now know that some atoms of the same element have different masses. We have had to include the existence of isotopes in our models.

6. Deflection of cathode rays by magnetic and electric fields led to the conclusion that they were negatively charged. The ray was produced at the negative electrode and repelled by the negative pole of the applied electric field.

7. β-particles are electrons. A cathode ray is a stream of electrons.

8. J. J. Thomson discovered the electrons. Henri Becquerel discovered radioactivity. Lord Rutherford proposed the nuclear model of the atom. Dalton's original model proposed that atoms were indivisible particles (that is, atoms had no internal structure). Thomson and Becquerel discovered sub-atomic particles and Rutherford's model attempted to describe the internal structure of the atom.

9. The atomic number of an element is equal to the number of protons in the nucleus of an atom of that element. The mass number is the sum of the number of protons plus neutrons in the nucleus. The atomic weight is the actual mass of a particular isotope (including electrons). As we will see in chapter three, the average mass of an atom is taken from a measurement made on a large number of atoms and it is this average value we see on the periodic table.

10. A family is a set of elements in the same vertical column. A family is also called a group. A period is a set of elements in the same horizontal row.

11. A compound will always contain the same numbers (and types) of atoms. A given amount of hydrogen will react only with a specific amount of oxygen. Any excess oxygen will remain unreacted.

12. The halogens have a high affinity for electrons and one important way they react is to form anions of the type X^-. The alkali metals tend to give up electrons easily and in most of their compounds exist as M^+ cations. Note: these two very reactive groups are only one electron away (in the periodic table) from the least reactive family of elements, the noble gases.

EXERCISES

Development of the Atomic Theory

13. a. The composition of a substance depends on the numbers of atoms of each element making up the compound (i.e., the formula of the compound) and not on the composition of the mixture from which it was formed.

    b. $H_2 + Cl_2 \rightarrow 2\ HCl$. The volume of HCl produced is twice the volume of H_2 (or Cl_2) used.

14. $Cl_2 + 3\ F_2 \rightarrow 2\ X$. Two molecules of X contain 6 atoms of F and two atoms of Cl. Therefore, the formula of X is ClF_3.

15. $\frac{1.188}{1.188} = 1.000;\ \frac{2.375}{1.188} = 1.999;\ \frac{3.563}{1.188} = 2.999$

    The masses of fluorine are simple ratios of whole numbers to each other, 1:2:3.

16. Hydrazine: 1.44×10^{-1} g H/g N

    Ammonia: 2.16×10^{-1} g H/g N

    Hydrogen azide: 2.40×10^{-2} g H/g N

    Let's try all of the ratios:

    $\frac{0.216}{0.144} = 1.50 = \frac{3}{2};\ \frac{0.144}{0.024} = 6.00;\ \frac{0.216}{0.024} = 9.00$

    They can all be expressed as simple whole number ratios. The g H/g N in hydrazine, ammonia, and hydrogen azide are in the ratios 6:9:1.

17. To get the atomic mass of H to be 1.00, we divide the mass that reacts with 1.00 g of oxygen by 0.1260. $\frac{0.1260}{0.1260} = 1.00$

    To get Na and Mg on the same scale, we do the same division.

    Na: $\frac{2.8750}{0.1260} = 22.8$; Mg: $\frac{1.5000}{0.1260} = 11.9$

    For O: $\frac{1.00}{0.1260} = 7.94$

	H	O	Na	Mg
Scale	1.00	7.94	22.8	11.9
Accepted Value	1.01 (1.008)	16.00	22.99	24.31

The atomic masses of O and Mg are incorrect. The atomic masses of H and Na are close. Something must be wrong about the assumed formulas of the compounds. It turns out the correct formulas are H_2O, Na_2O, and MgO. The smaller discrepancies result from the error in the atomic mass of H.

18. If the formula was Be_2O_3, then 2 times the atomic weight of Be would combine with three times the atomic weight of oxygen, or

$$\frac{2\text{ A}}{3(16.00)} = \frac{0.5633}{1.000}$$

Atomic weight of Be = 13.52

The Nature of the Atom

19. Density of nucleus:

$$V_{nucleus} = \frac{4}{3}\pi r^3 = \frac{4}{3}(3.14)(5 \times 10^{-14}\text{ cm})^3 = 5 \times 10^{-40}\text{ cm}^3$$

$$D = \frac{1.67 \times 10^{-24}\text{ g}}{5 \times 10^{-40}\text{ cm}^3} = 3 \times 10^{15}\text{ g/cm}^3$$

Density of H-atom:

$$V_{atom} = \frac{4}{3}(3.14)(1 \times 10^{-8}\text{ cm})^3 = 4 \times 10^{-24}\text{ cm}^3$$

$$D = \frac{1.67 \times 10^{-24} + 9 \times 10^{-28}\text{ g}}{4 \times 10^{-24}\text{ cm}^3} = 0.4\text{ g/cm}^3$$

20. Let's set up a ratio:

$$\frac{\text{diameter of nucleus}}{\text{diameter of atom}} = \frac{1\text{ mm}}{\text{diameter of model}} = \frac{1 \times 10^{-13}\text{ cm}}{1 \times 10^{-8}\text{ cm}} = 1 \times 10^{-5}$$

diameter of model = 1×10^5 mm = 1×10^2 m

21. $5.93 \times 10^{-18}\text{ C} \times \frac{1\text{ neg charge}}{1.602 \times 10^{-19}\text{ C}}$ = 37 negative charges on the oil drop

22. First: divide all charges by 2.56×10^{-12}.

$$\frac{2.56 \times 10^{-12}}{2.56 \times 10^{-12}} = 1.00 \qquad \frac{7.68}{2.56} = 3.00$$

$$\frac{3.84 \times 10^{-12}}{2.56 \times 10^{-12}} = 1.50 \qquad \frac{5.12}{2.56} = 2.00$$

Not all results are integers. So no one single drop carries a single negative charge. However, if the first drop contains 2 charges, then the electronic charge equals $2.56 \times 10^{-12}/2 = 1.28 \times 10^{-12}$ zirkombs. This number goes into all the other charges an integral number of times. But if the first drop contained 4 charges, $2.56 \times 10^{-12}/4 = 6.40 \times 10^{-13}$ zirkombs would also work. In fact, dividing 2.56×10^{-12} by any even number will give a set of consistent results.

23. gold - Au; silver - Ag; mercury - Hg; potassium - K: iron - Fe; antimony - Sb; tungsten - W

24. sodium - Na; beryllium - Be; manganese - Mn; chromium - Cr; uranium - U

25. fluorine - F; chlorine - Cl; bromine - Br; sulfur - S; oxygen - O; phosphorus -P

26. titanium - Ti; selenium - Se; plutonium - Pu; nitrogen - N; silicon - Si

27. Sn - tin; Pt - platinum; Co - cobalt; Ni - nickel; Mg - magnesium; Ba - barium; K - potassium

28. As - arsenic; I - iodine; Xe - xenon; He - helium; C - carbon; Si - silicon

29. The noble gases are He, Ne, Ar, Kr, Xe, and Rn (helium, neon, argon, krypton, xenon, and radon). Radon has only radioactive isotopes. In the periodic table the whole number enclosed in paranthesis is the mass number of the longest lived isotope of the element.

30. promethium (Pm) and technetium (Tc)

31. a. Eight, Li to Ne b. Eight, Na to Ar c. Eighteen, K to Kr d. Five, N, P, As, Sb, Bi

32. a. Six, Be, Mg, Ca, Sr, Ba, Ra
b. Five, O, S, Se, Te, Po
c. Three, Ni, Pd, Pt
d. Six, He, Ne, Ar, Kr, Xe, Rn

33. a. $^{238}_{94}Pu$; 94 protons, 238 - 94 = 144 neutrons
b. $^{65}_{29}Cu$; 29 protons, 65 - 29 = 36 neutrons
c. $^{52}_{24}Cr$; 24 protons, 28 neutrons
d. $^{4}_{2}He$; 2 protons, 2 neutrons
e. $^{60}_{27}Co$; 27 protons, 33 neutrons
f. $^{54}_{24}Cr$; 24 protons, 30 neutrons

34. a. ^{15}N; 7 protons, 8 neutrons
b. ^{3}H; 1 proton, 2 neutrons
c. ^{207}Pb; 82 protons, 125 neutrons
d. ^{151}Eu; 63 protons, 88 neutrons
e. ^{107}Ag; 47 protons, 60 neutrons
f. ^{109}Ag; 47 protons, 62 neutrons

35. 9 protons means the atomic number is 9. The mass number is 9 + 10 = 19; Symbol: $^{19}_{9}F$

36. a. P b. I c. K d. Yb

37. Atomic number = 63 (Eu); Charge = +63 - 60 = +3; Mass number = 63 + 88 = 151; Symbol: $^{151}_{63}Eu^{3+}$

38. 50 protons; The element is Sn; Mass number = 50 + 68 = 118; Net charge = +50 - 48 = +2; The symbol is: $^{118}_{50}Sn^{2+}$

39. Atomic number = 16 (S); Charge = +16 - 18 = -2; Mass number = 16 + 18 = 34; Symbol: $^{34}_{16}S^{2-}$

40. Atomic number = 16 (S); Charge = +16 - 18 = -2; Mass number = 16 + 16 = 32; Symbol: $^{32}_{16}S^{2-}$

41.

Symbol	Number of protons in nucleus	Number of neutrons in nucleus	Number of electrons	Net charge
$^{75}_{33}As^{3+}$	33	42	30	3+
$^{128}_{52}Te^{2-}$	52	76	54	2-
$^{32}_{16}S$	16	16	16	0
$^{204}_{81}Tl^{+}$	81	123	80	1+
$^{195}_{78}Pt$	78	117	78	0

42.

Symbol	Number of protons in nucleus	Number of neutrons in nucleus	Number of electrons	Net charge
$^{238}_{92}U$	92	146	92	0
$^{40}_{20}Ca^{2+}$	20	20	18	2+
$^{51}_{23}V^{3+}$	23	28	20	3+
$^{89}_{39}Y$	39	50	39	0
$^{79}_{35}Br^{-}$	35	44	36	1-
$^{31}_{15}P^{3-}$	15	16	18	3-

43. Metals: Mg, Ti, Au, Bi, Ge, Eu, Am Non-metals: Si, B, At, Rn, Br

44. Si, Ge, B, At

The elements at the boundary between the metals and the non-metals: B, Si, Ge, As, Sb, Te, Po, At.

45. a and d

A group is a vertical column of elements in the periodic table. Elements in the same family (group) have similar chemical properties.

46. b (Group 4A) and d (Group 2A)

47. Carbon is a non-metal. Silicon and germanium are metalloids. Tin and lead are metals. Thus, metallic character increases as one goes down a family in the periodic table.

48. The metallic character decreases from left to right.

49. a. Na^+ lose one e^-
b. Sr^{2+} lose two e^-
c. Ba^{2+} lose two e^-
d. I^- gain one e^-
e. Al^{3+} lose three e^-
f. S^{2-} gain two e^-

50. a. lose 3 e^- to form B^{3+}
b. lose 1 e^- to form Cs^+
c. gain 2 e^- to form Se^{2-}
d. gain 3 e^- to form N^{3-}
e. gain 2 e^- to form O^{2-}
f. lose 2 e^- to form Mg^{2+}

Nomenclature

51. a. chromium(VI) oxide
b. chromium(III) oxide
c. aluminum oxide
d. selenium dioxide
e. selenium trioxide
f. nitrogen triiodide
g. phosphorus trichloride
h. sulfur difluoride

52. a. nickel(II)oxide
b. iron(III) oxide
c. cerium(IV) oxide
d. cerium(III) oxide
e. silver(I) sulfide
f. manganese(IV) oxide
g. sodium hydride
h. dihydrogen sulfide or hydrosulfuric acid

53. a. sodium chloride
b. magnesium chloride
c. rubidium bromide
d. cesium fluoride
e. aluminum iodide
f. hydrogen iodide
g. nitrogen monoxide or nitric oxide (common name)
h. nitrogen trifluoride
i. dinitrogen tetrafluoride
j. dinitrogen dichloride
k. silicon tetrafluoride
l. dihydrogen selenide

54. a. potassium perchlorate
b. calcium phosphate
c. aluminum sulfate
d. lead(II) nitrate
e. barium sulfite
f. sodium nitrite
g. potassium permanganate
h. potassium dichromate
i. gold(III) chloride
j. iodous acid
k. titanium(IV) oxide
l. nickel(II) sulfide

55. a. nitric acid
b. nitrous acid
c. phosphoric acid
d. phosphorous acid
e. sodium hydrogen sulfate or sodium bisulfate (common name)
f. calcium hydrogen sulfite or calcium bisulfite (common name)
g. sodium bromate
h. iron(III) periodate
i. ruthenium(III) nitrate
j. vanadium(V) oxide
k. platinum(IV) chloride
l. iron(III) phosphate

56. a. copper(I) iodide or cuprous iodide
b. copper(II) iodide or cupric iodide
c. cobalt(II) iodide
d. sodium carbonate
e. sodium hydrogen carbonate or sodium bicarbonate
f. tetrasulfur tetranitride
g. sulfur hexafluoride
h. sodium hypochlorite
i. barium chromate
j. cerium(IV) nitrate

57. a. CsBr; b. $BaSO_4$; c. NH_4Cl; d. ClO; e. $SiCl_4$; f. ClF_3; g. BeO; h. MgF_2

58. a. $(NH_4)_2HPO_4$; b. Hg_2S; c. SiO_2; d. Na_2SO_3; e. $Al(HSO_4)_3$; f. NCl_3; g. HBr; h. $HBrO_2$; i. $HBrO_4$; j. KHS; k. CaI_2; l. $CsClO_4$

59. a. SF_2; b. SF_6; c. NaH_2PO_4; d. Li_3N; e. $Cr_2(CO_3)_3$; f. SnF_2; g. $(NH_4)(C_2H_3O_2)$; h. $(NH_4)(HSO_4)$; i. $Co(NO_3)_3$; j. Hg_2Cl_2: mercury(I) exists as Hg_2^{2+}; k. $KClO_3$; l. NaH

60. a. NaOH; b. $Al(OH)_3$; c. HCN; d. Na_2O_2; e. $Cu(C_2H_3O_2)_2$; f. CF_4; g. PbO; h. PbO_2; i. PbO_2; j. CuBr; k. GeO_2

l. GaAs If we treat this as an ionic compound we would predict the stable ions to be Ga^{3+} and As^{3-}.

ADDITIONAL EXERCISES:

61. There should be no difference. The composition of insulin from both sources will be the same and therefore, it will have the same activity regardless of the source. As a practical note, trace contaminants in the two types of insulin may be different. These trace components may be important.

62. a. $^{24}_{12}Mg$; 12 protons, 12 neutrons, 12 electrons

b. $^{24}_{12}Mg^{2+}$; 12 p, 12 n, 10 e^-

c. $^{59}_{27}Co^{2+}$; 27 p, 32 n, 25 e^-

d. $^{59}_{27}Co^{3+}$; 27 p, 32 n, 24 e^-

e. $^{59}_{27}Co$; 27 p, 32 n, 27 e^-

f. $^{79}_{34}Se$; 34 p, 45 n, 34 e^-

g. $^{79}_{34}Se^{2-}$; 34 p, 45 n, 36 e^-

h. $^{63}_{28}Ni$; 28 p, 35 n, 28 e^-

i. $^{59}_{28}Ni^{2+}$; 28 p, 31 n, 26 e^-

63. a. $Pb(C_2H_3O_2)_2$: lead(II) acetate

b. $CuSO_4$: copper(II) sulfate

c. CaO: calcium oxide

d. $MgSO_4$: magnesium sulfate

e. $Mg(OH)_2$: magnesium hydroxide

f. $CaSO_4$: calcium sulfate

g. N_2O: dinitrogen monoxide or nitrous oxide

64. a. Atomic number is 54. Xe

b. Atomic number is 34. Se

c. Atomic number is 20. Ca

d. Atomic number is 42. Mo

e. Atomic number is 94. Pu

65. Yes, 1.0 g H would react with 37.0 g ^{37}Cl and 1.0 g H would react with 35.0 g ^{35}Cl.

No, the ratio of H/Cl would always be 1 g H/37 g Cl for ^{37}Cl and 1 g H/35 g Cl for ^{35}Cl.

As long as we had pure ^{35}Cl and pure ^{37}Cl the above ratios will always hold. If we have a mixture (such as the natural abundance of chlorine) the ratio will also be constant as long as the composition of the mixture of the two isotopes does not change.

CHALLENGE PROBLEMS:

66. copper (Cu), silver (Ag), and gold (Au)

67. a. Ba^{2+} and O^{2-}: BaO, barium oxide

b. Li^+ and H^-: LiH, lithium hydride

c. In^{3+} and F^-: InF_3, indium(III) fluoride

d. As^{3+} and S^{2-}: As_2S_3, diarsenic trisulfide

e. B^{3+} and O^{2-}; B_2O_3, boron oxide or diboron trioxide

f. OF_2; oxygen difluoride

g. Al^{3+} and H^-; AlH_3, aluminum hydride

h. In^{3+} and P^{3-}; InP, indium(III) phosphide

68. ^{98}Tc; 43 protons and 55 neutrons; ^{99}Tc; 43 protons and 56 neutrons

Tc is in the same family as Mn. We would expect it to have similar properties:

permanganate: MnO_4^-; pertechnetate: TcO_4^-; ammonium pertechnetate: NH_4TcO_4

69. The solid residue must have come from the flask.

70. In the case of sulfur, SO_4^{2-} is sulfate; SO_3^{2-} is sulfite. By analogy:

SeO_4^{2-} selenate SeO_3^{2-} selenite TeO_4^{2-} tellurate TeO_3^{2-} tellurite

71. The PO_4^{3-} ion is phosphate.

Na_3AsO_4 contain Na^+ ions and AsO_4^{3-} ions. So the name would be sodium arsenate.

H_3AsO_4: arsenic acid

$Mg_3(SbO_4)_2$: Mg^{2+} ions and SbO_4^{3-} ions, magnesium antimonate

CHAPTER THREE: STOICHIOMETRY

QUESTIONS

1. The two major isotopes of boron are ^{10}B and ^{11}B. The listed mass of 10.81 is the average mass of a very large number of boron atoms.

2. No, it only means that the average mass is 52.00.

3. The molecular formula tells us the actual number of atoms of each element in a molecule (or formula unit) of a compound. The empirical formula tells only the simplest whole number ratio of atoms of each element in a molecule. The molecular formula is a whole number multiple of the empirical formula. If that multiplier is one, the molecular and empirical formulas are the same. For example, both the molecular and empirical formulas of water are H_2O. They are the same. For hydrogen peroxide, the empirical formula is OH; the molecular formula is H_2O_2.

4. Side reactions may occur. For example, in the combustion of CH_4 (methane) to CO_2 and H_2O, some CO is also formed. Also, some reactions only go part way to completion and reach a state of equilibrium with both reactants and products present (See Ch. 13).

EXERCISES

Atomic Masses and Mass Spectrometer

5. AW = 0.7899(23.9850 amu) + 0.1000(24.9858 amu) + 0.1101(25.9826 amu)
   = 18.95 amu + 2.499 amu + 2.861 amu = 24.31 amu

6. 0.3460(283.4 amu) + 0.2120(284.7 amu) + 0.4420(287.8 amu) = 285.6 amu

7. Let x = % of ^{151}Eu and y = % of ^{153}Eu

   $x + y = 100$ so $y = 100 - x$

   $$151.96 = \frac{x(150.9196) + (100 - x)(152.9209)}{100}$$

   $$15196 = 150.9196x + 15292.09 - 152.9209x$$

   $$-96 = -2.0013x$$

   $x = 48\%$; 48% ^{151}Eu and 52% ^{153}Eu

8. If Ag is 51.82% ^{107}Ag, then the remainder is ^{109}Ag or 48.18%. The average atomic mass is then:

   $$107.868 = \frac{51.82\ (106.905) + 48.18\ (A)}{100}$$

   10786.8 = 5540. + 48.18 A; A = 108.9 amu

9. There are three peaks in the mass spectrum, each 2 mass units apart. This is consistent with two isotopes, differing in mass by two mass units. The peak at 157.84 corresponds to a Br_2 molecule composed of two atoms of the lighter isotope. This isotope has mass equal to 157.84/2 or 78.92. This corresponds to ^{79}Br. The second isotope is ^{81}Br with mass 161.84/2 = 80.92. The peaks in the mass spectrum correspond to $^{79}Br_2$, $^{79}Br^{81}Br$ and $^{81}Br_2$ in order of increasing mass. The intensities of the highest and lowest mass tell us the two isotopes are present at about equal abundance. The actual abundance is 50.69% ^{79}Br and 49.31% ^{81}Br. The calculation of the abundance from the mass spectrum is beyond the scope of this text.

10.

Compound	Mass	Intensity	Scaled Intensity Largest Peak = 100
$H_2{}^{120}Te$	121.92	0.09	0.3
$H_2{}^{122}Te$	123.92	2.46	7.1
$H_2{}^{123}Te$	124.92	0.87	2.5
$H_2{}^{124}Te$	125.92	4.61	13.4
$H_2{}^{125}Te$	126.92	6.99	20.3
$H_2{}^{126}Te$	127.92	18.71	54.3
$H_2{}^{128}Te$	129.92	31.79	92.2
$H_2{}^{130}Te$	131.93	34.48	100.0

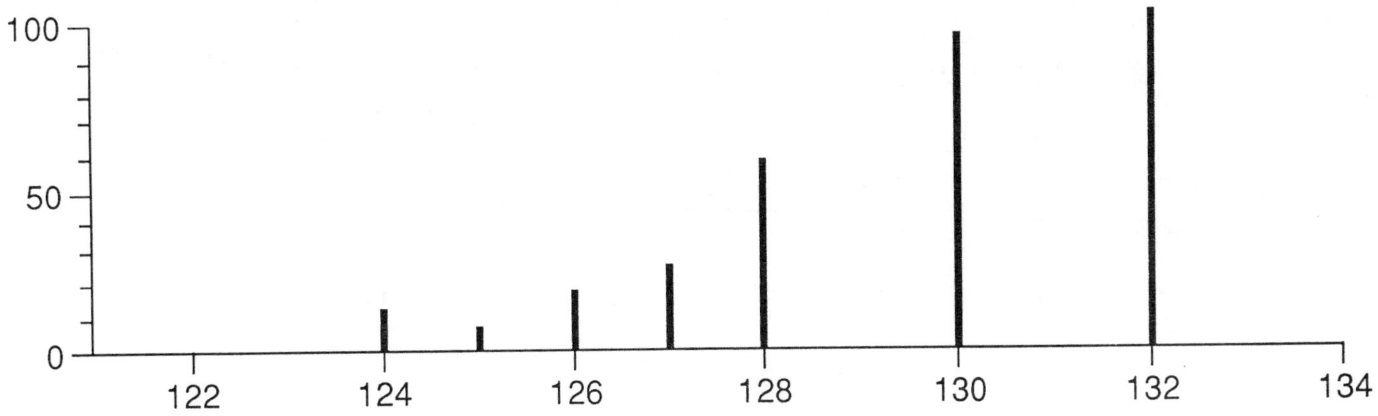

100
50
0
122 124 126 128 130 132 134

11. $$\frac{9.123 \times 10^{-23}\ \text{g}}{\text{atom}} \times \frac{6.022 \times 10^{23}\ \text{atom}}{\text{mol}} = \frac{54.94\ \text{g}}{\text{mol}}$$

The atomic mass is 54.94. The element is manganese (Mn).

12. $$\text{Atomic mass} = \frac{90.51(19.992) + 0.27(20.994) + 9.22(21.990)}{100} = 20.18\ \text{amu}$$ The element is neon.

Moles and Molar Masses

13. a. $1.0 \text{ mol } NH_3 \times \frac{1 \text{ mol N}}{\text{mol } NH_3} = 1.0 \text{ mol N}$ $\quad 1.0 \text{ mol } NH_3 \times \frac{3 \text{ mol H}}{\text{mol } NH_3} = 3.0 \text{ mol H}$

b. $1.0 \text{ mol } N_2H_4 \times \frac{2 \text{ mol N}}{\text{mol } N_2H_4} = 2.0 \text{ mol N}$ $\quad 1.0 \text{ mol } N_2H_4 \times \frac{4 \text{ mol H}}{\text{mol } N_2H_4} = 4.0 \text{ mol H}$

c. $1.0 \text{ mol } (NH_4)_2Cr_2O_7 \times \frac{2 \text{ mol N}}{\text{mol } (NH_4)_2Cr_2O_7} = 2.0 \text{ mol N}$

$1.0 \text{ mol } (NH_4)_2Cr_2O_7 \times \frac{8 \text{ mol H}}{\text{mol } (NH_4)_2Cr_2O_7} = 8.0 \text{ mol H}$

$1.0 \text{ mol } (NH_4)_2Cr_2O_7 \times \frac{2 \text{ mol Cr}}{\text{mol } (NH_4)_2Cr_2O_7} = 2.0 \text{ mol Cr}$

$1.0 \text{ mol } (NH_4)_2Cr_2O_7 \times \frac{7 \text{ mol O}}{\text{mol } (NH_4)_2Cr_2O_7} = 7.0 \text{ mol O}$

14. a. $1.0 \text{ mol } CoCl_2{\bullet}6\,H_2O \times \frac{1 \text{ mol Co}}{\text{mol } CoCl_2{\bullet}6\,H_2O} = 1.0 \text{ mol Co}$

$1.0 \text{ mol } CoCl_2{\bullet}6\,H_2O \times \frac{2 \text{ mol Cl}}{\text{mol } CoCl_2{\bullet}6\,H_2O} = 2.0 \text{ mol Cl}$

$1.0 \text{ mol } CoCl_2{\bullet}6\,H_2O \times \frac{6 \text{ mol } H_2O}{\text{mol } CoCl_2{\bullet}6\,H_2O} \times \frac{2 \text{ mol H}}{\text{mol } H_2O} = 12 \text{ mol H}$

$1.0 \text{ mol } CoCl_2{\bullet}6\,H_2O \times \frac{6 \text{ mol } H_2O}{\text{mol } CoCl_2{\bullet}6\,H_2O} \times \frac{1 \text{ mol O}}{\text{mol } H_2O} = 6.0 \text{ mol O}$

b. 1 mol of $Ca_3(PO_4)_2$ contains 3.0 mol of Ca, 2.0 mol of P, and 8.0 mol of O.

c. 1 mol of Na_2HPO_4 contains 2 mol of Na, 1.0 mol of H, 1.0 mol of P, and 4.0 mol of O.

15. a. 1.0 mol NH_3 contains 1.0 mol N and 3.0 mol H.

$1.0 \text{ mol N} \times \frac{6.02 \times 10^{23} \text{ atoms N}}{\text{mol N}} = 6.0 \times 10^{23} \text{ atoms N}$

$3.0 \text{ mol H} \times \frac{6.02 \times 10^{23} \text{ atoms H}}{\text{mol H}} = 1.8 \times 10^{24} \text{ atoms H}$

b. 1.0 mol N_2H_4 contains 2.0 mol N and 4.0 mol H.

$2.0 \text{ mol N} \times \frac{6.02 \times 10^{23} \text{ atoms N}}{\text{mol N}} = 1.2 \times 10^{24} \text{ atoms N}$

$4.0 \text{ mol H} \times \frac{6.02 \times 10^{23} \text{ atoms H}}{\text{mol H}} = 2.4 \times 10^{24} \text{ atoms H}$

c. 1.0 mol of $(NH_4)_2Cr_2O_7$ contains 2.0 mol N, 8.0 mol H, 2.0 mol Cr, and 7.0 mol O.

$2.0 \text{ mol N} \times \frac{6.02 \times 10^{23} \text{ atoms N}}{\text{mol N}} = 1.2 \times 10^{24} \text{ atoms N}$

$$8.0 \text{ mol H} \times \frac{6.02 \times 10^{23} \text{ atoms H}}{\text{mol H}} = 4.8 \times 10^{24} \text{ atoms H}$$

$$2.0 \text{ mol Cr} \times \frac{6.02 \times 10^{23} \text{ atoms Cr}}{\text{mol Cr}} = 1.2 \times 10^{24} \text{ atoms Cr}$$

$$7.0 \text{ mol O} \times \frac{6.02 \times 10^{23} \text{ atoms O}}{\text{mol O}} = 4.2 \times 10^{24} \text{ atoms O}$$

16. a. 1.0 mol $CoCl_2 \cdot 6\ H_2O$ contains 1.0 mol Co, 2.0 mol Cl, 12 mol H, and 6.0 mol O.

$$1.0 \text{ mol Co} \times \frac{6.02 \times 10^{23} \text{ atoms Co}}{\text{mol Co}} = 6.0 \times 10^{23} \text{ atoms Co}$$

$$2.0 \text{ mol Cl} \times \frac{6.02 \times 10^{23} \text{ atoms Cl}}{\text{mol Cl}} = 1.2 \times 10^{24} \text{ atoms Cl}$$

$$12 \text{ mol H} \times \frac{6.02 \times 10^{23} \text{ atoms H}}{\text{mol H}} = 7.2 \times 10^{24} \text{ atoms H}$$

$$6.0 \text{ mol O} \times \frac{6.02 \times 10^{23} \text{ atoms O}}{\text{mol O}} = 3.6 \times 10^{24} \text{ atoms O}$$

b.

$$1.0 \text{ mol } Ca_3(PO_4)_2 \times \frac{3 \text{ mol Ca}}{\text{mol } Ca_3(PO_4)_2} \times \frac{6.02 \times 10^{23} \text{ atoms Ca}}{\text{mol Ca}} = 1.8 \times 10^{24} \text{ atoms Ca}$$

$$1.0 \text{ mol } Ca_3(PO_4)_2 \times \frac{2 \text{ mol P}}{\text{mol } Ca_3(PO_4)_2} \times \frac{6.02 \times 10^{23} \text{ atoms P}}{\text{mol P}} = 1.2 \times 10^{24} \text{ atoms P}$$

$$1.0 \text{ mol } Ca_3(PO_4)_2 \times \frac{8 \text{ mol O}}{\text{mol } Ca_3(PO_4)_2} \times \frac{6.02 \times 10^{23} \text{ atoms O}}{\text{mol O}} = 4.8 \times 10^{24} \text{ atoms O}$$

c.

$$1.0 \text{ mol } Na_2HPO_4 \times \frac{2 \text{ mol Na}}{\text{mol } Na_2HPO_4} \times \frac{6.02 \times 10^{23} \text{ atoms Na}}{\text{mol Na}} = 1.2 \times 10^{24} \text{ atoms Na}$$

$$1.0 \text{ mol } Na_2HPO_4 \times \frac{1 \text{ mol H}}{\text{mol } Na_2HPO_4} \times \frac{6.02 \times 10^{23} \text{ atoms H}}{\text{mol H}} = 6.0 \times 10^{23} \text{ atoms H}$$

$$1.0 \text{ mol } Na_2HPO_4 \times \frac{1 \text{ mol P}}{\text{mol } Na_2HPO_4} \times \frac{6.02 \times 10^{23} \text{ atoms P}}{\text{mol P}} = 6.0 \times 10^{23} \text{ atoms P}$$

$$1.0 \text{ mol } Na_2HPO_4 \times \frac{4 \text{ mol O}}{\text{mol } Na_2HPO_4} \times \frac{6.02 \times 10^{23} \text{ atoms O}}{\text{mol O}} = 2.4 \times 10^{24} \text{ atoms O}$$

17. a. 1.0 mol NH_3 contains 1.0 mol N and 3.0 mol H.

$$1.0 \text{ mol N} \times \frac{14.01 \text{ g N}}{\text{mol N}} = 14 \text{ g N}$$

$$3.0 \text{ mol H} \times \frac{1.008 \text{ g H}}{\text{mol H}} = 3.0 \text{ g H}$$

b. 1.0 mol N_2H_4 contains 2.0 mol N and 4.0 mol H.

$$2.0 \text{ mol N} \times \frac{14.01 \text{ g N}}{\text{mol N}} = 28 \text{ g N}$$

$$4.0 \text{ mol H} \times \frac{1.008 \text{ g H}}{\text{mol H}} = 4.0 \text{ g H}$$

c. 1.0 mol $(NH_4)_2Cr_2O_7$ contains 2.0 mol N, 8.0 mol H, 2.0 mol Cr and 7.0 mol O.

$$2.0 \text{ mol N} \times \frac{14.01 \text{ g N}}{\text{mol N}} = 28 \text{ g N}$$

$$8.0 \text{ mol H} \times \frac{1.008 \text{ g H}}{\text{mol H}} = 8.0 \text{ g H}$$

$$2.0 \text{ mol Cr} \times \frac{52.00 \text{ g Cr}}{\text{mol Cr}} = 104 \text{ g Cr} = 1.0 \times 10^2 \text{ g Cr (2 S.F.)}$$

$$7.0 \text{ mol O} \times \frac{16.00 \text{ g O}}{\text{mol O}} = 112 \text{ g O} = 1.1 \times 10^2 \text{ g O (2 S.F.)}$$

18. a. 1.0 mol $CoCl_2{\bullet}6\ H_2O$ contains 1.0 mol Co, 2.0 mol Cl, 12 mol H, and 6.0 mol O.

$$1.0 \text{ mol Co} \times \frac{58.93 \text{ g Co}}{\text{mol Co}} = 59 \text{ g Co}$$

$$2.0 \text{ mol Cl} \times \frac{35.45 \text{ g Cl}}{\text{mol Cl}} = 71 \text{ g Cl}$$

$$12 \text{ mol H} \times \frac{1.008 \text{ g H}}{\text{mol H}} = 12 \text{ g H}$$

$$6.0 \text{ mol O} \times \frac{16.00 \text{ g O}}{\text{mol O}} = 96 \text{ g O}$$

b. $$1.0 \text{ mol } Ca_3(PO_4)_2 \times \frac{3 \text{ mol Ca}}{\text{mol } Ca_3(PO_4)_2} \times \frac{40.08 \text{ g Ca}}{\text{mol Ca}} = 1.2 \times 10^2 \text{ g Ca}$$

$$1.0 \text{ mol } Ca_3(PO_4)_2 \times \frac{2 \text{ mol P}}{\text{mol } Ca_3(PO_4)_2} \times \frac{30.97 \text{ g P}}{\text{mol P}} = 62 \text{ g P}$$

$$1.0 \text{ mol } Ca_3(PO_4)_2 \times \frac{8 \text{ mol O}}{\text{mol } Ca_3(PO_4)_2} \times \frac{16.00 \text{ g O}}{\text{mol O}} = 128 \text{ g} = 1.3 \times 10^2 \text{ g O}$$

c. $$1.0 \text{ mol } Na_2HPO_4 \times \frac{2 \text{ mol Na}}{\text{mol } Na_2HPO_4} \times \frac{22.99 \text{ g Na}}{\text{mol Na}} = 46 \text{ g Na}$$

$$1.0 \text{ mol } Na_2HPO_4 \times \frac{1 \text{ mol H}}{\text{mol } Na_2HPO_4} \times \frac{1.008 \text{ g H}}{\text{mol H}} = 1.0 \text{ g H}$$

$$1.0 \text{ mol } Na_2HPO_4 \times \frac{1 \text{ mol P}}{\text{mol } Na_2HPO_4} \times \frac{30.97 \text{ g P}}{\text{mol P}} = 3.1 \text{ g P}$$

$$1.0 \text{ mol } Na_2HPO_4 \times \frac{4 \text{ mol O}}{\text{mol } Na_2HPO_4} \times \frac{16.00 \text{ g O}}{\text{mol O}} = 64 \text{ g O}$$

19. a. For NH_3 the molar mass is 14.01 + 3(1.008) = 17.03 g/mol.

$$1.0\ g\ NH_3 \times \frac{1\ mol\ NH_3}{17.03\ g\ NH_3} \times \frac{1\ mol\ N}{mol\ NH_3} = 5.9 \times 10^{-2}\ mol\ N$$

$$1.0\ g\ NH_3 \times \frac{1\ mol\ NH_3}{17.03\ g\ NH_3} \times \frac{3\ mol\ H}{mol\ NH_3} = 1.8 \times 10^{-1}\ mol\ H$$

b. The molar mass of N_2H_4 is 2(14.01) + 4(1.008) = 32.05 g/mol.

$$1.0\ g\ N_2H_4 \times \frac{1\ mol\ N_2H_4}{32.05\ g\ N_2H_4} \times \frac{2\ mol\ N}{mol\ N_2H_4} = 6.2 \times 10^{-2}\ mol\ N$$

$$1.0\ g\ N_2H_4 \times \frac{1\ mol\ N_2H_4}{32.05\ g\ N_2H_4} \times \frac{4\ mol\ H}{mol\ N_2H_4} = 1.2 \times 10^{-1}\ mol\ H$$

c. The molar mass of $(NH_4)_2Cr_2O_7$ is 2(14.01) + 8(1.008) + 2(52.00) + 7(16.00) = 252.08 g/mol.

$$1.0\ g\ (NH_4)_2Cr_2O_7 \times \frac{1\ mol\ (NH_4)_2Cr_2O_7}{252.08\ g} \times \frac{2\ mol\ N}{mol\ (NH_4)_2Cr_2O_7} = 7.9 \times 10^{-3}\ mol\ N$$

$$1.0\ g\ (NH_4)_2Cr_2O_7 \times \frac{1\ mol\ (NH_4)_2Cr_2O_7}{252.08\ g} \times \frac{8\ mol\ H}{mol\ (NH_4)_2Cr_2O_7} = 3.2 \times 10^{-2}\ mol\ H$$

$$1.0\ g\ (NH_4)_2Cr_2O_7 \times \frac{1\ mol\ (NH_4)_2Cr_2O_7}{252.08\ g} \times \frac{2\ mol\ Cr}{mol\ (NH_4)_2Cr_2O_7} = 7.9 \times 10^{-3}\ mol\ Cr$$

$$1.0\ g\ (NH_4)_2Cr_2O_7 \times \frac{1\ mol\ (NH_4)_2Cr_2O_7}{252.08\ g} \times \frac{7\ mol\ O}{mol\ (NH_4)_2Cr_2O_7} = 2.8 \times 10^{-2}\ mol\ O$$

20. a. The molar mass of $CoCl_2{\bullet}6\ H_2O$ is 1(58.93) + 2(35.45) + 12(1.008) + 6(16.00) = 237.93 g/mol.

$$1.0\ g\ CoCl_2{\bullet}6\ H_2O \times \frac{1\ mol\ CoCl_2{\bullet}6\ H_2O}{237.93\ g\ CoCl_2{\bullet}6\ H_2O} \times \frac{1\ mol\ Co}{mol\ CoCl_2{\bullet}6\ H_2O} = 4.2 \times 10^{-3}\ mol\ Co$$

$$1.0\ g\ CoCl_2{\bullet}6\ H_2O \times \frac{1\ mol\ CoCl_2{\bullet}6\ H_2O}{237.93\ g} \times \frac{2\ mol\ Cl}{mol\ CoCl_2{\bullet}6\ H_2O} = 8.4 \times 10^{-3}\ mol\ Cl$$

$$1.0\ g\ CoCl_2{\bullet}6\ H_2O \times \frac{1\ mol\ CoCl_2{\bullet}6\ H_2O}{237.93\ g} \times \frac{12\ mol\ H}{mol\ CoCl_2{\bullet}6\ H_2O} = 5.0 \times 10^{-2}\ mol\ H$$

$$1.0\ g\ CoCl_2{\bullet}6\ H_2O \times \frac{1\ mol\ CoCl_2{\bullet}6\ H_2O}{237.93\ g} \times \frac{6\ mol\ O}{mol\ CoCl_2{\bullet}6\ H_2O} = 2.5 \times 10^{-2}\ mol\ O$$

b. Molar mass of $Ca_3(PO_4)_2$ = 3(40.08) + 2(30.97) + 8(16.00) = 310.18 g/mol

$$1.0\ g\ Ca_3(PO_4)_2 \times \frac{3\ mol\ Ca}{310.18\ g\ Ca_3(PO_4)_2} = 9.7 \times 10^{-3}\ mol\ Ca$$

$$1.0 \text{ g } Ca_3(PO_4)_2 \times \frac{2 \text{ mol P}}{310.18 \text{ g } Ca_3(PO_4)_2} = 6.4 \times 10^{-3} \text{ mol P}$$

$$1.0 \text{ g } Ca_3(PO_4)_2 \times \frac{8 \text{ mol O}}{310.18 \text{ g } Ca_3(PO_4)_2} = 2.6 \times 10^{-2} \text{ mol O}$$

c. Molar mass of Na_2HPO_4 = 2(22.99) + 1(1.008) + 1(30.97) + 4(16.00) = 141.96 g/mol.

$$1.0 \text{ g } Na_2HPO_4 \times \frac{2 \text{ mol Na}}{141.96 \text{ g } Na_2HPO_4} = 1.4 \times 10^{-2} \text{ mol Na}$$

$$1.0 \text{ g } Na_2HPO_4 \times \frac{1 \text{ mol H}}{141.96 \text{ g } Na_2HPO_4} = 7.0 \times 10^{-3} \text{ mol H}$$

$$1.0 \text{ g } Na_2HPO_4 \times \frac{1 \text{ mol P}}{141.96 \text{ g } Na_2HPO_4} = 7.0 \times 10^{-3} \text{ mol P}$$

$$1.0 \text{ g } Na_2HPO_4 \times \frac{4 \text{ mol O}}{141.96 \text{ g } Na_2HPO_4} = 2.8 \times 10^{-2} \text{ mol O}$$

21. a. $$1.0 \text{ g } NH_3 \times \frac{1 \text{ mol } NH_3}{17.03 \text{ g}} \times \frac{1 \text{ mol N}}{\text{mol } NH_3} \times \frac{6.02 \times 10^{23} \text{ atoms N}}{\text{mol N}} = 3.5 \times 10^{22} \text{ atoms N}$$

The first two conversions are the same ones that we did in Exercise 3.19. So, we can take our answers from the previous exercise and convert the moles of each element in 1.0 g into the number of atoms of each element in 1.0 g.

$$1.8 \times 10^{-1} \text{ mol H} \times \frac{6.02 \times 10^{23} \text{ atoms H}}{\text{mol H}} = 1.1 \times 10^{23} \text{ atoms H}$$

b. 1.0 g of N_2H_4 contains (from Exercise 3.19):

$$6.2 \times 10^{-2} \text{ mol N} \times \frac{6.02 \times 10^{23} \text{ atoms N}}{\text{mol N}} = 3.7 \times 10^{22} \text{ atoms N}$$

$$1.2 \times 10^{-1} \text{ mol H} \times \frac{6.02 \times 10^{23} \text{ atoms H}}{\text{mol H}} = 7.2 \times 10^{22} \text{ atoms H}$$

But, by doing the complete calculation and carrying an extra significant figure:

$$1.0 \text{ g } N_2H_4 \times \frac{1 \text{ mol } N_2H_4}{32.05 \text{ g}} \times \frac{4 \text{ mol H}}{\text{mol } N_2H_4} = 1.25 \times 10^{-1} \text{ mol H}$$

$$1.25 \times 10^{-1} \text{ mol H} \times \frac{6.02 \times 10^{23} \text{ atoms H}}{\text{mol H}} = 7.5 \times 10^{22} \text{ atoms H}$$

c. 1.0 g of $(NH_4)_2Cr_2O_7$ contains:

$$7.9 \times 10^{-3} \text{ mol N} \times \frac{6.02 \times 10^{23} \text{ atoms N}}{\text{mol N}} = 4.8 \times 10^{21} \text{ atoms N}$$

$$3.2 \times 10^{-2} \text{ mol H} \times \frac{6.02 \times 10^{23} \text{ atoms H}}{\text{mol H}} = 1.9 \times 10^{22} \text{ atoms H}$$

$$7.9 \times 10^{-3} \text{ mol Cr} \times \frac{6.02 \times 10^{23} \text{ atoms Cr}}{\text{mol Cr}} = 4.8 \times 10^{21} \text{ atoms Cr}$$

$$2.8 \times 10^{-2} \text{ mol O} \times \frac{6.02 \times 10^{23} \text{ atoms O}}{\text{mol O}} = 1.7 \times 10^{22} \text{ atoms O}$$

22. a. 1.0 g of $CoCl_2{\cdot}6\,H_2O$ contains (from Exercise 3.20):

$$4.2 \times 10^{-3} \text{ mol Co} \times \frac{6.02 \times 10^{23} \text{ atoms Co}}{\text{mol Co}} = 2.5 \times 10^{21} \text{ atoms Co}$$

$$8.4 \times 10^{-3} \text{ mol Cl} \times \frac{6.02 \times 10^{23} \text{ atoms Cl}}{\text{mol Cl}} = 5.1 \times 10^{21} \text{ atoms Cl}$$

$$5.0 \times 10^{-2} \text{ mol H} \times \frac{6.02 \times 10^{23} \text{ atoms H}}{\text{mol H}} = 3.0 \times 10^{22} \text{ atoms H}$$

$$2.5 \times 10^{-2} \text{ mol O} \times \frac{6.02 \times 10^{23} \text{ atoms O}}{\text{mol O}} = 1.5 \times 10^{22} \text{ atoms O}$$

b. From Exercise 3.20 we know that 1.0 g $Ca_3(PO_4)_2$ contains:

$$9.7 \times 10^{-3} \text{ mol Ca} \times \frac{6.02 \times 10^{23} \text{ atoms Ca}}{\text{mol Ca}} = 5.8 \times 10^{21} \text{ atoms Ca}$$

$$6.4 \times 10^{-3} \text{ mol P} \times \frac{6.02 \times 10^{23} \text{ atoms P}}{\text{mol P}} = 3.9 \times 10^{21} \text{ atoms P}$$

$$2.6 \times 10^{-2} \text{ mol O} \times \frac{6.02 \times 10^{23} \text{ atoms O}}{\text{mol O}} = 1.6 \times 10^{22} \text{ atoms O}$$

c. From Exercise 3.20 we know that 1.0 g Na_2HPO_4 contains:

$$1.4 \times 10^{-2} \text{ mol Na} \times \frac{6.02 \times 10^{23} \text{ atoms Na}}{\text{mol Na}} = 8.4 \times 10^{21} \text{ atoms Na}$$

$$7.0 \times 10^{-3} \text{ mol H} \times \frac{6.02 \times 10^{23} \text{ atoms H}}{\text{mol H}} = 4.2 \times 10^{21} \text{ atoms H}$$

$$7.0 \times 10^{-3} \text{ mol P} \times \frac{6.02 \times 10^{23} \text{ atoms P}}{\text{mol P}} = 4.2 \times 10^{21} \text{ atoms P}$$

$$2.8 \times 10^{-2} \text{ mol O} \times \frac{6.02 \times 10^{23} \text{ atoms O}}{\text{mol O}} = 1.7 \times 10^{22} \text{ atoms O}$$

23. Al_2O_3; 2(26.98) + 3(16.00) = 101.96 g/mol

Na_3AlF_6; 3(22.99) + 1(26.98) + 6(19.00) = 209.95 g/mol

24. HFC - 134a, CH_2FCF_3: 2(12.01) + 2(1.008) + 4(19.00) = 102.04 g/mol

HCFC-124, $CHClFCF_3$: 2(12.01) + 1(1.008) + 1(35.45) + 4(19.00) = 136.48 g/mol

25. Molar mass of $C_6H_8O_6 = 6(12.01) + 8(1.008) + 6(16.00) = 176.12$ g/mol

$$500.0 \text{ mg} \times \frac{1 \text{ g}}{1000 \text{ mg}} \times \frac{1 \text{ mol}}{176.12 \text{ g}} = 2.839 \times 10^{-3} \text{ mol}$$

$$2.839 \times 10^{-3} \text{ mol} \times \frac{6.022 \times 10^{23} \text{ molecules}}{\text{mol}} = 1.710 \times 10^{21} \text{ molecules}$$

26. a. $9(12.01) + 8(1.008) + 4(16.00) = 180.15$ g/mol

b. $$500. \text{ mg} \times \frac{1 \text{ g}}{1000 \text{ mg}} \times \frac{1 \text{ mol}}{180.15 \text{ g}} = 2.78 \times 10^{-3} \text{ mol}$$

$$2.78 \times 10^{-3} \text{ mol} \times \frac{6.022 \times 10^{23} \text{ molecules}}{\text{mol}} = 1.67 \times 10^{21} \text{ molecules}$$

27. a. $$100 \text{ molecules } H_2O \times \frac{1 \text{ mol}}{6.022 \times 10^{23} \text{ molecules}} = 1.661 \times 10^{-22} \text{ mol}$$

b. $$100.0 \text{ g } H_2O \times \frac{1 \text{ mol } H_2O}{18.02 \text{ g } H_2O} = 5.549 \text{ mol}$$

c. $$500 \text{ atoms Fe} \times \frac{1 \text{ mol Fe}}{6.022 \times 10^{23} \text{ atoms}} = 8.303 \times 10^{-22} \text{ mol}$$

d. $$500.0 \text{ g Fe} \times \frac{1 \text{ mol Fe}}{55.85 \text{ g Fe}} = 8.953 \text{ mol}$$

e. $$150 \text{ molecules } O_2 \times \frac{1 \text{ mol } O_2}{6.022 \times 10^{23} \text{ molecules } O_2} = 2.491 \times 10^{-22} \text{ mol}$$

28. a. $$150.0 \text{ g } Fe_2O_3 \times \frac{1 \text{ mol}}{159.70 \text{ g}} = 0.9393 \text{ mol}$$

b. $$10.0 \text{ mg } NO_2 \times \frac{1 \text{ g}}{10^3 \text{ mg}} \times \frac{1 \text{ mol}}{46.01 \text{ g}} = 2.17 \times 10^{-4} \text{ mol } NO_2$$

c. $$1.0 \text{ fmol } NO_2 \times \frac{1 \text{ mol}}{10^{15} \text{ fmol}} = 1.0 \times 10^{-15} \text{ mol } NO_2$$

d. $$1.5 \times 10^{16} \text{ molecules } BF_3 \times \frac{1 \text{ mol}}{6.02 \times 10^{23} \text{ molecules}} = 2.5 \times 10^{-8} \text{ mol } BF_3$$

e. $$2.6 \text{ mg } BF_3 \times \frac{1 \text{ g}}{10^3 \text{ mg}} \times \frac{1 \text{ mol}}{67.81 \text{ g}} = 3.8 \times 10^{-5} \text{ mol } BF_3$$

29. a. $$3.00 \times 10^{20} \text{ molecules } N_2 \times \frac{1 \text{ mol } N_2}{6.022 \times 10^{23} \text{ molecules}} \times \frac{28.02 \text{ g } N_2}{\text{mol } N_2} = 1.40 \times 10^{-2} \text{ g } N_2$$

b. $$3.00 \times 10^{-3} \text{ mol } N_2 \times \frac{28.02 \text{ g } N_2}{\text{mol } N_2} = 8.41 \times 10^{-2} \text{ g } N_2$$

c. $$1.5 \times 10^{2} \text{ mol } N_2 \times \frac{28.02 \text{ g } N_2}{\text{mol } N_2} = 4.2 \times 10^{3} \text{ g } N_2$$

d. $$1 \text{ molecule } N_2 \times \frac{1 \text{ mol } N_2}{6.022 \times 10^{23} \text{ molecules } N_2} \times \frac{28.02 \text{ g } N_2}{\text{mol } N_2} = 4.653 \times 10^{-23} \text{ g } N_2$$

e. $2.00 \times 10^{-15} \text{ mol } N_2 \times \frac{28.02 \text{ g } N_2}{\text{mol } N_2} = 5.60 \times 10^{-14} \text{ g } N_2 = 56.0 \text{ fg } N_2$

f. $18.0 \text{ pmol } N_2 \times \frac{1 \text{ mol } N_2}{10^{12} \text{ pmol}} \times \frac{28.02 \text{ g } N_2}{\text{mol } N_2} = 5.04 \times 10^{-10} \text{ g } N_2 = 504 \text{ pg } N_2$

g. $5.0 \text{ nmol } N_2 \times \frac{1 \text{ mol } N_2}{10^9 \text{ nmol}} \times \frac{28.02 \text{ g } N_2}{\text{mol } N_2} = 1.4 \times 10^{-7} \text{ g } N_2 = 140 \text{ ng}$

30. a. Molar mass of CH_4O = 1(12.01) + 4(1.008) + 1(16.00) = 32.04 g/mol

$$1.0 \text{ g } CH_4O \times \frac{1 \text{ mol } CH_4O}{32.04 \text{ g } CH_4O} \times \frac{6.02 \times 10^{23} \text{ molecules } CH_4O}{\text{mol } CH_4O} \times \frac{1 \text{ atom C}}{\text{molecule } CH_4O} = 1.9 \times 10^{22} \text{ atoms of C}$$

b. Molar mass of CH_3CO_2H = 2(12.01) + 4(1.008) + 2(16.00) = 60.05 g/mol

$$1.0 \text{ g} \times \frac{1 \text{ mol}}{60.05 \text{ g}} \times \frac{6.02 \times 10^{23} \text{ molecules}}{\text{mol}} \times \frac{2 \text{ atoms C}}{\text{molecule}} = 2.0 \times 10^{22} \text{ atoms C}$$

c. Molar mass of Na_2CO_3 = 2(22.99) + 1(12.01) + 3(16.00) = 105.99 g/mol

$$1.0 \text{ g } Na_2CO_3 \times \frac{1 \text{ mol } Na_2CO_3}{105.99 \text{ g } Na_2CO_3} \times \frac{1 \text{ mol C}}{\text{mol } Na_2CO_3} \times \frac{6.02 \times 10^{23} \text{ atoms C}}{\text{mol C}}$$

$= 5.7 \times 10^{21}$ atoms C

d. $$1.0 \text{ g } C_6H_{12}O_6 \times \frac{1 \text{ mol } C_6H_{12}O_6}{180.16 \text{ g } C_6H_{12}O_6} \times \frac{6 \text{ mol C}}{\text{mol } C_6H_{12}O_6} \times \frac{6.02 \times 10^{23} \text{ atoms C}}{\text{mol C}}$$

$= 2.0 \times 10^{22}$ atoms C

e. $$1.0 \text{ g } C_{12}H_{22}O_{11} \times \frac{1 \text{ mol } C_{12}H_{22}O_{11}}{342.30 \text{ g } C_{12}H_{22}O_{11}} \times \frac{12 \text{ mol C}}{\text{mol } C_{12}H_{22}O_{11}} \times \frac{6.02 \times 10^{23} \text{ atoms C}}{\text{mol C}}$$

$= 2.1 \times 10^{22}$ atoms C

f. $$1.0 \text{ g } CHCl_3 \times \frac{1 \text{ mol } CHCl_3}{119.37 \text{ g } CHCl_3} \times \frac{1 \text{ mol C}}{\text{mol } CHCl_3} \times \frac{6.02 \times 10^{23} \text{ atoms C}}{\text{mol C}}$$

$= 5.0 \times 10^{21}$ atoms C

31. a. $$14 \text{ mol C} \left(\frac{12.01 \text{ g}}{\text{mol C}}\right) + 18 \text{ mol H} \left(\frac{1.008 \text{ g}}{\text{mol H}}\right) + 2 \text{ mol N} \left(\frac{14.01 \text{ g}}{\text{mol N}}\right) + 5 \text{ mol O} \left(\frac{16.00 \text{ g}}{\text{mol O}}\right)$$

= 294.30 g/mol

b. $10.0 \text{ g aspartame} \times \frac{1 \text{ mol}}{294.30 \text{ g}} = 3.40 \times 10^{-2} \text{ mol}$

c. $1.56 \text{ mol} \times \frac{294.30 \text{ g}}{\text{mol}} = 459 \text{ g}$

d. $5.0 \text{ mg} \times \frac{1 \text{ g}}{1000 \text{ mg}} \times \frac{1 \text{ mol}}{294.30 \text{ g}} \times \frac{6.02 \times 10^{23} \text{ molecules}}{\text{mol}} = 1.0 \times 10^{19} \text{ molecules}$

e. $1.2 \text{ g aspartame} \times \frac{1 \text{ mol aspartame}}{294.30 \text{ g aspartame}} \times \frac{2 \text{ mol N}}{\text{mol aspartame}} \times \frac{6.02 \times 10^{23} \text{ atoms N}}{\text{mol N}}$

$= 4.9 \times 10^{21}$ atoms of nitrogen

f. $1.0 \times 10^{9} \text{ molecules} \times \frac{1 \text{ mol}}{6.02 \times 10^{23} \text{ molecules}} \times \frac{294.30 \text{ g}}{\text{mol}} = 4.9 \times 10^{-13} \text{ g or } 490 \text{ fg}$

g. $1 \text{ molecule aspartame} \times \frac{1 \text{ mol}}{6.022 \times 10^{23} \text{ molecules}} \times \frac{294.30 \text{ g}}{\text{mol}} = 4.887 \times 10^{-22} \text{ g}$

32. As we shall see later the formula written as $(CH_3)_2N_2O$ tries to tell us something about how the atoms are attached to each other. For our purposes in this problem we can write the formula as $C_2H_6N_2O$.

a. $2(12.01) + 6(1.008) + 2(14.01) + 1(16.00) = 74.09 \text{ g/mol}$

b. $250 \text{ mg} \times \frac{1 \text{ g}}{1000 \text{ mg}} \times \frac{1 \text{ mol}}{74.09 \text{ g}} = 3.4 \times 10^{-3} \text{ mol}$

c. $0.050 \text{ mol} \times \frac{74.09 \text{ g}}{\text{mol}} = 3.7 \text{ g}$

d. $1.0 \text{ mol } C_2H_6N_2O \times \frac{6.02 \times 10^{23} \text{ molecules } C_2H_6N_2O}{\text{mol } C_2H_6N_2O} \times \frac{6 \text{ atoms of H}}{\text{molecule } C_2H_6N_2O}$

$= 3.6 \times 10^{24}$ atoms of hydrogen

e. $1.0 \times 10^{6} \text{ molecules} \times \frac{1 \text{ mol}}{6.02 \times 10^{23} \text{ molecules}} \times \frac{74.09 \text{ g}}{\text{mol}}$

$= 1.2 \times 10^{-16} \text{ g} = 0.12 \text{ fg}$

f. $\frac{74.09 \text{ g}}{\text{mol}} \times \frac{1 \text{ mol}}{6.022 \times 10^{23} \text{ molecules}} = 1.230 \times 10^{-22} \text{ g/molecule}$

Percent Composition

33. From Exercise 2.15

i. 1.188 g for every 1.000 g S

$\%F = \frac{1.188 \text{ g}}{1.188 \text{ g} + 1.000 \text{ g}} \times 100 = 54.30\% \text{ F};$ $\%S = 100.00 - 54.30 = 45.70\% \text{ S}$

ii. 2.375 g F for every 1.000 g S

$\%F = \frac{2.375 \text{ g}}{3.375 \text{ g}} \times 100 = 70.37\% \text{ F};$ $\%S = 100.00 - 70.37 = 29.63\% \text{ S}$

iii. 3.563 g F for every 1.000 g S

$\%F = \frac{3.563\ g}{4.563\ g} \times 100 = 78.08\%\ F;$ $\%S = 100.00 - 78.08 = 21.92\%\ S$

34. From Exercise 2.16

Hydrazine: 1.00 g N and 1.44×10^{-1} g H

$\%H = \frac{0.144\ g}{1.00\ g + 0.144\ g} = 12.6\%\ H;$ $\%N = 100.0 - 12.6 = 87.4\%\ N$

Ammonia: 1.00 g N and 2.16×10^{-1} g H

$\%H = \frac{0.216\ g}{1.00\ g + 0.216\ g} \times 100 = 17.8\%\ H;$ $\%N = 100.0 - 17.8 = 82.2\%\ N$

Hydrogen azide: 1.00 g N and 2.40×10^{-2} g H

$\%H = \frac{2.40 \times 10^{-2}\ g}{1.00\ g + 0.0240\ g} \times 100 = 2.34\%\ H;$ $\%N = 100.00 - 2.34 = 97.66\%\ N$

35. In 1 mole of $YBa_2Cu_3O_7$ there are 1 mole of Y, 2 moles of Ba, 3 moles of Cu, and 7 moles of O.

$$\text{Molar mass} = 1\ \text{mol Y}\left(\frac{88.91\ \text{g Y}}{\text{mol Y}}\right) + 2\ \text{mol Ba}\left(\frac{137.3\ \text{g Ba}}{\text{mol Ba}}\right) + 3\ \text{mol Cu}\left(\frac{63.55\ \text{g Cu}}{\text{mol Cu}}\right) + 7\ \text{mol O}\left(\frac{16.00\ \text{g O}}{\text{mol O}}\right)$$

Molar mass = 88.91 + 274.6 + 190.65 + 112.00

Molar mass = 666.2 g/mol

$\%Y = \frac{88.91\ g}{666.2\ g} \times 100 = 13.35\%\ Y;$ $\%Ba = \frac{274.6\ g}{666.2\ g} \times 100 = 41.22\%\ Ba$

$\%Cu = \frac{190.65\ g}{666.2\ g} \times 100 = 28.62\%\ Cu;$ $\%O = \frac{112.0\ g}{666.2\ g} \times 100 = 16.81\%\ O$

or %O = 100.00 - (13.35 + 41.22 + 28.62) = 100.00 - (83.19) = 16.81% O

36. a. NO; $\%N = \frac{14.01\ \text{g N}}{30.01\ \text{g NO}} \times 100 = 46.68\%\ N$

b. NO_2; $\%N = \frac{14.01\ \text{g N}}{46.01\ \text{g } NO_2} \times 100 = 30.45\%\ N$

c. N_2O_4; $\%N = \frac{28.02\ \text{g N}}{92.02\ \text{g } N_2O_4} \times 100 = 30.45\%\ N$

d. N_2O; $\%N = \frac{28.02\ \text{g N}}{44.02\ \text{g } N_2O} \times 100 = 63.65\%\ N$

37. Na_3PO_4: molar mass = 3(22.99) + 30.97 + 4(16.00) = 163.94 g/mol

$$\%P = \frac{30.97 \text{ g}}{163.94 \text{ g}} \times 100 = 18.89\% \text{ P}$$

PH_3 : molar mass = 30.97 + 3(1.008) = 33.99 g/mol

$$\%P = \frac{30.97 \text{ g}}{33.99 \text{ g}} \times 100 = 91.12\% \text{ P}$$

P_4O_{10}: molar mass = 4(30.97) + 10(16.00) = 283.88 g/mol

$$\%P = \frac{123.88 \text{ g}}{283.88 \text{ g}} \times 100 = 43.638\%$$

$(NPCl_2)_3$: molar mass = 3(14.01) + 3(30.97) + 6(35.45) = 347.64 g/mol

$$\%P = \frac{92.91 \text{ g}}{347.64 \text{ g}} \times 100 = 26.73\%$$

The order from lowest to highest percentage of phosphorus is:

$$Na_3PO_4 < (NPCl_2)_3 < P_4O_{10} < PH_3$$

38. $NO_2 = N_2O_4 < NO < N_2O$

39. $\%Co = 4.34 = \frac{58.93 \text{ g Co}}{\text{molar mass}} \times 100$; molar mass = 1360 g/mol

40. 0.342 g Fe for every 100 g hemoglobin (Hb)

$$\frac{100 \text{ g Hb}}{0.342 \text{ g Fe}} \times \frac{55.85 \text{ g Fe}}{\text{mol Fe}} \times \frac{4 \text{ mol Fe}}{\text{mol Hb}} = \frac{63{,}500 \text{ g Hb}}{\text{mol Hb}}$$

41. Cds: $\%Cd = \frac{112.4 \text{ g Cd}}{144.5 \text{ g CdS}} \times 100 = 77.79\% \text{ Cd}$

CdSe: $\%Cd = \frac{112.4 \text{ g}}{191.4 \text{ g}} \times 100 = 58.73\% \text{ Cd}$

CdTe: $\%Cd = \frac{112.4 \text{ g}}{240.0 \text{ g}} \times 100 = 46.83\% \text{ Cd}$

42. XeF_2: $\%F = \frac{38.00 \text{ g}}{169.3 \text{ g}} \times 100 = 22.45\%$

XeF_4: $\%F = \frac{76.00 \text{ g}}{207.3 \text{ g}} \times 100 = 36.66\%$

XeF_6: $\%F = \frac{114.00 \text{ g}}{245.3 \text{ g}} \times 100 = 46.47\%$

43. a. $C_3H_4O_2$: Molar mass = 3(12.01) + 4(1.008) + 2(16.00) = 36.03 + 4.032 + 32.00 = 72.06 g/mol

$$\%C = \frac{36.03 \text{ g C}}{72.06 \text{ g compound}} \times 100 = 50.00\% \text{ C}$$

$$\%H = \frac{4.032 \text{ g H}}{72.06 \text{ g compound}} \times 100 = 5.595\% \text{ H}$$

$\%O = 100.00 - (50.00 + 5.595) = 44.41\%\ O$

or $\%O = \frac{32.00\ g}{72.06\ g} \times 100 = 44.41\%\ O$

b. $C_4H_6O_2$: Molar mass = 4(12.01) + 6(1.008) + 2(16.00) = 48.04 + 6.048 + 32.00 = 86.09 g/mol

$\%C = \frac{48.04\ g}{86.09\ g} \times 100 = 55.80\%\ C$

$\%H = \frac{6.048\ g}{86.09\ g} \times 100 = 7.025\%\ H$

$\%O = 100.00 - (55.80 + 7.025) = 37.18\%\ O$

c. C_3H_3N: Molar mass = 3(12.01) + 3(1.008) + 1(14.01) = 36.03 + 3.024 + 14.01 = 53.06 g/mol

$\%C = \frac{36.03\ g}{53.06\ g} \times 100 = 67.90\%\ C$; $\%H = \frac{3.024\ g}{53.06\ g} \times 100 = 5.699\%\ H$

$\%N = \frac{14.01\ g}{53.06\ g} \times 100 = 26.40\%\ N$; or $\%N = 100.00 - (67.90 + 5.699) = 26.40\%N$

44. Aluminum oxide: Al_2O_3

Molar mass = 2(26.98) + 3(16.00) = 101.96 g/mol

$\%Al = \frac{53.96\ g}{101.96\ g} \times 100 = 52.92\%\ Al$; $\%O = 100.00\% - 52.92\% = 47.08\%\ O$

Cryolite: Na_3AlF_6

Molar mass = 3(22.99) + 26.98 + 6(19.00) = 209.95 g/mol

$\%Na = \frac{68.97\ g}{209.95\ g} \times 100 = 32.85\%\ Na$; $\%Al = \frac{26.98\ g}{209.95\ g} \times 100 = 12.85\%\ Al$

$\%F = \frac{114.00\ g}{209.95\ g} \times 100 = 54.30\%\ F$

or $\%F = 100.00 - (32.85 + 12.85) = 100.00 - 45.70 = 54.30\%\ F$

HFC - 134a, CH_2FCF_3.

Molar mass = 2(12.01) + 2(1.008) + 4(19.00) = 102.04 g/mol

$\%C = \frac{24.02\ g}{102.04\ g} \times 100 = 23.54\%\ C$; $\%H = \frac{2.016\ g}{102.04\ g} \times 100 = 1.976\%\ H$

$\%F = \frac{76.00\ g}{102.04\ g} \times 100 = 74.48\%\ F$

HCFC - 124, $CHClFCF_3$

Molar mass = 2(12.01) + 1.008 + 35.45 + 4(19.00) = 136.48 g/mol

$\%C = \frac{24.02 \text{ g}}{136.48 \text{ g}} \times 100 = 17.60\%$ C; $\%H = \frac{1.008 \text{ g}}{136.48 \text{ g}} \times 100 = 0.7386\%$ H

$\%Cl = \frac{35.45 \text{ g}}{136.48 \text{ g}} \times 100 = 25.97\%$ Cl; $\%F = \frac{76.00 \text{ g}}{136.48 \text{ g}} \times 100 = 55.69\%$ F

Empirical and Molecular Formulas

45. From Exercise 2.15 we get

i. 1.188 g F for every 1.000 g S

$$1.188 \text{ g F} \times \frac{1 \text{ mol F}}{19.00 \text{ g F}} = 6.253 \times 10^{-2} \text{ mol F}$$

$$1.000 \text{ g S} \times \frac{1 \text{ mol S}}{32.07 \text{ g S}} = 3.118 \times 10^{-2} \text{ mol S}$$

$$\frac{6.253 \times 10^{-2} \text{ mol F}}{3.118 \times 10^{-2} \text{ mol S}} = \frac{2.005 \text{ mol F}}{\text{mol S}}$$; Empirical formula: SF_2

ii. $$2.375 \text{ g F} \times \frac{1 \text{ mol F}}{19.00 \text{ g F}} = 1.250 \times 10^{-1} \text{ mol F}$$

$$\frac{1.250 \times 10^{-1} \text{ mol F}}{3.118 \times 10^{-2} \text{ mol S}} = \frac{4 \text{ mol F}}{\text{mol S}}$$; Empirical formula: SF_4

iii. $$3.563 \text{ g F} \times \frac{1 \text{ mol F}}{19.00 \text{ g F}} = 1.875 \times 10^{-1} \text{ mol F}$$

$$\frac{1.875 \times 10^{-1} \text{ mol F}}{3.118 \times 10^{-2} \text{ mol S}} = \frac{6 \text{ mol F}}{\text{mol S}}$$; Empirical formula: SF_6

46. From Exercise 2.16 we get

Hydrazine: 1.00 g N and 1.44×10^{-1} g H

$$1.00 \text{ g N} \times \frac{1 \text{ mol N}}{14.01 \text{ g}} = 7.14 \times 10^{-2} \text{ mol N}$$

$$1.44 \times 10^{-1} \text{ H} \times \frac{1 \text{ mol H}}{1.008 \text{ g H}} = 1.43 \times 10^{-1} \text{ mol H}$$

$$\frac{1.43 \times 10^{-1} \text{ mol H}}{7.14 \times 10^{-2} \text{ mol N}} = \frac{2 \text{ mol H}}{\text{mol N}}$$; Empirical formula: NH_2

Ammonia:

$$0.216 \text{ g H} \times \frac{1 \text{ mol H}}{1.008 \text{ g H}} = 0.214 \text{ mol H}$$

$$\frac{0.214 \text{ mol H}}{0.0714 \text{ mol N}} = \frac{3 \text{ mol H}}{\text{mol N}}; \text{ Empirical formula: } NH_3$$

Hydrogen azide:

$$2.40 \times 10^{-2} \text{ g H} \times \frac{1 \text{ mol H}}{1.008 \text{ g H}} = 2.38 \times 10^{-2} \text{ mol H}$$

$$\frac{7.14 \times 10^{-2} \text{ mol N}}{2.38 \times 10^{-2} \text{ mol H}} = \frac{3 \text{ mol N}}{\text{mol H}}; \text{ Empirical formula: } N_3H$$

47. a. SNH: Mass of a single SNH is 32.07 + 14.01 + 1.008 = 47.09 g

$$\frac{188.35}{47.09} = 4; \text{ So the molecular formula is } (SNH)_4 \text{ or } S_4N_4H_4.$$

b. $NPCl_2$: Mass of an $NPCl_2$ unit is 14.01 + 30.97 + 2(35.45) = 115.88 g

$$\frac{347.64}{115.88} = 3; \text{ Molecular formula is } (NPCl_2)_3 \text{ or } N_3P_3Cl_6.$$

c. CoC_4O_4: 58.93 + 4(12.01) + 4(16.00) = 170.97 g

$$\frac{341.94}{170.97} = 2; \text{ Molecular formula: } Co_2C_8O_8$$

d. SN: 32.07 + 14.01 = 46.08 g

$$\frac{184.32}{46.08} = 4; \text{ Molecular formula: } S_4N_4$$

48. a. $C_3H_4O_3$ b. CH c. CH d. P_2O_5 e. CH_2O f. CH_2O

49. a.

$$\text{Molar mass of } CH_2O = 1 \text{ mol C}\left(\frac{12.01 \text{ g}}{\text{mol C}}\right) + 2 \text{ mol H}\left(\frac{1.008 \text{ g H}}{\text{mol H}}\right) + 1 \text{ mol O}\left(\frac{16.00 \text{ g}}{\text{mol O}}\right) = 30.03 \text{ g/mol}$$

$$\%C = \frac{12.01 \text{ g C}}{30.03 \text{ g } CH_2O} \times 100 = 39.99\% \text{ C}$$

$$\%H = \frac{2.016 \text{ g H}}{30.03 \text{ g } CH_2O} \times 100 = 6.713\% \text{ H}$$

$$\%O = \frac{16.00 \text{ g O}}{30.03 \text{ g } CH_2O} \times 100 = 53.28\% \text{ O} \qquad \text{or } \%O = 100 - (39.99 + 6.713) \doteq 53.30\%$$

b. Molar mass of $C_6H_{12}O_6$ = 6(12.01) + 12(1.008) + 6(16.00) = 180.16 g/mol

$$\%C = \frac{72.06 \text{ g C}}{180.16 \text{ g } C_6H_{12}O_6} \times 100 = 40.00\% \qquad \%H = \frac{12.096 \text{ g}}{180.16 \text{ g}} \times 100 = 6.7140\%$$

$$\%O = 100 - (40.00 + 6.714) = 53.29\%$$

c. Molar mass of $HC_2H_3O_2$ = 2(12.01) + 4(1.008) + 2(16.00) = 60.05 g/mol

$$\%C = \frac{24.02 \text{ g}}{60.05 \text{ g}} \times 100 = 40.00\% \quad \%H = \frac{4.032 \text{ g}}{60.05 \text{ g}} \times 100 = 6.714\%$$

$$\%O = 100 - (40.00 + 6.714) = 53.29\%$$

50. All three compounds have the same empirical formula, CH_2O, and different molecular formulas. The composition of all three in mass percent is also the same. Therefore, elemental analysis will give us only the empirical formula.

51. Out of 100 g of compound, there are:

$$48.38 \text{ g C} \times \frac{1 \text{ mol C}}{12.01 \text{ g C}} = 4.028 \text{ mol C}$$

$$8.12 \text{ g H} \times \frac{1 \text{ mol H}}{1.008 \text{ g H}} = 8.06 \text{ mol H}$$

$$\%O = 100 - 48.38 - 8.12 = 43.50\%$$

$$43.50 \text{ g O} \times \frac{1 \text{ mol O}}{16.00 \text{ g O}} = 2.719 \text{ mol O}$$

Look for common ratios.

$$\frac{4.028}{2.719} = 1.481 \approx \frac{3}{2}; \text{ So we try } \frac{2.719}{2} = 1.360$$

$$\frac{4.028}{1.36} \approx 3 \quad \frac{8.06}{1.36} \approx 6 \quad \frac{2.719}{1.36} \approx 2$$

So empirical formula is $C_3H_6O_2$.

52. Out of 100 g, there are:

$$69.6 \text{ g S} \times \frac{1 \text{ mol S}}{32.07 \text{ g S}} = 2.17 \text{ mol S}$$

$$30.4 \text{ g N} \times \frac{1 \text{ mol N}}{14.01 \text{ g N}} = 2.17 \text{ mol N}$$

Empirical formula is SN.

The mass of one SN unit is ~ 46 g.

Since $\frac{184}{46} = 4$, then the molecular formula is S_4N_4.

53. $0.979 \text{ g Na} \times \frac{1 \text{ mol Na}}{22.99 \text{ g Na}} = 4.26 \times 10^{-2} \text{ mol Na}$

$1.365 \text{ g S} \times \frac{1 \text{ mol S}}{32.07 \text{ g S}} = 4.256 \times 10^{-2} \text{ mol S}, \frac{\text{mol S}}{\text{mol Na}} = 1.00$

$1.021 \text{ g O} \times \frac{1 \text{ mol O}}{16.00 \text{ g O}} = 6.381 \times 10^{-2} \text{ mol O}$

$\frac{6.381 \times 10^{-2} \text{ mol O}}{4.256 \times 10^{-2} \text{ mol S}} = 1.499 \approx \frac{1.5 \text{ mol O}}{\text{mol S}} = \frac{3 \text{ mol O}}{2 \text{ mol S}}$

Empirical formula: $Na_2S_2O_3$

54. $1.121 \text{ g N} \times \frac{1 \text{ mol N}}{14.01 \text{ g N}} = 8.001 \times 10^{-2} \text{ mol N}$

$0.161 \text{ g H} \times \frac{1 \text{ mol H}}{1.008 \text{ g H}} = 1.60 \times 10^{-1} \text{ mol H}$

$0.480 \text{ g C} \times \frac{1 \text{ mol C}}{12.01 \text{ g C}} = 4.00 \times 10^{-2} \text{ mol C}$

$0.640 \text{ g O} \times \frac{1 \text{ mol O}}{16.00 \text{ g O}} = 4.00 \times 10^{-2} \text{ mol O}$

$\frac{8.001 \times 10^{-2}}{4.00 \times 10^{-2}} = 2$ and $\frac{1.60 \times 10^{-1}}{4.00 \times 10^{-2}} = 4$

So, empirical formula is: N_2H_4CO

55. Out of 100 g: $30.4 \text{ g N} \times \frac{1 \text{ mol N}}{14.01 \text{ g N}} = 2.17 \text{ mol N}$

$\%O = 100 - 30.4 = 69.6\% \text{ O}$; $69.6 \text{ g O} \times \frac{1 \text{ mol O}}{16.00 \text{ g O}} = 4.35 \text{ mol O}$

$\frac{2.17}{2.17} = 1$ $\frac{4.35}{2.17} = 2$; Empirical formula is NO_2

NO_2 unit has a mass of $14 + 2(16) \sim 46$ g

$\frac{92 \text{ g}}{46 \text{ g}} = 2$; Therefore, the molecular formula is N_2O_4.

56. Out of 100 g of the pigment, there are:

$59.9 \text{ g Ti} \times \frac{1 \text{ mol Ti}}{47.88 \text{ g Ti}} = 1.25 \text{ mol Ti}$

$40.1 \text{ g O} \times \frac{1 \text{ mol O}}{16.00 \text{ g O}} = 2.51 \text{ mol O}$

Empirical formula: TiO_2

57. First, get composition in mass percent.

We assume all of the carbon in 0.213 g CO_2 came from 0.157 g of the compound and that all of the hydrogen in the 0.0310 g H_2O came from the 0.157 g of the compound.

$$0.213 \text{ g } CO_2 \times \frac{12.01 \text{ g C}}{44.01 \text{ g } CO_2} = 0.0581 \text{ g C}$$

$$\%C = \frac{0.0581 \text{ g C}}{0.157 \text{ g compound}} \times 100 = 37.0\% \text{ C}$$

$$0.0310 \text{ g } H_2O \times \frac{2.016 \text{ g H}}{18.02 \text{ g } H_2O} = 3.47 \times 10^{-3} \text{ g H}$$

$$\%H = \frac{3.47 \times 10^{-3} \text{ g}}{0.157 \text{ g}} = 2.21\% \text{ H}$$

We get %N from the second experiment:

$$0.0230 \text{ g } NH_3 \times \frac{14.01 \text{ g N}}{17.03 \text{ g } NH_3} = 1.89 \times 10^{-2} \text{ g N}$$

$$\%N = \frac{1.89 \times 10^{-2} \text{ g}}{0.103 \text{ g}} \times 100 = 18.3\% \text{ N}$$

The mass percent of oxygen is obtained by difference:

$$\%O = 100 - (37.0 + 2.21 + 18.3) = 42.5\%$$

So out of 100 g of compound, there are:

$$37.0 \text{ g C} \times \frac{1 \text{ mol C}}{12.01 \text{ g C}} = 3.08 \text{ mol C}; \quad 2.21 \text{ g H} \times \frac{1 \text{ mol H}}{1.008 \text{ g H}} = 2.19 \text{ mol H}$$

$$18.3 \text{ g N} \times \frac{1 \text{ mol N}}{14.01 \text{ g N}} = 1.31 \text{ mol N}; \quad 42.5 \text{ g O} \times \frac{1 \text{ mol O}}{16.00 \text{ g O}} = 2.66 \text{ mol O}$$

Lastly, and often the hardest part, we need to find simple whole number ratios. We do this by trial and error.

$$\frac{2.19}{1.31} = 1.67 \approx 1\frac{2}{3} \approx \frac{5}{3}$$

So we try $\frac{1.31}{3} = 0.437$ as lowest common denominator.

$$\frac{3.08}{0.437} \approx 7 \quad \frac{2.19}{0.437} \approx 5 \quad \frac{1.31}{0.437} = 3 \quad \frac{2.66}{0.437} \approx 6$$

Empirical formula: $C_7H_5N_3O_6$.

58. First, determine composition by mass percent:

$$16.01 \text{ mg } CO_2 \times \frac{12.01 \text{ mg C}}{44.01 \text{ mg } CO_2} = 4.369 \text{ mg C}$$

$$\%C = \frac{4.369 \text{ mg C}}{10.68 \text{ mg compound}} \times 100 = 40.91\% \text{ C}$$

$$4.37 \text{ mg } H_2O \times \frac{2.016 \text{ mg H}}{18.02 \text{ mg } H_2O} = 0.489 \text{ mg H}$$

$$\%H = \frac{0.489 \text{ mg}}{10.68 \text{ mg}} \times 100 = 4.58\% \text{ H}$$

$$\%O = 100 - (40.91 + 4.58) = 54.51\% \text{ O}$$

So, if we have 100 g of the compound, we have

$$40.91 \text{ g C} \times \frac{1 \text{ mol C}}{12.01 \text{ g C}} = 3.406 \text{ mol C}$$

$$4.58 \text{ g H} \times \frac{1 \text{ mol H}}{1.008 \text{ g H}} = 4.54 \text{ mol H}$$

$$54.51 \text{ g O} \times \frac{1 \text{ mol O}}{16.00 \text{ g O}} = 3.407 \text{ mol O}$$

Look for common ratios $\frac{3.406}{4.54} = 0.750 = \frac{3}{4}$

Therefore, empirical formula is $C_3H_4O_3$.

The mass of one $C_3H_4O_3$ unit is $3(12) + 4(1) + 3(16) \approx 88$ g.

Since $\frac{176.1}{88} = 2$, then the molecular formula is $C_6H_8O_6$.

Balancing Chemical Equations

59. a. $Fe_2O_3 + CO \rightarrow Fe_3O_4 + CO_2$ Balance Fe atoms first:

$3\ Fe_2O_3(s) + CO(g) \rightarrow 2\ Fe_3O_4(s) + CO_2(g)$

The equation is balanced (1-C and 10-O atoms on each side).

b. $Fe_2O_4 + CO \rightarrow FeO + CO_2$ Balance Fe atoms first:

$Fe_3O_4(s) + CO(g) \rightarrow 3\ FeO(s) + CO_2(g)$

The equation is balanced (1-C and 5-O atoms on each side).

60. a. $3\ Ca(OH)_2(aq) + 2\ H_3PO_4(aq) \rightarrow 6\ H_2O(l) + Ca_3(PO_4)_2(s)$

b. $Al(OH)_3(s) + 3\ HCl(aq) \rightarrow AlCl_3(aq) + 3\ H_2O(l)$

c. $2\ AgNO_3(aq) + H_2SO_4(aq) \rightarrow Ag_2SO_4(s) + 2\ HNO_3(aq)$

61. a. $C_{12}H_{22}O_{11}(s) + 12\ O_2(g) \rightarrow 12\ CO_2(g) + 11\ H_2O(g)$

b. $C_6H_6(l) + \frac{15}{2}\ O_2(g) \rightarrow 6\ CO_2(g) + 3\ H_2O(g)$

or $2\ C_6H_6(l) + 15\ O_2(g) \rightarrow 12\ CO_2(g) + 6\ H_2O(g)$

c. $2\ Fe + \frac{3}{2}\ O_2 \rightarrow Fe_2O_3$

or $4\ Fe(s) + 3\ O_2(g) \rightarrow 2\ Fe_2O_3(s)$

d. $C_4H_{10} + \frac{13}{2}\ O_2 \rightarrow 4\ CO_2 + 5\ H_2O$

or $2\ C_4H_{10}(g) + 13\ O_2(g) \rightarrow 8\ CO_2(g) + 10\ H_2O(g)$

e. $2\ FeO(s) + \frac{1}{2}\ O_2(g) \rightarrow Fe_2O_3(s)$

or $4\ FeO(s) + O_2(g) \rightarrow 2\ Fe_2O_3(s)$

62. a. $16\ Cr(s) + 3\ S_8(s) \rightarrow 8\ Cr_2S_3(s)$

b. $2\ NaHCO_3(s) \rightarrow Na_2CO_3(s) + CO_2(g) + H_2O(g)$

c. $2\ KClO_3(s) \rightarrow 2\ KCl(s) + 3\ O_2(g)$

d. $2\ Eu(s) + 6\ HF(g) \rightarrow 2\ EuF_3(s) + 3\ H_2(g)$

63. a. $Cu(s) + 2\ AgNO_3(aq) \rightarrow 2\ Ag(s) + Cu(NO_3)_2(aq)$

b. $Zn(s) + 2\ HCl(aq) \rightarrow ZnCl_2(aq) + H_2(g)$

c. $Au_2S_3(s) + 3\ H_2(g) \rightarrow 2\ Au(s) + 3\ H_2S(g)$

d. $Ca(s) + 2\ H_2O(l) \rightarrow Ca(OH)_2(aq) + H_2(g)$

64. a. $HNO_3(l) + P_4O_{10}(s) \rightarrow (HPO_3)_3(l) + N_2O_5(g)$

Balance P atoms: $HNO_3 + 3\ P_4O_{10} \rightarrow 4\ (HPO_3)_3 + N_2O_5$

Balance H atoms: $12\ HNO_3 + 3\ P_4O_{10} \rightarrow 4\ (HPO_3)_3 + N_2O_5$

Balance N atoms: $12\ HNO_3 + 3\ P_4O_{10} \rightarrow 4\ (HPO_3)_3 + 6\ N_2O_5$

The oxygen atoms are also balanced.

$12\ HNO_3(l) + 3\ P_4O_{10}(s) \rightarrow 4\ (HPO_3)_3(l) + 6\ N_2O_5(g)$ Equation is balanced.

b. $Fe_2S_3(s) + HCl(g) \rightarrow FeCl_3(s) + H_2S(g)$

Balance Fe atoms: $Fe_2S_3 + HCl \rightarrow 2\ FeCl_3 + H_2S$

Balance S: $Fe_2S_3 + HCl \rightarrow 2\ FeCl_3 + 3\ H_2S$

There are 6 H and 6 Cl on right balance with 6 HCl on left:

$Fe_2S_3(s) + 6\ HCl(g) \rightarrow 2\ FeCl_3(s) + 3\ H_2S(g)$

c. $CS_2(l) + NH_3(g) \rightarrow H_2S(g) + NH_4SCN(s)$

C & S balanced; balance N

$CS_2 + 2\ NH_3 \rightarrow H_2S + NH_4SCN$

H is also balanced.

So: $CS_2(l) + 2\ NH_3(g) \rightarrow H_2S(g) + NH_4SCN(s)$

65. a. $SiO_2(s) + C(s) \rightarrow Si(s) + CO(g)$

Balance oxygen atoms

$SiO_2 + C \rightarrow Si + 2\ CO$

Balance carbon atoms

$SiO_2(s) + 2\ C(s) \rightarrow Si(s) + 2\ CO(g)$

b. $SiCl_4(l) + Mg(s) \rightarrow Si(s) + MgCl_2(s)$

Balance Cl atoms: $SiCl_4 + Mg \rightarrow Si + 2\ MgCl_2$

Balance Mg atoms: $SiCl_4(l) + 2\ Mg(s) \rightarrow Si(s) + 2\ MgCl_2(s)$

c. $Na_2SiF_6(s) + Na(s) \rightarrow Si(s) + NaF(s)$

Balance F atoms: $Na_2SiF_6 + Na \rightarrow Si + 6\ NaF$

Balance Na atoms: $Na_2SiF_6(s) + 4\ Na(s) \rightarrow Si(s) + 6\ NaF(s)$

66. Unbalanced equation:

$$CaF_2 \cdot 3Ca_3(PO_4)_2(s) + H_2SO_4(aq) \rightarrow H_3PO_4(aq) + HF(aq) + CaSO_4 \cdot 2H_2O(s)$$

Balancing Ca^{2+}, F^-, and PO_4^{3-}

$$CaF_2 \cdot 2Ca_3(PO_4)_2(s) + H_2SO_4(aq) \rightarrow 6\ H_3PO_4(aq) + 2\ HF(aq) + 10\ CaSO_4 \cdot 2H_2O(s)$$

On the right hand side there are 20 extra hydrogen atoms, 10 extra sulfates, and 20 extra water molecules. We can balance the hydrogen and sulfate with 10 sulfuric acid molecules. The extra waters came from the water in the sulfuric acid solution.

$CaF_2{\bullet}3Ca_3(PO_4)_2(s) + 10\ H_2SO_4(aq) + 20\ H_2O(l) \rightarrow 6\ H_3PO_4(aq) + 2\ HF(aq) + 10\ CaSO_4{\bullet}2\ H_2O(s)$

67. $Pb(NO_3)_2(aq) + H_3AsO_4(aq) \rightarrow PbHAsO_4(s) + 2\ HNO_3(aq)$

Note: The insecticide used is $PbHAsO_4$ and is commonly called lead arsenate. This is not the correct name, however. Correctly, lead arsenate would be $Pb_3(AsO_4)_2$ and $PbHAsO_4$ should be called lead hydrogen arsenate.

68. $2\ NaCl(aq) + 2\ H_2O(l) \rightarrow Cl_2(g) + H_2(g) + 2\ NaOH(aq)$

Reaction Stoichiometry

69. The balanced reaction is:

$(NH_4)_2Cr_2O_7(s) \rightarrow Cr_2O_3(s) + N_2(g) + 4\ H_2O(g)$

$$10.8\text{ g }(NH_4)_2Cr_2O_7 \times \frac{1\text{ mol }(NH_4)_2Cr_2O_7}{252.08\text{ g}} = 4.28 \times 10^{-2}\text{ mol }(NH_4)_2Cr_2O_7$$

$$4.28 \times 10^{-2}\text{ mol }(NH_4)_2Cr_2O_7 \times \frac{1\text{ mol }Cr_2O_3}{\text{mol }(NH_4)_2Cr_2O_7} \times \frac{152.00\text{ g }Cr_2O_3}{\text{mol }Cr_2O_3} = 6.51\text{ g }Cr_2O_3$$

$$4.28 \times 10^{-2}\text{ mol }(NH_4)_2Cr_2O_7 \times \frac{1\text{ mol }N_2}{\text{mol }(NH_4)_2Cr_2O_7} \times \frac{28.02\text{ g }N_2}{\text{mol }N_2} = 1.20\text{ g }N_2$$

$$4.28 \times 10^{-2}\text{ mol }(NH_4)_2Cr_2O_7 \times \frac{4\text{ mol }H_2O}{\text{mol }(NH_4)_2Cr_2O_7} \times \frac{18.02\text{ g }H_2O}{\text{mol }H_2O} = 3.09\text{ g }H_2O$$

70. $Fe_2O_3 + 2\ Al \rightarrow 2\ Fe + Al_2O_3$

$$15.0\text{ g Fe} \times \frac{1\text{ mol Fe}}{55.85\text{ g Fe}} = 0.269\text{ mol Fe}$$

$$0.269\text{ mol Fe} \times \frac{2\text{ mol Al}}{2\text{ mol Fe}} \times \frac{26.98\text{ g Al}}{\text{mol Al}} = 7.26\text{ g Al}$$

$$0.269\text{ mol Fe} \times \frac{1\text{ mol }Fe_2O_3}{2\text{ mol Fe}} \times \frac{159.70\text{ g }Fe_2O_3}{\text{mol }Fe_2O_3} = 21.5\text{ g }Fe_2O_3$$

$$0.269\text{ mol Fe} \times \frac{1\text{ mol }Al_2O_3}{2\text{ mol Fe}} \times \frac{101.96\text{ g }Al_2O_3}{\text{mol }Al_2O_3} = 13.7\text{ g }Al_2O_3$$

71. $$1.0\text{ kg Al} \times \frac{1000\text{ g Al}}{\text{kg Al}} \times \frac{1\text{ mol Al}}{26.98\text{ g Al}} \times \frac{3\text{ mol }NH_4ClO_4}{3\text{ mol Al}} \times \frac{117.49\text{ g }NH_4ClO_4}{\text{mol }NH_4ClO_4}$$

= 4355 g or 4.4 kg

72. $1.0 \times 10^6 \text{ kg HNO}_3 \times \frac{1000 \text{ g HNO}_3}{\text{kg HNO}_3} \times \frac{1 \text{ mol HNO}_3}{63.02 \text{ g HNO}_3} = 1.6 \times 10^7 \text{ mol HNO}_3$

We need to get the relationship between moles of HNO_3 and moles of NH_3. We have to use all 3 equations.

$$\frac{2 \text{ mol HNO}_3}{3 \text{ mol NO}_2} \times \frac{2 \text{ mol NO}_2}{2 \text{ mol NO}} \times \frac{4 \text{ mol NO}}{4 \text{ mol NH}_3} = \frac{8 \text{ mol HNO}_3}{12 \text{ mol NH}_3}$$

Thus, we can produce 8 mol HNO_3 for every 12 mol NH_3 that we begin with.

$$1.6 \times 10^7 \text{ mol HNO}_3 \times \frac{12 \text{ mol NH}_3}{8 \text{ mol HNO}_3} \times \frac{17.03 \text{ g NH}_3}{\text{mol NH}_3} = 4.1 \times 10^8 \text{ g} \quad \text{or } 4.1 \times 10^5 \text{ kg}$$

This is an oversimplified answer. In practice the NO produced in the 3rd step is recycled back into the process in the second step.

73. $1.0 \times 10^2 \text{ g Ca}_3(\text{PO}_4)_2 \times \frac{1 \text{ mol Ca}_3(\text{PO}_4)_2}{310.18 \text{ g Ca}_3(\text{PO}_4)_2} \times \frac{3 \text{ mol H}_2\text{SO}_4}{\text{mol Ca}_3(\text{PO}_4)_2} \times \frac{98.09 \text{ g H}_2\text{SO}_4}{\text{mol H}_2\text{SO}_4}$

$= 95 \text{ g H}_2\text{SO}_4$ are needed

$$95 \text{ g H}_2\text{SO}_4 \times \frac{100 \text{ g concentrated reagent}}{98 \text{ g H}_2\text{SO}_4} = 97 \text{ g of concentrated sulfuric acid}$$

74. a. $\text{Ba(OH)}_2\text{•8H}_2\text{O(s)} + 2\text{NH}_4\text{SCN(s)} \longrightarrow \text{Ba(SCN)}_2\text{(s)} + 10 \text{ H}_2\text{O(l)} + 2 \text{ NH}_3\text{(g)}$

b. $6.5 \text{ g Ba(OH)}_2\text{•8H}_2\text{O} \times \frac{1 \text{ mol Ba(OH)}_2\text{•8H}_2\text{O}}{315.4 \text{ g}} = 0.0206 \text{ mol} = 0.021 \text{ mol}$

$$0.021 \text{ mol Ba(OH)}_2\text{•8H}_2\text{O} \times \frac{2 \text{ mol NH}_4\text{SCN}}{1 \text{ mol Ba(OH)}_2\text{•8H}_2\text{O}} \times \frac{76.13 \text{ g NH}_4\text{SCN}}{\text{mol NH}_4\text{SCN}} = 3.2 \text{ g NH}_4\text{SCN}$$

c. $0.021 \text{ mol Ba(OH)}_2\text{•8H}_2\text{O} \times \frac{1 \text{ mol Ba(SCN)}_2}{1 \text{ mol Ba(OH)}_2\text{•8H}_2\text{O}} \times \frac{253.5 \text{ g Ba(SCN)}_2}{\text{mol Ba(SCN)}_2} = 5.3 \text{ g Ba(SCN)}_2$

$$0.021 \text{ mol Ba(OH)}_2\text{•8H}_2\text{O} \times \frac{10 \text{ mol H}_2\text{O}}{\text{mol Ba(OH)}_2\text{•8H}_2\text{O}} \times \frac{18.02 \text{ g H}_2\text{O}}{\text{mol H}_2\text{O}} = 3.8 \text{ g H}_2\text{O}$$

$$0.021 \text{ mol Ba(OH)}_2\text{•8H}_2\text{O} \times \frac{2 \text{ mol NH}_3}{\text{mol Ba(OH)}_2\text{•8H}_2\text{O}} \times \frac{17.03 \text{ g NH}_3}{\text{mol NH}_3} = 0.72 \text{ g NH}_3$$

75. a. $1.0 \times 10^2 \text{ mg NaHCO}_3 \times \frac{1 \text{ g}}{1000 \text{ mg}} \times \frac{1 \text{ mol NaHCO}_3}{84.01 \text{ g NaHCO}_3} \times \frac{1 \text{ mol C}_6\text{H}_8\text{O}_7}{3 \text{ mol NaHCO}_3}$

$$\times \frac{192.12 \text{ g C}_6\text{H}_8\text{O}_7}{\text{mol C}_6\text{H}_8\text{O}_7} = 0.076 \text{ g} \quad \text{or } 76 \text{ mg C}_6\text{H}_8\text{O}_7$$

b. $0.10 \text{ g NaHCO}_3 \times \frac{1 \text{ mol NaHCO}_3}{84.01 \text{ g NaHCO}_3} \times \frac{3 \text{ mol CO}_2}{3 \text{ mol NaHCO}_3} \times \frac{44.01 \text{ g CO}_2}{\text{mol CO}_2}$

$= 0.052 \text{ g or } 52 \text{ mg CO}_2$

76. a. $1.00 \times 10^2 \text{ g } C_7H_6O_3 \times \frac{1 \text{ mol } C_7H_6O_3}{138.12 \text{ g } C_7H_6O_3} \times \frac{1 \text{ mol } C_4H_6O_3}{1 \text{ mol } C_7H_6O_3} \times \frac{102.09 \text{ g } C_4H_6O_3}{\text{mol } C_4H_6O_3}$

$= 73.9 \text{ g } C_4H_6O_3$

b. $1.00 \times 10^2 \text{ g } C_7H_6O_3 \times \frac{1 \text{ mol } C_7H_6O_3}{138.12 \text{ g } C_7H_6O_3} \times \frac{1 \text{ mol } C_9H_8O_4}{1 \text{ mol } C_7H_6O_3} \times \frac{180.15 \text{ g } C_9H_8O_4}{\text{mol } C_9H_8O_4}$

$= 1.30 \times 10^2$ g aspirin

Limiting Reactants and Percent Yield

77. a. $Mg(s) + I_2(s) \rightarrow MgI_2(s)$

100 molecules of I_2 reacts completely with 100 atoms of Mg. We have a stoichiometric mixture. Neither is limiting.

b. $150 \text{ atoms Mg} \times \frac{1 \text{ molecule } I_2}{1 \text{ atom Mg}} = 150$ molecules I_2 needed.

We need 150 molecules I_2 to react completely with 150 atoms Mg; we only have 100. I_2 is limiting.

c. $200 \text{ atoms Mg} \times \frac{1 \text{ molecule } I_2}{1 \text{ atom Mg}} = 200$ molecules I_2. Mg is limiting.

d. $0.16 \text{ mol Mg} \times \frac{1 \text{ mol } I_2}{1 \text{ mol Mg}} = 0.16$ mol I_2 needed. Mg is limiting.

e. $0.14 \text{ mol Mg} \times \frac{1 \text{ mol } I_2}{1 \text{ mol Mg}} = 0.14$ mol I_2 needed.

Stoichiometric mixture. Neither is limiting.

f. $0.12 \text{ mol Mg} \times \frac{1 \text{ mol } I_2}{1 \text{ mol Mg}} = 0.12$ mol I_2 needed. I_2 is limiting.

g. $6.078 \text{ g Mg} \times \frac{1 \text{ mol Mg}}{24.31 \text{ g Mg}} \times \frac{1 \text{ mol } I_2}{1 \text{ mol Mg}} \times \frac{253.8 \text{ g } I_2}{\text{mol } I_2} = 63.46 \text{ g } I_2$.

Stoichiometric mixture. Neither is limiting.

h. $1.00 \text{ g Mg} \times \frac{1 \text{ mol Mg}}{24.31 \text{ g Mg}} \times \frac{1 \text{ mol } I_2}{1 \text{ mol Mg}} \times \frac{253.8 \text{ g } I_2}{\text{mol } I_2} = 10.4$ g I_2 needed.

10.4 g I_2 needed, we only have 2.00 g. I_2 is limiting.

i. From part h. above, we calculated that 10.4 g I_2 will react completely with 1.00 g Mg. We have 20.00 g I_2. I_2 is in excess. Mg is limiting.

78. $2\ H_2(g) + O_2(g) \longrightarrow 2\ H_2O(g)$

a. $50 \text{ molecules } H_2 \times \dfrac{1 \text{ molecule } O_2}{2 \text{ molecules } H_2} = 25 \text{ molecules } O_2.$

Stoichiometric mixture. Neither is limiting.

b. $100 \text{ molecules } H_2 \times \dfrac{1 \text{ molecule } O_2}{2 \text{ molecules } H_2} = 50 \text{ molecules } O_2.$ O_2 is limiting.

c. From b, 50 molecules of O_2 will react completely with 100 molecules of H_2. We have 100 molecules (an excess) of O_2. So, H_2 is limiting.

d. $0.50 \text{ mol } H_2 \times \dfrac{1 \text{ mol } O_2}{2 \text{ mol } H_2} = 0.25 \text{ mol } O_2.$ H_2 is limiting.

e. $0.80 \text{ mol } H_2 \times \dfrac{1 \text{ mol } O_2}{2 \text{ mol } H_2} = 0.40 \text{ mol } O_2.$ H_2 is limiting.

f. $1.0 \text{ g } H_2 \times \dfrac{1 \text{ mol } H_2}{2.016 \text{ g } H_2} \times \dfrac{1 \text{ mol } O_2}{2 \text{ mol } H_2} = 0.25 \text{ mol } O_2.$

Stoichiometric mixture, neither is limiting.

g. $5.00 \text{ g } H_2 \times \dfrac{1 \text{ mol } H_2}{2.016 \text{ g } H_2} \times \dfrac{1 \text{ mol } O_2}{2 \text{ mol } H_2} \times \dfrac{32.00 \text{ g } O_2}{\text{mol } O_2} = 39.7 \text{ g } O_2.$ H_2 is limiting.

79. $2\ Cu(s) + S(s) \longrightarrow Cu_2S(s)$

$1.50 \text{ g Cu} \times \dfrac{1 \text{ mol Cu}}{63.55 \text{ g Cu}} \times \dfrac{1 \text{ mol } Cu_2S}{2 \text{ mol Cu}} \times \dfrac{159.17 \text{ g } Cu_2S}{\text{mol } Cu_2S} = 1.88 \text{ g } Cu_2S$ is theoretical yield.

$\% \text{ yield} = \dfrac{\text{actual yield}}{\text{theoretical yield}} \times 100 = \dfrac{1.76 \text{ g}}{1.88 \text{ g}} \times 100 = 93.6\%$

80. $6.0 \text{ g Al} \times \dfrac{1 \text{ mol Al}}{26.98 \text{ mol Al}} \times \dfrac{2 \text{ mol } AlBr_3}{2 \text{ mol Al}} \times \dfrac{266.68 \text{ g } AlBr_3}{\text{mol } AlBr_3} = 59 \text{ g } AlBr_3$

$\% \text{ Yield} = \dfrac{50.3 \text{ g}}{59 \text{ g}} \times 100 = 85\%.$

81. a. $2.0 \text{ g Ag} \times \dfrac{1 \text{ mol Ag}}{107.9 \text{ g Ag}} \times \dfrac{8 \text{ mol } Ag_2S}{16 \text{ mol Ag}} = 9.3 \times 10^{-3} \text{ mol } Ag_2S$

$2.0 \text{ g } S_8 \times \dfrac{1 \text{ mol } S_8}{256.56 \text{ g } S_8} \times \dfrac{8 \text{ mol } Ag_2S}{\text{mol } S_8} = 6.2 \times 10^{-2} \text{ mol } Ag_2S$

When 9.3×10^{-3} mol Ag_2S is formed all of the Ag will be consumed. Ag is limiting.

$9.3 \times 10^{-3} \text{ mol } Ag_2S \times 247.9 \text{ g/mol} = 2.3 \text{ g } Ag_2S$

b. S_8 is left unreacted. 0.3 g S (2.3 g Ag_2S - 2.0 g Ag) is required to make 2.3 g of Ag_2S. Therefore, 1.7 g S_8 remains.

82. a. $1.00 \times 10^3 \text{ g } N_2 \times \frac{1 \text{ mol } N_2}{28.02 \text{ g } N_2} \times \frac{2 \text{ mol } NH_3}{\text{mol } N_2} = 71.4 \text{ mol } NH_3$

$5.00 \times 10^2 \text{ g } H_2 \times \frac{1 \text{ mol } H_2}{2.016 \text{ g } H_2} \times \frac{2 \text{ mol } NH_3}{3 \text{ mol } H_2} = 165 \text{ mol } NH_3$

Therefore, N_2 is limiting. The mass of NH_3 produced is:

$71.4 \text{ mol } NH_3 \times \frac{17.03 \text{ g } NH_3}{\text{mol } NH_3} = 1220 \text{ g } NH_3$

b. 220 g H_2 required to produce the 1220 g NH_3 (contains 1.00×10^3 g N).

Therefore, 500. - 220 = 280 g H_2 is left unreacted. If we carry all calculations to the nearest gram (one extra significant figure for ammonia), we get 1216 g NH_3 from 1.00×10^3 g N_2, requiring 216 g H_2, leaving 284 g H_2 unreacted.

83. a. $C_8H_4O_3 + C_6H_6 \rightarrow C_{14}H_8O_2 + H_2O$

$2.00 \times 10^3 \text{ g } C_8H_4O_3 \times \frac{1 \text{ mol } C_8H_4O_3}{148.11 \text{ g } C_8H_4O_3} \times \frac{1 \text{ mol } C_6H_6}{\text{mol } C_8H_8O_3} \times \frac{78.11 \text{ g } C_6H_6}{\text{mol } C_6H_6} = 1050 \text{ g } C_6H_6$

b. $2.00 \times 10^3 \text{ g } C_8H_4O_3 \times \frac{1 \text{ mol } C_8H_4O_3}{148.11 \text{ g } C_8H_4O_3} \times \frac{1 \text{ mol } C_{14}H_8O_2}{\text{mol } C_8H_4O_3} \times \frac{208.20 \text{ g } C_{14}H_8O_2}{\text{mol } C_{14}H_8O_2}$

$= 2810 \text{ g } C_{14}H_8O_2$

$2.00 \times 10^3 \text{ g } C_8H_4O_3 \times \frac{1 \text{ mol } C_8H_4O_3}{148.11 \text{ g } C_8H_4O_3} \times \frac{1 \text{ mol } H_2O}{\text{mol } C_8H_4O_3} \times \frac{18.02 \text{ g } H_2O}{\text{mol } H_2O} = 243 \text{ g } H_2O$

c. $\%\text{Yield} = \frac{\text{Actual Yield}}{\text{Theoretical Yield}} \times 100 = \frac{1960 \text{ g}}{2810 \text{ g}} \times 100 = 69.8\%$

84. $C_7H_6O_3 + C_4H_6O_3 \rightarrow C_9H_8O_4 + C_2H_4O_2$

(SA) (AA) (ASA)

$1.50 \text{ g SA} \times \frac{1 \text{ mol SA}}{138.12 \text{ g SA}} \times \frac{1 \text{ mol ASA}}{1 \text{ mol SA}} = 1.09 \times 10^{-2} \text{ mol ASA}$

$2.00 \text{ g AA} \times \frac{1 \text{ mol AA}}{102.09 \text{ g AA}} \times \frac{1 \text{ mol ASA}}{1 \text{ mol AA}} = 1.96 \times 10^{-2} \text{ mol ASA}$

Therefore salicylic acid, $C_7H_6O_3$, is the limiting reagent. The theoretical yield is:

$1.09 \times 10^{-2} \text{ mol ASA} \times \frac{180.15 \text{ g ASA}}{\text{mol ASA}} = 1.96 \text{ g aspirin}$ $\%\text{Yield} = \frac{1.50 \text{ g}}{1.96 \text{ g}} \times 100 = 76.5\%$

ADDITIONAL EXERCISES

85. a. $Cu(H_2O)_6Cl_2$; 1(63.55) + 12(1.008 g) + 6(16.00) + 2(35.45) = 242.55 g/mol

b. $NaBrO_3$; 1(22.99) + 1(79.90) + 3(16.00) = 150.89 g/mol

c. $(C_2F_4)_{500}$; $1000(12.01) + 2000(19.00) = 5 \times 10^4$ g/mol

d. H_3PO_4; $3(1.008) + 1(30.97) + 4(16.00) = 97.99$ g/mol

e. $CaCO_3$; $1(40.08) + 1(12.01) + 3(16.00) = 100.09$ g/mol

86. a. $2(12.01) + 3(1.008) + 3(35.45) + 2(16.00) = 165.39$ g/mol

b. $500.0 \text{ g} \times \frac{1 \text{ mol}}{165.39 \text{ g}} = 3.023 \text{ mol}$

c. $2.0 \times 10^{-2} \text{ mol} \times \frac{165.39 \text{ g}}{\text{mol}} = 3.3 \text{ g}$

d. $5.0 \text{ g } C_2H_3Cl_3O_2 \times \frac{1 \text{ mol}}{165.39 \text{ g}} \times \frac{6.02 \times 10^{23} \text{ molecules}}{\text{mol}} \times \frac{3 \text{ atoms Cl}}{\text{molecule}}$

$= 5.5 \times 10^{22}$ atoms of chlorine

e. $1.0 \text{ g Cl} \times \frac{1 \text{ mol Cl}}{35.45 \text{ g}} \times \frac{1 \text{ mol } C_2H_3Cl_3O_2}{3 \text{ mol Cl}} \times \frac{165.39 \text{ g } C_2H_3Cl_3O_2}{\text{mol } C_2H_3Cl_3O_2} = 1.6$ g chloral hydrate

f. $500 \text{ molecules} \times \frac{1 \text{ mol}}{6.022 \times 10^{23} \text{ molecules}} \times \frac{165.39 \text{ g}}{\text{mol}} = 1.373 \times 10^{-19} \text{ g}$

87. Mass of repeating unit is:

$$8 \text{ mol C}\left(\frac{12 \text{ g}}{\text{mol C}}\right) + 14 \text{ mol H}\left(\frac{1 \text{ g}}{\text{mol H}}\right) + 2 \text{ mol O}\left(\frac{16 \text{ g}}{\text{mol O}}\right) \approx 142 \text{ g}$$

$$\frac{100{,}000}{142} = 704 \text{ or about 700 monomer units}$$

88. a. $In + O_2 \rightarrow In_2O_3$

$4 In(s) + 3 O_2(g) \rightarrow 2 In_2O_3(s)$

b. $C_6H_{12}O_6 \rightarrow C_2H_5OH + CO_2$

Balance C-atoms first. 6-C on left, 3-C on right.

First, try multiplying both products by 2:

$C_6H_{12}O_6 \rightarrow 2 C_2H_5OH + 2 CO_2$

O and H are also balanced, and the balanced equation is:

$C_6H_{12}O_6(aq) \rightarrow 2 C_2H_5OH(aq) + 2 CO_2(g)$

c. $K + H_2O \rightarrow KOH + H_2$

$2 K(s) + 2 H_2O(l) \rightarrow 2 KOH(aq) + H_2(g)$

d. $BaO + HNO_3 \rightarrow Ba(NO_3)_2 + H_2O$

$BaO(s) + 2 HNO_3(aq) \rightarrow Ba(NO_3)_2(aq) + H_2O(l)$

89. $C_8H_{18}(l) + \frac{25}{2} O_2(g) \longrightarrow 8\ CO_2(g) + 9\ H_2O(g)$

or $2\ C_8H_{18}(l) + 25\ O_2(g) \longrightarrow 16\ CO_2(g) + 18\ H_2O(g)$

$$1.2 \times 10^{10} \text{ gallon} \times \frac{4 \text{ qt}}{\text{gal}} \times \frac{946 \text{ mL}}{\text{qt}} \times \frac{0.692 \text{ g}}{\text{mL}} = 3.1 \times 10^{13} \text{ g of gasoline}$$

$$3.1 \times 10^{13} \text{ g } C_8H_{18} \times \frac{1 \text{ mol } C_8H_{18}}{114.22 \text{ g } C_8H_{18}} \times \frac{16 \text{ mol } CO_2}{2 \text{ mol } C_8H_{18}} \times \frac{44.01 \text{ g } CO_2}{\text{mol } CO_2} = 9.6 \times 10^{13} \text{ g } CO_2$$

90. Let's first get the elemental composition of the confiscated substance:

$$150.0 \text{ mg } CO_2 \times \frac{12.01 \text{ mg C}}{44.01 \text{ mg } CO_2} = 40.93 \text{ mg C}$$

$$\%C = \frac{40.93 \text{ mg}}{50.86 \text{ mg}} \times 100 = 80.48\% \text{ C}$$

$$46.05 \text{ mg } H_2O \times \frac{2.016 \text{ mg H}}{18.02 \text{ mg } H_2O} = 5.152 \text{ mg H}$$

$$\%H = \frac{5.152 \text{ mg}}{50.86 \text{ mg}} \times 100 = 10.13\% \text{ H and } \%N = 9.39\%$$

$$\%O = 100 - (80.48 + 10.13 + 9.39) = 0.00\%$$

Composition of cocaine: $C_{17}H_{21}NO_4$

Molar mass of cocaine = 17(12.01) + 21(1.008) + 1(14.01) + 4(16.00) = 303.35 g/mol

$$\%C = \frac{204.17 \text{ g}}{303.35 \text{ g}} \times 100 = 67.31\% \qquad \%N = \frac{14.01 \text{ g}}{303.35 \text{ g}} \times 100 = 4.618\%$$

$$\%H = \frac{21.168 \text{ g}}{303.35 \text{ g}} \times 100 = 6.978\% \qquad \%O = \frac{64.00 \text{ g}}{303.35 \text{ g}} \times 100 = 21.10\%$$

Obviously, the composition by mass is not the same. The chemist can conclude the compound is not cocaine, assuming he analyzed a pure substance.

91. $156.8 \text{ mg } CO_2 \times \frac{12.01 \text{ mg C}}{44.01 \text{ mg } CO_2} = 42.79 \text{ mg C}$

$$42.8 \text{ mg } H_2O \times \frac{2.016 \text{ mg H}}{18.02 \text{ mg } H_2O} = 4.79 \text{ mg H}$$

$$\%C = \frac{42.79 \text{ mg}}{47.6 \text{ mg}} \times 100 = 89.9\% \text{ C}; \quad \%H = 100.0 - 89.9 = 10.1\% \text{ H}$$

Out of 100.0 g Cumene, we have

$$89.9 \text{ g C} \times \frac{1 \text{ mol C}}{12.01 \text{ g C}} = 7.49 \text{ mol C}$$

$$10.1 \text{ g H} \times \frac{1 \text{ mol H}}{1.008 \text{ g H}} = 10.0 \text{ mol H}$$

$$\frac{7.49}{10.0} = 0.749 \approx 0.75 \text{ which is a 3:4 ratio}$$

Empirical formula: C_3H_4

Mass of one empirical formula ≈ 3(12) + 4 = 40

So molecular formula is$(C_3H_4)_3$ or C_9H_{12}

92. Since 4.784 g In combines with 1.000 g O, then out of 5.784 g of the oxide there will be 4.784 g In and 1.000 g O.

$$\%In = \frac{4.784\text{ g}}{5.784\text{ g}} \times 100 = 82.71\%\text{ In};\ \ \%O = 100 - 82.71 = 17.29\%\text{ O}$$

Assume In_2O_3 is the formula and that we know that the atomic weight of oxygen is 16.00. This was known at the time.

Out of 100 g compound:

$$17.29\text{ g O} \times \frac{1\text{ mol O}}{16.00\text{ g O}} \times \frac{2\text{ mol In}}{3\text{ mol O}} = 0.7204\text{ mol In}$$

So, 0.7204 mol In has a mass of 82.71 g, or the atomic mass is:

$$\frac{82.71\text{ g}}{0.7204\text{ mol}} = 114.8\text{ g/mol}$$

which is in good agreement with the modern value.

If the formula is InO:

$$17.29\text{ g O} \times \frac{1\text{ mol O}}{16.00\text{ g O}} \times \frac{1\text{ mol In}}{\text{mol O}} = 1.081\text{ mol In}$$

Then, 1.081 mol of In has a mass of 82.71 g or an atomic mass of $\frac{82.71\text{ g}}{1.081\text{ mol}} = 76.51$ g/mol

Obviously, Mendeleev was correct.

93. Out of 100 g of compound there are:

$$83.53\text{ g Sb} \times \frac{1\text{ mol Sb}}{121.8\text{ g Sb}} = 0.6858\text{ mol Sb};\qquad 16.47\text{ g O} \times \frac{1\text{ mol O}}{16.00\text{ g O}} = 1.029\text{ mol O}$$

$\frac{0.6858}{1.029} = 0.6665 \approx \frac{2}{3}$; Empirical formula: Sb_2O_3, Empirical weight = 291.6

Mass of Sb_4O_6 is 583.2, which is in the correct range. Molecular formula: Sb_4O_6.

94. a. Mass of Zn in alloy; $0.0985\text{ g ZnCl}_2 \times \frac{65.38\text{ g Zn}}{136.28\text{ g ZnCl}_2} = 0.0473\text{ g Zn}$

$$\%Zn = \frac{0.0473\text{ g Zn}}{0.5065\text{ g brass}} \times 100 = 9.34\%\text{ Zn};\ \ \%Cu = 100 - 9.34 = 90.66\%\text{ Cu}$$

b. The Cu remains unreacted. After filtering, washing, and drying, the mass of the unreacted copper could be measured.

95. $41.98 \text{ mg } CO_2 \times \frac{12.01 \text{ mg C}}{44.01 \text{ mg } CO_2} = 11.46 \text{ mg C}$ $\%C = \frac{11.46 \text{ mg}}{19.81 \text{ mg}} \times 100 = 57.85\% \text{ C}$

$6.45 \text{ mg } H_2O \times \frac{2.016 \text{ mg H}}{18.02 \text{ mg } H_2O} = 0.722 \text{ mg H}$ $\%H = \frac{0.722 \text{ mg}}{19.81 \text{ mg}} \times 100 = 3.64\% \text{ H}$

$\%O = 100.00 - (57.85 + 3.64) = 38.51\% \text{ O}$

Out of 100.00 g terephthalic acid, there are:

$57.85 \text{ g C} \times \frac{1 \text{ mol C}}{12.01 \text{ g C}} = 4.817 \text{ mol C}$

$3.64 \text{ g H} \times \frac{1 \text{ mol H}}{1.008 \text{ g H}} = 3.61 \text{ mol H}$

$38.51 \text{ g O} \times \frac{1 \text{ mol O}}{16.00 \text{ g O}} = 2.407 \text{ mol O}$

$\frac{4.817}{2.407} = 2, \frac{3.61}{2.407} = 1.5, \frac{2.407}{2.407} = 1$

C:H:O ratio is 2:1.5:1 or 4:3:2

Empirical formula: $C_4H_3O_2$ Mass of $C_4H_3O_2 \approx 4(12) + 3(1) + 2(16) = 83$

$\frac{166}{83} = 2$; Molecular formula: $C_8H_6O_4$

96. a. $Pt(NH_3)_2Cl_2$

molar mass = 195.1 + 2(14.01) + 6(1.008) + 2(35.45) = 300.1 g/mol

$\%Pt = \frac{195.1 \text{ g}}{300.1 \text{ g}} \times 100 = 65.01\% \text{ Pt}$

$\%N = \frac{28.02 \text{ g}}{300.1 \text{ g}} \times 100 = 9.335\% \text{ N}$

$\%H = \frac{6.048 \text{ g}}{300.1 \text{ g}} \times 100 = 2.015\% \text{ H}$

$\%Cl = \frac{70.90 \text{ g}}{300.1 \text{ g}} \times 100 = 23.63\% \text{ Cl}$

65.01% Pt; 9.335% N; 2.015% H; 23.63% Cl

b. $65 \text{ g } K_2PtCl_4 \times \frac{1 \text{ mol } K_2PtCl_4}{415.1 \text{ g } K_2PtCl_4} \times \frac{1 \text{ mol cisplatin}}{\text{mol } K_2PtCl_4} \times \frac{300.1 \text{ g cisplatin}}{\text{mol cisplatin}} = 47 \text{ g cisplatin}$

$65 \text{ g } K_2PtCl_4 \times \frac{1 \text{ mol } K_2PtCl_4}{415.1 \text{ g } K_2PtCl_4} \times \frac{2 \text{ mol KCl}}{\text{mol } K_2PtCl_4} \times \frac{74.55 \text{ g KCl}}{\text{mol KCl}} = 23 \text{ g KCl}$

97. $4 \text{ Al} + 3 \text{ O}_2 \rightarrow 2 \text{ Al}_2\text{O}_3$

a. $1.0 \text{ mol Al} \times \frac{3 \text{ mol } O_2}{4 \text{ mol Al}} = 0.75 \text{ mol } O_2$; Al is limiting.

b. $2.0 \text{ mol Al} \times \frac{3 \text{ mol } O_2}{4 \text{ mol Al}} = 1.5 \text{ mol } O_2$; Al is limiting.

c. $0.50 \text{ mol Al} \times \frac{3 \text{ mol } O_2}{4 \text{ mol Al}} = 0.38 \text{ mol } O_2$; Al is limiting.

d. $64.75 \text{ g Al} \times \frac{1 \text{ mol Al}}{26.98 \text{ g Al}} \times \frac{3 \text{ mol } O_2}{4 \text{ mol Al}} \times \frac{32.00 \text{ g } O_2}{\text{mol } O_2} = 57.60 \text{ g } O_2$; Al is limiting.

e. $75.89 \text{ g Al} \times \frac{1 \text{ mol Al}}{26.98 \text{ g Al}} \times \frac{3 \text{ mol } O_2}{4 \text{ mol Al}} \times \frac{32.00 \text{ g } O_2}{\text{mol } O_2} = 67.51 \text{ g } O_2$; Al is limiting.

f. $51.28 \text{ g Al} \times \frac{1 \text{ mol Al}}{26.98 \text{ g Al}} \times \frac{3 \text{ mol } O_2}{4 \text{ mol Al}} \times \frac{32.00 \text{ g } O_2}{\text{mol } O_2} = 45.62 \text{ g } O_2$; Al is limiting.

98. $1.0 \times 10^3 \text{ g } Br_2 \times \frac{1 \text{ mol } Br_2}{159.80 \text{ g } Br_2} \times \frac{1 \text{ mol } Cl_2}{\text{mol } Br_2} \times \frac{70.90 \text{ g } Cl_2}{\text{mol } Cl_2} = 4.4 \times 10^2 \text{ g } Cl_2$

99. $2.00 \times 10^6 \text{ g } CaCO_3 \times \frac{1 \text{ mol } CaCO_3}{100.09 \text{ g } CaCO_3} \times \frac{1 \text{ mol CaO}}{\text{mol } CaCO_3} \times \frac{56.08 \text{ g CaO}}{\text{mol CaO}} = 1.12 \times 10^6 \text{ g CaO}$

100. $Ca_3(PO_4)_2 + 3\ H_2SO_4 \longrightarrow 3\ CaSO_4 + 2\ H_3PO_4$

$1.0 \times 10^3 \text{ g } Ca_3(PO_4)_2 \times \frac{1 \text{ mol } Ca_3(PO_4)_2}{310.18 \text{ g } Ca_3(PO_4)_2} \times \frac{3 \text{ mol } CaSO_4}{\text{mol } Ca_3(PO_4)_2} = 9.7 \text{ mol } CaSO_4$

$1.0 \times 10^3 \text{ g con } H_2SO_4 \times \frac{98 \text{ g } H_2SO_4}{100 \text{ g con } H_2SO_4} = 980 \text{ g } H_2SO_4$

$980 \text{ g } H_2SO_4 \times \frac{1 \text{ mol } H_2SO_4}{98.09 \text{ g } H_2SO_4} \times \frac{3 \text{ mol } CaSO_4}{3 \text{ mol } H_2SO_4} = 10.\text{ mol } CaSO_4$

The calcium phosphate is the limiting reagent.

$9.7 \text{ mol } CaSO_4 \times \frac{136.15 \text{ g } CaSO_4}{\text{mol } CaSO_4} = 1300 \text{ g } CaSO_4$

$9.7 \text{ mol } CaSO_4 \times \frac{2 \text{ mol } H_3PO_4}{3 \text{ mol } CaSO_4} \times \frac{97.99 \text{ g } H_3PO_4}{\text{mol } H_3PO_4} = 630 \text{ g } H_3PO_4$

101. Mass of Ni in sample:

$98.4 \text{ g Ni(CO)}_4 \times \frac{58.69 \text{ g Ni}}{170.73 \text{ g Ni(CO)}_4} = 33.8 \text{ g Ni}$

$\%Ni = \frac{33.8 \text{ g}}{94.2 \text{ g}} = 35.9\% \text{ Ni}$

102. $Hg + Br_2 \longrightarrow HgBr_2$

a. $10.0 \text{ g Hg} \times \frac{1 \text{ mol Hg}}{200.6 \text{ g Hg}} \times \frac{1 \text{ mol } HgBr_2}{\text{mol Hg}} = 4.99 \times 10^{-2} \text{ mol } HgBr_2$

$9.00 \text{ g } Br_2 \times \frac{1 \text{ mol } Br_2}{159.80 \text{ } Br_2} \times \frac{1 \text{ mol } HgBr_2}{\text{mol } Br_2} = 5.63 \times 10^{-2} \text{ mol } HgBr_2$

Hg is limiting. $HgBr_2$ produced is:

$$4.99 \times 10^{-2} \text{ mol } HgBr_2 \times \frac{360.4 \text{ g } HgBr_2}{\text{mol } HgBr_2} = 17.98 \text{ g} \approx 18.0 \text{ g } HgBr_2$$

18.0 g $HgBr_2$ with 1.0 g Br_2 left unreacted.

b. $$5.00 \text{ mL Hg} \times \frac{13.5 \text{ g Hg}}{\text{mL}} = 67.5 \text{ g Hg}$$

$$5.00 \text{ mL } Br_2 \times \frac{3.2 \text{ g } Br_2}{\text{mL } Br_2} = 16 \text{ g } Br_2$$

$$67.5 \text{ g Hg} \times \frac{1 \text{ mol Hg}}{200.6 \text{ g Hg}} \times \frac{1 \text{ mol } HgBr_2}{\text{mol Hg}} = 0.336 \text{ mol } HgBr_2$$

$$16 \text{ g } Br_2 \times \frac{1 \text{ mol } Br_2}{159.80 \text{ g } Br_2} \times \frac{1 \text{ mol } HgBr_2}{\text{mol } Br_2} = 0.10 \text{ mol } HgBr_2$$

Br_2 is limiting.

$$0.10 \text{ mol } HgBr_2 \times \frac{360.4 \text{ g } HgBr_2}{\text{mol } HgBr_2} = 36 \text{ g } HgBr_2$$

103. $C_6H_{10}O_4 + 2\ NH_3 + 4\ H_2 \rightarrow C_6H_{16}N_2 + 4\ H_2O$

Adip. (Adipic acid) HMD

a. $$1.00 \times 10^3 \text{ g Adip.} \times \frac{1 \text{ mol Adip.}}{146.14 \text{ g Adip}} \times \frac{1 \text{ mol HMD}}{\text{mol Adip.}} \times \frac{116.21 \text{ g HMD}}{\text{mol HMD}} = 795 \text{ g HMD}$$

b. $$\% \text{ Yield} = \frac{765 \text{ g}}{795 \text{ g}} \times 100 = 96.2\%$$

CHALLENGE PROBLEMS

104. $$1.0 \text{ lb flour} \times \frac{454 \text{ g flour}}{\text{lb flour}} \times \frac{30.0 \times 10^{-9} \text{ g EDB}}{\text{g flour}} \times \frac{1 \text{ mol EDB}}{187.85 \text{ g}}$$

$$\times \frac{6.02 \times 10^{23} \text{ molecules}}{\text{mol EDB}} = 4.4 \times 10^{16} \text{ molecules of EDB}$$

105. Consider the case of aluminum plus oxygen. Aluminum forms Al^{3+} ions; oxygen forms O^{2-} anions. The simplest compound of the two elements is Al_2O_3. Similarly we would expect the formula of any group VI element with Al to be Al_2X_3. Assuming this, out of 100 g of compound there are 18.56 g Al and 81.44 g of the unknown element.

$$18.56 \text{ g Al} \times \frac{1 \text{ mol Al}}{26.98 \text{ g Al}} \times \frac{3 \text{ mol X}}{2 \text{ mol Al}} = 1.032 \text{ mol X}$$

100 g of the compound must contain 1.032 mol of X, if the formula is Al_2X_3.

Therefore, $$\frac{81.44 \text{ g X}}{1.032 \text{ mol X}} = 78.91 \text{ g X/mol}$$

The unknown element is selenium, Se, and the formula is Al_2Se_3.

106. The reaction is:

$$BaX_2(aq) + H_2SO_4(aq) \longrightarrow BaSO_4(s) + 2\ HX(aq)$$

$$0.124\ g\ BaSO_4 \times \frac{137.3\ g\ Ba}{233.4\ g\ BaSO_4} = 0.0729\ g\ Ba$$

$$\%Ba = \frac{0.0729\ g\ Ba}{0.158\ g\ BaX_2} \times 100 = 46.1\%\ Ba$$

The formula is BaX_2 (from positions of the elements in the periodic table) and 100 g of compound contains 46.1 g Ba and 53.9 g of the unknown halogen. There must also be:

$$46.1\ g\ Ba \times \frac{1\ mol\ Ba}{137.3\ g\ Ba} \times \frac{2\ mol\ X}{mol\ Ba} = 0.672\ mol\ of\ the\ halogen\ in\ 100\ g\ of\ BaX_2$$

Therefore the atomic mass of the halogen is $\frac{53.9\ g}{0.672\ mol} = 80.2\ g/mol.$

This atomic mass is close to that of bromine. Thus the formula of the compound is $BaBr_2$.

107. $$5.00 \times 10^6\ g\ NH_3 \times \frac{1\ mol\ NH_3}{17.03\ g\ NH_3} \times \frac{2\ mol\ HCN}{2\ mol\ NH_3} = 2.94 \times 10^5\ mol\ HCN$$

$$5.00 \times 10^6\ g\ O_2 \times \frac{1\ mol\ O_2}{32.00\ g\ O_2} \times \frac{2\ mol\ HCN}{3\ mol\ O_2} = 1.04 \times 10^5\ mol\ HCN$$

$$5.00 \times 10^6\ g\ CH_4 \times \frac{1\ mol\ CH_4}{16.04\ g\ CH_4} \times \frac{2\ mol\ HCN}{2\ mol\ CH_4} = 3.12 \times 10^5\ mol\ HCN$$

O_2 is limiting. Therefore,

$$1.04 \times 10^5\ mol\ HCN \times \frac{27.03\ g\ HCN}{mol\ HCN} = 2.81 \times 10^6\ g\ HCN$$

$$1.04 \times 10^5\ mol\ HCN \times \frac{6\ mol\ H_2O}{2\ mol\ HCN} \times \frac{18.02\ g\ H_2O}{mol\ H_2O} = 5.62 \times 10^6\ g\ H_2O$$

108. $$2\ C_3H_6 + 2\ NH_3 + 3\ O_2 \longrightarrow 2\ C_3H_3N + 6\ H_2O$$

a. $$5.00 \times 10^2\ g\ C_3H_6 \times \frac{1\ mol\ C_3H_6}{42.08\ g\ C_3H_6} \times \frac{2\ mol\ C_3H_3N}{2\ mol\ C_3H_6} = 11.9\ mol\ C_3H_3N$$

$$5.00 \times 10^2\ g\ NH_3 \times \frac{1\ mol\ NH_3}{17.03\ g\ NH_3} \times \frac{2\ mol\ C_3H_3N}{2\ mol\ NH_3} = 29.4\ mol\ C_3H_3N$$

$$1.00 \times 10^3\ g\ O_2 \times \frac{1\ mol\ O_2}{32.00\ g\ O_2} \times \frac{2\ mol\ C_3H_3N}{3\ mol\ O_2} = 20.8\ mol\ C_3H_3N$$

Therefore, C_3H_6 is limiting and the mass of acrylonitrile produced is:

$$11.9\ mol \times \frac{53.06\ g\ C_3H_3N}{mol} = 631\ g$$

b. $11.9 \text{ mol } C_3H_3N \times \frac{6 \text{ mol } H_2O}{2 \text{ mol } C_3H_3N} \times \frac{18.02 \text{ g } H_2O}{\text{mol } H_2O} = 643 \text{ g } H_2O$

Amount of NH_3 needed:

$$11.9 \text{ mol } C_3H_3N \times \frac{2 \text{ mol } NH_3}{2 \text{ mol } C_3H_3N} \times \frac{17.03 \text{ g } NH_3}{\text{mol } NH_3} = 203 \text{ g } NH_3$$

Amount NH_3 left = 500. g - 203 g = 297 g

Amount O_2 needed:

$$11.9 \text{ mol } C_3H_3N \times \frac{3 \text{ mol } O_2}{2 \text{ mol } C_3H_3N} \times \frac{32.00 \text{ g } O_2}{\text{mol } O_2} = 571 \text{ g } O_2$$

Amount O_2 left = 1.00×10^3 g - 571 g = 430 g

297 g NH_3 and 430 g O_2 left unreacted.

109. $10.00 \text{ g } XCl_2 \rightarrow 12.55 \text{ g } XCl_4$

XCl_4 contains 2.55 g Cl and 10.00 g XCl_2

XCl_2 contains 2.55 g Cl and 7.45 g X

$$2.55 \text{ g Cl} \times \frac{1 \text{ mol Cl}}{35.45 \text{ g Cl}} \times \frac{1 \text{ mol } XCl_2}{2 \text{ mol Cl}} \times \frac{1 \text{ mol X}}{1 \text{ mol } XCl_2} = 3.60 \times 10^{-2} \text{ mol X}$$

$$\frac{7.45 \text{ g X}}{3.60 \times 10^{-2} \text{ mol X}} = \frac{207 \text{ g}}{\text{mol X}};$$ X is Pb.

110. $LaH_{2.90}$: If only La^{3+} present, LaH_3. If only La^{2+} present, LaH_2.

$(La^{2+})_x(La^{3+})_yH_{(2x + 3y)}$

$$\begin{aligned} x + y &= 1.00 \\ 2x + 3y &= 2.90 \end{aligned} \qquad \begin{aligned} -2x - 2y &= -2.00 \\ 2x + 3y &= 2.90 \\ \hline y = 0.90 &\text{ and } x = 0.10 \end{aligned}$$

$\frac{1}{10}$ La^{2+} or 10.% La^{2+} and $\frac{9}{10}$ La^{3+} or 90.% La^{3+}.

111. $1.252 \text{ g Cu} \times \frac{1 \text{ mol Cu}}{63.55 \text{ g Cu}} = 1.970 \times 10^{-2} \text{ mol Cu}$

Let x = g Cu_2O and y = g CuO; then x + y = 1.500 and

$$2\left(\frac{x}{143.10}\right) + \frac{y}{79.55} = 1.970 \times 10^{-2} \text{ total mol Cu}$$

$$\begin{array}{r} 1.112\ x + y = 1.567 \\ -x - y = -1.500 \\ \hline 0.112\ x = 0.067 \end{array}$$

x = 0.60 g Cu_2O; 40.% Cu_2O by mass. y = 0.90 g CuO; 60.% CuO by mass.

CHAPTER FOUR: TYPES OF CHEMICAL REACTIONS AND SOLUTION STOICHIOMETRY

QUESTIONS

1. "Slightly soluble" refers to substances that dissolve only to a small extent. A slightly soluble salt may still dissociate completely to ions and, hence, be a strong electrolyte. An example of such a substance is $Mg(OH)_2$. It is a strong electrolyte, but not very soluble. A weak electrolyte is a substance that doesn't dissociate completely to produce ions. A weak electrolyte may be very soluble in water, or it may not be very soluble. Acetic acid is an example of a weak electrolyte that is very soluble in water.

2. Measure the electrical conductivity of a solution and compare it to the conductivity of a solution of equal concentration of a strong electrolyte.

EXERCISES

Aqueous Solutions: Strong and Weak Electrolytes

3. a. $NaBr(s) \rightarrow Na^+(aq) + Br^-(aq)$

   b. $MgCl_2(s) \rightarrow Mg^{2+}(aq) + 2\ Cl^-(aq)$

   c. $Al(NO_3)_3(s) \rightarrow Al^{3+}(aq) + 3\ NO_3^-(aq)$

   d. $(NH_4)_2SO_4(s) \rightarrow 2\ NH_4^+(aq) + SO_4^{2-}(aq)$

   e. $HI(g) \rightarrow H^+(aq) + I^-(aq)$

   f. $FeSO_4(s) \rightarrow Fe^{2+}(aq) + SO_4^{2-}(aq)$

   g. $KMnO_4(s) \rightarrow K^+(aq) + MnO_4^-(aq)$

   h. $HClO_4(s) \rightarrow H^+(aq) + ClO_4^-(aq)$

   i. $NH_4C_2H_3O_2(s) \rightarrow NH_4^+(aq) + C_2H_3O_2^-(aq)$

4. a. $HCl(g) \rightarrow H^+(aq) + Cl^-(aq)$

   b. $HNO_3(l) \rightarrow H^+(aq) + NO_3^-(aq)$

   c. $Ca(OH)_2(s) \rightarrow Ca^{2+}(aq) + 2\ OH^-(aq)$

   d. $KOH(s) \rightarrow K^+(aq) + OH^-(aq)$

5. $CaCl_2(s) \rightarrow Ca^{2+}(aq) + 2\ Cl^-(aq)$

6. $MgSO_4(s) \rightarrow Mg^{2+}(aq) + SO_4^{2-}(aq)$ $\qquad NH_4NO_3(s) \rightarrow NH_4^+(aq) + NO_3^-(aq)$

Solution Concentration: Molarity

7. a. $1.0\ L \times \frac{0.10\ mol\ NaCl}{L} \times \frac{58.44\ g\ NaCl}{mol} = 5.8\ g\ NaCl$

   Place 5.8 g NaCl in a 1 L volumetric flask; add water to dissolve the NaCl and fill to the mark.

   b. $1.0\ L \times \frac{0.10\ mol\ NaCl}{L} \times \frac{1\ L\ stock}{2.5\ mol\ NaCl} = 4.0 \times 10^{-2}\ L$

   Add 40. mL of 2.5 *M* solution to a 1 L volumetric flask; fill to the mark with water.

c. $1.0\ L \times \frac{0.20\ mol\ NaIO_3}{L} \times \frac{197.9\ g\ NaIO_3}{mol\ NaIO_3} = 4.0 \times 10^1\ g\ NaIO_3$

As in a, using 40. g $NaIO_3$.

d. $1.0\ L \times \frac{0.010\ mol\ NaIO_3}{L} \times \frac{1\ L\ stock}{0.20\ mol\ NaIO_3} = 0.050\ L$

As in b, using 50. mL of the 0.20 *M* sodium iodate stock solution.

e. $1.0\ L \times \frac{0.050\ mol\ KHP}{L} \times \frac{204.22\ g\ KHP}{mol\ KHP} = 10.2\ g\ KHP = 10.\ g\ KHP$

As in a, using 10. g KHP

f. $1.0\ L \times \frac{0.040\ mol\ KHP}{L} \times \frac{1\ L\ stock}{0.50\ mol\ KHP} = 0.080\ L$

As in b, using 80. mL of the 0.50 *M* KHP stock solution.

8. a. $1.00\ L\ solution \times \frac{0.50\ mol\ H_2SO_4}{L} = 0.50\ mol\ H_2SO_4$

$0.50\ mol\ H_2SO_4 \times \frac{1\ L}{18\ mol\ H_2SO_4} = 2.8 \times 10^{-2}\ L\ conc.\ H_2SO_4$ or 28 mL

Dilute 28 mL of concentrated H_2SO_4 to a total volume of 1.00 L with water.

b. We will need 0.50 mol HCl.

$0.50\ mol\ HCl \times \frac{1\ L}{12\ mol\ HCl} = 4.2 \times 10^{-2}\ L = 42\ mL$

Dilute 42 mL of concentrated HCl to a final volume of 1.00 L.

c. We need 0.50 mol $NiCl_2$.

$0.50\ mol\ NiCl_2 \times \frac{1\ mol\ NiCl_2 \cdot 6\ H_2O}{mol\ NiCl_2} \times \frac{237.69\ g\ NiCl_2 \cdot 6\ H_2O}{mol\ NiCl_2 \cdot 6\ H_2O}$

$= 118.8\ g\ NiCl_2 \cdot 6\ H_2O \approx 120\ g$

Dissolve 120 g $NiCl_2 \cdot 6\ H_2O$ in water, and add water until the total volume of the solution is 1.00 L.

d. $1.00\ L \times \frac{0.50\ mol\ HNO_3}{L} = 0.50\ mol\ HNO_3$

$0.50\ mol\ HNO_3 \times \frac{1\ L}{16\ mol} = 0.031\ L = 31\ mL$

Dissolve 31 mL of concentrated HNO_3 in water. Dilute to a total volume of 1.00 L.

e. We need 0.50 mol Na_2CO_3

$$0.50 \text{ mol } Na_2CO_3 \times \frac{105.99 \text{ g } Na_2CO_3}{\text{mol}} = 53 \text{ g } Na_2CO_3$$

Dissolve 53 g of Na_2CO_3 in water, dilute to 1.00 L.

9. a. $$5.623 \text{ g } NaHCO_3 \times \frac{1 \text{ mol } NaHCO_3}{84.01 \text{ g } NaHCO_3} = 6.693 \times 10^{-2} \text{ mol}$$

$$M = \frac{6.693 \times 10^{-2} \text{ mol}}{250.0 \text{ mL}} \times \frac{1000 \text{ mL}}{\text{L}} = 0.2677\ M$$

b. $$0.1846 \text{ g } K_2Cr_2O_7 \times \frac{1 \text{ mol } K_2Cr_2O_7}{294.20 \text{ g } K_2Cr_2O_7} = 6.275 \times 10^{-4} \text{ mol}$$

$$M = \frac{6.275 \times 10^{-4} \text{ mol}}{500.0 \times 10^{-3} \text{ L}} = 1.255 \times 10^{-3}\ M$$

c. $$0.1025 \text{ g Cu} \times \frac{1 \text{ mol Cu}}{63.55 \text{ g Cu}} = 1.613 \times 10^{-3} \text{ mol Cu} = 1.613 \times 10^{-3} \text{ mol } Cu^{2+}$$

$$M = \frac{1.613 \times 10^{-3} \text{ mol } Cu^{2+}}{200.0 \text{ mL}} \times \frac{1000 \text{ mL}}{\text{L}} = 8.065 \times 10^{-3}\ M$$

10. a. $$\frac{16.45 \text{ g NaCl}}{1.000 \text{ L}} \times \frac{1 \text{ mol NaCl}}{58.44 \text{ g NaCl}} = 0.2815 \text{ mol/L}$$

b. $$853.5 \text{ mg } KIO_3 \times \frac{1 \text{ g}}{1000 \text{ mg}} \times \frac{1 \text{ mol } KIO_3}{214.0 \text{ g } KIO_3} = 3.988 \times 10^{-3} \text{ mol}$$

$$\frac{3.988 \times 10^{-3} \text{ mol}}{250.0 \text{ mL}} \times \frac{1000 \text{ mL}}{\text{L}} = 1.595 \times 10^{-2} \text{ mol/L}$$

c. $$0.4508 \text{ g Fe} \times \frac{1 \text{ mol Fe}}{55.85 \text{ g Fe}} = 8.072 \times 10^{-3} \text{ mol Fe} = 8.072 \times 10^{-3} \text{ mol } Fe^{3+}$$

$$\frac{8.072 \times 10^{-3} \text{ mol } Fe^{3+}}{500.0 \text{ mL}} \times \frac{1000 \text{ mL}}{\text{L}} = 1.614 \times 10^{-2} \text{ mol/L}$$

11. a. $CaCl_2(s) \rightarrow Ca^{2+}(aq) + 2\ Cl^-(aq)$

$M_{Ca^{2+}} = 0.15\ M$; $M_{Cl^-} = 2(0.15) = 0.30\ M$

b. $Al(NO_3)_3(s) \rightarrow Al^{3+}(aq) + 3\ NO_3^-(aq)$

$M_{Al^{3+}} = 0.26\ M$; $M_{NO_3^-} = 3(0.26) = 0.78\ M$

c. $K_2Cr_2O_7(s) \rightarrow 2\ K^+(aq) + Cr_2O_7^{2-}(aq)$

$M_{K^+} = 0.50\ M$; $M_{Cr_2O_7^{2-}} = 0.25\ M$

d. $Al_2(SO_4)_3(s) \rightarrow 2\ Al^{3+}(aq) + 3\ SO_4^{2-}(aq)$

$$M_{Al^{3+}} = \frac{2.0 \times 10^{-3}\ \text{mol}\ Al_2(SO_4)_3}{L} \times \frac{2\ \text{mol}\ Al^{3+}}{\text{mol}\ Al_2(SO_4)_3} = 4.0 \times 10^{-3}\ M$$

$$M_{SO_4^{2-}} = \frac{2.0 \times 10^{-3}\ \text{mol}\ Al_2(SO_4)_3}{L} \times \frac{3\ \text{mol}\ SO_4^{2-}}{\text{mol}\ Al_2(SO_4)_3} = 6.0 \times 10^{-3}\ M$$

12. a. $0.100\ \text{g}\ MgCl_2 \times \frac{1\ \text{mol}\ MgCl_2}{95.21\ \text{g}\ MgCl_2} = 1.05 \times 10^{-3}\ \text{mol}$

$$M = \frac{1.05 \times 10^{-3}\ \text{mol}\ MgCl_2}{0.100\ L} = 1.05 \times 10^{-2}\ M\ MgCl_2$$

$MgCl_2(s) \rightarrow Mg^{2+}(aq) + 2\ Cl^-(aq)$

$M_{Mg^{2+}} = 1.05 \times 10^{-2}\ M$; $M_{Cl^-} = 2.10 \times 10^{-2}\ M$

b. $55.1 \times 10^{-3}\ \text{g}\ NH_4Br \times \frac{1\ \text{mol}}{97.94\ \text{g}} = 5.63 \times 10^{-4}\ \text{mol}\ NH_4Br$

$$M = \frac{5.63 \times 10^{-4}\ \text{mol}}{0.500\ L} = 1.13 \times 10^{-3}\ M\ NH_4Br$$

$NH_4Br(s) \rightarrow NH_4^+(aq) + Br^-(aq)$; $M_{NH_4^+} = M_{Br^-} = 1.13 \times 10^{-3}\ M$

c. $$\frac{5.47\ \text{g}\ Na_2S \times \frac{1\ \text{mol}\ Na_2S}{78.05\ \text{g}\ Na_2S}}{1.00\ L} = 7.01 \times 10^{-2}\ M\ Na_2S$$

$Na_2S(s) \rightarrow 2\ Na^+(aq) + S^{2-}(aq)$

$M_{Na^+} = 1.40 \times 10^{-1}\ M$; $M_{S^{2-}} = 7.01 \times 10^{-2}\ M$

d. $0.208\ \text{g}\ AlCl_3 \times \frac{1\ \text{mol}}{133.33\ \text{g}} = 1.56 \times 10^{-3}\ \text{mol}\ AlCl_3$

$$M_{Al^{3+}} = \frac{1.56 \times 10^{-3}\ \text{mol}}{250.0\ \text{mL}} \times \frac{1000\ \text{mL}}{L} = 6.24 \times 10^{-3}\ \text{M}\ AlCl_3$$

$AlCl_3(s) \rightarrow Al^{3+}(aq) + 3\ Cl^-(aq)$

$M_{Al^{3+}} = 6.24 \times 10^{-3}\ M$; $M_{Cl^-} = 1.87 \times 10^{-2}\ M$

13. $10.8\ \text{g}\ (NH_4)_2SO_4 \times \frac{1\ \text{mol}}{132.15\ \text{g}} = 8.17 \times 10^{-2}\ \text{mol}$

$$\text{Molarity} = \frac{8.17 \times 10^{-2}\ \text{mol}}{100.0\ \text{mL}} \times \frac{1000\ \text{mL}}{L} = \frac{0.817\ \text{mol}}{L}$$

Moles of $(NH_4)_2SO_4$ in final solution:

$$\frac{0.817\ \text{mol}}{\text{L}} \times 10.00 \times 10^{-3}\ \text{L} = 8.17 \times 10^{-3}\ \text{mol}$$

$$\text{Molarity of final solution} = \frac{8.17 \times 10^{-3}\ \text{mol}}{(10.00 + 50.00)\ \text{mL}} \times \frac{1000\ \text{mL}}{\text{L}} = \frac{0.136\ \text{mol}}{\text{L}}$$

$M_{NH_4^+} = 2(0.136) = 0.272\ M$; $M_{SO_4^{2-}} = 0.136\ M$

14. $10.0\ \text{mL} \times \frac{0.79\ \text{g}}{\text{mL}} \times \frac{1\ \text{mol}}{46.07\ \text{g}} = 0.17\ \text{mol ethanol}$

$$\text{Molarity} = \frac{0.17\ \text{mol}}{0.250\ \text{L}} = \frac{0.68\ \text{mol}}{\text{L}}$$

15. $\frac{30.\ \text{mg}\ Mg^{2+}}{\text{L}} \times \frac{1\ \text{g}}{1000\ \text{mg}} \times \frac{1\ \text{mol}\ Mg^{2+}}{24.31\ \text{g}\ Mg^{2+}} = \frac{1.2 \times 10^{-3}\ \text{mol}}{\text{L}}$

$$\frac{75\ \text{mg}\ Ca^{2+}}{\text{L}} \times \frac{1\ \text{g}}{1000\ \text{mg}} \times \frac{1\ \text{mol}\ Ca^{2+}}{40.08\ \text{g}\ Ca^{2+}} = \frac{1.9 \times 10^{-3}\ \text{mol}}{\text{L}}$$

$$\frac{1250\ \text{mg}\ Mg^{2+}}{\text{L}} \times \frac{1\ \text{g}}{1000\ \text{mg}} \times \frac{1\ \text{mol}\ Mg^{2+}}{24.31\ \text{g}} = \frac{0.0514\ \text{mol}}{\text{L}}$$

$$\frac{200.\ \text{mg}\ Ca^{2+}}{\text{L}} \times \frac{1\ \text{g}}{1000\ \text{mg}} \times \frac{1\ \text{mol}\ Ca^{2+}}{40.08\ \text{g}} = \frac{4.99 \times 10^{-3}\ \text{mol}}{\text{L}}$$

16. $\frac{50.0 \times 10^{-3}\ \text{g Myocrisin}}{0.500 \times 10^{-3}\ \text{L}} \times \frac{1\ \text{mol Myocrisin}}{390.1\ \text{g}} = 0.256\ M$

$$\frac{300. \times 10^{-6}\ \text{g Au}}{100.0\ \text{mL}} \times \frac{1\ \text{mol Au}}{197.0\ \text{g Au}} \times \frac{1\ \text{mol Myocrisin}}{\text{mol Au}} \times \frac{1000\ \text{mL}}{\text{L}} = 1.523 \times 10^{-5}\ M$$

17. $0.5842\ \text{g} \times \frac{1\ \text{mol}}{90.04\ \text{g}} = 6.488 \times 10^{-3}\ \text{mol}$

$\frac{6.488 \times 10^{-3}\ \text{mol}}{100.0\ \text{mL}} \times \frac{1000\ \text{mL}}{\text{L}} = 6.488 \times 10^{-2}\ M$ which is the concentration of the initial solution.

Consider, next, the dilution step:

$$\frac{6.488 \times 10^{-2}\ \text{mol}}{\text{L}} \times 10.00 \times 10^{-3}\ \text{L} = 6.488 \times 10^{-4}\ \text{mol}$$

The final solution contains 6.488×10^{-4} mol of oxalic acid in 250.0 mL of solution; or

$$M = \frac{6.488 \times 10^{-4}\ \text{mol}}{0.2500\ \text{L}} = 2.595 \times 10^{-3}\ M$$

18. Stock solution:

$$1.584 \text{ g Mn} \times \frac{1 \text{ mol Mn}}{54.94 \text{ g Mn}} = 2.883 \times 10^{-2} \text{ mol Mn}; \quad \frac{2.883 \times 10^{-2} \text{ mol Mn}}{1.00 \text{ L}} = 2.88 \times 10^{-2}\ M$$

Solution A contains:

$$50.0 \text{ mL} \times \frac{1 \text{ L}}{1000 \text{ mL}} \times \frac{2.88 \times 10^{-2} \text{ mol}}{\text{L}} = 1.44 \times 10^{-3} \text{ mol}$$

$$\text{Molarity} = \frac{1.44 \times 10^{-3} \text{ mol}}{1000.0 \text{ mL}} \times \frac{1000 \text{ mL}}{\text{L}} = 1.44 \times 10^{-3}\ M$$

Solution B contains:

$$10.0 \text{ mL} \times \frac{1 \text{ L}}{1000 \text{ mL}} \times \frac{1.44 \times 10^{-3} \text{ mol}}{\text{L}} = 1.44 \times 10^{-5} \text{ mol}$$

$$\text{Molarity} = \frac{1.44 \times 10^{-5} \text{ mol}}{0.250 \text{ L}} = 5.76 \times 10^{-5}\ M$$

Solution C contains:

$$10.0 \times 10^{-3} \text{ L} \times \frac{5.76 \times 10^{-5} \text{ mol}}{\text{L}} = 5.76 \times 10^{-7} \text{ mol}$$

$$\text{Molarity} = \frac{5.76 \times 10^{-7} \text{ mol}}{0.500 \text{ L}} = 1.15 \times 10^{-6}\ M$$

Precipitation Reactions

19. a. $BaCl_2(aq) + Na_2SO_4(aq) \rightarrow BaSO_4(s) + 2\ NaCl(aq)$

$Ba^{2+}(aq) + 2\ Cl^-(aq) + 2\ Na^+(aq) + SO_4^{2-}(aq) \rightarrow BaSO_4(s) + 2\ Na^+(aq) + 2Cl^-(aq)$

$Ba^{2+}(aq) + SO_4^{2-}(aq) \rightarrow BaSO_4(s)$

b. $Pb(NO_3)_2(aq) + 2\ HCl(aq) \rightarrow PbCl_2(s) + 2\ HNO_3(aq)$

$Pb^{2+}(aq) + 2\ NO_3^-(aq) + 2\ H^+(aq) + 2\ Cl^-(aq) \rightarrow PbCl_2(s) + 2\ H^+(aq) + 2\ NO_3^-(aq)$

$Pb^{2+}(aq) + 2Cl^-(aq) \rightarrow PbCl_2(s)$

c. $3\ AgNO_3(aq) + Na_3PO_4(aq) \rightarrow Ag_3PO_4(s) + 3\ NaNO_3(aq)$

$3\ Ag^+(aq) + 3\ NO_3^-(aq) + 3\ Na^+(aq) + PO_4^{3-}(aq) \rightarrow Ag_3PO_4(s) + 3\ Na^+(aq) + 3\ NO_3^-(aq)$

$3\ Ag^+(aq) + PO_4^{3-}(aq) \rightarrow Ag_3PO_4(s)$

d. $3\ NaOH(aq) + Fe(NO_3)_3(aq) \rightarrow Fe(OH)_3(s) + 3\ NaNO_3(aq)$

$3\ Na^+(aq) + 3OH^-(aq) + Fe^{3+}(aq) + 3\ NO_3^-(aq) \rightarrow Fe(OH)_3(s) + 3\ Na^+(aq) + 3\ NO_3^-(aq)$

$Fe^{3+}(aq) + 3\ OH^-(aq) \rightarrow Fe(OH)_3(s)$

20. a. $CuCl_2(aq) + Na_2S(aq) \rightarrow CuS(s) + 2\ NaCl(aq)$

$Cu^{2+}(aq) + 2Cl^-(aq) + 2\ Na^+(aq) + S^{2-}(aq) \rightarrow CuS(s) + 2\ Na^+(aq) + 2Cl^-(aq)$

$Cu^{2+}(aq) + S^{2-}(aq) \rightarrow CuS(s)$

b. $NiSO_4(aq) + 2\ KOH(aq) \rightarrow Ni(OH)_2(s) + K_2SO_4(aq)$

$Ni^{2+}(aq) + SO_4^{2-}(aq) + 2\ K^+(aq) + 2\ OH^-(aq) \rightarrow Ni(OH)_2(s) + 2\ K^+(aq) + SO_4^{2-}(aq)$

$Ni^{2+}(aq) + 2\ OH^-(aq) \rightarrow Ni(OH)_2(s)$

c. $KOH(aq) + NaNO_3(aq) \rightarrow$ No reaction

d. $2\ NaOH(aq) + MnSO_4(aq) \rightarrow Mn(OH)_2(s) + Na_2SO_4(aq)$

$2\ Na^+(aq) + 2\ OH^-(aq) + Mn^{2+}(aq) + SO_4^{2-}(aq) \rightarrow Mn(OH)_2(s) + 2\ Na^+(aq) + SO_4^{2-}(aq)$

$Mn^{2+}(aq) + 2\ OH^-(aq) \rightarrow Mn(OH)_2(s)$

21. a. Silver iodide is insoluble. $AgNO_3(aq) + KI(aq) \rightarrow AgI(s) + KNO_3(aq)$

$Ag^+(aq) + I^-(aq) \rightarrow AgI(s)$

b. Copper(II) sulfide is insoluble. $CuSO_4(aq) + Na_2S(aq) \rightarrow CuS(s) + Na_2SO_4(aq)$

$Cu^{2+}(aq) + S^{2-}(aq) \rightarrow CuS(s)$

c. $CoCl_2(aq) + 2\ NaOH(aq) \rightarrow Co(OH)_2(s) + 2\ NaCl(aq)$

$Co^{2+}(aq) + 2\ OH^-(aq) \rightarrow Co(OH)_2(s)$

d. The potential products are $Ni(NO_3)_2$ and HCl. Both are soluble in water. Thus, no reaction occurs.

22. a. AgCl is insoluble. $Ag^+(aq) + Cl^-(aq) \rightarrow AgCl(s)$

b. FeS is insoluble. $Fe^{2+}(aq) + S^{2-}(aq) \rightarrow FeS(s)$

c. No reaction

d. $Hg_2(NO_3)_2$ is made up of Hg_2^{2+} and NO_3^- ions.

Hg_2Cl_2, mercury(I) chloride or mercurous chloride, is insoluble.

$Hg_2^{2+}(aq) + 2\ Cl^-(aq) \rightarrow Hg_2Cl_2(s)$

23. a. $(NH_4)_2SO_4(aq) + Ba(NO_3)_2(aq) \rightarrow 2\ NH_4NO_3(aq) + BaSO_4(s)$

$Ba^{2+}(aq) + SO_4^{2-}(aq) \rightarrow BaSO_4(s)$

b. $Pb(NO_3)_2(aq) + 2\ NaCl(aq) \rightarrow PbCl_2(s) + 2\ NaNO_3(aq)$

$Pb^{2+}(aq) + 2\ Cl^-(aq) \rightarrow PbCl_2(s)$

c. Potassium phosphate and sodium nitrate are both soluble in water. No reaction occurs.

d. No reaction occurs.

e. $CuCl_2(aq) + 2\ NaOH(aq) \rightarrow Cu(OH)_2(s) + 2\ NaCl(aq)$

$Cu^{2+}(aq) + 2\ OH^-(aq) \rightarrow Cu(OH)_2(s)$

24. a. $Fe(NO_3)_3(aq) + 3\ NaOH(aq) \rightarrow Fe(OH)_3\ (s) + 3\ NaNO_3(aq)$

$Fe^{3+}(aq) + 3\ OH^-(aq) \rightarrow Fe(OH)_3(s)$

b. $CdCl_2(aq) + Na_2S(aq) \rightarrow CdS(s) + 2\ NaCl(aq)$

$Cd^{2+}(aq) + S^{2-}(aq) \rightarrow CdS(s)$

c. $AgNO_3(aq) + RbBr(aq) \rightarrow AgBr(s) + RbNO_3(aq)$

$Ag^+(aq) + Br^-(aq) \rightarrow AgBr(s)$

d. $CuCl_2(aq) + Ca(OH)_2(aq) \rightarrow Cu(OH)_2(s) + CaCl_2(aq)$

$Cu^{2+}(aq) + 2OH^-(aq) \rightarrow Cu(OH)_2(s)$

25. Three possibilities are:

Addition of H_2SO_4 solution to give a white ppt. of $PbSO_4$.

Addition of HCl solution to give a white ppt. of $PbCl_2$.

Addition of K_2CrO_4 solution to give a bright yellow ppt. of $PbCrO_4$.

26. Since no precipitates formed upon addition of NaCl or Na_2SO_4, we can conclude that Hg_2^{2+} and Ba^{2+} are not present. The precipitate with NaOH is $Mn(OH)_2$. Therefore, Mn^{2+} was present.

27. The reaction is:

$$AgNO_3(aq) + NaBr(aq) \longrightarrow AgBr(s) + NaNO_3(aq)$$

$$100.0\text{ mL } AgNO_3 \times \frac{1\text{ L}}{1000\text{ mL}} \times \frac{0.150\text{ mol } AgNO_3}{\text{L } AgNO_3} \times \frac{1\text{ mol AgBr}}{\text{mol } AgNO_3} = 1.50 \times 10^{-2}\text{ mol AgBr}$$

$$20.0\text{ mL NaBr} \times \frac{1\text{ L}}{1000\text{ mL}} \times \frac{1.00\text{ mol NaBr}}{\text{L NaBr}} \times \frac{1\text{ mol AgBr}}{\text{mol NaBr}} = 2.00 \times 10^{-2}\text{ mol AgBr}$$

Therefore, $AgNO_3$ is the limiting reagent.

$$1.50 \times 10^{-2}\text{ mol AgBr} \times \frac{187.8\text{ g AgBr}}{\text{mol AgBr}} = 2.82\text{ g AgBr}$$

28. The unbalanced reaction is:

$$BaCl_2(aq) + Fe_2(SO_4)_3(aq) \longrightarrow BaSO_4(s) + FeCl_3(aq)$$

Balancing the equation, we get

$$3\ BaCl_2(aq) + Fe_2(SO_4)_3(aq) \longrightarrow 3\ BaSO_4(s) + 2\ FeCl_3(aq)$$

$$100.0\text{ mL } BaCl_2 \times \frac{1\text{ L}}{1000\text{ mL}} \times \frac{0.100\text{ mol } BaCl_2}{\text{L}} \times \frac{3\text{ mol } BaSO_4}{3\text{ mol } BaCl_2} = 1.00 \times 10^{-2}\text{ mol } BaSO_4$$

$$100.0\text{ mL } Fe(SO_4)_3 \times \frac{1\text{ L}}{1000\text{ mL}} \times \frac{0.100\text{ mol } Fe_2(SO_4)_3}{\text{L } Fe_2(SO_4)_3} \times \frac{3\text{ mol } BaSO_4}{\text{mol } Fe_2(SO_4)_3}$$

$$= 3.00 \times 10^{-2}\text{ mol } BaSO_4$$

Therefore, the barium chloride solution is the limiting reagent and the mass of barium sulfate produced is:

$$1.00 \times 10^{-2}\text{ mol } BaSO_4 \times \frac{233.4\text{ g}}{\text{mol}} = 2.33\text{ g } BaSO_4$$

29. The reaction is: $Ni(NO_3)_2(aq) + 2\ NaOH(aq) \longrightarrow Ni(OH)_2(s) + 2\ NaNO_3(aq)$

$$150.0\text{ mL } Ni(NO_3)_2 \times \frac{1\text{ L}}{1000\text{ mL}} \times \frac{0.250\text{ mol } Ni(NO_3)_2}{\text{L } Ni(NO_3)_2} \times \frac{2\text{ mol NaOH}}{1\text{ mol } Ni(NO_3)_2} \times \frac{1\text{ L NaOH}}{0.100\text{ mol NaOH}}$$

$$= 0.750\text{ L or } 750.\text{ mL}$$

30. The reaction is: $AgNO_3(aq) + HCl(aq) \longrightarrow AgCl(s) + HNO_3(aq)$

$$50.0 \times 10^{-3}\text{ L } AgNO_3 \times \frac{0.0500\text{ mol } AgNO_3}{\text{L}} \times \frac{1\text{ mol HCl}}{1\text{ mol } AgNO_3} \times \frac{1\text{ L}}{0.204\text{ mol HCl}}$$

$$= 1.23 \times 10^{-2}\text{ L} = 12.3\text{ mL}$$

Acids and Bases

31. a. $2HClO_4(aq) + Mg(OH_2)(s) \rightarrow 2\ H_2O(l) + Mg(ClO_4)_2(aq)$

$2\ H^+(aq) + 2ClO_4^-(aq) + Mg(OH)_2(s) \rightarrow 2\ H_2O(l) + Mg^{2+}(aq) + 2ClO_4^-(aq)$

$2H^+(aq) + Mg(OH)_2(s) \rightarrow 2H_2O(l) + Mg^{2+}(aq)$

b. $HCN(aq) + NaOH(aq) \rightarrow H_2O(l) + NaCN(aq)$

$HCN(aq) + Na^+(aq) + OH^-(aq) \rightarrow H_2O(l) + Na^+(aq) + CN^-(aq)$

$HCN(aq) + OH^-(aq) \rightarrow H_2O(l) + CN^-(aq)$

c. $HCl(aq) + NaOH(aq) \rightarrow H_2O(l) + NaCl(aq)$

$H^+(aq) + Cl^-(aq) + Na^+(aq) + OH^-(aq) \rightarrow H_2O(l) + Na^+(aq) + Cl^-(aq)$

$H^+(aq) + OH^-(aq) \rightarrow H_2O(l)$

32. a. $3\ HNO_3(aq) + Al(OH)_3(s) \rightarrow 3\ H_2O(l) + Al(NO_3)_3(aq)$

$3\ H^+(aq) + 3\ NO_3^-(aq) + Al(OH)_3(s) \rightarrow 3\ H_2O(l) + Al^{3+}(aq) + 3\ NO_3^-(aq)$

$3\ H^+(aq) + Al(OH)_3(s) \rightarrow 3\ H_2O(l) + Al^{3+}(aq)$

b. $HC_2H_3O_2(aq) + KOH(aq) \rightarrow H_2O(l) + KC_2H_3O_2(aq)$

$HC_2H_3O_2(aq) + K^+(aq) + OH^-(aq) \rightarrow H_2O(l) + K^+(aq) + C_2H_3O_2^-(aq)$

$HC_2H_3O_2(aq) + OH^-(aq) \rightarrow H_2O(l) + C_2H_3O_2^-(aq)$

c. $Ca(OH)_2(aq) + 2\ HCl(aq) \rightarrow 2\ H_2O(l) + CaCl_2(aq)$

$Ca^{2+}(aq) + 2\ OH^-(aq) + 2\ H^+(aq) + 2\ Cl^-(aq) \rightarrow 2\ H_2O(l) + Ca^{2+}(aq) + 2\ Cl^-(aq)$

$2\ H^+(aq) + 2\ OH^-(aq) \rightarrow 2\ H_2O(l)$ or $H^+(aq) + OH^-(aq) \rightarrow H_2O(l)$

33. a. $KOH(aq) + HNO_3(aq) \rightarrow H_2O(l) + KNO_3(aq)$

$K^+(aq) + OH^-(aq) + H^+(aq) + NO_3^-(aq) \rightarrow H_2O(l) + K^+(aq) + NO_3^-(aq)$

$OH^-(aq) + H^+(aq) \rightarrow H_2O(l)$

b. $Ba(OH)_2(aq) + 2\ HCl(aq) \rightarrow 2\ H_2O(l) + BaCl_2(aq)$

$Ba^{2+}(aq) + 2\ OH^-(aq) + 2\ H^+(aq) + 2\ Cl^-(aq) \rightarrow Ba^{2+}(aq) + 2\ Cl^-(aq) + 2\ H_2O(l)$

$2\ OH^-(aq) + 2\ H^+(aq) \rightarrow 2\ H_2O(l)$ or $OH^-(aq) + H^+(aq) \rightarrow H_2O(l)$

c. $3\ HClO_4(aq) + Fe(OH)_3(s) \rightarrow 3\ H_2O(l) + Fe(ClO_4)_3(aq)$

$3\ H^+(aq) + 3\ ClO_4^-(aq) + Fe(OH)_3(s) \rightarrow 3\ H_2O(l) + Fe^{3+}(aq) + 3ClO_4^-(aq)$

$3\ H^+(aq) + Fe(OH)_3(s) \rightarrow 3\ H_2O(l) + Fe^{3+}(aq)$

34. a. $AgOH(s) + HBr(aq) \rightarrow AgBr(s) + H_2O(l)$

$AgOH(s) + H^+(aq) + Br^-(aq) \rightarrow AgBr(s) + H_2O(l)$

$AgOH(s) + H^+(aq) + Br^-(aq) \rightarrow AgBr(s) + H_2O(l)$

b. $Sr(OH)_2(aq) + 2\ HI(aq) \rightarrow 2\ H_2O(l) + SrI_2(aq)$

$Sr^{2+}(aq) + 2\ OH^-(aq) + 2\ H^+(aq) + 2\ I^-(aq) \rightarrow 2\ H_2O(l) + Sr^{2+}(aq) + 2\ I^-(aq)$

$2\ OH^-(aq) + 2\ H^+(aq) \rightarrow 2\ H_2O(l)$ or $OH^-(aq) + H^+(aq) \rightarrow H_2O(l)$

c. $Fe(OH)_3(s) + 3\ HNO_3(aq) \rightarrow 3\ H_2O(l) + Fe(NO_3)_3(aq)$

$Fe(OH)_3(s) + 3\ H^+(aq) + 3\ NO_3^-(aq) \rightarrow 3\ H_2O(l) + Fe^{3+}(aq) + 3\ NO_3^-(aq)$

$Fe(OH_3)(s) + 3\ H^+(aq) \rightarrow 3\ H_2O(l) + Fe^{3+}(aq)$

35. If we begin with 50.00 mL of 0.200 *M* NaOH, then

$$50.00 \times 10^{-3}\ L \times \frac{0.200\ mol}{L} = 1.00 \times 10^{-2}\ mol\ NaOH \text{ to be neutralized.}$$

a. $NaOH(aq) + HCl(aq) \rightarrow NaCl(aq) + H_2O(l)$

$$1.00 \times 10^{-2}\ mol\ NaOH \times \frac{1\ mol\ HCl}{mol\ NaOH} \times \frac{1\ L\ soln}{0.100\ mol} = 0.100\ L \text{ or } 100.\ mL$$

b. $3\ NaOH(aq) + H_3PO_4(aq) \rightarrow Na_3PO_4(aq) + 3\ H_2O(l)$

$$1.00 \times 10^{-2}\ mol\ NaOH \times \frac{1\ mol\ H_3PO_4}{3\ mol\ NaOH} \times \frac{1\ L\ soln}{0.200\ mol\ H_3PO_4} = 1.67 \times 10^{-2}\ L \text{ or } 16.7\ mL$$

c. $$50.00\ mL \times \frac{1\ L}{1000\ mL} \times \frac{0.200\ mol\ NaOH}{L} = 1.00 \times 10^{-2}\ mol\ NaOH$$

$HNO_3(aq) + NaOH(aq) \rightarrow H_2O(l) + NaNO_3(aq)$

$$1.00 \times 10^{-2}\ mol\ NaOH \times \frac{1\ mol\ HNO_3}{mol\ NaOH} \times \frac{1\ L}{0.150\ mol\ HNO_3} = 6.67 \times 10^{-2}\ L \text{ or } 66.7\ mL$$

d. $HC_2H_3O_2(aq) + NaOH(aq) \rightarrow H_2O(l) + NaC_2H_3O_2(aq)$

$$1.00 \times 10^{-2} \text{ mol NaOH} \times \frac{1 \text{ mol } HC_2H_3O_2}{\text{mol NaOH}} \times \frac{1 \text{ L}}{0.200 \text{ mol } HC_2H_3O_2} = 5.00 \times 10^{-2} \text{ L or } 50.0 \text{ mL}$$

36. We begin with 25.00 mL of 0.200 M HCl or 25.00×10^{-3} L $\times$ 0.200 mol/L = 5.00×10^{-3} mol HCl.

a. $HCl(aq) + NaOH(aq) \rightarrow H_2O(l) + NaCl(aq)$

$$5.00 \times 10^{-3} \text{ mol HCl} \times \frac{1 \text{ mol NaOH}}{\text{mol HCl}} \times \frac{1 \text{ L}}{0.100 \text{ mol NaOH}} = 5.00 \times 10^{-2} \text{ L} = 50.0 \text{ mL}$$

b. $2\,HCl(aq) + Ba(OH)_2(aq) \rightarrow 2\,H_2O(l) + BaCl_2(aq)$

$$5.00 \times 10^{-3} \text{ mol HCl} \times \frac{1 \text{ mol } Ba(OH)_2}{2 \text{ mol HCl}} \times \frac{1 \text{ L}}{0.0500 \text{ mol } Ba(OH)_2} = 5.00 \times 10^{-2} \text{ L} = 50.0 \text{ mL}$$

c. $HCl(aq) + KOH(aq) \rightarrow H_2O(l) + KCl(aq)$

$$5.00 \times 10^{-3} \text{ mol HCl} \times \frac{1 \text{ mol KOH}}{\text{mol HCl}} \times \frac{1 \text{ L}}{0.250 \text{ mol KOH}} = 2.00 \times 10^{-2} \text{ L} = 20.0 \text{ mL}$$

37. $$15 \text{ g NaOH} \times \frac{1 \text{ mol NaOH}}{40.00 \text{ g}} = 0.38 \text{ mol NaOH}$$

$$0.15 \text{ L} \times \frac{0.25 \text{ mol } HNO_3}{\text{L}} = 0.038 \text{ mol } HNO_3$$

We have added more moles of NaOH. The reaction is 1:1

$HNO_3(aq) + NaOH(aq) \rightarrow NaNO_3(aq) + H_2O(l)$

We have excess NaOH, the solution will be basic.

In the final solution there will be 0.038 mol NO_3^-, and (0.38 - 0.038) = 0.34 mol of OH^-, and 0.38 mol Na^+.

$$C_{NO_3^-} = \frac{0.038 \text{ mol}}{0.15 \text{ L}} = \frac{0.25 \text{ mol } NO_3^-}{\text{L}}; \quad C_{Na^+} = \frac{0.38 \text{ mol}}{0.15 \text{ L}} = \frac{2.5 \text{ mol } Na^+}{\text{L}}$$

$$C_{OH^-} = \frac{0.34 \text{ mol}}{0.15 \text{ L}} = \frac{2.3 \text{ mol } OH^-}{\text{L}}$$

38. $Ba(OH)_2(aq) + 2\,HCl(aq) \rightarrow BaCl_2(aq) + H_2O(l)$

$H^+(aq) + OH^-(aq) \rightarrow H_2O(l)$

$$75 \times 10^{-3} \text{ L} \times \frac{0.25 \text{ mol HCl}}{\text{L}} = 1.9 \times 10^{-2} \text{ mol HCl}$$

$$225 \times 10^{-3} \text{ L} \times \frac{0.055 \text{ mol } Ba(OH)_2}{\text{L}} \times \frac{2 \text{ mol } OH^-}{\text{mol } Ba(OH)_2} = 2.5 \times 10^{-2} \text{ mol } OH^-$$

We have an excess of OH^- .

Since 1.9×10^{-2} mol OH^- will react, we will have $(2.5 - 1.9) \times 10^{-2} = 0.6 \times 10^{-2}$ mol OH^- left unreacted.

$$C_{OH^-} = \frac{6 \times 10^{-3}\ \text{mol}}{300.\ \text{mL}} \times \frac{1000\ \text{mL}}{\text{L}} = \frac{2 \times 10^{-2}\ \text{mol OH}^-}{\text{L}}$$

39. $HCl(aq) + NaOH(aq) \longrightarrow H_2O(l) + NaCl(aq)$

$$24.16 \times 10^{-3}\ \text{L NaOH soln} \times \frac{0.106\ \text{mol NaOH}}{\text{L}} \times \frac{1\ \text{mol HCl}}{\text{mol NaOH}} = 2.56 \times 10^{-3}\ \text{HCl}$$

$$\text{Molarity of HCl} = \frac{2.56 \times 10^{-3}\ \text{mol}}{25.00 \times 10^{-3}\ \text{L}} = \frac{0.102\ \text{mol}}{\text{L}}$$

40. $2\ HNO_3(aq) + Ca(OH)_2(aq) \longrightarrow 2\ H_2O(l) + Ca(NO_3)_2(aq)$

$$34.66 \times 10^{-3}\ \text{L HNO}_3\ \text{soln.} \times \frac{0.0980\ \text{mol HNO}_3}{\text{L}} \times \frac{1\ \text{mol Ca(OH)}_2}{2\ \text{mol HNO}_3}$$

$$= 1.70 \times 10^{-3}\ \text{mol Ca(OH)}_2$$

$$\text{Molarity of Ca(OH)}_2 = \frac{1.70 \times 10^{-3}\ \text{mol}}{50.00 \times 10^{-3}\ \text{L}} = 0.0340\ M$$

Oxidation-Reduction Reactions

41. a. K, +1: O, -2: Mn, +7

b. Ni, +4: O, -2

c. $K_4Fe(CN)_6$: Fe, +2

K^+ ions and $Fe(CN)_6^{4-}$ anions; $Fe(CN)_6^{4-}$ composed of Fe^{2+} and CN^-

d. $(NH_4)_2HPO_4$ is made of NH_4^+ cations and HPO_4^{2-} anions. Assign +1 as oxidation number of H and -2 as oxidation number of O. Then we get N, -3: P, +5

e. P, +3: O, -2

f. O, -2: Fe, +8/3

g. O, -2: F, -1: Xe, +6

h. S, +4: F, -1

i. C, +2: O, -2

j. Na, +1: O, -2: C, +3

42. a. UO_2^{2+}: O, -2: U, $x + 2(-2) = +2$, $x = \underline{+6}$

b. As_2O_3: O, -2: As, $2(x) + 3(-2) = 0$, $x = \underline{+3}$

c. $NaBiO_3$: Na, +1: O, -2: Bi, $+1 + x + 3(-2) = 0$, $x = \underline{+5}$

d. As_4: As, O

e. $HAsO_2$: assign H = +1 and O = -2: As, $+1 + x + 2(-2) = 0$; $x = \underline{+3}$

f. $Mg_2P_2O_7$: composed of Mg^{2+} ions and $P_2O_7^{4-}$ ions. Oxidation numbers are:

Mg, +2: O, -2: P, +5 .

g. $Na_2S_2O_3$: Na^+ ions and $S_2O_3^{2-}$ ions Na, +1: O, -2: S, +2

h. Hg_2Cl_2: Hg, +1: Cl, -1

i. $Ca(NO_3)_2$: Ca^{2+} ions and NO_3^- ions Ca, +2: O, -2: N, +5

43. OCl^-: oxidation number of oxygen is (-2).

$-2 + x = -1$: $x = +1$: The oxidation number of Cl in OCl^- is +1.

ClO_2^-: $2(-2) + x = -1$: $x = +3$ ClO_3^-: $3(-2) + x = -1$: $x = +5$ ClO_4^-: $4(-2) + x = -1$: $x = +7$

44. a. -3 d. +2 g. +3
b. -3 e +1 h. +5
c. $2(x) + 4 = 0$ f. +4 i. 0
$x = -2$

45.

	Redox?	Oxidizing Agent	Reducing Agent	Substance Oxidized	Substance Reduced
a.	Yes	O_2	CH_4	CH_4 (C)	O_2 (O)
b.	Yes	HCl	Zn	Zn	HCl (H)
c.	No	---	---	---	---
d.	Yes	O_3	NO	NO (N)	O_3 (O)
e.	Yes	H_2O_2	H_2O_2	H_2O_2 (O)	H_2O_2 (O)
f.	Yes	CuCl	CuCl	CuCl (Cu)	CuCl (Cu)

In c, no oxidation numbers change from reactants to products.

46.

	Redox?	Oxidizing Agent	Reducing Agent	Substance Oxidized	Substance Reduced
a.	Yes	Ag^+	Cu	Cu	Ag^+
b.	No	---	---	---	---
c.	No	---	---	---	---
d.	Yes	$SiCl_4$	Mg	Mg	$SiCl_4$ (Si)
e.	No	---	---	---	---

In b, c, and e, no oxidation numbers change.

47. a. $Zn \rightarrow Zn^{2+} + 2\,e^-$ $\qquad 2\,e^- + 2\,HCl \rightarrow H_2 + 2Cl^-$

Adding the two balanced half reactions, $Zn(s) + 2\,HCl(aq) \rightarrow H_2(g) + Zn^{2+}(aq) + 2Cl^-(aq)$

b. $3\,I^- \rightarrow I_3^- + 2e^-$ $\qquad ClO^- \rightarrow Cl^-$

$2e^- + 2H^+ + ClO^- \rightarrow Cl^- + H_2O$

Adding the two balanced half reactions so electrons cancel:

$$3\,I^-(aq) + 2\,H^+(aq) + ClO^-(aq) \rightarrow I_3^-(aq) + Cl^-(aq) + H_2O(l)$$

c. $As_2O_3 \rightarrow H_3AsO_4$ $\qquad NO_3^- \rightarrow NO + 2\,H_2O$

$As_2O_3 \rightarrow 2\,H_3AsO_4$ $\qquad 4\,H^+ + NO_3^- \rightarrow NO + 2\,H_2O$

left 3 - O $\qquad (3\,e^- + 4\,H^+ + NO_3^- \rightarrow NO + 2\,H_2O) \times 4$

right 6 H and 8 - O

right hand side has 6 extra H and 5 extra O

Balance the oxygen atoms first using H_2O, then balance H using H^+, and finally balance charge using electrons.

$$(5\,H_2O + As_2O_3 \rightarrow 2\,H_3AsO_4 + 4\,H^+ + 4\,e^-) \times 3$$

Common factor is a transfer of 12 e^-. Add half reactions so electrons cancel.

$$12\,e^- + 16\,H^+ + 4\,NO_3^- \rightarrow 4\,NO + 8\,H_2O$$
$$15\,H_2O + 3\,As_2O_3 \rightarrow 6\,H_3AsO_4 + 12\,H^+ + 12\,e^-$$

$$7\,H_2O(l) + 4\,H^+(aq) + 3\,As_2O_3(s) + 4\,NO_3^-(aq) \rightarrow 4\,NO(g) + 6\,H_3AsO_4(aq)$$

d. $(2\,Br^- \rightarrow Br_2 + 2\,e^-) \times 5$ $\qquad MnO_4^- \rightarrow Mn^{2+} + 4\,H_2O$

$(5\,e^- + 8\,H^+ + MnO_4^- \rightarrow Mn^{2+} + 4\,H_2O) \times 2$

$$10\,Br^- \rightarrow 5\,Br_2 + 10\,e^-$$
$$10\,e^- + 16\,H^+ + 2\,MnO_4^- \rightarrow 2\,Mn^{2+} + 8\,H_2O$$

$$16\,H^+(aq) + 2\,MnO_4^-(aq) + 10\,Br^-(aq) \rightarrow 5\,Br_2(l) + 2\,Mn^{2+}(aq) + 8\,H_2O(l)$$

e. $CH_3OH + Cr_2O_7^{2-} \rightarrow CH_2O + Cr^{3+}$

$CH_3OH \rightarrow CH_2O$

$(CH_3OH \rightarrow CH_2O + 2\,H^+ + 2\,e^-) \times 3$

$$Cr_2O_7^{2-} \rightarrow Cr^{3+}$$
$$14\ H^+ + Cr_2O_7^{2-} \rightarrow 2\ Cr^{3+} + 7\ H_2O$$
$$6\ e^- + 14\ H^+ + Cr_2O_7^{2-} \rightarrow 2\ Cr^{3+} + 7\ H_2O$$

$$3\ CH_3OH \rightarrow 3\ CH_2O + 6\ H^+ + 6\ e^-$$
$$6\ e^- + 14\ H^+ + Cr_2O_7^{2-} \rightarrow 2\ Cr^{3+} + 7\ H_2O$$

$$8\ H^+(aq) + 3\ CH_3OH(l) + Cr_2O_7^{2-}(aq) \rightarrow 2\ Cr^{3+}(aq) + 3\ CH_2O(l) + 7\ H_2O(l)$$

48. a. $(Cu \rightarrow Cu^{2+} + 2e^-) \times 3$

$$HNO_3 \rightarrow NO + 2\ H_2O$$
$$(3\ e^- + 3\ H^+ + HNO_3 \rightarrow NO + 2\ H_2O) \times 2$$

$$3\ Cu \rightarrow 3\ Cu^{2+} + 6\ e^-$$
$$6\ e^- + 6\ H^+ + 2\ HNO_3 \rightarrow 2\ NO + 4\ H_2O$$

$$3\ Cu(s) + 6\ H^+(aq) + 2\ HNO_3(aq) \rightarrow 3\ Cu^{2+}(aq) + 2\ NO(g) + 4\ H_2O(l)$$

or $3\ Cu(s) + 8\ HNO_3(aq) \rightarrow 3\ Cu(NO_3)_2(aq) + 2\ NO(g) + 4\ H_2O(l)$

b. $(2\ Cl^- \rightarrow Cl_2 + 2\ e^-) \times 3$

$$Cr_2O_7^{2-} \rightarrow 2\ Cr^{3+} + 7\ H_2O$$
$$6\ e^- + 14\ H^+ + Cr_2O_7^{2-} \rightarrow 2\ Cr^{3+} + 7\ H_2O$$

Add the two half reactions with six electrons being transferred:

$$6\ Cl^- \rightarrow 3\ Cl_2 + 6\ e^-$$
$$6\ e^- + 14\ H^+ + Cr_2O_7^{2-} \rightarrow 2\ Cr^{3+} + 7\ H_2O$$

$$14\ H^+(aq) + Cr_2O_7^{2-}\ (aq) + 6\ Cl^-(aq) \rightarrow 3\ Cl_2(g) + 2\ Cr^{3+}(aq) + 7\ H_2O\ (l)$$

c.

$$Pb \rightarrow PbSO_4$$
$$Pb + H_2SO_4 \rightarrow PbSO_4 + 2\ H^+$$
$$Pb + H_2SO_4 \rightarrow PbSO_4 + 2\ H^+ + 2\ e^-$$

$$PbO_2 \rightarrow PbSO_4$$
$$PbO_2 + H_2SO_4 \rightarrow PbSO_4 + 2\ H^+$$
$$2\ e^- + 2\ H^+ + PbO_2 + H_2SO_4 \rightarrow PbSO_4 + 2\ H_2O$$

Add the two half reactions:

$$2\,e^- + 2\,H^+ + PbO_2 + H_2SO_4 \rightarrow PbSO_4 + 2\,H_2O$$
$$Pb + H_2SO_4 \rightarrow PbSO_4 + 2\,H^+ + 2\,e^-$$

$$Pb(s) + 2\,H_2SO_4(aq) + PbO_2(s) \rightarrow 2\,PbSO_4(s) + 2\,H_2O(l)$$

This is the reaction that occurs in an automobile lead storage battery.

d.

$$Mn^{2+} \rightarrow MnO_4^-$$
$$(4\,H_2O + Mn^{2+} \rightarrow MnO_4^- + 8\,H^+ + 5\,e^-) \times 2$$

$$NaBiO_3 \rightarrow Bi^{3+} + Na^+$$
$$6\,H^+ + NaBiO_3 \rightarrow Bi^{3+} + Na^+ + 3\,H_2O$$
$$(2\,e^- + 6\,H^+ + NaBiO_3 \rightarrow Bi^{3+} + Na^+ + 3\,H_2O) \times 5$$

$$8\,H_2O + 2\,Mn^{2+} \rightarrow 2\,MnO_4^- + 16\,H^+ + 10\,e^-$$
$$10\,e^- + 30\,H^+ + 5\,NaBiO_3 \rightarrow 5\,Bi^{3+} + 5\,Na^+ + 15\,H_2O$$

$$8\,H_2O + 30\,H^+ + 2\,Mn^{2+} + 5\,NaBiO_3 \rightarrow 2\,MnO_4^- + 5\,Bi^{3+} + 5\,Na^+ + 15\,H_2O + 16\,H^+$$

Simplifying to:

$$14\,H^+(aq) + 2\,Mn^{2+}(aq) + 5\,NaBiO_3(s) \rightarrow 2\,MnO_4^-(aq) + 5\,Bi^{3+}(aq) + 5\,Na^+(aq) + 7\,H_2O(l)$$

e.

$$H_3AsO_4 \rightarrow AsH_3$$
$$H_3AsO_4 \rightarrow AsH_3 + 4\,H_2O$$
$$8\,e^- + 8\,H^+ + H_3AsO_4 \rightarrow AsH_3 + 4\,H_2O$$

$$(Zn \rightarrow Zn^{2+} + 2\,e^-) \times 4$$

$$8\,e^- + 8\,H^+ + H_3AsO_4 \rightarrow AsH_3 + 4\,H_2O$$
$$4\,Zn \rightarrow 4\,Zn^{2+} + 8\,e^-$$

$$8\,H^+(aq) + H_3AsO_4(aq) + 4\,Zn(s) \rightarrow 4\,Zn^{2+}(aq) + AsH_3(g) + 4\,H_2O(l)$$

49. a.

$$Al \rightarrow Al(OH)_4^-$$
$$4\,OH^- + Al \rightarrow Al(OH)_4^-$$
$$4\,OH^- + Al \rightarrow Al(OH)_4^- + 3\,e^-$$

$$MnO_4^- \rightarrow MnO_2$$
$$4\,OH^- + 4\,H^+ + MnO_4^- \rightarrow MnO_2 + 2\,H_2O + 4\,OH^-$$
$$2\,H_2O + MnO_4^- \rightarrow MnO_2 + 4\,OH^-$$
$$3\,e^- + 2\,H_2O + MnO_4^- \rightarrow MnO_2 + 4\,OH^-$$

$$4\ OH^- + Al \rightarrow Al(OH)_4^- + 3\ e^-$$
$$3\ e^- + 2\ H_2O + MnO_4^- \rightarrow MnO_2 + 4\ OH^-$$

$$2\ H_2O(l) + Al(s) + MnO_4^-(aq) \rightarrow Al(OH)_4^-(aq) + MnO_2(s)$$

b.

$$Cl_2 \rightarrow Cl^-$$
$$2\ e^- + Cl_2 \rightarrow 2\ Cl^-$$

$$Cl_2 \rightarrow ClO^-$$
$$2\ H_2O + Cl_2 \rightarrow 2\ ClO^- + 4\ H^+$$
$$4\ OH^- + 2\ H_2O + Cl_2 \rightarrow 2ClO^- + 4\ H^+ + 4\ OH^-$$
$$4\ OH^- + Cl_2 \rightarrow 2\ ClO^- + 2\ H_2O + 2\ e^-$$

$$2\ e^- + Cl_2 \rightarrow 2\ Cl^-$$
$$4\ OH^- + Cl_2 \rightarrow 2\ ClO^- + 2\ H_2O + 2\ e^-$$

$$4\ OH^- + 2\ Cl_2 \rightarrow 2\ Cl^- + 2\ ClO^- + 2\ H_2O$$

or $$2\ OH^-(aq) + Cl_2(g) \rightarrow Cl^-(aq) + ClO^-(aq) + H_2O(l)$$

c.

$$NO_2^- \rightarrow NH_3$$
$$7\ H^+ + NO_2^- \rightarrow NH_3 + 2\ H_2O$$
$$7\ OH^- + 7\ H^+ + NO_2^- \rightarrow NH_3 + 2\ H_2O + 7\ OH^-$$
$$5\ H_2O + NO_2^- \rightarrow NH_3 + 7\ OH^-$$
$$6\ e^- + NO_2^- + 5\ H_2O \rightarrow NH_3 + 7\ OH^-$$

$$Al \rightarrow AlO_2^-$$
$$4\ OH^- + 2\ H_2O + Al \rightarrow AlO_2^- + 4\ H^+ + 4\ OH^-$$
$$4\ OH^- + Al \rightarrow AlO_2^- + 2\ H_2O$$
$$(4\ OH^- + Al \rightarrow AlO_2^- + 2\ H_2O + 3\ e^-) \times 2$$

$$6\ e^- + NO_2^- + 5\ H_2O \rightarrow NH_3 + 7\ OH^-$$
$$8\ OH^- + 2\ Al \rightarrow 2\ AlO_2^- + 4\ H_2O + 6\ e^-$$

$$OH^-(aq) + H_2O(l) + NO_2^-(aq) + 2\ Al(s) \rightarrow NH_3(g) + 2\ AlO_2^-(aq)$$

50. a.

$$Cr \rightarrow Cr(OH)_3$$
$$3\ OH^- + Cr \rightarrow Cr(OH)_3 + 3\ e^-$$

$$CrO_4^- \rightarrow Cr(OH)_3$$
$$5\ H^+ + CrO_4^{2-} \rightarrow Cr(OH)_3 + H_2O$$
$$5\ OH^- + 5\ H^+ + CrO_4^{2-} \rightarrow Cr(OH)_3 + H_2O + 5\ OH^-$$
$$4\ H_2O + CrO_4^{2-} \rightarrow Cr(OH)_3 + 5\ OH^-$$
$$3\ e^- + 4\ H_2O + CrO_4^{2-} \rightarrow Cr(OH)_3 + 5\ OH^-$$

$$3\ OH^- + Cr \rightarrow Cr(OH)_3 + 3\ e^-$$
$$3\ e^- + 4\ H_2O + CrO_4^{2-} \rightarrow Cr(OH)_3 + 5\ OH^-$$

$$4\ H_2O(l) + Cr(s) + CrO_4^{2-}(aq) \rightarrow 2\ Cr(OH)_3(s) + 2\ OH^-(aq)$$

b. $$MnO_4^- + S^{2-} \rightarrow MnS + S$$

$$S^{2-} \rightarrow S$$
$$(S^{2-} \rightarrow S + 2\ e^-) \times 5$$

$$MnO_4^- \rightarrow MnS$$
$$MnO_4^- + S^{2-} \rightarrow MnS$$
$$8\ OH^- + 8\ H^+ + MnO_4^- + S^{2-} \rightarrow MnS + 4\ H_2O + 8\ OH^-$$
$$4\ H_2O + MnO_4^- + S^{2-} \rightarrow MnS + 8\ OH^-$$
$$(5\ e^- + 4\ H_2O + MnO_4^- + S^{2-} \rightarrow MnS + 8\ OH^-) \times 2$$

Common factor is a transfer of 10 e^-.

$$5\ S^{2-} \rightarrow 5\ S + 10\ e^-$$
$$10\ e^- + 8\ H_2O + 2\ MnO_4^- + 2\ S^{2-} \rightarrow 2\ MnS + 16\ OH^-$$

$$8\ H_2O(l) + 2\ MnO_4^-(aq) + 7\ S^{2-}(aq) \rightarrow 5\ S(s) + 2\ MnS(s) + 16\ OH^-(aq)$$

c. $$CN^- \rightarrow CNO^-$$
$$H_2O + CN^- \rightarrow CNO^- + 2\ H^+$$
$$2\ OH^- + H_2O + CN^- \rightarrow CNO^- + 2\ H^+ + 2\ OH^-$$
$$(2\ OH^- + CN^- \rightarrow CNO^- + H_2O + 2\ e^-) \times 3$$

$$MnO_4^- \rightarrow MnO_2$$
$$4\ H^+ + MnO_4^- \rightarrow MnO_2 + 2\ H_2O$$
$$4\ H^+ + 4\ OH^- + MnO_4^- \rightarrow MnO_2 + 2\ H_2O + 4\ OH^-$$
$$(3\ e^- + MnO_4^- + 2\ H_2O \rightarrow MnO_2 + 4\ OH^-) \times 2$$

Common factor is a transfer of 6 electrons.

$$6\ OH^- + 3\ CN^- \rightarrow 3\ CNO^- + 3\ H_2O + 6\ e^-$$

$$6\ e^- + 2\ MnO_4^- + 4\ H_2O \rightarrow 2\ MnO_2 + 8\ OH^-$$

$$3\ CN^-(aq) + 2\ MnO_4^-(aq) + H_2O(l) \rightarrow 3\ CNO^-(aq) + 2\ MnO_2(s) + 2\ OH^-(aq)$$

51. a.

$$\overset{+1\ -1}{NaCl} + \overset{+1\ +6\ -2}{H_2SO_4} + \overset{+4\ -2}{MnO_2} \rightarrow \overset{+1\ +6\ -2}{Na_2SO_4} + \overset{+2\ -1}{MnCl_2} + \overset{0}{Cl_2} + \overset{+1\ -2}{H_2O}$$

$$2\ NaCl + H_2SO_4 + MnO_2 \rightarrow Na_2SO_4 + MnCl_2 + Cl_2 + H_2O$$

To balance Cl, need two more NaCl for Cl in $MnCl_2$.

$$4\ NaCl + H_2SO_4 + MnO_2 \rightarrow Na_2SO_4 + MnCl_2 + Cl_2 + H_2O$$

Balance Na and SO_4^{2-}

$$4\ NaCl + 2\ H_2SO_4 + MnO_2 \rightarrow 2\ Na_2SO_4 + MnCl_2 + Cl_2 + H_2O$$

Left: 4-H and 10-O

Right: 8-O not counting H_2O

Need 2 H_2O on right

$$4\ NaCl(aq) + 2\ H_2SO_4(aq) + MnO_2(s) \rightarrow 2\ Na_2SO_4(aq) + MnCl_2(aq) + Cl_2(g) + 2\ H_2O(l)$$

52. $Au + HNO_3^- + HCl \rightarrow AuCl_4^- + NO$

Only deal with ions that are reacting: $Au + NO_3^- + Cl^- \rightarrow AuCl_4^- + NO$

$$Au + 4\ Cl^- \rightarrow AuCl_4^- + 3\ e^-$$

$$NO_3^- \rightarrow NO$$

$$3\ e^- + 4\ H^+ + NO_3^- \rightarrow NO + 2\ H_2O$$

Adding the two balanced half reactions:

$$Au(s) + 4\ Cl^-(aq) + 4\ H^+(aq) + NO_3^-(aq) \rightarrow AuCl_4^-(aq) + NO(g) + 2\ H_2O(l)$$

53. $H_2C_2O_4 \rightarrow 2\ CO_2 + 2\ H^+$

$$(H_2C_2O_4 \rightarrow 2\ CO_2 + 2\ H^+ + 2\ e^-) \times 5$$

$$(5\ e^- + 8\ H^+ + MnO_4^- \rightarrow Mn^{2+} + 4\ H_2O) \times 2$$

$$5\ H_2C_2O_4 \rightarrow 10\ CO_2 + 10\ H^+ + 10\ e^-$$
$$10\ e^- + 16\ H^+ + 2\ MnO_4^- \rightarrow 2\ Mn^{2+} + 8\ H_2O$$

$$6\ H^+ + 5\ H_2C_2O_4 + 2\ MnO_4^- \rightarrow 10\ CO_2 + 2\ Mn^{2+} + 8\ H_2O$$

$$0.1058\ \text{g oxalic acid} \times \frac{1\ \text{mol oxalic acid}}{90.04\ \text{g}} \times \frac{2\ \text{mol}\ MnO_4^-}{5\ \text{mol oxalic acid}} = 4.700 \times 10^{-4}\ \text{mol}$$

$$\text{Molarity} = \frac{4.700 \times 10^{-4}\ \text{mol}}{28.97\ \text{mL}} \times \frac{1000\ \text{mL}}{\text{L}} = 1.622 \times 10^{-2}\ M$$

54.

$$MnO_4^- + H_2C_2O_4 \rightarrow CO_2 + Mn^{2+}$$

$$(5\ e^- + 8\ H^+ + MnO_4^- \rightarrow Mn^{2+} + 4\ H_2O) \times 2$$
$$(H_2C_2O_4 \rightarrow 2\ CO_2 + 2\ H^+ + 2\ e^-) \times 5$$

$$2\ MnO_4^- + 6\ H^+ + 5\ H_2C_2O_4 \rightarrow 10\ CO_2 + 2\ Mn^{2+} + 8\ H_2O$$

$$0.500\ \text{g}\ H_2C_2O_4 \times \frac{1\ \text{mol}\ H_2C_2O_4}{90.04\ \text{g}\ H_2C_2O_4} \times \frac{10\ \text{mol}\ CO_2}{5\ \text{mol}\ H_2C_2O_4} = 1.11 \times 10^{-2}\ \text{mol}\ CO_2$$

$$50.00 \times 10^{-3}\ \text{L}\ MnO_4^-\ \text{soln.} \times \frac{0.0200\ \text{mol}\ MnO_4^-}{\text{L}} \times \frac{10\ \text{mol}\ CO_2}{2\ \text{mol}\ MnO_4^-} = 5.00 \times 10^{-3}\ \text{mol}\ CO_2$$

The $KMnO_4$ solution is the limiting reagent.

$$5.00 \times 10^{-3}\ \text{mol}\ CO_2 \times \frac{44.01\ \text{g}\ CO_2}{\text{mol}\ CO_2} = 0.220\ \text{g}\ CO_2$$

55. a. $Fe^{2+} \rightarrow Fe^{3+} + e^-$

$5\ e^- + 8\ H^+ + MnO_4^- \rightarrow Mn^{2+} + 4\ H_2O$

Balanced equation is:

$$8\ H^+ + MnO_4^- + 5\ Fe^{2+} \rightarrow 5\ Fe^{3+} + Mn^{2+} + 4\ H_2O$$

$$20.62 \times 10^{-3}\ \text{L soln} \times \frac{0.0216\ \text{mol}\ MnO_4^-}{\text{L soln}} \times \frac{5\ \text{mol}\ Fe^{2+}}{1\ \text{mol}\ MnO_4^-} = 2.23 \times 10^{-3}\ \text{mol}\ Fe^{2+}$$

$$\text{Molarity} = \frac{2.23 \times 10^{-3}\ \text{mol}\ Fe^{2+}}{50.0 \times 10^{-3}\ \text{L}} = 4.46 \times 10^{-2}\ M$$

b. $Fe^{2+} \rightarrow Fe^{3+} + e^-$

$Cr_2O_7^{2-} \rightarrow 2\ Cr^{3+} + 7\ H_2O$

$6\ e^- + 14\ H^+ + Cr_2O_7^{2-} \rightarrow 2\ Cr^{3+} + 7\ H_2O$

The balanced equation is:

$14\ H^+ + Cr_2O_7^{2-} + 6\ Fe^{2+} \rightarrow 6\ Fe^{3+} + 2\ Cr^{3+} + 7\ H_2O$

$$50.00 \times 10^{-3}\ L \times \frac{4.46 \times 10^{-2}\ mol\ Fe^{2+}}{L} \times \frac{1\ mol\ Cr_2O_7^{2-}}{6\ mol\ Fe^{2+}} \times \frac{1\ L}{0.0150\ mol\ Cr_2O_7^{2-}}$$

$= 2.48 \times 10^{-2}$ L or 24.8 mL

56. This is the same titration as in problem 4.55. The stoichiometry of the reaction depends on the reaction:

$$8\ H^+ + MnO_4^- + 5\ Fe^{2+} \rightarrow 5\ Fe^{3+} + Mn^{2+} + 4\ H_2O$$

From the titration data we can get the number of moles of Fe^{2+}. We then convert this to a mass of iron and calculate the mass percent of iron in the sample.

$$41.95 \times 10^{-3}\ L\ MnO_4^- \times \frac{0.0205\ mol\ MnO_4^-}{L} \times \frac{5\ mol\ Fe^{2+}}{mol\ MnO_4^-} = 4.30 \times 10^{-3}\ mol\ Fe^{2+}$$

$$4.30 \times 10^{-3}\ mol\ Fe \times \frac{55.85\ g\ Fe}{mol\ Fe} = 0.240\ g\ Fe$$

$\%Fe = \frac{0.240\ g}{0.6128\ g} \times 100 = 39.2\%$ Fe since the concentration of the permanganate solution is only accurate to three significant figures.

ADDITIONAL EXERCISES

57. $$4.25\ g\ Ca \times \frac{1\ mol\ Ca}{40.08\ g\ Ca} \times \frac{1\ mol\ Ca(OH)_2}{mol\ Ca} \times \frac{2\ mol\ OH^-}{mol\ Ca(OH)_2} = 0.212\ mol\ OH^-$$

$$Molarity = \frac{0.212\ mol}{225 \times 10^{-3}\ L} = 0.942\ M$$

58. a. $4\ NH_3(g) + 5\ O_2(g) \rightarrow 4\ NO(g) + 6\ H_2O(g)$

-3 +1   0   +2 -2   +1 -2   oxidation numbers

$2\ NO(g) + O_2(g) \rightarrow 2\ NO_2(g)$

+2 -2   0   +4 -2

$3\ NO_2(g) + H_2O(l) \rightarrow 2\ HNO_3(aq) + NO(g)$

+4 -2   +1 -2   +1+5 -2   +2 -2

All three reactions are oxidation reduction reactions.

b. $4\ NH_3 + 5\ O_2 \rightarrow 4\ NO + 6\ H_2O$

O_2 is the oxidizing agent.

NH_3 is the reducing agent.

$2\ NO + O_2 \rightarrow 2\ NO_2$

O_2 is the oxidizing agent.

NO is the reducing agent.

$3\ NO_2 + H_2O \rightarrow 2\ HNO_3 + NO$

NO_2 is both the oxidizing and reducing agent.

c. $$5.0 \times 10^6\ g\ NH_3 \times \frac{1\ mol\ NH_3}{17.03\ g\ NH_3} \times \frac{4\ mol\ NO}{4\ mol\ NH_3} = 2.9 \times 10^5\ mol\ NO$$

$$5.0 \times 10^7\ g\ O_2 \times \frac{1\ mol\ O_2}{32.00\ g\ O_2} \times \frac{4\ mol\ NO}{5\ mol\ O_2} = 1.3 \times 10^6\ mol\ NO$$

The NH_3 will be consumed completely before the O_2 will. NH_3 is limiting.

$$2.9 \times 10^5\ mol\ NO \times \frac{30.01\ g\ NO}{mol\ NO} = 8.7 \times 10^6\ g\ NO$$

59. $Mn + HNO_3 \rightarrow Mn^{2+} + NO_2$

$Mn \rightarrow Mn^{2+} + 2\ e^-$

$HNO_3 \rightarrow NO_2$

$HNO_3 \rightarrow NO_2 + H_2O$

$(e^- + H^+ + HNO_3 \rightarrow NO_2 + H_2O) \times 2$

$$Mn \rightarrow Mn^{2+} + 2\ e^-$$

$$2\ e^- + 2\ H^+ + 2\ HNO_3 \rightarrow 2\ NO_2 + 2\ H_2O$$

$$2\ H^+(aq) + Mn(s) + 2\ HNO_3(aq) \rightarrow Mn^{2+}(aq) + 2\ NO_2(g) + 2\ H_2O(l)$$

$Mn^{2+} + IO_4^- \rightarrow MnO_4^- + IO_3^-$

$(4\ H_2O + Mn^{2+} \rightarrow MnO_4^- + 8\ H^+ + 5\ e^-) \times 2$

$(2\ e^- + 2\ H^+ + IO_4^- \rightarrow IO_3^- + H_2O) \times 5$

$$8\ H_2O + 2\ Mn^{2+} \longrightarrow 2\ MnO_4^- + 16\ H^+ + 10\ e^-$$

$$10\ e^- + 10\ H^+ + 5\ IO_4^- \longrightarrow 5\ IO_3^- + 5\ H_2O$$

$$3\ H_2O(l) + 2\ Mn^{2+}(aq) + 5\ IO_4^-(aq) \longrightarrow 2\ MnO_4^-(aq) + 5\ IO_3^-(aq) + 6\ H^+(aq)$$

60. $1.50\text{ g Cu} \times \frac{1\text{ mol Cu}}{63.55\text{ g}} \times \frac{2\text{ mol Ag}^+}{\text{mol Cu}} = 4.72 \times 10^{-2}\text{ mol Ag}^+$ required to dissolve all of the Cu.

We have: $250 \times 10^{-3}\text{ L} \times \frac{0.20\text{ mol Ag}^+}{\text{L}} = 5.0 \times 10^{-2}\text{ mol Ag}^+$

Yes, all of the copper will dissolve.

61. $1.00\text{ L} \times \frac{0.200\text{ mol Na}_2\text{S}_3\text{O}_3}{\text{L}} \times \frac{1\text{ mol AgBr}}{2\text{ mol Na}_2\text{S}_2\text{O}_3} \times \frac{187.8\text{ g AgBr}}{\text{mol AgBr}} = 18.8\text{ g}$

62. a. The unbalanced reaction is:

$$KOH(aq) + Mg(NO_3)_2(aq) \longrightarrow Mg(OH)_2(s) + KNO_3(aq)$$

Balancing the equation gives:

$$2\ KOH(aq) + Mg(NO_3)_2(aq) \longrightarrow Mg(OH)_2(s) + 2\ KNO_3(aq)$$

b. The precipitate is magnesium hydroxide.

c. $0.1000\text{ L soln} \times \frac{0.200\text{ mol KOH}}{\text{L soln}} \times \frac{1\text{ mol Mg(OH)}_2}{2\text{ mol KOH}} = 1.00 \times 10^{-2}\text{ mol}$

$0.1000\text{ L soln} \times \frac{0.200\text{ mol Mg(NO}_3\text{)}_2}{\text{L soln}} \times \frac{1\text{ mol Mg(OH)}_2}{\text{mol Mg(NO}_3\text{)}_2} = 2.00 \times 10^{-2}\text{ mol}$

Therefore, the potassium hydroxide is the limiting reagent and the mass of magnesium hydroxide produced is:

$$1.00 \times 10^{-2}\text{ mol} \times \frac{58.33\text{ g}}{\text{mol}} = 0.583\text{ g Mg(OH)}_2$$

63. The molecular weight of $Al(C_9H_6NO)_3$ is

$$1\text{ mol Al}\left(\frac{26.98\text{ g}}{\text{mol Al}}\right) + 27\text{ mol C}\left(\frac{12.01\text{ g}}{\text{mol C}}\right) + 3\text{ mol N}\left(\frac{14.01\text{ g}}{\text{mol N}}\right)$$

$$+ 3\text{ mol O}\left(\frac{16.00\text{ g}}{\text{mol O}}\right) + 18\text{ mol H}\left(\frac{1.008\text{ g}}{\text{mol H}}\right) = \frac{459.42\text{ g}}{\text{mol}}$$

$$0.1248\text{ g Al(C}_9\text{H}_6\text{NO)}_3 \times \frac{26.98\text{ g Al}}{459.42\text{ g Al(C}_9\text{H}_6\text{NO)}_3} = 7.329 \times 10^{-3}\text{ g Al}$$

$$\%\text{Al} = \frac{7.329 \times 10^{-3}\text{ g}}{1.8571\text{ g}} \times 100 = 0.3947\ \%\text{ Al}$$

64. $0.1824 \text{ g TlI} \times \frac{204.4 \text{ g Tl}}{331.3 \text{ g TlI}} \times \frac{504.9 \text{ g Tl}_2\text{SO}_4}{408.8 \text{ g Tl}} = 0.1390 \text{ g Tl}_2\text{SO}_4$

$\frac{0.1390 \text{ g Tl}_2\text{SO}_4}{9.486 \text{ g pesticide}} \times 100 = 1.465\% \text{ Tl}_2\text{SO}_4$

65. $CH_3CO_2H(aq) + NaOH(aq) \rightarrow H_2O(l) + CH_3CO_2Na(aq)$

a. $16.58 \times 10^{-3} \text{ L soln} \times \frac{0.5062 \text{ mol NaOH}}{\text{L soln}} \times \frac{1 \text{ mol acetic acid}}{\text{mol NaOH}} = 8.393 \times 10^{-3}$ mol acetic acid

Concentration of acetic acid $= \frac{8.393 \times 10^{-3} \text{ mol}}{0.01000 \text{ L}} = 0.8393\ M$

b. If we have 1.000 L of solution:

Total mass = $1000. \text{ mL} \times \frac{1.006 \text{ g}}{\text{mL}} = 1006 \text{ g}$

Mass of acetic acid = $0.8393 \text{ mol} \times \frac{60.05 \text{ g}}{\text{mol}} = 50.40 \text{ g}$

% acetic acid = $\frac{50.40 \text{ g}}{1006 \text{ g}} \times 100 = 5.010\%$

66. $39.47 \times 10^{-3} \text{ L HCl soln} \times \frac{0.0984 \text{ mol HCl}}{\text{L}} \times \frac{1 \text{ mol NH}_3}{\text{mol HCl}} = 3.88 \times 10^{-3} \text{ mol NH}_3$

Molarity of $NH_3 = \frac{3.88 \times 10^{-3} \text{ mol}}{50.00 \times 10^{-3} \text{ L}} = \frac{0.0776 \text{ mol}}{\text{L}}$

67. Since KHP is a monoprotic acid, the reaction is

$$NaOH + HX \rightarrow NaX + H_2O$$

where HX is short hand for potassium hydrogen phthalate.

$0.1082 \text{ g KHP} \times \frac{1 \text{ mol KHP}}{204.22 \text{ g KHP}} \times \frac{1 \text{ mol NaOH}}{\text{mol KHP}} = 5.298 \times 10^{-4} \text{ mol NaOH}$

There is 5.298×10^{-4} mol of sodium hydroxide in 20.46 mL of solution. Therefore, the concentration of sodium hydroxide is:

$$\frac{5.298 \times 10^{-4} \text{ mol}}{20.46 \times 10^{-3} \text{ L}} = 2.589 \times 10^{-2}\ M$$

68. $Na_2CrO_4(aq) + Pb(NO_3)_2(aq) \rightarrow PbCrO_4(s) + 2\ NaNO_3(aq)$

$0.1000 \text{ L} \times \frac{0.4100 \text{ mol Na}_2\text{CrO}_4}{\text{L}} \times \frac{1 \text{ mol chrome yellow}}{\text{mol Na}_2\text{CrO}_4} = 4.100 \times 10^{-2} \text{ mol}$

$0.1000 \text{ L} \times \frac{0.3200 \text{ mol Pb(NO}_3)_2}{\text{L}} \times \frac{1 \text{ mol chrome yellow}}{\text{mol Pb(NO}_3)_2} = 3.200 \times 10^{-2} \text{ mol}$

$Pb(NO_3)_2$ is limiting. Therefore,

$$3.200 \times 10^{-2} \text{ mol } PbCrO_4 \times \frac{323.2 \text{ g } PbCrO_4}{\text{mol}} = 10.34 \text{ g } PbCrO_4$$

69. a. No, no element shows a change in oxidation number.

b. $$1.0 \text{ L oxalic acid} \times \frac{0.14 \text{ mol oxalic acid}}{\text{L}} \times \frac{1 \text{ mol } Fe_2O_3}{6 \text{ mol oxalic acid}} \times \frac{159.70 \text{ g } Fe_2O_3}{\text{mol } Fe_2O_3}$$

$$= 3.7 \text{ g } Fe_2O_3$$

70. $Fe(NO_3)_3 + 3\ NaOH \longrightarrow Fe(OH)_3 + 3\ NaNO_3$

$$75.0 \times 10^{-3} \text{ L soln} \times \frac{0.105 \text{ mol } Fe(NO_3)_3}{\text{L soln}} \times \frac{1 \text{ mol } Fe(OH)_3}{\text{mol } Fe(NO_3)_3} = 7.88 \times 10^{-3} \text{ mol } Fe(OH)_3$$

$$125 \times 10^{-3} \text{ L soln} \times \frac{0.150 \text{ mol NaOH}}{\text{L soln}} \times \frac{1 \text{ mol } Fe(OH)_3}{3 \text{ mol NaOH}} = 6.25 \times 10^{-3} \text{ mol } Fe(OH)_3$$

The NaOH is limiting.

$$6.25 \times 10^{-3} \text{ mol } Fe(OH)_3 \times \frac{106.87 \text{ g } Fe(OH)_3}{3 \text{ mol NaOH}} = 0.668 \text{ g or } 668 \text{ mg}$$

71. Fe^{2+} will react with MnO_4^- (purple) producing Fe^{3+} and Mn^{2+} (almost colorless). There is no reaction between MnO_4^- and Fe^{3+}. Therefore, add a few drops of the potassium permanganate solution. If the purple color is present the solution contains iron(III) sulfate. If the color disappears, iron(II) sulfate is present.

72. a. The balanced reaction is:

$$16\ H^+(aq) + 2\ Cr_2O_7^{2-}(aq) + C_2H_5OH(l) \longrightarrow 4\ Cr^{3+}(aq) + 11\ H_2O(l) + 2\ CO_2(g)$$

$$35.48 \times 10^{-3} \text{ L soln} \times \frac{0.05182 \text{ mol } K_2Cr_2O_7}{\text{L soln}} \times \frac{1 \text{ mol } C_2H_5OH}{2 \text{ mol } K_2Cr_2O_7} \times \frac{46.07 \text{ g } C_2H_5OH}{\text{mol } C_2H_5OH}$$

$$= 4.235 \times 10^{-2} \text{ g } C_2H_5OH$$

$$\%C_2H_5OH = \frac{4.235 \times 10^{-2} \text{ g}}{50.02 \text{ g}} \times 100 = 0.08467\%$$

The person is not legally intoxicated.

b. $$48.02 \times 10^{-3} \text{ L } K_2Cr_2O_7 \times \frac{0.05182 \text{ mol } K_2Cr_2O_7}{\text{L } K_2Cr_2O_7} \times \frac{1 \text{ mol } C_2H_5OH}{2 \text{ mol } K_2Cr_2O_7}$$

$$\times \frac{46.07 \text{ g } C_2H_5OH}{\text{mol } C_2H_5OH} = 5.732 \times 10^{-2} \text{ g } C_2H_5OH$$

$$\%C_2H_5OH = \frac{5.732 \times 10^{-2} \text{ g}}{48.91 \text{ g}} \times 100 = 0.1172\%$$

Yes, in this case the person is legally intoxicated.

73. $KMnO_4$ added to Sb(III) solution.

$$25.00 \times 10^{-3}\ L \times \frac{0.0233\ mol\ KMnO_4}{L} = 5.83 \times 10^{-4}\ mol$$

$KMnO_4$ left unreacted:

$$8\ H^+ + MnO_4^- + 5\ Fe^{2+} \longrightarrow Mn^{2+} + 5\ Fe^{3+} + 4\ H_2O$$

$$2.58 \times 10^{-3}\ L\ Fe^{2+}\ soln \times \frac{0.0843\ mol\ Fe^{2+}}{L} \times \frac{1\ mol\ MnO_4^-}{5\ mol\ Fe^{2+}} = 4.35 \times 10^{-5}\ mol\ MnO_4^-$$

Thus, 4.35×10^{-5} mol MnO_4^- was unreacted.

The amount of $KMnO_4$ that reacted with Sb(III) was:

$$5.83 \times 10^{-4}\ mol - 0.435 \times 10^{-4}\ mol = 5.40 \times 10^{-4}\ mol$$

$MnO_4^- \longrightarrow Mn^{2+}$ (change by 5 e^-); Sb(III) $\longrightarrow$ Sb(V) (change by 2 e^-)

2 mol MnO_4^- will oxidize 5 mol Sb(III).

$$5.40 \times 10^{-4}\ mol\ MnO_4^- \times \frac{5\ mol\ Sb(III)}{2\ mol\ MnO_4^-} = 1.35 \times 10^{-3}\ mol\ Sb$$

$$1.35 \times 10^{-3}\ mol\ Sb \times \frac{1\ mol\ Sb_2S_3}{2\ mol\ Sb} \times \frac{339.8\ g\ Sb_2S_3}{mol\ Sb_2S_3} = 0.229\ g\ Sb_2S_3$$

$$\%Sb_2S_3 = \frac{0.229\ g\ Sb_2S_3}{0.506\ g\ ore} \times 100 = 45.3\%\ Sb_2S_3$$

74. $$0.5032\ g\ BaSO_4 \times \frac{32.07\ g\ S}{233.4\ g\ BaSO_4} \times \frac{183.19\ g\ saccharin}{32.07\ g\ S} = 0.3949\ g\ saccharin$$

$$\frac{\text{Avg. Mass}}{\text{Tablet}} = \frac{0.3949\ g}{10\ tablets} = \frac{3.949 \times 10^{-2}\ g}{tablet} = \frac{39.39\ mg}{tablet}$$

$$\text{Avg. Mass \%} = \frac{0.3949\ g\ saccharin}{0.5894\ g} \times 100 = 67.00\%\ \text{saccharin by mass}$$

75. $$0.104\ g\ AgCl \times \frac{35.45\ g\ Cl^-}{143.4\ g\ AgCl} = 2.57 \times 10^{-2}\ g\ Cl^-$$

$$\text{Chlorisondiamine} = 14(12.01) + 18(1.008) + 6(35.45) + 2(14.01) = \frac{427.00\ g}{mol}$$

There are 6(35.45) = 212.70 g chlorine for every mole (427.00 g) of chlorisondiamine.

$$2.57 \times 10^{-2}\ g\ Cl^- \times \frac{427.00\ g\ drug}{212.70\ g\ Cl^-} = 5.16 \times 10^{-2}\ g\ drug$$

$$\%\ drug = \frac{5.16 \times 10^{-2}\ g}{1.28\ g} \times 100 = 4.03\%$$

76. $Zn_2P_2O_7$: $2(65.38) + 2(30.97) + 7(16.00) = \frac{304.69\ g}{mol}$

$$0.4089\ g\ Zn_2P_2O_7 \times \frac{130.76\ g\ Zn}{304.69\ g\ Zn_2P_2O_7} = 0.1755\ g\ Zn$$

$$\%Zn = \frac{0.1755\ g}{1.200\ g} \times 100 = 14.63\%\ Zn$$

CHALLENGE PROBLEMS

77. a. HCl(aq) dissociates to $H^+(aq) + Cl^-(aq)$. For simplicity let's use H^+ and Cl^- separately.

$$H^+ \rightarrow H_2 \qquad\qquad Fe \rightarrow HFeCl_4$$

$$(2\ H^+ + 2e^- \rightarrow H_2) \times 3 \qquad\qquad (H^+ + 4\ Cl^- + Fe \rightarrow HFeCl_4 + 3e^-) \times 2$$

$$6\ H^+ + 6\ e^- \rightarrow 3\ H_2$$

$$2\ H^+ + 8\ Cl^- + 2\ Fe \rightarrow 2\ HFeCl_4 + 6\ e^-$$

$$8\ H^+ + 8\ Cl^- + 2\ Fe \rightarrow 2\ HFeCl_4 + 3\ H_2$$

or $8\ HCl(aq) + 2\ Fe(s) \rightarrow 2\ HFeCl_4(aq) + 3\ H_2(g)$

b.

$$IO_3^- \rightarrow I_3^- \qquad\qquad I^- \rightarrow I_3^-$$

$$3\ IO_3^- \rightarrow I_3^- \qquad\qquad (3\ I^- \rightarrow I_3^- + 2\ e^-) \times 8$$

$$3\ IO_3^- \rightarrow I_3^- + 9\ H_2O$$

$$16\ e^- + 18\ H^+ + 3\ IO_3^- \rightarrow I_3^- + 9\ H_2O$$

$$16\ e^- + 18\ H^+ + 3\ IO_3^- \rightarrow I_3^- + 9\ H_2O$$

$$24\ I^- \rightarrow 8\ I_3^- + 16\ e^-$$

$$18\ H^+ + 24\ I^- + 3\ IO_3^- \rightarrow 9\ I_3^- + 9\ H_2O$$

or $6\ H^+(aq) + 8\ I^-(aq) + IO_3^-(aq) \rightarrow 3\ I_3^-(aq) + 3\ H_2O(l)$

c. $Ce^{4+} + e^- \rightarrow Ce^{3+}$

$$Cr(NCS)_6^{4-} \rightarrow Cr^{3+} + NO_3^- + CO_2 + SO_4^{2-}$$

$$54\ H_2O + Cr(NCS)_6^{4-} \rightarrow Cr^{3+} + 6\ NO_3^- + 6\ CO_2 + 6\ SO_4^{2-} + 108\ H^+$$

Charge on left -4

Charge on right (+3) + 6(-1) + 6(-2) + 108(+1) = +93

Add 97 e^- to the right then add two balanced half reactions.

$$54\ H_2O + Cr(NCS)_6^{4-} \rightarrow Cr^{3+} + 6\ NO_3^- + 6\ CO_2 + 6\ SO_4^{2-} + 108\ H^+ + 97\ e^-$$

$$97\ e^- + 97\ Ce^{4+} \rightarrow 97\ Ce^{3+}$$

$$97\ Ce^{4+} + 54\ H_2O + Cr(NCS)_6^{4-} \rightarrow 97\ Ce^{3+} + Cr^{3+} + 6\ NO_3^- + 6\ CO_2 + 6\ SO_4^{2-} + 108\ H^+$$

This is very complicated. A check of the net charge is a good check to see if the equation is balanced. Note: See question for states, i.e., aq, l, or g.

Left: charge = 97 (+4) - 4 = +384

Right: charge = 97 (+3) + 3 + 6(-1) + 6(-2) + 108(+1) = +384

d.

$$Cl_2 \rightarrow Cl^-$$

$$(2\ e^- + Cl_2 \rightarrow 2\ Cl^-) \times 27$$

$$CrI_3 \rightarrow CrO_4^{2-} + IO_4^-$$

$$32\ OH^- + 16\ H_2O + CrI_3 \rightarrow CrO_4^{2-} + 3\ IO_4^- + 32\ H^+ + 32\ OH^-$$

$$32\ OH^- + CrI_3 \rightarrow CrO_4^{2-} + 3\ IO_4^- + 16\ H_2O$$

Net charge on left: -32; Net charge on right: -5

Add 27 e^- to the right.

$$(32\ OH^- + CrI_3 \rightarrow CrO_4^{2-} + 3\ IO_4^- + 16\ H_2O + 27\ e^-) \times 2$$

Common factor is a transfer of 54 e^-

$$54\ e^- + 27\ Cl_2 \rightarrow 54\ Cl^-$$

$$64\ OH^- + 2\ CrI_3 \rightarrow 2\ CrO_4^{2-} + 6\ IO_4^- + 32\ H_2O + 54\ e^-$$

$$64\ OH^-(aq) + 2\ CrI_3(s) + 27\ Cl_2(g) \rightarrow 54\ Cl^-(aq) + 2\ CrO_4^{2-}(aq) + 6\ IO_4^-(aq) + 32\ H_2O(l)$$

e.

$$Ce^{4+} \rightarrow Ce(OH)_3$$

$$(e^- + 3\ OH^- + Ce^{4+} \rightarrow Ce(OH)_3) \times 61$$

$$Fe(CN)_6^{4-} \rightarrow Fe(OH)_3 + CO_3^{2-} + NO_3^-$$

$$3\ OH^- + Fe(CN)_6^{4-} \rightarrow Fe(OH)_3 + 6\ CO_3^{2-} + 6\ NO_3^-$$

36 extra O atoms on right

Add 36 H_2O to left, 72 H^+ to right, then neutralize with 72 OH^- on each side.

$$72\ OH^- + 36\ H_2O + 3\ OH^- + Fe(CN)_6^{4-} \rightarrow Fe(OH)_3 + 6\ CO_3^{2-} + 6\ NO_3^- + 72\ H^+ + 72\ OH^-$$

$$75\ OH^- + Fe(CN)_6^{4-} \rightarrow Fe(OH)_3 + 6\ CO_3^{2-} + 6\ NO_3^- + 36\ H_2O$$

net charge = -79 net charge = -18

Add 61 e^- to the right then add two balanced half reactions.

$$75\ OH^- + Fe(CN)_6^{4-} \rightarrow Fe(OH)_3 + 6\ CO_3^{2-} + 6\ NO_3^- + 36\ H_2O + 61\ e^-$$

$$61\ e^- + 183\ OH^- + 61\ Ce^{4+} \rightarrow 61\ Ce(OH)_3$$

$$258\ OH^-(aq) + Fe(CN)_6^{4-}(aq) + 61\ Ce^{4+}(aq) \rightarrow Fe(OH)_3(s) + 61\ Ce(OH)_3(s) + 6\ CO_3^{2-}(aq) + 6\ NO_3^- + 36\ H_2O(l)$$

f. $Fe(OH)_2 + H_2O_2 \rightarrow Fe(OH)_3$

$$Fe(OH)_2 \rightarrow Fe(OH)_3 \qquad H_2O_2 \rightarrow OH^-$$

$$(OH^- + Fe(OH)_2 \rightarrow Fe(OH)_3 + e^-) \times 2 \qquad 2\ e^- + H_2O_2 \rightarrow 2\ OH^-$$

$$2\ e^- + H_2O_2 \rightarrow 2\ OH^-$$

$$2\ OH^- + 2\ Fe(OH)_2 \rightarrow 2\ Fe(OH)_3 + 2\ e^-$$

$$2\ Fe(OH)_2(s) + H_2O_2(aq) \rightarrow 2\ Fe(OH)_3(s)$$

78. a. $H_3PO_4(aq) + 3\ NaOH(aq) \rightarrow 3\ H_2O(l) + Na_3PO_4(aq)$

b. $2\ Al(OH)_3(s) + 3\ H_2SO_4(aq) \rightarrow 6\ H_2O(l) + Al_2(SO_4)_3(aq)$

c. $H_2Se(aq) + Ba(OH)_2(s) \rightarrow 2\ H_2O(l) + BaSe(s)$

d. $H_2C_2O_4(aq) + 2\ NaOH(aq) \rightarrow 2\ H_2O(l) + Na_2C_2O_4(aq)$

79. $CaCO_3(s) + H_2SO_4(aq) \rightarrow CaSO_4(aq) + H_2O(l) + CO_2(g)$

80. $41.28 \times 10^{-3}\ \text{L NaOH} \times \frac{0.250\ \text{mol NaOH}}{\text{L NaOH}} \times \frac{1\ \text{mol citric acid}}{3\ \text{mol NaOH}} = 3.44 \times 10^{-3}\ \text{mol citric acid}$

The concentration of citric acid is: $\frac{3.44 \times 10^{-3}\ \text{mol}}{0.0100\ \text{L}} = 0.344\ M$

81. $$35.08 \text{ mL NaOH} \times \frac{1 \text{ L}}{1000 \text{ mL}} \times \frac{2.12 \text{ mol NaOH}}{\text{L NaOH}} \times \frac{1 \text{ mol H}_2\text{SO}_4}{2 \text{ mol NaOH}} = 3.72 \times 10^{-2} \text{ mol H}_2\text{SO}_4$$

$$\text{Molarity} = \frac{3.72 \times 10^{-2} \text{ mol}}{10.00 \text{ mL}} \times \frac{1000 \text{ mL}}{\text{L}} = 3.72\ M$$

82. a. $MgO(s) + 2\ HCl(aq) \rightarrow MgCl_2(aq) + H_2O(l)$

$Mg(OH)_2(s) + 2\ HCl(aq) \rightarrow MgCl_2(aq) + 2\ H_2O(l)$

$Al(OH)_3(s) + 3\ HCl(aq) \rightarrow AlCl_3(aq) + 3\ H_2O(l)$

b. Let's calculate the number of moles of HCl neutralized per gram. We can get these directly from the balanced equation.

$$\frac{2 \text{ mol HCl}}{\text{mol MgO}} \times \frac{1 \text{ mol MgO}}{40.\text{ g MgO}} = \frac{5.0 \times 10^{-2} \text{ mol HCl}}{\text{g MgO}}$$

$$\frac{2 \text{ mol HCl}}{\text{mol Mg(OH)}_2} \times \frac{1 \text{ mol Mg(OH)}_2}{58 \text{ g Mg(OH)}_2} = \frac{3.4 \times 10^{-2} \text{ mol HCl}}{\text{g Mg(OH)}_2}$$

$$\frac{3 \text{ mol HCl}}{\text{mol Al(OH)}_3} \times \frac{1 \text{ mol Al(OH)}_3}{78 \text{ g Al(OH)}_3} = \frac{3.8 \times 10^{-2} \text{ mol HCl}}{\text{g Al(OH)}_3}$$

Therefore, one gram of magnesium oxide would neutralize the most 0.1 M HCl.

83. The pertinent reactions are:

$$2\ NaOH(aq) + H_2SO_4(aq) \rightarrow Na_2SO_4(aq) + 2\ H_2O(l)$$

$$HCl(aq) + NaOH(aq) \rightarrow NaCl(aq) + H_2O(l)$$

Amount of NaOH added:

$$0.0500 \text{ L} \times \frac{0.213 \text{ mol}}{\text{L}} = 1.07 \times 10^{-2} \text{ mol NaOH}$$

Amount of NaOH neutralized by HCl:

$$0.01321 \text{ L HCl} \times \frac{0.103 \text{ mol HCl}}{\text{L HCl}} \times \frac{1 \text{ mol NaOH}}{\text{mol HCl}} = 1.36 \times 10^{-3} \text{ mol NaOH}$$

The difference, 9.3×10^{-3} mol, is the amount of NaOH neutralized by the sulfuric acid.

$$9.3 \times 10^{-3} \text{ mol NaOH} \times \frac{1 \text{ mol H}_2\text{SO}_4}{2 \text{ mol NaOH}} = 4.7 \times 10^{-3} \text{ mol H}_2\text{SO}_4$$

The concentration of H_2SO_4 is:

$$\frac{4.7 \times 10^{-3} \text{ mol}}{0.100 \text{ L}} = 4.7 \times 10^{-2}\ M$$

84. We get the empirical formula from the elemental analysis. Out of 100 g carminic acid there are:

$$53.66 \text{ g C} \times \frac{1 \text{ mol C}}{12.01 \text{ g C}} = 4.468 \text{ mol C}$$

$$4.09 \text{ g H} \times \frac{1 \text{ mol H}}{1.008 \text{ g H}} = 4.06 \text{ mol H}$$

$$42.25 \text{ g O} \times \frac{1 \text{ mol O}}{16.00 \text{ g O}} = 2.641 \text{ mol O}$$

Taking ratios in the usual way: $\frac{4.468}{4.06} = 1.10 = \frac{11}{10}$

So let's try $\frac{4.06}{10} = 0.406$ as a common factor.

$$\frac{4.468}{0.406} = 11.0; \quad \frac{4.06}{0.406} = 10.0; \quad \frac{2.641}{0.406} = 6.50$$

$C_{22}H_{20}O_{13}$ is the empirical formula, to make all units whole numbers.

We can get molecular weight from the titration data:

$$18.02 \times 10^{-3} \text{ L soln} \times \frac{0.0406 \text{ mol NaOH}}{\text{L soln}} \times \frac{1 \text{ mol carminic acid}}{\text{mol NaOH}} = 7.32 \times 10^{-4} \text{ mol carminic acid}$$

$$\text{Molar mass} = \frac{0.3602 \text{ g}}{7.32 \times 10^{-4} \text{ mol}} = \frac{492 \text{ g}}{\text{mol}}$$

The molar mass of $C_{22}H_{20}O_{13} \approx 22(12) + 20(1) + 13(16) = 492$ g.

Therefore, the molecular formula of carminic acid is $C_{22}H_{20}O_{13}$.

85. a. $IO_3^- + I^- \rightarrow I_3^-$ See 4.77b

$$IO_3^- \rightarrow I_3^- \qquad (3\,I^- \rightarrow I_3^- + 2\,e^-) \times 8$$

$$16\,e^- + 18\,H^+ + 3\,IO_3^- \rightarrow I_3^- + 9\,H_2O$$

$$24\,I^- \rightarrow 8\,I_3^- + 16\,e^-$$

$$16\,e^- + 18\,H^+ + 3\,IO_3^- \rightarrow I_3^- + 9\,H_2O$$

$$18\,H^+ + 24\,I^- + 3\,IO_3^- \rightarrow 9\,I_3^- + 9\,H_2O$$ which simplifies to

$$6\,H^+(aq) + 8\,I^-(aq) + IO_3^-(aq) \rightarrow 3\,I_3^-(aq) + 3\,H_2O(l)$$

b. $S_2O_3^{2-} + I_3^- \rightarrow I^- + S_4O_6^{2-}$

$I_3^- + 2\,e^- \rightarrow 3\,I^-$

$S_2O_3^{2-} \rightarrow S_4O_6^{2-}$

$2\,S_2O_3^{2-} \rightarrow S_4O_6^{2-} + 2\,e^-$

Adding the balanced half reactions gives:

$$2\,S_2O_3^{2-}(aq) + I_3^-(aq) \rightarrow 3\,I^-(aq) + S_4O_6^{2-}(aq)$$

c. $25.00 \times 10^{-3}\ \text{L}\ IO_3^- \times \dfrac{0.0100\ \text{mol}\ IO_3^-}{\text{L}} \times \dfrac{3\ \text{mol}\ I_3^-}{\text{mol}\ IO_3^-} = 7.50 \times 10^{-4}\ \text{mol}\ I_3^-$

$7.50 \times 10^{-4}\ \text{mol}\ I_3^- \times \dfrac{2\ \text{mol}\ S_2O_3^{2-}}{\text{mol}\ I_3^-} = 1.50 \times 10^{-3}\ \text{mol}\ S_2O_3^{2-}$

$M_{S_2O_3^{2-}} = \dfrac{1.50 \times 10^{-3}\ \text{mol}}{32.04 \times 10^{-3}\ \text{L}} = 0.0468\ M$

d. $0.5000\ \text{L} \times \dfrac{0.0100\ \text{mol}\ KIO_3}{\text{L}} \times \dfrac{214.0\ \text{g}\ KIO_3}{\text{mol}\ KIO_3} = 1.07\ \text{g}\ KIO_3$

Place 1.07 g KIO_3 in a 500 mL volumetric flask; add water to dissolve KIO_3; continue adding water to the mark.

86. a. $0.308\ \text{g AgCl} \times \dfrac{35.45\ \text{g Cl}}{143.4\ \text{g AgCl}} = 0.0761\ \text{g Cl}$

$\%Cl = \dfrac{0.0761\ \text{g}}{0.256\ \text{g}} \times 100 = 29.7\%\ Cl$

Cobalt(III) oxide, Co_2O_3 $2(58.93) + 3(16.00) = \dfrac{165.86\ \text{g}}{\text{mol}}$

$0.145\ \text{g}\ Co_2O_3 \times \dfrac{117.86\ \text{g Co}}{165.86\ \text{g}\ Co_2O_3} = 0.103\ \text{g Co}$

$\%Co = \dfrac{0.103\ \text{g}}{0.416\ \text{g}} \times 100 = 24.8\%\ Co$

The remainder, 100 - (29.7 + 24.8) = 45.5%, is water.

24.8% Co; 29.7% Cl; 45.5% H_2O

b. Out of 100 g of compound, there are:

$$24.8\ \text{g Co} \times \frac{1\ \text{mol}}{58.93\ \text{g Co}} = 0.421\ \text{mol Co}$$

$$29.7 \text{ g Cl} \times \frac{1 \text{ mol}}{35.45 \text{ g Cl}} = 0.838 \text{ mol Cl}$$

$$45.5 \text{ g } H_2O \times \frac{1 \text{ mol}}{18.02 \text{ g } H_2O} = 2.52 \text{ mol } H_2O$$

Dividing all results by 0.421, we get $CoCl_2 \cdot 6H_2O$

c. $CoCl_2 \cdot 6H_2O(aq) + 2\ AgNO_3(aq) \rightarrow 2\ AgCl(s) + Co(NO_3)_2(aq) + 6\ H_2O(l)$

$CoCl_2 \cdot 6H_2O(aq) + 2\ NaOH(aq) \rightarrow Co(OH)_2(s) + 2\ NaCl(aq) + 6\ H_2O(l)$

$Co(OH)_2 \rightarrow Co_2O_3$ This is an oxidation. Thus, we also need to include an oxidizing agent. The obvious choice is O_2.

$4\ Co(OH)_2 + O_2 \rightarrow 2\ Co_2O_3 + 4\ H_2O$

87. a. $Fe^{3+}(aq) + 3\ OH^-(aq) \rightarrow Fe(OH)_3(s)$

$Fe(OH)_3$: 55.85 + 3(16.00) + 3(1.008) = 106.87 g/mol

$$0.107 \text{ g } Fe(OH)_3 \times \frac{55.85 \text{ g Fe}}{106.87 \text{ g } Fe(OH)_3} = 0.0559 \text{ g Fe}$$

b. $Fe(NO_3)_3$: 55.85 + 3(14.01) + 9(16.00) = 241.86 g/mol

$$0.559 \text{ g Fe} \times \frac{241.86 \text{ g } Fe(NO_3)_3}{55.85 \text{ g Fe}} = 0.242 \text{ g } Fe(NO_3)_3$$

c. $$\%Fe(NO_3)_3 = \frac{0.242 \text{ g}}{0.456 \text{ g}} \times 100 = 53.1\%$$

88. Molar masses: KCl, 39.10 + 35.45 = 74.55 g/mol

KBr, 39.10 + 79.90 = 119.00 g/mol

AgCl, 107.9 + 35.45 = 143.4 g/mol

AgBr, 107.90 + 79.90 = 187.8 g/mol

Let x = number of moles of KCl in mixture and y = number of moles of KBr in mixture.

Since $Ag^+ + Cl^- \rightarrow AgCl$ and $Ag^+ + Br^- \rightarrow AgBr$

then, x = moles AgCl and y = moles AgBr

0.1024 g = 74.55 x + 119.0 y

0.1889 g = 143.4 x + 187.8 y

Multiply the first equation by $\frac{187.8}{119.0}$, and subtract from the second.

$$\begin{array}{rl} 0.1889 = & 143.4\text{ x} + 187.8\text{ y} \\ -0.1616 = & -117.7\text{ x} - 187.8\text{ y} \\ \hline 0.0273 = & 25.7\text{ x} \end{array} \qquad x = 1.06 \times 10^{-3}\text{ mol KCl}$$

$$1.06 \times 10^{-3}\text{ mol KCl} \times \frac{74.55\text{ g KCl}}{\text{mol KCl}} = 0.0790\text{ g KCl}$$

$$\%\text{KCl} = \frac{0.0790}{0.1024} \times 100 = 77.1\%;\ \ \%\text{KBr} = 100 - 77.1 = 22.9\%$$

Alternatively, we can calculate the mass of AgX, X = Cl or Br, assuming sample was pure KCl or pure KBr and set up a ratio.

If pure KCl: $0.1024\text{ g KCl} \times \frac{143.4\text{ g AgCl}}{74.55\text{ g KCl}} = 0.1970\text{ g AgCl}$

If pure KBr: $0.1024\text{ g KBr} \times \frac{187.8\text{ g AgBr}}{119.0\text{ g KBr}} = 0.1616\text{ g AgBr}$

So, 0% KCl will give 0.1616 g AgX. 100% KCl will give 0.1970 g AgX.

The mass in excess of 0.1616 g will give us % KCl:

$$\frac{0.1889 - 0.1616}{0.1970 - 0.1616} \times 100 = \%\text{ KCl} = 77.1\% \qquad \text{and } \%\text{ KBr} = 22.9\%$$

89. $0.298\text{ g BaSO}_4 \times \frac{96.07\text{ g SO}_4^{2-}}{233.4\text{ g BaSO}_4} = 0.123\text{ g sulfate}$

$$\%\text{ sulfate} = \frac{0.123\text{ g SO}_4^{2-}}{0.205\text{ g}} = 60.0\%$$

Assume we have 100 g of the mixture of Na_2SO_4 and K_2SO_4. There is:

$$60.0\text{ g SO}_4^{2-} \times \frac{1\text{ mol}}{96.07\text{ g}} = 0.625\text{ mol SO}_4^{2-}$$

There must be $2 \times 0.625 = 1.25$ mol of cations.

Let x = number of moles of K^+ and y = number of moles of Na^+

Then, x + y = 1.25

The total mass of Na^+ and K^+ must be 40.0 g.

$$\text{x mol K}^+ \left(\frac{39.10\text{ g}}{\text{mol}}\right) + \text{y mol Na}^+ \left(\frac{22.99\text{ g}}{\text{mol}}\right) = 40.0\text{ g}$$

So, we have two equations in two unknowns:

$$x + y = 1.25 \text{ and } 39.10x + 22.99y = 40.0$$

Since $x = 1.25 - y$, $39.10(1.25 - y) + 22.99y = 40.0$

$48.9 - 39.10y + 22.99y = 40.0$; $-16.11y = -8.9$

$y = 0.55$ mol Na^+ and $x = 1.25 - 0.55 = 0.70$ mol K^+

Therefore:

$$0.70 \text{ mol } K^+ \times \frac{1 \text{ mol } K_2SO_4}{2 \text{ mol } K^+} = 0.35 \text{ mol } K_2SO_4$$

$$0.35 \text{ mol } K_2SO_4 \times \frac{174.27 \text{ g}}{\text{mol}} = 61 \text{ g } K_2SO_4$$

Since we assumed 100 g then the mixture is 61% K_2SO_4 and 39% Na_2SO_4.

90. a. $5.0 \text{ ppb Hg in water} = \frac{5.0 \text{ ng Hg}}{\text{mL } H_2O} = \frac{5.0 \times 10^{-9} \text{ g Hg}}{\text{mL } H_2O}$

$$\frac{5.0 \times 10^{-9} \text{ g Hg}}{\text{mL } H_2O} \times \frac{1 \text{ mol Hg}}{200.6 \text{ g Hg}} \times \frac{1000 \text{ mL}}{\text{L}} = 2.5 \times 10^{-8} \ M$$

b. $$\frac{1.0 \times 10^{-9} \text{ g } CHCl_3}{\text{mL}} \times \frac{1 \text{ mol } CHCl_3}{119.37 \text{ g } CHCl_3} \times \frac{1000 \text{ mL}}{\text{L}} = 8.4 \times 10^{-9} \ M$$

c. $10.0 \text{ ppm As} = \frac{10.0 \ \mu\text{g As}}{\text{mL}} = \frac{10.0 \times 10^{-6} \text{ g As}}{\text{mL}}$

$$\frac{10.0 \times 10^{-6} \text{ g As}}{\text{mL}} \times \frac{1 \text{ mol As}}{74.92 \text{ g As}} \times \frac{1000 \text{ mL}}{\text{L}} = 1.33 \times 10^{-4} \ M$$

d. $$\frac{0.10 \times 10^{-6} \text{ g DDT}}{\text{mL}} \times \frac{1 \text{ mol DDT}}{354.46 \text{ g DDT}} \times \frac{1000 \text{ mL}}{\text{L}} = 2.8 \times 10^{-7} \ M$$

91. a. $\text{ppm Na} = \frac{\mu\text{g Na}}{\text{mL}}$

$$150 \times 10^{-3} \text{ g } Na_2CO_3 \times \frac{45.98 \text{ g } Na^+}{105.99 \text{ g } Na_2CO_3} = 6.5 \times 10^{-2} \text{ g } Na^+$$

$$\frac{6.5 \times 10^{-2} \text{ g } Na^+}{1.0 \text{ L}} \times \frac{1 \text{ L}}{1000 \text{ mL}} = \frac{6.5 \times 10^{-5} \text{ g } Na^+}{\text{mL}} = \frac{65 \times 10^{-6} \text{ g } Na^+}{\text{mL}} = 65 \text{ ppm}$$

b. $\text{ppb} = \frac{\text{ng dioctylphthalate}}{\text{mL}}$

$$\frac{2.5 \times 10^{-3} \text{ g}}{500.0 \text{ mL}} = \frac{5.0 \times 10^{-6} \text{ g}}{\text{mL}} \times \frac{10^9 \text{ ng}}{\text{g}} = \frac{5.0 \times 10^3 \text{ ng}}{\text{mL}} = 5.0 \times 10^3 \text{ ppb}$$

92. $\text{Stock solution} = \frac{10.0 \text{ mg}}{500.0 \text{ mL}} = \frac{10.0 \times 10^{-3} \text{ g}}{500.0 \text{ mL}} = \frac{2.00 \times 10^{-5} \text{ g}}{\text{mL}}$

$$100.0 \times 10^{-6} \text{ L stock} \times \frac{1000 \text{ mL}}{\text{L}} \times \frac{2.00 \times 10^{-5} \text{ g steroid}}{\text{mL}} = 2.00 \times 10^{-6} \text{ g steroid}$$

This is diluted to a final volume of 100.0 mL.

$$\text{ppb steroid} = \frac{\text{g steroid}}{\text{g solution}} \times 10^9 = \frac{\text{ng steroid}}{\text{mL aqueous solution}}$$

$$\text{ppb steroid} = \frac{2.00 \times 10^{-6}\text{ g}}{100.0 \text{ mL}} = \frac{2.00 \times 10^{-8}\text{ g}}{\text{mL}} = \frac{20.0 \times 10^{-9}\text{ g}}{\text{mL}} = 20.0 \text{ ppb}$$

$$\frac{20.0 \times 10^{-9}\text{ g steroid}}{\text{mL}} \times \frac{1000 \text{ mL}}{\text{L}} \times \frac{1 \text{ mol steroid}}{336.43 \text{ g steroid}} = 5.95 \times 10^{-8}\ M$$

93. We want 100.0 mL of each standard. To make the 100. ppm standard:

$$\frac{100.\ \mu\text{g Cu}}{\text{mL}} \times 100.0 \text{ mL solution} = 1.00 \times 10^4\ \mu\text{g Cu needed}$$

$$1.00 \times 10^4\ \mu\text{g Cu} \times \frac{1 \text{ mL stock}}{1000.0\ \mu\text{ g}} = 10.0 \text{ mL of stock solution}$$

Therefore, to make 100.0 mL of 100. ppm solution, transfer 10.0 mL of the 1000.0 ppm stock solution to a 100 mL volumetric flask and dilute to the mark.

Similarly for the:

75.0 ppm standard, dilute 7.50 mL of the 1000.0 ppm stock to 100.0 mL

50.0 ppm standard, dilute 5.00 mL of the 1000.0 ppm stock to 100.0 mL

25.0 ppm standard, dilute 2.50 mL of the 1000.0 ppm stock to 100.0 mL

10.0 ppm standard, dilute 1.00 mL of the 1000.0 ppm stock to 100.0 mL

94. First we will calculate the molarity while ignoring the uncertainty.

$$0.150 \text{ g} \times \frac{1 \text{ mol}}{58.44 \text{ g}} = 2.57 \times 10^{-3} \text{ moles}$$

$$\text{Molarity} = \frac{2.57 \times 10^{-3} \text{ mol}}{0.1000 \text{ L}} = \frac{2.57 \times 10^{-2} \text{ mol}}{\text{L}}$$

The maximum value for the molarity is:

$$\frac{0.153 \text{ g} \times \frac{1 \text{ mol}}{58.44 \text{ g}}}{0.0995 \text{ L}} = \frac{2.63 \times 10^{-2} \text{ mol}}{\text{L}}$$

The minimum value for the molarity is:

$$\frac{0.147 \text{ g} \times \frac{1 \text{ mol}}{58.44 \text{ g}}}{0.1005 \text{ L}} = \frac{2.50 \times 10^{-2} \text{ mol}}{\text{L}}$$

The range of the molarity is 0.0250 M to 0.0263 M or we can express this range as 0.0257 $M \pm$ 0.0007 M.

95. The amount of KHP used is:

$$0.4016 \text{ g} \times \frac{1 \text{ mol}}{204.22 \text{ g}} = 1.967 \times 10^{-3} \text{ mol KHP}$$

Since one mole of NaOH reacts completely with one mole of KHP, the NaOH solution contains 1.967×10^{-3} mol NaOH.

$$\text{Molarity of NaOH} = \frac{1.967 \times 10^{-3} \text{ mol}}{25.06 \times 10^{-3} \text{ L}} = \frac{7.849 \times 10^{-2} \text{ mol}}{\text{L}}$$

$$\text{Maximum molarity} = \frac{1.967 \times 10^{-3} \text{ mol}}{25.01 \times 10^{-3} \text{ L}} = \frac{7.865 \times 10^{-2} \text{ mol}}{\text{L}}$$

$$\text{Minimum molarity} = \frac{1.967 \times 10^{-3} \text{ mol}}{25.11 \times 10^{-3} \text{ L}} = \frac{7.834 \times 10^{-2} \text{ mol}}{\text{L}}$$

We can express this as 0.07849 $M \pm 0.00016$ M.

An alternate is to express the molarity as 0.0785 $M \pm 0.0002$ M.

This second way is consistent with our convention on significant figures. The advantage of the first method is that it shows that we made all of our individual measurements to four significant figures.

96. Desired uncertainty is 1% of 0.02 or $\pm$ 0.0002. So we want the solution to be 0.0200 $\pm$ 0.0002 M or the concentration should be between 0.0198 and 0.0202 M. We should use a volumetric flask to make the solution. They are good to $\pm$ 0.1%. We want to weigh out between 0.0198 mol and 0.0202 mol of KIO_3.

$$\text{Molar mass of } KIO_3 = 39.10 + 126.9 + 3(16.00) = \frac{214.0 \text{ g}}{\text{mol}}$$

$$0.0198 \text{ mol} \times \frac{214.0 \text{ g}}{\text{mol}} = 4.2372 \text{ g (For now we will carry extra digits.)}$$

$$0.0202 \text{ mol} \times \frac{214.0 \text{ g}}{\text{mol}} = 4.3228 \text{ g}$$

We should weigh out between 4.24 and 4.32 g of KIO_3. To be safe we should weigh to the nearest mg or 0.1 mg. Dissolve the KIO_3 in water and dilute to the mark in a one liter volumetric flask.

97. The first procedure would give the most accurate concentration. The uncertainty in pipetting three times from a 10 mL volumetric pipet would still be less than trying to make a single transfer of 0.050 mL. Each volume measurement with the 10 mL volumetric pipet has an uncertainty of $0.01/10.00 \times 100 = 0.1\%$. The uncertainty in measuring the 0.050 mL aliquot is $0.001/0.050 \times 100 = 2\%$.

CHAPTER FIVE: GASES

QUESTIONS

1. a. Heating the can will increase the pressure of the gas inside the can. $P \propto T$, V & n const. As the pressure increases it may be enough to rupture the can.

   b. As you draw a vacuum in your mouth, atmospheric pressure pushes the liquid up the straw.

   c. The external atmospheric pressure pushes on the can. Since there is no opposing pressure from the air in the inside, the can collapses.

2. How "hard" the tennis ball is depends on the difference between the pressure of the air inside the tennis ball and atmospheric pressure. A "sea level" ball will be much "harder" at high altitude. A high altitude ball will be "soft" at sea level.

3. $PV = nRT$ = constant at constant n & T. At two sets of conditions, P_1V_1 = const. = P_2V_2.

   $P_1V_1 = P_2V_2$ (Boyle's Law).

   $\frac{V}{T} = \frac{nR}{P}$ = constant at constant n & P. At two sets of conditions, $\frac{V_1}{T_1}$ = const. = $\frac{V_2}{T_2}$

   $\frac{V_1}{T_1} = \frac{V_2}{T_2}$ (Charles's Law)

4. Boyle's Law: $P \propto 1/V$ at constant n and T

   In the kinetic molecular theory (kmt) P is proportional to the collision frequency which is proportional to 1/V. As the volume increases there will be fewer collisions with the walls of the container and pressure decreases (Boyle's Law).

   Charles's Law: $V \propto T$ at constant n and P

   Pressure is proportional to collision frequency. If pressure is constant, then the collision frequency of the gas molecules with the walls of the container is constant. Volume is inversely proportional to collision frequency and temperature is directly proportional. If the temperature increases, to keep the pressure (and collision frequency) constant, the volume of the container must increase. Therefore, volume and temperature are directly related at constant n and P (Charles's Law).

   Dalton's Law: $P_{tot} = P_1 + P_2 + ...$

   One of the postulates of the kmt is that gas particles do not interact with each other. If this is so, then we can treat each gas in a mixture of gases independently.

EXERCISES

Pressure

5. $1 \text{ atm} \times \frac{760 \text{ mm Hg}}{1 \text{ atm}} \times \frac{1 \text{ cm}}{10 \text{ mm}} \times \frac{1 \text{ in}}{2.54 \text{ cm}} = 29.92$ in of Hg

6. a. $2200 \text{ psi} \times \frac{1 \text{ atm}}{14.7 \text{ psi}} = 150 \text{ atm}$

b. $150 \text{ atm} \times \frac{1.01 \times 10^5 \text{ Pa}}{\text{atm}} \times \frac{1 \text{ MPa}}{10^6 \text{ Pa}} = 15 \text{ MPa}$

c. $150 \text{ atm} \times \frac{760 \text{ torr}}{\text{atm}} = 1.1 \times 10^5 \text{ torr}$

7. $6.5 \text{ cm} \times \frac{10 \text{ mm}}{\text{cm}} = 65 \text{ mm Hg or } 65 \text{ torr}$

$65 \text{ torr} \times \frac{1 \text{ atm}}{760 \text{ torr}} = 8.6 \times 10^{-2} \text{ atm}$

$8.6 \times 10^{-2} \text{ atm} \times \frac{1.01 \times 10^5 \text{ Pa}}{\text{atm}} = 8.7 \times 10^3 \text{ Pa}$

8. $70. \text{ psi} \times \frac{1 \text{ atm}}{14.7 \text{ psi}} = 4.8 \text{ atm}$

$4.8 \text{ atm} \times \frac{760 \text{ mm Hg}}{\text{atm}} = 3.6 \times 10^3 \text{ mm Hg}$

$4.8 \text{ atm} \times 1.01 \times 10^5 \frac{\text{Pa}}{\text{atm}} = 4.8 \times 10^5 \text{ Pa} = 4.8 \times 10^2 \text{ kPa} = 0.48 \text{ MPa}$

9. If the levels of Hg in each arm of the manometer are equal, then the pressure in the flask is equal to atmospheric pressure. When they are unequal, the difference in height in mm will be equal to the difference in pressure in mm Hg (torr) between the flask and the atmosphere. Which level is higher will tell us whether the pressure in the flask is less than or greater than atmospheric pressure.

a. $P \text{ flask} < P \text{ atm}$; $P \text{ flask} = 760. - 140. = 620. \text{ torr}$

$620. \text{ torr} \times \frac{1 \text{ atm}}{760 \text{ torr}} = 0.816 \text{ atm}$

$0.816 \text{ atm} \times \frac{1.013 \times 10^5 \text{ Pa}}{\text{atm}} = 8.27 \times 10^4 \text{ Pa}$

b. $P \text{ flask} > P \text{ atm}$; $P \text{ flask} = 760. + 175 \text{ torr} = 935 \text{ torr}$

$935 \text{ torr} \times \frac{1 \text{ atm}}{760 \text{ torr}} = 1.23 \text{ atm}$

$1.23 \text{ atm} \times \frac{1.013 \times 10^5 \text{ Pa}}{\text{atm}} = 1.25 \times 10^5 \text{ Pa}$

c. $P \text{ flask} = 635 - 140. = 495 \text{ torr.}$; $P \text{ flask} = 635 + 175 = 810. \text{ torr}$

10. a. The pressure is proportional to the mass of the fluid. The mass is proportional to the volume of the column of fluid (or the height of the column assuming the area of the column of fluid is constant).

$D = \frac{\text{mass}}{\text{volume}}$ In this case the volume of silicon oil will be the same as the volume of Hg in problem 5.9.

$$V = \frac{m}{D}, \; V_1 = V_2, \; \frac{m_1}{D_1} = \frac{m_2}{D_2}; \; m_2 = \frac{m_1 D_2}{D_1}$$

Since P is proportional to the mass:

$$P_{oil} = P_{Hg}\left(\frac{D_{oil}}{D_{Hg}}\right) = P_{Hg}\left(\frac{1.30}{13.6}\right) = 0.0956\ P_{Hg}$$

This conversion applies only to the column of liquid.

$$P_{flask} = 760.\ \text{torr} - (140. \times 0.0956)\ \text{torr} = 760. - 13.4 = 747\ \text{torr}$$

$$747\ \text{torr} \times \frac{1\ \text{atm}}{760\ \text{torr}} = 0.983\ \text{atm}$$

$$0.983\ \text{atm} \times \frac{1.013 \times 10^5\ \text{Pa}}{\text{atm}} = 9.96 \times 10^4\ \text{Pa}$$

$$P_{flask} = 760.\ \text{torr} + (175 \times 0.0956)\ \text{torr} = 760. + 16.7 = 777\ \text{torr}$$

$$777\ \text{torr} \times \frac{1\ \text{atm}}{760\ \text{torr}} = 1.02\ \text{atm}$$

$$1.02\ \text{atm} \times \frac{1.013 \times 10^5\ \text{Pa}}{\text{atm}} = 1.03 \times 10^5\ \text{Pa}$$

b. If we are measuring the same pressure, the height of the silicon oil column would be 13.6 ÷ 1.30 = 10.5 times the height of a mercury column. The advantage of using a less dense fluid than mercury is in measuring small pressures. The quantity measured (length) will be larger for the less dense fluid. Thus, the measurement will be more precise.

Gas Laws

11. a. PV = nRT

$$V = \frac{nRT}{P} = \frac{(2.00\ \text{mol})\left(\frac{0.08206\ \text{L atm}}{\text{mol K}}\right)(155 + 273)\ \text{K}}{5.00\ \text{atm}} = 14.0\ \text{L}$$

b. PV = nRT

$$n = \frac{PV}{RT} = \frac{0.300\ \text{atm} \times 2.00\ \text{L}}{\frac{0.08206\ \text{L atm}}{\text{mol K}} \times 155\ \text{K}} = 4.72 \times 10^{-2}\ \text{mol}$$

c. PV = nRT

$$T = \frac{PV}{nR} = \frac{4.47\ \text{atm} \times 25.0\ \text{L}}{2.01\ \text{mol} \times \frac{0.08206\ \text{L atm}}{\text{mol K}}} = 678\ \text{K}$$

d. PV = nRT

$$P = \frac{nRT}{V} = \frac{(10.5\ \text{mol})\left(\frac{0.08206\ \text{L atm}}{\text{mol K}}\right)(273 + 75)\ \text{K}}{2.25\ \text{L}} = 133\ \text{atm}$$

12. a. $PV = nRT$

$$T = \frac{PV}{nR} = \frac{875 \text{ torr} \times 275 \times 10^{-3} \text{ L}}{0.0105 \text{ mol} \times \frac{0.08206 \text{ L atm}}{\text{mol K}}} \times \frac{1 \text{ atm}}{760 \text{ torr}} = 367 \text{ K} = 94° \text{ C}$$

b. $$P = \frac{nRT}{V} = \frac{0.200 \text{ mol} \times \frac{0.08206 \text{ L atm}}{\text{mol K}} \times 311 \text{ K}}{0.100 \text{ L}} = 51.0 \text{ atm}$$

c. $$V = \frac{nRT}{P} = \frac{3.00 \text{ mol} \times \frac{0.08206 \text{ L atm}}{\text{mol K}} \times 838 \text{ K}}{2.50 \text{ atm}} = 82.5 \text{ L}$$

d. $$n = \frac{PV}{RT} = \frac{688 \text{ torr} \times 986 \times 10^{-3} \text{ L}}{\frac{0.08206 \text{ L atm}}{\text{mol K}} \times 565 \text{ K}} \times \frac{1 \text{ atm}}{760 \text{ torr}} = 1.93 \times 10^{-2} \text{ mol}$$

13. $$R = 0.08206 \frac{\text{L atm}}{\text{mol K}} \times 1.01325 \times 10^5 \frac{\text{Pa}}{\text{atm}} = 8.315 \times 10^3 \frac{\text{L Pa}}{\text{mol K}}$$

14. From 5.13, $R = 8.315 \times 10^3 \frac{\text{L Pa}}{\text{mol K}}$

$$8.315 \times 10^3 \frac{\text{L Pa}}{\text{mol K}} \times \frac{1 \text{ dm}^3}{\text{L}} \times \left(\frac{1 \text{ m}}{10 \text{ dm}}\right)^3 = 8.315 \frac{\text{m}^3 \text{ Pa}}{\text{mol K}}$$

$$\text{m}^3 \text{ Pa} = \text{m}^3 \times \frac{\text{N}}{\text{m}^2} = \text{N m} = \text{J (Joules)}; \; R = 8.315 \frac{\text{N m}}{\text{mol K}} \text{ or } 8.315 \text{ J/mol}\bullet\text{K}$$

15. Use the relationship $P_1V_1 = P_2V_2$

For H_2: $P_2 = \frac{P_1 V_1}{V_2} = 475 \text{ torr} \times \frac{1.00 \text{ L}}{1.50 \text{ L}} = 317 \text{ torr}$

For N_2: $P_2 = 45 \text{ kPa} \times \frac{0.50 \text{ L}}{1.50 \text{ L}} = 15 \text{ kPa}$

$$15 \text{ kPa} \times \frac{1000 \text{ Pa}}{\text{kPa}} \times \frac{760 \text{ torr}}{1.013 \times 10^5 \text{ Pa}} = 110 \text{ torr}$$

$P_{total} = P_{H_2} + P_{N_2} = 317 + 110 = 430$ torr

16. $P_1V_1 = P_2V_2$; $P_2 = \frac{P_1 V_1}{V_2}$

For H_2: $P_2 = 360 \text{ torr} \times \left(\frac{1.00 \text{ L}}{1.50 \text{ L}}\right) = 240 \text{ torr}$

For N_2: $P_2 = 240 \text{ torr} \times \left(\frac{0.50 \text{ L}}{1.50 \text{ L}}\right) = 80. \text{ torr}$

$P_{total} = P_{H_2} + P_{N_2} = 240 + 80. = 320$ torr

17. PV = nRT, Assume n is constant.

$$\frac{PV}{T} = nR = \text{constant}; \quad \frac{P_1 V_1}{T_1} = \frac{P_2 V_2}{T_2}$$

$$\frac{V_2}{V_1} = \frac{T_2 P_1}{T_1 P_2} = \frac{(273 + 15)\text{ K} \times 720.\text{ torr}}{(273 + 25)\text{ K} \times 605\text{ torr}} = 1.15$$

$V_2 = 1.15\ V_1$ or the volume has increased by 15%

18. PV = nRT, n constant

$$\frac{PV}{T} = nR = \text{constant}; \quad \frac{P_1 V_1}{T_1} = \frac{P_2 V_2}{T_2}$$

$$P_2 = \frac{P_1 V_1 T_2}{V_2 T_1} = 710.\text{ torr} \times \frac{5.0 \times 10^2\text{ mL}}{25\text{ mL}} \times \frac{(273 + 820)\text{ K}}{(273 + 30.)\text{ K}} = 5.1 \times 10^4\text{ torr}$$

19. $$PV = nRT;\ n = \frac{PV}{RT} = \frac{145\text{ atm} \times 75 \times 10^{-3}\text{ L}}{\frac{0.08206\text{ L atm}}{\text{mol K}} \times 295\text{ K}} = 0.45\text{ mol } O_2$$

20. PV = nRT

$$P = \frac{nRT}{V} = \frac{\left(0.60\text{ g} \times \frac{1\text{ mol}}{32.00\text{ g}}\right) \times \frac{0.08206\text{ L atm}}{\text{mol K}} \times (273 + 22)\text{K}}{5.0\text{ L}} = 0.091\text{ atm}$$

21. $$1.00\text{ g } H_2 \times \frac{1\text{ mol } H_2}{2.016\text{ g } H_2} = 0.496\text{ mol } H_2$$

$$1.00\text{ g He} \times \frac{1\text{ mol He}}{4.003\text{ g He}} = 0.250\text{ mol He} \qquad P_{H_2} = \chi_{H_2} P_{tot}; \quad \chi_{H_2} = \frac{n_{H_2}}{n_{tot}}$$

$$P_{H_2} = \left(\frac{0.496}{0.250 + 0.496}\right)(0.480\text{ atm}) = 0.319\text{ atm}$$

$$P_{H_2} + P_{He} = P_{tot} = 0.480\text{ atm}$$

$$P_{He} = 0.480 - 0.319 = 0.161\text{ atm}$$

22. a. $$\chi_{CH_4} = \frac{P_{CH_4}}{P_{tot}} = \frac{0.175}{0.175 + 0.250} = 0.412; \quad \chi_{O_2} = 1.000 - 0.412 = 0.588$$

b. $$PV = nRT;\ n = \frac{PV}{RT} = \frac{0.425\text{ atm} \times 10.5\text{ L}}{\frac{0.08206\text{ L atm}}{\text{mol K}} \times 338\text{ K}} = 0.161\text{ mol}$$

c. $$\chi_{CH_4} = \frac{n_{CH_4}}{n_{tot}}; \quad n_{CH_4} = \chi_{CH_4}(n_{tot}) = 0.412(0.161\text{ mol}) = 6.63 \times 10^{-2}\text{ mol } CH_4$$

$$6.63 \times 10^{-2}\text{ mol } CH_4 \times \frac{16.04\text{ g } CH_4}{\text{mol } CH_4} = 1.06\text{ g } CH_4$$

$$(0.588)(0.161 \text{ mol}) = 9.47 \times 10^{-2} \text{ mol } O_2$$

$$9.47 \times 10^{-2} \text{ mol } O_2 \times \frac{32.00 \text{ g } O_2}{\text{mol } O_2} = 3.03 \text{ g } O_2$$

23. PV = nRT, P and n constant

$$\frac{V}{T} = \frac{nR}{P} = \text{constant}, \ \frac{V_1}{T_1} = \frac{V_2}{T_2}, \ V_2 = \frac{V_1 T_2}{T_1}; \ T_1 = 273 + 20. = 293 \text{ K}$$

$$V_2 = 700. \text{ mL} \times \frac{100. \text{ K}}{293 \text{ K}} = 239 \text{ mL}$$

24. At constant n and P, $\frac{V_1}{T_1} = \frac{V_2}{T_2}$ and $V_2 = 125\%$ of $V_1 = 1.25\ V_1$

$$T_2 = \frac{T_1 V_2}{V_1} = \frac{(273 + 19) \text{ K } (1.25\ V_1)}{V_1} = 365 \text{ K or } 92° \text{ C}$$

25. a. $PV = nRT; \ 175 \text{ g Ar} \times \frac{1 \text{ mol Ar}}{39.95 \text{ g Ar}} = 4.38 \text{ mol Ar}$

$$T = \frac{PV}{nR} = \frac{2.50 \text{ L} \times 10.0 \text{ atm}}{4.38 \text{ mol} \times \frac{0.08206 \text{ L atm}}{\text{mol K}}} = 69.6 \text{ K}$$

b. $PV = nRT; \ P = \frac{nRT}{V} = \frac{4.38 \text{ mol} \times \frac{0.08206 \text{ L atm}}{\text{mol K}} \times 225 \text{ K}}{2.50 \text{ L}} = 32.3 \text{ atm}$

26. PV = nRT, V and n are constant

$$\frac{P}{T} = \frac{nR}{V} = \text{constant}, \ \frac{P_1}{T_1} = \frac{P_2}{T_2}, \text{ or } P_2 = \frac{P_1 T_2}{T_1}$$

a. $P_2 = 685 \text{ torr} \times \frac{(273 + 150.) \text{ K}}{273 \text{ K}} = 1060 \text{ torr}$

b. $P_2 = 685 \text{ torr} \times \frac{77 \text{ K}}{273 \text{ K}} = 193 \text{ torr} \approx 190 \text{ torr}$

27. PV = nRT, V and n constant; $\frac{P}{T} = \frac{nR}{V} = \text{constant}, \ \frac{P_1}{T_1} = \frac{P_2}{T_2}$,

$$P_2 = \frac{P_1 T_2}{T_1} = 13.7 \text{ MPa} \times \frac{(273 + 450) \text{ K}}{(273 + 23) \text{ K}} = 33 \text{ MPa}$$

28. Begin with the ideal gas law, PV = nRT.

From the information given, we know that the volume is constant, i.e., the gas is in a steel cylinder, and P, n and T are changing. So, we can say that

$$\frac{nT}{P} = \frac{V}{R} = \text{a constant}; \text{ or } \frac{n_1 T_1}{P_1} = \frac{n_2 T_2}{P_2}$$

Thus, $n_2 = \frac{n_1 T_1 P_2}{T_2 P_1}$; $= 150 \text{ mol} \times \frac{298 \text{ K}}{292 \text{ K}} \times \frac{2.0 \text{ MPa}}{7.5 \text{ MPa}} = 41 \text{ mol Ar}$

$$41 \text{ mol Ar} \times \frac{39.95 \text{ g}}{\text{mol}} = 1600 \text{ g}$$

29. $P_{total} = 1.00 \text{ atm} = 760. \text{ torr}$

$760. = P_{N_2} + P_{H_2O} = P_{N_2} + 17.5;\ P_{N_2} = 743 \text{ torr}$

$$PV = nRT;\ n = \frac{PV}{RT} = \frac{(743 \text{ torr})(2.50 \times 10^2 \text{ mL})}{\left(\frac{0.08206 \text{ L atm}}{\text{mol K}}\right)(293 \text{ K})} \times \frac{1 \text{ atm}}{760 \text{ torr}} \times \frac{1 \text{ L}}{1000 \text{ mL}}$$

$n = 1.02 \times 10^{-2} \text{ mol } N_2$

$$1.02 \times 10^{-2} \text{ mol } N_2 \times \frac{28.02 \text{ g } N_2}{\text{mol } N_2} = 0.286 \text{ g } N_2$$

30. $P_{total} = P_{O_2} + P_{H_2O}$

$P_{O_2} = P_{tot} - P_{H_2O} = 641 \text{ torr} - 23.8 \text{ torr} = 617 \text{ torr}$

$$PV = nRT;\ n = \frac{PV}{RT} = \frac{\frac{617}{760} \text{ atm} \times 0.5000 \text{ L}}{\frac{0.08206 \text{ L atm}}{\text{mol K}} \times 298 \text{ K}} = 1.66 \times 10^{-2} \text{ mol}$$

$$1.66 \times 10^{-2} \text{ mol} \times \frac{32.00 \text{ g}}{\text{mol}} = 0.531 \text{ g } O_2$$

Gas Density, Molar Mass, and Reaction Stoichiometry

31. $PV = nRT;\ \frac{n}{V} = \frac{P}{RT}$

$\frac{nM}{V} = \frac{PM}{RT}$ and $\frac{nM}{V} = d$, M = molar mass, so $d = \frac{PM}{RT}$

For $SiCl_4$, $M = 28.09 + 4(35.45) = \frac{169.89 \text{ g}}{\text{mol}}$

$$d = \frac{758 \text{ torr} \times \frac{169.89 \text{ g}}{\text{mol}}}{\frac{0.08206 \text{ L atm}}{\text{mol K}} \times 358 \text{ K}} \times \frac{1 \text{ atm}}{760 \text{ torr}} = 5.77 \text{ g/L for } SiCl_4$$

For $SiHCl_3$, $M = 28.09 + 1.008 + 3(35.45) = \frac{135.45 \text{ g}}{\text{mol}}$

$$d = \frac{PM}{RT} = \frac{758 \text{ torr} \times \frac{135.45 \text{ g}}{\text{mol}}}{\frac{0.08206 \text{ L atm}}{\text{mol K}} \times 358 \text{ K}} \times \frac{1 \text{ atm}}{760 \text{ torr}} = 4.60 \text{ g/L for } SiHCl_3$$

32. $PV = nRT; \ \frac{n}{V} = \frac{P}{RT}$

$$\frac{nM}{V} = d = \frac{PM}{RT} = \frac{635 \text{ torr} \times \frac{17.03 \text{ g}}{\text{mol}}}{\frac{0.08206 \text{ L atm}}{\text{mol K}} \times 300. \text{ K}} \times \frac{1 \text{ atm}}{760 \text{ torr}} = 0.578 \text{ g/L}$$

33. Out of 100 g of compound, there are:

$$87.4 \text{ g N} \times \frac{1 \text{ mol N}}{14.01 \text{ g N}} = 6.24 \text{ mol N}; \ \frac{6.24}{6.24} = 1$$

$$12.6 \text{ g H} \times \frac{1 \text{ mol H}}{1.008 \text{ g H}} = 12.5 \text{ mol H}; \ \frac{12.5}{6.24} = 2$$

Empirical formula, NH_2

$$PV = nRT \quad n = \frac{g}{M} \text{ where g = mass in g, M = molar mass in } \frac{g}{\text{mol}}$$

$$PV = \frac{g}{M} RT, \ PM = \frac{g}{V} RT = dRT$$

$$M = \frac{dRT}{P} = \frac{\frac{0.977 \text{ g}}{\text{L}} \times \frac{0.08206 \text{ L atm}}{\text{mol K}} \times 373 \text{ K}}{710. \text{ torr}} \times \frac{760 \text{ torr}}{\text{atm}} = 32.0 \text{ g/mol}$$

Mass of NH_2 = 16.0 g. Therefore, molecular formula is N_2H_4.

34. $PV = \frac{gRT}{M}$

$$M = \frac{gRT}{PV} = \frac{0.80 \text{ g} \times \frac{0.08206 \text{ L atm}}{\text{mol K}} \times 373 \text{ K}}{750 \text{ torr} \times 0.256 \text{ L}} \times \frac{760 \text{ torr}}{\text{atm}} = 97 \text{ g/mol}$$

Mass of CHCl ≈ 12 + 1 + 35.5 = 48.5, so molecular formula is $C_2H_2Cl_2$.

35. $27.37 \times 10^9 \text{ lb} \times \frac{1 \text{ kg}}{2.2046 \text{ lb}} \times \frac{1000 \text{ g}}{\text{kg}} \times \frac{1 \text{ mol } NH_3}{17.03 \text{ g}} = 7.290 \times 10^{11} \text{ mol of } NH_3$

$$7.290 \times 10^{11} \text{ mol } NH_3 \times \frac{1 \text{ mol } N_2}{2 \text{ mol } NH_3} = 3.645 \times 10^{11} \text{ mol } N_2$$

$$PV = nRT; \ V = \frac{nRT}{P} = \frac{3.645 \times 10^{11} \text{ mol} \times \frac{0.08206 \text{ L atm}}{\text{mol K}} \times 273.2 \text{ K}}{1.000 \text{ atm}}$$

$V = 8.172 \times 10^{12}$ L of N_2 at STP

$$7.290 \times 10^{11} \text{ mol } NH_3 \times \frac{3 \text{ mol } H_2}{2 \text{ mol } NH_3} = 1.094 \times 10^{12} \text{ mol } H_2$$

N_2 and H_2 are measured at the same pressure and temperature.

PV = nRT, P and T constant

$$\frac{V}{n} = \frac{RT}{P} = \text{constant}, \quad \frac{V_1}{n_1} = \frac{V_2}{n_2}$$

$$V_{H_2} = \frac{V_{N_2} \times n_{H_2}}{n_{N_2}} = 8.172 \times 10^{12}\ \text{L} \times \frac{1.094 \times 10^{12}\ \text{mol}}{3.645 \times 10^{11}\ \text{mol}}$$

$V_{H_2} = 2.453 \times 10^{13}$ L of H_2 at STP

There are several other ways to do this problem.

36. $P_{total} = P_{NH_3} + P_{CO_2}$; $P_{CO_2} = 14.0\ \text{atm} - 9.0\ \text{atm} = 5.0\ \text{atm}$

$$PV = nRT;\ n = \frac{PV}{RT};\ n_{NH_3} = \frac{9.0\ \text{atm} \times 5.0\ \text{L}}{\frac{0.08206\ \text{L atm}}{\text{mol K}} \times 296\ \text{K}} = 1.9\ \text{mol } NH_3$$

$$1.9\ \text{mol } NH_3 \times \frac{1\ \text{mol urea}}{2\ \text{mol } NH_3} = 0.95\ \text{mol urea}$$

$$n_{CO_2} = \frac{5.0\ \text{atm} \times 5.0\ \text{L}}{\frac{0.08206\ \text{L atm}}{\text{mol K}} \times 293\ \text{K}} = 1.0\ \text{mol } CO_2;\ 1.0\ \text{mol } CO_2 \times \frac{1\ \text{mol urea}}{1\ \text{mol } CO_2} = 1.0\ \text{mol urea}$$

Therefore, NH_3 is the limiting reagent and $0.95\ \text{mol urea} \times \frac{60.06\ \text{g urea}}{\text{mol urea}} = 57$ g urea

37. The unbalanced reaction is: $C_8H_{18} + O_2 \rightarrow CO_2 + H_2O$

The balanced equation is: $2\ C_8H_{18} + 25\ O_2 \rightarrow 16\ CO_2 + 18\ H_2O$

$$125\ \text{g } C_8H_{18} \times \frac{1\ \text{mol } C_8H_{18}}{114.22\ \text{g } C_8H_{18}} \times \frac{25\ \text{mol } O_2}{2\ \text{mol } C_8H_{18}} = 13.7\ \text{mol } O_2$$

$$PV = nRT,\ V = \frac{nRT}{P} = \frac{13.7\ \text{mol} \times \frac{0.08206\ \text{L atm}}{\text{mol K}} \times 273\ \text{K}}{1.00\ \text{atm}} = 307\ \text{L of } O_2$$

Or we can make use of the fact that at STP one mole of an ideal gas occupies a volume of 22.42 L. This can be calculated using the ideal gas law.

$$\text{So: } 13.7\ \text{mol } O_2 \times \frac{22.42\ \text{L}}{\text{mol}} = 307\ \text{L}$$

38. $$4.1\ \text{g HgO} \times \frac{1\ \text{mol HgO}}{216.6\ \text{g HgO}} \times \frac{1\ \text{mol } O_2}{2\ \text{mol HgO}} = 9.5 \times 10^{-3}\ \text{mol } O_2$$

$$V = \frac{nRT}{P} = \frac{9.5 \times 10^{-3}\ \text{mol} \times \frac{0.08206\ \text{L atm}}{\text{mol K}} \times 303\ \text{K}}{725\ \text{torr}} \times \frac{760\ \text{torr}}{\text{atm}} = 0.25\ \text{L}$$

39. a. $CH_4(g) + NH_3(g) + O_2(g) \rightarrow HCN(g) + H_2O(g)$

$CH_4 + NH_3 + O_2 \rightarrow HCN + 3\ H_2O$; $CH_4 + NH_3 + \frac{3}{2}\ O_2 \rightarrow HCN + 3\ H_2O$

$2\ CH_4(g) + 2\ NH_3(g) + 3\ O_2(g) \rightarrow 2\ HCN(g) + 6\ H_2O(g)$

b. $PV = nRT$; $n \propto V$ since P & T are constant.

$$20.0 \text{ L } CH_4 \times \frac{2 \text{ L HCN}}{2 \text{ L } CH_4} = 20.0 \text{ L HCN}; \; 20.0 \text{ L } NH_3 \times \frac{2 \text{ L HCN}}{2 \text{ L } NH_3} = 20.0 \text{ L HCN}$$

$$20.0 \text{ L } O_2 \times \frac{2 \text{ L HCN}}{3 \text{ L } O_2} = 13.3 \text{ L HCN};$$ O_2 is limiting, 13.3 L of HCN is produced.

40. a. $PV = nRT$, P & T constant. The number of moles is directly proportional to the volume.

From the reaction we get 4 moles of NO for every 4 moles of NH_3. Since the moles are equal, the volume of gas should be equal. We will get 10.0 L of NO.

b. $$10.0 \times 10^3 \text{ g } NH_3 \times \frac{1 \text{ mol } NH_3}{17.03 \text{ g } NH_3} \times \frac{5 \text{ mol } O_2}{4 \text{ mol } NH_3} = 734 \text{ mol of } O_2$$

$$PV = nRT; \; V = \frac{nRT}{P} = \frac{(734 \text{ mol})\left(\frac{0.08206 \text{ L atm}}{\text{mol K}}\right)(273 \text{ K})}{1.00 \text{ atm}} = 1.64 \times 10^4 \text{ L } O_2$$

c. $$PV = nRT; \; n = \frac{PV}{RT} = \frac{(3.00 \text{ atm})(5.00 \times 10^2 \text{ L})}{\left(\frac{0.08206 \text{ L atm}}{\text{mol K}}\right)(273 \text{ K} + 250.) \text{ K}} = 35.0 \text{ mol } NH_3$$

$$35.0 \text{ mol } NH_3 \times \frac{4 \text{ mol NO}}{4 \text{ mol } NH_3} \times \frac{30.01 \text{ g NO}}{\text{mol NO}} = 1.05 \times 10^3 \text{ g NO}$$

d. $$4 \; NH_3(g) + 5 \; O_2(g) \longrightarrow 4 \; NO(g) + 6 \; H_2O(g)$$

$$65 \text{ L } NH_3 \times \frac{6 \text{ L } H_2O}{4 \text{ L } NH_3} = 98 \text{ L } H_2O; \; 75 \text{ L } O_2 \times \frac{6 \text{ L } H_2O}{5 \text{ L } O_2} = 90. \text{ L } H_2O$$

The O_2 is limiting, 90. L H_2O is produced.

At STP, the volume of 1.0 mol of an ideal gas is 22.42 L.

$$90. \text{ L } H_2O \times \frac{1 \text{ mol } H_2O}{22.42 \text{ L } H_2O} \times \frac{18.02 \text{ g } H_2O}{\text{mol } H_2O} = 72 \text{ g } H_2O$$

e. $$90. \text{ L } H_2O \times \frac{4 \text{ L NO}}{6 \text{ L } H_2O} \times \frac{1 \text{ mol NO}}{22.42 \text{ L NO}} = 2.7 \text{ mol NO}$$

41. $P_{total} = P_{N_2} + P_{H_2O}$

$$P_{N_2} = 726 \text{ torr} - 23.8 \text{ torr} = 702 \text{ torr} \times \frac{1 \text{ atm}}{760 \text{ torr}} = 0.924 \text{ atm}$$

$$PV = nRT; \; n = \frac{PV}{RT} = \frac{0.924 \text{ atm} \times 31.8 \times 10^{-3} \text{ L}}{\frac{0.08206 \text{ L atm}}{\text{mol K}} \times 298 \text{ K}} = 1.20 \times 10^{-3} \text{ mol } N_2$$

$$\text{Mass of N in compound} = 1.20 \times 10^{-3} \text{ mol} \times \frac{28.02 \text{ g } N_2}{\text{mol}} = 3.36 \times 10^{-2} \text{ g}$$

$$\%N = \frac{3.36 \times 10^{-2} \text{ g}}{0.253 \text{ g}} \times 100 = 13.3\% \text{ N}$$

42. $0.2766 \text{ g } CO_2 \times \frac{12.01 \text{ g C}}{44.01 \text{ g } CO_2} = 7.548 \times 10^{-2} \text{ g C}$

$$\%C = \frac{7.548 \times 10^{-2} \text{ g}}{0.1023 \text{ g}} \times 100 = 73.78\% \text{ C}$$

$$0.0991 \text{ g } H_2O \times \frac{2.016 \text{ g H}}{18.02 \text{ g } H_2O} = 1.11 \times 10^{-2} \text{ g H}$$

$$\%H = \frac{1.11 \times 10^{-2} \text{ g}}{0.1023 \text{ g}} \times 100 = 10.9\% \text{ H}$$

$$PV = nRT;\ n = \frac{PV}{RT} = \frac{1.00 \text{ atm} \times 27.6 \times 10^{-3} \text{ L}}{\frac{0.08206 \text{ L atm}}{\text{mol K}} \times 273 \text{ K}} = 1.23 \times 10^{-3} \text{ mol } N_2$$

$$1.23 \times 10^{-3} \text{ mol } N_2 \times \frac{28.02 \text{ g } N_2}{\text{mol } N_2} = 3.45 \times 10^{-2} \text{ g nitrogen}$$

$$\%N = \frac{3.45 \times 10^{-2} \text{ g}}{0.4831 \text{ g}} \times 100 = 7.14\% \text{ N}$$

$$\%O = 100 - (73.78 + 10.9 + 7.14) = 8.2\% \text{ O}$$

Out of 100 g of compound, there are:

$$73.78 \text{ g C} \times \frac{1 \text{ mol}}{12.01 \text{ g}} = 6.143 \text{ mol C};\ 7.14 \text{ g N} \times \frac{1 \text{ mol}}{14.01 \text{ g}} = 0.510 \text{ mol N}$$

$$10.9 \text{ g H} \times \frac{1 \text{ mol}}{1.008 \text{ g}} = 10.8 \text{ mol H};\ 8.2 \text{ g O} \times \frac{1 \text{ mol}}{16.00 \text{ g}} = 0.51 \text{ mol O}$$

Divide all values by 0.51, and we get the empirical formula $C_{12}H_{21}NO$.

$$M = \frac{dRT}{P} = \frac{\frac{4.02 \text{ g}}{\text{L}} \times \frac{0.08206 \text{ L atm}}{\text{mol K}} \times 400. \text{ K}}{256 \text{ torr}} \times \frac{760 \text{ torr}}{\text{atm}} = 392 \text{ g/mol}$$

Mass of $C_{12}H_{21}NO$ is ~ 195 g and $\frac{392}{195} \approx 2$. Thus, the molecular formula is $C_{24}H_{42}N_2O_2$.

43. $3.7 \text{ g } KClO_3 \times \frac{1 \text{ mol } KClO_3}{122.55 \text{ g } KClO_3} \times \frac{3 \text{ mol } O_2}{2 \text{ mol } KClO_3} = 4.5 \times 10^{-2} \text{ mol } O_2$

$$P_{total} = P_{O_2} + P_{H_2O}$$

$$P_{O_2} = P_{total} - P_{H_2O} = 735 - 26.7 = 708 \text{ torr} \times \frac{1 \text{ atm}}{760 \text{ torr}} = 0.932 \text{ atm}$$

$$V = \frac{nRT}{P} = \frac{4.5 \times 10^{-2} \text{ mol} \times \frac{0.08206 \text{ L atm}}{\text{mol K}} \times 300. \text{ K}}{0.932 \text{ atm}} = 1.2 \text{ L} = 1200 \text{ mL}$$

44. The unbalanced equation is: $CaC_2 + H_2O \rightarrow C_2H_2 + Ca(OH)_2$

Balancing the O - atoms, we get: $CaC_2 + 2\,H_2O \rightarrow C_2H_2 + Ca(OH)_2$

$$5.20\text{ g } CaC_2 \times \frac{1\text{ mol } CaC_2}{64.10\text{ g } CaC_2} \times \frac{1\text{ mol } C_2H_2}{\text{mol } CaC_2} = 8.11 \times 10^{-2}\text{ mol } C_2H_2$$

$$P_{C_2H_2} = 715 - 23.8 = 691\text{ torr};\quad \chi_{C_2H_2} = \frac{P_{C_2H_2}}{P_{tot}} = \frac{691}{715} = 0.966$$

$$\chi_{C_2H_2} = 0.966 = \frac{n_{C_2H_2}}{n_{tot}} = \frac{8.11 \times 10^{-2}\text{ mol}}{n_{tot}};\quad n_{tot} = 8.40 \times 10^{-2}\text{ mol}$$

$$V_{tot} = V_{wet} = \frac{n_{tot}RT}{P_{tot}} = \frac{8.40 \times 10^{-2} \times 0.08206 \times 298}{715/760} = 2.18\text{ L}$$

Kinetic Molecular Theory and Real Gases

45. $(KE)_{avg} = 3/2\ RT$

at 273 K: $(KE)_{avg} = \frac{3}{2} \times \frac{8.3145\text{ J}}{\text{mol K}} \times 273\text{ K} = 3.40 \times 10^3\text{ J/mol}$

at 546 K: $(KE)_{avg} = \frac{3}{2} \times \frac{8.3145\text{ J}}{\text{mol K}} \times 546\text{ K} = 6.81 \times 10^3\text{ J/mol}$

46. $(KE)_{avg} = 3/2\ RT$. Since the kinetic energy depends only on temperature, methane and nitrogen at the same temperature will have the same average kinetic energy. So for N_2 the average kinetic energy is 3.40×10^3 J/mol (273 K) and 6.81×10^3 J/mol (546 K). See Exercise 5.45 for calculations.

47. $u_{rms} = \left(\frac{3RT}{M}\right)^{1/2}$, where $R = \frac{8.3145\text{ J}}{\text{mol K}}$ and M = molar mass in kg; 1.604×10^{-2} kg/mol for CH_4

For CH_4 at 273 K, $u_{rms} = \left(\frac{\frac{3 \times 8.3145\text{ J}}{\text{mol K}} \times 273\text{ K}}{1.604 \times 10^{-2}\text{ kg/mol}}\right)^{1/2} = 652\text{ m/s}$

Similarly u_{rms} for CH_4 at 546 K is 921 m/s.

48. $u_{rms} = \left(\frac{3RT}{M}\right)^{1/2}$, where R = 8.3145 J/mol•K and M = 2.802×10^{-2} kg/mol for N_2.

$$u_{rms} = \left(\frac{\frac{3 \times 8.3145\text{ J}}{\text{mol K}} \times 273\text{ K}}{2.802 \times 10^{-2}\text{ kg/mol}}\right)^{1/2} = 493\text{ m/s}$$

Similarly for N_2 at 546 K, u_{rms} = 697 m/s.

49. No, the number calculated in 5.45 is the average energy. There is a distribution of energies.

50. No, the number calculated in 5.47 is an average velocity. There is a distribution of velocities.

51. a. They will all have the same average kinetic energy since they are all at the same temperature.

b. Flask C, H_2 has the smallest molar mass. Lightest molecules are fastest.

52.

	a	b	c	d
avg. KE	inc	dec	same $KE \propto T$	same
avg. velocity	inc	dec	same $\frac{1}{2}mv^2 \propto T$	same
coll. freq wall	inc	dec	inc	inc
impact E	inc	dec	same impact $E \propto KE \propto T$	same

Collision frequency is proportional to the average velocity (as velocity increases it takes less time to move to the next collision) and proportional to the quantity n/V (as molecules per volume increases, collision frequency increases).

53. $\frac{R_1}{R_2} = \left(\frac{M_2}{M_1}\right)^{1/2}$; $\frac{31.50}{30.50} = \left(\frac{32.00}{M}\right)^{1/2} = 1.033$, M = molar mass of the unknown;

$\frac{32.00}{M} = 1.067$, so M = 29.99

Of the choices, the gas would be NO, nitric oxide.

54. $\frac{R_1}{R_2} = \left(\frac{M_2}{M_1}\right)^{1/2}$; $R_1 = \frac{24.0 \text{ mL}}{\text{min}}$, $R_2 = \frac{47.8 \text{ mL}}{\text{min}}$, $M_2 = \frac{16.04 \text{ g}}{\text{mol}}$ and $M_1 = ?$

$\frac{24.0}{47.8} = \left(\frac{16.04}{M_1}\right)^{1/2} = 0.502$; $16.04 = (0.502)^2 M_1$; $M_1 = \frac{16.04}{0.252} = \frac{63.7 \text{ g}}{\text{mol}}$

55. a. PV = nRT

$$P = \frac{nRT}{V} = \frac{0.5000 \text{ mol}\left(\frac{0.08206 \text{ L atm}}{\text{mol K}}\right)(25.0 + 273.2) \text{ K}}{1.000 \text{ L}} = 12.24 \text{ atm}$$

b. $\left[P + a\left(\frac{n}{V}\right)^2\right](V - nb) = nRT$; For N_2: $a = 1.39$ atm L^2/mol^2 and $b = 0.0391$ L/mol

$$\left[P + 1.39\left(\frac{0.5000}{1.000}\right)^2 \text{atm}\right](1.000 \text{ L} - 0.5000 \times 0.0391 \text{ L}) = 12.24 \text{ L atm}$$

(P + 0.348 atm)(0.9805 L) = 12.24 L atm (carry extra significant figure)

$$P = \frac{12.24 \text{ L atm}}{0.9805 \text{ L}} - 0.348 \text{ atm} = 12.48 - 0.35 = 12.13 \text{ atm}$$

c. The ideal gas is high by 0.11 atm or $\frac{0.11}{12.13} \times 100 = 0.91\%$ if we carry extra digits through the entire calculation.

56. a. $PV = nRT$

$$P = \frac{nRT}{V} = \frac{0.5000 \text{ mol} \times \frac{0.08206 \text{ L atm}}{\text{mol K}} \times 298.2 \text{ K}}{10.00 \text{ L}} = 1.224 \text{ atm}$$

b. $\left[P + a\left(\frac{n}{V}\right)^2\right](V - nb) = nRT$; $a = 1.39$ atm L^2/mol^2 and $b = 0.0391$ L/mol

$$\left[P + 1.39\left(\frac{0.5000}{10.00}\right)^2 \text{atm}\right](10.00 \text{ L} - 0.5000 \times 0.0391 \text{ L}) = 12.24 \text{ L atm}$$

$$(P + 0.00348 \text{ atm})(10.00 - 0.0196) \text{ L} = 12.24 \text{ L atm}$$

$$P + 0.00348 \text{ atm} = \frac{12.24 \text{ L atm}}{9.980 \text{ L}} = 1.226 \text{ atm (carry an extra significant figure)}$$

$$P = 1.226 - 0.00348 = 1.223 \text{ atm}$$

c. The results agree to ± 0.001 (0.08%), if we carry an extra significant figure.

d. In 5.55 the pressure is relatively high and there is a significant disagreement. In 5.56 the pressure is around 1 atm and there is better agreement between the two gas laws. The ideal gas law is valid at relatively low pressures.

Atmospheric Chemistry

57. $\chi_{NO} = 5 \times 10^{-7}$ from Table 5.4.

$$P_{NO} = \chi_{NO} P_{total} = 5 \times 10^{-7} \times 1.0 \text{ atm} = 5 \times 10^{-7} \text{ atm}$$

$$PV = nRT; \ \frac{n}{V} = \frac{P}{RT} = \frac{5 \times 10^{-7} \text{ atm}}{\left(\frac{0.08206 \text{ L atm}}{\text{mol K}}\right)(273 \text{ K})} = 2 \times 10^{-8} \text{ mol/L}$$

$$\frac{2 \times 10^{-8} \text{ mol}}{\text{L}} \times \frac{1 \text{ L}}{1000 \text{ cm}^3} \times \frac{6.022 \times 10^{23} \text{ molecules}}{\text{mol}} = 1 \times 10^{13} \text{ molecules/cm}^3$$

58. $\chi_{He} = 5.24 \times 10^{-6}$ from Table 5.4. $P_{He} = \chi_{He} P_{tot} = 5.24 \times 10^{-6}$ (1.0 atm) $= 5.2 \times 10^{-6}$ atm

$$\frac{n}{V} = \frac{P}{RT} = \frac{5.2 \times 10^{-6} \text{ atm}}{\left(\frac{0.08206 \text{ L atm}}{\text{mol K}}\right)(298 \text{ K})} = 2.1 \times 10^{-7} \text{ mol/L}$$

$$\frac{2.1 \times 10^{-7} \text{ mol}}{\text{L}} \times \frac{1 \text{ L}}{1000 \text{ cm}^3} \times \frac{6.022 \times 10^{23} \text{ molecules}}{\text{mol}} = 1.3 \times 10^{14} \text{ molecules/cm}^3$$

59. At 100. km, $T \approx -90°$ C and $P \approx 10^{-5.5} \approx 3 \times 10^{-6}$ atm

$PV = nRT$; $\frac{PV}{T} = nR = \text{Const.}$; $\frac{P_1 V_1}{T_1} = \frac{P_2 V_2}{T_2}$

$$V_2 = \frac{V_1 P_1 T_2}{T_1 P_2} = \frac{10.0 \text{ L} \times 3 \times 10^{-6} \text{ atm} \times 273 \text{ K}}{183 \text{ K} \times 1.0 \text{ atm}} = 4 \times 10^{-5} \text{ L} = 0.04 \text{ mL}$$

60. At 15 km, $T \approx -50°$ C and $P \approx 0.1$ atm. Use $\frac{P_1 V_1}{T_1} = \frac{P_2 V_2}{T_2}$

$$V_2 = \frac{V_1 P_1 T_2}{T_1 P_2} = \frac{1.0 \text{ L} \times 1.00 \text{ atm} \times 223 \text{ K}}{298 \text{ K} \times 0.1 \text{ atm}} = 7 \text{ L}$$

61. $N_2(g) + O_2(g) \rightarrow 2\ NO(g)$ (automobile combustion or caused by lightning)

$2\ NO(g) + O_2(g) \rightarrow 2\ NO_2(g)$ (reaction with atmospheric O_2)

$2\ NO_2(g) + H_2O(l) \rightarrow HNO_3(aq) + HNO_2(aq)$ (reaction with atmospheric H_2O)

$S(s) + O_2(g) \rightarrow SO_2(g)$ (combustion of coal)

$2\ SO_2(g) + O_2(g) \rightarrow 2\ SO_3(g)$ (reaction with atmospheric O_2)

$H_2O(l) + SO_3(g) \rightarrow H_2SO_4(aq)$ (reaction with atmospheric H_2O)

62. $HNO_3 + CaCO_3 \rightarrow Ca(NO_3)_2 + H_2CO_3$ (unbalanced)

$2\ HNO_3(aq) + CaCO_3(s) \rightarrow Ca(NO_3)_2(aq) + H_2O(l) + CO_2(g)$

$H_2SO_4(aq) + CaCO_3(s) \rightarrow CaSO_4(aq) + H_2O(l) + CO_2(g)$

ADDITIONAL EXERCISES

63. a. $PV = nRT$, n and T Const

$PV = \text{Constant}$

b. $PV = nRT$

$P = \left(\frac{nR}{V}\right)T = \text{Const} \times T$

c. $PV = nRT$

$T = \left(\frac{P}{nR}\right)V = \text{Const} \times V$

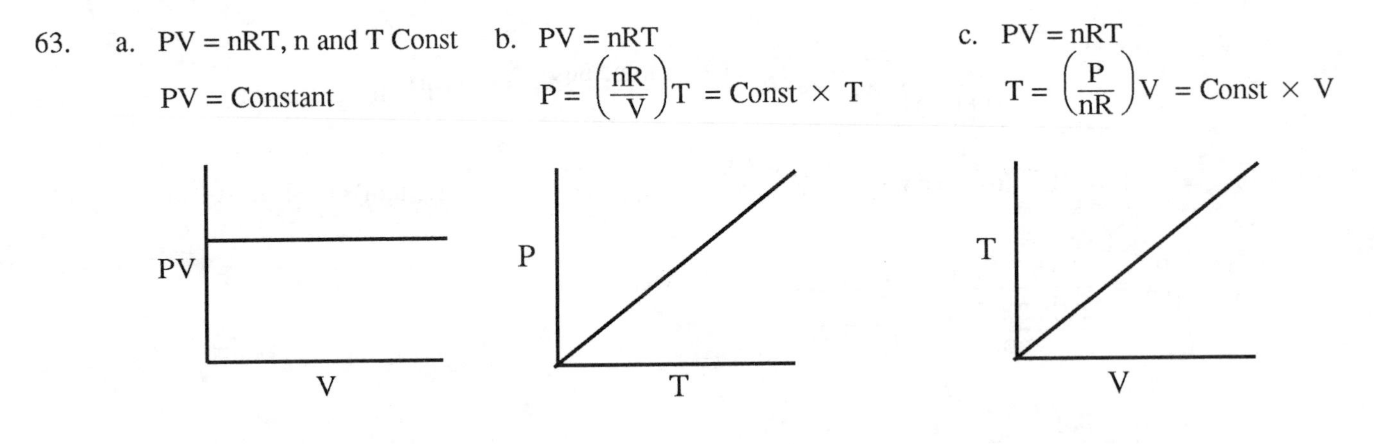

d. $PV = nRT$

$P = \frac{nRT}{V} = \frac{Const}{V}$

e. $P = \frac{nRT}{V} = \frac{Const}{V} = Const\left(\frac{1}{V}\right)$

f. $PV = nRT$

$\frac{PV}{T} = nR = Const$

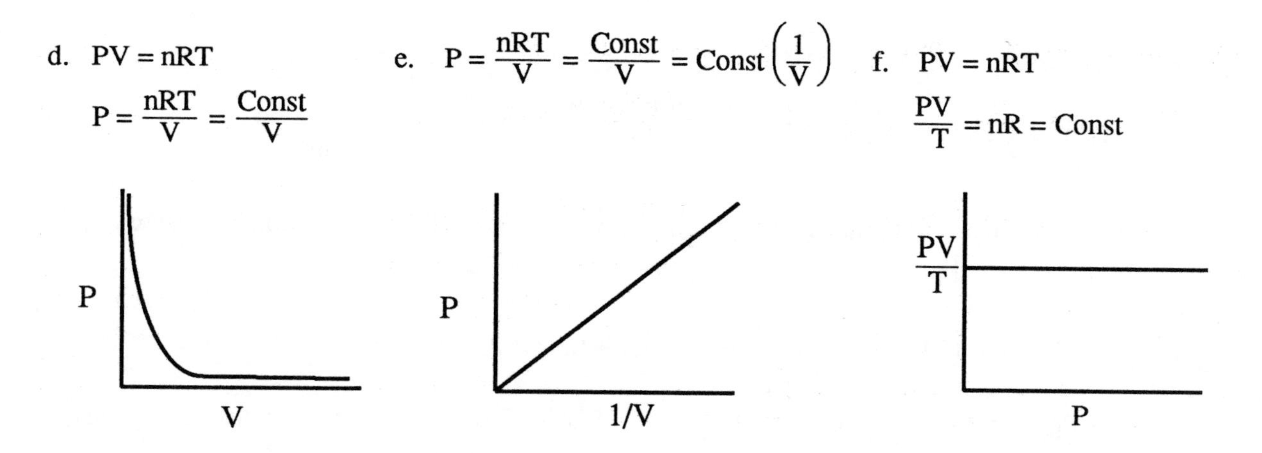

64. We can use the ideal gas law to calculate the partial pressure of each gas or to calculate the total pressure. There will be less math if we calculate the total pressure from the ideal gas law.

$$n_{O_2} = 1.5 \times 10^2 \text{ mg } O_2 \times \frac{1 \text{ g}}{1000 \text{ mg}} \times \frac{1 \text{ mol } O_2}{32.00 \text{ g } O_2} = 4.7 \times 10^{-3} \text{ mol } O_2$$

$$n_{NH_3} = 1.0 \times 10^3 \text{ cm}^3 \times \frac{5.0 \times 10^{18} \text{ molecules } NH_3}{\text{cm}^3} \times \frac{1 \text{ mol } NH_3}{6.022 \times 10^{23} \text{ molecules } NH_3}$$

$$= 8.3 \times 10^{-3} \text{ mol } NH_3$$

$$n_{total} = n_{N_2} + n_{O_2} + n_{NH_3} = 5.0 \times 10^{-2} + 4.7 \times 10^{-3} + 8.3 \times 10^{-3} = 6.3 \times 10^{-2} \text{ mol}$$

$$P_{total} = \frac{nRT}{V} = \frac{6.3 \times 10^{-2} \text{ mol} \times \frac{0.08206 \text{ L atm}}{\text{mol K}} \times 273 \text{ K}}{1.0 \text{ L}} = 1.4 \text{ atm}$$

$$P_{N_2} = \chi_{N_2} P_{tot}, \quad \chi_{N_2} = \frac{n_{N_2}}{n_{tot}}; \qquad P_{N_2} = \frac{5.0 \times 10^{-2}}{6.3 \times 10^{-2}} \times 1.4 \text{ atm} = 1.1 \text{ atm}$$

$$P_{O_2} = \frac{4.7 \times 10^{-3}}{6.3 \times 10^{-2}} \times 1.4 \text{ atm} = 0.10 \text{ atm} \qquad P_{NH_3} = \frac{8.3 \times 10^{-3}}{6.3 \times 10^{-2}} \times 1.4 \text{ atm} = 0.18 \text{ atm}$$

65. a. $PV = nRT$, n and V are constant.

$$\frac{P}{T} = \frac{nR}{V} \text{ or } \frac{P_1}{T_1} = \frac{P_2}{T_2}; \; P_2 = \frac{P_1 T_2}{T_1} = 40.0 \text{ atm} \times \frac{318 \text{ K}}{273 \text{ K}} = 46.6 \text{ atm}$$

b. $$\frac{P_1}{T_1} = \frac{P_2}{T_2}; \; T_2 = \frac{T_1 P_2}{P_1} = 273 \text{ K} \times \frac{150. \text{ atm}}{40.0 \text{ atm}} = 1.02 \times 10^3 \text{ K}$$

c. $$T_2 = \frac{T_1 P_2}{P_1} = 273 \text{ K} \times \frac{25.0 \text{ atm}}{40.0 \text{ atm}} = 171 \text{ K}$$

66. $PV = nRT$

$$P = \frac{nRT}{V} = \frac{\left(7.8 \text{ g} \times \frac{1 \text{ mol}}{44.01 \text{ g}}\right) \times \frac{0.08206 \text{ L atm}}{\text{mol K}} \times 300. \text{ K}}{4.0 \text{ L}} = 1.1 \text{ atm}$$

With air present, the partial pressure of CO_2 will still be 1.1 atm. The total pressure will be the sum of the partial pressures.

$$P_{tot} = 1.1 \text{ atm} + \left(740 \text{ torr} \times \frac{1 \text{ atm}}{760 \text{ torr}}\right) = 1.1 + 0.97 = 2.1 \text{ atm}$$

67. $P_{He} + P_{H_2O} = 1.0 \text{ atm} = 760 \text{ torr}$, $P_{H_2O} = 23.8 \text{ torr}$; $P_{He} = 736 \text{ torr} = 740 \text{ torr}$ (2 sig. figs.)

$P_{tot}V_{tot} = n_{tot}RT$, $V_{tot} = V_{He} + V_{H_2O} = V_{wet}$

$$P_{He} = \chi_{He}P_{tot} = \frac{n_{He}}{n_{tot}}(760) = 740; \quad n_{He} = 0.586 \text{ g} \times \frac{1 \text{ mol}}{4.003 \text{ g}} = 0.146 \text{ mol}$$

$$740 = \frac{0.146 \times 760}{n_{tot}}, \quad n_{tot} = 0.15 \text{ mol}; \quad V_{wet} = \frac{0.15 \times 0.08206 \times 298}{1.0} = 3.7 \text{ L}$$

68. $$P_{total} = 94 \text{ kPa} = 94 \times 10^3 \text{ Pa} \times \frac{1 \text{ atm}}{1.013 \times 10^5 \text{ Pa}} \times \frac{760 \text{ torr}}{\text{atm}} = 710 \text{ torr}$$

$$P_{total} = P_{N_2O} + P_{H_2O}; \quad P_{N_2O} = 710 - 21 = 690 \text{ torr} \times \frac{1 \text{ atm}}{760 \text{ torr}} = 0.91 \text{ atm}$$

$$2.6 \text{ g } NH_4NO_3 \times \frac{1 \text{ mol } NH_4NO_3}{80.05 \text{ g } NH_4NO_3} \times \frac{1 \text{ mol } N_2O}{\text{mol } NH_4NO_3} = 3.2 \times 10^{-2} \text{ mol}$$

$$V = \frac{nRT}{P} = \frac{3.2 \times 10^{-2} \text{ mol} \times \frac{0.08206 \text{ L atm}}{\text{mol K}} \times 295 \text{ K}}{0.91 \text{ atm}} = 0.85 \text{ L or } 850 \text{ mL}$$

69. $Br_2 + 3 F_2 \rightarrow 2 X$; Two moles of X must contain two moles of Br and 6 moles of F; X must have the formula, BrF_3.

70. $$750. \text{ mL juice} \times \frac{12 \text{ mL alcohol}}{100 \text{ mL juice}} = 90. \text{ mL alcohol present}$$

$$90. \text{ mL alcohol} \times \frac{0.79 \text{ g}}{\text{mL}} = 71 \text{ g } C_2H_5OH$$

$$71 \text{ g } C_2H_5OH \times \frac{1 \text{ mol } C_2H_5OH}{46.07 \text{ g } C_2H_5OH} \times \frac{2 \text{ mol } CO_2}{2 \text{ mol } C_2H_5OH} = 1.5 \text{ mol } CO_2$$

The CO_2 will occupy the (825 - 750.) or 75 mL not occupied by the liquid (headspace).

$$P = \frac{nRT}{V} = \frac{1.5 \text{ mol} \times \frac{0.08206 \text{ L atm}}{\text{mol K}} \times 298 \text{ K}}{75 \times 10^{-3} \text{ L}} = 490 \text{ atm}$$

Actually, CO_2 will dissolve in the wine to lower the pressure of CO_2 to a much more reasonable value.

71. $PV = nRT;\ \frac{nT}{P} = \frac{V}{R} = \text{Const.};\ \frac{n_1 T_1}{P_1} = \frac{n_2 T_2}{P_2};$

$\text{mol} \times \frac{g}{\text{mol}} = g;\ \frac{n_1 M T_1}{P_1} = \frac{n_2 M T_2}{P_2};\ \frac{g_1 T_1}{P_1} = \frac{g_2 T_2}{P_2}$

$$g_2 = \frac{g_1 T_1 P_2}{T_2 P_1} = \frac{500.\ g \times 291\ K \times 980.\ psi}{299\ K \times 2050.\ psi} = 233\ g$$

72. $P_1V_1 = P_2V_2$, the total volume is 1.0 L + 1.0 L + 2.0 L = 4.0 L

For He: $P_2 = \frac{P_1 V_1}{V_2} = 180\ \text{torr} \times \frac{1.0\ L}{4.0\ L} = 45\ \text{torr}$

For Ne: $P_2 = 0.45\ \text{atm} \times \frac{1.0\ L}{4.0\ L} = 0.11\ \text{atm}$

$0.11\ \text{atm} \times \frac{760\ \text{torr}}{\text{atm}} = 84\ \text{torr}$

For Ar: $P_2 = 25\ \text{kPa} \times \frac{2.0\ L}{4.0\ L} = 12.5\ \text{kPa} = 13\ \text{kPa}$

$13\ \text{kPa} \times \frac{1\ \text{atm}}{101.3\ \text{kPa}} \times \frac{760\ \text{torr}}{\text{atm}} = 98\ \text{torr}$ (94 torr without rounding)

$P_{total} = 45 + 84 + 98 = 227$ torr (223 torr without rounding P_{Ar})

73. If Be^{3+}, formula is $Be(C_5H_7O_2)_3$ and $M \approx 13.5 + 15(12) + 21(1) + 6(16) = 311$ g/mol

If Be^{2+}, formula is $Be(C_5H_7O_2)_2$ and $M \approx 9.0 + 10(12) + 14(1) + 4(16) = 207$ g/mol

Data Set I:

$$M = \frac{gRT}{PV} = \frac{0.2022\ g \times \frac{0.08206\ L\ atm}{mol\ K} \times 286\ K \times 760\ \frac{torr}{atm}}{22.6 \times 10^{-3}\ L \times 765.2\ torr} = 209\ g/mol$$

Data Set II:

$$M = \frac{gRT}{PV} = \frac{0.2224\ g \times \frac{0.08206\ L\ atm}{mol\ K} \times 290.\ K \times 760\ \frac{torr}{atm}}{26.0 \times 10^{-3}\ L \times 764.6\ torr} = 202\ g/mol$$

These results are close to the expected value of 207 g/mol for $Be(C_5H_7O_2)_2$. Thus, we conclude from these data that beryllium is a divalent element with an atomic weight of 9.0 g/mol.

74. $$n = \frac{PV}{RT} = \frac{\left(1.00 \times 10^{-6}\ torr \times \frac{1\ atm}{760\ torr}\right) \times 1.00\ L}{\frac{0.08206\ L\ atm}{mol\ K} \times 295\ torr} = 5.44 \times 10^{-11}\ mol$$

$$5.44 \times 10^{-11}\ mol \times \frac{6.022 \times 10^{23}\ molecules}{mol} = 3.28 \times 10^{13}\ molecules$$

$$\frac{3.28 \times 10^{13}\ molecules}{1.00 \times 10^{3}\ cm^3} = 3.28 \times 10^{10}\ molecules/cm^3$$

75. Out of 100.0 g compounds there are:

$$58.51 \text{ g C} \times \frac{1 \text{ mol C}}{12.01 \text{ g C}} = 4.872 \text{ mol C}; \ \frac{4.872}{2.435} = 2$$

$$7.37 \text{ g H} \times \frac{1 \text{ mol H}}{1.008 \text{ g H}} = 7.31 \text{ mol H}; \ \frac{7.31}{2.435} = 3$$

$$34.12 \text{ g N} \times \frac{1 \text{ mol N}}{14.01 \text{ g N}} = 2.435 \text{ mol N}$$

Empirical formula: C_2H_3N

$$\frac{\text{Rate (1)}}{\text{Rate (2)}} = \left(\frac{M_2}{M_1}\right)^{1/2}; \text{ Let Gas (1) = He}; \qquad 3.20 = \left(\frac{M_2}{4.003}\right)^{1/2}$$

$M_2 = 41.0$, Mass of C_2H_3N: $2(12.0) + 3(1.0) + 1(14.0) = 41.0$

So molecular formula is also C_2H_3N

76. $$33.5 \text{ mg } CO_2 \times \frac{12.01 \text{ mg C}}{44.01 \text{ mg } CO_2} = 9.14 \text{ mg C}; \ \% \text{ C} = \frac{9.14 \text{ mg}}{35.0 \text{ mg}} \times 100 = 26.1\% \text{ C}$$

$$41.1 \text{ mg } H_2O \times \frac{2.016 \text{ mg H}}{18.02 \text{ mg } H_2O} = 4.60 \text{ mg H}; \ \% \text{ H} = \frac{4.60 \text{ mg}}{35.0 \text{ mg}} \times 100 = 13.1\% \text{ H}$$

$$n_{N_2} = \frac{PV}{RT} = \frac{\frac{740.}{760} \text{ atm} \times 35.6 \times 10^{-3} \text{ L}}{\frac{0.08206 \text{ L atm}}{\text{mol K}} \times 298 \text{ K}} = 1.42 \times 10^{-3} \text{ mol } N_2$$

$$1.42 \times 10^{-3} \text{ mol } N_2 \times \frac{28.02 \text{ g } N_2}{\text{mol } N_2} = 3.98 \times 10^{-2} \text{ g N} = 39.8 \text{ mg N}$$

$$\%\text{N} = \frac{39.8 \text{ mg}}{65.2 \text{ mg}} \times 100 = 61.0\% \text{ N}$$

or we can get %N by difference: % N = 100 - (26.1 + 13.1) = 60.8%

Out of 100 g:

$$26.1 \text{ g C} \times \frac{1 \text{ mol}}{12.01 \text{ mg}} = 2.17 \text{ mol}; \ \frac{2.17}{2.17} = 1$$

$$13.1 \text{ g H} \times \frac{1 \text{ mol}}{1.008 \text{ g}} = 13.0 \text{ mol}; \ \frac{13.0}{2.17} \approx 6$$

$$60.8 \text{ g N} \times \frac{1 \text{ mol}}{14.01 \text{ g}} = 4.34 \text{ mol}; \ \frac{4.34}{2.17} = 2$$

Empirical formula is CH_6N_2

$$\left(\frac{M}{39.95}\right)^{1/2} = \frac{26.4}{24.6} = 1.07; \ M = (1.07)^2 (39.95) = 45.7;$$

Mass of $CH_6N_2 \approx 12 + 6 + 28 = 46$. Thus, molecular formula is CH_6N_2.

77. $PV = nRT = \text{Const.}; \; P_1V_1 = P_2V_2$

Let Set (1) correspond to He from the tank that can be used to fill balloons.

$P_1 = 199$ atm (200. - 1) = 199. We must leave 1.0 atm of He in the tank.

$V_1 = 15.0$ L

Set (2) will correspond to the filled balloon. $P_2 = 1.00$ atm

$V_2 = N(2.00\text{ L})$ where N is the number of filled balloons.

(199)(15.0) = (1.00)(2.00) N; N = 1492.5; We can't fill 0.5 of a balloon.

So N = 1492 balloons or to 3 significant figures 1490 balloons.

78. The pressure will increase because H_2 will effuse into container A faster than air will escape.

79. The van der Waals' constant, *b*, is a measure of the size of the molecule. Thus, C_3H_8 should have the largest value of *b*.

80. The values of *a* are:

$$H_2 \; \frac{0.244\text{ L}^2\text{ atm}}{\text{mol}^2}, \; CO_2 \; 3.59, \; N_2 \; 1.39, \; CH_4 \; 2.25$$

Since *a* is a measure of interparticle attractions, the attractions are greatest for CO_2.

CHALLENGE PROBLEMS

81. $M = \frac{dRT}{P}$, P & M constant.

$dT = \frac{PM}{R} = \text{const}$ or $d_1T_1 = d_2T_2$, where T is the absolute temperature.

$T = °C + x; \; 1.2930(0.0 + x) = 0.9460(100.0 + x)$

$1.2930x = 0.9460x + 94.60; \; 0.3470x = 94.60; \; x = 272.6$

From these data absolute zero would be -272.6 °C. Actual value: -273.15 °C.

82. $$\frac{^{12}C^{17}O}{^{12}C^{18}O} = \left(\frac{30.0}{29.0}\right)^{1/2} = 1.02; \quad \frac{^{12}C^{16}O}{^{12}C^{18}O} = \left(\frac{30.0}{28.0}\right)^{1/2} = 1.04$$

The relative rates of effusion of $^{12}C^{16}O$: $^{12}C^{17}O$: $^{12}C^{18}O$ are 1.04: 1.02: 1.00.

Advantages: CO_2 isn't as toxic as CO.

Major disadvantages of using CO_2 instead of CO:

1. Differences in rates of effusion are not as large if only one ^{18}O is present.
2. Same species, e.g., $^{12}C^{16}O^{18}O$ and $^{12}C^{17}O_2$ would effuse at about the same rate. Thus, at some points no separation occurs.

83. $\frac{\text{Rate (1)}}{\text{Rate (2)}} = \left(\frac{M_2}{M_1}\right)^{1/2}$; Let Gas (1) = He, Gas (2) = Cl_2

$$\frac{\frac{1.0\ \text{L}}{4.5\ \text{min}}}{\frac{1.0\ \text{L}}{t}} = \left(\frac{70.90}{4.003}\right)^{1/2};\quad \frac{t}{4.5\ \text{min}} = 4.21;\ t = 19\ \text{min}$$

84. From the van der Waals' equation:

$$P = \frac{nRT}{V - nb} = a\left(\frac{n}{V}\right)^2$$

where P is the measured pressure and V is the volume of the container.

For NH_3: $a = 4.17\ L^2\ atm/mol^2$ and $b = 0.0371$ L/mol

For the first experiment:

$$P = \frac{nRT}{V - nb} - a\left(\frac{n}{V}\right)^2 = \frac{1.0 \times 0.08206 \times 273.15}{172.1 - 0.0371} - 4.17\left(\frac{1.0}{172.1}\right)^2 \quad \text{(carry extra sig. figs.)}$$

P = 0.1303 - 0.0001 = 0.1302 atm. In sample Exercise 5.3, P_{obs} = 0.1300 atm. The difference is less than 0.1%. The ideal gas law also gives 0.1302 atm. At low pressures the van der Waals equation agrees with the ideal gas law.

For the last experiment:

$$P = \frac{nRT}{V - nb} - a\left(\frac{n}{V}\right)^2 = \frac{1.0 \times 0.08206 \times 273.15}{22.08 - 0.0371} - 4.17\left(\frac{1.0}{22.08}\right)^2$$

Note: carry extra significant figures.

P = 1.016 - 0.0086 = 1.007 atm

P_{obs} = 1.000 atm from sample Exercise 5.3. The ideal gas equation gives P = 1.015 atm. Therefore, the van der Waals equation accounts for more than half of the deviation from ideal behavior. Note: as pressure increases, deviation from the ideal gas equation increases.

85. $\left(P + \frac{an^2}{V^2}\right)(V\text{-}nb) = nRT;\ PV + \frac{an^2 V}{V^2} - nbP - \frac{an^3 b}{V^2} = nRT;\quad PV + \frac{an^2}{V^2} - nbP - \frac{an^3 b}{V^2} = nRT$

At low P and high T, the molar volume of a gas will be relatively large. $\frac{an^2}{V}$ and $\frac{an^3 b}{V^2}$ become negligible because V is large.

Since nb is the actual volume of the gas molecules themselves: nb << V and -nbP is negligible compared to PV. Thus PV = nRT.

86. $PV = nRT$, n constant. $\frac{PV}{T} = nR = \text{constant}, \frac{P_1 V_1}{T_1} = \frac{P_2 V_2}{T_2}$

$V_2 = 1.040\ V_1$, so $\frac{V_1}{V_2} = \frac{1.000}{1.040}$

$$P_2 = \frac{P_1 V_1 T_2}{V_2 T_1} = 610 \text{ kPa} \times \frac{1.000}{1.040} \times \frac{(273 + 58)\text{ K}}{(273 + 19)\text{ K}} = 660 \text{ kPa}$$

87. $PV = nRT$, P constant; $\frac{nT}{V} = \frac{P}{R}, \frac{n_1 T_1}{V_1} = \frac{n_2 T_2}{V_2}$

$$\frac{n_2}{n_1} = \frac{T_1 V_2}{T_2 V_1} = \frac{294 \text{ K}}{335 \text{ K}} \times \frac{4.20 \times 10^3 \text{ m}^3}{4.00 \times 10^3 \text{ m}^3} = 0.921$$

88. a. Volume of hot air: $V = \frac{4}{3}\pi r^3 = \frac{4}{3}\pi(2.5 \text{ m})^3 = 65 \text{ m}^3$

(Note: radius = diameter/2 = 5.0/2 = 2.5 m)

$$65 \text{ m}^3 \times \left(\frac{10 \text{ dm}}{\text{m}}\right)^3 \times \frac{1 \text{ L}}{\text{dm}^3} = 6.5 \times 10^4 \text{ L}$$

$$n = \frac{PV}{RT} = \frac{\left(745 \text{ torr} \times \frac{1 \text{ atm}}{760 \text{ torr}}\right) \times 6.5 \times 10^4 \text{ L}}{\frac{0.08206 \text{ L atm}}{\text{mol K}} \times (273 + 65) \text{ K}} = 2.3 \times 10^3 \text{ mol}$$

$$\text{Mass of hot air} = 2.3 \times 10^3 \text{ mol} \times \frac{29.0 \text{ g}}{\text{mol}} = 6.7 \times 10^4 \text{ g}$$

Mass of air displaced:

$$n = \frac{PV}{RT} = \frac{\frac{745}{760} \text{ atm} \times 6.5 \times 10^4 \text{ L}}{\frac{0.08206 \text{ L atm}}{\text{mol K}} \times (273 + 21) \text{ K}} = 2.6 \times 10^3 \text{ mol}$$

$$\text{Mass of air displaced} = 2.6 \times 10^3 \text{ mol} \times \frac{29.0 \text{ g}}{\text{mol}} = 7.5 \times 10^4 \text{ g}$$

Lift = 7.5×10^4 g - 6.7×10^4 g = 8×10^3 g

b. Mass displaced is the same, 7.5×10^4 g.

Moles of He in balloon will be the same as moles of air displaced, 2.6×10^3 mol.

$$\text{Mass of He} = 2.6 \times 10^3 \text{ mol} \times \frac{4.003 \text{ g}}{\text{mol}} = 1.0 \times 10^4 \text{ g}$$

Lift = 7.5×10^4 g - 1.0×10^4 g = 6.5×10^4 g

c. Mass of hot air:

$$n = \frac{PV}{RT} = \frac{\frac{630.}{760}\text{ atm} \times 6.5 \times 10^4\text{ L}}{\frac{0.08206\text{ L atm}}{\text{mol K}} \times 338\text{ K}} = 1.9 \times 10^3\text{ mol}$$

$$1.9 \times 10^3\text{ mol} \times \frac{29.0\text{ g}}{\text{mol}} = 5.5 \times 10^4\text{ g of hot air}$$

Mass of air displaced:

$$n = \frac{PV}{RT} = \frac{\frac{630.}{760}\text{ atm} \times 6.5 \times 10^4\text{ L}}{\frac{0.08206\text{ L atm}}{\text{mol K}} \times 294\text{ K}} = 2.2 \times 10^3\text{ mol}$$

$$2.2 \times 10^3\text{ mol} \times \frac{29.0\text{ g}}{\text{mol}} = 6.4 \times 10^4\text{ g}$$

$$\text{Lift} = 6.4 \times 10^4\text{ g} - 5.5 \times 10^4\text{ g} = 0.9 \times 10^4\text{ g} = 9 \times 10^3\text{ g}$$

89. 10.10 atm - 7.62 atm = 2.48 atm is a pressure of the amount of F_2 reacted.

PV = nRT, V and T are constant. $\frac{P}{n}$ = constant, $\frac{P_1}{n_1} = \frac{P_2}{n_2}$ or $\frac{P_1}{P_2} = \frac{n_1}{n_2}$

$$\frac{\text{moles } F_2 \text{ reacted}}{\text{moles Xe reacted}} = \frac{2.48\text{ atm}}{1.24\text{ atm}} = 2;\ \text{So Xe} + 2\,F_2 \longrightarrow XeF_4$$

90. We assume that 28.01 g/mol is the true value for the molar mass of N_2. The value of 28.15 g/mol is the average of the amount of N_2 and Ar in air.

Let x = % of the number of moles that are N_2 molecules.

Then 100 - x = % of the number of moles that are Ar atoms.

$$28.15 = \frac{x(28.01) + (100 - x)(39.95)}{100};\ 2815 = 28.01x + 3995 - 39.95x$$

$11.94x = 1180.$; $x = 98.83\%\ N_2$; % Ar = 100 - x = 1.17% Ar

Ratio of moles of Ar to moles of N_2 will be equal to $\frac{1.17}{98.83} = 1.18 \times 10^{-2}$

91. PV = nRT, V and T are constant. $\frac{P_1}{n_1} = \frac{P_2}{n_2}$, $\frac{P_2}{P_1} = \frac{n_2}{n_1}$

Let's calculate the partial pressure of C_3H_3N that can be produced from each of the starting materials.

$$P = 0.50\text{ MPa} \times \frac{2\text{ mol } C_3H_3N}{2\text{ mol } C_3H_6} = 0.50\text{ MPa if } C_3H_6 \text{ is limiting.}$$

$$P = 0.80\text{ MPa} \times \frac{2\text{ mol } C_3H_3N}{2\text{ mol } NH_3} = 0.80\text{ MPa if } NH_3 \text{ is limiting.}$$

$$P = 1.50 \text{ MPa} \times \frac{2 \text{ mol } C_3H_3N}{3 \text{ mol } O_2} = 1.00 \text{ MPa if } O_2 \text{ is limiting.}$$

Thus, C_3H_6 is limiting. The partial pressure of C_3H_3N after the reaction is:

$$0.50 \times 10^6 \text{ Pa} \times \frac{1 \text{ atm}}{1.013 \times 10^5 \text{ Pa}} = 4.9 \text{ atm}$$

$$n = \frac{PV}{RT} = \frac{4.9 \text{ atm} \times 150. \text{ L}}{\frac{0.08206 \text{ L atm}}{\text{mol K}} \times 298 \text{ K}} = 30. \text{ mol } C_3H_3N$$

$$30. \text{ mol} \times \frac{53.06 \text{ g}}{\text{mol}} = 1.6 \times 10^3 \text{ g}$$

92. The partial pressure of CO_2 that reacted is 740 - 390 = 350 torr. Thus, the number of moles of CO_2 that reacts is given by:

$$n = \frac{PV}{RT} = \frac{\frac{350}{760} \text{ atm} \times 3.0 \text{ L}}{\frac{0.08206 \text{ L atm}}{\text{mol K}} \times 293 \text{ K}} = 5.7 \times 10^{-2} \text{ mol } CO_2$$

$$5.7 \times 10^{-2} \text{ mol } CO_2 \times \frac{1 \text{ mol MgO}}{1 \text{ mol } CO_2} \times \frac{40.31 \text{ g MgO}}{\text{mol MgO}} = 2.3 \text{ g MgO}$$

$$\% \text{ MgO} = \frac{2.3 \text{ g}}{2.85 \text{ g}} \times 100 = 81\% \text{ MgO}$$

93. a. If we have 10^6 L of air, there are 3.0×10^2 L of CO.

$$P_{CO} = \chi_{CO}P_{tot}; \quad \chi_{CO} = \frac{V_{CO}}{V_{tot}} \text{ since } V \propto n.$$

$$P_{CO} = \frac{3.0 \times 10^2}{1.0 \times 10^6} \times 628 = 0.19 \text{ torr}$$

b. $n_{CO} = \frac{P_{CO} V}{RT}$ Assume 1.0 m^3 air, 1.0 m^3 = 1.0×10^3 L

$$n_{CO} = \frac{\frac{0.19}{760} \text{ atm} \times 1.0 \times 10^3 \text{ L}}{\frac{0.08206 \text{ L atm}}{\text{mol K}} \times 273 \text{ K}} = 1.1 \times 10^{-2} \text{ mol}$$

$$1.1 \times 10^{-2} \text{ mol} \times \frac{6.022 \times 10^{23} \text{ molecules}}{\text{mol}} = 6.6 \times 10^{21} \text{ molecules in the 1.0 } m^3 \text{ of air}$$

c. $$\frac{6.6 \times 10^{21} \text{ molecules}}{m^3} \times \left(\frac{1 \text{ m}}{100 \text{ cm}}\right)^3 = \frac{6.6 \times 10^{15} \text{ molecules}}{cm^3}$$

94. Total volume = 5.0×10^3 ft. $\left(5.0 \text{ mi} \times \dfrac{5280 \text{ ft}}{\text{mi}}\right)\left(10.0 \text{ mi} \times \dfrac{5280 \text{ ft}}{\text{mi}}\right) = 7.0 \times 10^{12} \text{ ft}^3$

$$7.0 \times 10^{12} \text{ ft}^3 \left(\frac{12 \text{ in}}{\text{ft}} \times \frac{2.54 \text{ cm}}{\text{in}}\right)^3 \left(\frac{1 \text{ L}}{1000 \text{ cm}^3}\right) = 2.0 \times 10^{14} \text{ L}$$

$$\text{ppmv NO}_2 = 0.2 = \frac{V_{NO_2}}{2.0 \times 10^{14} \text{ L}} \times 10^6$$

$$V_{NO_2} = \frac{(0.2)(2.0 \times 10^{14} \text{ L})}{10^6} = 4 \times 10^7 \text{ L NO}_2$$

$$PV = nRT; \; n = \frac{PV}{RT} = \frac{(1.0 \text{ atm})\,(4 \times 10^7 \text{ L})}{\left(\dfrac{0.08206 \text{ L atm}}{\text{mol K}}\right)(273 \text{ K})} = 2 \times 10^6 \text{ mol NO}_2$$

$$2 \times 10^6 \text{ mol NO}_2 \times \frac{46.01 \text{ g NO}_2}{\text{mol NO}_2} = 9 \times 10^7 \text{ g NO}_2$$

95. For benzene: $101 \times 10^{-9} \text{ g} \times \dfrac{1 \text{ mol}}{78.11 \text{ g}} = 1.29 \times 10^{-9}$ mol

$$V = \frac{nRT}{P} = \frac{1.29 \times 10^{-9} \text{ mol} \times \dfrac{0.08206 \text{ L atm}}{\text{mol K}} \times 296 \text{ K}}{748 \text{ torr}} \times \frac{760 \text{ torr}}{\text{atm}} = 3.18 \times 10^{-8} \text{ L}$$

$$\text{Mixing ratio} = \frac{3.18 \times 10^{-8} \text{ L}}{2.0 \text{ L}} \times 10^6 = 1.6 \times 10^{-2} \text{ ppmv}$$

or better yet: Use ppbv, defined similarly to ppmv in problem 5.93.

$$\text{ppbv} = \frac{\text{vol. of } x}{\text{total vol.}} \times 10^9$$

In this case Mixing ratio = 16 ppbv

$$\frac{1.29 \times 10^{-9} \text{ mol benzene}}{2.0 \text{ L}} \times \frac{1 \text{ L}}{1000 \text{ cm}^3} \times \frac{6.022 \times 10^{23} \text{ molecules}}{\text{mol}}$$

$= 3.9 \times 10^{11}$ molecules benzene/cm^3

For toluene: $274 \times 10^{-9} \text{ g } C_7H_8 \times \dfrac{1 \text{ mol}}{92.13 \text{ g}} = 2.97 \times 10^{-9}$ mol toluene

$$V_{tol} = \frac{nRT}{P} = \frac{2.97 \times 10^{-9} \text{ mol} \times \dfrac{0.08206 \text{ L atm}}{\text{mol K}} \times 296 \text{ K}}{748 \text{ torr}} \times \frac{760 \text{ torr}}{\text{atm}} = 7.33 \times 10^{-8} \text{ L}$$

$$\text{Mixing ratio} = \frac{7.33 \times 10^{-8} \text{ L}}{2.0 \text{ L}} \times 10^9 = 37 \text{ ppbv}$$

$$\frac{2.97 \times 10^{-9} \text{ mol toluene}}{2.0 \text{ L}} \times \frac{1 \text{ L}}{1000 \text{ cm}^3} \times \frac{6.022 \times 10^{23} \text{ molecules}}{\text{mol}}$$

$= 8.9 \times 10^{11}$ molecules toluene/cm^3

96. A concentration of 1.0 ppbv means there is 1.0 L of CH_2O for every 1.0×10^9 L of air measured at the same temperature and pressure.

The molar volume of an ideal gas is 22.42 at STP. Assume 1.0×10^9 L air:

$$\frac{1.0\ \text{L}\ CH_2O}{1.0 \times 10^9\ \text{L}} \times \frac{1\ \text{mol}\ CH_2O}{22.42\ \text{L}} \times \frac{6.022 \times 10^{23}\ \text{molecules}}{\text{mol}} \times \frac{1\ \text{L}}{1000\ \text{cm}^3}$$

$$= \frac{2.7 \times 10^{10}\ \text{molecules}}{\text{cm}^3}$$

$$V = \left(18\ \text{ft} \times \frac{12\ \text{in}}{\text{ft}} \times \frac{2.54\ \text{cm}}{\text{in}}\right)\left(24\ \text{ft} \times \frac{12\ \text{in}}{\text{ft}} \times \frac{254\ \text{cm}}{\text{in}}\right)$$

$$\times \left(8\ \text{ft} \times \frac{12\ \text{in}}{\text{ft}} \times \frac{2.54\ \text{cm}}{\text{in}}\right) = 9.8 \times 10^7\ \text{cm}^3 = 1 \times 10^8\ \text{cm}^3$$

$$1 \times 10^8\ \text{cm}^3 \times \frac{2.7 \times 10^{10}\ \text{molecules}}{\text{cm}^3} \times \frac{1\ \text{mol}}{6.022 \times 10^{23}\ \text{molecules}} \times \frac{30.03\ \text{g}}{\text{mol}}$$

$$= 1 \times 10^{-4}\ \text{g}\ CH_2O$$

CHAPTER SIX: THERMOCHEMISTRY

QUESTIONS

1. A solution calorimeter is at constant (atmospheric) pressure. The heat at constant P is ΔH. A bomb calorimeter is at constant volume. The heat at constant volume is ΔE.

2. If there are fewer forests, then less CO_2 will be removed from the atmosphere by photosynthesis. The greenhouse effect is a result of increased levels of CO_2 in the atmosphere, and thus will be more severe with forest destruction.

3. A state function is a function whose change depends only on the initial and final states and not on how one got from the initial to the final state. If H and E were not state functions, the law of conservation of energy (first law) would not be true.

4. If Hess's law were not true it would be possible to create energy by reversing a reaction using a different series of steps. This violates the law of conservation of energy (first law). Thus, Hess's law is another statement of the law of conservation of energy.

5. We can only measure differences in enthalpy. Thus, in order for different workers to compare measured values of ΔH to each other, a common reference (or zero) point must be chosen. The definition of ΔH_f^o establishes the pure elements in their standard states as that common reference point.

6. Advantages: H_2 burns cleanly (less pollution) and gives a lot of energy per gram of fuel.

   Disadvantages: Expense and storage.

EXERCISES

Potential and Kinetic Energy

7. $KE = \frac{1}{2}mv^2$, convert mass and velocity to SI units. $1\ J = \frac{1\ kg\ m^2}{s^2}$

$$5.25\ oz \times \frac{1\ lb}{16\ oz} \times \frac{1\ kg}{2.205\ lb} = 0.149\ kg = mass$$

$$Velocity = \frac{1.0 \times 10^2\ mi}{hr} \times \frac{1\ hr}{60\ min} \times \frac{1\ min}{60\ s} \times \frac{1760\ yd}{mi} \times \frac{1\ m}{1.094\ yd} = \frac{45\ m}{s}$$

$$KE = \frac{1}{2}mv^2 = \frac{1}{2}\ (0.149\ kg)\left(\frac{45\ m}{s}\right)^2 = 150\ J$$

8. $$KE = \frac{1}{2}mv^2 = \frac{1}{2}\left(1.0 \times 10^{-5}\ g \times \frac{1\ kg}{1000\ g}\right)\left(\frac{2.0 \times 10^5\ cm}{sec} \times \frac{1\ m}{100\ cm}\right)^2 = 2.0 \times 10^{-2}\ J$$

9. $$KE = \frac{1}{2}mv^2 = \frac{1}{2}\ (2.0\ kg)\left(\frac{1.0\ m}{s}\right)^2 = 1.0\ J$$

$$KE = \frac{1}{2}mv^2 = \frac{1}{2}\ (1.0\ kg)\left(\frac{2.0\ m}{s}\right)^2 = 2.0\ J$$

The 1.0 kg object with a velocity of 2.0 m/s has the greater kinetic energy.

10. Ball A: $PE = mgz = 5.0\ kg \times \frac{9.8\ m}{s^2} \times 5.0\ m = \frac{245\ kg\ m^2}{s^2} = 245\ J = 250\ J$

At Point I: All of the energy is transferred to Ball B. All of B's energy is kinetic energy at this point. $E_{tot} = KE = 250\ J$

At Point II: $PE = mgz = 1.0\ kg \times \frac{9.8\ m}{s^2} \times 2.0\ m = 19.6\ J \approx 20.\ J$

$KE = E_{tot} - PE = 250\ J - 20.\ J = 230\ J$

Heat and Work

11. a. $\Delta E = q + w = 51\ kJ + (-15\ kJ) = 36\ kJ$

b. $\Delta E = 100.\ kJ + (-65\ kJ) = 35\ kJ$ c. $\Delta E = -65 - 20. = -85\ kJ$

d. When the system delivers work to the surroundings, $w < 0$. This is the case in all these examples, a, b and c.

12. a. $\Delta E = q + w = -47 + 88 = 41\ kJ$

b. $\Delta E = 82 + 47 = 129\ kJ$

c. $\Delta E = 47 + 0 = 47\ kJ$

d. When the surroundings deliver work to the system, $w > 0$. This is the case for a and b.

13. $\Delta E = q + w = +45 - 29 = +16\ kJ$

14. $\Delta E = q + w = -125 + 104 = -21\ kJ$

15. $w = -P\Delta V = -P(V_f - V_i) = -2.0\ atm\ (5.0 \times 10^{-3}\ L - 5.0\ L)$

$w = -2.0\ atm\ (-5.0\ L) = 10.\ L\ atm$

We can also calculate the work in Joules.

$1\ atm = 1.013 \times 10^5\ Pa = 1.013 \times 10^5\ \frac{N}{m^2}$

$1\ m^3 = (10\ dm)^3 = 1000\ dm^3 = 1000\ L$

$1\ L\ atm = 10^{-3}\ m^3 \times 1.013 \times 10^5\ \frac{N}{m^2} = 101.3\ N\ m = 101.3\ J$

$w = 10.\ L\ atm \times \frac{101.3\ J}{L\ atm} = 1013\ J = 1.0 \times 10^3\ J$

16. In this problem $q = w = -950.\ J$

$-950.\ J \times \frac{1\ L\ atm}{101.3\ J} = -9.38\ L\ atm$ of work done by the gases.

$w = -P\Delta V;\ -9.38\ L\ atm = \frac{-650.}{760}\ atm\ (V_f - 0.040\ L)$

$V_f - 0.040 = 11.0\ L;\ V_f = 11.0\ L$

17. $q = \frac{20.8\text{ J}}{°\text{C mol}} \times 39.1\text{ mol} \times (38.0 - 0.0)\ °\text{C}$

$q = 30{,}900\text{ J} = 30.9\text{ kJ}$

$w = -P\Delta V = -1.00\text{ atm}(998\text{ L} - 875\text{ L}) = -123\text{ L atm} \times \frac{101.3\text{ J}}{\text{L atm}}$

$w = -12{,}500\text{ J} = -12.5\text{ kJ}$

$\Delta E = q + w = 30.9\text{ kJ} + (-12.5\text{ kJ}) = 18.4\text{ kJ}$

18. $313\text{ g He} \times \frac{1\text{ mol He}}{4.003\text{ g He}} = 78.2\text{ mol He}$

$q = 78.2\text{ mol He} \times \frac{20.8\text{ J}}{\text{mol °C}} \times (-15\ °\text{C}) = -2.4 \times 10^4\text{ J or } -24\text{ kJ}$

$w = -P\Delta V = -1.00\text{ atm }(1840 - 1910)\text{ L} = 70\text{ L atm}$

There are 101.3 J/L•atm. (An easy way to get this is from tabulated values of R.)

$$\frac{\frac{8.3145\text{ J}}{\text{mol K}}}{\frac{0.08206\text{ L atm}}{\text{mol K}}} = 101.3\text{ J/L}\cdot\text{atm}$$

$w = 70\text{ L atm} \times \frac{101.3\text{ J}}{\text{L atm}} = 7.1 \times 10^3\text{ J} = 7\text{ kJ}$

$\Delta E = q + w = -24\text{ kJ} + 7\text{ kJ} = -17\text{ kJ}$

Properties of Enthalpy

19. a. endothermic b. exothermic c. exothermic d. endothermic

20. Sign of ΔH is negative; the reaction is exothermic.

Heat is evolved by the system to the surroundings.

$$\Delta H = \frac{-67\text{ kJ}}{\text{mol } C_6H_{12}O_6} \times \frac{1\text{ mol } C_6H_{12}O_6}{180.16\text{ g } C_6H_{12}O_6} = -0.37\text{ kJ/g}$$

21. $S(s) + O_2(g) \longrightarrow SO_2(g)\quad \Delta H° = \frac{-296\text{ kJ}}{\text{mol}}$

a. $275\text{ g S} \times \frac{1\text{ mol S}}{32.07\text{ g S}} \times \frac{296\text{ kJ}}{\text{mol S}} = 2.54 \times 10^3\text{ kJ}$

b. $25\text{ mol S} \times \frac{296\text{ kJ}}{\text{mol S}} = 7.4 \times 10^3\text{ kJ}$

c. $150.\text{ g } SO_2 \times \frac{1\text{ mol } SO_2}{64.07\text{ g } SO_2} \times \frac{296\text{ kJ}}{\text{mol } SO_2} = 693\text{ kJ}$

22. $B_2H_6(g) + 3\,O_2(g) \longrightarrow B_2O_3(s) + 3\,H_2O(g) \quad \Delta H° = -2035$ kJ

Molar mass of $B_2H_6 = 2(10.81) + 6(1.008) = 27.67$ g/mol

a. $1.0\text{ g } B_2H_6 \times \frac{1\text{ mol } B_2H_6}{27.67\text{ g } B_2H_6} \times \frac{2035\text{ kJ}}{\text{mol } B_2H_6} = 74\text{ kJ}$

b. $1.0 \times 10^3\text{ g } B_2H_6 \times \frac{1\text{ mol } B_2H_6}{27.67\text{ g } B_2H_6} \times \frac{2035\text{ kJ}}{\text{mol } B_2H_6} = 7.4 \times 10^4\text{ kJ}$

c. $1.0\text{ mol } B_2H_6 \times \frac{2035\text{ kJ}}{\text{mol } B_2H_6} = 2.0 \times 10^3\text{ kJ}$

d. $1.0 \times 10^2\text{ mol } B_2H_6 \times \frac{2035\text{ kJ}}{\text{mol}} = 2.0 \times 10^5\text{ kJ}$

e. $2.6 \times 10^{22}\text{ molecules } B_2H_6 \times \frac{1\text{ mol } B_2H_6}{6.02 \times 10^{23}\text{ molecules}} \times \frac{2035\text{ kJ}}{\text{mol } B_2H_6} = 88\text{ kJ}$

f. $10.0\text{ g } B_2H_6 \times \frac{1\text{ mol}}{27.67\text{ g}} = 0.361\text{ mol } B_2H_6$

$10.0\text{ g } O_2 \times \frac{1\text{ mol } O_2}{32.00\text{ g } O_2} \times \frac{1\text{ mol } B_2H_6}{3\text{ mol } O_2} = 0.104\text{ mol } B_2H_6$

We only have enough O_2 to completely react with 0.104 mol B_2H_6. We have 0.361 mol B_2H_6, so the O_2 is the limiting reactant.

$0.104\text{ mol } B_2H_6\text{ reacts} \times \frac{2035\text{ kJ}}{\text{mol } B_2H_6} = 212\text{ kJ}$

or we get the same answer by:

$10.0\text{ g } O_2 \times \frac{1\text{ mol } O_2}{32.00\text{ g } O_2} \times \frac{2035\text{ kJ}}{3\text{ mol } O_2} = 212\text{ kJ}$

Calorimetry and Heat Capacity

23. a. $\Delta H = q = 850.\text{ g} \times \frac{0.900\text{ J}}{\text{g °C}} \times (94.6 - 22.8)\text{ °C} = 5.49 \times 10^4\text{ J or } 54.9\text{ kJ}$

b. $\frac{0.900\text{ J}}{\text{g °C}} \times \frac{26.98\text{ g}}{\text{mol}} = \frac{24.3\text{ J}}{\text{mol °C}}$

24. a. $1.0\text{ mol C} \times \frac{12.01\text{ g C}}{\text{mol C}} \times \frac{0.71\text{ J}}{\text{g °C}} \times 1.0\text{ °C} = 8.5\text{ J}$

b. $850\text{ g C} \times \frac{0.71\text{ J}}{\text{g °C}} \times 150\text{ °C} = 9.1 \times 10^4\text{ J} = 91\text{ kJ}$

c. $75\text{ kg C} \times \frac{1000\text{ g}}{\text{kg}} \times \frac{0.71\text{ J}}{\text{g °K}} \times (348 - 294)\text{K} = 2.9 \times 10^6\text{ J} = 2.9 \times 10^3\text{ kJ}$

Note: $\Delta T(\text{K}) = \Delta T(\text{°C})$

25. Specific heat capacity (C): J/g•°C

$$C = \frac{78.2\ J}{45.6\ g \times 13.3\ °C} = \frac{0.129\ J}{g\ °C}$$

$$\text{Molar heat capacity} = \frac{0.129\ J}{g\ °C} \times \frac{207.2\ g}{mol} = \frac{26.7\ J}{mol\ °C}$$

26. $$C = \frac{585\ J}{125.6\ g\ (53.5 - 20.0)\ °C} = 0.139\ J/g•°C$$

$$\text{Molar heat capacity} = \frac{0.139\ J}{g\ °C} \times \frac{200.6\ g}{mol\ Hg} = \frac{27.9\ J}{mol\ °C}$$

27. Heat gained by water = Heat lost by nickel

$$\text{Heat gain} = \frac{4.18\ J}{g\ °C} \times 150.\ g \times (25.0\ °C - 23.5\ °C) = 940\ J$$

Note: A temperature change of one Kelvin is the same as a temperature change of one degree Celsius.

$$\text{Heat loss} = C \times m \times \Delta T$$

$$940\ J = C \times 28.2\ g \times (99.8 - 25.0)\ °C$$

$$C = \frac{940\ J}{28.2\ g \times 74.8\ °C} = \frac{0.45\ J}{g\ °C}$$

28. Heat gained by water = Heat lost by copper

$$\frac{4.18\ J}{g\ °C} \times 75.0\ g \times 2.2°C = C \times 46.2\ g \times 73.6\ °C$$

$$C = 0.20\ J/g•°C$$

29. Heat lost by solution = Heat gained by KBr

$$\text{Heat lost by solution} = \frac{4.18\ J}{g\ °C} \times 136\ g \times 3.1\ °C = 1800\ J$$

$$\text{Heat gained by KBr} = \frac{1800\ J}{10.5\ g} = \frac{170\ J}{g}$$

$$\frac{170\ J}{g\ KBr} \times \frac{119.00\ g\ KBr}{mol\ KBr} \times \frac{1\ kJ}{1000\ J} = \frac{20.2\ kJ}{mol} = \frac{2.0 \times 10^1\ kJ}{mol} = \Delta H$$

30. $NaOH(aq) + HCl(aq) \longrightarrow NaCl(aq) + H_2O(l)$

We have a stoichiometric mixture. All of the NaOH and HCl will react.

$$0.10\ L \times \frac{1.0\ mol}{L} = 0.10 \text{ mol of HCl is being neutralized.}$$

Heat lost by chemicals = Heat gained by solution

$$\text{Heat Gain} = \frac{4.18\ J}{g\ °C}\left(200.\ mL \times \frac{1.0\ g}{mL}\right)(31.3 - 24.6)°C = 5.6 \times 10^3\ J = 5.6\ kJ$$

Heat Loss = 5.6 kJ. This is the heat released by the neutralization of 0.10 mol HCl (0.100 L HCl $\times$ 1.0 mol/L). The reaction is exothermic.

$$\Delta H = \frac{-5.6 \text{ kJ}}{0.10 \text{ mol}} = -56 \text{ kJ/mol}$$

31. Heat lost by camphor = Heat gained by calorimeter

Heat lost by combustion of camphor = $\frac{5903.6 \text{ kJ}}{\text{mol}} \times \frac{1 \text{ mol}}{152.23 \text{ g}} \times 0.1204 \text{ g} = 4.669 \text{ kJ}$

Heat gained by calorimeter = C $\times$ ΔT; C $\times$ (2.28 °C) = 4.669 kJ; C = 2.05 kJ/°C

32. Heat Loss = Heat Gain

Heat Gain by calorimeter = $\frac{1.56 \text{ kJ}}{°\text{C}} \times 3.2\ °\text{C} = 5.0 \text{ kJ}$

Heat Loss = 5.0 kJ which is the heat evolved (exothermic reaction) by the combustion of 0.1964 g of quinone.

$$q = \frac{-5.0 \text{ kJ}}{0.1964 \text{ g}} = -25 \text{ kJ/g}$$

$$q = \frac{-25 \text{ kJ}}{\text{g}} \times \frac{108.09 \text{ g}}{\text{mol}} = -2700 \text{ kJ/mol}$$

33. $50.0 \times 10^{-3} \text{ L} \times \frac{0.100 \text{ mol}}{\text{L}} = 5.00 \times 10^{-3}$ mol of both $AgNO_3$ and HCl is reacted.

Thus, 5.00×10^{-3} mol of AgCl will be produced.

Heat lost by chemicals = Heat gain by solution

Heat Gain = $\frac{4.18 \text{ J}}{\text{g}\ °\text{C}} \times 100.\ \text{g} \times 0.8°\text{C} = 300 \text{ J}$

Heat Loss = 300 J: this is the heat evolved (exothermic reaction) when 5.00×10^{-3} mol of AgCl is produced.

Thus, $\Delta H = \frac{-300 \text{ J}}{5.00 \times 10^{-3} \text{ mol}} \times \frac{1 \text{ kJ}}{1000 \text{ J}} = -60 \text{ kJ/mol}$

Since, we only know the temperature difference to one significant figure, ΔH = -60 kJ/mol (without intermediate rounding we get ΔH = -66.9 kJ/mol $\approx$ -70 kJ/mol).

34. $0.100 \text{ L} \times \frac{0.500 \text{ mol HCl}}{\text{L}} \times \frac{1 \text{ mol } BaCl_2}{2 \text{ mol HCl}} = 2.50 \times 10^{-2} \text{ mol } BaCl_2$

$0.300 \text{ L} \times \frac{0.500 \text{ mol } Ba(OH)_2}{\text{L}} \times \frac{1 \text{ mol } BaCl_2}{1 \text{ mol } Ba(OH)_2} = 1.50 \times 10^{-1} \text{ mol } BaCl_2$

Thus, the HCl is limiting.

$\frac{118 \text{ kJ}}{\text{mol } BaCl_2} \times 2.50 \times 10^{-2} \text{ mol } BaCl_2$ = 2.95 kJ of heat is evolved by reaction.

Heat gained by solution $= 2.95 \times 10^3\ J = \frac{4.18\ J}{°C\ g} \times 400.0\ g \times \Delta T$

$\Delta T = 1.76° = T_f - T_i = T_f - 25°C;\ T_f = 26.76° = 27°C$

Hess's Law

35.

$S + 3/2\ O_2 \rightarrow SO_3$	$\Delta H° = -395.2$ kJ	
$SO_3 \rightarrow SO_2 + 1/2\ O_2$	$\Delta H° = +198.2$ kJ/2 = +99.1 kJ	÷(-2)
$S(s) + O_2(g) \rightarrow SO_2(g)$	$\Delta H° = -296.1$ kJ	

36. We want $\Delta H°$ for $2\ NO_2 \rightarrow N_2O_4$

$2\ NO_2 \rightarrow N_2 + 2\ O_2$	$\Delta H° = -67.7$ kJ	(-)
$N_2 + 2\ O_2 \rightarrow N_2O_4$	$\Delta H° = 9.7$ kJ	
$2\ NO_2(g) \rightarrow N_2O_4(g)$	$\Delta H° = -58.0$ kJ	

37.

$4\ HNO_3 \rightarrow 2\ N_2O_5 + 2\ H_2O$	$\Delta H° = 2(+76.6$ kJ)	×(-2)
$2\ N_2 + 6\ O_2 + 2\ H_2 \rightarrow 4\ HNO_3$	$\Delta H° = 4(-174.1$ kJ)	×4
$2\ H_2O \rightarrow 2\ H_2 + O_2$	$\Delta H° = 2(+285.8$ kJ)	×(-2)
$2\ N_2(g) + 5\ O_2(g) \rightarrow 2\ N_2O_5(g)$	$\Delta H° = 28.4$ kJ	

38.

$2\ C + 2\ O_2 \rightarrow 2\ CO_2$	$\Delta H° = 2(-394$ kJ)	×2
$H_2 + 1/2\ O_2 \rightarrow H_2O$	$\Delta H° = -286$ kJ	
$2\ CO_2 + H_2O \rightarrow C_2H_2 + 5/2\ O_2$	$\Delta H° = +1300.$ kJ	(-)
$2\ C(s) + H_2(g) \rightarrow C_2H_2(g)$	$\Delta H° = 226$ kJ	

39.

$NO + O_3 \rightarrow NO_2 + O_2$	$\Delta H° = -199$ kJ	
$3/2\ O_2 \rightarrow O_3$	$\Delta H° = +427$ kJ/2	÷(-2)
$O \rightarrow 1/2\ O_2$	$\Delta H° = -495$ kJ/2	÷(-2)
$NO(g) + O(g) \rightarrow NO_2(g)$	$\Delta H° = -233$ kJ	

40.

$C_6H_4(OH)_2(aq) \rightarrow C_6H_4O_2(aq) + H_2(s)$	$\Delta H° = +177.4$ kJ	
$H_2O_2(aq) \rightarrow H_2(g) + O_2(g)$	$\Delta H° = +191.2$ kJ	(-)
$2\ H_2(g) + O_2(g) \rightarrow 2\ H_2O(g)$	$\Delta H° = 2(-241.8$ kJ)	×2
$2\ H_2O(g) \rightarrow 2\ H_2O(l)$	$\Delta H° = 2(-43.8$ kJ)	×2

$C_6H_4(OH)_2(aq) + H_2O_2(aq) \rightarrow C_6H_4O_2(aq) + 2\ H_2O(l)$ $\Delta H° = -202.6$ kJ

41.

$1/2\ O_2 + 1/2\ H_2 \rightarrow OH$	$\Delta H° = 77.9$ kJ/2	÷2
$O \rightarrow 1/2\ O_2$	$\Delta H° = -495$ kJ/2	÷(-2)
$H \rightarrow 1/2\ H_2$	$\Delta H° = -435.9$ kJ/2	÷(-2)

$O(g) + H(g) \rightarrow OH(g)$ $\Delta H° = -427$ kJ

42. We want $\Delta H°$ for $N_2H_4(l) + O_2(g) \rightarrow N_2(g) + 2\ H_2O(l)$

It will be easier to calculate $\Delta H°$ for combustion of four moles of N_2H_4.

$9\ H_2(g) + 9/2\ O_2(g) \rightarrow 9\ H_2O(l)$	$\Delta H° = 9(-286$ kJ)	×9
$3\ N_2H_4(l) + 3\ H_2O(l) \rightarrow 3\ N_2O(g) + 9\ H_2(g)$	$\Delta H° = 3(317$ kJ)	×(-3)
$2\ NH_3(g) + 3\ N_2O(g) \rightarrow 4\ N_2(g) + 3\ H_2O(l)$	$\Delta H° = -1010.$ kJ	
$N_2H_4(l) + H_2O(l) \rightarrow 2\ NH_3(g) + 1/2\ O_2(g)$	$\Delta H° = +\ 143$ kJ	(-)

$4\ N_2H_4(l) + 4\ O_2(g) \rightarrow 4\ N_2(g) + 8\ H_2O(l)$ $\Delta H° = -2490$ kJ

So for:

$N_2H_4(l) + O_2(g) \rightarrow N_2(g) + 2\ H_2O(l)$ $\Delta H° = \frac{-2490\text{ kJ}}{4} = -623$ kJ

Standard Enthalpies of Formation

43. The enthalpy change for the formation of one mole of a compound at 25°C from its elements, with all substances in their standard states.

$Na(s) + 1/2\ Cl_2(g) \rightarrow NaCl(s)$

$H_2(g) + 1/2\ O_2(g) \rightarrow H_2O(l)$

$6\ C(graphite) + 6\ H_2(g) + 3\ O_2(g) \rightarrow C_6H_{12}O_6(s)$

$Pb(s) + S(s) + 2\ O_2(g) \rightarrow PbSO_4(s)$

44. a. aluminum oxide Al_2O_3

$2\ Al(s) + 3/2\ O_2(g) \rightarrow Al_2O_3(s)$

b. $C_2H_5OH(l) + O_2(g) \rightarrow CO_2(g) + H_2O(l)$

$C_2H_5OH(l) + 3\ O_2 \rightarrow 2\ CO_2(g) + 3\ H_2O(l)$

c. $Ba(OH)_2(aq) + 2\ HCl(aq) \rightarrow 2\ H_2O(l) + BaCl_2(aq)$

d. $2\ C(graphite) + 3/2\ H_2(g) + 1/2\ Cl_2(g) \rightarrow C_2H_3Cl(g)$

e. $C_6H_6(l) + O_2(g) \rightarrow 6\ CO_2(g) + 3\ H_2O(l)$

15 - O atoms on right

$C_6H_6(l) + 15/2\ O_2(g) \rightarrow 6\ CO_2(g) + 3\ H_2O(l)$

or $2\ C_6H_6(l) + 15\ O_2(g) \rightarrow 12\ CO_2(g) + 6\ H_2O(l)$

f. $NH_4Br(s) \rightarrow NH_4^+(aq) + Br^-(aq)$

45. In general:

$$\Delta H° = \Sigma \Delta H_f^\circ \text{ (products)} - \Sigma \Delta H_f^\circ \text{ (reactants)}$$

a. $2\ NH_3(g) + 3\ O_2(g) + 2\ CH_4(g) \rightarrow 2\ HCN(g) + 6\ H_2O(g)$

$\Delta H° = 2 \text{ mol HCN} \times \Delta H_f^\circ (HCN) + 6 \text{ mol } H_2O(g) \times \Delta H_f^\circ (H_2O,g)$

$- [2 \text{ mol } NH_3 \times \Delta H_f^\circ (NH_3) + 2 \text{ mol } CH_4 \times \Delta H_f^\circ (CH_4)]$

$\Delta H° = 2(135.1) + 6(-242) - [2(-46) + 2(-75)] = -940. \text{ kJ}$

b. $\Delta H° = 3 \text{ mol } CaSO_4 \text{ (-1433 kJ/mol)} + 2 \text{ mol } H_3PO_4(l) \text{ (-1267 kJ/mol)}$

$-[1 \text{ mol } Ca_3(PO_4)_2 \text{ (-4126 kJ/mol)} + 3 \text{ mol } H_2SO_4(l) \text{ (-814 kJ/mol)}]$

$\Delta H° = -6833 \text{ kJ} - (-6568 \text{ kJ}) = -265 \text{ kJ}$

c. $\Delta H° = 1 \text{ mol } NH_4Cl \times \Delta H_f^\circ (NH_4Cl)$

$- [1 \text{ mol } NH_3 \times \Delta H_f^\circ (NH_3) + 1 \text{ mol HCl} \times \Delta H_f^\circ (HCl)]$

$\Delta H° = 1 \text{ mol (-314 kJ/mol)} - [1 \text{ mol (-46 kJ/mol)} + 1 \text{ mol (-92 kJ/mol)}]$

$\Delta H° = -314 \text{ kJ} + 138 \text{ kJ} = -176 \text{ kJ}$

46. a. $C_2H_5OH(l) + 3\ O_2(g) \rightarrow 2\ CO_2(g) + 3\ H_2O(g)$

$\Delta H° = [2 \text{ mol (-393.5 kJ/mol)} + 3 \text{ mol (-242 kJ/mol)}] - [1 \text{ mol (-278 kJ/mol)}]$

$= -1513 \text{ kJ} - (-278 \text{ kJ}) = -1235 \text{ kJ}$

b. $SiCl_4(l) + 2\ H_2O(l) \rightarrow SiO_2(s) + 4\ HCl(aq)$

Since HCl(aq) is $H^+(aq) + Cl^-(aq)$, $\Delta H_f^\circ = 0 - 167 = -167$ kJ/mol

$\Delta H^\circ = [4 \text{ mol } (-167 \text{ kJ/mol}) + 1 \text{ mol}(-911 \text{ kJ/mol})]$

$- [1 \text{ mol } (-687 \text{ kJ/mol}) + 2 \text{ mol}(-286 \text{ kJ/mol})]$

$\Delta H^\circ = -1579 \text{ kJ} - (-1259 \text{ kJ}) = -320. \text{ kJ}$

c. $MgO(s) + H_2O(l) \rightarrow Mg(OH)_2(s)$

$\Delta H^\circ = 1 \text{ mol}(-925 \text{ kJ/mol}) - [1 \text{ mol } (-602 \text{ kJ/mol}) + 1 \text{ mol } (-286 \text{ kJ/mol})]$

$\Delta H^\circ = -925 \text{ kJ} - (-888 \text{ kJ}) = -37 \text{ kJ}$

47. a. $4\ NH_3(g) + 5\ O_2(g) \rightarrow 4\ NO(g) + 6\ H_2O(g)$

$\Delta H^\circ = 4 \text{ mol } (90 \text{ kJ/mol}) + 6 \text{ mol } (-242 \text{ kJ/mol}) - 4 \text{ mol } (-46 \text{ kJ/mol}) = -908 \text{ kJ}$

$2\ NO(g) + O_2(g) \rightarrow 2\ NO_2(g)$

$\Delta H^\circ = 2 \text{ mol } (34 \text{ kJ/mol}) - 2 \text{ mol } (90 \text{ kJ/mol}) = -112 \text{ kJ}$

$3\ NO_2(g) + H_2O(l) \rightarrow 2\ HNO_3(aq) + NO(g)$

$\Delta H^\circ = 2 \text{ mol } (-207 \text{ kJ/mol}) + 1 \text{ mol } (90 \text{ kJ/mol})$

$-[3 \text{ mol } (34 \text{ kJ/mol}) + 1 \text{ mol } (-286 \text{ kJ/mol})] = -140. \text{ kJ}$

b.

$12\ NH_3(g) + 15\ O_2(g) \rightarrow 12\ NO(g) + 18\ H_2O(g)$

$12\ NO(g) + 6\ O_2(g) \rightarrow 12\ NO_2(g)$

$12\ NO_2(g) + 4\ H_2O(l) \rightarrow 8\ HNO_3(aq) + 4\ NO(g)$

$4\ H_2O(g) \rightarrow 4\ H_2O(l)$

$12\ NH_3(g) + 21\ O_2(g) \rightarrow 8\ HNO_3(aq) + 4\ NO(g) + 14\ H_2O(g)$

The overall reaction is exothermic, since each step is exothermic.

48. $4\ Na(s) + O_2(g) \rightarrow 2\ Na_2O(s)$

$\Delta H^\circ = 2 \text{ mol } (-416 \text{ kJ/mol}) = -832 \text{ kJ}$

$2\ Na(s) + 2\ H_2O(l) \rightarrow 2\ NaOH(aq) + H_2(g)$

$\Delta H^\circ = 2 \text{ mol } (-470 \text{ kJ/mol}) - 2 \text{ mol } (-286 \text{ kJ/mol}) = -368 \text{ kJ}$

$2\ Na(s) + CO_2(g) \rightarrow Na_2O(s) + CO(g)$

$\Delta H^\circ = [1 \text{ mol } (-416 \text{ kJ/mol}) + 1 \text{ mol } (-110.5 \text{ kJ/mol})] - 1 \text{ mol } (-393.5 \text{ kJ/mol}) = -133 \text{ kJ}$

In both cases sodium metal reacts with the "extinguishing agent." Both reactions are exothermic, and each reaction produces a flammable gas, H_2 and CO respectively.

49. $3\ Al(s) + 3\ NH_4ClO_4(s) \rightarrow Al_2O_3(s) + AlCl_3(s) + 3\ NO(g) + 6\ H_2O(g)$

$\Delta H° = 6\text{ mol }(-242\text{ kJ/mol}) + 3\text{ mol }(90\text{ kJ/mol}) + 1\text{ mol }(-704\text{ kJ/mol})$

$+ 1\text{ mol }(-1676\text{ kJ/mol}) - [3\text{ mol }(-295\text{ kJ/mol})] = -2677\text{ kJ}$

50. $5\ N_2O_4(l) + 4\ N_2H_3CH_3(l) \rightarrow 12\ H_2O(g) + 9\ N_2(g) + 4\ CO_2$

$\Delta H° = 12\text{ mol }(-242\text{ kJ/mol}) + 4\text{ mol }(-393.5\text{ kJ/mol})$

$- [5\text{ mol }(-20\text{ kJ/mol}) + 4\text{ mol }(54\text{ kJ/mol})] = -4594\text{ kJ}$

51. $C(s) + H_2O(g) \rightarrow H_2(g) + CO(g)$

$\Delta H° = -110.5\text{ kJ} - (-242\text{ kJ}) = +132\text{ kJ}$

52. $TiCl_4(g) + 2\ H_2O(g) \rightarrow TiO_2(s) + 4\ HCl(g)$

$\Delta H° = [-945\text{ kJ} + 4(-92)\text{ kJ}] - [-763\text{ kJ} + 2(-242)\text{ kJ}] = -66\text{ kJ}$

53. $2\ ClF_3(g) + 2\ NH_3(g) \rightarrow N_2(g) + 6\ HF(g) + Cl_2(g)$

$\Delta H° = 6\ \Delta H_f^\circ\ (HF) - [2\ \Delta H_f^\circ\ (ClF_3) + 2\ \Delta H_f^\circ\ (NH_3)]$

$-1196\text{ kJ} = 6\text{ mol }(-271\text{ kJ/mol}) - 2\ \Delta H_f^\circ - 2\text{ mol }(-46\text{ kJ/mol})$

$-1196\text{ kJ} = -1626\text{ kJ} - 2\ \Delta H_f^\circ + 92\text{ kJ}$

$$\Delta H_f^\circ = \frac{(-1626 + 92 + 1196)\text{ kJ}}{2\text{ mol}} = \frac{-169\text{ kJ}}{\text{mol}}$$

54. $C_2H_4(g) + 3\ O_2(g) \rightarrow 2\ CO_2(g) + 2\ H_2O(l)$

$\Delta H° = -1411.1\text{ kJ} = 2(-393.5) + 2(-285.9) - \Delta H_f^\circ$

$-1411.1\text{ kJ} = -1358.8\text{ kJ} - \Delta H_f^\circ$

$\Delta H_f^\circ = 52.3\text{ kJ/mol}$

Energy Consumption and Sources

55. $$4.19 \times 10^6\text{ kJ} \times \frac{1\text{ mol }CH_4}{891\text{ kJ}} \times \frac{22.42\text{ L }CH_4\text{ @ STP}}{\text{mol }CH_4} = 1.05 \times 10^5\text{ L}$$

56. $C_2H_5OH(l) + 3\ O_2(g) \rightarrow 2\ CO_2(g) + 3\ H_2O(l)$

$\Delta H° = 2(-393.5\text{ kJ}) + 3(-286\text{ kJ}) - (-278\text{ kJ})$

$$\Delta H° = \frac{-1367\text{ kJ}}{\text{mol ethanol}}$$

$$\frac{-1367\text{ kJ}}{\text{mol}} \times \frac{1\text{ mol}}{46.07\text{ g}} = -29.67\text{ kJ/g}$$

57. $CO(g) + 2\ H_2(g) \longrightarrow CH_3OH(l)$

$\Delta H° = (-239\ kJ) - (-110.5\ kJ) = -129\ kJ$

58. $C_3H_8(g) + 5\ O_2(g) \longrightarrow 3\ CO_2(g) + 4\ H_2O(l)$

$\Delta H° = 3(-393.5\ kJ) + 4(-286\ kJ) - (-104\ kJ)$

$\Delta H° = -2221$ kJ per mole of C_3H_8

$\frac{-2221\ kJ}{mol} \times \frac{1\ mol}{44.09\ g} = \frac{-50.37\ kJ}{g}$ vs. -47.8 kJ/g for octane (example 6.11)

The fuel values of the fuels are very close. An advantage of propane is that it burns more cleanly. The boiling point of propane is -42°C. Thus, it is more difficult to store propane and there are extra safety hazards associated in using the necessary high pressure compressed gas tanks.

ADDITIONAL EXERCISES

59. a. $1.00\ g\ CH_4 \times \frac{1\ mol\ CH_4}{16.04\ g\ CH_4} \times \frac{(-891\ kJ)}{mol\ CH_4} = -55.5\ kJ$

b. $PV = nRT$

$$n = \frac{PV}{RT} = \frac{\frac{740.}{760}\ atm \times 1.00 \times 10^3\ L}{\frac{0.08206\ L\ atm}{mol\ K} \times 298\ K} = 39.8\ mol$$

$$\frac{-891\ kJ}{mol} \times 39.8\ mol = -3.55 \times 10^4\ kJ$$

60. $PV = nRT$

$$n = \frac{PV}{RT} = \frac{125\ atm \times 50.0 \times 10^{-3}\ L}{\frac{0.08206\ L\ atm}{mol\ K} \times 295\ K} = 0.258\ mol\ O_2$$

The balanced reaction for the combustion of benzoic acid is:

$$2\ C_7H_6O_2(s) + 15\ O_2(g) \longrightarrow 14\ CO_2(g) + 6\ H_2O(l)$$

If we want 25 times as much oxygen as needed, only 0.258/25 moles of oxygen can be burned.

$$\frac{0.258\ mol\ O_2}{25} \times \frac{2\ mol\ C_7H_6O_2}{15\ mol\ O_2} \times \frac{122.12\ g\ C_7H_6O_2}{mol\ C_7H_6O_2} = 0.168\ g = 0.17\ g$$

61. The specific heat capacities are: 0.89 J/g•°C (Al) and 0.45 J/g•°C (Fe)
Al would be the better choice. It has a higher heat capacity and a lower density than Fe. Using Al, the same amount of heat could be dissipated by a smaller mass, keeping the mass of the amplifier down.

62. First, we need to get the heat capacity of the calorimeter from the combustion of benzoic acid.

Heat lost by combustion = Heat gained by calorimeter

$$\text{Heat Loss} = \frac{26.42 \text{ kJ}}{\text{g}} \times 0.1584 \text{ g} = 4.185 \text{ kJ}$$

$$\text{Heat Gain} = 4.185 \text{ kJ} = C_{cal} \times \Delta T$$

$$C_{cal} = \frac{4.185 \text{ kJ}}{2.54°\text{C}} = 1.65 \text{ kJ/°C}$$

Now we can calculate the heat of combustion of vanillin.

Heat Loss = Heat Gain

$$\text{Heat Gain by calorimeter} = \frac{1.65 \text{ kJ}}{°\text{C}} \times 3.25°\text{C} = 5.36 \text{ kJ}$$

Heat Loss = 5.36 kJ which is the heat evolved by the combustion of the vanillin.

$$\text{Heat of Combustion} = \frac{-5.36 \text{ kJ}}{0.2130 \text{ g}} = -25.2 \text{ kJ/g}$$

$$\frac{-25.2 \text{ kJ}}{\text{g}} \times \frac{152.14 \text{ g}}{\text{mol}} = -3830 \text{ kJ/mol}$$

63. $V = 10.0 \text{ m} \times 4.0 \text{ m} \times 3.0 \text{ m} = 1.2 \times 10^2 \text{ m}^3 \times (100 \text{ cm/m})^3 = 1.2 \times 10^8 \text{ cm}^3$

Mass of water = 1.2×10^8 g since the density of water is 1.0 g/cm^3

$$\text{Heat} = \frac{4.18 \text{ J}}{\text{g °C}} \times 1.2 \times 10^8 \text{ g} \times 9.8°\text{C} = 4.9 \times 10^9 \text{ J or } 4.9 \times 10^6 \text{ kJ}$$

64. The specific heat of water is 4.18 J/g•°C, which is equal to 4.18 kJ/kg•°C.

We have 1.00 kg of H_2O, so

$$1.00 \text{ kg} \times \frac{4.18 \text{ kJ}}{\text{kg °C}} = 4.18 \text{ kJ/°C}$$

This is the portion of the heat capacity that can be attributed to H_2O.

Total heat Cap. = $C_{cal} + C_{H_2O}$

$C_{cal} = 10.84 - 4.18 = 6.66$ kJ/°C

65. $2\text{ K(s)} + 2\text{ H}_2\text{O(l)} \longrightarrow 2\text{ KOH(aq)} + \text{H}_2\text{(g)}$

$\Delta H° = 2(-481 \text{ kJ}) - 2(-286 \text{ kJ}) = -390. \text{ kJ}$

$$5.00 \text{ g K} \times \frac{1 \text{ mol K}}{39.10 \text{ g K}} \times \frac{390. \text{ kJ}}{2 \text{ mol K}} = 24.9 \text{ kJ}$$ of heat released upon reaction of 5.00 g of potassium.

$$24{,}900 \text{ J} = \frac{4.18 \text{ J}}{\text{g °C}} \times 1.00 \times 10^3 \text{ g} \times \Delta T$$

$$\Delta T = \frac{24{,}900}{4.18 \times 1.00 \times 10^3} = 5.96°\text{C}$$

Final Temperature = 24.0 + 5.96 = 30.0°C

66. $N_2H_4(l) + O_2(g) \longrightarrow N_2(g) + 2\ H_2O(g)$

$\Delta H° = 2(-242 \text{ kJ}) - (51 \text{ kJ}) = -535 \text{ kJ}$

$2\ N_2H_4(l) + N_2O_4(l) \longrightarrow 3\ N_2(g) + 4\ H_2O(g)$

$\Delta H° = 4(-242 \text{ kJ}) - [2(51 \text{ kJ}) + (-20 \text{ kJ})] = -1050. \text{ kJ}$

For hydrazine plus oxygen the stoichiometric reactant mixture contains 1 mol N_2H_4 for every mol O_2. This is a total mass of 64.05 g or 0.06405 kg.

$$\Delta H = \frac{-535 \text{ kJ}}{0.06405 \text{ kg}} = -8.35 \times 10^3 \text{ kJ/kg for } N_2H_4(l) + O_2(g)$$

For hydrazine plus N_2O_4 the optimum mixture contains 2 mol of N_2H_4 for every mol N_2O_4. This is a total mass of = 156.12 g or 0.15612 kg

$$\Delta H = \frac{-1050. \text{ kJ}}{0.15612 \text{ kg}} = -6.726 \times 10^3 \text{ kJ/kg for } 2\ N_2H_4(l) + N_2O_4(l)$$

From Exercise 49 a mixture of 3 mol Al and 3 mol NH_4ClO_4 yields 2677 kJ of energy.

The mass of the stoichiometric reactant mixture is:

$$3 \text{ mol} \times \frac{26.98 \text{ g}}{\text{mol}} + 3 \text{ mol} \times \frac{117.49 \text{ g}}{\text{mol}} = 433.41 \text{ g}$$

For 1.000 kg we get: $1.000 \times 10^3 \text{ g} \times \frac{-2677 \text{ kJ}}{433.38 \text{ g}} = -6177 \text{ kJ}$

In exercise 50 we get 4594 kJ of energy from 5 mol of N_2O_4 and 4 mol of $N_2H_3CH_3$.

The mass is: $5 \text{ mol} \times \frac{92.02 \text{ g}}{\text{mol}} + 4 \text{ mol} \times \frac{46.08 \text{ g}}{\text{mol}} = 644.42 \text{ g}$

For 1.000 kg of fuel mixture we get: $1.000 \times 10^3 \text{ g} \times \frac{-4594 \text{ kJ}}{644.42 \text{ g}} = -7129 \text{ kJ}$

The hydrazine plus oxygen mixture is the most efficient fuel.

67. $C_2H_2(g) + 5/2\ O_2(g) \longrightarrow 2\ CO_2(g) + H_2O(g)$

$\Delta H° = 2(-393.5 \text{ kJ}) + (-242 \text{ kJ}) - (227 \text{ kJ}) = -1256 \text{ kJ}$

If one mole of acetylene is burned with a stoichiometric amount of oxygen, the products will be 2 mol CO_2 and 1 mol H_2O. The heat capacity of this mixture is:

$$C = 2.00 \text{ mol } CO_2 \times \frac{37.1 \text{ J}}{\text{mol °C}} + 1.00 \text{ mol } H_2O \times \frac{33.6 \text{ J}}{\text{mol °C}}$$

$$C = 108 \text{ J/°C} = 0.108 \text{ kJ/°C}$$

$$1256 \text{ kJ} = C \times \Delta T$$

$$\Delta T = \frac{1256 \text{ kJ}}{\frac{0.108 \text{ kJ}}{\text{°C}}} = 11{,}600\text{°C}$$

or final temperature is $11{,}600 + 25 \approx 12{,}000$°C

Note: The actual temperature of an oxyacetylene torch is about 3000°C. The assumption of no heat loss is not a very good one.

68. $w = -P\Delta V$

$\Delta V > 0$, $w < 0$ and system does work on surroundings (e and f)

$\Delta V < 0$, $w > 0$ and surroundings does work on system (a, c, and d)

$\Delta V = 0$, $w = 0$ (b)

69. We need 1.3×10^8 J of energy or 1.3×10^5 kJ.

For:

$$C_3H_8(g) + 5\ O_2(g) \longrightarrow 3\ CO_2(g) + 4\ H_2O(l)$$

$$\Delta H° = 4(-286 \text{ kJ}) + 3(-393.5 \text{ kJ}) - (-104 \text{ kJ}) = -2221 \text{ kJ}$$

So the mass of C_3H_8 is:

$$1.3 \times 10^5 \text{ kJ} \times \frac{1 \text{ mol } C_3H_8}{2221 \text{ kJ}} \times \frac{44.09 \text{ g } C_3H_8}{\text{mol}} = 2600 \text{ g of } C_3H_8$$

70. Information given:

$C(s) + O_2(g) \longrightarrow CO_2(g)$ $\Delta H° = -393.7$ kJ

$CO(g) + 1/2\ O_2(g) \longrightarrow CO_2(g)$ $\Delta H° = -283.3$ kJ

So:

$2\ C + 2\ O_2 \longrightarrow 2\ CO_2$ $\Delta H° = 2(-393.7)$ kJ $\times 2$

$2\ CO_2 \longrightarrow 2\ CO + O_2$ $\Delta H° = 2(+283.3)$ KJ $\times(-2)$

$2\ C(s) + O_2(g) \longrightarrow 2\ CO(g)$ $\Delta H° = -220.8$ kJ

71. a. $C_2H_4(g) + O_3(g) \longrightarrow CH_3CHO(g) + O_2(g)$

$\Delta H° = -166 \text{ kJ} - [143 \text{ kJ} + 52 \text{ kJ}] = -361 \text{ kJ}$

b. $O_3 + NO \longrightarrow NO_2 + O_2$

$\Delta H° = 1 \text{ mol}(34 \text{ kJ/mol}) - [1 \text{ mol } (90 \text{ kJ/mol}) + 1 \text{ mol } (143 \text{ kJ/mol})] = -199 \text{ kJ}$

c. $SO_3(g) + H_2O(l) \longrightarrow H_2SO_4(aq)$

$\Delta H° = 1 \text{ mol } (-909 \text{ kJ/mol}) - [1 \text{ mol } (-396 \text{ kJ/mol}) + 1 \text{ mol } (-286 \text{ kJ/mol})] = -227 \text{ kJ}$

d. $2 NO(g) + O_2(g) \longrightarrow 2 NO_2(g)$

$\Delta H° = 2(34) - 2(90) = 68 \text{ kJ} - 180. \text{ kJ} = -112 \text{ kJ}$

CHALLENGE PROBLEMS

72.

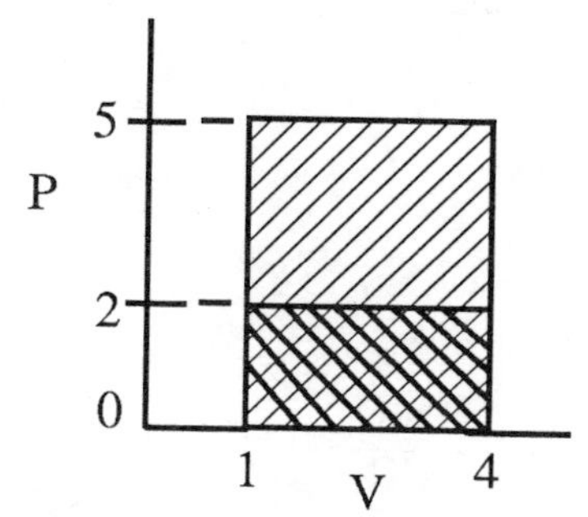

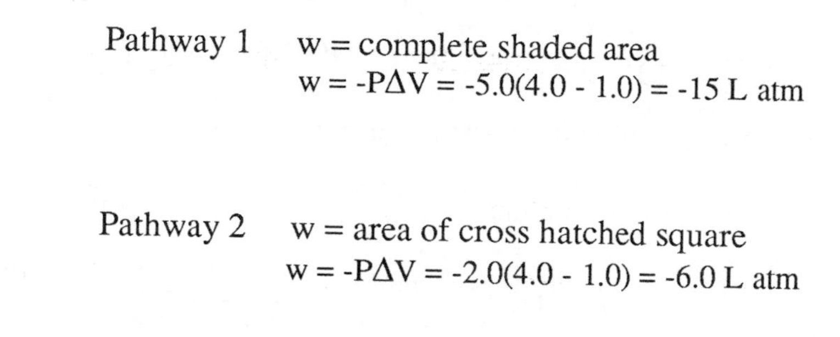

Pathway 1 w = complete shaded area
w = -PΔV = -5.0(4.0 - 1.0) = -15 L atm
Pathway 2 w = area of cross hatched square
w = -PΔV = -2.0(4.0 - 1.0) = -6.0 L atm

Sign is (-) because system is doing work on the surroundings.

Since w depends on pathway it cannot be a state function.

73. One should try to cool the reaction mixture or provide some means of removing heat. The reaction is very exothermic. The $H_2SO_4(aq)$ will get very hot and possibly boil.

74. $KE = \frac{1}{2}\left(7.8 \text{ g } \frac{1 \text{ kg}}{1000 \text{ g}}\right)\left(\frac{5.0 \times 10^4 \text{ cm}}{\text{sec}} \times \frac{1 \text{ m}}{100 \text{ cm}}\right)^2 = 980 \text{ J}$

Heat gained by bullet and wood = 980 J

$980 \text{ J} = 7.8 \text{ g} \times \frac{0.13 \text{ J}}{\text{g °C}} \times \Delta T + 1.00 \times 10^3 \text{ g} \times \frac{2.1 \text{ J}}{\text{g °C}} \times \Delta T$

$980 \text{ J} = \frac{1.0 \text{ J}}{\text{g °C}} \Delta T + \frac{2100 \text{ J}}{\text{g °C}} \Delta T$

$\Delta T = 0.47°C;\ \Delta T = T_f - T_i;\ 0.47 = T_f - 25.0;\ T_f = 25.5°C$

75. To avoid fractions, let's first calculate ΔH for the reaction:

$$6\,FeO(s) + 6\,CO(g) \rightarrow 6\,Fe(s) + 6\,CO_2(g)$$

$6\,FeO + 2\,CO_2 \rightarrow 2\,Fe_3O_4 + 2\,CO$	$\Delta H^\circ = 2(-18\text{ kJ})$	$\times(-2)$
$2\,Fe_3O_4 + CO_2 \rightarrow 3\,Fe_2O_3 + CO$	$\Delta H^\circ = +39\text{ kJ}$	(-)
$3\,Fe_2O_3 + 9\,CO \rightarrow 6\,Fe + 9\,CO_2$	$\Delta H^\circ = 3(-23\text{ kJ})$	$\times 3$
$6\,FeO(s) + 6\,CO(g) \rightarrow 6\,Fe(s) + 6\,CO_2(g)$	$\Delta H^\circ = -66\text{ kJ}$	

So for

$$FeO(s) + CO(g) \rightarrow Fe(s) + CO_2(g) \qquad \Delta H^\circ = \frac{-66\text{ kJ}}{6} = -11\text{ kJ}$$

76. a. $2\,HNO_3(aq) + Na_2CO_3(s) \rightarrow 2\,NaNO_3(aq) + H_2O(l) + CO_2(g)$

$$\Delta H^\circ = 2(-467) + (-286\text{ kJ}) + (-393.5\text{ kJ}) - [2(-207\text{ kJ}) + (-1131\text{ kJ})] = -69\text{ kJ}$$

$$2.0 \times 10^4\text{ gallons} \times \frac{4\text{ qt}}{\text{gal}} \times \frac{946\text{ mL}}{\text{qt}} \times \frac{1.42\text{ g}}{\text{mL}}$$

$= 1.1 \times 10^8$ g of concentrated nitric acid solution.

$$1.1 \times 10^8\text{ g solution} \times \frac{70.0\text{ g }HNO_3}{100\text{ g solution}} = 7.7 \times 10^7\text{ g }HNO_3$$

$$7.7 \times 10^7\text{ g }HNO_3 \times \frac{1\text{ mol}}{63.02\text{ g}} \times \frac{1\text{ mol }Na_2CO_3}{2\text{ mol }HNO_3} \times \frac{105.99\text{ g }Na_2CO_3}{\text{mol }Na_2CO_3}$$

$$= 6.5 \times 10^7\text{ g }Na_2CO_3$$

There are $7.7 \times 10^7/63.02$ mol of HNO_3 from previous calculation. There are 69 kJ of heat evolved for every two moles of nitric acid neutralized.

Combining those two results:

$$7.7 \times 10^7\text{ g }HNO_3 \times \frac{1\text{ mol }HNO_3}{63.02\text{ g }HNO_3} \times \frac{-69\text{ kJ}}{2\text{ mol }HNO_3} = -4.2 \times 10^7\text{ kJ}$$

b. They feared the heat generated by the neutralization reaction would vaporize unreacted nitric acid, causing widespread airborne contamination.

77. $400\text{ kcal} \times \frac{4.18\text{ kJ}}{\text{kcal}} = 1.67 \times 10^3\text{ kJ} \approx 2 \times 10^3\text{ kJ}$

$$E = mgz = \left(180\text{ lb} \times \frac{1\text{ kg}}{2.205\text{ lb}}\right)\left(\frac{9.8\text{ m}}{s^2}\right)\left(8\text{ in} \times \frac{2.54\text{ cm}}{\text{in}} \times \frac{1\text{ m}}{100\text{ cm}}\right) = 160\text{ J} \approx 200\text{ J}$$

200 J of energy are needed to climb one step.

$$2 \times 10^6\text{ J} \times \frac{1\text{ step}}{200\text{ J}} = 1 \times 10^4\text{ steps}$$

CHAPTER SEVEN: ATOMIC STRUCTURE AND PERIODICITY

QUESTIONS

1. Planck found that heated bodies only give off certain frequencies of light. Einstein's studies of the photoelectric effect.

2. When something is quantized, it can only have certain discrete values. In the Bohr model of the H-atom, the energy of the electron is quantized.

3. Very small particles (small mass) at large velocities.

4. 1) Electrons can be diffracted like light.

   2) The electron microscope uses electrons in a fashion similar to the way in which light is used in a light microscope.

5. a. A discrete bundle of light energy.

   b. A number describing a discrete state of an electron.

   c. The lowest energy state of the electron(s) in an atom or ion.

   d. An allowed energy state that is higher in energy than the ground state.

6. n gives the energy (it completely specifies the energy only for the H-atom or ions with one electron) and the relative size of the orbitals.

   l gives the type (shape) of orbital.

   m_l gives information about the direction in which the orbital is pointing.

7. The 2p orbitals differ from each other in the direction in which they point in space.

8. The 2p and 3p orbitals differ from each other in their size and number of nodes.

9. A nodal surface in an atomic orbital is a surface in which the probability of finding an electron is zero.

10. ψ^2 gives the probability of finding the electron at that point.

11. The electrostatic energy of repulsion, from Coulomb's Law will be of the form e^2/r where r is the distance between the two electrons. From the Heisenberg Uncertainty Principle, we cannot know precisely the position of each electron. Thus, we cannot precisely know the distance between the electrons nor the value of the electrostatic repulsions.

12. No, the spin is a convenient model. Since we cannot locate or "see" the electron, we cannot see if it is spinning.

13. There is a higher probability of finding the 4s electron very close to the nucleus than for the 3d electron.

14. When atoms interact with each other, it will be the outermost electrons that are involved in those interactions.

15. If one more electron is added to a half-filled subshell, electron-electron repulsions will increase.

16. As we remove succeeding electrons, it takes more energy, because each succeeding electron is being removed from an increasingly positively charged species.

17. Each element has a characteristic spectrum. Thus, the presence of the characteristic spectral lines of an element confirms its presence in any particular sample.

18. Yes, the maximum number of unpaired electrons in any configuration corresponds to a minimum in electron-electron repulsions.

19. The electron is no longer part of that atom. The proton and electron are completely separated.

20. Elements in the same group have similar chemical properties.

EXERCISES

Light and Matter

21. $99.5 \text{ MHz} = 99.5 \times 10^6 \text{ Hz} = 99.5 \times 10^6 \text{ s}^{-1}$

$$\lambda = \frac{c}{\nu} = \frac{2.998 \times 10^8 \text{ m/s}}{99.5 \times 10^6 \text{ s}^{-1}} = 3.01 \text{ m}$$

22. $$\nu = \frac{c}{\lambda} = \frac{2.998 \times 10^8 \text{ m/s}}{780. \times 10^{-9} \text{ m}} = 3.84 \times 10^{14} \text{ s}^{-1}; \text{ E} = h\nu = 2.54 \times 10^{-19} \text{ J}$$

23. $$\nu = \frac{c}{\lambda} = \frac{3.00 \times 10^8 \text{ m/s}}{1.0 \times 10^{-2} \text{ m}} = 3.0 \times 10^{10} \text{ s}^{-1}$$

$$E = h\nu = 6.63 \times 10^{-34} \text{ J s} \times 3.0 \times 10^{10} \text{ s}^{-1} = 2.0 \times 10^{-23} \text{ J}$$

$$2.0 \times 10^{-23} \text{ J/photon} \times 6.02 \times 10^{23} \text{ photons/mol} = 12 \text{ J/mol}$$

24. For 404.7 nm light:

$$\nu = \frac{c}{\lambda} = \frac{2.9979 \times 10^8 \text{ m/s}}{404.7 \times 10^{-9} \text{ m}} = 7.408 \times 10^{14} \text{ s}^{-1}$$

$$E = h\nu = (6.626 \times 10^{-34} \text{ J s})(7.408 \times 10^{14} \text{ s}^{-1}) = 4.909 \times 10^{-19} \text{ J}$$

$$\frac{4.909 \times 10^{-19} \text{ J}}{\text{photon}} \times \frac{6.022 \times 10^{23} \text{ photon}}{\text{mol}} = 2.956 \times 10^5 \text{ J/mol} = 295.6 \text{ kJ/mol}$$

For 435.8 nm light:

$$\nu = \frac{c}{\lambda} = \frac{2.9979 \times 10^8 \text{ m/s}}{435.8 \times 10^{-9} \text{ m}} = 6.879 \times 10^{14} \text{ s}^{-1}$$

$E = h\nu = (6.626 \times 10^{-34}\ \text{J s})(6.879 \times 10^{14}\ \text{s}^{-1}) = 4.558 \times 10^{-19}\ \text{J}$

$$\frac{4.558 \times 10^{-19}\ \text{J}}{\text{photon}} \times \frac{6.022 \times 10^{23}\ \text{photon}}{\text{mol}} = 2.745 \times 10^{5}\ \text{J/mol} = 274.5\ \text{kJ/mol}$$

25. $\Delta E = 7.21 \times 10^{-19}\ \text{J} = E_{photon}$

$E_{photon} = h\nu$ and $\lambda\nu = c$. So, $\nu = \frac{c}{\lambda}$ and $E = \frac{hc}{\lambda}$

$$\lambda = \frac{hc}{E_{photon}} = \frac{(6.626 \times 10^{-34}\ \text{J s})(2.998 \times 10^{8}\ \text{m/s})}{7.21 \times 10^{-19}\ \text{J}}$$

$\lambda = 2.76 \times 10^{-7}\ \text{m} = 276 \times 10^{-9}\ \text{m} = 276\ \text{nm}$

26. The energy needed to remove a single electron is:

$$\frac{279.7\ \text{kJ}}{\text{mol}} \times \frac{1\ \text{mol}}{6.022 \times 10^{23}} = 4.645 \times 10^{-22}\ \text{kJ}$$

$$4.645 \times 10^{-22}\ \text{kJ} \times \frac{1000\ \text{J}}{\text{kJ}} = 4.645 \times 10^{-19}\ \text{J}$$

$E = \frac{hc}{\lambda}$ for a single photon.

$$\lambda = \frac{hc}{E} = \frac{(6.626 \times 10^{-34}\ \text{J s})(2.9979 \times 10^{8}\ \text{m/s})}{4.645 \times 10^{-19}\ \text{J}} = 4.276 \times 10^{-7}\ \text{m or } 427.6\ \text{nm}$$

27. $$\frac{492\ \text{kJ}}{\text{mol}} \times \frac{1\ \text{mol}}{6.022 \times 10^{23}} = 8.17 \times 10^{-22}\ \text{kJ} = 8.17 \times 10^{-19}\ \text{J}$$

$E = \frac{hc}{\lambda}$

$$\lambda = \frac{hc}{E} = \frac{(6.626 \times 10^{-34}\ \text{J s})(2.998 \times 10^{8}\ \text{m/s})}{(8.17 \times 10^{-19}\ \text{J})} = 2.43 \times 10^{-7}\ \text{m} = 243\ \text{nm}$$

28. The energy to remove a single electron is:

$$\frac{208.4\ \text{kJ}}{\text{mol}} \times \frac{1\ \text{mol}}{6.022 \times 10^{23}} = 3.461 \times 10^{-22}\ \text{kJ} = 3.461 \times 10^{-19}\ \text{J}$$

Energy of 254 nm light is:

$$E = \frac{hc}{\lambda} = \frac{(6.626 \times 10^{-34}\ \text{J s})(2.998 \times 10^{8}\ \text{m/s})}{254 \times 10^{-9}\ \text{m}} = 7.82 \times 10^{-19}\ \text{J}$$

$E_{photon} = E_K + E_w$

$E_K = 7.82 \times 10^{-19}\ \text{J} - 3.461 \times 10^{-19}\ \text{J} = 4.36 \times 10^{-19}\ \text{J}$

29. $\frac{890.1 \text{ kJ}}{\text{mol}} \times \frac{1 \text{ mol}}{6.022 \times 10^{23} \text{ atoms}} = \frac{1.478 \times 10^{-21} \text{ kJ}}{\text{atom}} = \frac{1.478 \times 10^{-18} \text{ J}}{\text{atom}}$

$$E = \frac{hc}{\lambda}$$

$$\lambda = \frac{hc}{E} = \frac{(6.626 \times 10^{-34} \text{ J s})(2.9979 \times 10^8 \text{ m/s})}{1.478 \times 10^{-18} \text{ J}} = 1.344 \times 10^{-7} \text{ m} = 134.4 \text{ nm}$$

No, it will take light with a wavelength of 134.4 nm or less to ionize gold. A photon of light with a wavelength of 225 nm is a longer wavelength and, thus, less energy than 134.4 nm light.

30. $E_{photon} = \frac{hc}{\lambda} = \frac{6.626 \times 10^{-34} \text{ J s} \times 2.998 \times 10^8 \text{ m/s}}{589 \times 10^{-9} \text{ m}}$

$$= 3.37 \times 10^{-19} \text{ J} \times \frac{1 \text{ kJ}}{1000 \text{ J}} \times \frac{6.022 \times 10^{23}}{\text{mol}} = 203 \text{ kJ/mol}$$

No, since $E_{photon} < IE$ this light cannot ionize sodium.

Hydrogen Atom: Bohr Model

31. For the H-atom: $E_n = \frac{-R_H}{n^2}$, $R_H = 2.178 \times 10^{-18}$ J

For a spectral transition $\Delta E = E_f - E_i$

$$\Delta E = \left(\frac{-R_H}{n_f^2}\right) - \left(\frac{-R_H}{n_i^2}\right) = R_H \left(\frac{1}{n_i^2} - \frac{1}{n_f^2}\right)$$

Where n_i and n_f are the quantum numbers of the initial and final states, respectively. If we follow this convention a positive value of ΔE will correspond to absorption of light and a negative value of ΔE will correspond to emission of light. However, we will know whether we are observing emission or absorption of light. For the remainder, all examples will be worked so that the value of ΔE is always positive.

a. $\Delta E = 2.178 \times 10^{-18} \text{ J} \left(\frac{1}{2^2} - \frac{1}{3^2}\right) = 2.178 \times 10^{-18} \text{ J} \left(\frac{1}{4} - \frac{1}{9}\right)$

$$= 2.178 \times 10^{-18} \text{ J} (0.2500 - 0.1111) = 3.025 \times 10^{-19} \text{ J}$$

The photon of light must have precisely this energy.

$\Delta E = E_{photon} = h\nu = \frac{hc}{\lambda}$ or $\lambda = \frac{hc}{\Delta E} = \frac{(6.626 \times 10^{-34} \text{ J s})(2.9979 \times 10^8 \text{ m/s})}{3.025 \times 10^{-19} \text{ J}}$

$$= 6.567 \times 10^{-7} \text{ m or } 656.7 \text{ nm}$$

b. $\Delta E = 2.178 \times 10^{-18}\ J \left(\frac{1}{2^2} - \frac{1}{4^2}\right) = 4.084 \times 10^{-19}\ J$

$$\lambda = \frac{hc}{\Delta E} = \frac{(6.626 \times 10^{-34}\ J\ s)(2.9979 \times 10^{8}\ m/s)}{4.084 \times 10^{-19}\ J} = 4.864 \times 10^{-7}\ m \text{ or } 486.4\ nm$$

c. $\Delta E = 2.178 \times 10^{-18}\ J \left(\frac{1}{1^2} - \frac{1}{2^2}\right) = 1.634 \times 10^{-18}\ J$

$$\lambda = \frac{(6.626 \times 10^{-34}\ J\ s)(2.9979 \times 10^{8}\ m/s)}{1.634 \times 10^{-18}\ J} = 1.216 \times 10^{-7}\ m \text{ or } 121.6\ nm$$

32. a. $\Delta E = 2.178 \times 10^{-18}\ J \left(\frac{1}{3^2} - \frac{1}{4^2}\right) = 1.059 \times 10^{-19}\ J$

$$\lambda = \frac{(6.626 \times 10^{-34}\ J\ s)(2.9979 \times 10^{8}\ m/s)}{1.059 \times 10^{-19}\ J} = 1.886 \times 10^{-6}\ m \text{ or } 1886\ nm$$

b. $\Delta E = 2.178 \times 10^{-18}\ J \left(\frac{1}{4^2} - \frac{1}{5^2}\right) = 4.901 \times 10^{-20}\ J$

$$E_{photon} = \frac{hc}{\lambda} = \Delta E$$

$$\lambda = \frac{hc}{\Delta E} = \frac{(6.626 \times 10^{-34}\ J\ s)(2.9979 \times 10^{8}\ m/s)}{4.901 \times 10^{-20}\ J}$$

$$\lambda = 4.053 \times 10^{-6}\ m = 4053\ nm$$

c. $\Delta E = 2.178 \times 10^{-18}\ J \left(\frac{1}{3^2} - \frac{1}{5^2}\right) = 1.549 \times 10^{-19}\ J$

$$\Delta E = E_{photon} = \frac{hc}{\lambda}$$

$$\lambda = \frac{hc}{\Delta E} = \frac{(6.626 \times 10^{-34}\ J\ s)(2.9979 \times 10^{8}\ m/s)}{1.549 \times 10^{-19}\ J} = 1.282 \times 10^{-6}\ m = 1282\ nm$$

33. Ionization from n = 1 corresponds to the transition n = 1 $\longrightarrow$ n = ∞ and $E_\infty = 0$

$$\Delta E = E_\infty - E_1 = -E_1 = 2.178 \times 10^{-18} \left(\frac{1}{1^2}\right) = 2.178 \times 10^{-18}\ J$$

$$\lambda = \frac{hc}{\Delta E} = \frac{(6.626 \times 10^{-34}\ J\ s)(2.9979 \times 10^{8}\ m/s)}{2.178 \times 10^{-18}\ J} = 9.120 \times 10^{-8}\ m \text{ or } 91.20\ nm$$

To ionize from n = 2, $\Delta E = E_\infty - E_2 = 0 - E_2$

$$\Delta E = 2.178 \times 10^{-18} \left(\frac{1}{2^2}\right) = 5.445 \times 10^{-19}\ \text{J}$$

$$\lambda = \frac{(6.626 \times 10^{-34}\ \text{J s})(2.9979 \times 10^8\ \text{m/s})}{5.445 \times 10^{-19}\ \text{J}} = 3.648 \times 10^{-7}\ \text{m or } 364.8\ \text{nm}$$

34. To ionize from n = 4, $\Delta E = E_\infty - E_4 = 0 - E_4$

$$\Delta E = 2.178 \times 10^{-18} \left(\frac{1}{4^2}\right) = 1.361 \times 10^{-19}\ \text{J}$$

$$\lambda = \frac{hc}{\Delta E} = \frac{(6.626 \times 10^{-34}\ \text{J s})(2.9979 \times 10^8\ \text{m/s})}{1.361 \times 10^{-19}\ \text{J}} = 1.460 \times 10^{-6}\ \text{m} = 1460.\ \text{nm}$$

To ionize from n = 10, $\Delta E = 2.178 \times 10^{-18} \left(\frac{1}{10^2}\right) = 2.178 \times 10^{-20}\ \text{J}$

$$\lambda = \frac{hc}{\Delta E} = \frac{(6.626 \times 10^{-34}\ \text{J s})(2.9979 \times 10^8\ \text{m/s})}{2.178 \times 10^{-20}\ \text{J}} = 9.120 \times 10^{-6}\ \text{m} = 9120.\ \text{nm}$$

35.

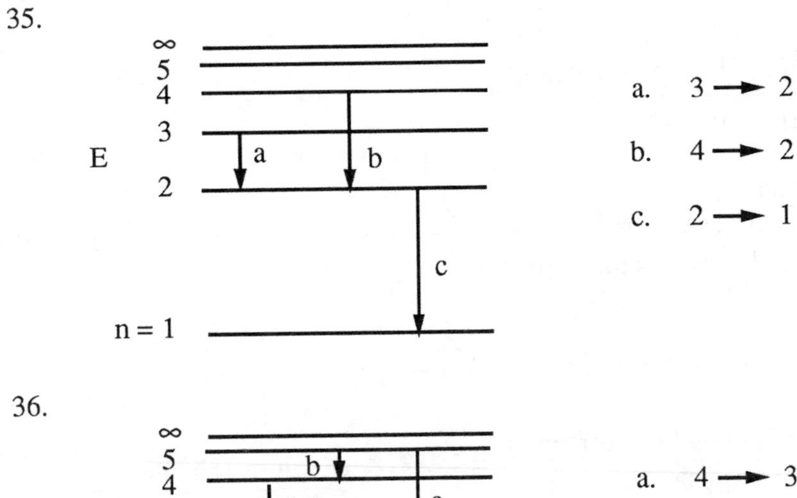

36.

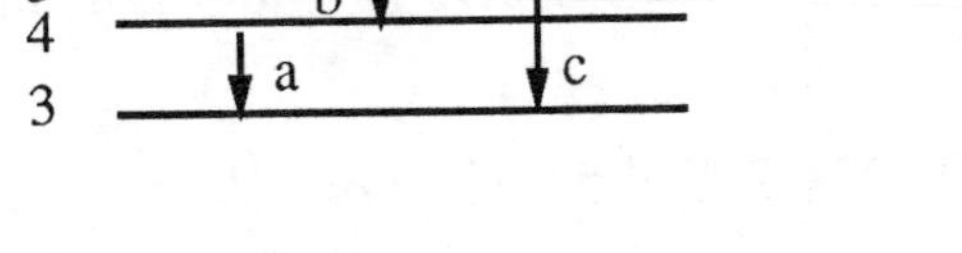

E

2

n = 1

a. 4 → 3

b. 5 → 4

c. 5 → 3

37. The longest wavelength light emitted will correspond to the transition with the lowest energy change. This is the transition from n = 6 to n = 5.

$$\Delta E = -2.178 \times 10^{-18} \text{ J} \left(\frac{1}{5^2} - \frac{1}{6^2}\right) = -2.662 \times 10^{-20} \text{ J}$$

$$\lambda = \frac{hc}{\Delta E} = \frac{(6.626 \times 10^{-34} \text{ J s})(2.9979 \times 10^8 \text{ m/s})}{2.662 \times 10^{-20} \text{ J}} = 7.462 \times 10^{-6} \text{ m} = 7462 \text{ nm}$$

The shortest wavelength emitted will correspond to the largest ΔE; this is $n = 6 \longrightarrow n = 1$.

$$\Delta E = -2.178 \times 10^{-18} \left(\frac{1}{1^2} - \frac{1}{6^2}\right) = -2.118 \times 10^{-18} \text{ J}$$

$$\lambda = \frac{hc}{\Delta E} = \frac{(6.626 \times 10^{-34} \text{ J s})(2.9979 \times 10^8 \text{ m/s})}{2.118 \times 10^{-18} \text{ J}} = 9.379 \times 10^{-8} \text{ m} = 93.79 \text{ nm}$$

38. $\Delta E = h\nu = (6.626 \times 10^{-34} \text{ J s})(1.141 \times 10^{14} \text{ s}^{-1}) = 7.560 \times 10^{-20} \text{ J}$

$\Delta E = -7.560 \times 10^{-20}$ J since light is emitted.

$$\Delta E = E_4 - E_n;\ -7.560 \times 10^{-20} \text{ J} = -2.178 \times 10^{-18} \text{ J} \left(\frac{1}{4^2} - \frac{1}{n^2}\right)$$

$$3.471 \times 10^{-2} = 6.250 \times 10^{-2} - \frac{1}{n^2};\ \frac{1}{n^2} = 2.779 \times 10^{-2}$$

$n^2 = 35.98;\ n = 6$

Wave Mechanics, Quantum Numbers, and Orbitals

39. $\lambda = \dfrac{h}{mv}$

a. 5.0% of speed of light = $0.050 \times 3.00 \times 10^8$ m/s = 1.5×10^7 m/s

$$\lambda = \frac{6.63 \times 10^{-34} \text{ J s}}{1.67 \times 10^{-27} \text{ kg} \times (1.5 \times 10^7 \text{ m/s})} = 2.6 \times 10^{-14} \text{ m} = 2.6 \times 10^{-5} \text{ nm}$$

Note: for units to come out the mass must be in kg since $J = \dfrac{\text{kg m}^2}{\text{s}^2}$

b. $$\lambda = \frac{6.63 \times 10^{-34} \text{ J s}}{9.11 \times 10^{-31} \text{ kg} \times (0.15 \times 3.00 \times 10^8 \text{ m/s})} = 1.6 \times 10^{-11} \text{ m} = 1.6 \times 10^{-2} \text{ nm}$$

c. $$m = 5.2 \text{ oz} \times \frac{1 \text{ lb}}{16 \text{ oz}} \times \frac{1 \text{ kg}}{2.205 \text{ lb}} = 0.15 \text{ kg}$$

$$v = \frac{100.8 \text{ mi}}{\text{hr}} \times \frac{1 \text{ hr}}{3600 \text{ s}} \times \frac{1760 \text{ yd}}{\text{mi}} \times \frac{0.9144 \text{ m}}{\text{yd}} = 45.06 \text{ m/s}$$

$$\lambda = \frac{h}{mv} = \frac{6.63 \times 10^{-34}\ \text{J s}}{0.15\ \text{kg} \times 45.06\ \text{m/s}} = 9.8 \times 10^{-35}\ \text{m} = 9.8 \times 10^{-26}\ \text{nm}$$

This number is so small it is essentially zero. We cannot detect a wavelength this small. The meaning of this number is that we do not have to worry about wave properties of large objects at low velocities.

40. a. $$\lambda = \frac{h}{mv} = \frac{6.626 \times 10^{-34}\ \text{J s}}{(1.675 \times 10^{-27}\ \text{kg})(0.0100 \times 2.998 \times 10^{8}\ \text{m/s})} = 1.32 \times 10^{-13}\ \text{m}$$

b. $$\lambda = \frac{h}{mv},\ \lambda mv = h,\ v = \frac{h}{\lambda m}$$

$$v = \frac{6.626 \times 10^{-34}\ \text{J s}}{(75 \times 10^{-12}\ \text{m})(1.675 \times 10^{-27}\ \text{kg})} = 5.3 \times 10^{3}\ \text{m/s}$$

41. $$\lambda = \frac{h}{mv},\ v = \frac{h}{\lambda m}$$

for $\lambda = 1.0 \times 10^{2}$ nm $= 1.0 \times 10^{-7}$ m

$$v = \frac{6.63 \times 10^{-34}\ \text{J s}}{(9.11 \times 10^{-31}\ \text{kg})(1.0 \times 10^{-7}\ \text{m})} = 7.3 \times 10^{3}\ \text{m/s}$$

for $\lambda = 1.0$ nm

$$v = \frac{h}{m\lambda} = \frac{6.63 \times 10^{-34}\ \text{J s}}{(9.11 \times 10^{-31}\ \text{kg})(1.0 \times 10^{-9}\ \text{m})} = 7.3 \times 10^{5}\ \text{m/s}$$

42. $$\lambda = \frac{h}{mv},\ \text{or } h = \lambda mv$$

If $\lambda = 5.0$ cm, m = 0.15 kg, and v = 45.06 m/s then

$$h = 5.0 \times 10^{-2}\ \text{m} \times 0.15\ \text{kg} \times 45.06\ \text{m/s} = 0.34\ \text{J s}$$

For $\lambda = 5.0 \times 10^{2}$ nm

$$h = 5.0 \times 10^{-7}\ \text{m} \times 0.15\ \text{kg} \times 45.06\ \text{m/s} = 3.4 \times 10^{-6}\ \text{J s}$$

43. $\Delta p = m\Delta v$

$$\Delta p = m\Delta v = 9.11 \times 10^{-31}\ \text{kg} \times 0.100\ \text{m/s} = \frac{9.11 \times 10^{-32}\ \text{kg m}}{\text{s}}$$

$$\Delta p\Delta x \geq \frac{h}{4\pi};\ \Delta x = \frac{h}{4\pi\Delta p} = \frac{6.626 \times 10^{-34}\ \text{J s}}{4 \times 3.1416 \times 9.11 \times 10^{-32}\ \text{kg m/s}}$$

$$\Delta x = 5.79 \times 10^{-4}\ \text{m} = 0.579\ \text{mm} = 5.79 \times 10^{5}\ \text{nm}$$

Diameter of H atom is roughly 1.0×10^{-8} cm. The uncertainty in position is much larger than the size of the atom.

44. $\Delta x = \frac{h}{4\pi\Delta p} = \frac{6.626 \times 10^{-34} \text{ J s}}{4 \times 3.1416 \times 0.145 \text{ kg} \times 0.100 \text{ m/s}} = 3.64 \times 10^{-33} \text{ m}$

The uncertainty is insignificant compared to the size of a baseball.

Polyelectronic Atoms

45. $n = 1, 2, 3, ...$

$l = 0, 1, 2, ... (n - 1)$

$m_l = -l ... -2, -1, 0, 1, 2, ... +l$

46. lp: $n = 1,\ l = 1$ not possible

3f: $n = 3, l = 3$ not possible

2d: $n = 2, l = 2$ not possible

In all three incorrect cases $n = l$. The maximum value l can have is $n - 1$, not n.

47. b. l must be smaller than n.

d. For $l = 0$, $m_l = 0$ only allowed value.

48. b. For $l = 3$, m_l can range from -3 to +3; thus +4 is not allowed.

c. n cannot equal zero

d. l cannot be a negative number

49. $E_n = \frac{-R_H Z^2}{n^2}$, $R_H = 2.178 \times 10^{-18}$ J

Normal ionization energy is for transition $n = 1 \rightarrow n = \infty$, $E = 0$. Thus, the I.E. is given by the energy of the state $n = 1$. Ionization energies are positive.

a. $\text{IE} = 2.178 \times 10^{-18} \text{ J} \left(\frac{1^2}{1^2}\right) = 2.178 \times 10^{-18} \text{ J}$

$$\text{IE} = \frac{2.178 \times 10^{-18} \text{ J}}{\text{atom}} \times \frac{1 \text{ kJ}}{1000 \text{ J}} \times \frac{6.022 \times 10^{23} \text{ atoms}}{\text{mol}} = 1311.6 \text{ kJ/mol}$$

IE = 1312 kJ/mol

For brevity, the I.E. of heavier one electron species can be given as:

$$\text{IE} = \frac{1311.6 \text{ kJ}}{\text{mol}} \left(\frac{Z^2}{n^2}\right)$$ (We will carry an extra significant figure.)

We get this by combining $\text{IE(J)} = 2.178 \times 10^{-18} \text{ J} \left(\frac{Z^2}{n^2}\right)$

and $\frac{2.178 \times 10^{-18} \text{ J}}{\text{atom}} \times \frac{6.022 \times 10^{23} \text{ atoms}}{\text{mol}} \times \frac{1 \text{ kJ}}{1000 \text{ J}} = \frac{1311.6 \text{ kJ}}{\text{mol}}$

b. He^+, $Z = 2$ IE = 5246 kJ/mol

c. Li^{2+}, Z = 3 IE = 1311.6 kJ/mol × 3^2 = 1.180 × 10^4 kJ/mol

The size of the 1s orbitals would be proportional to 1/Z, that is as Z increases, the electrons are more strongly attracted to the nucleus and will be drawn in closer. Thus the relative sizes would be:

H : He^+ : Li^{2+} $1 : \frac{1}{2} : \frac{1}{3}$ 1 : 0.50 : 0.33

50. IE = 1311.6 kJ/mol $\left(\frac{Z^2}{n^2}\right)$, See Exercise 7.49 for the derivation of this equation.

a. Be^{3+}, Z = 4. IE = 1311.6 kJ/mol$(4)^2$ = 2.099 × 10^4 kJ/mol

b. C^{5+}, Z = 6, IE = 1311.6 kJ/mol × 6^2 = 4.722 × 10^4 kJ/mol

c. Fe^{25+}, Z = 26, IE = 1311.6 kJ/mol × 26^2 = 8.866 × 10^5 kJ/mol

51. 5p: three $3d_{z^2}$: one 4d: five

n = 5: l = 0 (1 orbital), l = 1 (3 orbitals), l = 2 (5 orbitals)

l = 3 (7 orbitals), l = 4 (9 orbitals)

Total for n = 5 is 25 orbitals.

n = 4: l = 0 (1), l = 1 (3), l = 2 (5), l = 3 (7)

Total for n = 4 is 16 orbitals.

52. 1s - 2 electrons, 2p - 6 electrons, $3p_x$ - 2 electrons, 6f - 14 electrons, $2d_{xy}$ - none (2d orbitals are not possible).

53. a. For n = 4, l can be 0, 1, 2, or 3. Thus we have s(2 e^-), p(6 e^-), d(10 e^-) and f(14 e^-) orbitals present. Total number of electrons to fill these orbitals is 32.

b. n = 5, m_l = +1 For n = 5, l = 0, 1, 2, 3, 4.

l = 1, 2, 3, 4 can have m_l = +1. Four distinct orbitals, thus 8 electrons.

c. n = 5, l = 0, 1, 2, 3, 4

number of orbitals 1, 3, 5, 7, 9 for each value of l respectively.

There are 25 orbitals with n = 5. They can hold 50 electrons, 25 electrons can have m_s = +1/2.

d. n = 3, l = 2 defines a set of 3d orbitals.

5 orbitals, 10 electrons

e. n = 2, l = 1 defines a set of 2p orbitals.

3 orbitals, 6 electrons

54. a. It is impossible for n = 0. Thus, no electrons can have this set of quantum numbers.

b. The four quantum numbers completely specify a single electron.

c. n = 3: This corresponds to the 3s, 3p and 3d orbitals which can hold up to 18 electrons.

d. $n = 2, l = 2$: This combination is not possible; thus, zero electrons in an atom can have these quantum numbers.

e. $n = 1, l = 0, m_l = 0$: This defines a 1s orbital which can hold 2 electrons.

55. Sc : $1s^22s^22p^63s^23p^64s^23d^1$

Fe : $1s^22s^22p^63s^23p^64s^23d^6$

S : $1s^22s^22p^63s^23p^4$

P : $1s^22s^22p^63s^23p^3$

Cs : $1s^22s^22p^63s^23p^64s^23d^{10}4p^65s^24d^{10}5p^66s^1$

Eu : $1s^22s^22p^63s^23p^64s^23d^{10}4p^65s^24d^{10}5p^66s^24f^65d^1$*

Pt : $1s^22s^22p^63s^23p^64s^23d^{10}4p^65s^24d^{10}5p^66s^24f^{14}5d^8$*

Xe : $1s^22s^22p^63s^23p^64s^23d^{10}4p^65s^24d^{10}5p^6$

Br : $1s^22s^22p^63s^23p^64s^23d^{10}4p^5$

Se : $1s^22s^22p^63s^23p^64s^23d^{10}4p^4$

*Note: These electron configurations were written down using only the periodic table. Actual electron configurations are:

Eu: $[Xe]6s^24f^7$ and Pt: $[Xe]6s^14f^{14}5d^9$

56. K : $1s^22s^22p^63s^23p^64s^1$

Rb : $1s^22s^22p^63s^23p^64s^23d^{10}4p^65s^1$

Fr : $1s^22s^22p^63s^23p^64s^23d^{10}4p^65s^24d^{10}5p^66s^24f^{14}5d^{10}6p^67s^1$ or $[Rn]7s^1$

Pu : $[Rn]7s^26d^15f^5$ (expected from periodic table)

Sb : $1s^22s^22p^63s^23p^64s^23d^{10}4p^65s^24d^{10}5p^3$

Os : $[Xe]6s^24f^{14}5d^6$

Pd : $1s^22s^22p^63s^23p^64s^23d^{10}4p^65s^24d^8$ (expected from periodic table)

Cd : $1s^22s^22p^63s^23p^64s^23d^{10}4p^65s^24d^{10}$

Pb : $[Xe]6s^24f^{14}5d^{10}6p^2$

I : $1s^22s^22p^63s^23p^64s^23d^{10}4p^65s^24d^{10}5p^5$

57. Exceptions: Cr, Cu, Nb, Mo, Tc, Ru, Rh, Pd, Ag, Pt, Au

Tc, Ru, Rh, Pd and Pt do not correspond to the supposed extra stability of half-filled and filled subshells.

58. No, in the solid and liquid state the electrons may interact (bonding occurs) and the experimental configuration will not be valid. That is, the experimental electron configurations are for isolated atoms; in condensed phases, the atoms are no longer isolated.

59. a. The smallest halogen is fluorine: $1s^22s^22p^5$

b. K: $1s^22s^22p^63s^23p^64s^1$

c. Be: $1s^22s^2$ Mg: $1s^22s^22p^63s^2$ Ca: $1s^22s^22p^63s^23p^64s^2$

d. In: $[Kr]5s^24d^{10}5p^1$

e. C: $1s^22s^22p^2$ Si: $1s^22s^22p^63s^23p^2$

f. This will be element #118: $[Rn]7s^25f^{14}6d^{10}7p^6$

60. a. This atom has 10 electrons. Ne

b. S

c. The ground state configuration is $[Kr]5s^24d^9$. The element is Ag.

d. Bi

61. B : $1s^22s^22p^1$

	n	l	m_l	m_s
1s	1	0	0	+1/2
1s	1	0	0	-1/2
2s	2	0	0	+1/2
2s	2	0	0	-1/2
2p*	2	1	-1	+1/2

C : $1s^22s^22p^2$

	n	l	m_l	m_s	
1s	1	0	0	+1/2	
1s	1	0	0	-1/2	
2s	2	0	0	+1/2	
2s	2	0	0	-1/2	
2p*	2	1	-1	+1/2	or
2p	2	1	0	+1/2	-1/2

*This is only one of several possibilities for 2p electrons. For example, the 2p electron in B could have m_l = -1, 0, +1, and m_s = +1/2 or -1/2, a total of six possibilities.

N : $1s^2 2s^2 2p^3$

	n	l	m_l	m_s
1s	1	0	0	+1/2
1s	1	0	0	-1/2
2s	2	0	0	+1/2
2s	2	0	0	-1/2
2p	2	1	-1	+1/2
2p	2	1	0	+1/2
2p	2	1	+1	+1/2

(or all 2p electrons could have m_s = -1/2.)

62. Al : $1s^2 2s^2 2p^6 3s^2 3p^1$

	n	l	m_l	m_s
3s	3	0	0	+1/2
3s	3	0	0	-1/2
3p*	3	1	1	+1/2

* only one of six possibilities for $3p^1$

Sb : $[Kr]4d^{10} 5s^2 5p^3$

	n	l	m_l	m_s	
5s	5	0	0	+1/2	
5s	5	0	0	-1/2	
5p	5	1	-1	+1/2	} or -1/2
5p	5	1	0	+1/2	
5p	5	1	+1	+1/2	

Se : $1s^2 2s^2 2p^6 3s^2 3p^6 3d^{10} 4s^2 4p^4$

	n	l	m_l	m_s	
4s	4	0	0	+1/2	
4s	4	0	0	-1/2	
4p*	4	1	-1	+1/2	
4p	4	1	-1	-1/2	
4p	4	1	0	+1/2	} or -1/2
4p	4	1	+1	+1/2	

*one of several possibilities for $4p^4$

O : $1s^2 2s^2 2p^4$

	n	l	m_l	m_s	
1s	1	0	0	+1/2	
1s	1	0	0	-1/2	
2s	2	0	0	+1/2	
2s	2	0	0	-1/2	
2p*	2	1	-1	+1/2	
2p	2	1	-1	-1/2	
2p	2	1	0	+1/2	} or -1/2
2p	2	1	+1	+1/2	

* only one of several possibilities for $2p^4$

As : $1s^2 2s^2 3s^2 3p^6 4s^2 3d^{10} 4p^3$

	n	l	m_l	m_s	
4s	4	0	0	+1/2	
4s	4	0	0	-1/2	
4p	4	1	-1	+1/2	} or -1/2
4p	4	1	0	+1/2	
4p	4	1	+1	+1/2	

S : $1s^2 2s^2 2p^6 3s^2 3p^4$

	n	l	m_l	m_s	
3s	3	0	0	+1/2	
3s	3	0	0	-1/2	
3p*	3	1	-1	+1/2	
3p	3	1	-1	-1/2	
3p	3	1	0	+1/2	} or -1/2
3p	3	1	+1	+1/2	

*one of several possibilities for $3p^4$

63. O: $1s^22s^22p_x^{\ 2}2p_y^{\ 2}$ ↑↓ ↑↓ __

There are no unpaired electrons in this oxygen atom. This configuration would be an excited state and in going to the ground state ↑↓ ↑ ↑ energy would be released.

64. The number of unpaired electrons are in parentheses.

a. excited state of boron (1)

ground state $1s^22s^22p^1$ (1)

b. ground state of neon (0)

c. excited state of fluorine (3)

ground state: $1s^22s^22p^5$ (1)

d. excited state of iron (6)

ground state: $[Ar]4s^23d^6$ (4)

65. We get the number of unpaired electrons by looking at the incompletely filled subshells.

Sc:	$[Ar]4s^23d^1$:	$3d^1$:	↑ __ __ __ __	one unpaired e^-
Ti:	$[Ar]4s^23d^2$	$3d^2$:	↑ ↑ __ __ __	two unpaired e^-
Al:	$[Ne]3s^23p^1$	$3p^1$:	↑ __ __	one unpaired e^-
Sn:	$[Kr]5s^24d^{10}5p^2$	$5p^2$	↑ ↑ __	two unpaired e^-
Te:	$[Kr]5s^24d^{10}5p^4$	$5p^4$	↑↓ ↑ ↑	two unpaired e^-
Br:	$[Ar]4s^23d^{10}4p^5$	$4p^5$	↑↓ ↑↓ ↑	one unpaired e^-

66. We get the number of unpaired electrons by looking at the incompletely filled subshells.

O:	$[He]2s^22p^4$:	$2p^4$:	↑↓ ↑ ↑	two unpaired e^-
O^+:	$[He]2s^22p^3$	$2p^3$:	↑ ↑ ↑	three unpaired e^-
O^-:	$[He]2s^22p^5$	$2p^5$:	↑↓ ↑↓ ↑	one unpaired e^-
Fe:	$[Ar]4s^23d^6$	$3d^6$	↑↓ ↑ ↑ ↑ ↑	four unpaired e^-
Mn:	$[Ar]4s^23d^5$	$3d^5$	↑ ↑ ↑ ↑ ↑	five unpaired e^-
S:	$[Ne]3s^23p^4$	$3p^4$	↑↓ ↑ ↑	two unpaired e^-
F:	$[He]2s^22p^5$	$2p^5$	↑↓ ↑↓ ↑	one unpaired e^-
Ar:	$[Ne]3s^23p^6$	$3p^6$	↑↓ ↑↓ ↑↓	zero unpaired e^-

The Periodic Table and Periodic Properties

67. a. Be < Mg < Ca b. Xe < I < Te c. Ge < Ga < In

68. a. F < N < As b. F < Cl < S c. Li < K < Cs

69. a. Ca < Mg < Be b. Te < I < Xe c. In < Ga < Ge

70. a. As < N < F b. S < Cl < F c. Cs < K < Li

71. $IE \propto \frac{Z_{eff}^2}{n^2}$ or $Z_{eff} \propto n\,(IE)^{1/2}$

The effect of n is greater than the effect of the ionization energy.

a. Be < Mg < Ca b. Te < I < Xe c. Ga < Ge < In

72. a. N < F < As b. F < S < Cl c. Li < K < Cs

73. Ge: $[Ar]3d^{10}4s^24p^2$; As: $[Ar]3d^{10}4s^24p^3$; Se: $[Ar]3d^{10}4s^24p^4$

There are extra electron-electron repulsions in Se because two electrons are in the same 4p orbital, resulting in a lower ionization energy.

74. No, our generalization was that the first ionization energy decreases as we go down a group. For the coinage metals the first ionization energy decreases and then increases as we go down the group.

Sc, 549.5 kJ/mol	Ti, 658	Fe, 759.4	Ga, 578.8	As, 944
Y, 616	Zr, 660	Ru, 711	In, 558.3	Sb, 831.6
La, 538.1	Hf, 654	Os, 840	Tl, 589.3	Bi, 703.3

(Fe, Ru, Os) and (Ga, In, Tl) follow the same trend as the coinage metals. Only (As, Sb, Bi) follow the trend we would predict.

75. a. Li b. P c. O^+

d. Ar < Cl < S and Kr > Ar. Since variation in size down a family is greater than the variation across a period, we would predict Cl to be the smallest of the three.

e. Cu

76. a. Cs b. Ga c. Tl d. Tl e. O^{2-}

77. See graphs. mp ≈ 22°C, IE ≈ 340 kJ/mol

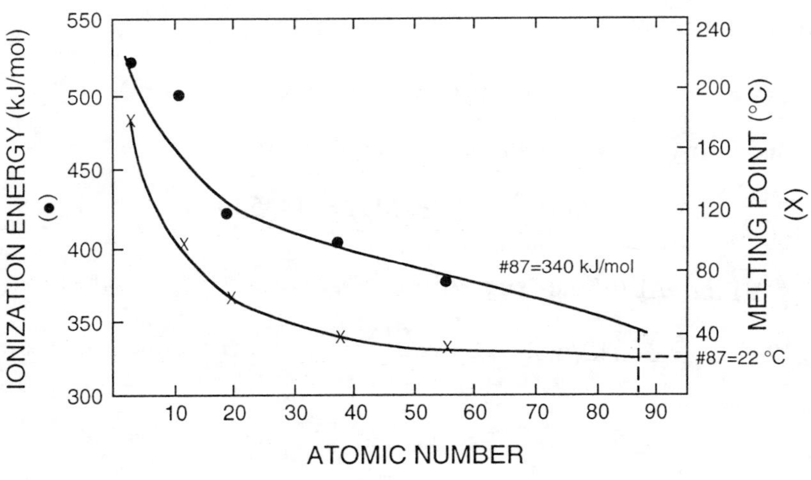

IONIZATION ENERGY (kJ/mol) (•)
550
500
450
400
350
300
MELTING POINT (°C) (X)
240
200
160
120
80
40
#87=340 kJ/mol
#87=22 °C
10 20 30 40 50 60 70 80 90
ATOMIC NUMBER

78. Uus. Atomic number 117

a. $[Rn]7s^2 5f^{14} 6d^{10} 7p^5$

b. It will be in the halogen family and most similar to astatine, At.

c. NaUus, $Mg(Uus)_2$, $C(Uus)_4$, $O(Uus)_2$

d. Assuming Uus is like the other halogens:

$UusO^-$, $UusO_2^-$, $UusO_3^-$, $UusO_4^-$

79. a $O < S$, S more exothermic

b. $I < Br < F < Cl$, Cl most exothermic

80. a. $N < O < F$ F most exothermic

b. $N < B < C$, C most exothermic

81. a. more favorable EA: Li and S

b. higher IE: Li and S

c. larger radius: K and Sc

82. a. more favorable EA: B and Cl

b. higher IE: N and F

c. larger radius: B and Cl

83. Yes, since it is an exothermic EA there are conditions in which we might expect Na^- to exist. Such compounds were first synthesized by James L. Dye at Michigan State University.

84. O, because electron-electron repulsions will be much more severe for $O^- + e^- \rightarrow O^{2-}$ than for $O + e \rightarrow O^-$.

85. $P(g) \rightarrow P^+(g) + e^-$

86. $P(g) + e^- \rightarrow P^-(g)$

87. a. The electron affinity of Mg^{2+} is ΔH for

$$Mg^{2+}(g) + e^- \rightarrow Mg^+(g)$$

This is just the reverse of the second ionization energy, or

$$EA(Mg^{2+}) = -IE_2(Mg) = -1445 \text{ kJ/mol}$$

b. EA of Al^+ is ΔH for: $Al^+(g) + e^- \rightarrow Al(g)$

$EA(Al^+) = -IE_1(Al) = -580$ kJ/mol

88. a. IE of Cl^- is ΔH for : $Cl^-(g) \rightarrow Cl(g) + e^-$

$IE(Cl^-) = -EA(Cl) = +348.7$ kJ/mol

b. $Cl(g) \rightarrow Cl^+(g) + e^-$ IE = 1255 kJ/mol (Table 7.5)

c. $Cl^+(g) + e^- \rightarrow Cl(g)$ $\Delta H = -IE_1 = -1255$ kJ/mol = EA (Cl^+)

The Alkali Metals

89. Li^+ ions will be the smallest of the alkali metal cations and will be most strongly attracted to the water molecules.

90. a. Carbonate ion is CO_3^{2-}. Lithium forms Li^+ ions. Thus, lithium carbonate is Li_2CO_3.

b. $$\frac{1 \times 10^{-3} \text{ mol Li}}{\text{L}} \times \frac{6.9 \text{ g Li}}{\text{mol Li}} = \frac{7 \times 10^{-3} \text{ g Li}}{\text{L}}$$

91. It should be potassium peroxide, K_2O_2. K^{2+} ions are not stable; the second ionization energy of K is very large compared to the first.

92. It should be magnesium(II) oxide. Mg^+ ions are not stable; the O_2^{2-} ion would oxidize Mg^+ to Mg^{2+} with O^{2-} as the other product.

93. $$\nu = \frac{c}{\lambda} = \frac{2.9979 \times 10^8 \text{ m/s}}{455.5 \times 10^{-9} \text{ m}} = 6.582 \times 10^{14} \text{ s}^{-1}$$

$$E = h\nu = (6.626 \times 10^{-34} \text{ J s})(6.582 \times 10^{14} \text{ s}^{-1}) = 4.361 \times 10^{-19} \text{ J}$$

94. $\nu = \frac{c}{\lambda}$ $E = h\nu$

589.0 nm $$\nu = \frac{2.9979 \times 10^8 \text{ m/s}}{589.0 \times 10^{-9} \text{ m}} = 5.090 \times 10^{14} \text{ s}^{-1}$$

$$E = 6.626 \times 10^{-34} \text{ J s} \times 5.090 \times 10^{14} \text{ s}^{-1} = 3.373 \times 10^{-19} \text{ J}$$

589.6 nm $\nu = 5.085 \times 10^{14} \text{ s}^{-1}$ $E = 3.369 \times 10^{-19}$ J

In kJ/mol:

589.0 nm: $$3.373 \times 10^{-19} \text{ J} \times \frac{1 \text{ kJ}}{1000 \text{ J}} \times \frac{6.022 \times 10^{23}}{\text{mol}} = 203.1 \text{ kJ/mol}$$

589.6 nm: $$3.369 \times 10^{-19} \text{ J} \times \frac{1 \text{ kJ}}{1000 \text{ J}} \times \frac{6.022 \times 10^{23}}{\text{mol}} = 202.9 \text{ kJ/mol}$$

95. a. Li_3N lithium nitride b. NaBr sodium bromide c. K_2S potassium sulfide

96. a. Li_3P lithium phosphide b. RbH rubidium hydride c. NaH sodium hydride

97. a. $4\,Li(s) + O_2(g) \rightarrow 2\,Li_2O(s)$ b. $2\,K(s) + S(s) \rightarrow K_2S(s)$

98. a. $2\,Cs(s) + 2\,H_2O(l) \rightarrow 2\,CsOH(aq) + H_2(g)$

b. $2\,Na(s) + Cl_2(g) \rightarrow 2\,NaCl(s)$

ADDITIONAL EXERCISES

99. It should be element #119 with ground state electron configuration: $[Rn]\,7s^2 5f^{14} 6d^{10} 7p^6 8s^1$

100. The IE is for removal of the electron from the atom in the gas phase. The work function is for the removal of an electron from the solid.

$M(g) \rightarrow M^+(g) + e^-$ IE

$M(s) \rightarrow M^+(s) + e^-$ work function

101. $E = \frac{310.\text{ kJ}}{\text{mol}} \times \frac{1\text{ mol}}{6.022 \times 10^{23}} = 5.15 \times 10^{-22}\text{ kJ} = 5.15 \times 10^{-19}\text{ J}$

$$E = \frac{hc}{\lambda};\ \lambda = \frac{hc}{E} = \frac{6.626 \times 10^{-34}\text{ J s} \times 2.998 \times 10^8\text{ m/s}}{5.15 \times 10^{-19}} = 3.86 \times 10^{-7}\text{ m} = 386\text{ nm}$$

102. a. n b. n and l

103. a. n = 3 We can have 3s, 3p, and 3d orbitals. Nine orbitals can hold 18 electrons.

b. n = 2, $l = 0$ specifies a 2s orbital. 2 electrons

c. n = 2, $l = 2$ is not possible. No electrons can have this combination of quantum numbers.

d. These four quantum numbers completely specify a single electron.

104. a. False, It takes less energy to ionize an electron from n = 3 than the ground state.

b. True

c. False, The energy difference from n = 3 $\rightarrow$ n = 2 is less than the energy difference from n = 3 $\rightarrow$ n = 1, thus, the wavelength is larger for n = 3 $\rightarrow$ n = 2 than for n = 3 $\rightarrow$ n = 1.

d. True

e. False, n = 2 is the first excited state, n = 3 is the second exited state.

105. No, for n = 2, the allowed values of l are 0 and 1; for n = 3 the allowed values of l are 0, 1, and 2.

106. The diagrams of orbitals give probabilities. We can never be 100% certain of the location of the electrons.

107. He: $1s^2$; Ne: $1s^2 2s^2 2p^6$; Ar: $1s^2 2s^2 2p^6 3s^2 3p^6$

Each peak in the diagram corresponds to a subshell with different values of n. Corresponding subshells are closer to the nucleus for heavier elements because of the increased nuclear charge.

108. $n = 5$; $m_l = -4, -3, -2, -1, 0, 1, 2, 3, 4$; 18 electrons

109. b and f are the only possible sets of quantum numbers.

a. For $l = 0$, m_l can only be 0.

c. m_s can only be +1/2 or -1/2.

d. For $n = 1$, l can only be 0.

e. For $l = 2$, m_l cannot be -3; lowest allowed value is -2.

110. a. As: $1s^22s^22p^63s^23p^64s^23d^{10}4p^3$

b. #116 will be below Po in the periodic table: $[Rn]7s^25f^{14}6d^{10}7p^4$

c. Ta: $[Xe]6s^24f^{14}5d^3$ or Ir: $[Xe]6s^24f^{14}5d^7$

d. Ti: $[Ar]\,4s^23d^2$ Ni: $[Ar]\,4s^23d^8$ Os: $[Xe]6s^24f^{14}5d^6$

111. Sb: $1s^22s^22p^63s^23p^64s^23d^{10}4p^65s^24d^{10}5p^3$

a. $l = 1$ designates p orbitals. There are 21 electrons in p orbitals ($l = 1$).

b. $m_l = 0$ All s electrons, 2 out of each set of 2p, 3p, 4p electrons, 2 out of each set of 3d and 4d electrons, and one of the 5p electrons have $m_l = 0$. 10 + 6 + 4 + 1 = 21 electrons with $m_l = 0$.

c. $m_l = 1$:

2 out of each set of 2p, 3p amd 4 p (6)
2 out of each set of 3d and 4d (4)
1 of the 5p (1)

11 electrons have $m_l = 1$.

112. a. The 4+ ion contains 20 electrons. Thus, the electrically neutral atom will contain 24 electrons. The atomic number is 24.

b. From the information the electron configuration of the ion must be: $1s^22s^22p^63s^23p^63d^2$. So there are 6 electrons in s orbitals.

c. 12 d. 2 e. This is an isotope of $^{50}_{24}Cr$. There are 26 neutrons in the nucleus.

f. $$3.01 \times 10^{23} \text{ atoms} \times \frac{49.9 \text{ amu}}{\text{atom}} \times \frac{1 \text{ g}}{6.022 \times 10^{23} \text{ amu}} = 24.94 \text{ g} = 24.9 \text{ g}$$

g. $1s^22s^22p^63s^23p^64s^13d^5$

113. Yes, the electron configuration is $1s^22s^22p^63s^23p^2$. There should be another big jump when the thirteenth electron is removed, i.e. when the 1s electrons begin to be removed.

114. A large IE is usually the result of a large Z_{eff}. A large Z_{eff} can also result in a large, exothermic EA. In both cases, a large Z_{eff} indicates a strong nuclear attraction for the valence electrons. The noble gases are an exception. The noble gases have a large IE but an endothermic EA. Noble gases have a completely filled shell of electrons. Added electrons must go to an unoccupied, higher energy shell. EA is not favorable for noble gases.

115. Expected order:

$$Li < Be < B < C < N < O < F < Ne$$

B and O are out of order. The IE of O (and also F and Ne) is lower because of the extra electron-electron repulsions present when two electrons are paired in the same orbital. B is out of order because of the different penetrating abilities of the 2p electron in B compared to the 2s electrons in Be.

116. Ti: $[Ar]4s^2 3d^2$

	n	l	m_l	m_s
4s	4	0	0	+1/2
4s	4	0	0	-1/2
3d*	3	2	-2	+1/2
3d	3	2	-1	+1/2

* Only one of 10 possible combinations of m_l and m_s.

For the ground state, the second d electron should be in a different orbital with parallel spin (4 possibilities).

117. Electron-electron repulsions become more important when we try to add electrons to an atom. From the standpoint of electron-electron repulsions, large atoms would have more favorable (more exothermic) electron affinities. Considering only electron-nucleus attractions, smaller atoms would be expected to have more favorable (more exothermic) EA's. These trends are exactly the opposite of each other. Thus, the overall variation is not as great as ionization energy in which attractions to the nucleus dominate.

118. Al (-44), Si(-120), P (-74), S (-200.4), Cl (-348.7)

Based on effective nuclear charge, we would expect the EA to become more exothermic as we go from left to right in the period. Phosphorus is out of line. The reaction for the EA of P is:

$$P(g) + e^- \longrightarrow P^-(g)$$

$$[Ne]3s^2 3p^3 \qquad [Ne]3s^2 3p^4$$

The additional electron in P^- will have to go into an orbital that already has one electron. There will be greater repulsion between electrons in P^-, causing the EA of P to be less favorable than predicted based solely on attractions to the nucleus.

119.

$1s^2 2s^2 2p^6 3s^2 3p^6$	:	Ar	:	1.5205 MJ/mol	:	0.94 Å
$1s^2 2s^2 2p^6 3s^2$	:	Mg	:	0.7377 MJ/mol	:	1.60 Å
$1s^2 2s^2 2p^6 3s^2 3p^6 4s^1$	:	K	:	0.4189 MJ/mol	:	1.97 Å

Size: Ar < Mg < K; IE: K < Mg < Ar

120. Electron-electron repulsions are much greater in O^- than S^- because the electron goes into a smaller 2p orbital vs. the larger 3p orbital on sulfur, resulting in a more favorable (more exothermic) EA for sulfur.

121. a.

$$Cu^+(g) + e^- \rightarrow Cu(g) \qquad -I_1 = -746 \text{ kJ}$$
$$Cu^+(g) \rightarrow Cu^{2+}(g) + e^- \qquad I_2 = 1958 \text{ kJ}$$
$$2\,Cu^+(g) \rightarrow Cu(g) + Cu^{2+}(g) \qquad \Delta H = 1212 \text{ kJ}$$

b.

$$Na^-(g) \rightarrow Na(g) + e^- \qquad -EA = 52 \text{ kJ}$$
$$Na^+(g) + e^- \rightarrow Na(g) \qquad -I_1 = -495 \text{ kJ}$$
$$Na^-(g) + Na^+(g) \rightarrow 2\,Na(g) \qquad \Delta H = -443 \text{ kJ}$$

c.

$$Mg^{2+}(g) + e^- \rightarrow Mg^+(g) \qquad -I_2 = -1445 \text{ kJ}$$
$$K(g) \rightarrow K^+(g) + e^- \qquad I_1 = 419 \text{ kJ}$$
$$Mg^{2+}(g) + K(g) \rightarrow Mg^+(g) + K^+(g) \qquad \Delta H = -1026 \text{ kJ}$$

d.

$$Na(g) \rightarrow Na^+(g) + e^- \qquad I_1 = 495 \text{ kJ}$$
$$Cl(g) + e^- \rightarrow Cl^-(g) \qquad EA = -349 \text{ kJ}$$
$$Na(g) + Cl(g) \rightarrow Na^+(g) + Cl^-(g) \qquad \Delta H = 146 \text{ kJ}$$

e.

$$Mg(g) \rightarrow Mg^+(g) + e^- \qquad \Delta H = I_1 = 735 \text{ kJ}$$
$$F(g) + e^- \rightarrow F^-(g) \qquad \Delta H = EA = -328 \text{ kJ}$$
$$Mg(g) + F(g) \rightarrow Mg^+ + F^-(g) \qquad \Delta H = 407 \text{ kJ}$$

f.

$$Mg^+(g) \rightarrow Mg^{2+}(g) + e^- \qquad \Delta H = I_2 = 1445 \text{ kJ}$$
$$F(g) + e^- \rightarrow F^-(g) \qquad \Delta H = EA = -328 \text{ kJ}$$
$$Mg^+(g) + F(g) \rightarrow Mg^{2+} + F^-(g) \qquad \Delta H = 1117 \text{ kJ}$$

g. From parts e and f we get:

$$Mg(g) + F(g) \rightarrow Mg^+(g) + F^-(g) \qquad \Delta H = 407 \text{ kJ}$$
$$Mg^+(g) + F(g) \rightarrow Mg^{2+} + F^-(g) \qquad \Delta H = 1117 \text{ kJ}$$
$$Mg(g) + 2\,F(g) \rightarrow Mg^{2+}(g) + 2\,F^-(g) \qquad \Delta H = 1524 \text{ kJ}$$

122. a. $Se^{3+}(g) \rightarrow Se^{4+}(g) + e^-$

b. $S^-(g) + e^- \rightarrow S^{2-}(g)$

c. $Fe^{3+}(g) + e^- \rightarrow Fe^{2+}(g)$

d. $Mg(g) \rightarrow Mg^+(g) + e^-$

e. $Mg(s) \rightarrow Mg^+(s) + e^-$

CHALLENGE PROBLEMS

123. Size also decreases going across a period. Sc & Ti and Y & Zr are adjacent elements. There are 14 elements (the Lanthanides) between La and Hf, making Hf considerably smaller.

124. a. As we remove succeeding electrons, the electron being removed is closer to the nucleus and there are fewer electrons left repelling it. In addition, we are removing a negatively charged particle from a positively charged particle. All of these factors go in the direction of requiring increasingly more energy to remove successive electrons.

b. Al : $1s^22s^22p^63s^23p^1$

For I_4, we begin removing an electron with n = 2. For I_3, we removed an electron with n = 3. In going from n = 3 to n = 2 there is a big jump in ionization energy because the n = 2 electrons are much closer to the nucleus than n = 3 electrons. Also, the n = 3 electrons are effectively shielded from the nuclear charge by all of the n = 2 electrons while the n = 2 electrons are not as effective in shielding each other from the nuclear charge.

c. Al^{4+}, the electron affinity for Al^{4+} is ΔH for the reaction

$$Al^{4+}(g) + e^- \rightarrow Al^{3+}(g) \quad \Delta H = -I_4 = -11{,}600 \text{ kJ/mol}$$

d. The greater the number of electrons, the greater will be the size. So:

$$Al^{4+} < Al^{3+} < Al^{2+} < Al^+ < Al$$

125. a. Each orbital could hold 3 electrons.

b. First period: 3. Second period: 12.

c. 15

d. 21

126. a. 1st period: p = 1, q = 1, r = 0, s = ± 1/2 (2 elements)

2nd period: p = 2, q = 1, r = 0, s = ± 1/2 (2 elements)

3rd period: $p = 3, q = 1, r = 0, s = \pm 1/2$ (2 elements)

$p = 3, q = 3, r = -2, s = \pm 1/2$ (2 elements)

$p = 3, q = 3, r = 0, s = \pm 1/2$ (2 elements)

$p = 3, q = 3, r = +2, s = \pm 1/2$ (2 elements)

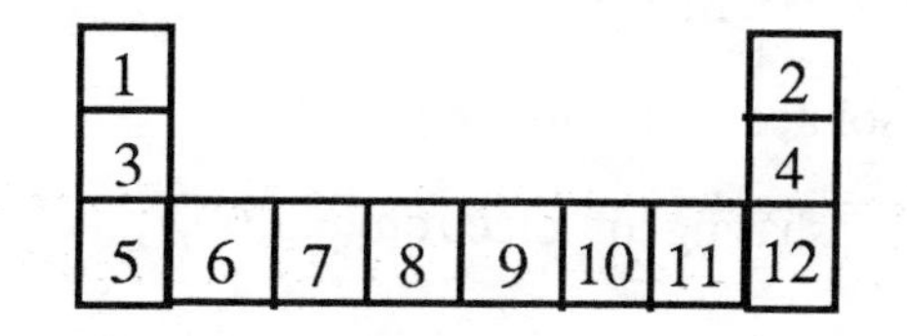

b. From (a), we can see that eight electrons can have $p = 3$.

c. $p = 4, q = 3, r = 2, s = \pm 1/2$ (2 electrons)

d. $p = 4, q = 3, r = -2, s = \pm 1/2$ (2 electrons)

$p = 4, q = 3, r = 0, s = \pm 1/2$ (2 electrons)

$p = 4, q = 3, r = +2, s = \pm 1/2$ (2 electrons)

A total of 6 electrons can have $p = 4$ and $q = 3$.

e. $p = 3, q = 0, r = 0$: This is not allowed; q must be odd. Zero electrons can have these quantum numbers.

f. $p = 5, q = 1, r = 0$

$q = 3, r = -2, 0, +2$

$q = 5; r = -4, -2, 0, +2, +4$

g. $p = 6, q = 1, r = 0, s = \pm 1/2$ (2 electrons)

$q = 3; r = -2, 0, +2; s = \pm 1/2$ (6 electrons)

$q = 5; r = -4, -2, 0, +2, 4; s = \pm 1/2$ (10 electrons)

Eighteen electrons can have $p = 6$.

127. None of the noble gases and no subatomic particles had been discovered when Mendeleev published his periodic table. Thus, there was not an element out of place in terms of reactivity. There was no reason to predict an entire family of elements. Mendeleev ordered his table by mass, he had no way of knowing there were gaps in atomic numbers (they hadn't been invented yet).

CHAPTER EIGHT: BONDING - GENERAL CONCEPTS

QUESTIONS

1. a. electronegativity: The ability of an atom in a molecule to attract electrons to itself.

   electron affinity: The energy change for the reaction:

   $$M(g) + e^- \rightarrow M^-(g)$$

   EA deals with isolated atoms in the gas phase.

   b. Covalent bond: sharing of electron pair(s); polar covalent bond: unequal sharing of electron pair(s).

   c. Ionic bond: electrons are no longer shared. Completely unequal sharing (i.e. transfer) of electrons from one atom to another.

2. Isoelectronic: same number of electrons.

   There are two variables, number of protons and number of electrons, that will determine the size of an ion. Keeping the number of electrons constant we only have to consider the number of protons to predict trends in size.

3. No, we would expect the more highly charged ions to have a greater attraction for electrons. Thus, the electronegativity of an element does depend on its oxidation state.

4. The extra electron-electron repulsions are much greater than the attraction of the electron for the nucleus.

5. The two requirements for a polar molecular are:

   1. polar bonds.

   2. a structure such that the bond dipoles do not cancel.

   In addition some molecules that have no polar bonds, but contain unsymmetrical lone pairs are polar. For example, PH_3 and AsH_3 are slightly polar.

6. If the valence shell contains a set of ns and np orbitals, there are 4 orbitals and, hence, room for only eight electrons.

EXERCISES

Chemical Bonds and Electronegativity

7. Using the periodic table we expect the general trend for electronegativity to be: 1) increase as we go from left to right across a period and 2) decrease as we go down a group.

   a. $C < N < O$ b. $Se < S < Cl$ c. $Sn < Ge < Si$ d. $Tl < Ge < S$

8. a. $Rb < K < Na$ b. $Ga < B < O$ c. $Br < Cl < F$ d. $S < O < F$

9. The greater the difference in electronegativity between two atoms will result in a more polar bond.

   a. Ge—F b. P—Cl c. S—F d. Ti—Cl

10. a. Sn—H b. Tl—Br c. Si—O d. O—F

11. The general trends in electronegativity used on Exercises 8.7 and 8.9 are only rules of thumb. In this exercise we use experimental values of electronegativities and can begin to see several exceptions.

Order of EN from Figure 8.3.

a. C (2.5) < N (3.0) < O (3.5) same as predicted.

b. Se (2.4) < S (2.5) < Cl (3.0) same

c. Si = Ge = Sn (1.8) different

d. Tl (1.8) = Ge (1.8) < S (2.5) different

Most polar bonds using actual EN values

a. Si—F and Ge—F equal polarity (Ge—F predicted)

b. P—Cl (same as predicted)

c. S—F (same as predicted)

d. Ti—Cl (correctly predicted) Si—Cl and Ge—Cl equal polarity

12. a. Rb (0.8) = K(0.8) < Na (0.9) different b. Ga (1.6) < B (2.0) < O (3.5) same

c. Br (2.8) < Cl (3.0) < F (4.0) same d. S (2.5) < O (3.5) < F (4.0) same

Most polar bonds using actual EN values.

a. C—H most polar (Sn—H predicted).

b. Al—Br most polar (Tl—Br predicted). Note: the order of polarity is precisely the opposite of what we might have predicted.

c. Si—O (same as predicted).

d. Same polarity, but in different directions. Oxygen is positive end of O—F dipole and negative end of O—Cl dipole (O—F predicted).

Ionic Compounds

13. a. $Cu > Cu^+ > Cu^{2+}$ b. $Pt^{2+} > Pd^{2+} > Ni^{2+}$ c. $Se^{2-} > S^{2-} > O^{2-}$

d. $La^{3+} > Eu^{3+} > Gd^{3+} > Yb^{3+}$ e. $Te^{2-} > I^- > Xe > Cs^+ > Ba^{2+} > La^{3+}$

14. a. $Co > Co^+ > Co^{2+} > Co^{3+}$ b. $N^{3-} > N^{2-} > N^- > N$ c. $Br^- > Cl^- > F^-$

d. $S^{2-} > Cl^- > Ar > K^+ > Ca^{2+}$ e. $Ca^{2+} > Mg^{2+} > Be^{2+}$

15. a. Mg^{2+}: $1s^22s^22p^6$; Sn^{2+}: $[Kr]\ 5s^24d^{10}$

K^+: $1s^22s^22p^63s^23p^6$; Al^{3+}: $1s^22s^22p^6$

Tl^+: $[Xe]6s^24f^{14}5d^{10}$; As^{3+}: $[Ar]4s^23d^{10}$

b. N^{3-}, O^{2-} and F^-: $1s^2 2s^2 2p^6$; Te^{2-}: $[Kr]5s^2 4d^{10} 5p^6$

c. Be^{2+}: $1s^2$; Rb^+: $[Ar]4s^2 3d^{10} 4p^6$; Ba^{2+}: $[Kr]5s^2 4d^{10} 5p^6$

Se^{2-}: $[Ar]4s^2 3d^{10} 4p^6$; I^-: $[Kr]5s^2 4d^{10} 5p^6$

16. a. Sn^{4+}: $[Kr]4d^{10}$ Cs^+: [Xe] or [Kr] $5s^2 4d^{10} 5p^6$

Ga^{3+} and As^{5+}: $[Ar]3d^{10}$ Tl^{3+}: $[Xe]4f^{14} 5d^{10}$

b. P^{3-} and S^{2-}: [Ar] or $1s^2 2s^2 2p^6 3s^2 3p^6$ Br^-: [Kr] or [Ar] $4s^2 3d^{10} 4p^6$

c. Ca^{2+} and Cl^-: [Ar] or $1s^2 2s^2 2p^6 3s^2 3p^6$ Sr^{2+}: [Kr] or $[Ar]4s^2 3d^{10} 4p^6$

Li^+: [He] or $1s^2$

17. a. Sc^{3+} b. Te^{2-} c. Ce^{4+} and Ti^{4+} d. Ba^{2+}

18. a. none b. none c. F^- d. Cs^+

19. Se^{2-}, Br^-, Rb^+, Sr^{2+}, Y^{3+}, Zr^{4+} are all isoelectronic with Kr.

20. P^{3-}, S^{2-}, Cl^-, K^+, Ca^{2+}, Sc^{3+} and Ti^{4+} are all isoelectronic with Ar.

21. Lattice energy is proportional to $\frac{q_1 q_2}{r}$

In general, charge effects are much greater than size effects.

a. NaCl, Na^+ smaller than K^+.

b. LiF, F^- smaller than Cl^-.

c. MgO, O^{2-} greater charge than OH^-.

d. $Fe(OH)_3$, Fe^{3+} greater charge than Fe^{2+}.

e. Na_2O, greater negative charge on O.

f. MgO, smaller ions.

22. a. LiF, Li^+ smaller than Cs^+.

b. NaBr, Br^- smaller than I^-.

c. BaO, O^{2-} greater charge than Cl^-.

d. $CaSO_4$, Ca^{2+} greater charge than Na^+.

e. K_2O, O^{2-} greater charge than F^-.

f. Li_2O, Li^+ smaller than Na^+ and O^{2-} smaller than S^{2-}

23.

$Na(s) \rightarrow Na(g)$	ΔH = 108 kJ	(sublimation)
$Na(g) \rightarrow Na^+(g) + e^-$	ΔH = 495 kJ	(ionization energy)
$1/2\ Cl_2(g) \rightarrow Cl(g)$	ΔH = 239/2 kJ	(bond energy)
$Cl(g) + e^- \rightarrow Cl^-(g)$	ΔH = -348 kJ	(electron affinity)
$Na^+(g) + Cl^-(g) \rightarrow NaCl(s)$	ΔH = -786 kJ	(lattice energy)
$Na(s) + 1/2\ Cl_2(g) \rightarrow NaCl(s)$	ΔH_f° = -412 kJ/mol	

24.

$Ba(s) \rightarrow Ba(g)$	$\Delta H_{sub} = 178$ kJ
$Ba(g) \rightarrow Ba^+(g) + e^-$	$I_1 = 503$ kJ
$Ba^+(g) \rightarrow Ba^{2+}(g) + e^-$	$I_2 = 965$ kJ
$Cl_2(g) \rightarrow 2\ Cl(g)$	D = 239 kJ
$2\ Cl(g) + 2\ e^- \rightarrow 2\ Cl^-(g)$	2 EA = 2(-348) kJ
$Ba^{2+}(g) + 2\ Cl^-(g) \rightarrow BaCl_2(s)$	Latt. En. = -2056 kJ
$Ba(s) + Cl_2(g) \rightarrow BaCl_2(s)$	$\Delta H_f^\circ = -867$ kJ/mol

25. a. From the data given, it costs less energy to produce $Mg^+(g) + O^-(g)$ than to produce $Mg^{2+}(g) + O^{2-}(g)$. However, the lattice energy for $Mg^{2+}O^{2-}$ will be much larger than for Mg^+O^-.

b. Mg^+ and O^- both have unpaired electrons. In Mg^{2+} and O^{2-} there are no unpaired electrons. Hence, Mg^+O^- would be paramagnetic; Mg^{2+} and O^{2-} would be diamagnetic. Paramagnetism can be detected by measuring the mass of a sample in the presence and absence of a magnetic field. The apparent mass of a paramagnetic substance will be larger in a magnetic field because of a force between the unpaired electrons and the field.

26.

$O(g) + e^- \rightarrow O^-(g)$	$\Delta H = -141$ kJ/mol
$O^-(g) + e^- \rightarrow O^{2-}(g)$	$\Delta H = +878$ kJ/mol
$O(g) + 2\ e^- \rightarrow O^{2-}(g)$	$\Delta H = +737$ kJ/mol

27. Ca^{2+} has greater charge than Na^+ and Se^{2-} is smaller than Te^{2-}.

The effect of charge on the lattice energy is greater than the effect of size. We expect the trend from most exothermic to least exothermic to be:

CaSe	>	CaTe	>	Na_2Se	>	Na_2Te	
(-2862)		(-2721)		(-2130)		(-2095)	This is what we observe.

28. The compounds are FeO, Fe_2O_3, $FeCl_2$ and $FeCl_3$. The lattice energy is proportional to the charge of the cation and the charge of the anion, q_1q_2.

Compound	q_1q_2	Lattice Energy
$FeCl_2$	(+2) (-1)	-2631 kJ/mol
FeO	(+2) (-2)	-5359 kJ/mol
$FeCl_3$	(+3) (-1)	-3865 kJ/mol
Fe_2O_3	(+3) (-2)	-14,744 kJ/mol

29. a. Li^+ and N^{3-} are expected ions. The formula of the compound would be Li_3N (lithium nitride).

b. Ga^{3+} and O^{2-}, Ga_2O_3, gallium(III) oxide

c. Rb^+ and Cl^-, RbCl, rubidium chloride

d. Ba^{2+} and S^{2-}, BaS, barium sulfide

30. a. Al^{3+} and Cl^-, $AlCl_3$, aluminum chloride

b. Na^+ and O^{2-}, Na_2O, sodium oxide

c. Sr^{2+} and F^-, SrF_2, strontium fluoride

d. Ca^{2+} and S^{2-}, CaS, calcium sulfide

Bond Energies (Assume all bond energies are ±1 kJ)

31.

```
      H                       H
      |                       |
  H—C—N≡C  ——▶  H—C—C≡N
      |                       |
      H                       H
```

Break C—N (305 kJ/mol)
Make C—C (347 kJ/mol)
$\Delta H° = 305 - 347 = -42$ kJ

32. a. H—H + Cl—Cl ——▶ 2 H—Cl

Bonds broken:

1 H—H (432 kJ/mol)
1 Cl—Cl (239 kJ/mol)

Bonds made:

2 H—Cl (427 kJ/mol)

$\Delta H = 432 \text{ kJ} + 239 \text{ kJ} - 2(427) \text{ kJ} = -183 \text{ kJ}$

b.

```
N≡N + 3 H—H ——▶ 2 H—N—H
                      |
                      H
```

Bonds broken:

1 N≡N (941 kJ/mol)
3 H—H (432 kJ/mol)

Bonds made:

6 N—H (391 kJ/mol)

$\Delta H = 941 \text{ kJ} = 3(432) \text{ kJ} - 6(391) \text{ kJ} = -109 \text{ kJ}$

c.

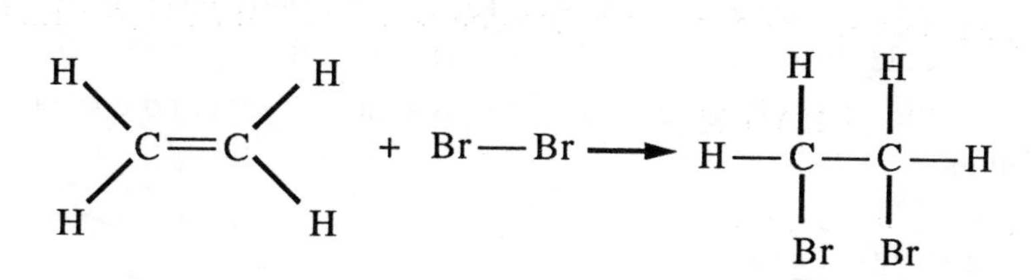

Bonds broken:

1 C═C (614 kJ/mol)
1 Br—Br (193 kJ/mol)

Bonds made:

1 C—C (347 kJ/mol)
2 C—Br (276 kJ/mol)

$\Delta H = 614 \text{ kJ} + 193 \text{ kJ} - [347 \text{ kJ} + 2(276 \text{ kJ})] = -92 \text{ kJ}$

d.

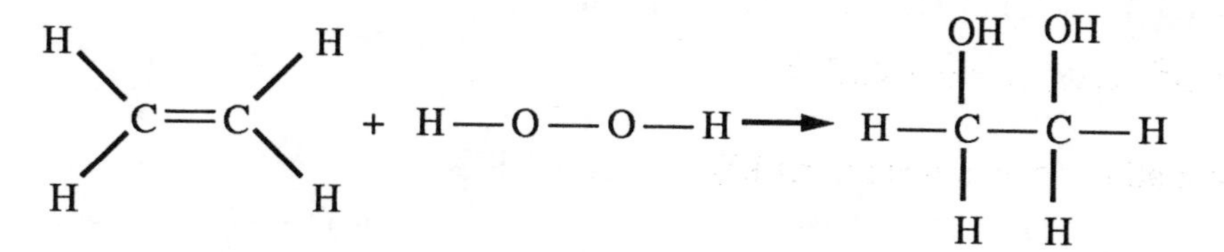

Bonds broken:

1 C═C (614 kJ/mol)
1 O—O (146 kJ/mol)

Bonds made:

1 C—C (347 kJ/mol)
2 C—O (358 kJ/mol)

$\Delta H = 614 \text{ kJ} + 146 \text{ kJ} - [347 \text{ kJ} + 2(358 \text{ kJ})] = -303 \text{ kJ}$

33.

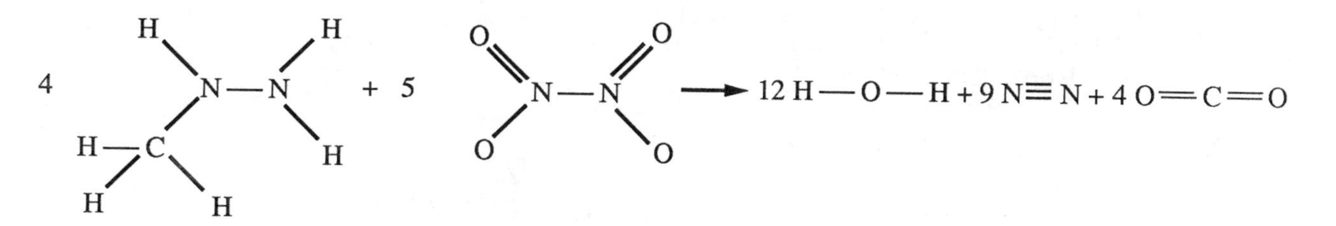

Bonds broken:

9 N—N (160 kJ/mol)
4 N—C (305 kJ/mol)
12 C—H (413 kJ/mol)
12 N—H (391 kJ/mol)
10 N═O (607 kJ/mol)
10 N—O (201 kJ/mol)

Bonds made:

24 O—H (467 kJ/mol)
9 N≡N (941 kJ/mol)
8 C═O (799 kJ/mol)

(Assume all ±1 kJ)

$\Delta H = \Sigma D(\text{broken}) - \Sigma D(\text{made})$

$= 9(160) + 4(305) + 12(413) + 12(391) + 10(607) + 10(201)$

$- [24(467) + 9(941) + 8(799)]$

$= 20{,}388 \text{ kJ} - 26{,}069 \text{ kJ} = -5681 \text{ kJ}$

From standard enthalpies of formation we got ΔH = -4594 kJ. The bond energy gives us an estimate that is about 24% high. We could do better. In 6.71 we used ΔH_f° for N_2O_4(l) and $CH_3N_2H_3$(l). It takes heat to convert liquids to the gas. ΔH for the gas phase reaction using bond energies will be more exothermic than for the liquids.

34. a. $\Delta H = 2\ \Delta H_f^\circ\ (HCl) = 2\ \text{mol}\ (-92\ \text{kJ/mol})$

$\Delta H = -184$ kJ/mol (-183 from bond energies)

b. $\Delta H = 2\ \Delta H_f^\circ\ (NH_3) = 2\ \text{mol}\ (-46\ \text{kJ/mol})$

$\Delta H = -92$ kJ/mol (-109 from bond energies)

Comparing the values for each reaction, bond energies seem to give a reasonably good estimate of the enthalpy change for a reaction. The estimate is especially good for gas phase reactions.

35.

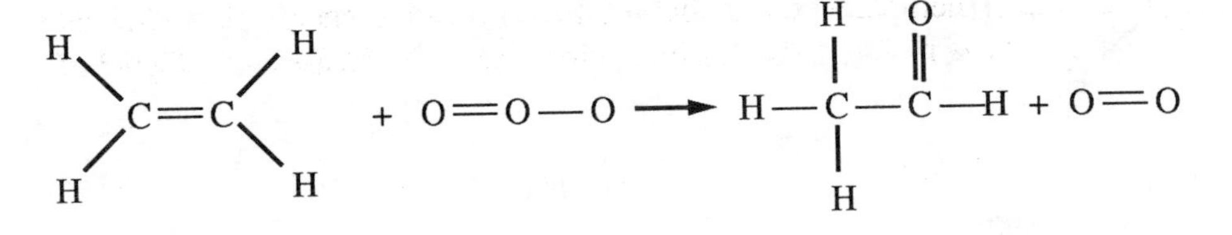

Bonds broken:

1 C═C (614 kJ/mol)
1 O—O (146 kJ/mol)

Bonds made:

1 C—C (347 kJ/mol)
1 C═O (799 kJ/mol)

$\Delta H = 614 + 146 - (347 + 799) = -386$ kJ

From Ex. 6.71, $\Delta H^\circ = -361$ kJ

36.

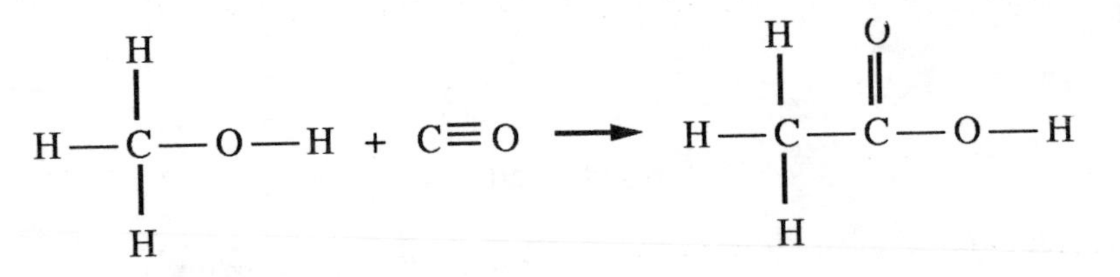

Bonds broken:

1 C≡O 1072 kJ/mol

Bonds made:

1 C—C (347 kJ/mol)
1 C═O (799 kJ/mol)

$\Delta H = 1072 - [347 + 799] = -74$ kJ

37. a. Using SF_4 data: $SF_4(g) \longrightarrow S(g) + 4\ F(g)$

$\Delta H^\circ = 4\ D_{SF} = 278.8 + 4(79.0) - (-775) = 1370.\ \text{kJ}$

$$D_{SF} = \frac{1370.\ \text{kJ}}{4\ \text{mol SF bonds}} = 342.5\ \text{kJ/mol}$$

Using SF_6 data: $SF_6(g) \rightarrow S(g) + 6\ F(g)$

$\Delta H° = 6\ D_{SF} = 278.8 + 6(79.0) - (-1209) = 1962\ kJ$

$$D_{SF} = \frac{1962\ kJ}{6} = 327.0\ kJ/mol$$

b. The S—F bond energy in the table is 327 kJ/mol. The value in the table was based on the S—F bond in SF_6.

c. S(g) and F(g) are not the most stable form of the element at 25°C. The most stable forms are $S_8(s)$ and $F_2(g)$; $\Delta H_f^\circ = 0$ for these two species.

38. $NH_3(g) \rightarrow N(g) + 3\ H(g)$

$\Delta H° = 3\ D_{NH} = 472.7 + 3(216.0) - (-46.1) = 1166.8\ kJ$

$$D_{NH} = \frac{1166.8\ kJ}{3\ mol\ NH\ bonds} = 388.93\ kJ/mol \approx 389\ kJ/mol$$

D_{calc} = 389 kJ/mol as compared to 391 kJ/mol in the table. There is good agreement.

Lewis Structures and Resonance

39. a. HCN has 1 + 4 + 5
= 10 valence electrons.

H—C—N   H—C≡N:

uses 4: (6 e^- left, 3 pairs)

b. PH_3 has 5 + 3(1)
= 8 valence electrons

uses 6: (2 e^- left, 1 pair)

c. $CHCl_3$ has 4 + 1 + 3(7)
= 26 valence electrons.

H, Cl—C—Cl, Cl   H, :Cl—C—Cl:, :Cl:

uses 8 e^-: (18 e^- left, 9 pairs)

d. NH_4^+ has 5 + 4(1) - 1
= 8 valence electrons

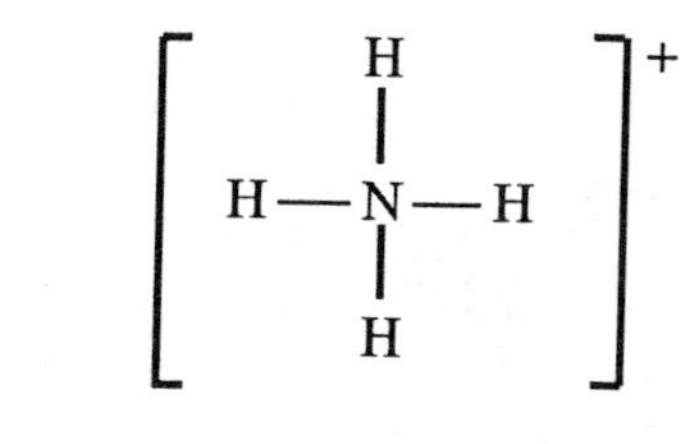

e. BF_4^- has 3 + 4(7) + 1
= 32 electrons.

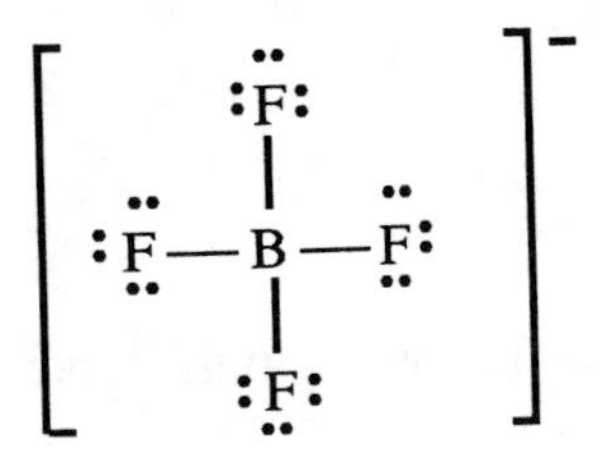

f. SeF_2 has 6 + 2(7)
= 20 valence electrons

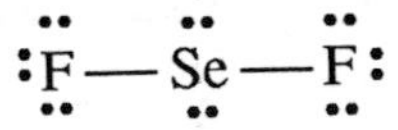

40. a. $POCl_3$ has 5 + 6 + 3(7) = 32 valence electrons.

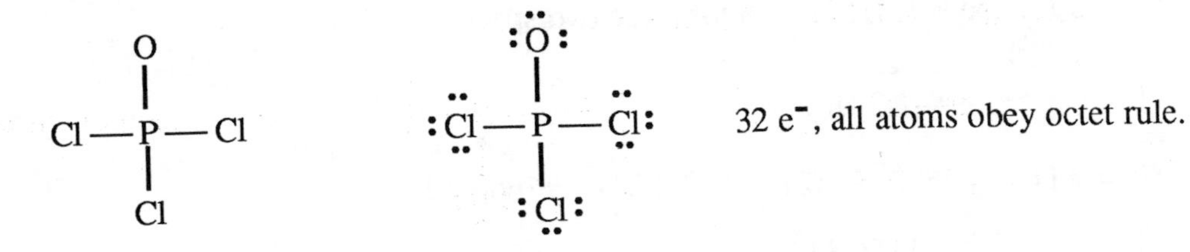

skeletal structure

complete octets

32 e^-, all atoms obey octet rule.

If the structure we draw contains all of the valence electrons and all of the atoms satisfy the octet rule then we have a valid Lewis structure.

SO_4^{2-} has 6 + 4(6) + 2 = 32 valence electrons.

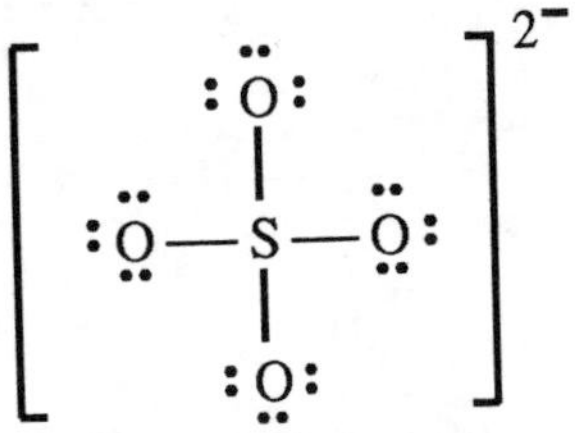

Note: A negatively charged ion will have additional electrons to those that come from the valence shells of the atoms.

XeO_4, 8 + 4(6) = 32 e^-

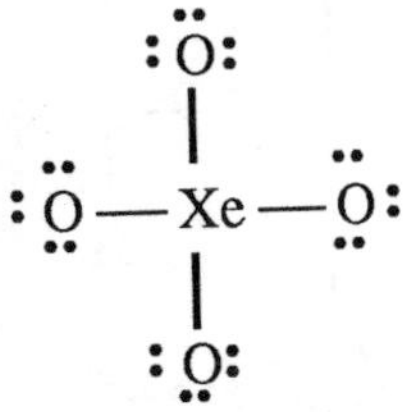

PO_4^{3-}, 5 + 4(6) + 3 = 32 e^-

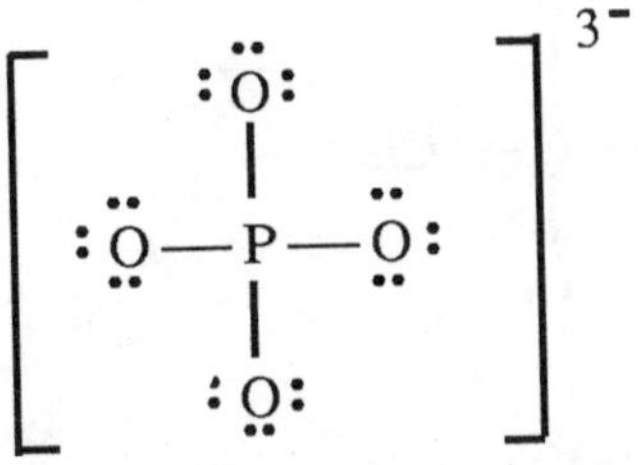

ClO_4^- has 7 + 4(6) + 1 = 32 valence electrons.

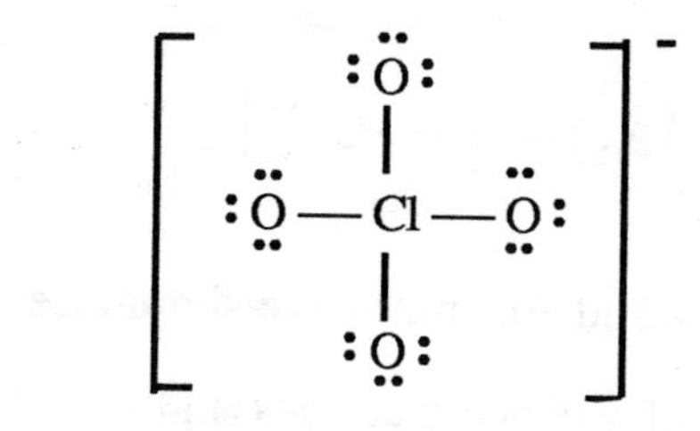

Note: All of these species have the same number of atoms and the same number of valence electrons. They also have the same Lewis structure.

b. NF_3 has 5 + 3(7) = 26 valence electrons.

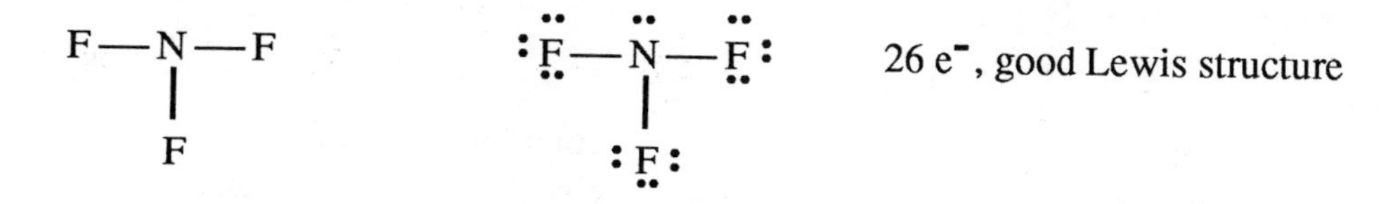

skeletal structure complete octets

SO_3^{2-}, 6 + 3(6) + 2 = 26 e$^-$

PO_3^{3-}, 5 + 3(6) + 3 = 26 e$^-$

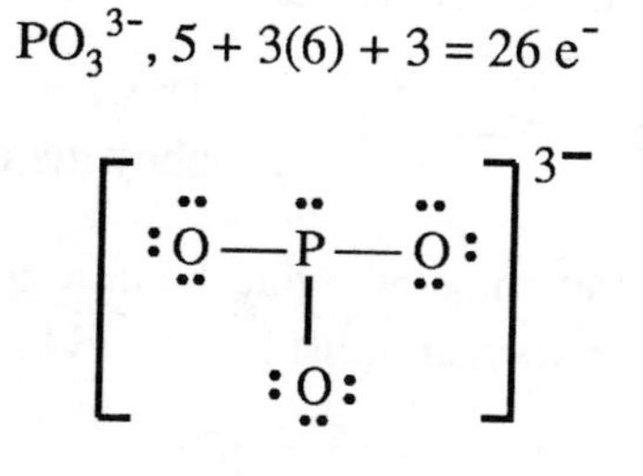

ClO_3^-, 7 + 3(6) + 1 = 26 e$^-$

Note: Species with the same number of atoms and valence electrons have similar Lewis structures.

c. ClO_2^-, has 7 + 2(6) + 1 = 20 valence electrons.

O—Cl—O

20 e$^-$, good Lewis structure

skeletal structure complete octets

All atoms obey the octet rule and we have used all the valence electrons.

SCl_2, 6 + 2(7) = 20 e^- PCl_2^-, 5 + 2(7) + 1 = 20 e^-

:C̤l̤—S̤—C̤l̤: [:C̤l̤—P̤—C̤l̤:]⁻

Note: Species with the same number of atoms and electrons have similar Lewis structures.

d. Molecules/ions that have the same number of valence electrons and the same number of atoms will have similar Lewis structures.

41. a. NO_2^- has 5 + 2(6) + 1 = 18 valence electrons.

O—N—O :Ö—N—Ö:

skeletal structure

complete octet of more electronegative oxygen (16 e^- used)

We have 1 more pair of e^-; putting this pair on the nitrogen we get:

:Ö—N̈—Ö: This accounts for all 18 electrons, but N does not obey the octet rule

To get an octet about the nitrogen and not use any more electrons, one of the unshared pairs on an oxygen must be shared between N and O instead.

[O=N̈—Ö:]⁻ (0 0 -1) ⟷ [:Ö—N̈=O]⁻ (-1 0 0)

Since, there is no prior reason to have the double bond to either oxygen atom, we can draw two resonance structures. Formal charges are shown.

HNO_2 has 1 + 5 + 2(6) = 18 valence electrons.

O=N̈—Ö—H (0 0 0 0)

NO_3^- has 5 + 3(6) + 1 = 24 valence electrons.

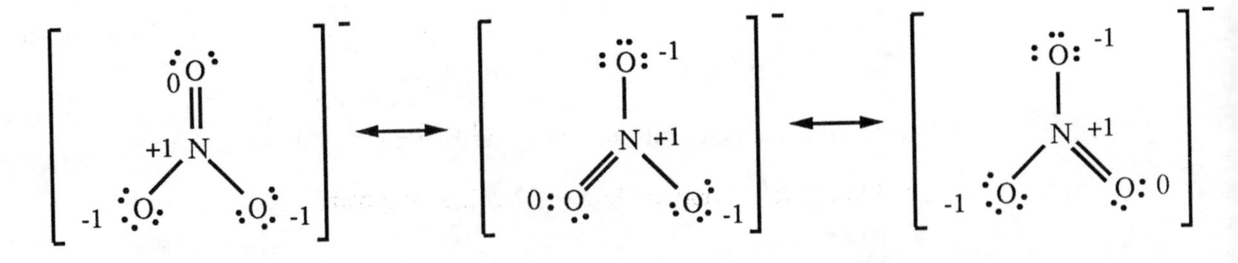

HNO_3 has 1 + 5 + 3(6) = 24 valence electrons.

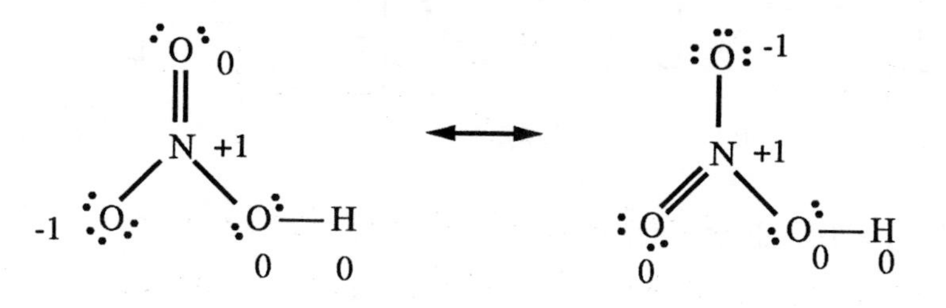

b. SO_4^{2-} see Exercise 8.40a.

HSO_4^- has 1 + 6 + 4(6) + 1 = 32 valence electrons.

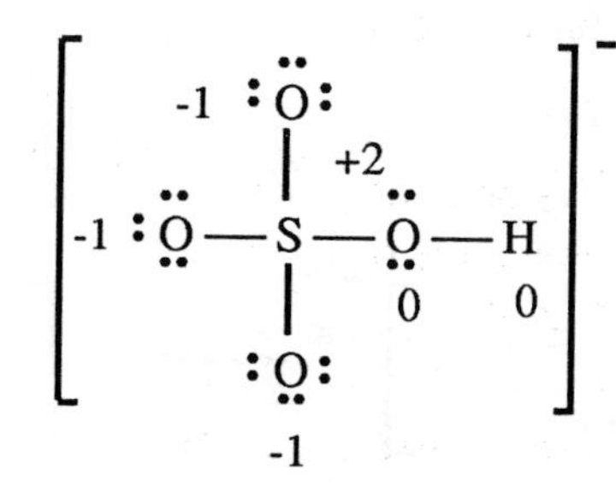

H_2SO_4 has 2 + 6 + 4(6) = 32 valence electrons.

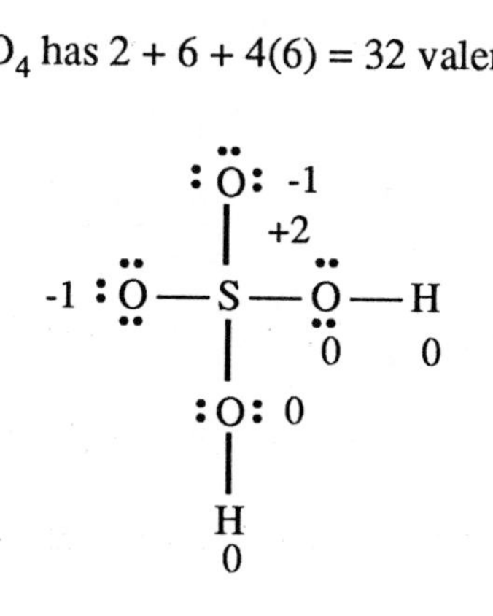

c. CN^-, 4 + 5 + 1 = 10 e^-

[:C≡N:]⁻
-1 0

HCN, 1 + 4 + 5 = 10 e^-

H—C≡N:
0 0 0

d. OCN^- has 6 + 4 + 5 + 1 = 16 valence electrons.

O—C—N

skeletal structure

:Ö—C—N̈:

complete octets of O and N

16 e^- but no octet about C.

We need to convert 2 nonbonding pairs to bonding pairs.

$$\left[:\ddot{\underset{..}{O}}-C\equiv N:\right]^- \longleftrightarrow \left[\dot{:}O=C=N\dot{:}\right]^- \longleftrightarrow \left[:O\equiv C-\ddot{\underset{..}{N}}:\right]^-$$

Formal charges: -1, 0, 0 ; 0, 0, -1 ; +1, 0, -2

Third structure not likely because of formal charge.

SCN^- has 6 + 4 + 5 + 1 = 16 valence electrons.

$$\left[:\ddot{\underset{..}{S}}-C\equiv N:\right]^- \longleftrightarrow \left[\dot{:}S=C=N\dot{:}\right]^- \longleftrightarrow \left[:S\equiv C-\ddot{\underset{..}{N}}:\right]^-$$

Formal charges: -1, 0, 0 ; 0, 0, -1 ; +1, 0, -2

First two structures, and especially the second, best from formal charge.

N_3^- has 3(5) + 1 = 16 valence electrons.

$$\left[:\ddot{\underset{..}{N}}-N\equiv N:\right]^- \longleftrightarrow \left[\dot{:}N=N=N\dot{:}\right]^- \longleftrightarrow \left[:N\equiv N-\ddot{\underset{..}{N}}:\right]^-$$

Formal charges: -2, +1, 0 ; -1, +1, -1 ; 0, +1, -2

Second structure best from formal charge.

e. NCCN has 2(5) + 2(4) = 18 valence electrons.

$$:N\equiv C-C\equiv N:$$

Formal charges: 0, 0, 0, 0

42. Ozone: O_3 has 3(6) = 18 valence electrons.

$$\dot{:}O=\ddot{O}-\ddot{\underset{..}{O}}: \longleftrightarrow :\ddot{\underset{..}{O}}-\ddot{O}=O\dot{:}$$

Sulfur dioxide: SO_2 has 6 + 2(6) = 18 valence electrons.

$$\dot{:}O=\ddot{S}-\ddot{\underset{..}{O}}: \longleftrightarrow :\ddot{\underset{..}{O}}-\ddot{S}=O\dot{:}$$

Sulfur trioxide: SO_3 has 6 + 3(6) = 24 valence electrons.

$$\dot{:}O=\underset{\underset{:\underset{..}{O}:}{|}}{S}-\ddot{\underset{..}{O}}: \longleftrightarrow :\ddot{\underset{..}{O}}-\underset{\underset{:\underset{..}{O}:}{|}}{S}=O\dot{:} \longleftrightarrow :\ddot{\underset{..}{O}}-\underset{\underset{\dot{:}O\dot{:}}{||}}{S}-\ddot{\underset{..}{O}}:$$

43. PAN ($H_3C_2NO_5$) has 3 + 2(4) + 5 + 5(6) = 46 valence electrons.

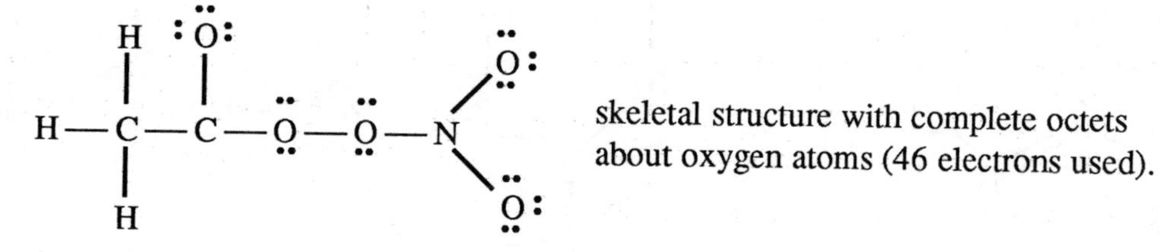

skeletal structure with complete octets about oxygen atoms (46 electrons used).

The skeletal structure has used all 46 electrons, but there are only six electrons around one of the carbon atoms and the nitrogen atom. Two unshared pairs must become shared, that is we form two double bonds.

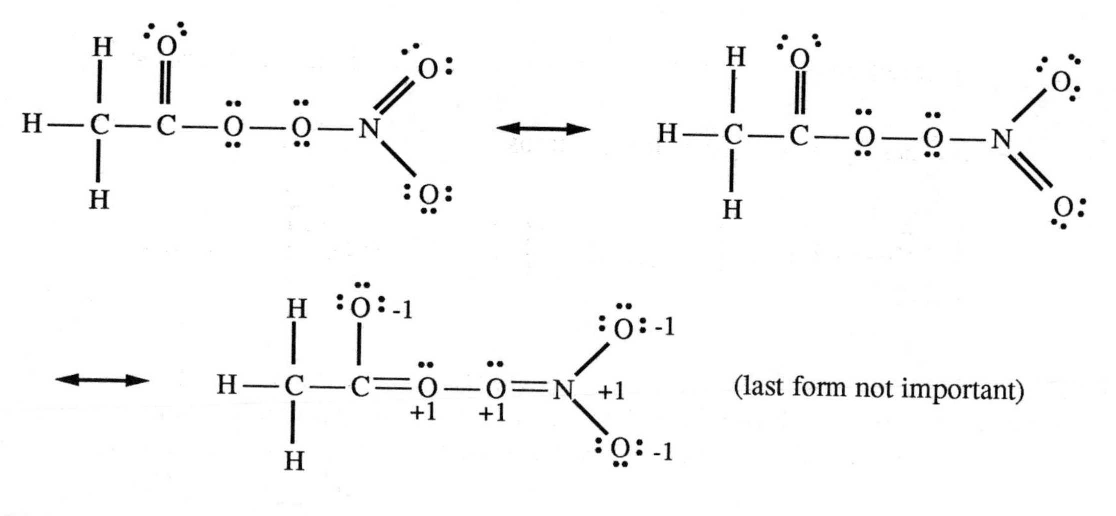

(last form not important)

44. CH_3NCO has 4 + 3(1) + 5 + 4 + 6 = 22 valence electrons.

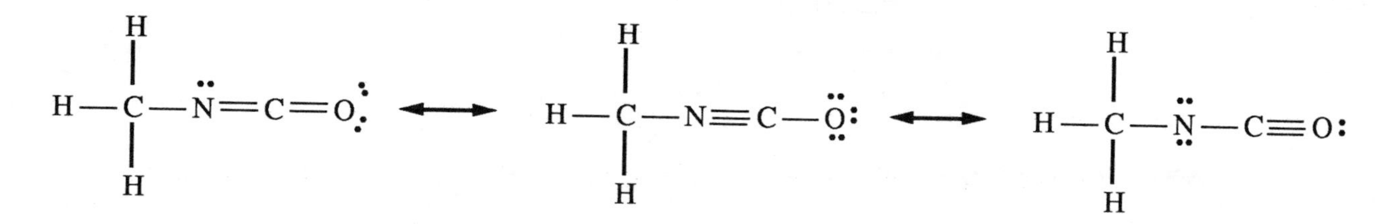

45.

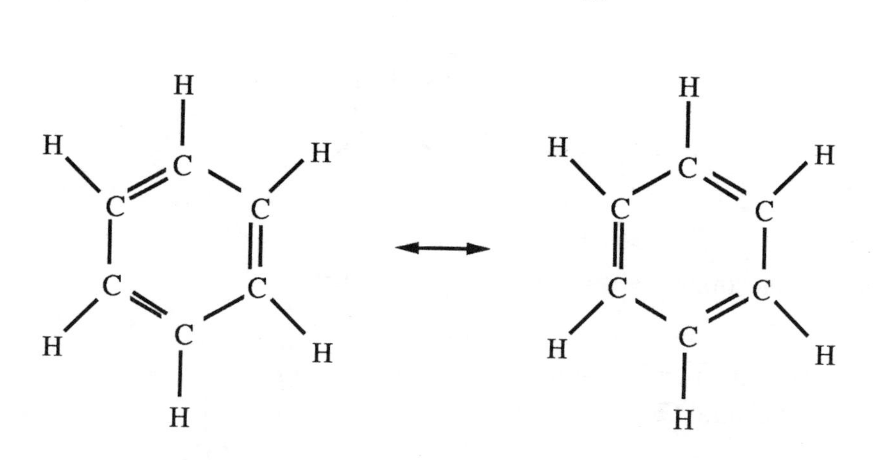

Benzene has 6(4) + 6(1) = 30 valence electrons.

46. Borazine ($B_3N_3H_6$) has 3(3) + 3(5) + 6(1) = 30 valence electrons.

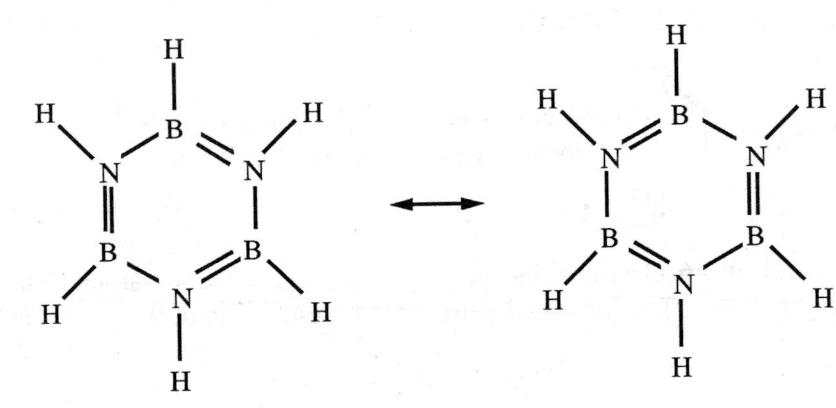

47. PF_5, 5 + 5(7) = 40 e^-

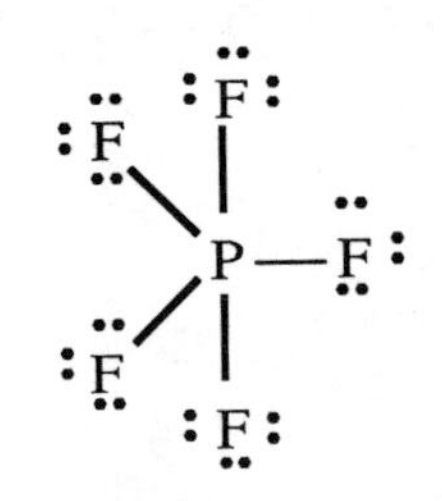

$Be(CH_3)_2$, 2 + 2[4 + 3(1)] = 16 e^-

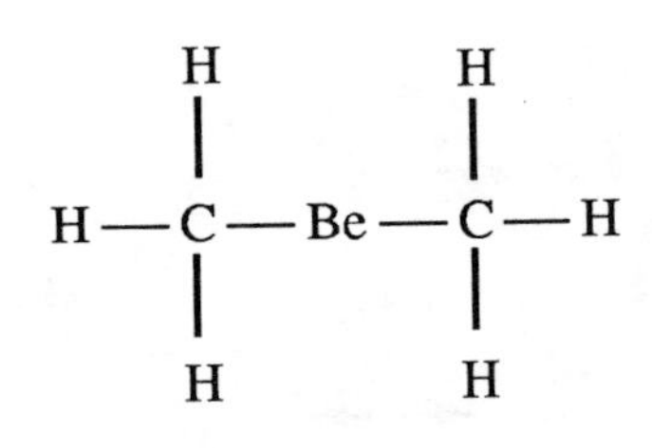

BCl_3, 3 + 3(7) = 24 e^-

$XeOF_4$, 8 + 6 + 4(7) = 42 e^-

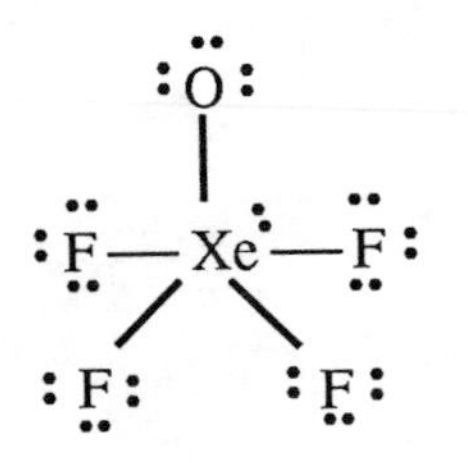

XeF_6, 8 + 6(7) = 50 e^-

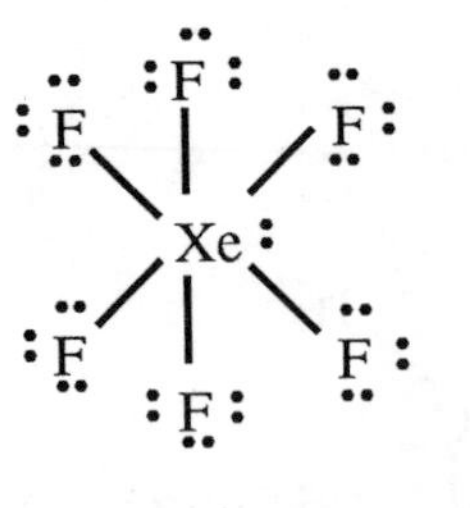

48. ClF_3 has 7 + 3(7) = 28 valence electrons.

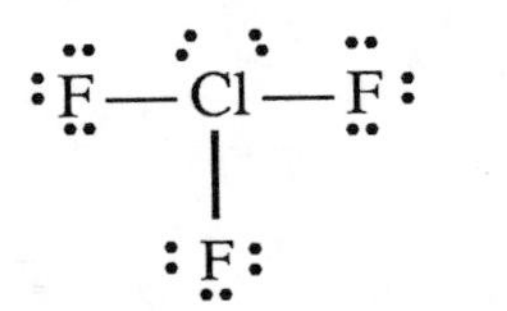

We expand the octet of the central Cl atom.

BrF_3: 28 e^-

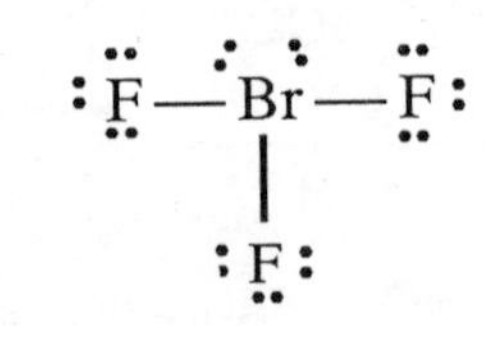

49.

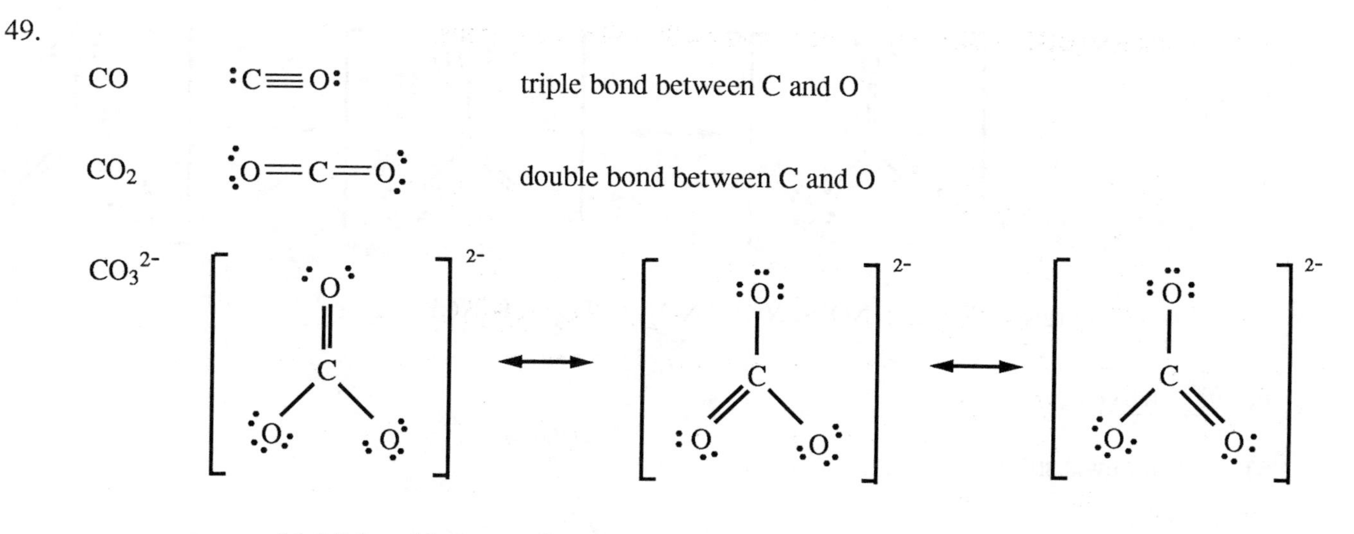

average of 1 1/3 bond between C and O

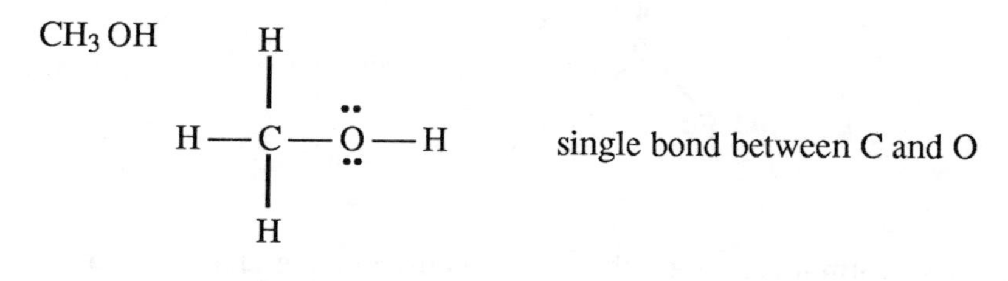

Longest ⟶ shortest

$CH_3OH > CO_3^{2-} > CO_2 > CO$

Weakest ⟶ strongest

$CH_3OH < CO_3^{2-} < CO_2 < CO$

50. H_2NOH:

H—N(—H)—O—H

N_2O: N=N=O ⟷ :N≡N—O: ⟷ :N—N≡O:

NO^+: $[:N\equiv O:]^+$

NO_2^-: $[O=N—O:]^-$ ⟷ $[:O—N=O]^-$

NO_3^-:

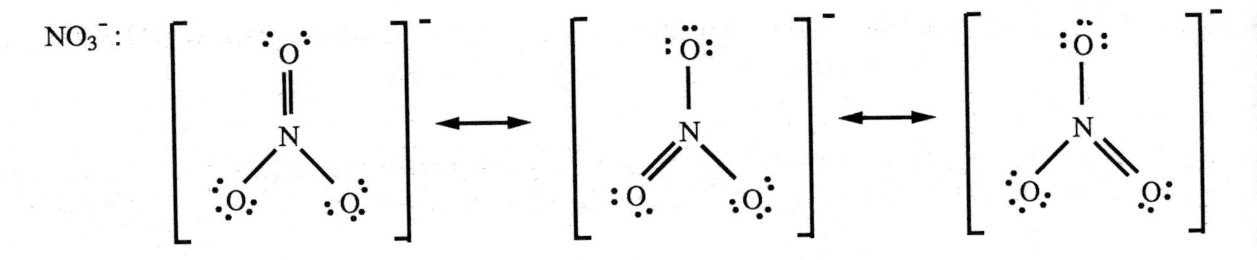

Shortest to longest: $NO^+ < N_2O < NO_2^- < NO_3^- < H_2NOH$

Formal Charge

51. The Lewis structure

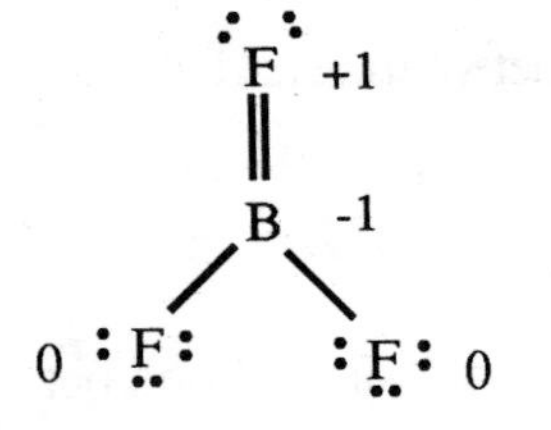

obeys the octet rule, but has a +1 formal charge on the most electronegative element there is, fluorine, and a negative formal charge on a much less electronegative element (boron). This is just the opposite of what we expect; negative formal charge on F and positive formal charge on B. Therefore, BF_3 does not follow the octet rule and there are only six electrons around the boron atom.

52. :C≡O:

Carbon: FC = 4 - 2 - 1/2(6) = -1

Oxygen: FC = 6 - 2 - 1/2(6) = +1

Electronegativity predicts the opposite polarization. The two opposing effects seem to cancel to give a much less polar molecular than expected.

53. See Exercise 8.40a for Lewis structures of a, b, c, and d.

a. $POCl_3$: P, FC = 5 - 1/2(8) = +1

b. SO_4^{2-}: S, FC = 6 - 1/2(8) = +2

c. ClO_4^-: Cl, FC = 7 - 1/2(8) = +3

d. PO_4^{3-}: P, FC = 5 - 1/2(8) = +1

e. SO_2Cl_2, 6 + 2(6) + 2(7) = 32 e^-

O
Cl — S — Cl
O

S, FC = 6 - 1/2(8) = +2

f. XeO_4, 8 + 4(6) = 32 e^-

O
O — Xe — O
O

Xe, FC = 8 - 1/2(8) = +4

g. ClO_3^-, 7 + 3(6) + 1 = 26 e^-

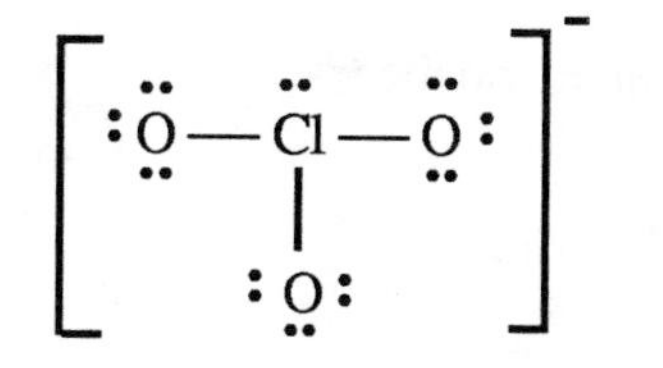

Cl, FC = 7 - 2 - 1/2(6) = +2

h. NO_4^{3-}, 5 + 4(6) + 3 = 32 e^-

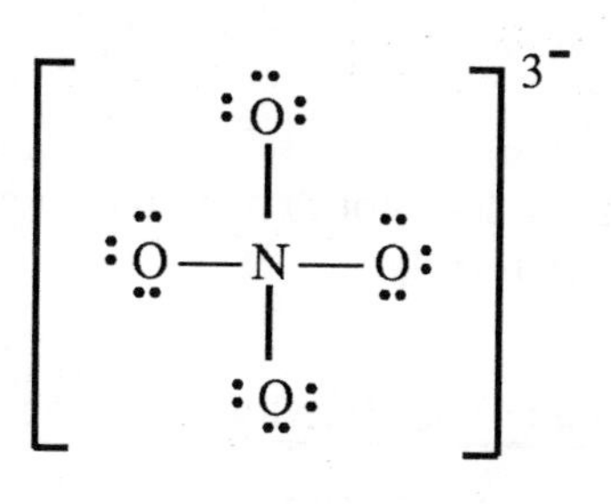

N, FC = 5 - 1/2(8) = +1

54. Note: only one resonance form is shown.

a.

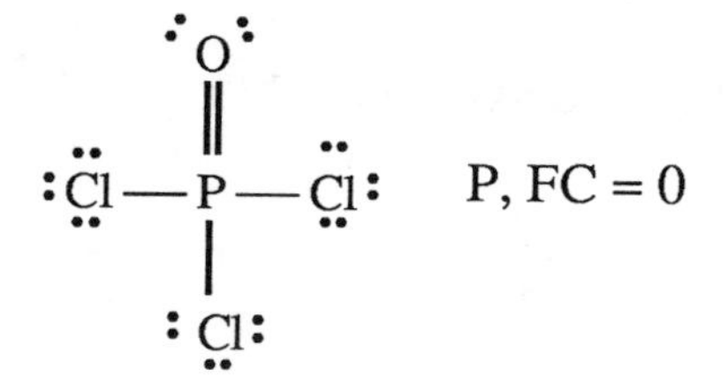

P, FC = 0

b.

O
O — S — O
O
2−

S, FC = 0

c.

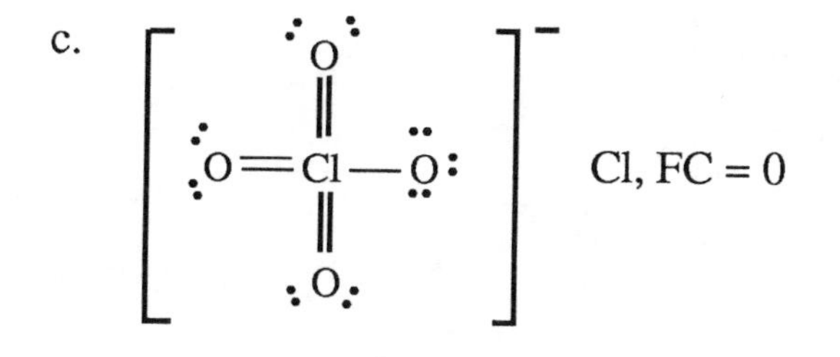

Cl, FC = 0

d.

O
O — P — O
O
3−

P, FC = 0

e.

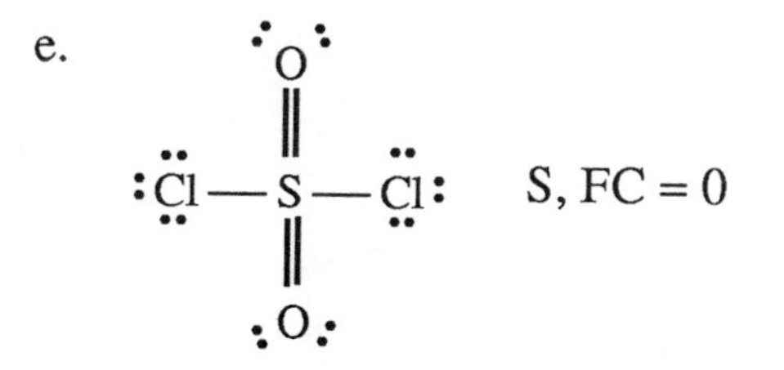

S, FC = 0

f.

O
O = Xe = O
O

Xe, FC = 0

g. $\left[:\ddot{O}=\ddot{Cl}=\ddot{O}: \; | \; :\ddot{O}: \right]^-$ Cl, FC = 0

h. We can't. $\left[\ddot{O}=N(-\ddot{O}:)_3 \right]^{3-}$ has lower formal charge,

but N does not expand its octet. We wouldn't expect this resonance form to exist.

Molecular Geometry and Polarity

55. 8.39 a. linear, 180°

b. trigonal pyramid, < 109.5°

c. tetrahedral, ≈ 109.5°

d. tetrahedral, 109.5°

e. tetrahedral, 109.5°

f. V-shaped or bent, < 109.5°

8.41 a. NO_2^-, bent, < 120°

HNO_2, bent about O and N, HON angle < 109.5°, ONO angle < 120°

NO_3^-, trigonal planar, 120°

HNO_3, trigonal planar about N, 120°

bent about N—O—H, angle < 109.5°

b. tetrahedral about S in all three, angle ≈ 109.5°

bent about S—O—H, angle < 109.5°

c. HCN is linear, 180°

d. all are linear, 180°

e. linear about both carbons, 180°

56. 8.40 a. all are tetrahedral, ≈ 109.5°

b. all are trigonal pyramidal, < 109.5°

c. all are bent (V-shaped), < 109.5°

57. a. 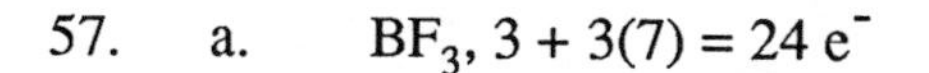

BF_3, $3 + 3(7) = 24\ e^-$

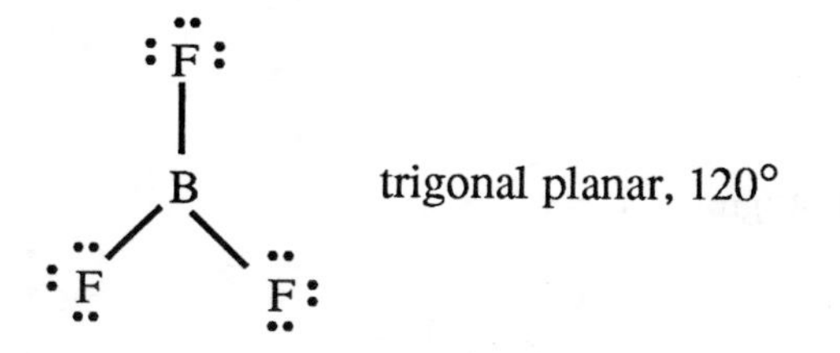

trigonal planar, 120°

b. BH_2^-, $3 + 2(1) + 1 = 6\ e^-$

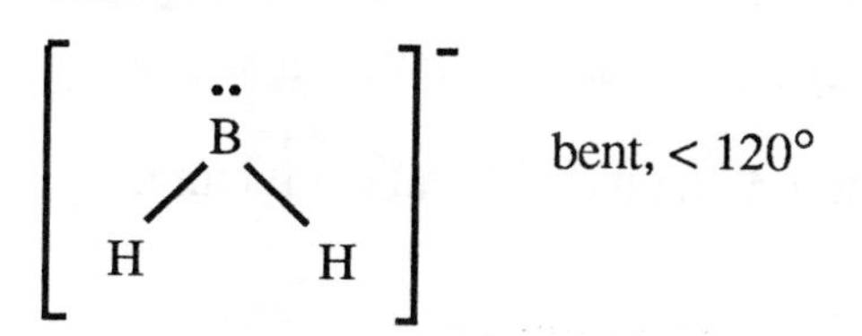

bent, < 120°

c. $COCl_2$, $4 + 6 + 2(7) = 24\ e^-$

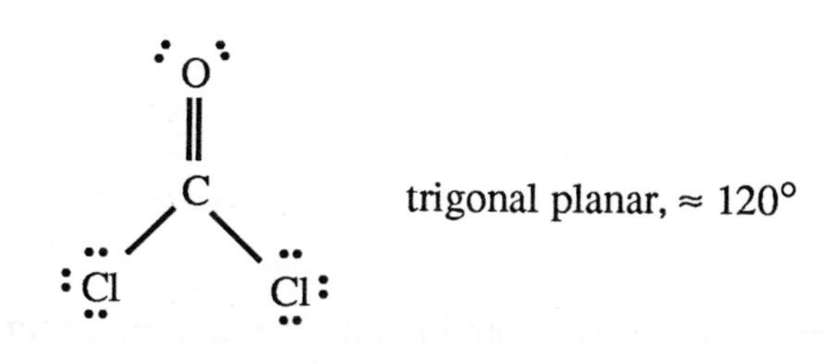

trigonal planar, ≈ 120°

Note: All of these structures have three effective pairs of electrons (a double bond counts as one effective pair for geometry) about the central atom. All of the structures are based on a trigonal planar geometry, but only a and c are described as having a trigonal planar structure.

58. SeO_3^{2-} has $6 + 3(6) + 2 = 26$ valence electrons.

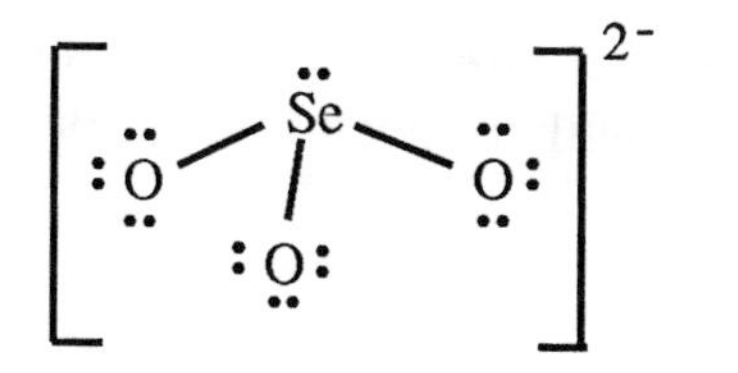

trigonal pyramid, angles < 109.5°

SeH_2 has $6 + 2(1) = 8$ valence electrons.

H—Se—H bent, < 109.5°

H | :Se—H < 109.5°

SeO_4^{2-} has 6 + 4(6) + 2 = 32 valence electrons.

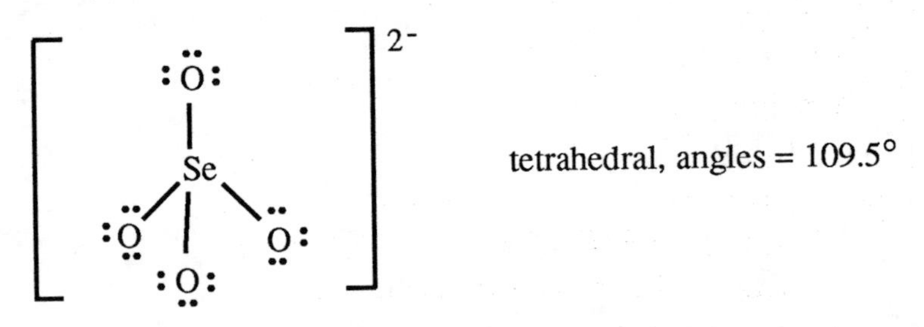

tetrahedral, angles = 109.5°

Note: There are 4 pairs of electrons about the Se atom in each case in this exercise. All of the structures are based on a tetrahedral geometry, but only SeO_4^{2-} has a tetrahedral structure.

We consider only the positions of the atoms in describing the molecular structure.

59. a. I_3^- has 3(7) + 1 = 22 valence electrons.

:I—I—I: Complete octets, but only 20 electrons used.

Expand octet of Central I.

[:I—I—I:]$^-$

In general we add extra pairs of electrons to the central atom. We can see a more specific reason when we look at how I_3^- forms in solution:

$$I_2(aq) + I^-(aq) \rightarrow I_3^-(aq)$$

:I—I: + [:I:]$^-$ → [:I—I—I:]$^-$

There are 5 pairs of electrons about the central ion. The structure will be based on a trigonal bipyramid geometry. The lone pairs require more room and will occupy the equatorial positions; the bonding pairs will occupy the axial positions.

[:I: | :I: | :I:]$^-$

axial: apex of the trigonal pyramid

equatorial: around the middle, the corners of the trigonal pyramid

Thus, I_3^- is linear with a 180° bond angle.

b. ClF_3 has 7 + 3(7) = 28 valence electrons.

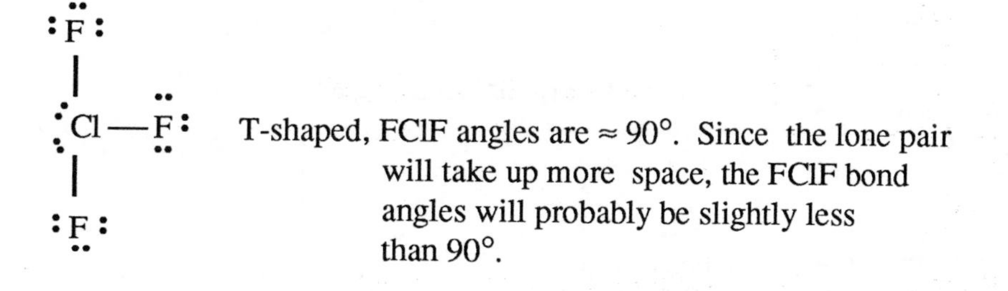

T-shaped, FClF angles are ≈ 90°. Since the lone pair will take up more space, the FClF bond angles will probably be slightly less than 90°.

c. IF_4^+ has 7 + 4(7) - 1 = 34 valence electrons.

< 120°

< 90°

See-saw or teeter totter

Also called bisphenoid.

d. SF_5^+ has 6 + 5(7) - 1 = 40 valence electrons.

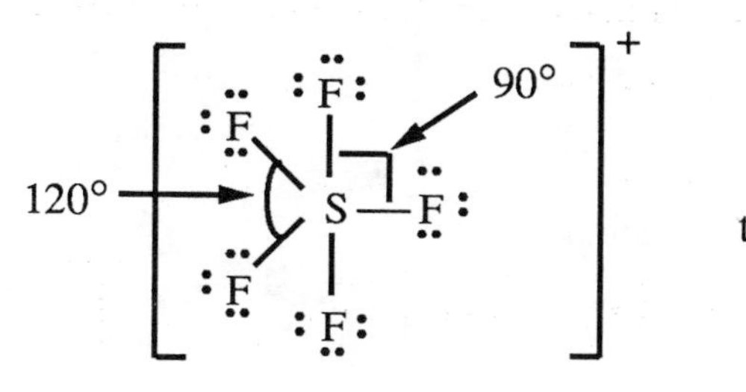

trigonal bipyramid

Note: All of the species in this exercise have 5 pairs of electrons around the central atom. All of the structures are based on a trigonal bipyramid, but only in SF_5^+ are all of the pairs bonding pairs. Thus, SF_5^+ is the only one we describe as being trigonal bipyramidal. Still, we had to begin with the trigonal bipyramid geometry to get to the structures of the others.

60. BrF_5 has 7 + 5(7) = 42 valence electrons.

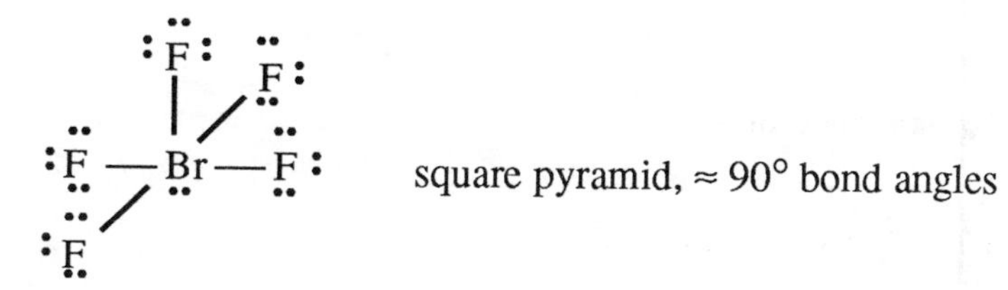

square pyramid, ≈ 90° bond angles

KrF_4 has 8 + 4(7) = 36 valence electrons.

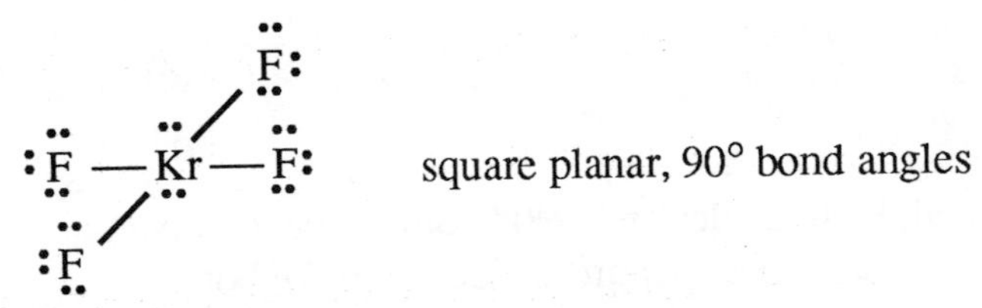

IF_6^+ has 7 + 6(7) - 1 = 48 valence electrons.

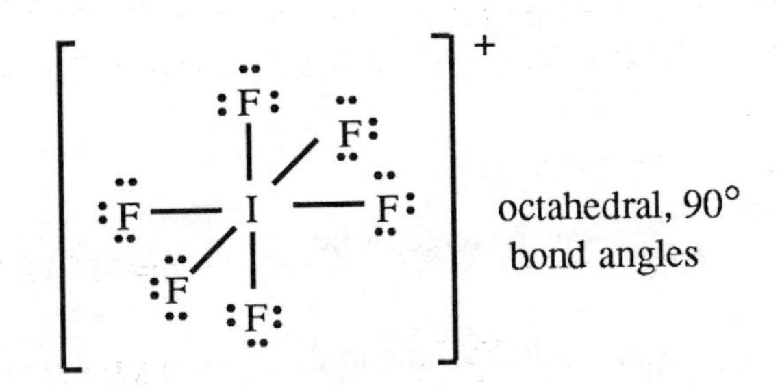

Note: All these species have 6 pairs of electrons around the central atom. All three structures are based on the octahedron, but only IF_6^+ has an octahedral structure.

61. SbF_5 has 5 + 5(7) = 40 valence electrons.

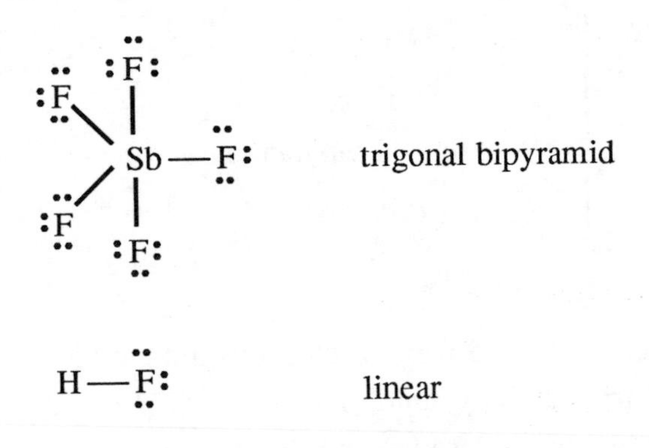

SbF_6^- has 5 + 6(7) + 1 = 48 valence electrons.

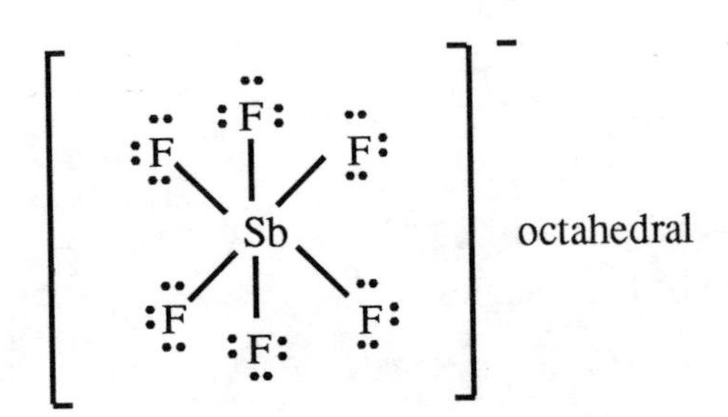

H_2F^+ has 2 + 7 - 1 = 8 valence electrons.

$$\left[H - \ddot{F} - H \right]^+$$

bent or V-shaped

62. a. CrO_4^{2-} has 6 + 4(6) + 2 = 32 valence electrons.

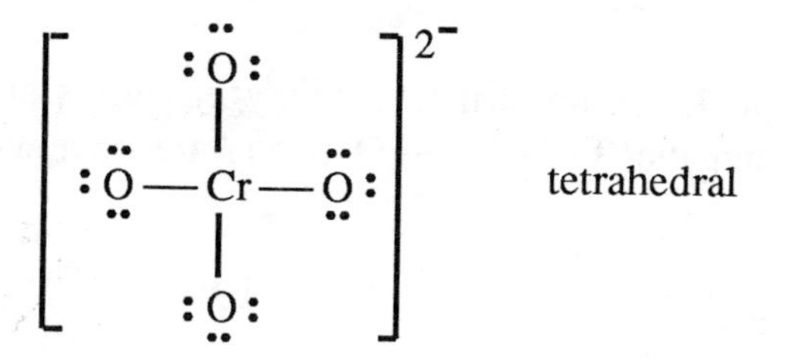

tetrahedral

Cr has electron configuration $[Ar]4s^2 3d^4$. We count all six electrons beyond the argon core.

b. Dichromate ion ($Cr_2O_7^{2-}$) has 2(6) + 7(6) + 2 = 56 valence electrons.

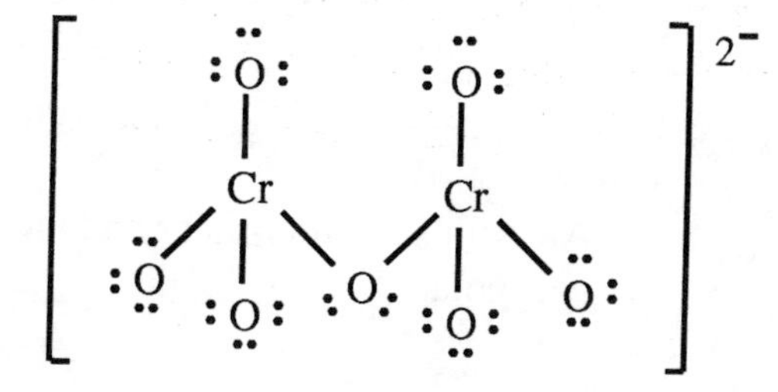

There is a no simple way to describe the structure. We would expect there to be a roughly tetrahedral arrangement of O - atoms about each Cr and the Cr—O—Cr bond to be bent.

c. $S_2O_3^{2-}$ has 2(6) + 3(6) + 2 = 32 valence electrons.

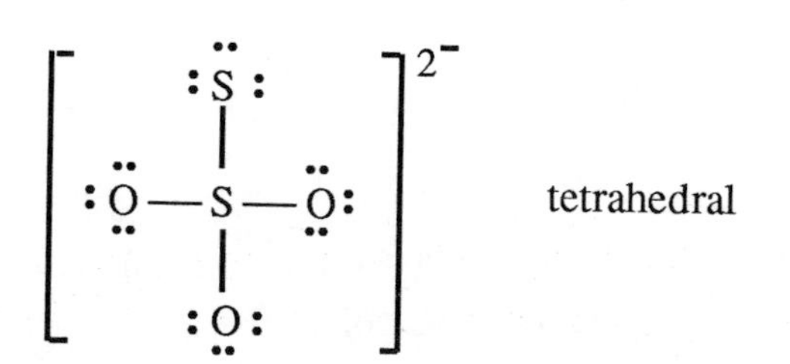

tetrahedral

d. $S_2O_8^{2-}$ has 2(6) + 8(6) + 2 = 62 valence electrons.

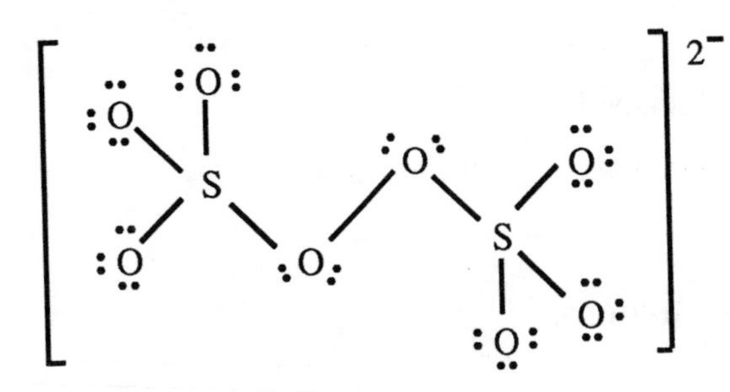

Again, there is no simple description of the entire structure. Oxygen atoms are roughly tetrahedrally arranged around each S and the S — O — O bonds are bent with angles slightly less than 109.5°.

63. a. OCl_2 has 6 + 2(7) = 20 valence electrons.

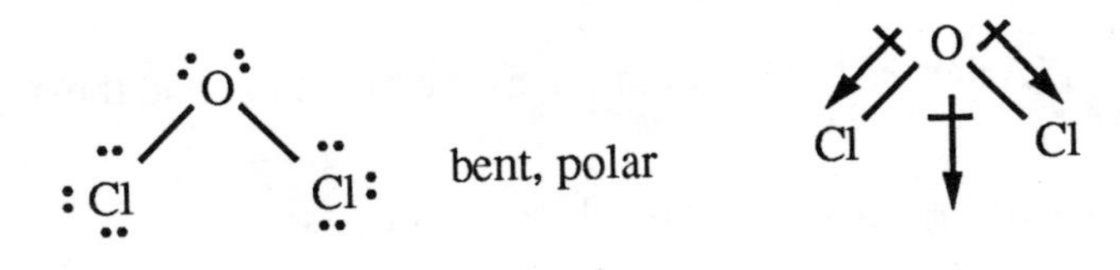

Molecule is polar because the O — Cl bond dipoles don't cancel. The resultant dipole moment is shown in the drawing.

Br_3^- has 3(7) + 1 = 22 valence electrons.

$[:\ddot{Br}—\ddot{Br}—\ddot{Br}:]^-$ linear, non-polar

All bonds are non-polar and molecule is non-polar.

BeH_2 has 2 + 2(1) = 4 valence electrons.

H — Be — H linear, non-polar

Be — H bond dipoles are equal and point in opposite directions. They cancel each other. BeH_2 is non-polar.

BH_2^- has 3 + 2 + 1 = 6 valence electrons.

bent, polar
Bond dipoles do not cancel.
BH_2^- is polar.

Note: All four species contain three atoms. They have different structures because the number of lone pairs of electrons around the central atom are different in each case.

b. BCl_3 has 3 + 3(7) = 24 valence electrons.

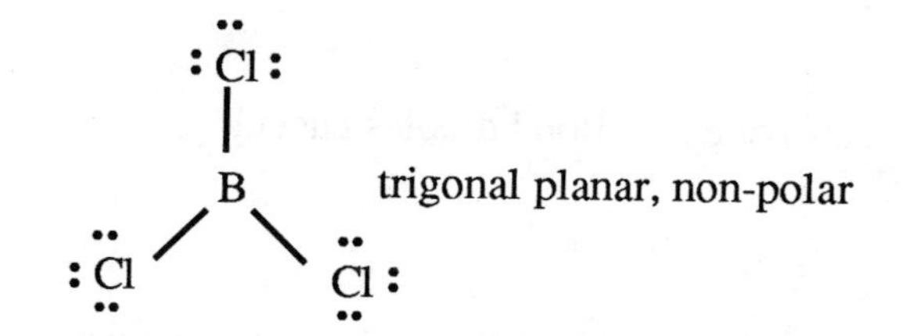

trigonal planar, non-polar   Bond dipoles cancel

NF_3 has 5 + 3(7) = 26 valence electrons.

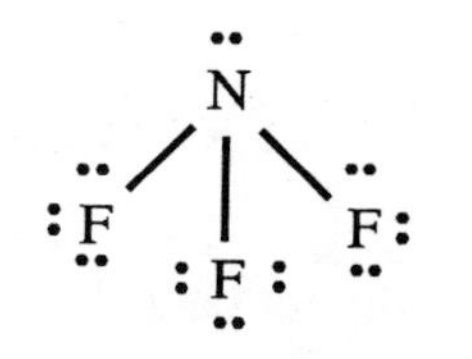

trigonal pyramid, polar   Bond dipoles do not cancel.

IF_3 has 7 + 3(7) = 28 valence electrons.

T-shaped, polar   Bond dipoles do not cancel.

Note: Each molecule contains the same number of atoms, but the structures are different because of differing numbers of lone pairs around each cental atom.

c. CF_4 has 4 + 4(7) = 32 valence electrons.

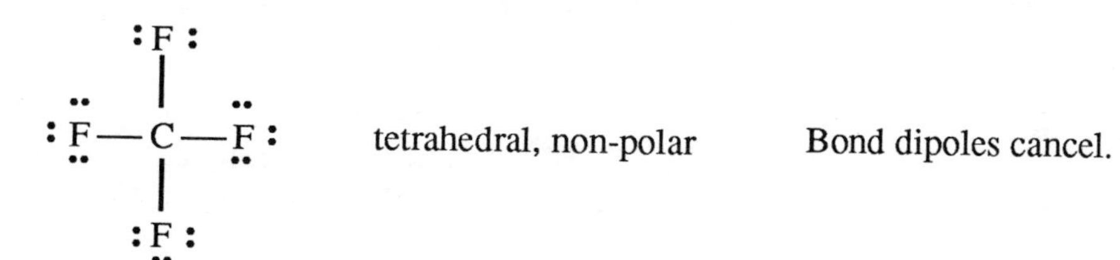

tetrahedral, non-polar   Bond dipoles cancel.

SeF_4 has 6 + 4(7) = 34 valence electrons.

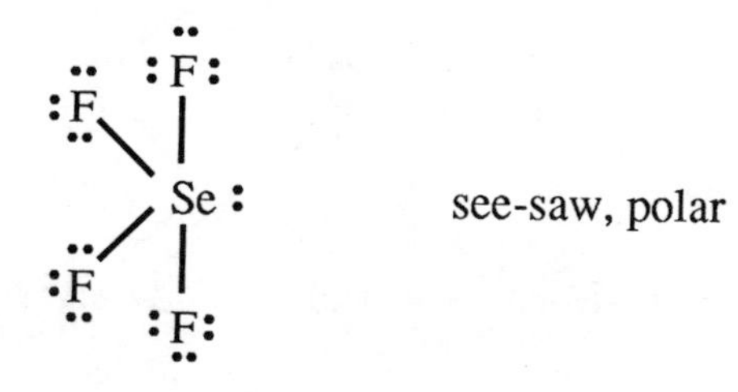

see-saw, polar   Bond dipoles do not cancel.

XeF_4 has 8 + 4(7) = 36 valence electrons.

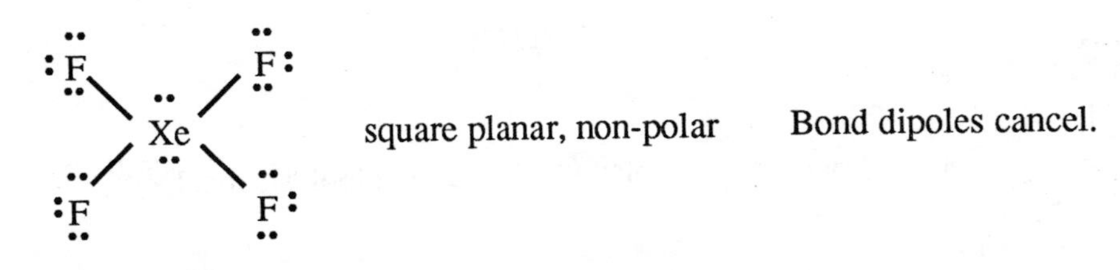

square planar, non-polar   Bond dipoles cancel.

Again, each molecule has the same number of atoms, but a different structure because of differing numbers of electrons around the central atom.

d. IF_5 has 7 + 5(7) = 42 valence electrons.

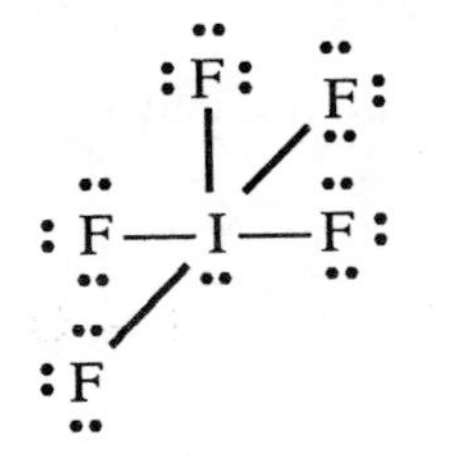

square pyramid, polar   Bond dipoles do not cancel.

AsF_5 has 5 + 5(7) = 40 valence electrons.

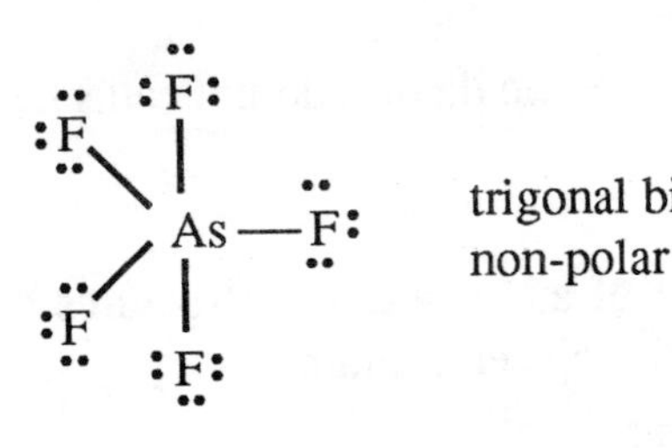

trigonal bipyramid, non-polar   Bond dipoles cancel.

Yet again, the molecules have the same number of atoms, but different structures because of the presence of differing numbers of lone pairs.

64. a.

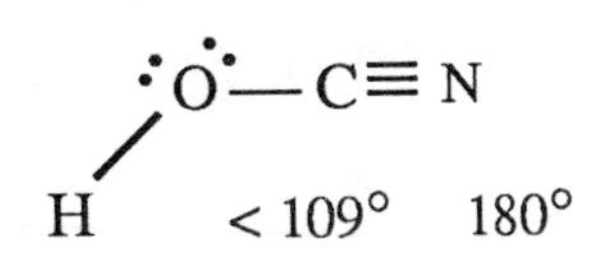

polar

b.

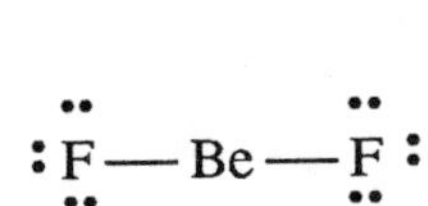

linear, non-polar

c.

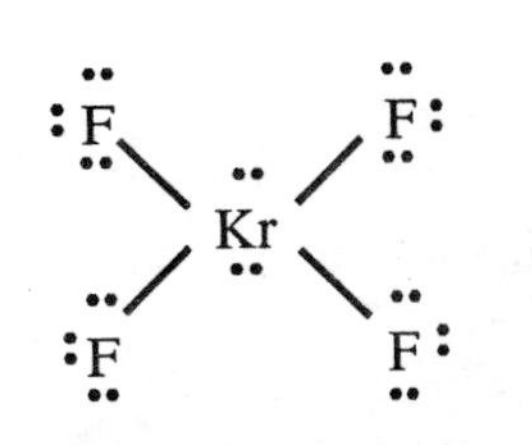

square planar, non-polar

d.

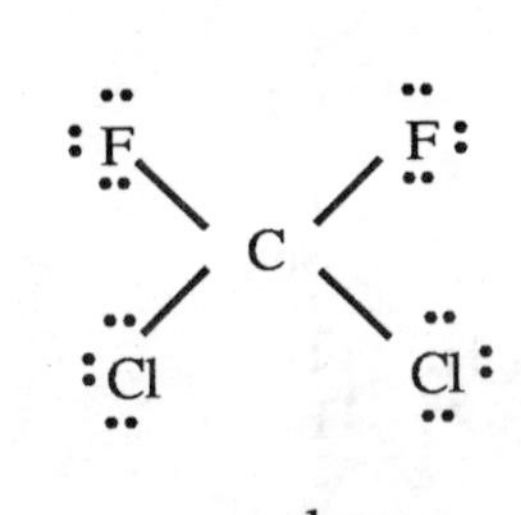

polar

e. H_2NNH_2

```
  ••    ••
H—N——N—H
  |    |
  H    H
```

polar

f. H_2CO

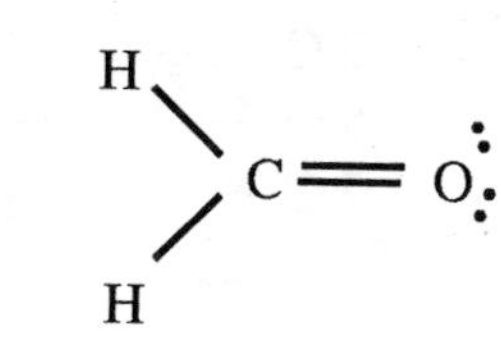

polar

ADDITIONAL EXERCISES

65. a.

```
   H   H   H
   |   |   |
H—C—C—C—H + 5 O═O ⟶ 3 O═C═O + 4 H—O—H
   |   |   |
   H   H   H
```

Bonds broken:

2 C—C	(347 kJ/mol)
8 C—H	(413 kJ/mol)
5 O═O	(495 kJ/mol)

Bonds made:

3 × 2 C═O	(799 kJ/mol)
4 × 2 H—O	(467 kJ/mol)

$\Delta H = 2(347) + 8(413) + 5(495) - [6(799) + 8(467)] = -2057$ kJ

b.

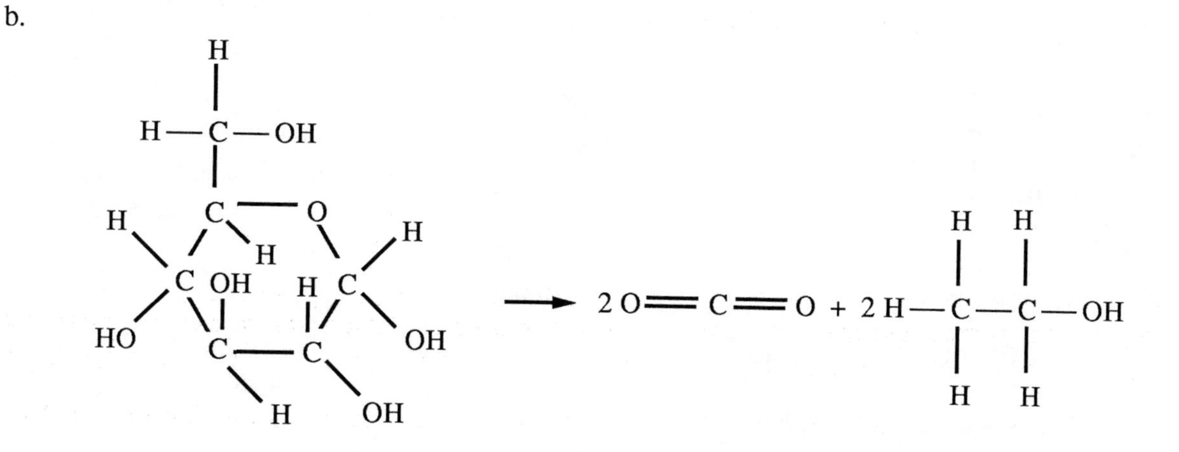

The molecules are complicated enough that it will be easier to break all bonds in glucose and make all the bonds in CO_2 and the alcohol.

Bonds broken:

5 C—C	(347 kJ/mol)
7 C—O	(358 kJ/mol)
5 O—H	(467 kJ/mol)
7 C—H	(413 kJ/mol)

Bonds made:

2 × 2 C=O	(799 kJ/mol)
2 × 5 C—H	(413 kJ/mol)
2 C—O	(358 kJ/mol)
2 O—H	(467 kJ/mol)
2 C—C	(347 kJ/mol)

$\Delta H = 5(347) + 7(358) + 5(467) + 7(413)$

$- [4(799) + 10(413) + 2(358) + 2(467) + 2(347)] = -203$ kJ

66. a. I.

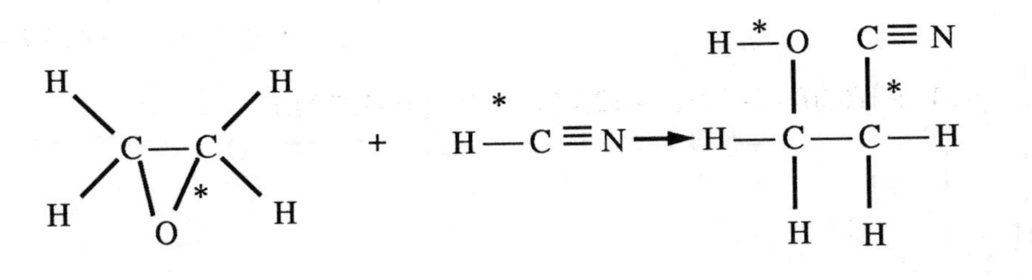

Bonds broken: (*)

1 C—C	(358 kJ)
1 H—C	(413 kJ)

Bonds made: (*)

1 O—H	(467 kJ)
1 C—C	(347 kJ)

$\Delta H_I = 358$ kJ + 413 kJ - [467 kJ + 347 kJ] = -43 kJ

II.

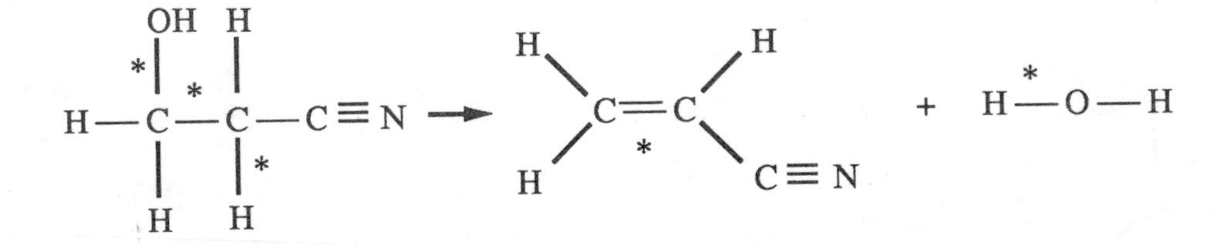

Bonds broken: (*)

1 C—O	(358 kJ/mol)
1 C—H	(413 kJ/mol)
1 C—C	(347 kJ/mol)

Bonds made: (*)

1 H—O	(467 kJ/mol)
1 C=C	(614 kJ/mol)

$\Delta H_{II} = 358$ kJ + 413 kJ + 347 kJ - [467 kJ + 614 kJ] = +37 kJ

$\Delta H = \Delta H_I + \Delta H_{II} = -43$ kJ + 37 kJ = -6 kJ

b.

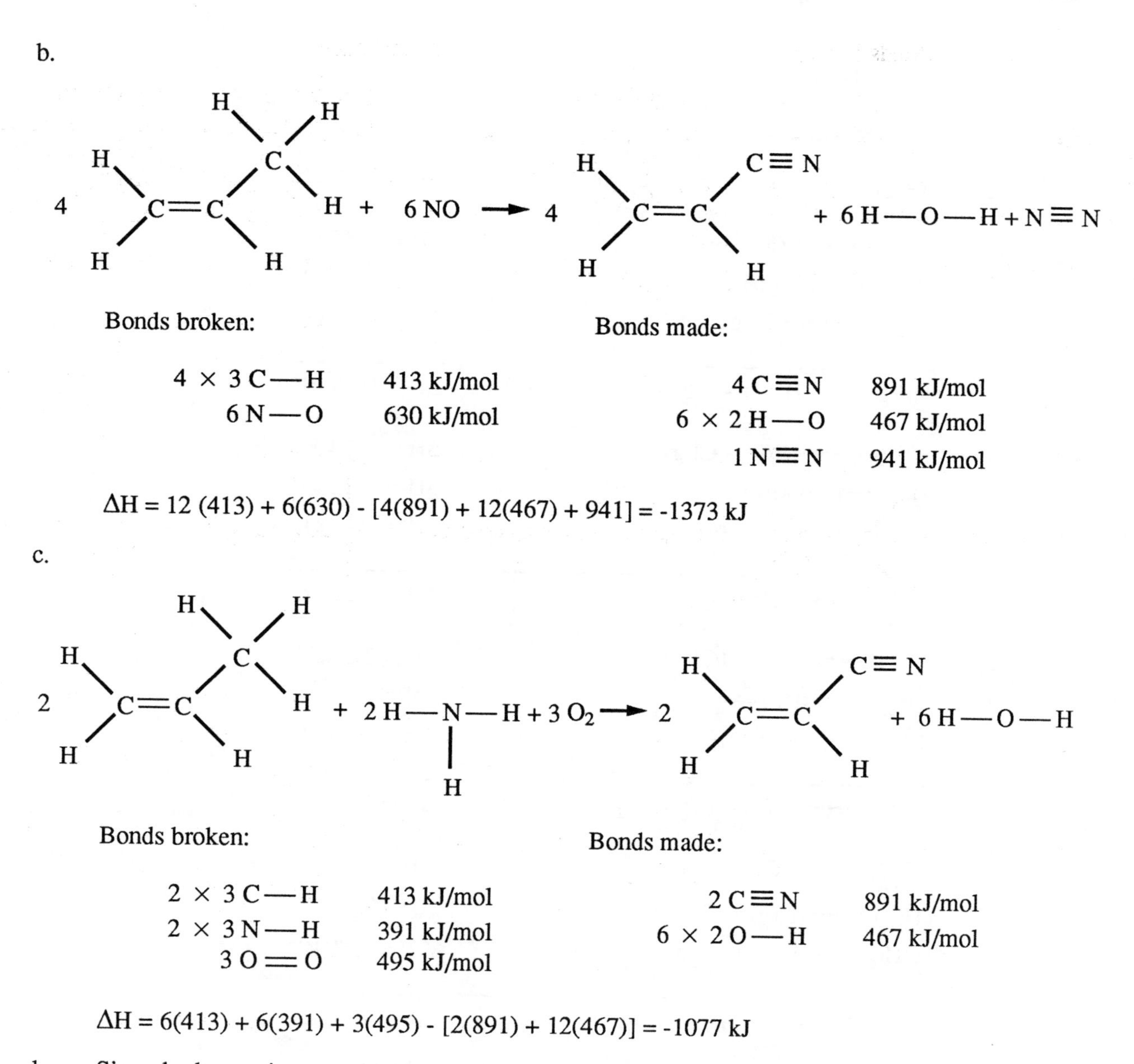

Bonds broken:

4×3 C—H	413 kJ/mol
6 N—O	630 kJ/mol

Bonds made:

4 C≡N	891 kJ/mol
6×2 H—O	467 kJ/mol
1 N≡N	941 kJ/mol

$\Delta H = 12\,(413) + 6(630) - [4(891) + 12(467) + 941] = -1373$ kJ

c.

Bonds broken:

2×3 C—H	413 kJ/mol
2×3 N—H	391 kJ/mol
3 O=O	495 kJ/mol

Bonds made:

2 C≡N	891 kJ/mol
6×2 O—H	467 kJ/mol

$\Delta H = 6(413) + 6(391) + 3(495) - [2(891) + 12(467)] = -1077$ kJ

d. Since both reactions are highly exothermic the high temperature is not needed to provide energy. It must be necessary for some other reason. This will be discussed in Chapter 12.

67.

$H_2N—NH_2 \longrightarrow 2\ N(g) + 4\ H(g)$

$\Delta H° = D_{N\text{-}N} + 4\ D_{N\text{-}H} = D_{N\text{-}N} + 4(388.9)$

$\Delta H° = 2\Delta H_f°(N) + 4\ \Delta H_f°(H) - \Delta H_f°(N_2H_4) = 2(472.7) + 4(216.0) - 95.4$

$\Delta H° = 1714.0\ \text{kJ} = D_{N\text{-}N} + 4(388.9)$

$D_{N\text{-}N} = 158.4$ kJ/mol (160 kJ/mol in Table 8.4)

68. ΔH_f° for H(g) is ΔH° for the reaction:

$1/2\ H_2(g) \longrightarrow H(g)$

$\Delta H_f^\circ(H) = D_{H\text{-}H}/2 = 432/2 = 216$ kJ/mol

69. a.

$HF(g) \longrightarrow H(g) + F(g)$	$\Delta H = 565$ kJ
$H(g) \longrightarrow H^+(g) + e^-$	$\Delta H = 1312$ kJ
$F(g) + e^- \longrightarrow F^-(g)$	$\Delta H = -328$ kJ
$HF(g) \longrightarrow H^+(g) + F(g)^-$	$\Delta H = 1549$ kJ

b.

$HCl(g) \longrightarrow H(g) + Cl(g)$	$\Delta H = 427$ kJ
$H(g) \longrightarrow H^+(g) + e^-$	$\Delta H = 1312$ kJ
$Cl(g) + e^- \longrightarrow Cl^-(g)$	$\Delta H = -349$ kJ
$HCl(g) \longrightarrow H^+(g) + Cl^-(g)$	$\Delta H = 1390.$ kJ

c.

$HI(g) \longrightarrow H(g) + I(g)$	$\Delta H = 295$ kJ
$H(g) \longrightarrow H^+(g) + e^-$	$\Delta H = 1312$ kJ
$I(g) + e^- \longrightarrow I^-(g)$	$\Delta H = -295$ kJ
$HI(g) \longrightarrow H^+(g) + I^-(g)$	$\Delta H = 1312$ kJ

d.

$H_2O(g) \longrightarrow OH(g) + H(g)$	$\Delta H = 467$ kJ
$H(g) \longrightarrow H^+(g) + e^-$	$\Delta H = 1312$ kJ
$OH(g) + e^- \longrightarrow OH^-(g)$	$\Delta H = -180$ kJ
$H_2O(g) \longrightarrow H^+(g) + OH^-(g)$	$\Delta H = 1599$ kJ

70. a. $Na^+(g) + Cl^-(g) \longrightarrow NaCl(s)$

b. $NH_4^+(g) + Br^-(g) \longrightarrow NH_4Br(s)$

c. $Mg^{2+}(g) + S^{2-}(g) \longrightarrow MgS(s)$

d. $1/2\ O_2(g) \longrightarrow O(g)$

e. $O_2(g) \longrightarrow 2\ O(g)$

71. Ionic solids can be characterized as being held together by strong-omnidirectional forces.

I. For electrical conductivity, charged species must be free to move. In ionic solids the charged ions are held rigidly in place. Once the forces are disrupted (melting or dissolution) the ions can move about (conduct).

II. Melting and boiling disrupts the attractions of the ions for each other. If the forces are strong it will take a lot of energy (high temp.) to accomplish this.

III. If we try to bend a piece of material, the atoms/ions must slide about each other. For an ionic solid the following might happen:

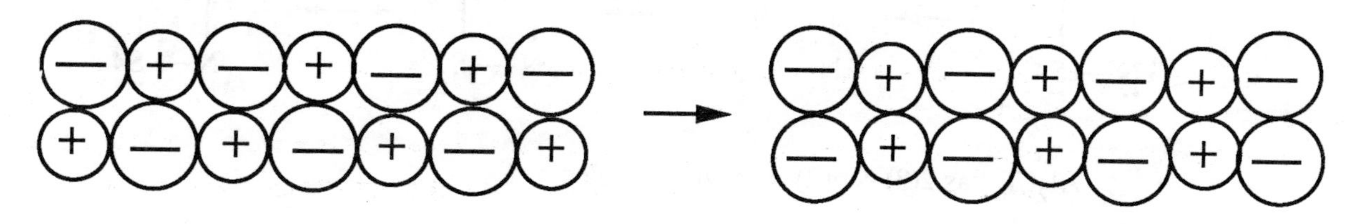

Just as the layers begin to slide, there will be very strong repulsions causing the solid to snap across a fairly clean plane.

IV. Polar molecules are attracted to ions and can break up the lattice.

These properties and their correlation to chemical forces will be discussed in detail in Chapter 10 and 11.

72. The stable species are:

a. SO_4^{2-}: We can't draw a Lewis structure that obeys the octet rule for SO_4. The two extra electrons (-2 charge) complete octets.

b. PF_5: N is too small and doesn't have low energy d-orbitals to expand its octet.

c. SF_6: O is too small and doesn't have low-energy d-orbitals to expand its octet.

d. BH_4^-: BH_3 doesn't obey octet rule.

e. MgO: In MgF we would have to have Mg^+F^- or Mg^{2+} F^{2-}. Neither Mg^+ nor F^{2-} are stable.

f. CsCl: Cs doesn't form Cs^{2+} ion.

g. KBr: Br doesn't form Br^{2-} ion.

73. S_2Cl_2 has 2(6) + 2(7) = 26 valence electrons.

: Cl — S — S — Cl :

SCl_2 has 6 + 2(7) = 20 valence electrons.

: Cl — S — Cl :

74. S_2N_2 has 2(6) + 2(5) = 22 valence electrons.

S — N
| |
N — S

Ring of alternating S and N atoms.
We've used 8 e^-; 14 e^- left over

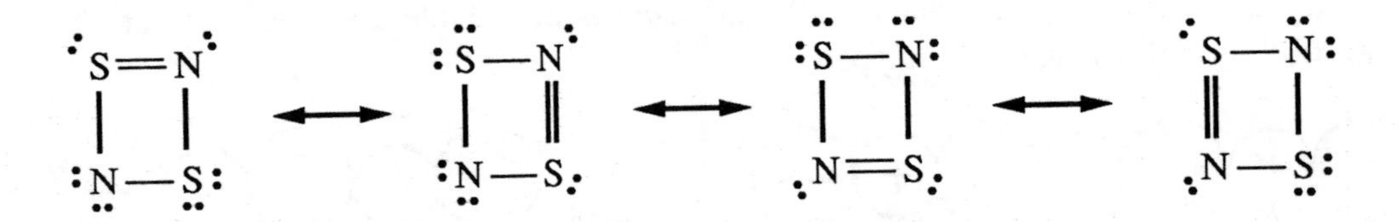

75. a. Al_2Cl_6 has 2(3) + 6(7) = 48 valence electrons.

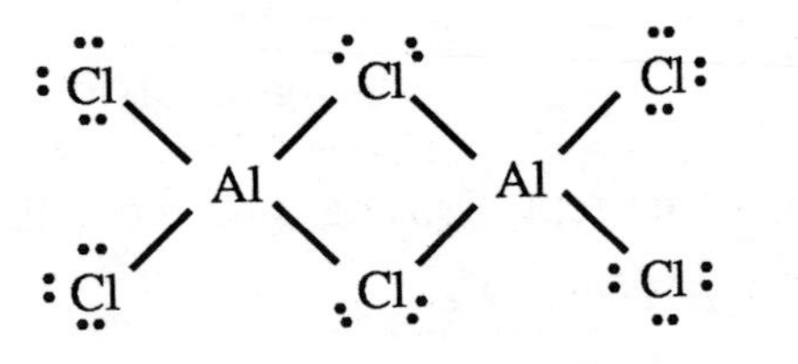

b. There are 4 pairs of electrons about each Al, we would predict the bond angles to be close to a tetrahedral angle of 109.5°.

c. non-polar

76. CO_3^{2-} has 4 + 3(6) + 2 = 24 valence electrons.

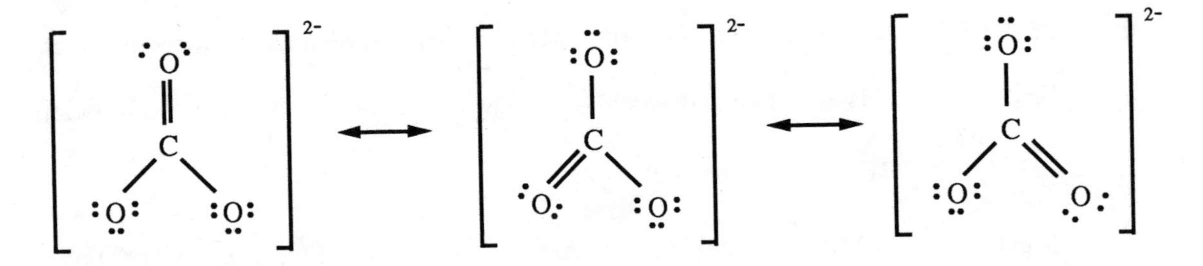

HCO_3^- has 1 + 4 + 3(6) + 1 = 24 valence electrons.

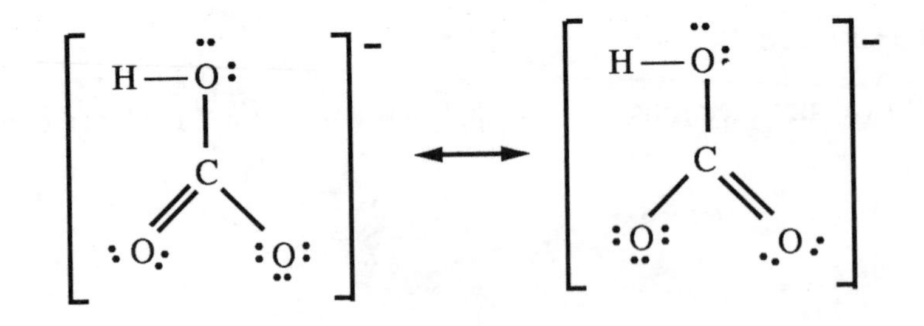

H_2CO_3 has 2(1) + 4 + 3(6) = 24 valence electrons.

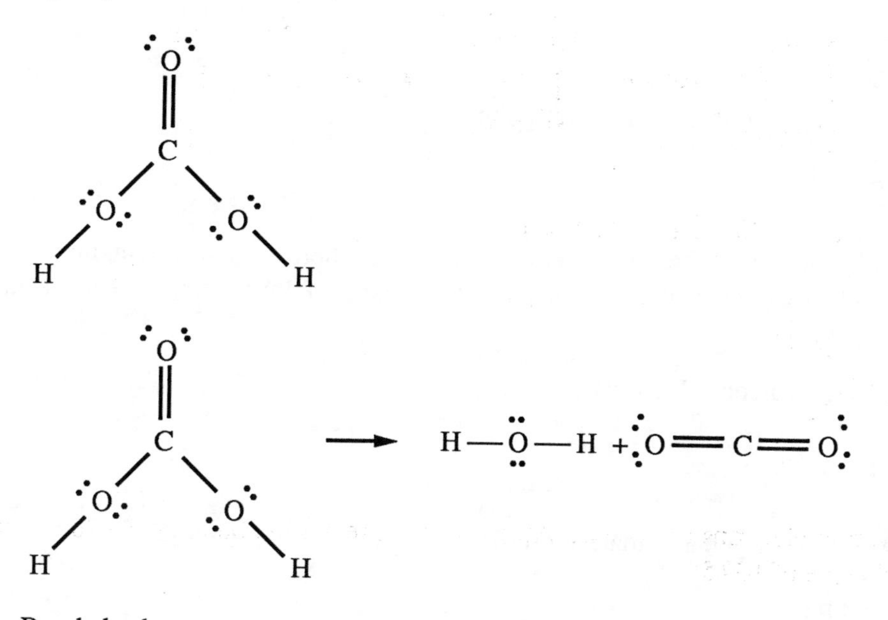

Bonds broken:

2 C—O 358 kJ/mol

Bonds made:

1 C=O 799 kJ/mol

ΔH = 2(358) - (799) = -83 kJ

The carbon-oxygen double bond is stronger than two carbon-oxygen single bonds.

77. Resonance is possible in CO_3^{2-}. (See previous exercise) We would expect all C—O bonds to be equal and intermediate in strength and length to a single and a double bond. The bond length of 136 pm is consistent with this view of the structure.

78. a. SiF_4 has 4 + 4(7) = 32 valence electrons.

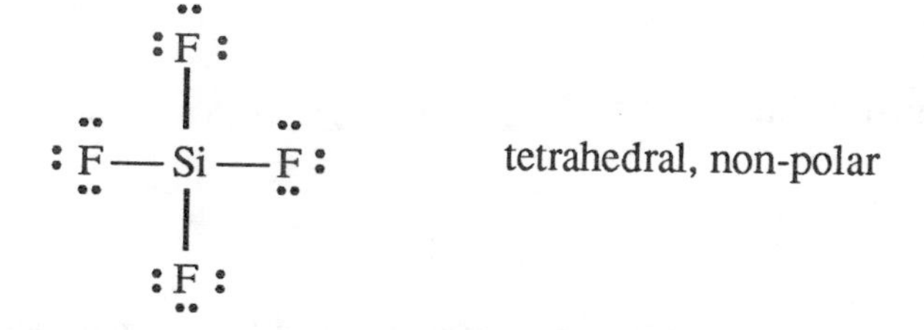

tetrahedral, non-polar

b. SeF_4 has 6 + 4(7) = 34 valence electrons.

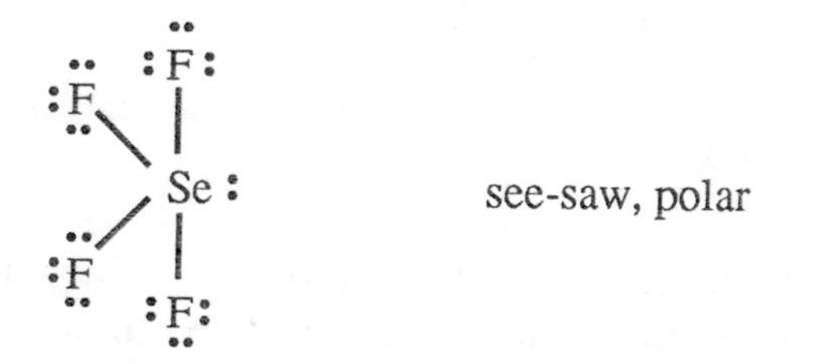

see-saw, polar

c. KrF_4 has 8 + 4(7) = 36 valence electrons.

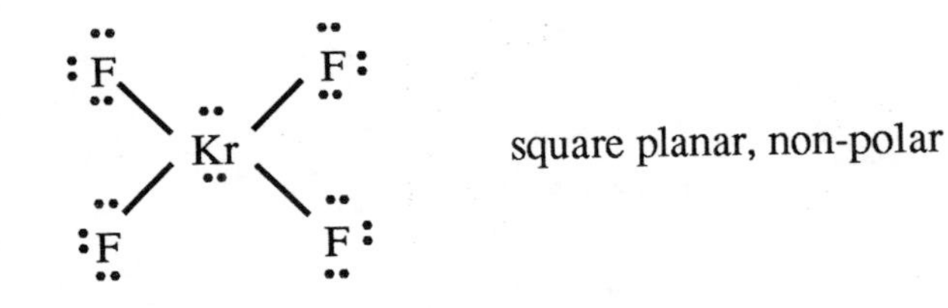

square planar, non-polar

Although all three molecules have the same number of atoms, there is a different number of valence electrons in each one. This results in differing numbers of lone pairs on each central atom; hence, different structures.

79. a. BF_3 has 3 + 3(7) = 24 valence electrons.

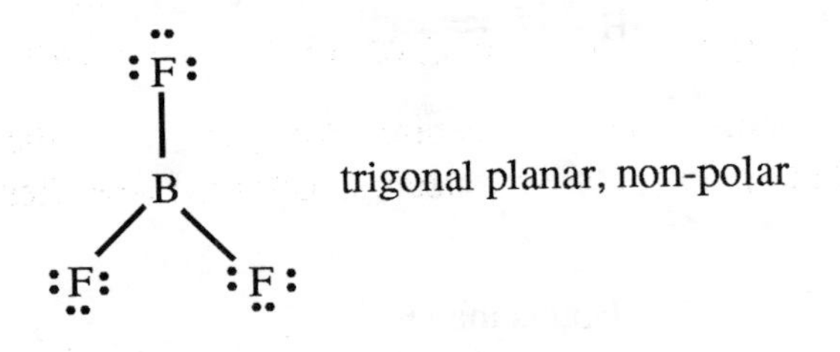

trigonal planar, non-polar

b. PF_3 has 5 + 3(7) = 26 valence electrons.

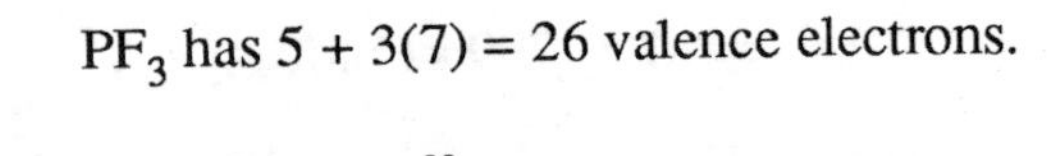

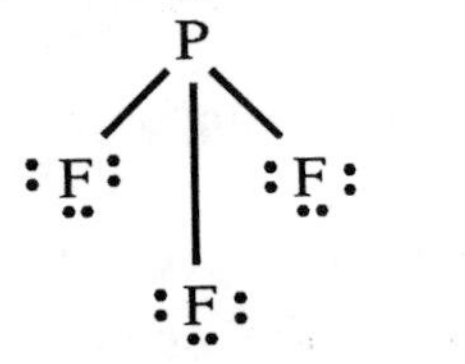

trigonal pyramid, polar

c. BrF_3 has 7 + 3(7) = 28 valence electrons.

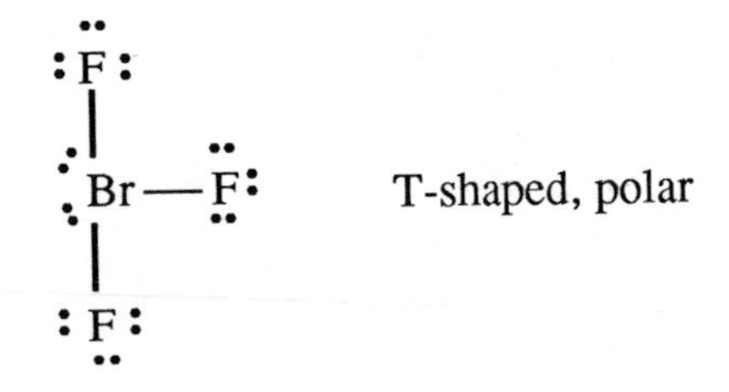

T-shaped, polar

Again we see a series of molecules with the same number of atoms, but different numbers of lone pairs: They have different structures.

80. The general structure of the trihalide ions is:

:X—X—X:

Bromine and iodine are large enough and have low energy, empty d-orbitals to accommodate the expanded octet. Flourine is small and its valence shell contains only 2s and 2p orbitals (4 orbitals) and cannot expand its octet. The d orbitals in F are 3d; they are too high in energy compared to 2s and 2p to be used.

81.

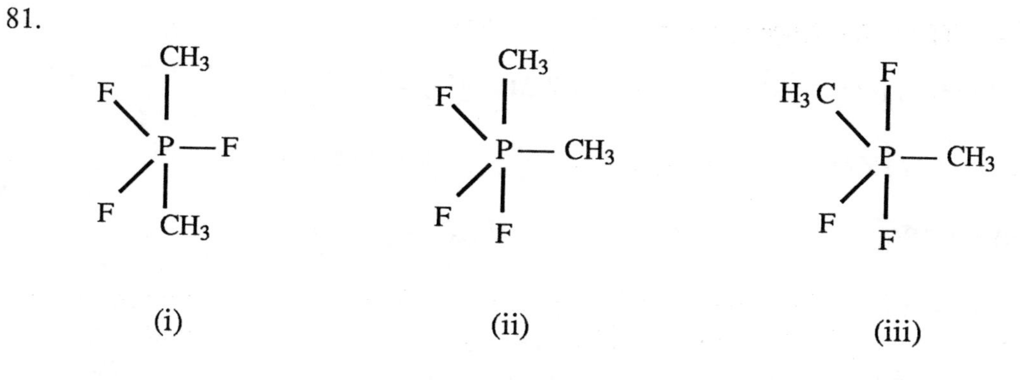

(i) (ii) (iii)

(i) is non-polar

(ii) and (iii) are polar with (ii) probably the more polar. Thus, dipole moments should help distinguish the three isomers; particularly in distinguishing (i) from the other two.

82. The polar bonds are symmetrically arranged about the central atoms and all of the individual bond dipoles cancel to give no net dipole moment for each molecule, i.e., the molecules are non-polar.

83. a.

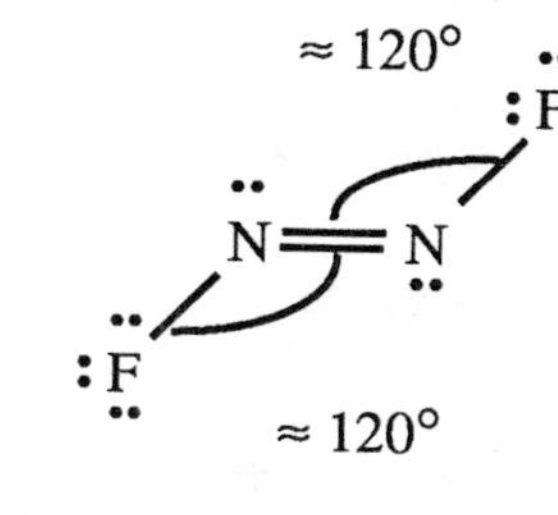

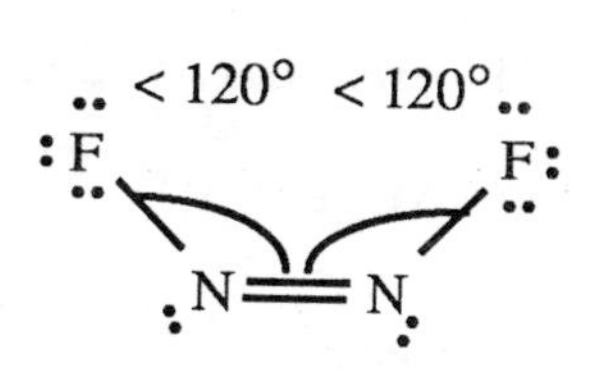

b. non-polar polar

84. Yes, each structure has the same number of effective pairs around the central atom. (We count a multiple bond as a single group of electrons.)

85. This is a very weak bond. The structure

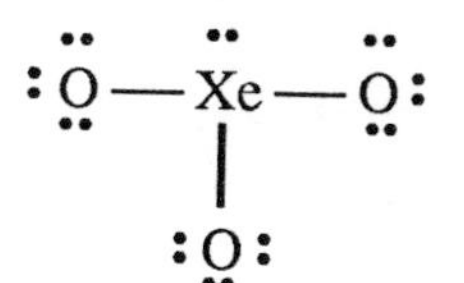

might be more accurate. It is difficult to think of a double bond as being so weak. This illustrates that no model is totally effective in describing the bonding in all molecules. We need to recognize that all models have limitations and that we can't apply all models to all molecules.

CHALLENGE PROBLEMS

86. C≡O 1072 kJ/mol and N≡N 941 kJ/mol

CO is polar while N_2 is non-polar. This may lead to a great reactivity.

87. Let us look at the complete cycle for Li_2S. (Assume all values are ±1 kJ).

$2\ Li(s) \rightarrow 2\ Li(g)$	$2\ \Delta H_{sub}(Li) = 2(162)$ kJ
$2\ Li(g) \rightarrow 2\ Li^+(g) + 2\ e^-$	$2\ I = 2(520)$ kJ
$S(s) \rightarrow S(g)$	$\Delta H_{sub}(S) = 277$ kJ
$S(g) + e^- \rightarrow S^-(g)$	$EA_1 = -200$ kJ
$S^-(g) + e^- \rightarrow S^{2-}(g)$	$EA_2 = ?$
$2\ Li^+(g) + S^{2-}(g) \rightarrow Li_2S(s)$	$LE = -2472$ kJ
$2\ Li(s) + S(s) \rightarrow Li_2S(s)$	$\Delta H_f^\circ = -500$ kJ

$\Delta H_f^\circ = 2\ \Delta H_{sub}(Li) + 2I + \Delta H_{sub}(S) + EA_1 + EA_2 + LE$

$-500 = -1031 + EA_2$; $EA_2 = +531$ kJ

For each salt: $\Delta H_f^\circ = 2\ \Delta H_{sub}(M) + 2I + 277 - 200 + EA_2 + LE$

Na_2S: $-365 = 2(108) + 2(496) + 277 - 200 - 2203 + EA_2$; $EA_2 = +553$ kJ

K_2S: $-381 = 2(90) + 2(419) + 277 - 200 - 2052 + EA_2$; $EA_2 = +576$ kJ

Rb_2S: $-361 = 2(82) + 2(409) + 277 - 200 - 1949 + EA_2$; $EA_2 = +529$ kJ

Cs_2S: $-360 = 2(78) + 2(382) + 277 - 200 - 1850 + EA_2$; $EA_2 = +493$ kJ

We get values from 493 to 576 kJ

The mean value is $\dfrac{531 + 553 + 576 + 529 + 493}{5} = 536$ kJ

We can represent the results as $EA_2 = 540$ kJ ± 50 kJ

88. The Si—Cl bonds are more polar than the C—Cl bonds.

89. If no resonance:

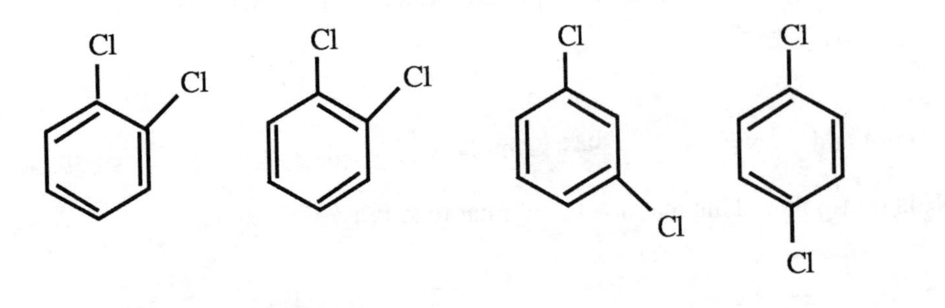

4 different molecules

With resonance:

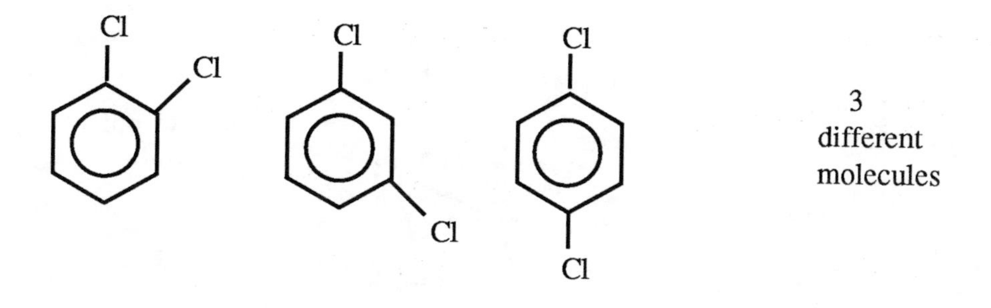

If resonance is present, we can't distinguish between a single and double bond between adjacent carbons that have a chlorine attached, since all carbon-carbon bonds are equivalent. That only 3 isomers are observed provides evidence for the existence of resonance.

90. If we can draw resonance forms for the anion after loss of H^+, we can argue that the extra stability of the anion causes the proton to be more readily lost (more acidic).

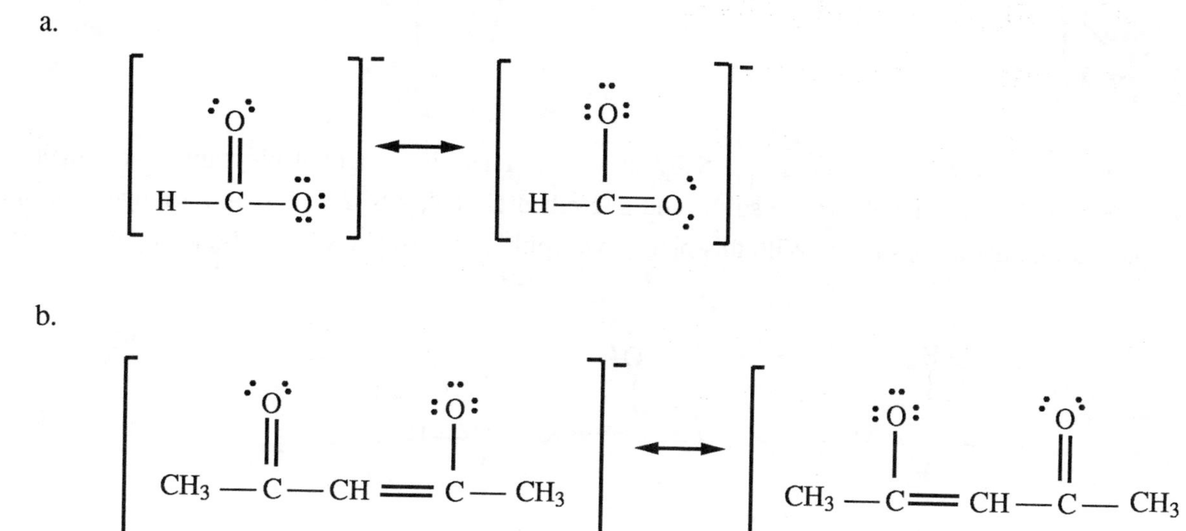

c.

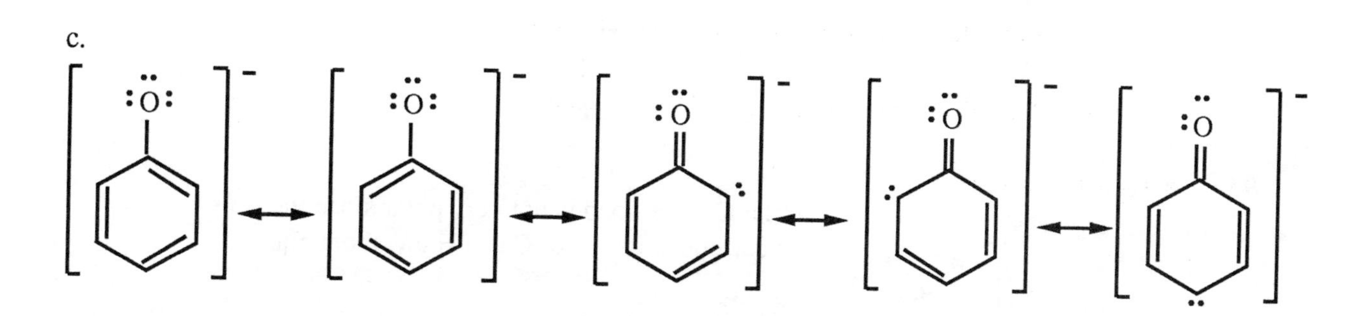

In all 3 cases, extra resonance forms can be drawn for the anion that are not possible when the H^+ is present.

91. SF_2 has 6 + 2(7) = 20 valence electrons.

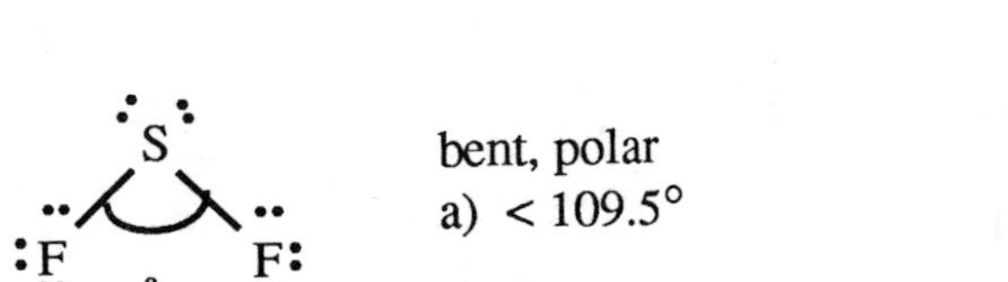

SF_4 has 6 + 4(7) = 34 valence electrons.

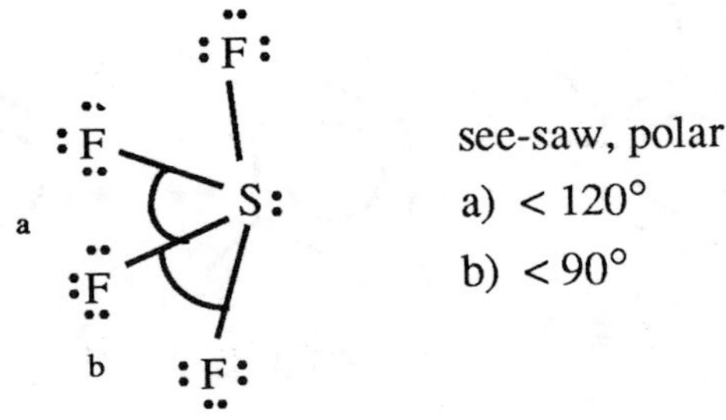

SF_6 has 6 + 6(7) = 48 valence electrons.

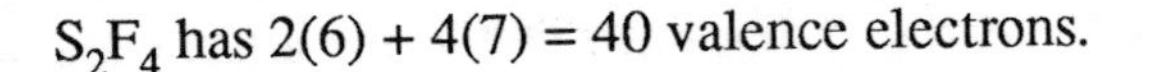

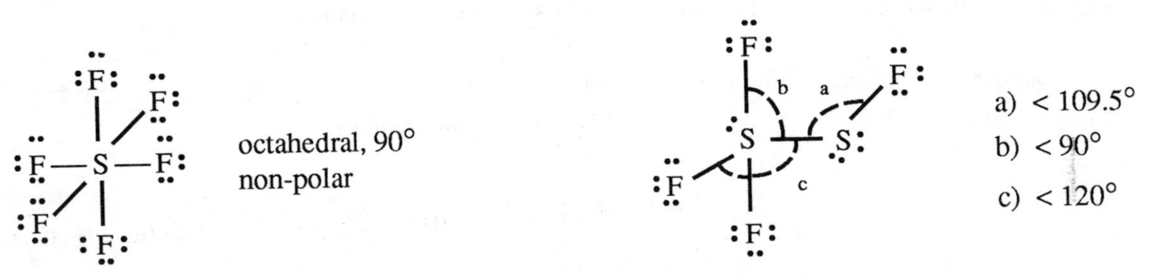

There is no easy description for the S_2F_4 structure. There is a trigonal bipyramidal arrangement (3 F, 1 S, 1 lone pair) about one sulfur and a tetrahedral (1 F, 1 S, 2 lone pairs) arrangement of e^- pairs about the other sulfur. With this picture we could predict values for all bond angles. Polar.

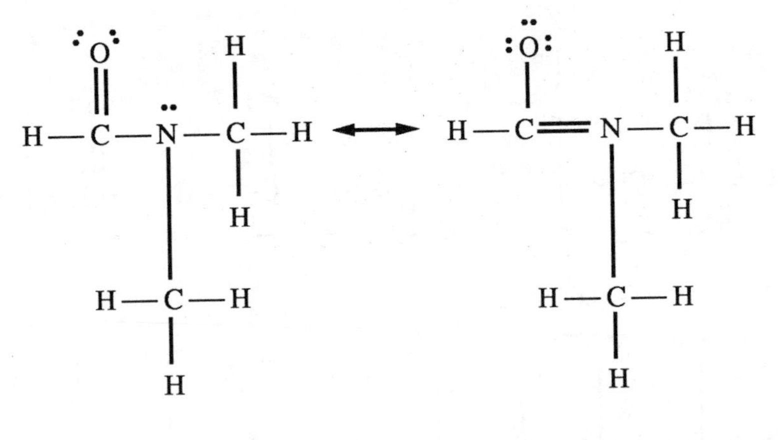

93. a.

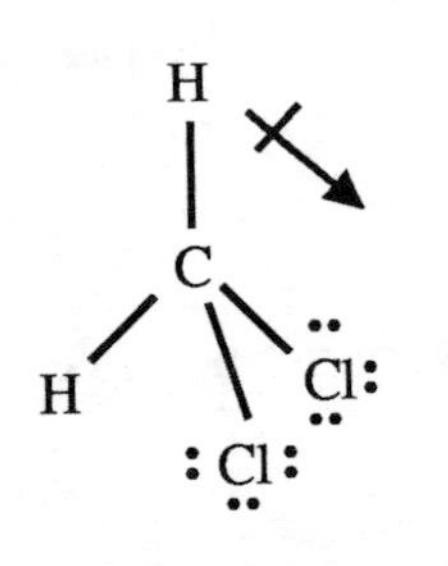

The C—H bond is not very polar since the electronegativities of C and H are about equal.

δ+ δ-
C—Cl is the charge distribution for each C—Cl bond. The two individual C—Cl bond dipoles add together to give an overall dipole moment for the molecule. The overall dipole will point from C (positive end) to the mid-point of the two Cl atoms (negative end).

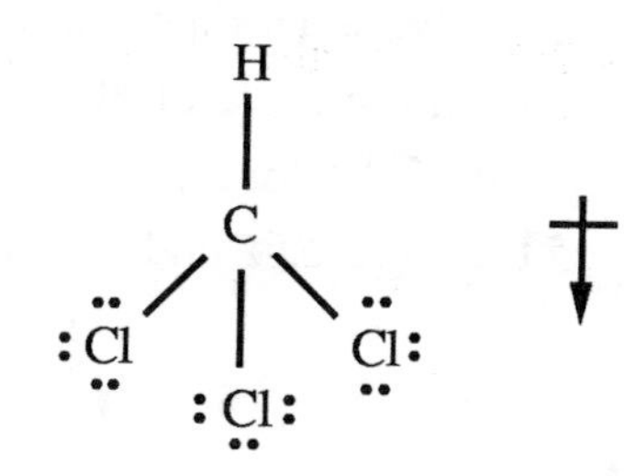

The C — H bond is essentially non-polar. The three C — Cl bond dipoles add together to give an overall dipole moment for the molecule. The overall dipole will have the negative end at the midpoint of the three chlorines and the positive end at the carbon.

CCl_4 is non-polar. CCl_4 is a tetrahedral molecule where all four C — Cl bond dipoles cancel when added together. Let's consider just the C and two of the Cl atoms. There will be a net dipole pointing in the direction of the middle of the two Cl atoms.

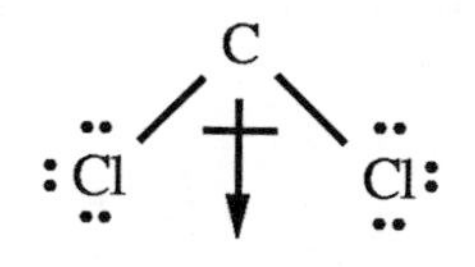

There will be an equal and opposite dipole arising from the other two Cl atoms. Combining:

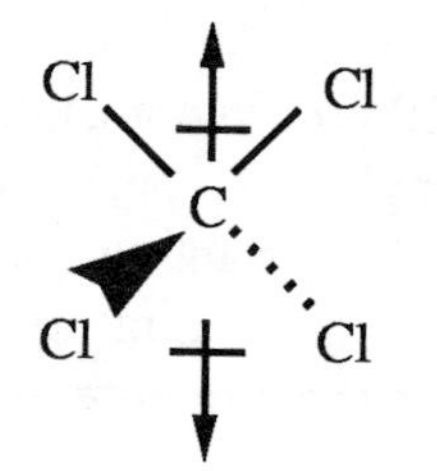

The two dipoles will cancel and CCl_4 is non-polar.

b. CO_2 is non-polar. CO_2 is a linear molecule with two equivalent bond dipoles that cancel. N_2O is polar since the bond dipoles do not cancel.

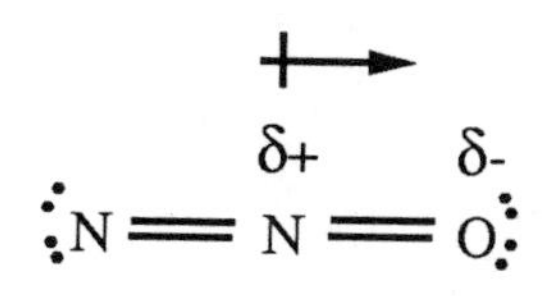

c. NH_3 is polar. The 3 N—H bond dipoles add together to give a net dipole in the direction of the lone pair. We would predict PH_3 to be non-polar on the basis of electronegativity, i.e., P—H bonds are non-polar. However, the presence of the lone pair makes the PH_3 molecule polar. The net dipole is in the direction of the lone pair and has a magntiude about one third that of the NH_3 dipole.

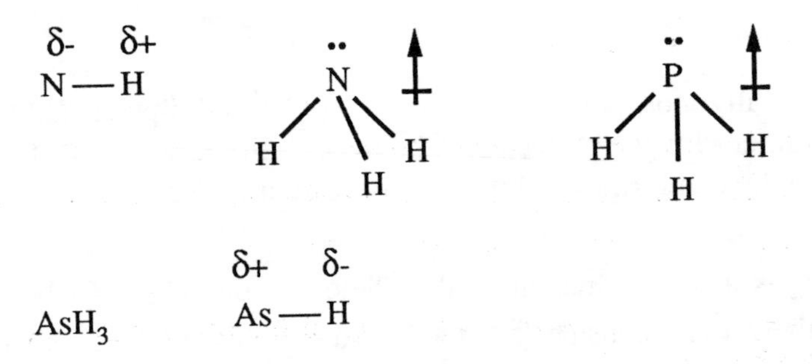

For AsH_3, the polarity arising from the lone pair is in the opposite direction as the bond dipoles. AsH_3 is slightly polar with a dipole moment one third that of PH_3 and in the direction of the lone pair.

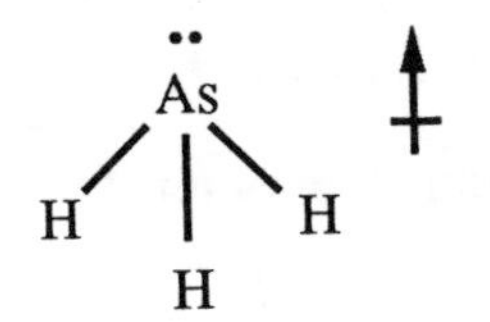

94. The nitrogen-nitrogen bond length of 112 pm is between a double (120 pm) and a triple (110 pm) bond. The nitrogen-oxygen bond length of 119 pm is between a single (147 pm) and a double bond (115 pm). The last resonance structure doesn't appear to be as important as the other two. There is no evidence from bond lengths for a nitrogen-oxygen triple bond or a nitrogen-nitrogen single bond as in the third resonance form. We can adequately describe the structure of N_2O using the resonance forms:

:N═N═O: ⟷ :N≡N—O:

Assigning formal charges for all 3 resonance forms:

:N═N═O:	⟷	:N≡N—O:	⟷	:N—N≡O:
-1 +1 0		0 +1 -1		-2 +1 +1

For: (:N═) FC = 5 - 4 - 1/2(4) = -1

(═N═) FC = 5 - 1/2(8) = +1 Same for ≡N— and —N≡

(:N—) FC = 5 - 6 - 1/2(2) = -2

(:N≡) FC = 5 - 2 - 1/2(6) = 0

(═ O:) FC = 6 - 4 - 1/2(4) = 0

(—Ö:) FC = 6 - 6 - 1/2(2) = -1

(≡ O:) FC = 6 - 2 - 1/2(6) = +1

We should eliminate N—N≡O since it has a formal charge of +1 on the most electronegative element. This is consistent with the observation that the N—N bond is between a double and triple bond and that the N—O bond is between a single and double bond.

95. a. EN is proportional to (IE - EA). The ionization energies of the noble gases are very high. This causes the EN to also be very high.

b. The electronegativity of F is greater than the electronegativity of Xe; it is also higher than the electronegativity of Kr. Thus, we might also expect F_2 to react with Kr and Rn.

CHAPTER NINE: COVALENT BONDING - ORBITALS

QUESTIONS

1. Bond energy is proportional to bond order. Bond length is inversely proportional to bond order. Bond energy and length can be measured.

2. We can determine molecular structures; sometimes directly by x-ray diffraction (Chapter 10). Our view is that if electrons are in orbitals in atoms then they must be in orbitals in molecules. However, atomic orbitals don't point in the right direction for some molecules. We had to invent the concept of hybrid orbitals. We decide what hybrid orbitals are used, once we know the structure. Structures are the facts. Hybrid orbitals come from a model to rationalize the facts.

3. Paramagnetic: Unpaired electrons are present.

   Measure the mass of a substance in the presence and absence of a magnetic field. A substance with unpaired electrons will be attracted by the magnetic field, giving an apparent increase in mass in the presence of the field. Greater number of unpaired electrons will give greater attraction and greater observed mass increase.

4. Bonding molecular orbitals have electron density between the bonded atoms and are lower in energy than the atomic orbitals from which they are formed. Anti-bonding molecular orbitals have minimal electron density between the bonded atoms and are higher in energy than the atomic orbitals from which they are formed.

EXERCISES

Localized Electron Model: Hybrid Orbitals

5. a.

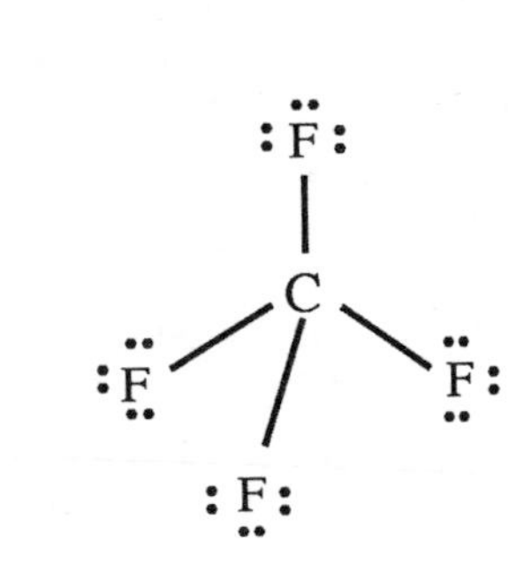

tetrahedral   109.5°
sp^3 non-polar

b.

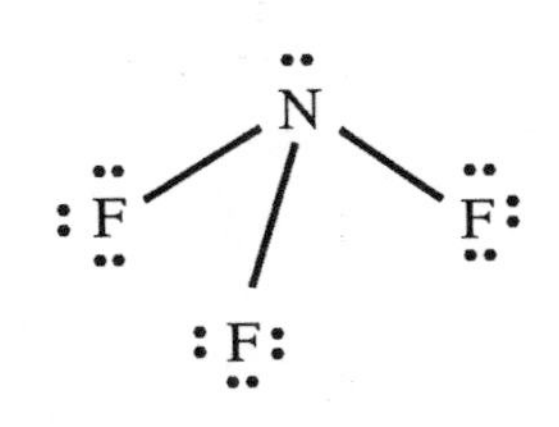

trigonal pyramid   < 109.5°
sp^3 polar

The angle should be slightly less than 109.5° because the lone pair requires more room than the bonding pairs.

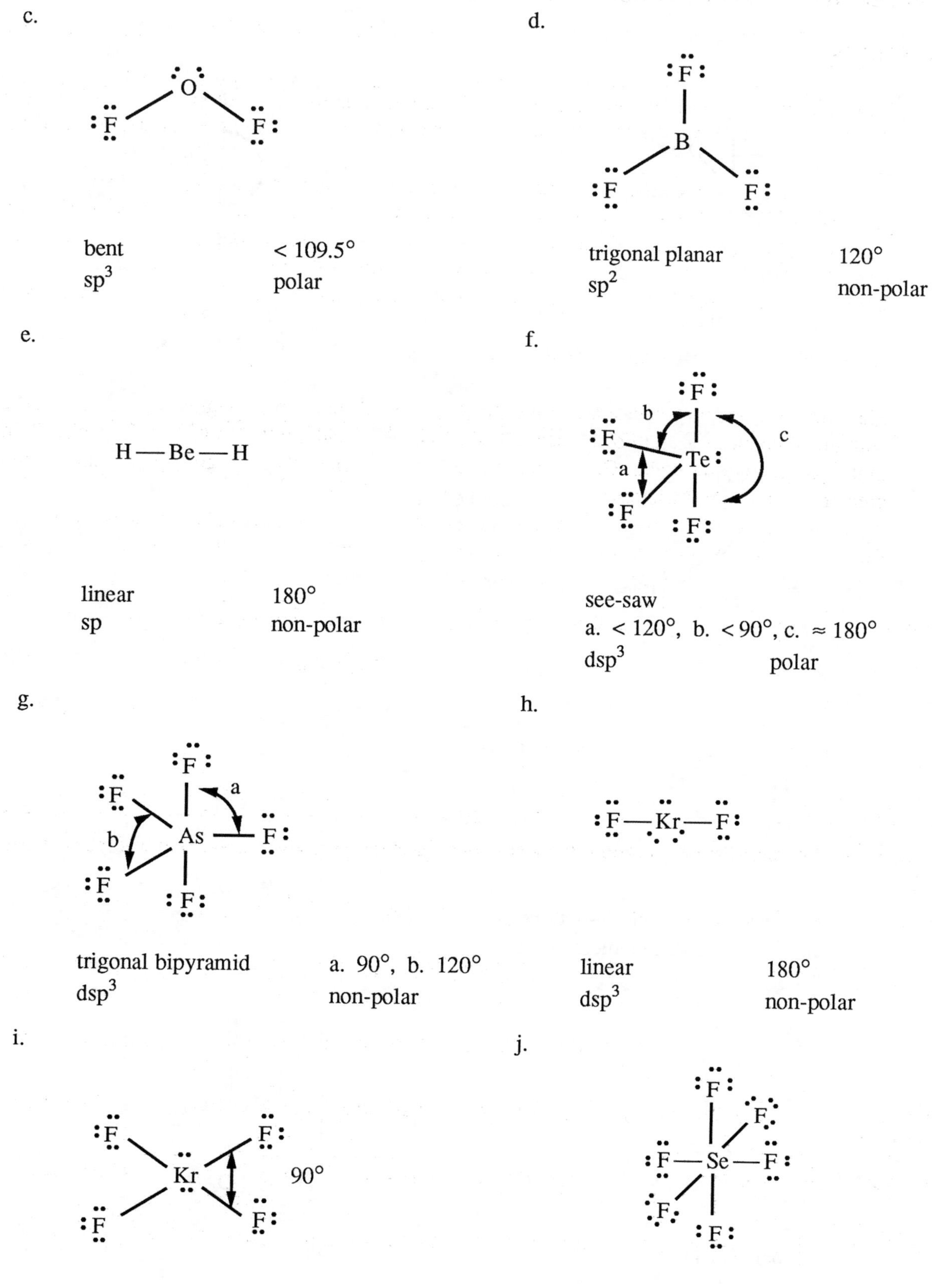
c.
O
F
F
bent
sp3
< 109.5°
polar
d.
F
B
F
F
trigonal planar
sp2
120°
non-polar
e.
H—Be—H
linear
sp
180°
non-polar
f.
F
b
F
c
Te
a
F
F
see-saw
a. < 120°, b. < 90°, c. ≈ 180°
dsp3
polar
g.
F
a
F
As
F
b
F
F
trigonal bipyramid
dsp3
a. 90°, b. 120°
non-polar
h.
F—Kr—F
linear
dsp3
180°
non-polar
i.
F
F
Kr
90°
F
F
square planar
d2sp3
90°
non-polar
j.
F
F
F—Se—F
F
F
octahedral
d2sp3
90°
non-polar

k.

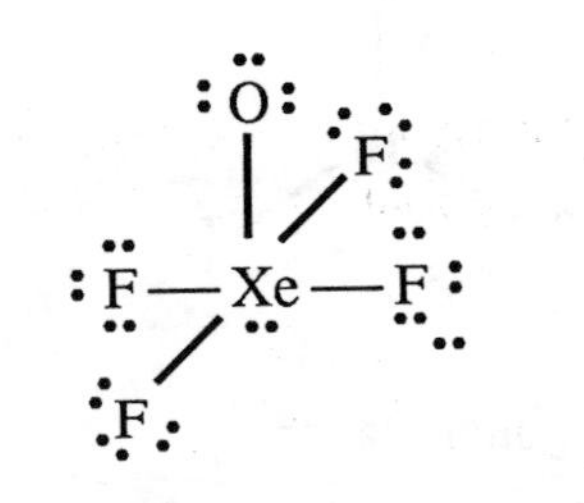

square pyramid $\approx 90°$
d^2sp^3 polar

l.

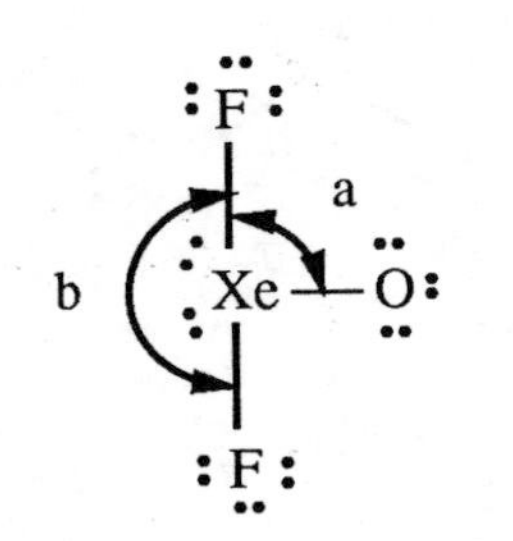

T-shaped
a. $< 90°$, b. $\approx 180°$
dsp^3 polar

m.

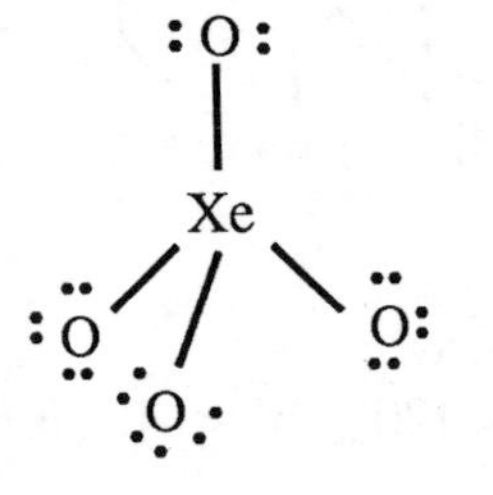

tetrahedral
109.5°
sp^3
non-polar

6. a.

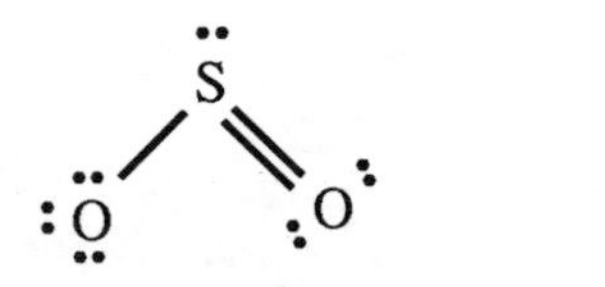

bent
120°
sp^2

Only one resonance form is shown.

Resonance does not change the position of the atoms. Realizing that the presence of resonance makes both S — O bond lengths equal, we can predict the geometry from only one of the resonance structures.

b. c.

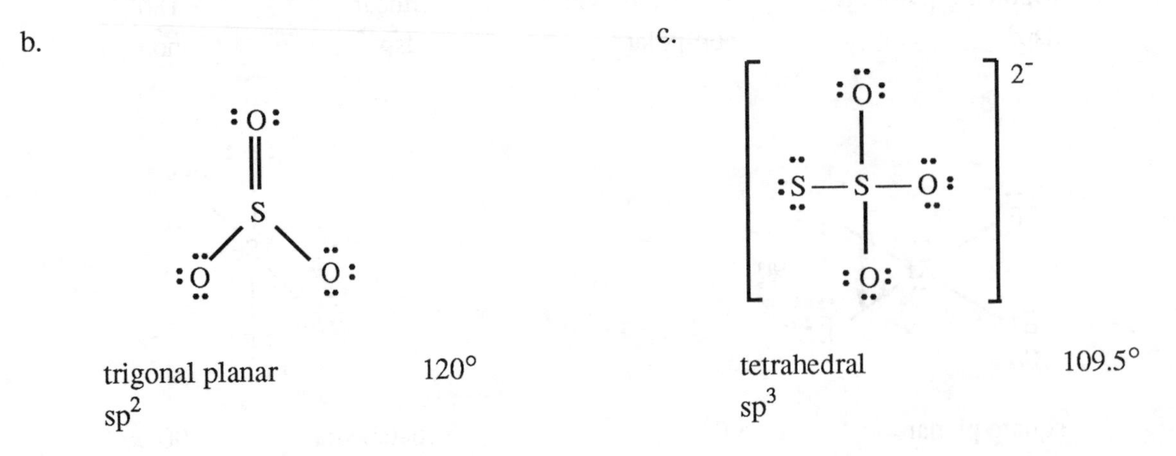

trigonal planar 120°
sp^2

Only one resonance form is shown.

tetrahedral 109.5°
sp^3

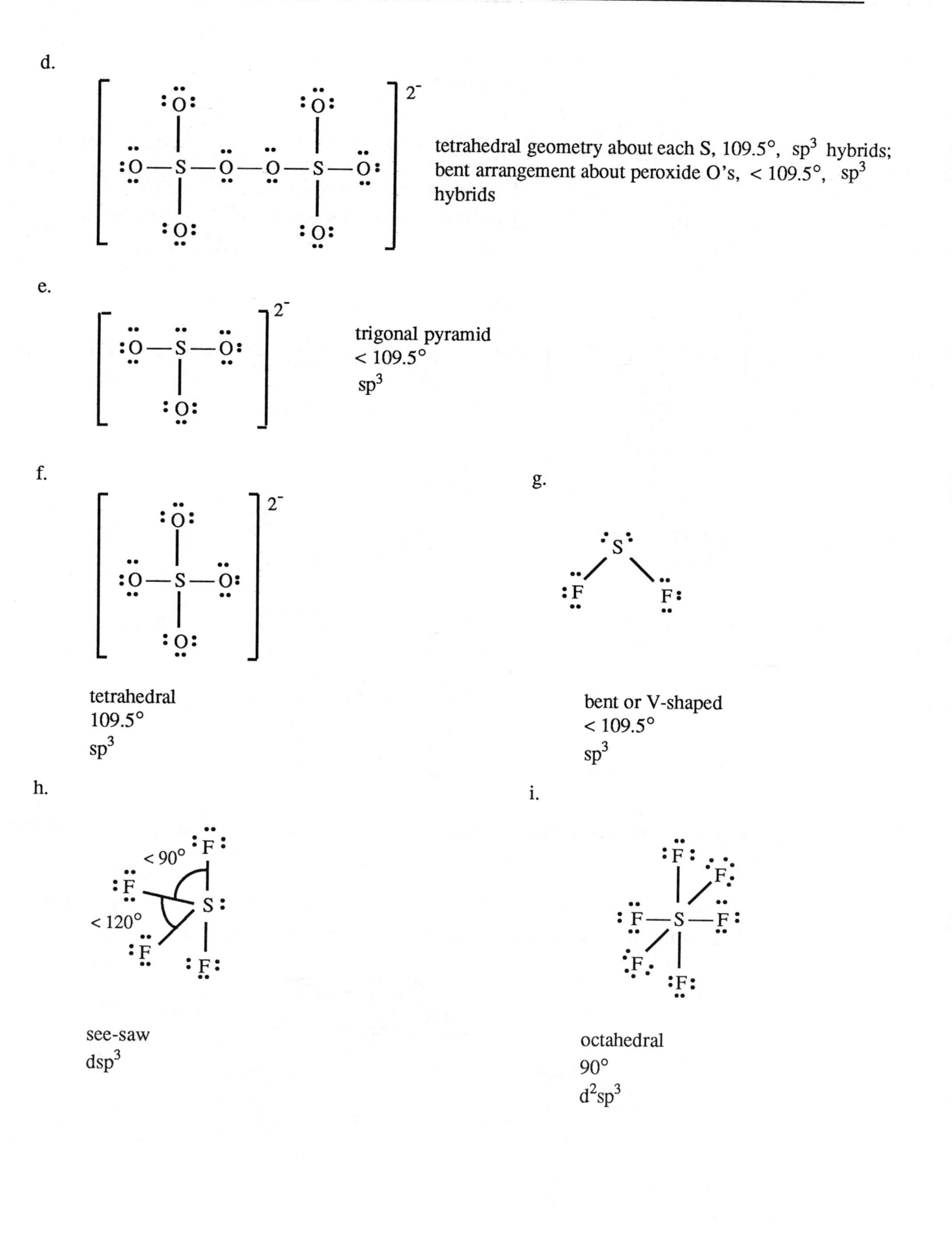
d.
2⁻
tetrahedral geometry about each S, 109.5°, sp³ hybrids; bent arrangement about peroxide O's, < 109.5°, sp³ hybrids
e.
2⁻
trigonal pyramid
< 109.5°
sp³
f.
2⁻
tetrahedral
109.5°
sp³
g.
bent or V-shaped
< 109.5°
sp³
h.
< 90°
< 120°
see-saw
dsp³
i.
octahedral
90°
d²sp³

j.

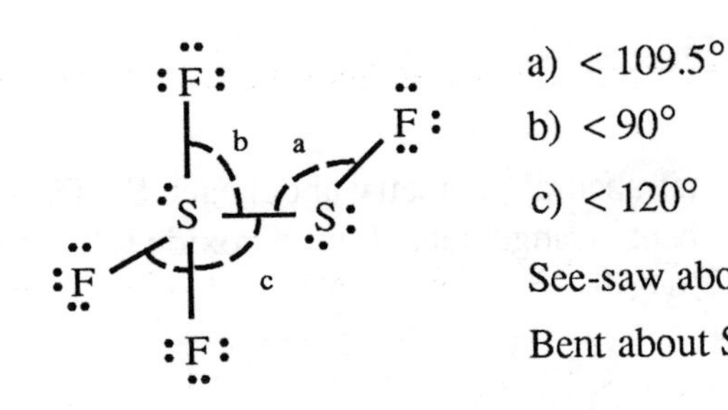

a) < 109.5°

b) < 90°

c) < 120°

See-saw about S atom with one lone pair (dsp^3)

Bent about S atom with two lone pairs (sp^3)

k.

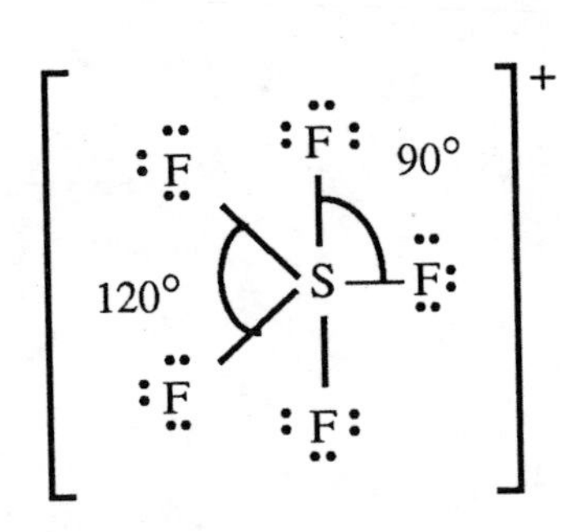

trigonal bipyramid,
90° and 120°, dsp^3

7.

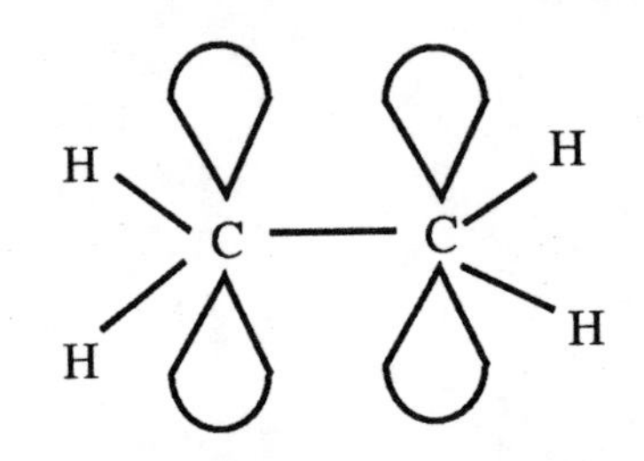

For the p-orbitals to properly line up to form the π bond, all six atoms are forced into the same plane.

8. No, the $= CH_2$ planes are mutually perpendicular. The center C-atom is involved in two π-bonds, thus, the p-orbitals used in each bond are perpendicular to the other.

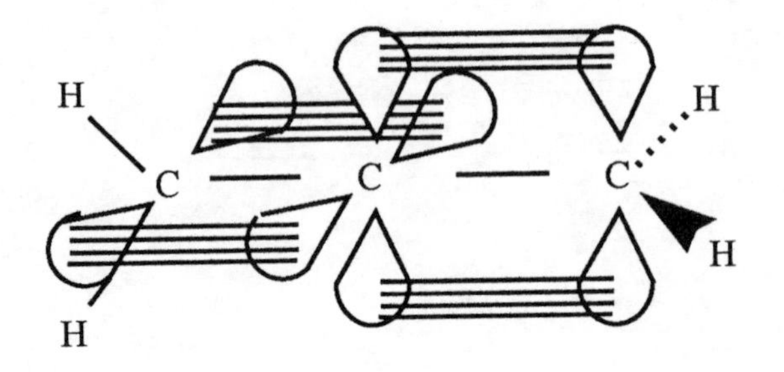

The result is the two 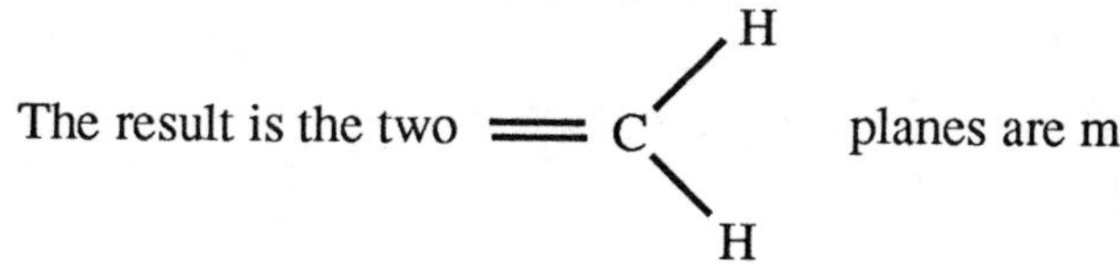 planes are mutually perpendicular.

9. Biacetlyl ($C_4H_6O_2$) has 4(4) + 6(1) + 2(6) = 34 valence electrons.

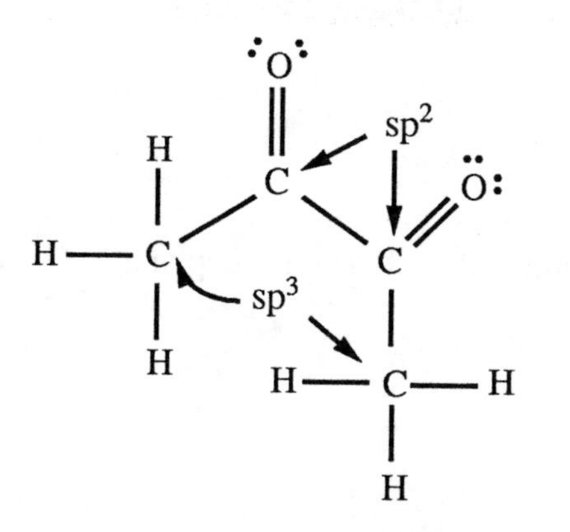

All CCO angles are 120°. The six atoms are not in the same plane because of free rotation about the carbon-carbon single bonds.

11 σ and 2 π bonds

Acetoin ($C_4H_8O_2$) has 4(4) + 8(1) + 2(6) = 36 valence electrons.

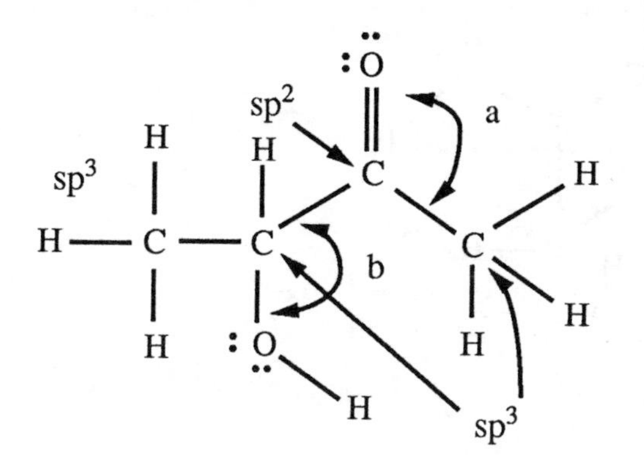

The carbon with doubly bonded O is sp^2 hybridized. The other 3 C-atoms are sp^3 hybridized.

Angle a) 120°

Angle b) 109.5°

13 σ and 1 π bonds

10. acrylonitrile: C_3H_3N has 3(4) + 3(1) + 5 = 20 valence electrons.

a. 120°

b. 120°

c. 180°

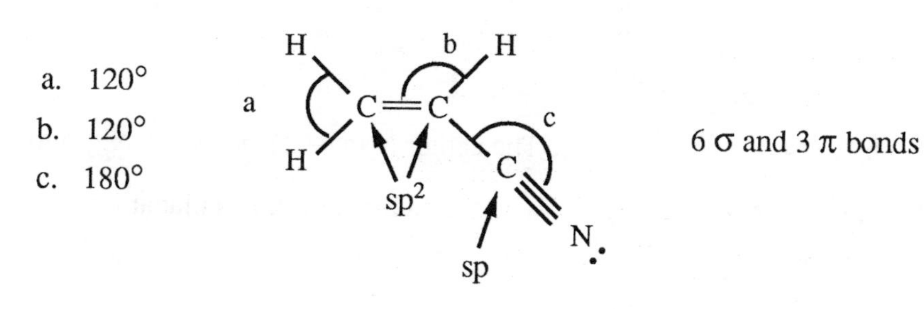

6 σ and 3 π bonds

methylmethacrylate: $C_5H_8O_2$

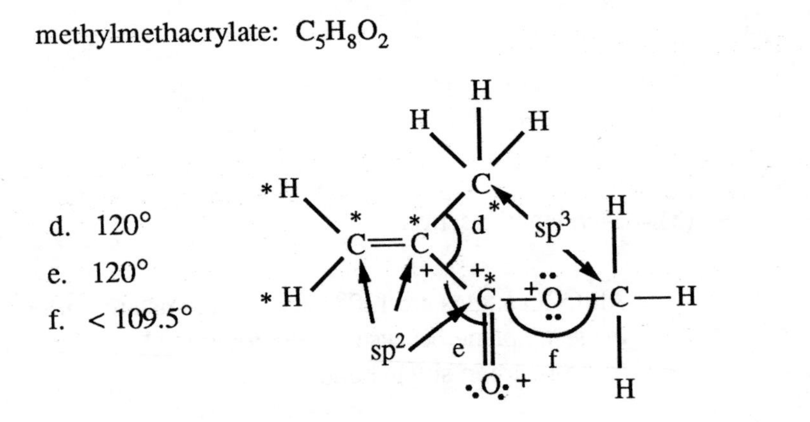

d. 120°

e. 120°

f. < 109.5°

14 σ and 2 π bonds

The atoms marked with an * are coplanar and the atoms marked with a + are coplanar. The two planes, however, do not have to coincide.

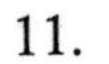

11.

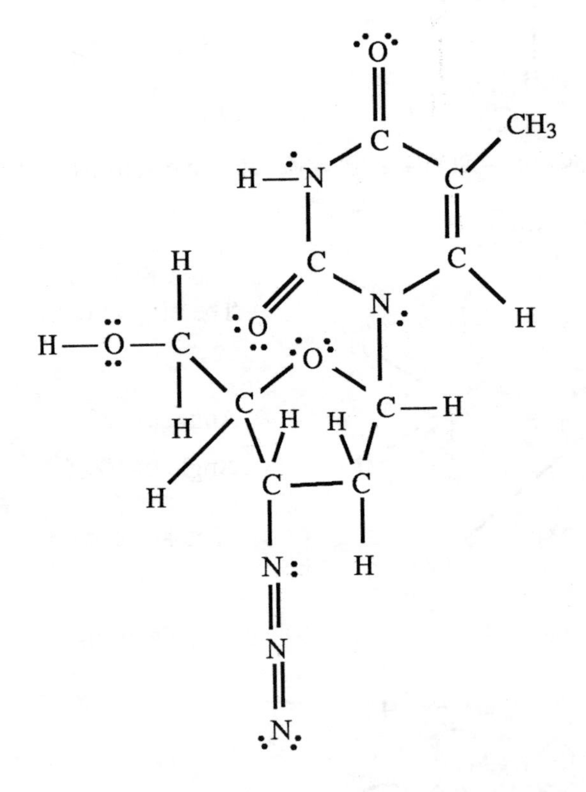

a.	6	b.	4	c.	The center N in —N=N=N group
d.	33 σ	e.	5 π bonds	f.	The six membered ring is planar.
g.	180°	h.	< 109.5°	i.	sp^3

12. a. NCN^{2-} has 5 + 4 + 5 + 2 = 16 valence electrons.

$$\left[\,:\ddot{N}=C=\ddot{N}:\,\right]^{2-} \longleftrightarrow \left[\,:N\equiv C-\ddot{\underset{..}{N}}:\,\right]^{2-} \longleftrightarrow \left[\,:\ddot{\underset{..}{N}}-C\equiv N:\,\right]^{2-}$$

H_2NCN has 2 + 5 + 4 + 5 = 16 valence electrons.

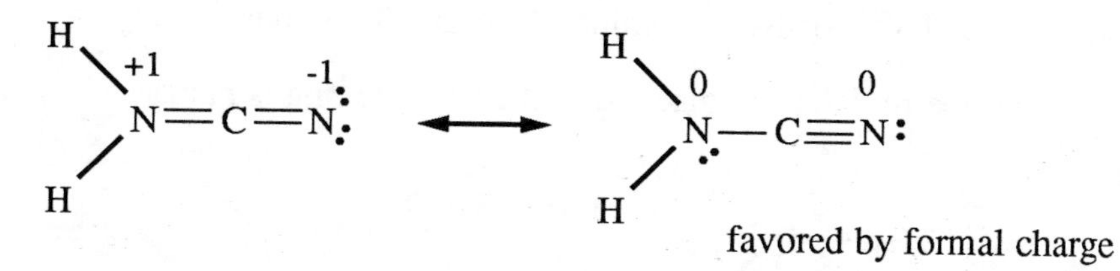

$NCNC(NH_2)_2$ has 5 + 4 + 5 + 4 + 2(5) + 4(1) = 32 valence electrons.

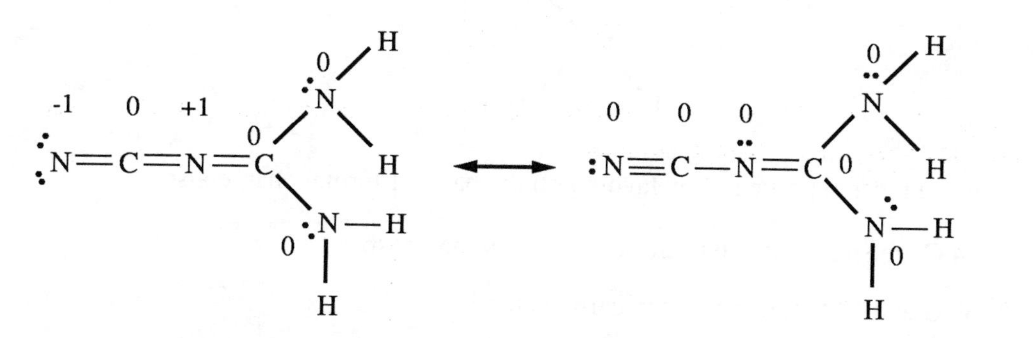

melamine: $C_3N_6H_6$ has 3(4) + 6(5) + 6(1) = 48 valence electrons

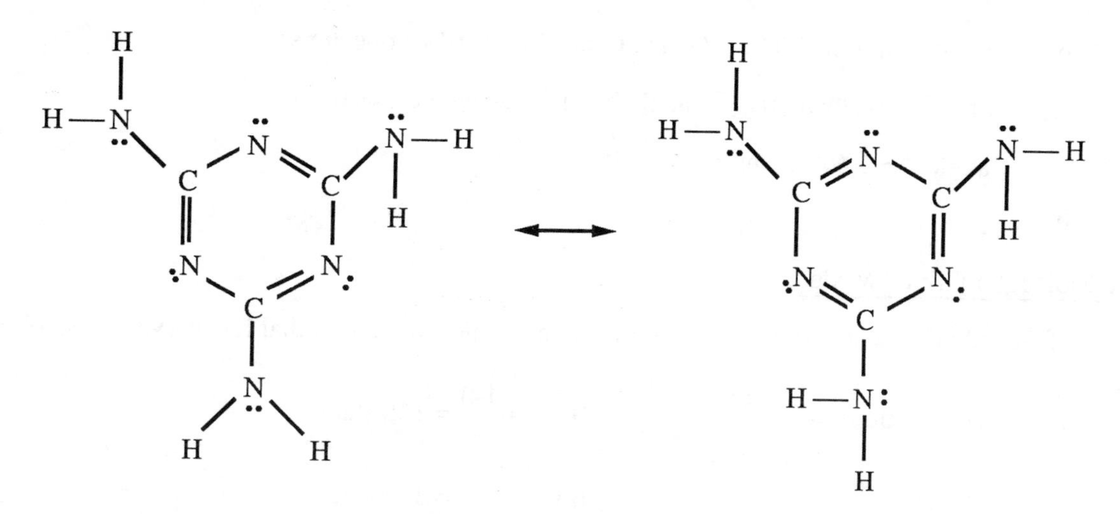

b. NCN^{2-}: C is sp hybridized. Cannot predict hybrids for N's (different for each resonance structure). For the remaining compounds, we will predict hybrids for the favored resonance structures only.

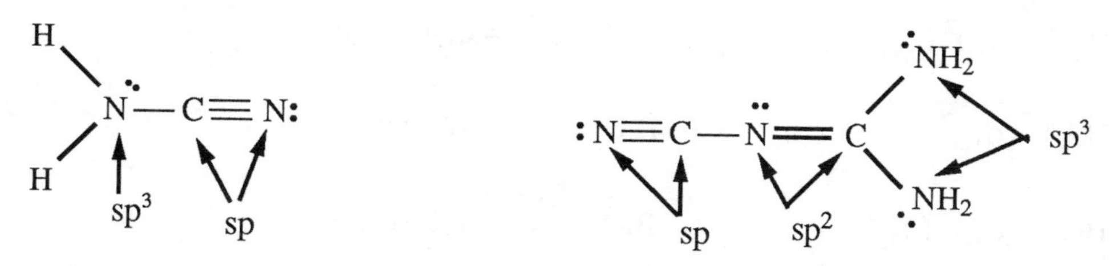

melamine: N in NH_2 groups are all sp^3 hybridized. Atoms in ring are all sp^2 hybridized.

c. NCN^{2-}: 2 σ and 2 π bonds; H_2NCN: 4 σ and 2 π bonds

dicyandiamide: 9 σ and 3 π bonds; melamine: 15 σ and 3 π bonds

d. The π-system forces the ring to be planar just as the benzene ring is planar.

e. The structure:

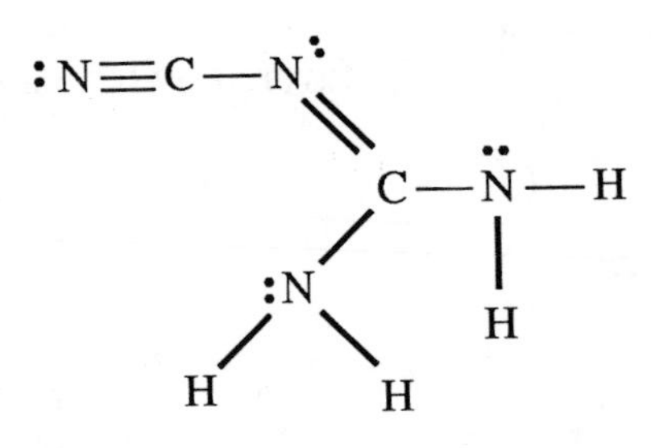

is the most important. It is favored on the basis of formal charge also.

13. a. 14 C are sp^3. 26 C are sp^2. None are sp.

b. 39 σ and 13 π bonds between carbon atoms.

14. a. sp^2

b. 9 sp^2 (6 in ring, NC═O, and C═C); 9 sp^3; None are sp

c. angle a less than 109.5°; angle b = 120°; angle c = 120°

angle d = 109.5°; angle e = 120°

d. five e. No

The Molecular Orbital Model

15. If we calculate a non-zero bond order for a molecule, we predict that it can exist (is stable).

a. H_2^+ $(\sigma_{1s})^1$ B.O. = $\frac{1-0}{2}$ = 1/2, stable

H_2 $(\sigma_{1s})^2$ B.O. = $\frac{2-0}{2}$ = 1, stable

H_2^- $(\sigma_{1s})^2(\sigma_{1s}^*)^1$ B.O. = $\frac{2-1}{2}$ = 1/2, stable

H_2^{2-} $(\sigma_{1s})^2(\sigma_{1s}^*)^2$ B.O. = $\frac{2-2}{2}$ = 0, not stable

b. He_2^{2+} $(\sigma_{1s})^2$ B.O. = $\frac{2-0}{2}$ = 1, stable

He_2^+ $(\sigma_{1s})^2(\sigma_{1s}^*)^1$ B.O. = $\frac{2-1}{2}$ = 1/2, stable

He_2 $(\sigma_{1s})^2(\sigma_{1s}^*)^2$ B.O. = $\frac{2-2}{2}$ = 0, not stable

16. a. N_2^{2-} $(\sigma_{2s})^2 (\sigma_{2s}*)^2 (\pi_{2p})^4 (\sigma_{2p})^2 (\pi_{2p}*)^2$ B.O. = $\frac{8\text{-}4}{2}$ = 2, stable

O_2^{2-} $(\sigma_{2s})^2 (\sigma_{2s}*)^2 (\pi_{2p})^4 (\sigma_{2p})^2 (\pi_{2p}*)^4$ B.O. = $\frac{8\text{-}6}{2}$ = 1, stable

F_2^{2-} $(\sigma_{2s})^2 (\sigma_{2s}*)^2 (\pi_{2p})^4 (\sigma_{2p})^2 (\pi_{2p}*)^4 (\sigma_{2p}*)^2$ B.O. = $\frac{8\text{-}8}{2}$ = 0, not stable

b. Be_2 $(\sigma_{2s})^2 (\sigma_{2s}*)^2$ B.O. = $\frac{2\text{-}2}{2}$ = 0, not stable

B_2 $(\sigma_{2s})^2 (\sigma_{2s}*)^2 (\pi_{2p})^2$ B.O. = $\frac{4\text{-}2}{2}$ = 1, stable

Li_2 $(\sigma_{2s})^2$ B.O. = 1, stable

17. The electron configurations are:

a. H_2: $(\sigma_{1s})^2$, B.O. = 1, diamagnetic

b. B_2: $(\sigma_{2s})^2 (\sigma_{2s}*)^2 (\pi_{2p})^2$, B.O. = 1, paramagnetic

c. F_2: $(\sigma_{2s})^2 (\sigma_{2s}*)^2 (\pi_{2p})^4 (\sigma_{2p})^2 (\pi_{2p}*)^4$, B.O. = 1, diamagnetic

d. CN^+: $(\sigma_{2s})^2 (\sigma_{2s}*)^2 (\pi_{2p})^4$, B.O. = $\frac{6 - 2}{2}$ = 2 diamagnetic

e. CN: $(\sigma_{2s})^2 (\sigma_{2s}*)^2 (\pi_{2p})^4 (\sigma_{2p})^1$, B.O. $\frac{7 - 2}{2}$ = 2.5 paramagnetic

f. CN^-: $(\sigma_{2s})^2 (\sigma_{2s}*)^2 (\pi_{2p})^4 (\sigma_{2p})^2$, B.O. = 3, diamagnetic

g. N_2: $(\sigma_{2s})^2 (\sigma_{2s}*)^2 (\pi_{2p})^4 (\sigma_{2p})^2$, B.O. = 3, diamagnetic

h. N_2^+: $(\sigma_{2s})^2 (\sigma_{2s}*)^2 (\pi_{2p})^4 (\sigma_{2p})^1$, B.O. = 2.5, paramagnetic

i. N_2^-: $(\sigma_{2s})^2 (\sigma_{2s}*)^2 (\pi_{2p})^4 (\sigma_{2p})^2 (\pi_{2p}*)^1$, B.O. = 2.5, paramagnetic

18. The complete electron configurations are:

O_2^+: $(\sigma_{2s})^2 (\sigma_{2s}*)^2 (\pi_{2p})^4 (\sigma_{2p})^2 (\pi_{2p}*)^1$

O_2: $(\sigma_{2s})^2 (\sigma_{2s}*)^2 (\pi_{2p})^4 (\sigma_{2p})^2 (\pi_{2p}*)^2$

O_2^-: $(\sigma_{2s})^2 (\sigma_{2s}*)^2 (\pi_{2p})^4 (\sigma_{2p})^2 (\pi_{2p}*)^3$

O_2^{2-}: $(\sigma_{2s})^2 (\sigma_{2s}*)^2 (\pi_{2p})^4 (\sigma_{2p})^2 (\pi_{2p}*)^4$

	O_2^+		O_2		O_2^-		O_2^{2-}
Bond order	2.5		2		1.5		1
# of unpaired electrons	1		2		1		0
Bond energy	O_2^{2-}	<	O_2^-	<	O_2	<	O_2^+
Bond length	O_2^+	<	O_2	<	O_2^-	<	O_2^{2-}

19.

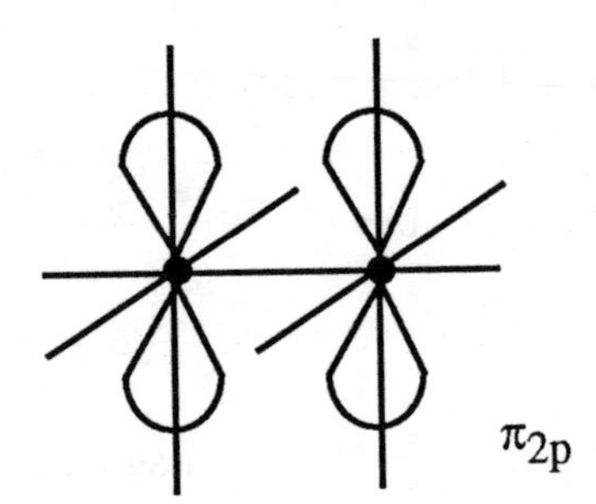

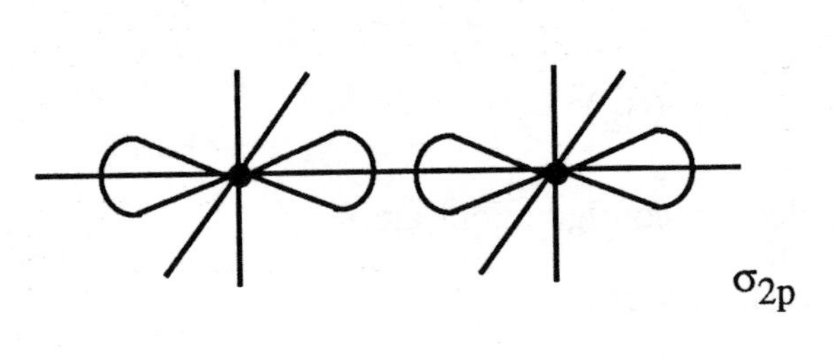

20.

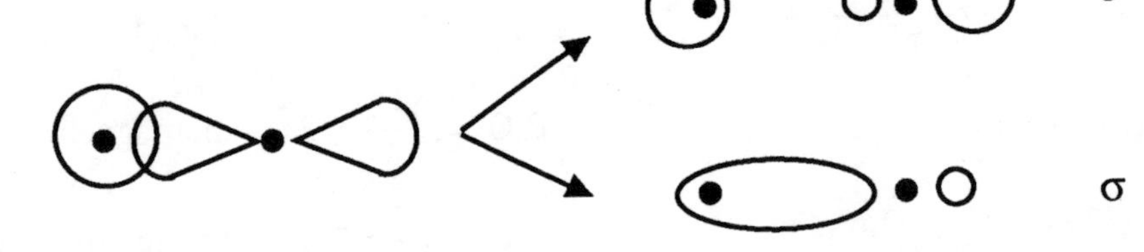

21. The bond orders are: CN^+, 2; CN, 2.5; CN^-, 3

Shortest ⟶ Longest bond length $CN^- < CN < CN^+$

Lowest ⟶ Highest bond energy $CN^+ < CN < CN^-$

22. Bond orders: N_2, 3; N_2^+, 2.5; N_2^-, 2.5

Bond length: $N_2 < N_2^+ \approx N_2^-$

Bond strength: $N_2 > N_2^+ \approx N_2^-$

We would expect the strengths and lengths of the bonds in N_2^+ and N_2^- to be close. Since there are more electrons on N_2^- than N_2^+, electron repulsions might be greater in N_2^- than N_2^+ causing the bond in N_2^- to be a little longer and a little weaker than the bond in N_2^+.

23. C_2^{2-} has 10 valence electrons.

$[:C{\equiv}C:]^{2-}$ sp hybrid orbitals used.

$(\sigma_{2s})^2 (\sigma_{2s}^*)^2 (\pi_{2p})^4 (\sigma_{2p})^2$ B.O. = 3

Both give the same picture, a triple bond composed of a σ and two π-bonds. Both predict the ion to be diamagnetic. Lewis structures deal well with diamagnetic (all electrons paired) species. The Lewis model cannot really predict magnetic properties. C_2^{2-} is isoelectronic with the CO molecule.

24.

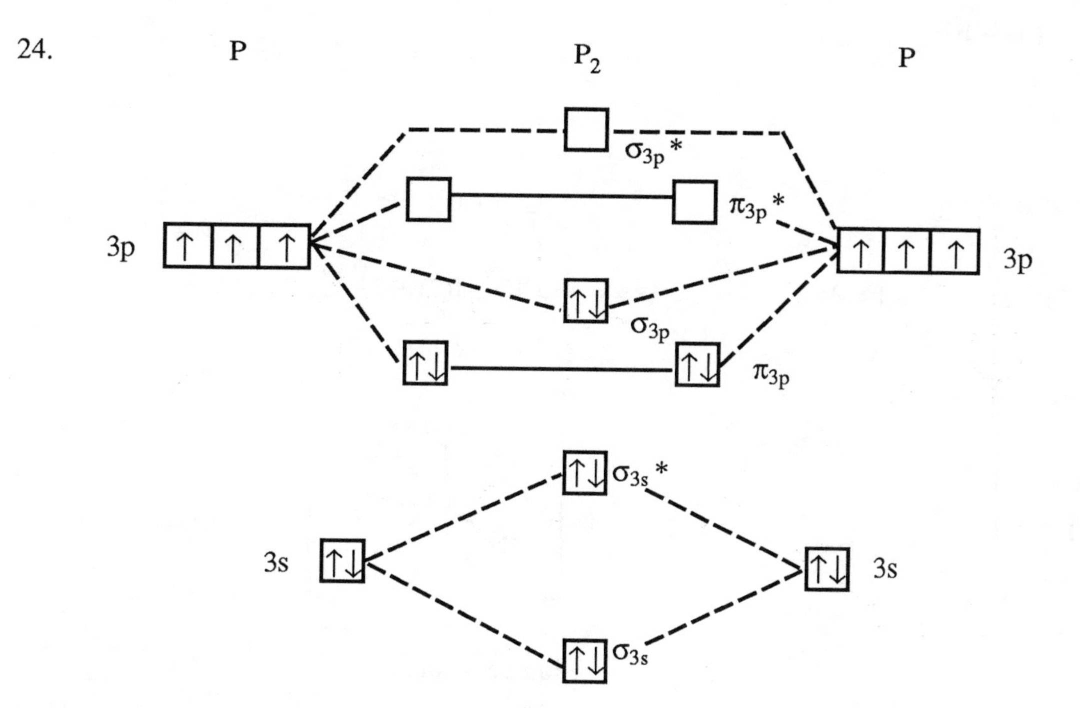

ADDITIONAL EXERCISES

25. a. FClO, 7 + 7 + 6 = 20 e^-

bent, sp^3 hybrid

b. $FClO_2$, 7 + 7 + 2(6) = 26 e^-

trigonal pyramid, sp^3

c. $FClO_3$, 7 + 7 + 3(6) = 32 e^-

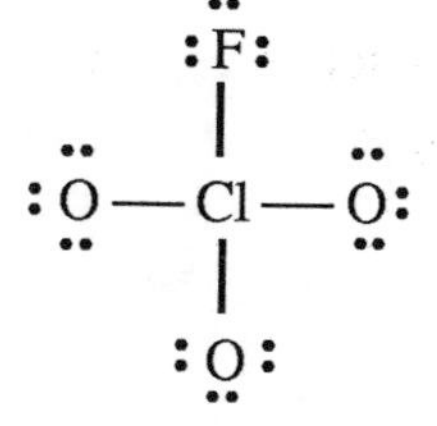

tetrahedral
sp^3

d. F_3ClO, 3(7) + 7 + 6 = 34 e^-

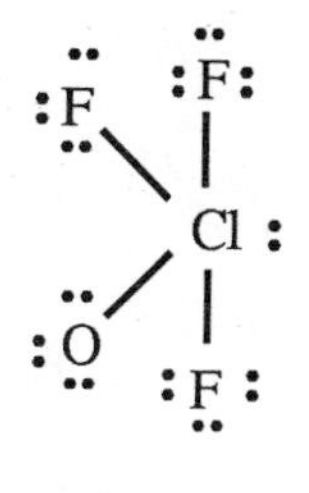

see-saw
dsp^3

e. F_3ClO_2: 3(7) + 7 + 2(6) = 40 valence e^-

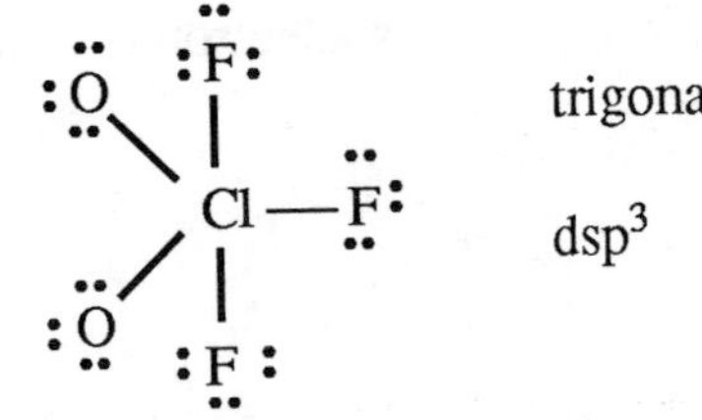

trigonal bipyramid

dsp^3

26. $FClO_2 + F^- \rightarrow F_2ClO_2^-$

$F_2ClO_2^-$, 2(7) + 2(6) + 7 + 1 = 34 e^-

see-saw
dsp^3

$F_3ClO \rightarrow F^- + F_2ClO^+$

F_2ClO^+, 2(7) + 7 + 6 - 1 = 26 e^-

trigonal pyramid
sp^3

$F_3ClO + F^- \rightarrow F_4ClO^-$

F_4ClO^-, 4(7) + 7 + 6 + 1 = 42 e^-

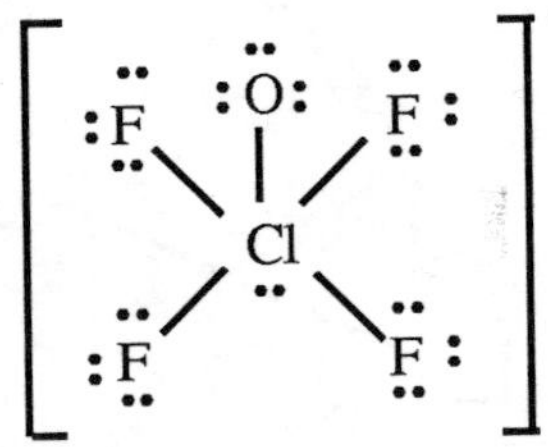
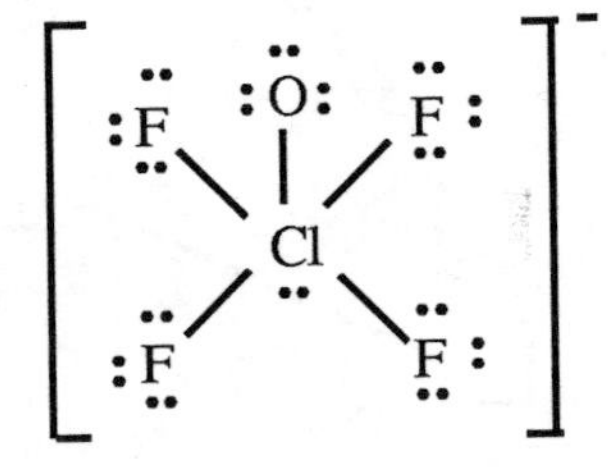

square pyramid
d^2sp^3

$F_3ClO_2 \rightarrow F^- + F_2ClO_2^+$

$F_2ClO_2^+$, 2(7) + 7 + 2(6) - 1 = 32 e^-

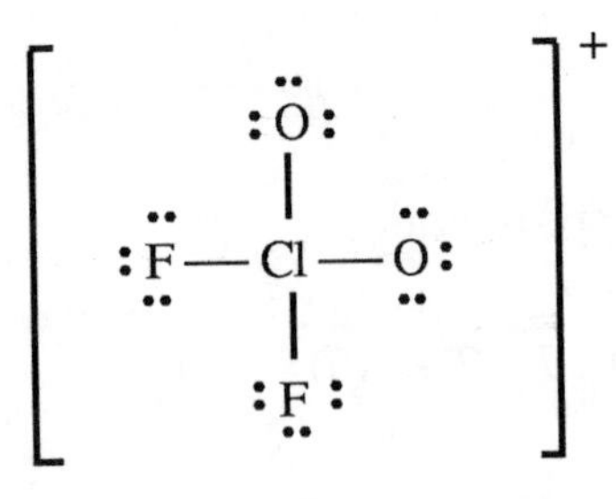

tetrahedral
sp^3

27. SbF_5 monomer:

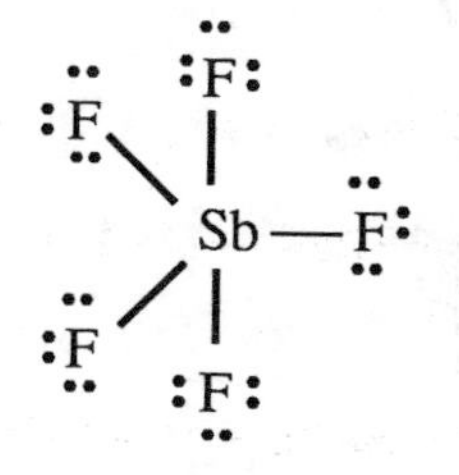

trigonal bipyramid, dsp^3

SbF_5 polymer: Essentially there is an octahedral arrangement of fluorines about each antimony atom (d^2sp^3):

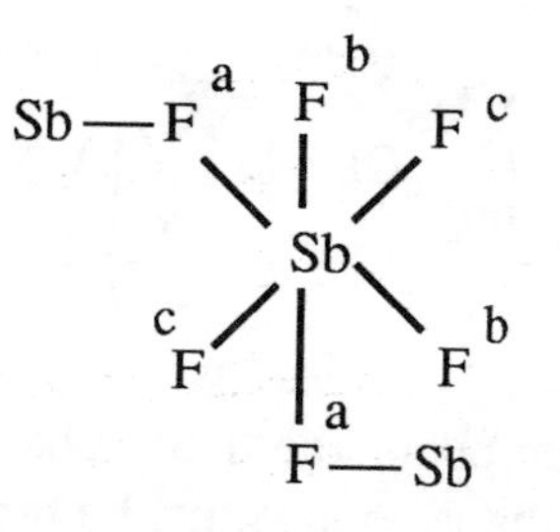

The three types of fluorines are:

a. bridging, bonded to two Sb

b. 90° away from one type (a) and 180° away from the other type (a)

c. 90° away from both type (a)

28. a.

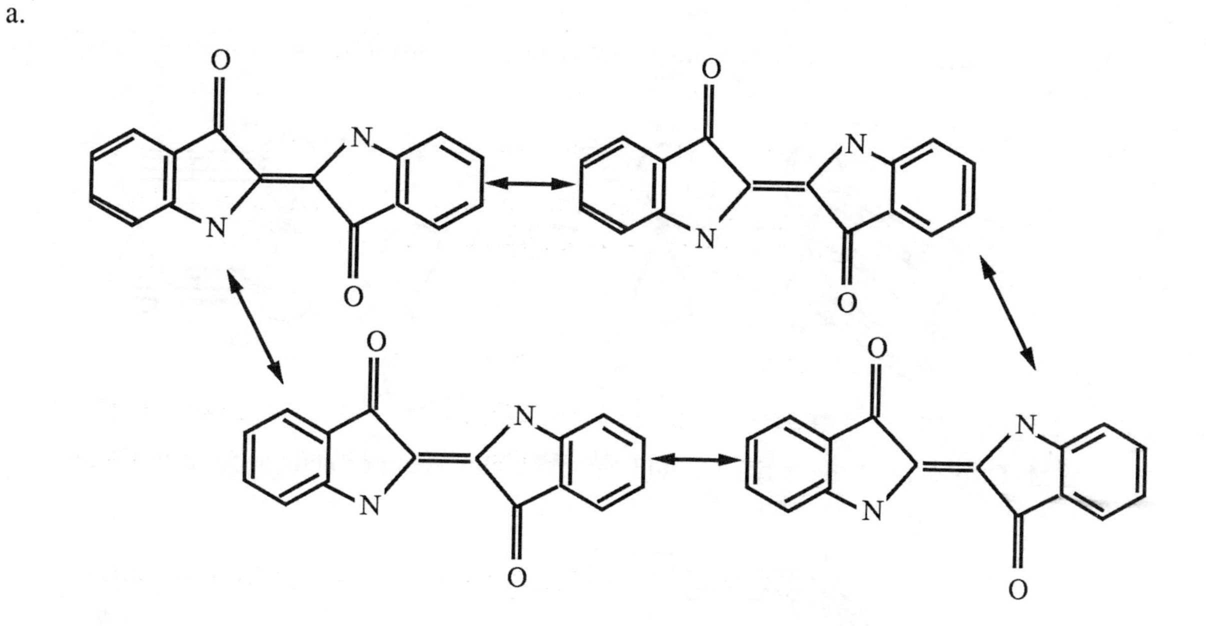

Plus other less likely forms.

b. 33 σ and 9 π bonds

c. The whole molecule is planar.

d. All C-atoms are sp^2 hybridized.

29.

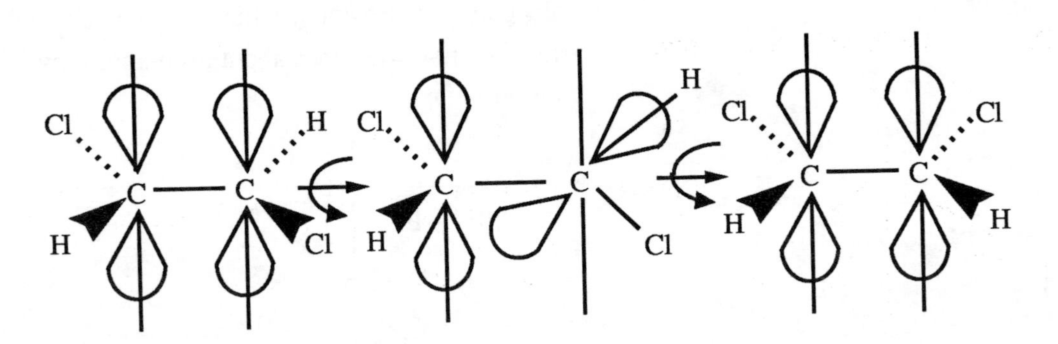

In order to rotate about the double bond, the molecule must go through the intermediate shown above. Hence, the π bonds must be broken, while the sigma bond remains intact. Bond energies are 347 kJ/mol for a C — C and 614 kJ/mol for a C $=$ C. If we take the single bond as the strength of the σ bond, then the strength of the π bond is (614 - 347) = 267 kJ/mol. Thus, 267 kJ/mol must be supplied to rotate about a carbon-carbon double bond.

30. O_3 and NO_2^- are isoelectronic, so we only need consider one of them.

$$:\ddot{O}=\ddot{O}-\ddot{O}: \longleftrightarrow :\ddot{O}-\ddot{O}=\ddot{O}:$$

Two resonance forms: central atom is sp^2 hybridized. Resonance view of the two resonance forms is:

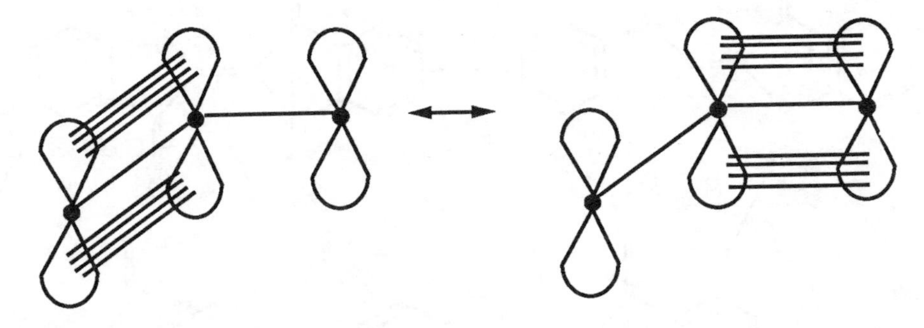

The M. O. picture of the π-bond would comprise all three p-orbitals to form three new M. O.'s. The lowest energy M. O. would look like:

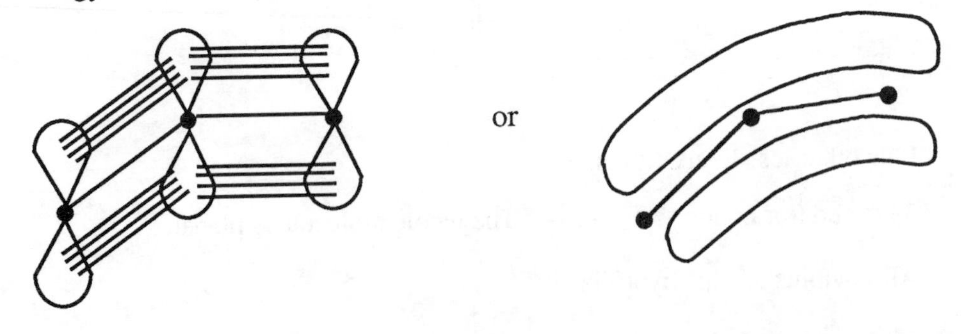

In the M. O. view, there would be two localized σ-bonds, and a π-bond delocalized over all 3 atoms in the molecule.

31. a. P_4 4(5) = 20 valence electrons

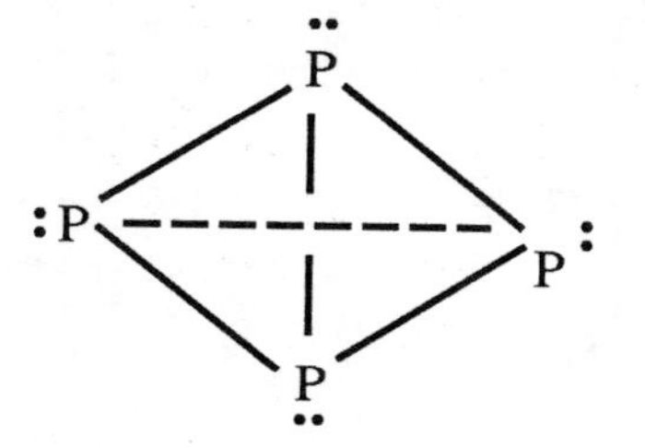

All P - sp^3

P—P—P bond angle 60° (Each P is at the corner of a tetrahedron. The faces of a tetrahedron are equilateral triangles.)

Note: We would normally predict 109.5° bond angles because there are 4 pair of e^- about each P. There is considerable strain in this molecule since each P is forced into 60° bond angles.

b. P_4O_6 4(5) + 6(6) = 56 valence electrons

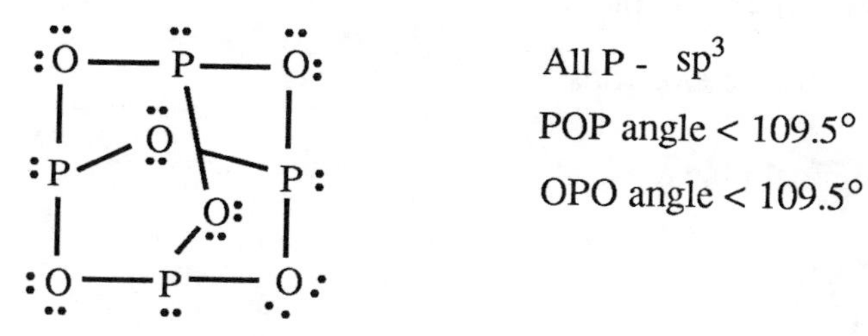

All P - sp^3

POP angle < 109.5°

OPO angle < 109.5°

c. P_4O_{10} 4(5) + 10(6) = 80 valence electrons

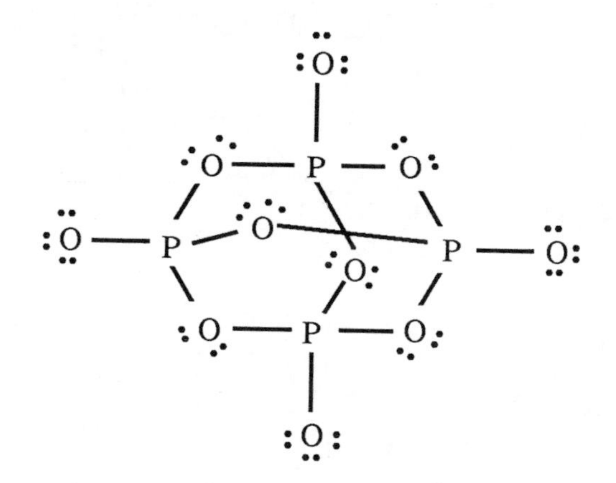

All P - sp^3

POP angle < 109.5°

OPO angle = 109.5°

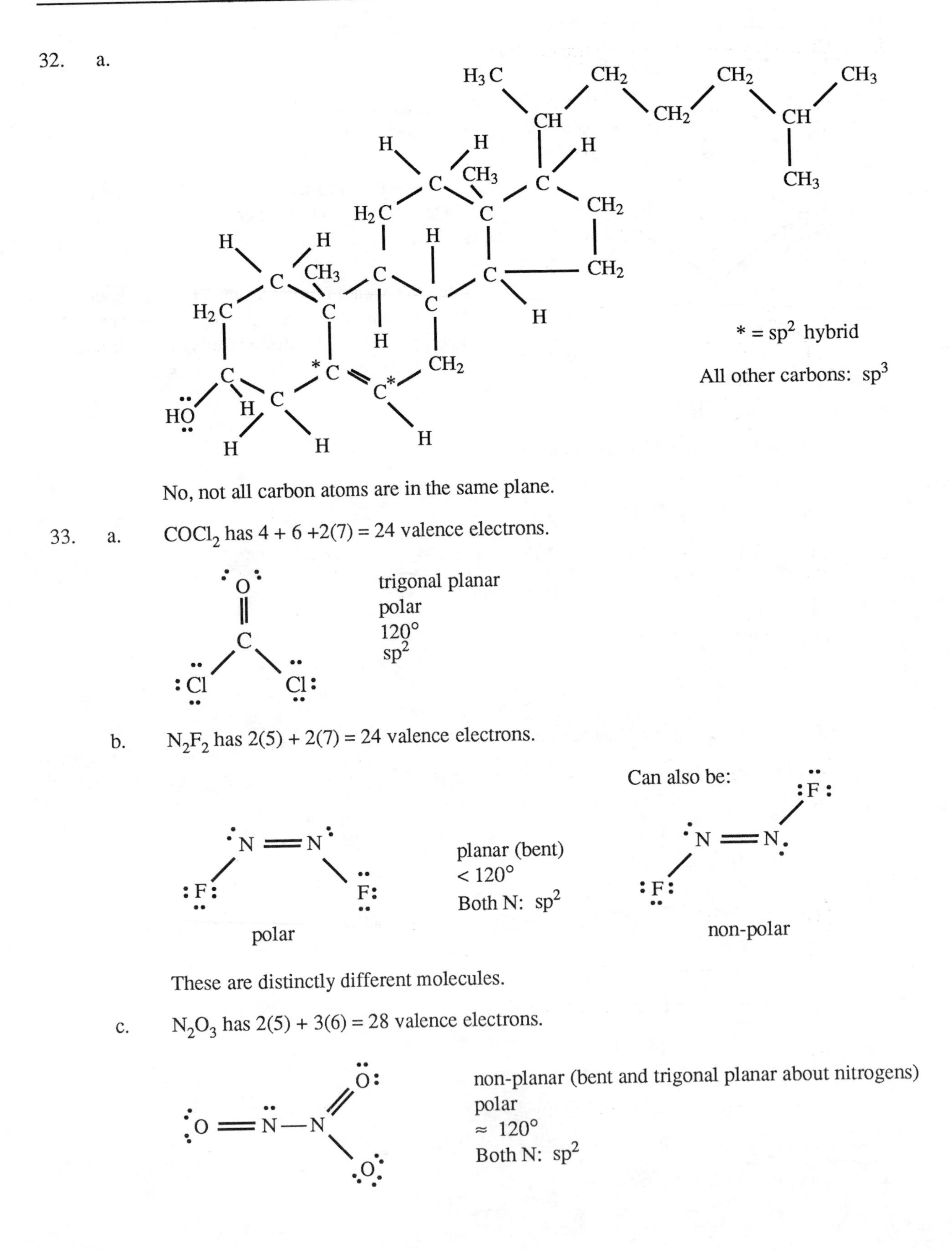

32. a.

$* = sp^2$ hybrid

All other carbons: sp^3

No, not all carbon atoms are in the same plane.

33. a. $COCl_2$ has 4 + 6 +2(7) = 24 valence electrons.

trigonal planar
polar
120°
sp^2

b. N_2F_2 has 2(5) + 2(7) = 24 valence electrons.

planar (bent)
< 120°
Both N: sp^2

polar

Can also be:

non-polar

These are distinctly different molecules.

c. N_2O_3 has 2(5) + 3(6) = 28 valence electrons.

non-planar (bent and trigonal planar about nitrogens)
polar
≈ 120°
Both N: sp^2

d. COS has 4 + 6 + 6 = 16 valence electrons.

:O═C═S:

linear
polar
180°
sp

e. ICl_3 has 7 + 3(7) = 28 valence electrons.

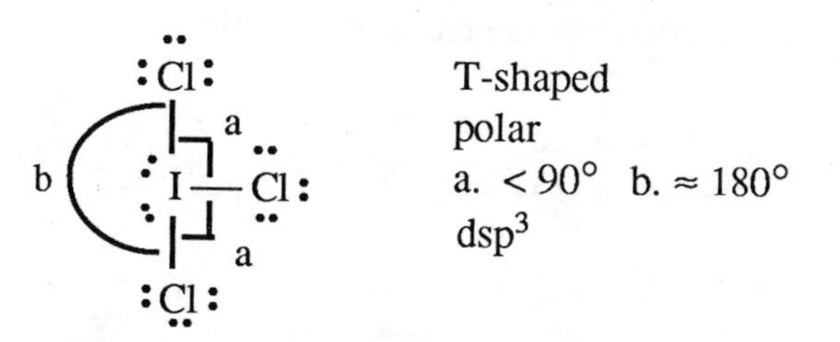

T-shaped
polar
a. < 90° b. ≈ 180°
dsp^3

34. c.

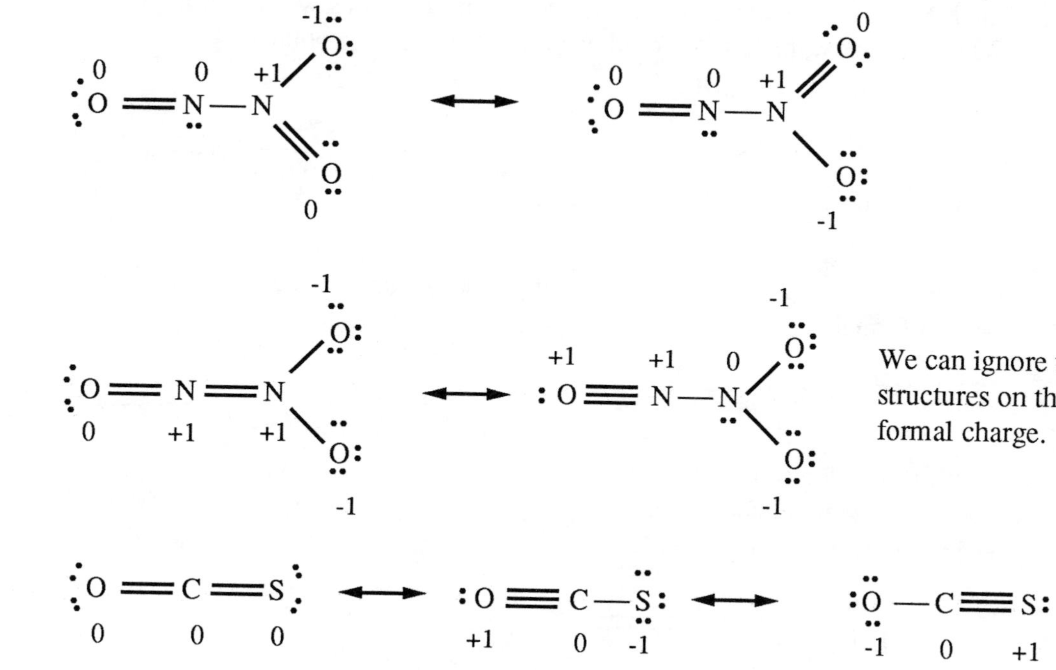

We can ignore these last two structures on the basis of formal charge.

d.

35. a. Yes, both have 4 effective pairs about the P.

b. sp^3 in each

c. P has to use one of its d orbitals to form the π bond.

d.

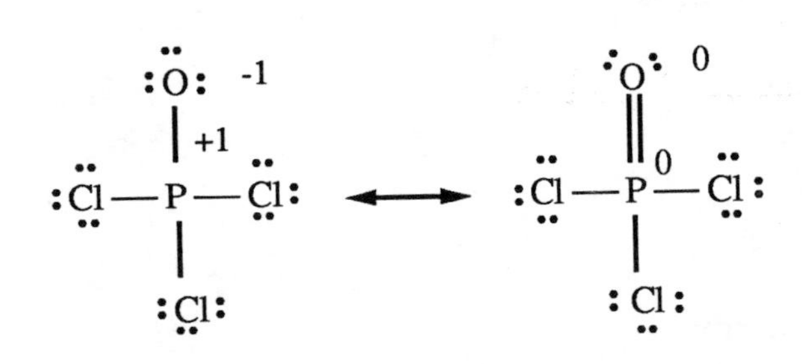

The structure with the P$=$O bond is favored on the basis of formal charge.

36. a.

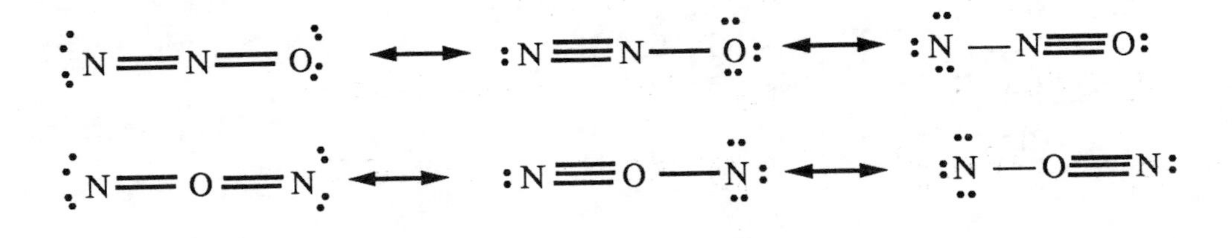

The NNO structure is correct. From the Lewis structures we would predict both NNO and NON to be linear, but NON would be non-polar. NNO is polar.

b.

$:\!N{=}N{=}O\!:$ $\longleftrightarrow$ $:\!N{\equiv}N{-}O\!:$ $\longleftrightarrow$ $:\!N{-}N{\equiv}O\!:$

-1	+1	0	0	+1	-1	-2	+1	+1	Formal charges

The central N is sp hybridized. We can probably ignore the 3rd resonance structure on the basis of formal charge.

c. $:\!N{\equiv}N{-}O\!:$

sp hybrid orbitals on the center N overlap with atomic orbitals (or hybrid orbitals) on the other two atoms to form two sigma bonds. The remaining p orbitals on the center N overlap with p orbitals on the other N to form two π-bonds.

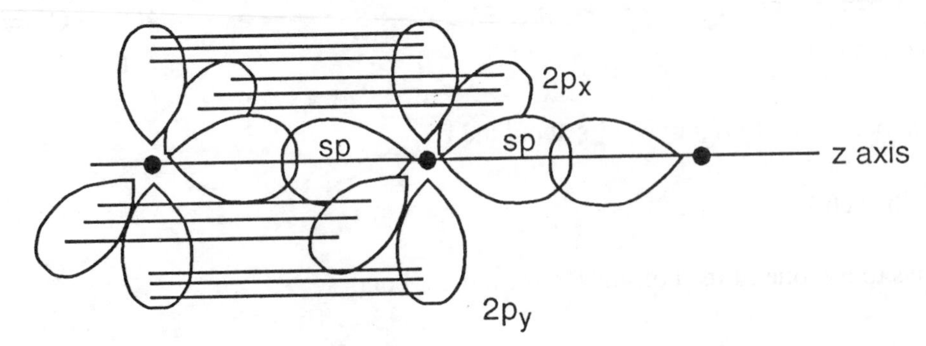

37. NO^+ has 5 + 6 - 1 = 10 valence electrons.

$[:\!N{\equiv}O\!:]^+$

NO has 5 + 6 = 11 valence electrons.

$:\dot{N}=\ddot{O}: \longleftrightarrow :\dot{N}=\ddot{O}: \longleftrightarrow \dot{N}=\ddot{O}:$

NO^- has 12 valence electrons.

$[:\ddot{N}=\ddot{O}:]^-$

M.O. model:

NO^+ $(\sigma_{2s})^2 (\sigma_{2s}^*)^2 (\pi_{2p})^4 (\sigma_{2p})^2$, B.O. = 3, diamagnetic

NO $(\sigma_{2s})^2 (\sigma_{2s}^*)^2 (\pi_{2p})^4 (\sigma_{2p})^2 (\pi_{2p}^*)^1$, B.O. = 2.5, 1 unpaired e^-

NO^- $(\sigma_{2s})^2 (\sigma_{2s}^*)^2 (\pi_{2p})^2 (\sigma_{2p})^2 (\pi_{2p}^*)^2$, B.O. = 2, 2 unpaired e^-

Bond Energies: $NO^- < NO < NO^+$

Bond Lengths: $NO^+ < NO < NO^-$

The two models only give the same results for NO^+. The MO view is the correct one for NO and NO^-.

38. For NO: $(\sigma_{2s})^2 (\sigma_{2s}^*)^2 (\pi_{2p})^4 (\sigma_{2p})^2 (\pi_{2p}^*)^1$

$$B.O. = \frac{8-3}{2} = 2.5$$

MO model does a better job. It predicts a bond order of 2.5. Measured bond energies and bond lengths are consistent with the bond in NO being stronger than a double bond.

	Bond Length	Bond Energy
$>N-O$	140 pm	201 kJ/mol
$-N=O$	121 pm	607 kJ/mol
NO (nitric oxide)	115 pm	678 kJ/mol

It is impossible to draw a Lewis structure satisfying the octet rule for any odd electron species. The most reasonable attempt is to draw:

$:\dot{N}=\ddot{O}: \longleftrightarrow :\dot{N}=\ddot{O}: \longleftrightarrow \dot{N}=\ddot{O}:$

None of which makes the correct prediction for the bond strength or number of unpaired electrons.

39. N_2: $(\sigma_{2s})^2 (\sigma_{2s}*)^2 (\pi_{2p})^4 (\sigma_{2p})^2$ in ground state;

B.O. = 3; diamagnetic

1st excited state: $(\sigma_{2s})^2 (\sigma_{2s}*)^2 (\pi_{2p})^4 (\sigma_{2p})^1 (\pi_{2p}*)^1$

B.O. = $\frac{7 - 3}{2}$ = 2; paramagnetic (2 unpaired e^-)

40. Be: $1s^22s^2$; each Be has 2 valence electrons.

We can draw a Lewis structure:

Be ═ Be

This doesn't obey the octet rule, but is consistent with structures we've drawn earlier for Be compounds. There are two pairs of electrons around each Be atom just as in earlier drawings. From this Lewis structure we would predict Be_2 to be stable.

The MO model (see Exercise 9.16b) predicts a bond order of zero; Be_2 shouldn't exist. Be_2 has not yet been observed. Be is very small; electron-electron repulsions will be much too great in the Lewis structure drawn.

41. Considering only the twelve valence electrons in O_2 we can put them in an energy level diagram.

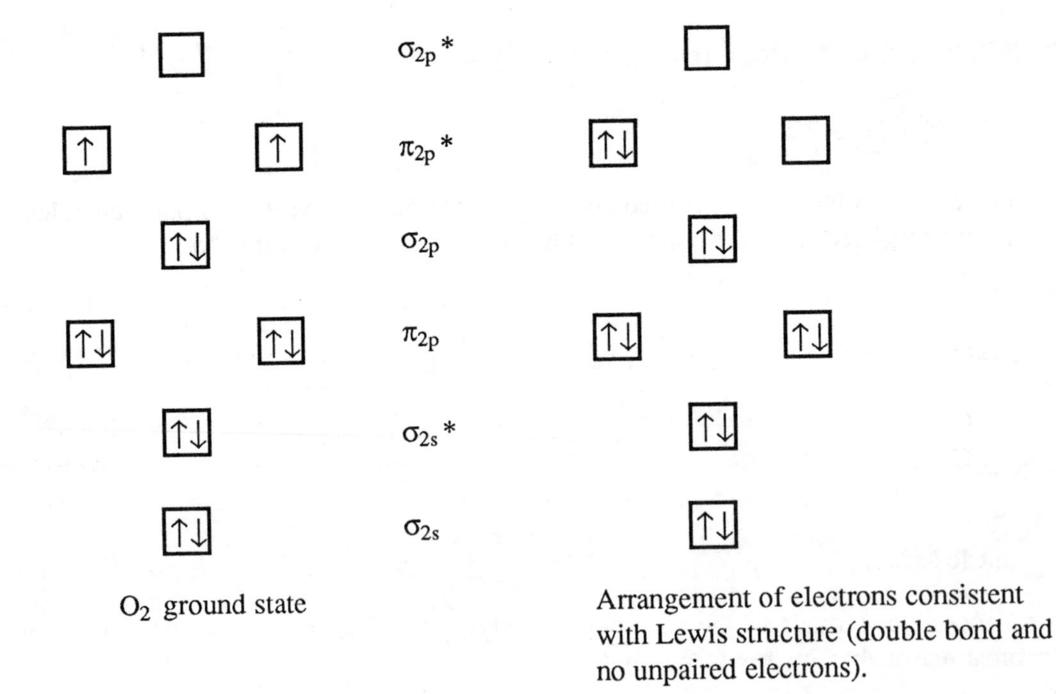

O_2 ground state

Arrangement of electrons consistent with Lewis structure (double bond and no unpaired electrons).

It takes energy to pair electrons in the same orbital. Thus, the structure with no unpaired electrons is at a higher energy; it is in an excited state.

CHALLENGE PROBLEMS

42. a. No, some atoms are in different places. Thus, these are not resonance structures.

b.

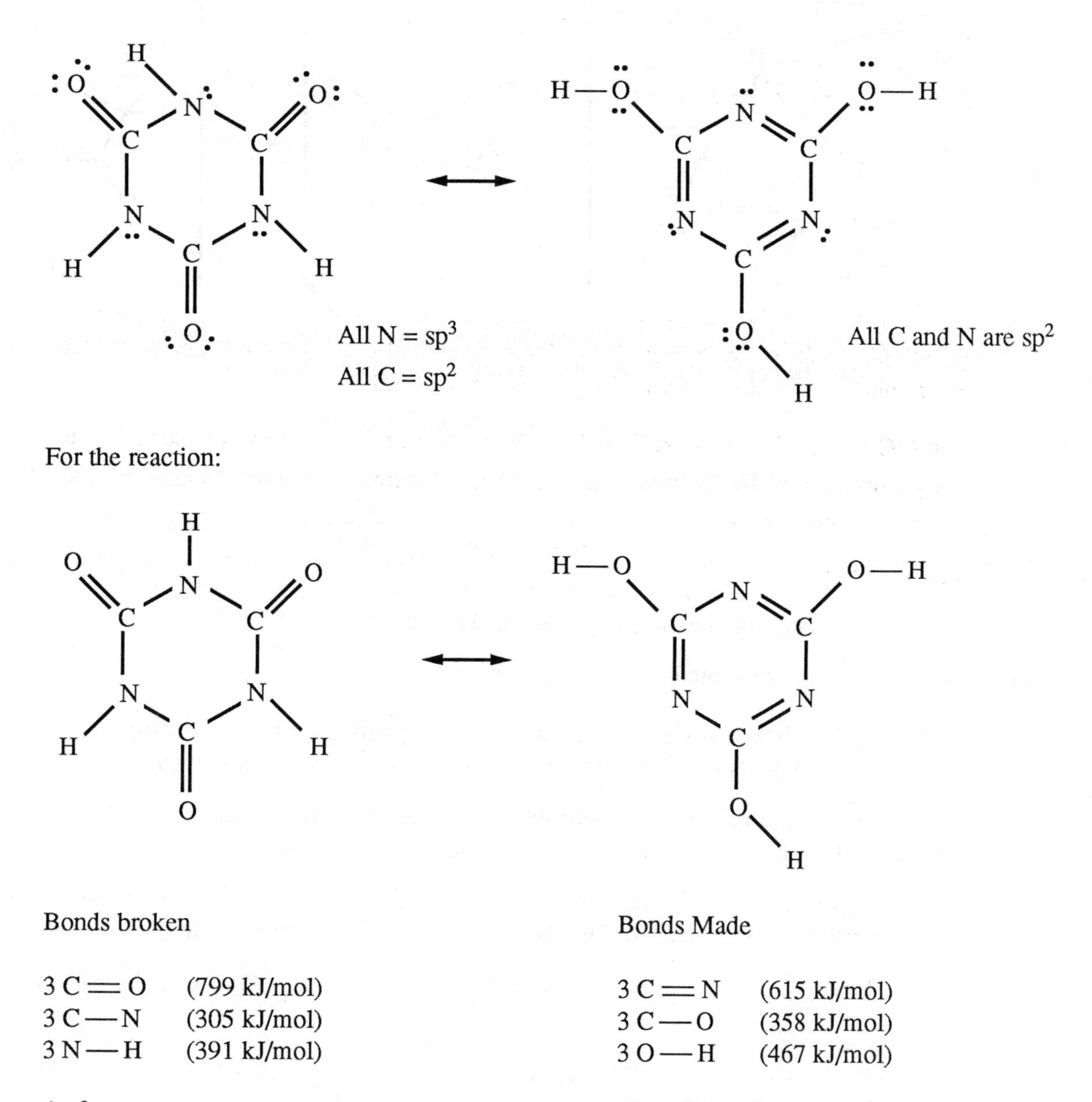

All N = sp3
All C = sp2
All C and N are sp2
For the reaction:
Bonds broken
3 C=O (799 kJ/mol)
3 C—N (305 kJ/mol)
3 N—H (391 kJ/mol)
Bonds Made
3 C=N (615 kJ/mol)
3 C—O (358 kJ/mol)
3 O—H (467 kJ/mol)

$\Delta H° = 3(799) + 3(305) + 3(391) - [3(615) + 3(358) + 3(467)]$

$\Delta H° = 4485 \text{ kJ} - 4320 \text{ kJ} = + 165 \text{ kJ}$

The bonds are stronger in the first structure with the carbon-oxygen double bonds.

43.

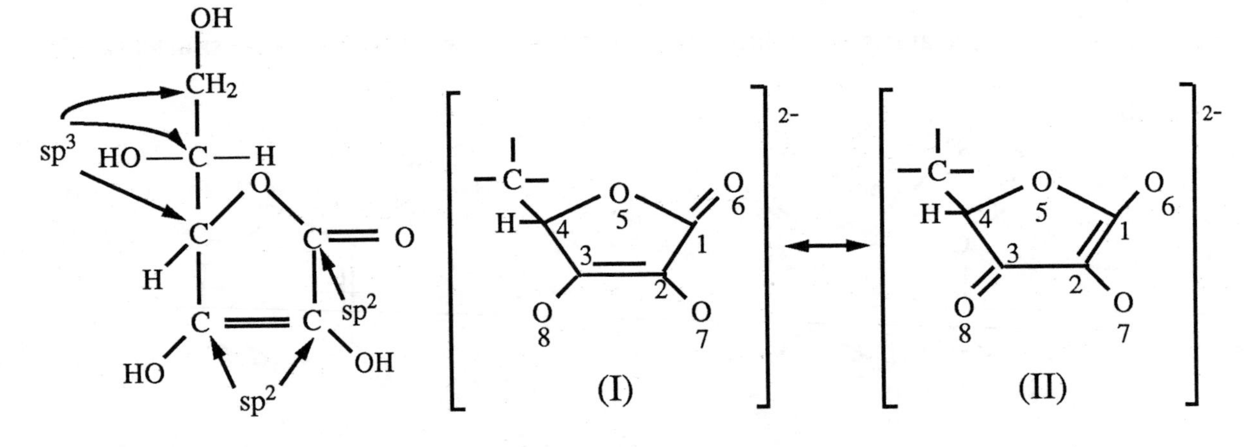

Two resonance forms can be drawn for the ring portion of the anion. In (I) the following groups of atoms must be coplanar because of the position of the double bonds:

$$C_1C_2O_5O_6 \text{ and } C_1C_2C_3C_4O_7O_8$$

In resonance form (II) the following two groups of atoms must be coplanar:

$$C_1C_2C_3O_5O_6O_7 \text{ and } C_2C_3C_4O_8$$

The only way to satisfy all of these requirements is for all eight atoms (the four carbons and one oxygen in the ring and the three oxygens bonded to the ring) to lie in the same plane.

44. N_2: The π and π^* orbitals are symmetrical.

CO: The π orbitals would place more electron density near the more electronegative oxygen. The π^* orbitals would place more electron density near the carbon atom.

45. The π bonds between S atoms and between C and S atoms are not as strong. The orbitals do not overlap with each other as well as the atomic orbitals of C and O.

46. a. The electron density would be closer to F on the average. The F atom is more electronegative than the H atom. The 2p orbital of F is lower in energy than the 1s orbital of H.

b. The bonding MO would have more fluorine 2p character. The 2p is closer in energy to the bonding MO.

c. The anti-bonding MO would place more electron density closer to H and would have a greater contribution from the hydrogen 1s atomic orbital.

47. a. The electron removed from N_2 is in the σ_{2p} molecular orbital which is lower in energy than the 2p atomic orbital from which the electron in atomic nitrogen is removed. Since the electron removed from N_2 is lower in energy than the electron in N, the ionization energy of N_2 should be greater than for N.

b. F_2 should have a lower first ionization energy than F. The electron removed from F_2 is in a π_{2p}^* anti-bonding molecular orbital which is higher in energy than the 2p atomic orbitals. Thus, it is easier to remove an electron from F_2 than from F.

48. a.

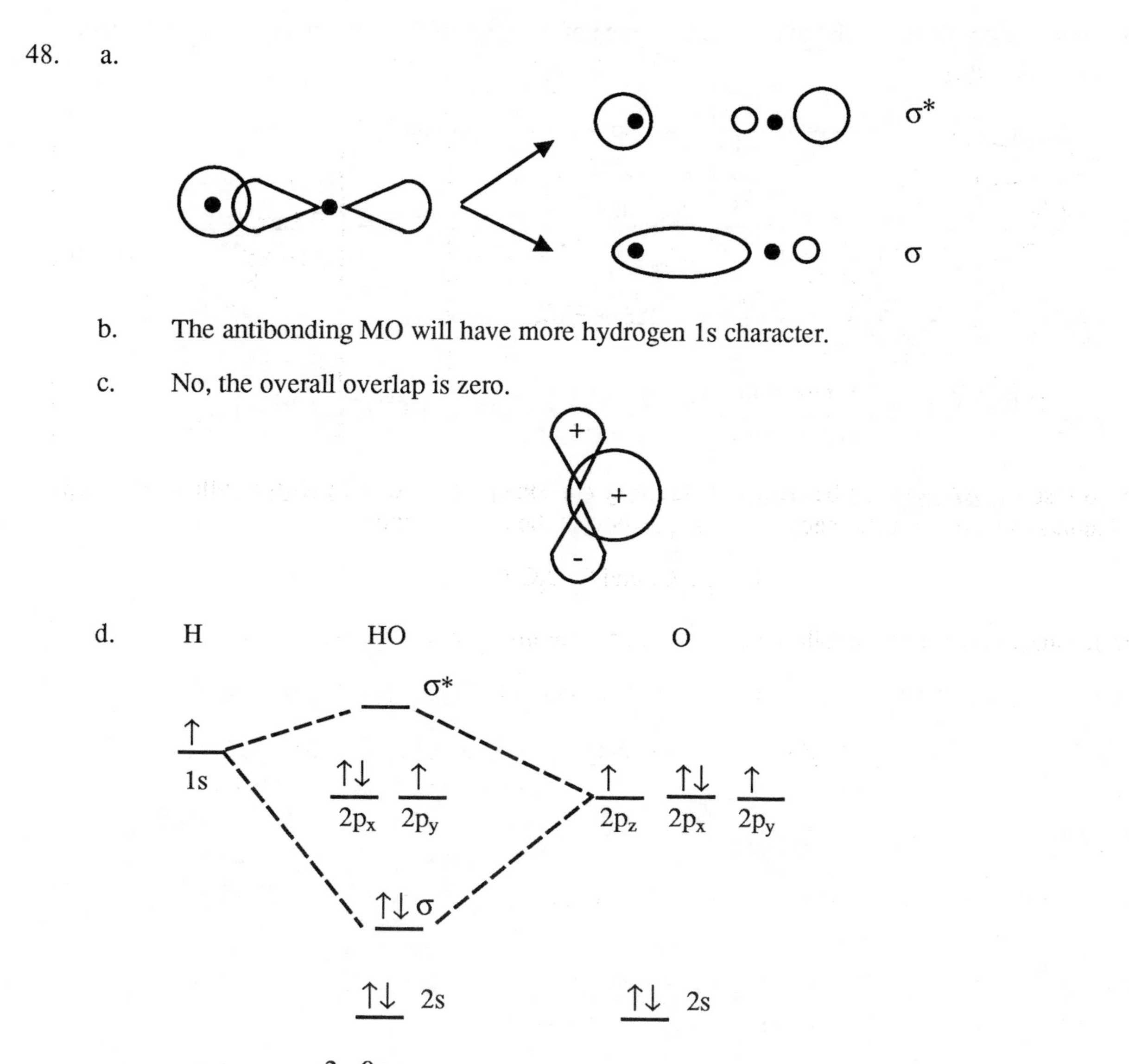

b. The antibonding MO will have more hydrogen 1s character.

c. No, the overall overlap is zero.

d.

e. Bond order = $\frac{2 - 0}{2} = 1$ Note: $2p_x$ and $2p_y$ electrons have no effect on the bond order.

f. To form OH^+ a non-bonding electron is removed from OH, which has no effect on the bond order.

Bond order = $\frac{2 - 0}{2} = 1$

49. $E = \frac{hc}{\lambda} = \frac{(6.626 \times 10^{-34}\ \text{J s})(2.998 \times 10^{8}\ \text{m/s})}{25 \times 10^{-9}\ \text{m}} = 7.9 \times 16^{-18}\ \text{J}$

$7.9 \times 10^{-18}\ \text{J} \times \frac{6.022 \times 10^{23}}{\text{mol}} \times \frac{1\ \text{kJ}}{1000\ \text{J}} = 4800\ \text{kJ/mol}$

25 nm light has sufficient energy to ionize N_2 and N, and to break the triple bond. Thus, N_2, N_2^+, N, and N^+ will all be present.

To produce atomic nitrogen but no ions the range of energies of the light must be from 941 kJ/mol to 1402 kJ/mol.

$$941 \text{ kJ/mol} \times \frac{1 \text{ mol}}{6.022 \times 10^{23}} \times \frac{1000 \text{ J}}{\text{kJ}} = 1.56 \times 10^{-18} \text{ J/photon}$$

$$\lambda = \frac{hc}{E} = \frac{(6.626 \times 10^{-34} \text{ J s})(2.998 \times 10^{8} \text{ m/s})}{1.56 \times 10^{-18} \text{ J}} = 1.27 \times 10^{-7} \text{ m} = 127 \text{ nm}$$

$$\frac{1402 \text{ kJ}}{\text{mol}} \times \frac{1 \text{ mol}}{6.022 \times 10^{23}} \times \frac{1000 \text{ J}}{\text{kJ}} = 2.328 \times 10^{-18} \text{ J/photon}$$

$$\lambda = \frac{hc}{E} = \frac{(6.626 \times 10^{-34} \text{ J s})(2.9979 \times 10^{8} \text{ m/s})}{2.328 \times 10^{-18} \text{ J}} = 8.533 \times 10^{-8} \text{ m} = 85.33 \text{ nm}$$

Light with wavelengths in the range 85.33 nm $< \lambda <$ 127 nm will produce N but no ions.

CHAPTER TEN: LIQUIDS AND SOLIDS

QUESTIONS

1. There is an electrostatic attraction between the permanent dipoles of the polar molecules. The greater the polarity, the greater the attraction among molecules, the stronger the intermolecular forces.

2. London dispersion (LDF) < dipole-dipole < H-bonding < metallic bonding, covalent network, ionic.

   Yes, there is considerable overlap. Consider some of the examples in Exercise 10.30. Benzene (only LDF) has a higher boiling point than acetone (dipole-dipole). Also, there is even more overlap of the stronger forces (metallic, covalent, and ionic).

3. As the size of the molecules increases, the strength of the London dispersion forces also increase. As the electron cloud gets larger it is easier for the electrons to be drawn away from the nucleus (more polarizable).

4. Yes, there are some substances in which only dispersion forces are present such as naphthalene, $C_{10}H_{18}$, and polyethylene that are solids at room temperature. That these substances are solids at room temperature tells us that their interparticle forces are stronger than those that are liquids at room temperature, such as water in which there are hydrogen bonds.

5. As the strengths of interparticle forces increase; surface tension, viscosity, melting point, and boiling point increase, while the vapor pressure decreases. See discussion in text.

6. The nature of the forces stays the same. As the temperature increases and the phase changes, solid $\rightarrow$ liquid $\rightarrow$ gas, occur, a greater fraction of the forces are overcome by the increased thermal (kinetic) energy of the particles.

7. Dipole forces are generally weaker than hydrogen bonding. They are similar in that they arise from an unequal sharing of electrons. We can look at hydrogen bonding as a particularly strong dipole force.

8. a. Polarizability of an atom refers to the ease of distorting the electron cloud. It can also refer to distorting the electron clouds in molecules or ions. Polarity refers to the presence of a permanent dipole moment in a molecule.

   b. London dispersion forces are present in all substances. LDF can be referred to as accidental dipole-induced dipole forces. Dipole-dipole forces involve the attraction of molecules with permanent dipoles for each other.

   c. inter: between

      intra: within

      For example, in H_2 the covalent bond is an intramolecular force, holding the two H-atoms together in the molecule. The much weaker London dispersion forces are intermolecular forces of attraction.

9. Liquids and solids both have characteristic volume and are not very compressible. Liquids and gases flow and assume the shape of their container.

10. Critical temperature: The temperature above which a liquid cannot exist, i.e. the gas cannot be liquified by increased pressure.

Critical pressure: The pressure that must be applied to a substance at its critical temperature to produce a liquid.

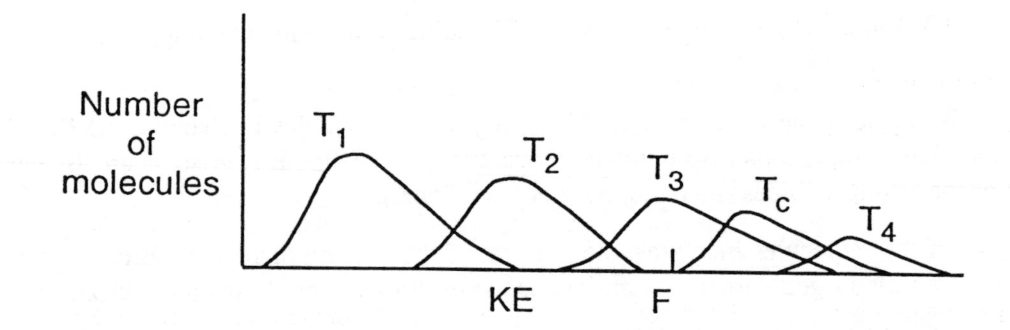

The kinetic energy distribution changes as one raises the temperature ($T_4 > T_c > T_3 > T_2 > T_1$). At the critical temperature, T_c, all molecules have kinetic energies greater than the interparticle forces, (F) and a liquid can't form.

11. As the interparticle forces increase the critical temperature increases.

12. When a liquid evaporates, the molecules that escape have larger kinetic energies. The average kinetic energy of the remaining molecules is lower, thus, the temperature of the liquid is lower.

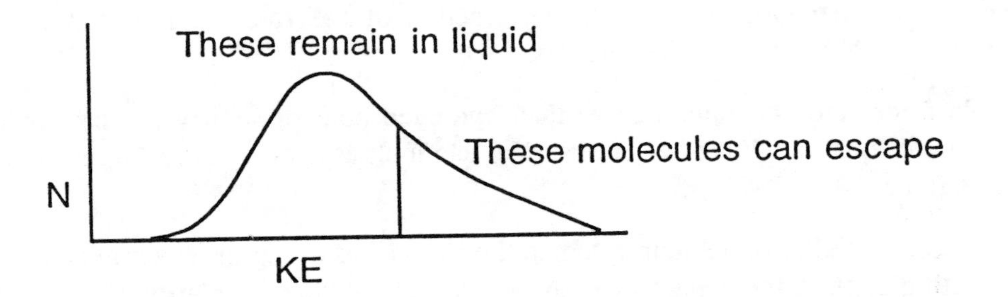

13. a. crystalline solid: regular, repeating structure.

amorphous solid: irregular arrangement of atoms or molecules.

b. ionic solid: made up of ions held together by ionic bonding.

molecular solid: made up of discrete covalently bonded molecules held together in the solid by weaker forces (LDF, dipole, or hydrogen bonds).

c. molecular solid: discrete molecules.

covalent network solid: no discrete molecules. A covalent network is like a large polymer. The interparticle forces are the covalent bonds between atoms.

d. metallic solid: completely delocalized electrons, conductor of electricity (ions in a sea of electrons).

covalent network: localized electrons, insulator or semiconductor.

14. A crystalline solid because a regular, repeating arrangement is necessary to produce planes of atoms that will diffract the x-rays.

15. No, an example is common glass which is primarily amorphous SiO_2 (a covalent network solid) as compared to ice (a crystalline solid held together by weaker H-bonds).

The interparticle forces in the amorphous solid in this case are stronger than those in the crystalline solid. Whether or not a solid is amorphous or crystalline depends on the long range order in the solid and not on the strengths of the interparticle forces.

16. conductor: partially filled valence band: Electrons can move through valence band.

insulator: filled valence band, large band gap: Electrons cannot move in valence band and cannot jump to next band.

semiconductor: filled valence band, small band gap: Electrons cannot move in valence band but can jump to next band.

17. a. As the temperature is increased, more electrons in the valence band have sufficient kinetic energy, KE, to jump from the valence band to the conduction band.

b. A photon of light is absorbed by an electron which then goes from the valence band to the conduction band.

c. An impurity either adds electrons at an energy near that of the conduction band (n-type) or adds holes at an energy near that of the valence band (p-type).

18. In metallic conductors the electrical conductivity is inversely proportional to the absolute temperature. Increases in temperature causes motions of atoms in the metal which gives rise to increased resistance. In a semiconductor the electrical conductivity is proportional to the temperature. An increase in temperature gives some electrons enough kinetic energy to jump from the valence band to the conduction band, increasing conductivity.

19. a. condensation: vapor $\rightarrow$ liquid

b. evaporation: liquid $\rightarrow$ vapor

c. sublimation: solid $\rightarrow$ vapor

d. A supercooled liquid is a liquid which is at a temperature below its freezing point.

20. equilibrium: There is no change in composition; the vapor pressure is constant.

dynamic: Two processes, vapor $\rightarrow$ liquid and liquid $\rightarrow$ vapor, are both occurring with equal rates.

21. a. As intermolecular forces increase, the rate of evaporation decreases.

    b. increase T: increase rate

    c. increase surface area: increase rate

22. Although liquid water cannot exist, sublimation may still occur.

23. The phase change, $H_2O(g) \rightarrow H_2O(l)$ releases heat that can cause additional damage. Also steam can be at a temperature greater than 100°C.

24. Only a fraction of the hydrogen bonds are broken in going from solid to liquid. Most of the hydrogen bonds are still present in the liquid and must be broken during the liquid to gas phase change.

EXERCISES

Interparticle Forces and Physical Properties

25. a. ionic b. LDF, dipole c. LDF only

    d. LDF mostly (For all practical purposes, we consider a C — H bond to be non-polar even though there is a small difference in electronegativity.)

    e. metallic f. metallic g. LDF only

    h. H-bonding, LDF

26. a. ionic b. metallic

    c. LDF mostly (C — F bonds are polar, but polymers like teflon are so large the LDF are the predominant interparticle forces.)

    d. LDF e. LDF, dipole f. covalent network

    g. LDF, dipole h. LDF

27. a. OCS: OCS has dipole forces in addition to LDF. CO_2 is non-polar (only has LDF forces). The same reasoning is true for b-d answers.

    b. PF_3 c. SF_2 d. SO_2

28. a. $H_2NCH_2CH_2NH_2$: More extensive hydrogen bonding is possible.

    b. $B(OH)_3$: No, hydrogen bonding in BH_3. For hydrogen bonding an H-atom must be covalently bonded to O, N, or F.

    c. CH_3OH d. HF

29. a. Neopentane is more compact than n-pentane. Thus, there is less area of contact between neopentane molecules: hence, weaker interparticle forces and a lower boiling point.

    b. Ethanol is capable of H-bonding; dimethylether is not.

    c. HF is capable of H-bonding.

d. LiCl, ionic: Ionic forces are much stronger than the intermolecular forces present in molecular solids.

$TiCl_4$: $4 + 4(7) = 32\ e^-$

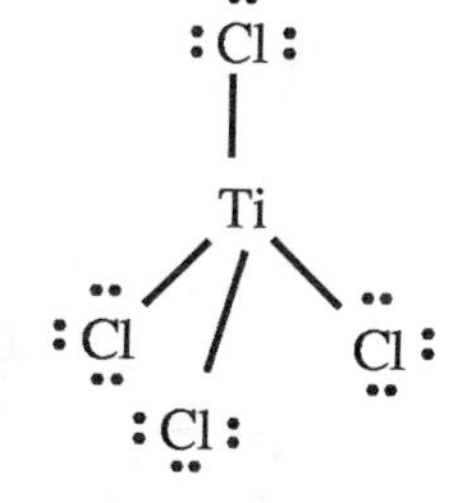

$TiCl_4$ is a non-polar molecular substance with relatively weak London dispersion forces.

e. LiCl, ionic: HCl, molecular; Only dipole forces and LDF between HCl molecules. Ionic forces are much stronger than the molecular forces.

f. Both LiCl and CsCl are ionic. The lattice energy of LiCl is greater than that of CsCl because Li^+ ion is smaller than Cs^+ ion. Thus, stronger forces in LiCl.

30.

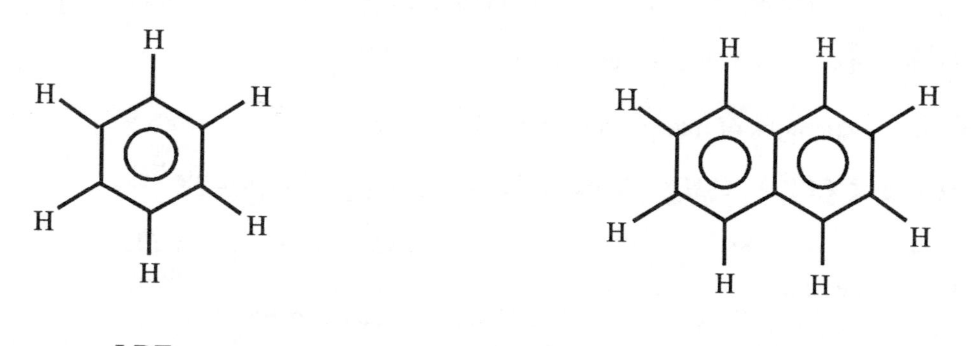

LDF LDF

Note: LDF in molecules like benzene and naphthalene can be fairly large. The molecules are flat and there is efficient contact between molecules.

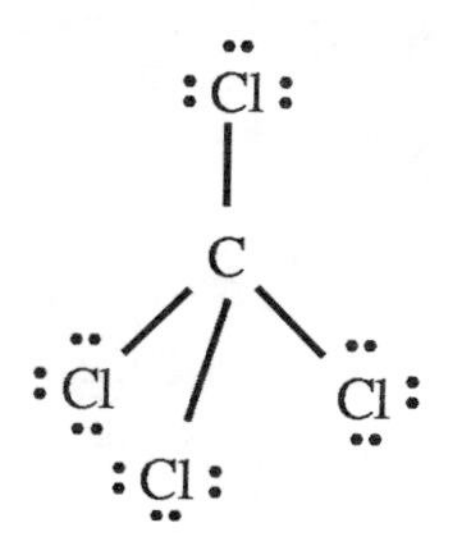

LDF (polar bonds but a non-polar molecule)

In terms of size: $CCl_4 < C_6H_6 < C_{10}H_8$

The strengths of the LDF are proportional to size. The strengths of LDF will be in the same order. The physical properties are consistent with this order. Each of the physical properties will increase with an increase in interparticle forces.

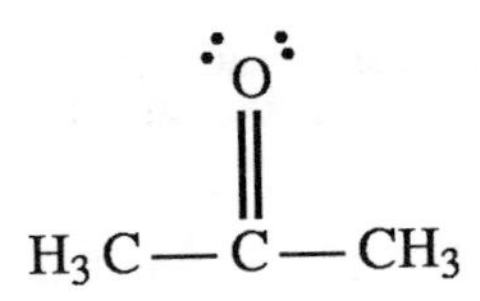

LDF, dipole

$H_3C—C(=O)—O—H$

LDF, dipole, H-bonding

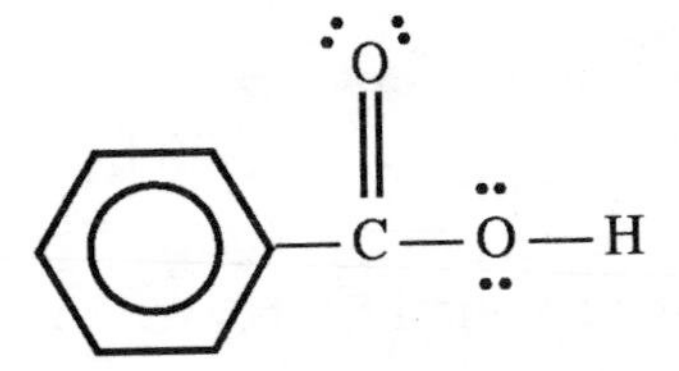

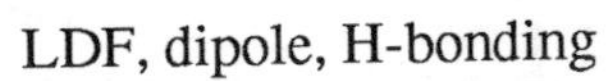

LDF, dipole, H-bonding

We would predict the strength of interparticle forces of the last three molecules to be:

acetone < acetic acid < benzoic acid

polar H-bond H-bond, but large LDF because of greater size and ring.

This is consistent with the values given for bp, mp, and ΔH_{vap}.

The order of the strengths of interparticle forces based on physical properties is:

acetone < CCl_4 < C_6H_6 < acetic acid < naphthalene < benzoic acid

The order seems reasonable except for acetone and naphthalene. Acetone is lowest. We would not predict this because it is polar. The LDF in acetone must be very small compared to the other molecules. Naphthalene has very strong LDF because of its size and shape.

31. As the electronegativity of the atoms bonded to H in a hydrogen bond increases, the strength of the hydrogen bond increases.

N····· H— N< < N····· H— O— < O····· H— O— etc.

weakest

32. FHF^- can be looked upon as forming from an HF molecule and a F^- ion to give equal F — H distances.

$:\ddot{F}—H + :\ddot{F}: \longrightarrow [:\ddot{F}—H—\ddot{F}:]^-$

linear

33. a. NaCl strong ionic bonding in lattice

b. CH_3CN polar but no H-bonding

c. H_2 non-polar like CH_4, smaller than CH_4, weaker LDF

d. SiO_2 covalent network solid vs. a gas and a liquid

e. $HOCH_2CH_2OH$ (ethylene glycol) greatest amount of H-bonding since two —OH groups are present.

f. NH_3 Need N—H, O—H, or F—H for hydrogen bonding.

g. predict HF because of H-bonding - actually HI (see Exercise 10.77)

h. CO_2 non-polar molecular substance, only weak LDF

34. a. CH_3CH_3 smaller size, weaker LDF

b. NaCl strong ionic bonds

c. gasoline non-polar, weakest interparticle forces gives highest vapor pressure

d. N_2 non-polar, smallest, weakest forces

e. H_2O O—H bond more polar than N—H bond

f. CH_4 smallest molecule, weakest LDF

g. MgO ionic bonds stronger than Li_2O because of charge of cations

h. $CH_3CH_2CH_2OH$ H-bonding

Properties of Liquids

35. The attraction of H_2O for glass is stronger than H_2O — H_2O attraction. The miniscus is concave to increase the area of contact between glass and H_2O.

The Hg — Hg attraction is greater than the Hg — glass attraction. The miniscus is convex to minimize the Hg — glass contact. Polyethylene is a non-polar substance. The H_2O — H_2O attraction is stronger than the H_2O — polyethylene attraction; thus, the miniscus will have a convex shape.

36. H_2O will rise higher in a glass capillary because of a greater attraction for glass than polyethylene.

37. The structure of H_2O_2 is H—O—O—H and produces greater hydrogen bonding than water. Long chains of hydrogen bonded H_2O_2 molecules get tangled together.

38. CO_2 is a gas at room temperature. Strengths of forces: $CO_2 < CS_2 < CSe_2$

As mp and bp increase, the strength of interparticle forces increases. The above order is the same as the order of mp and bp. From a structural standpoint this is expected. All three are linear non-polar molecules. Thus, only dispersion forces are present. Since the molecules increase in size, $CO_2 < CS_2 < CSe_2$, the interparticle forces will be in the same order.

Structures and Properties of Solids

39. a. CO_2: molecular
b. SiO_2: covalent network
c. Si: atomic, covalent network
d. CH_4: molecular
e. Ru: atomic, metallic
f. I_2: molecular
g. KBr: ionic
h. H_2O: molecular
i. NaOH: ionic
j. U: atomic, metallic
k. $CaCO_3$: ionic
l. PH_3: molecular

40. a. C: atomic, covalent network
b. CO: molecular
c. P_4: molecular
d. S_8: molecular
e. Mo: atomic, metallic
f. Pt: atomic, metallic
g. NH_4Cl: ionic
h. BaS: ionic
i. Li_2O: ionic
j. Xe: atomic, molecular
k. H_2S: molecular
l. $NaHSO_4$: ionic

41. $n\lambda = 2d \sin\theta$

$$\lambda = \frac{2d \sin\theta}{n} = \frac{2 \times 201 \times 10^{-12} \text{ m} \sin 34.68°}{1}$$

$\lambda = 2.29 \times 10^{-10}$ m = 229 pm = 0.229 nm

42. $n\lambda = 2d \sin\theta$ and 1 Å = 10^{-8} cm = 10^{-10} m = 100 pm

$$\sin\theta = \frac{n\lambda}{2d} = \frac{1 \times 71.2 \text{ pm}}{1993 \text{ pm}} = 0.0357$$

$\theta = 2.05°$

43.

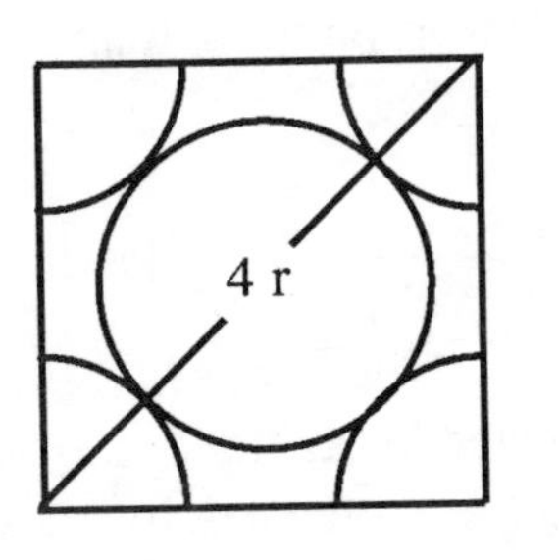

3.92×10^{-8} cm = length of cube edge

4r = length of diagonal

$(4r)^2 = (3.92 \times 10^{-8})^2 + (3.92 \times 10^{-8})^2$

$r = 1.39 \times 10^{-8}$ cm = 139 pm = 1.39 Å

In a face centered cubic unit cell:

$$8 \text{ corners} \times \frac{1/8 \text{ atom}}{\text{corner}} + 6 \text{ faces} \times \frac{1/2 \text{ atom}}{\text{face}} = 4 \text{ atoms}$$

$$\text{Density} = \frac{\text{mass}}{\text{volume}}$$

$$\text{Mass} = 4 \text{ atoms} \times \frac{\text{AW}}{\text{atom}} \times \frac{1 \text{ g}}{6.022 \times 10^{23} \text{ amu}}$$

$$\text{Volume} = (\text{edge})^3 = (3.92 \times 10^{-8} \text{ cm})^3$$

$$21.45 \text{ g/cm}^3 = \frac{4 \times \text{AW} \times \dfrac{1\text{g}}{6.022 \times 10^{23} \text{ amu}}}{(3.92 \times 10^{-8} \text{ cm})^3}$$

AW = 194.5 amu ≈ 195 g/mol. The metal is platinum.

44. There are 4 Ir atoms in the unit cell. (See Exercise 10.43.)

$$22.61 \text{ g/cm}^3 = \frac{4 \times 192.2 \text{ amu} \times \dfrac{1 \text{ g}}{N_o \text{ amu}}}{(3.833 \times 10^{-8} \text{ cm})^3}$$

$$N_o = \frac{4 \times 192.2}{22.61 \times (3.833 \times 10^{-8})^3} = 6.038 \times 10^{23}$$

45. Body centered unit cell:

$$8 \text{ corners} \times \frac{1/8 \text{ Ti}}{\text{corner}} + 1 \text{ Ti at body center} = 2 \text{ Ti atoms}$$

$$4.50 \text{ g/cm}^3 = \frac{2(47.88 \text{ amu}) \times \dfrac{1 \text{ g}}{6.022 \times 10^{23} \text{ amu}}}{l^3}$$

l = length of unit cell = 3.28×10^{-8} cm = 328 pm

Assume Ti atoms just touch along the body diagonal of the cube:

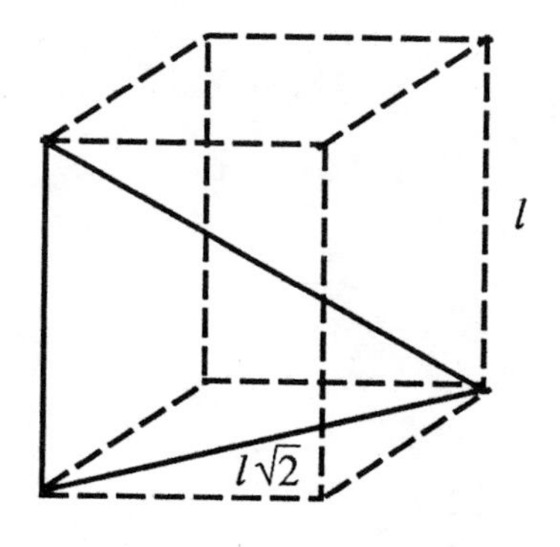

The triangle we need to solve is:

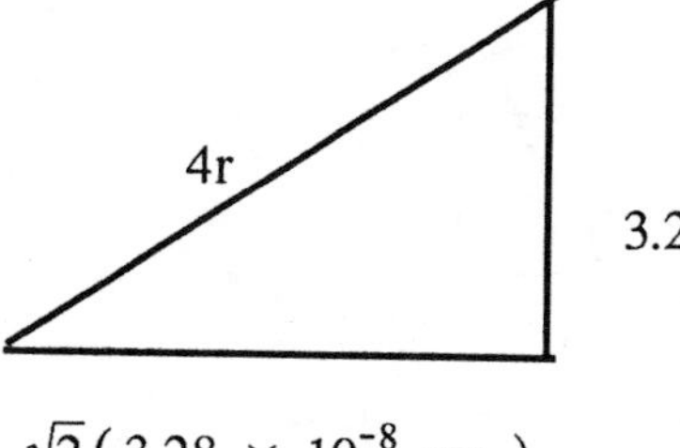

$$(4r)^2 = (3.28 \times 10^{-8})^2 + [\sqrt{2}(3.28 \times 10^{-8})]^2$$

$$r = 1.42 \times 10^{-8} \text{ cm} = 142 \text{ pm} = 1.42 \text{ Å}$$

46. From Exercise 10.45:

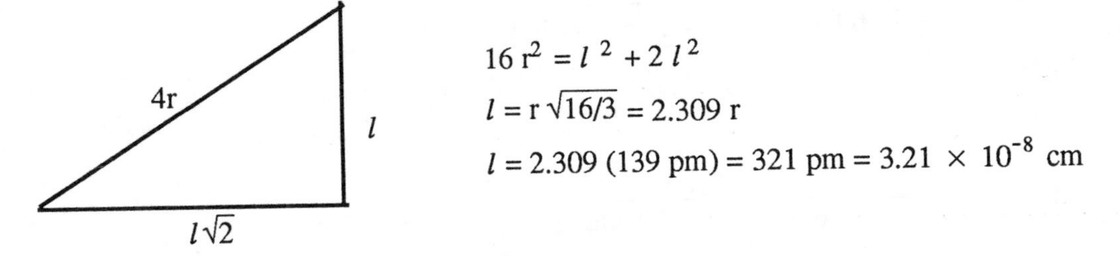

$$16\,r^2 = l^2 + 2\,l^2$$

$$l = r\sqrt{16/3} = 2.309\ r$$

$$l = 2.309\ (139\ \text{pm}) = 321\ \text{pm} = 3.21 \times 10^{-8}\ \text{cm}$$

In bcc, there are 2 atoms/unit cell.

$$D = \frac{2(183.9\ \text{amu})}{(3.21 \times 10^{-8}\ \text{cm})^3} \times \frac{1\ \text{g}}{6.022 \times 10^{23}\ \text{amu}} = \frac{18.5\ \text{g}}{\text{cm}^3}$$

47. There are 4 Ni atoms in the unit cell.

$$\text{Density} = \frac{\text{Mass}}{\text{Volume}}$$

$$6.84\ \text{g/cm}^3 = \frac{4 \times 58.69\ \text{amu} \times \dfrac{1\ \text{g}}{6.022 \times 10^{23}\ \text{amu}}}{l^3}$$

$$l = 3.85 \times 10^{-8}\ \text{cm}$$

For a face centered cube:

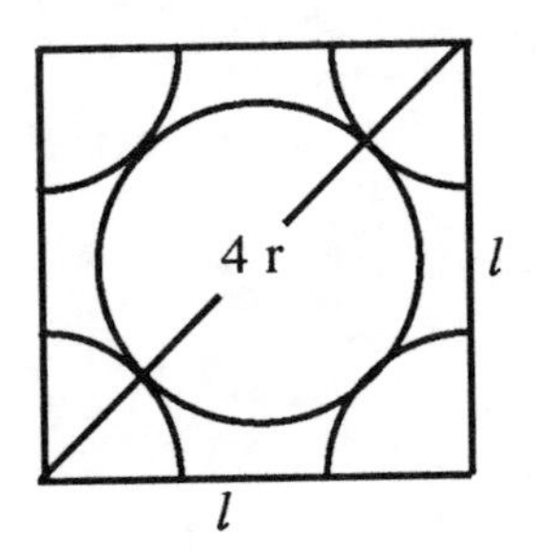

$$(4r)^2 = l^2 + l^2 = 2\,l^2$$

$$16\,r^2 = 2(3.85 \times 10^{-8}\ \text{cm})^2$$

$$r = 1.36 \times 10^{-8}\ \text{cm} = 1.36\ \text{Å} = 136\ \text{pm}$$

48. If fcc structure: 4 atoms/unit cell

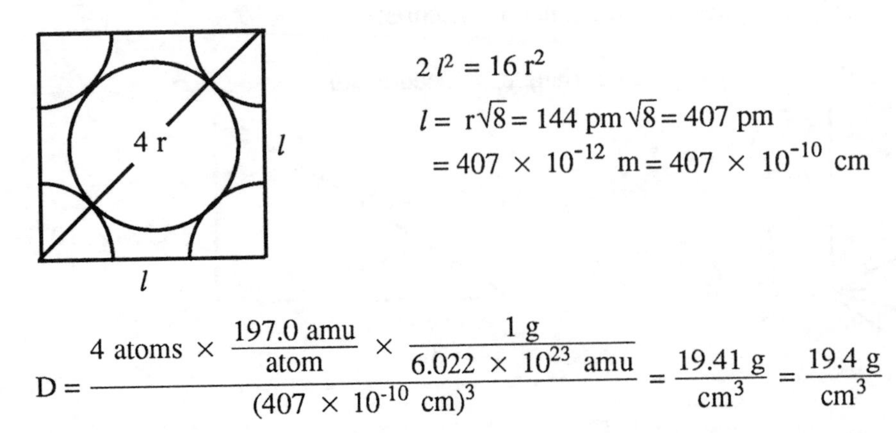

$$2\,l^2 = 16\,r^2$$

$$l = r\sqrt{8} = 144\ \text{pm}\sqrt{8} = 407\ \text{pm}$$

$$= 407 \times 10^{-12}\ \text{m} = 407 \times 10^{-10}\ \text{cm}$$

$$D = \frac{4\ \text{atoms} \times \dfrac{197.0\ \text{amu}}{\text{atom}} \times \dfrac{1\ \text{g}}{6.022 \times 10^{23}\ \text{amu}}}{(407 \times 10^{-10}\ \text{cm})^3} = \frac{19.41\ \text{g}}{\text{cm}^3} = \frac{19.4\ \text{g}}{\text{cm}^3}$$

If bcc: 2 atoms/unit cell

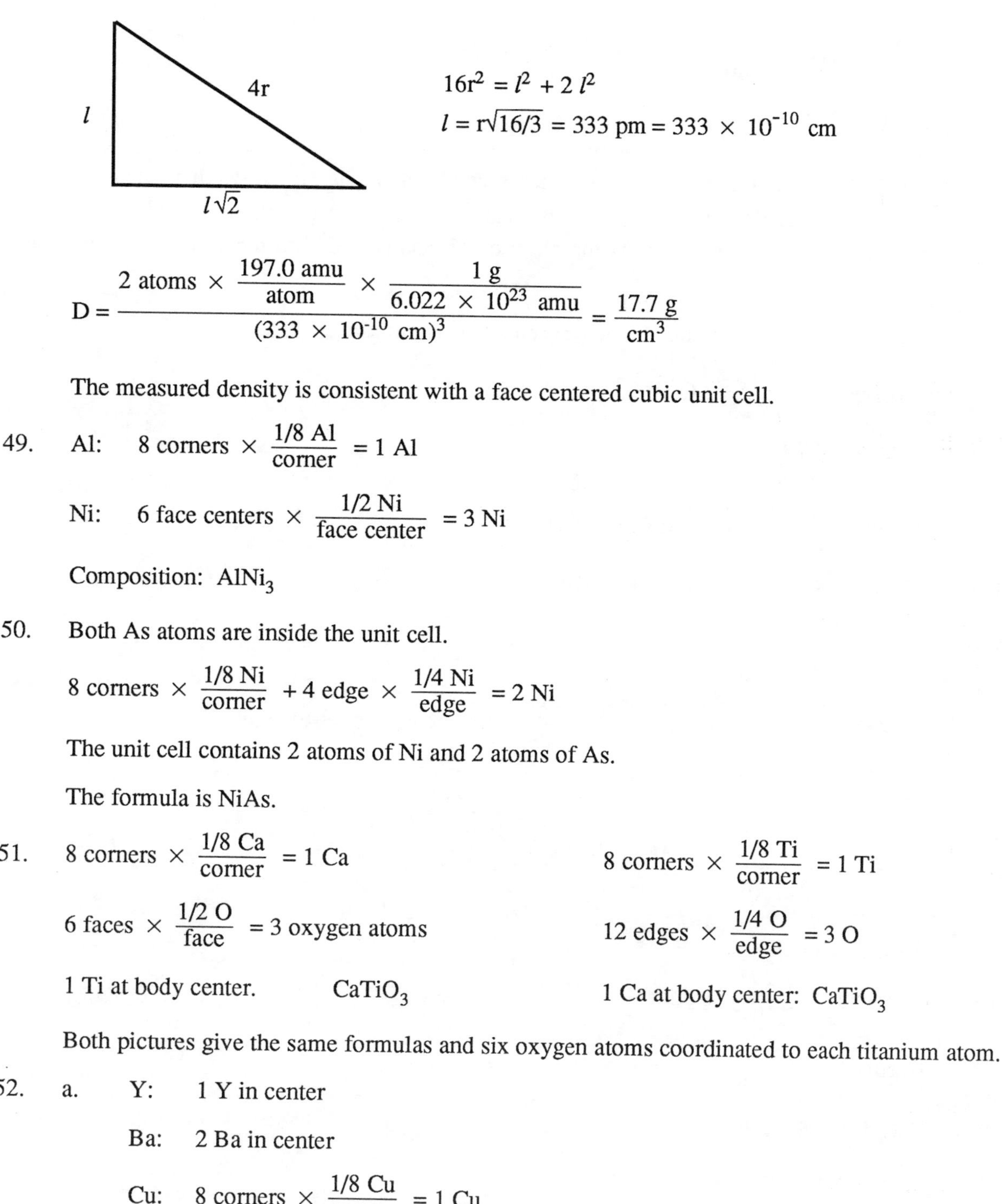

$$16r^2 = l^2 + 2\,l^2$$

$$l = r\sqrt{16/3} = 333 \text{ pm} = 333 \times 10^{-10} \text{ cm}$$

$$D = \frac{2 \text{ atoms} \times \frac{197.0 \text{ amu}}{\text{atom}} \times \frac{1 \text{ g}}{6.022 \times 10^{23} \text{ amu}}}{(333 \times 10^{-10} \text{ cm})^3} = \frac{17.7 \text{ g}}{\text{cm}^3}$$

The measured density is consistent with a face centered cubic unit cell.

49. Al: $8 \text{ corners} \times \frac{1/8 \text{ Al}}{\text{corner}} = 1 \text{ Al}$

Ni: $6 \text{ face centers} \times \frac{1/2 \text{ Ni}}{\text{face center}} = 3 \text{ Ni}$

Composition: $AlNi_3$

50. Both As atoms are inside the unit cell.

$$8 \text{ corners} \times \frac{1/8 \text{ Ni}}{\text{corner}} + 4 \text{ edge} \times \frac{1/4 \text{ Ni}}{\text{edge}} = 2 \text{ Ni}$$

The unit cell contains 2 atoms of Ni and 2 atoms of As.

The formula is NiAs.

51. $8 \text{ corners} \times \frac{1/8 \text{ Ca}}{\text{corner}} = 1 \text{ Ca}$

$6 \text{ faces} \times \frac{1/2 \text{ O}}{\text{face}} = 3$ oxygen atoms

1 Ti at body center. $CaTiO_3$

$8 \text{ corners} \times \frac{1/8 \text{ Ti}}{\text{corner}} = 1 \text{ Ti}$

$12 \text{ edges} \times \frac{1/4 \text{ O}}{\text{edge}} = 3 \text{ O}$

1 Ca at body center: $CaTiO_3$

Both pictures give the same formulas and six oxygen atoms coordinated to each titanium atom.

52. a. Y: 1 Y in center

Ba: 2 Ba in center

Cu: $8 \text{ corners} \times \frac{1/8 \text{ Cu}}{\text{corner}} = 1 \text{ Cu}$

$8 \text{ edges} \times \frac{1/4 \text{ Cu}}{\text{edge}} = 2 \text{ Cu}$

O: $20 \text{ edges} \times \frac{1/4 \text{ O}}{\text{edge}} = 5 \text{ oxygen}$

$8 \text{ faces} \times \frac{1/2 \text{ O}}{\text{face}} = 4 \text{ oxygen}$

Formula: $YBa_2Cu_3O_9$

b. The $YBa_2Cu_3O_9$ structure is based on the second perovskite structure. It is composed of three of these cubic perovskite unit cells stacked on top of each other. The oxygen atoms are in the same places, Cu takes the place of Ti, and two Ca are replaced by Ba and one replaced by Y.

c. Y, Ba, and Cu are the same. Some oxygen atoms are missing.

$12 \text{ edges} \times \frac{1/4 \text{ O}}{\text{edge}} = 3 \text{ O}$

$8 \text{ faces} \times \frac{1/2 \text{ O}}{\text{face}} = 4 \text{ O}$

Superconductor is $YBa_2Cu_3O_7$

53. $E = 2.5 \text{ eV} \times 1.6 \times 10^{-19} \text{ J/eV} = 4.0 \times 10^{-19} \text{ J}$

$E = \frac{hc}{\lambda}$

$$\lambda = \frac{hc}{E} = \frac{(6.63 \times 10^{-34} \text{ J s})(3.00 \times 10^8 \text{ m/s})}{4.0 \times 10^{-19} \text{ J}} = 4.97 \times 10^{-7} \text{ m} = 497 \text{ nm}$$

$\approx 5.0 \times 10^2 \text{ nm}$

54. $$E = \frac{hc}{\lambda} = \frac{(6.63 \times 10^{-34} \text{ J s})(3.00 \times 10^8 \text{ m/s})}{730 \times 10^{-9} \text{ m}} = 2.7 \times 10^{-19} \text{ J}$$

= band gap energy

55. a. $8 \text{ corners} \times \frac{1/8 \text{ Cl}}{\text{corner}} + 6 \text{ faces} \times \frac{1/2 \text{ Cl}}{\text{face}} = 4 \text{ Cl}$

$12 \text{ edges} \times \frac{1/4 \text{ Na}}{\text{edge}} + 1 \text{ Na at body center} = 4 \text{ Na}$

NaCl is the formula.

b. 1 Cs at body center

$8 \text{ corners} \times \frac{1/8 \text{ Cl}}{\text{corner}} = 1 \text{ Cl}$

CsCl is the formula.

c. There are 4 Zn inside the cube.

$$8 \text{ corners} \times \frac{1/8 \text{ S}}{\text{corner}} + 6 \text{ faces} \times \frac{1/2 \text{ S}}{\text{face}} = 4 \text{ S}$$

ZnS is the formula.

d. $$8 \text{ corners} \times \frac{1/8 \text{ Ti}}{\text{corner}} + 1 \text{ Ti at body center} = 2 \text{ Ti}$$

$$4 \text{ faces} \times \frac{1/2 \text{ O}}{\text{face}} + 2 \text{ O inside cube} = 4 \text{ O}$$

TiO_2 is the formula.

56. $$8 \text{ corners} \times \frac{1/8 \text{ Xe}}{\text{corner}} + 1 \text{ center} \times \frac{1 \text{ Xe}}{\text{center}} = 2 \text{ Xe}$$

$$8 \text{ edges} \times \frac{1/4 \text{ F}}{\text{edge}} + 2 \text{ F inside cell} = 4 \text{ F}$$

Formula is XeF_2.

$$\text{mass} = [2(131.3 \text{ amu}) + 4(19.00 \text{ amu})] \times \frac{1 \text{ g}}{6.022 \times 10^{23} \text{ amu}} = 5.62 \times 10^{-22} \text{ g}$$

$$\text{volume} = (702 \times 10^{-12} \text{ m})(432 \times 10^{-12} \text{ m})(432 \times 10^{-12} \text{ m})$$

$$= 1.31 \times 10^{-28} \text{ m}^3 \times \left(\frac{100 \text{ cm}}{\text{m}}\right)^3 = 1.31 \times 10^{-22} \text{ cm}^3$$

$$D = \frac{\text{mass}}{\text{volume}} = \frac{5.62 \times 10^{-22} \text{ g}}{1.31 \times 10^{-22} \text{ cm}^3} = 4.29 \text{ g/cm}^3$$

57. In has fewer valence electrons than Se, thus, Se doped with In would be a p-type semiconductor.

58. To make a p-type semiconductor we need to dope the material with atoms with fewer valence electrons. The average number of valence electrons is four. We could dope with more of the Group 3A elements or with atoms of Zn or Cd. Cadmium is the most common impurity used to produce p-type GaAs semiconductors. To make an n-type GaAs semiconductor, dope with an excess Group 5 element or dope with a Group 6 element such as sulfur.

Phase Changes and Phase Diagrams

59. At 100°C (373 K) the vapor pressure of H_2O is 1.00 atm.

$$\ln\left(\frac{P_2}{P_1}\right) = \frac{\Delta H}{R}\left(\frac{1}{T_1} - \frac{1}{T_2}\right)$$

$$\ln\left(\frac{P_2}{1.00}\right) = \frac{40.7 \times 10^3 \text{ J/mol}}{8.3145 \text{ J/K•mol}}\left(\frac{1}{373 \text{ K}} - \frac{1}{388 \text{ K}}\right)$$

$\ln P_2 = 0.51$, $P_2 = e^{0.51} = 1.7$ atm (See Appendix One for a review of natural logarithms)

$$\ln\left(\frac{3.5}{1.0}\right) = \frac{40.7 \times 10^3 \text{ J/mol}}{8.3145 \text{ J/K•mol}}\left(\frac{1}{373} - \frac{1}{T_2}\right) = 1.253$$

$$2.6 \times 10^{-4} = \left(\frac{1}{373} - \frac{1}{T_2}\right)$$

$$2.6 \times 10^{-4} = 2.68 \times 10^{-3} - \frac{1}{T_2};\ \frac{1}{T_2} = 2.42 \times 10^{-3}$$

$$T_2 = \frac{1}{2.42 \times 10^{-3}} = 413 \text{ K or } 140.°\text{C}$$

60. $$\ln\left(\frac{P_2}{1.00}\right) = \frac{40.7 \times 10^3 \text{ J/mol}}{8.3145 \text{ J/K•mol}}\left(\frac{1}{373 \text{ K}} - \frac{1}{623 \text{ K}}\right)$$

$$\ln P_2 = 5.27$$

$$P_2 = e^{5.27} = 194 \text{ atm} = 190 \text{ atm}$$

61. $H_2O(s, -20°C) \rightarrow H_2O(s, 0°C)$

$$q_1 = C_{ice} \times M \times \Delta T = 2.1 \frac{\text{J}}{\text{g °C}} \times 5.0 \times 10^2 \text{ g} \times 20.°\text{C} = 2.1 \times 10^4 \text{ J} = 21 \text{ kJ}$$

$H_2O(s, 0°C) \rightarrow H_2O(l, 0°C)$

$$q_2 = 5.0 \times 10^2 \text{ g } H_2O \times \frac{1 \text{ mol}}{18.02 \text{ g}} \times 6.0 \frac{\text{kJ}}{\text{mol}} = 170 \text{ kJ}$$

$H_2O(l, 0°C) \rightarrow H_2O(l, 100°C)$

$$q_3 = 4.2 \frac{\text{J}}{\text{g °C}} \times 5.0 \times 10^2 \text{ g} \times 100.°\text{C} = 2.1 \times 10^5 \text{ J} = 210 \text{ kJ}$$

$H_2O(l, 100°C) \rightarrow H_2O(g, 100°C)$

$$q_4 = 5.0 \times 10^2 \text{ g} \times \frac{1 \text{ mol}}{18.02 \text{ g}} \times \frac{40.7 \text{ kJ}}{\text{mol}} = 1100 \text{ kJ}$$

$H_2O(g, 100°C) \rightarrow H_2O(g, 250°C)$

$$q_5 = 1.8 \frac{\text{J}}{\text{g °C}} \times 5.0 \times 10^2 \text{ g} \times 150.°\text{C} = 1.4 \times 10^5 \text{ J} = 140 \text{ kJ}$$

$$q_{total} = q_1 + q_2 + q_3 + q_4 + q_5 = 21 + 170 + 210 + 1100 + 140 \approx 1600 \text{ kJ}$$

62. To melt the 10.0 g of ice at 0°C it will take:

$$10.0 \text{ g} \times \frac{1 \text{ mol}}{18.02 \text{ g}} \times 6.02 \frac{\text{kJ}}{\text{mol}} = 3.34 \text{ kJ} = 3340 \text{ J}$$

Only 850. J of heat is added. Some, but not all, of the ice will melt and the temperature will still be 0°C.

63. $\ln\left(\frac{P_1}{P_2}\right) = \frac{\Delta H_{vap}}{R}\left(\frac{1}{T_2} - \frac{1}{T_1}\right)$

At normal boiling point (1), P_1 = 760. torr, T_1 = 56.5 °C = 329.7 K

T_2 = 25°C = 298.2 K

$$\ln\left(\frac{760.}{P_2}\right) = \frac{32.0 \times 10^3 \text{ J/mol}}{8.3145 \text{ J/K}\cdot\text{mol}}\left(\frac{1}{298.2} - \frac{1}{329.7}\right)$$

$6.633 - \ln P_2 = 1.23$

$\ln P_2 = 5.40$

$P_2 = e^{5.40} = 221$ torr

64. $\ln\left(\frac{P_1}{P_2}\right) = \frac{\Delta H_{vap}}{R}\left(\frac{1}{T_2} - \frac{1}{T_1}\right)$

P_1 = 760. torr $\quad T_1$ = 56.5°C $\quad$ the normal bp from previous exercise

P_2 = 630. torr $\quad T_2$ = ?

$$\ln\left(\frac{760.}{630.}\right) = \frac{32.0 \times 10^3 \text{ J/mol}}{8.3145 \text{ J/K}\cdot\text{mol}}\left(\frac{1}{T_2} - \frac{1}{329.7}\right)$$

$$0.188 = 3.85 \times 10^3\left(\frac{1}{T_2} - 3.033 \times 10^{-3}\right)$$

$$\frac{1}{T_2} - 3.033 \times 10^{-3} = 4.88 \times 10^{-5}$$

$$\frac{1}{T_2} = 3.082 \times 10^{-3}$$

T_2 = 324.5 K = 51.3°C

$$\ln\left(\frac{630.}{P_2}\right) = \frac{32.0 \times 10^3}{8.3145}\left(\frac{1}{298} - \frac{1}{324.5}\right)$$

$6.446 - \ln P_2 = 1.05, \quad \ln P_2 = 5.40 \quad P_2 = e^{5.40} = 220$ torr

65. We want to graph ln P vs 1/T. The slope of the resulting straight line will be $-\Delta H/R$. Use Kelvin temperature.

P	ln P	T(Li)	1/T	T(Mg)	1/T
1	0	1023	9.775×10^{-4}	893	11.2×10^{-4}
10.	2.30	1163	8.598×10^{-4}	1013	9.872×10^{-4}
100.	4.61	1353	7.391×10^{-4}	1173	8.525×10^{-4}
400.	5.99	1513	6.609×10^{-4}	1313	7.616×10^{-4}
760.	6.63	1583	6.314×10^{-4}	1383	7.231×10^{-4}

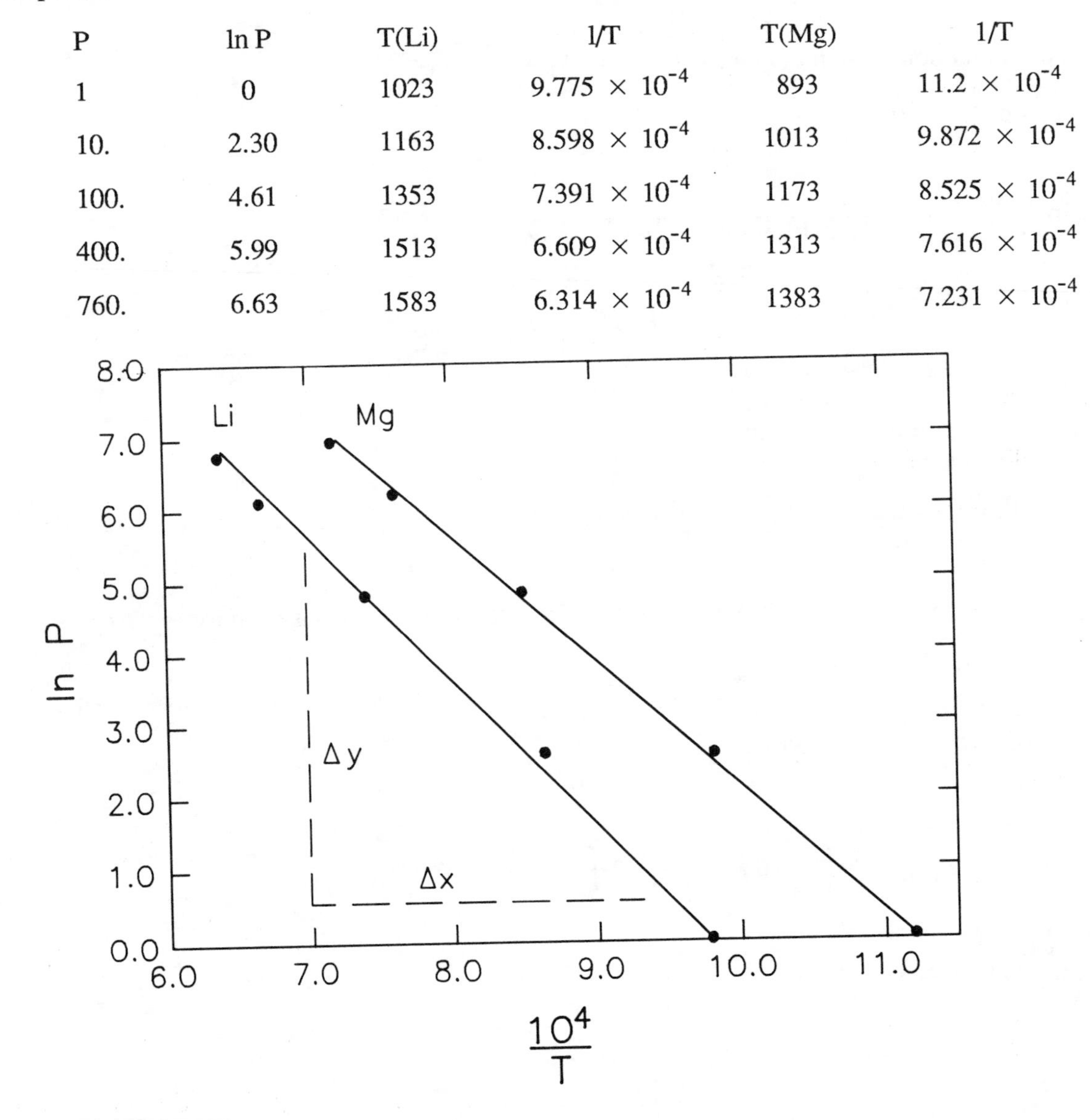

For Li, we get the slope by taking two points (*x*, *y*) that are on the line we draw, not data points. There may be experimental error and individual data points may not fall directly on the line. For a line:

$$\text{Slope} = \frac{\Delta y}{\Delta x} = \frac{y_2 - y_1}{x_2 - x_1}$$

or we can fit the straight line using a computer or calculator.

The equation of this line is:

$$\ln P = -1.90 \times 10^4 (1/T) + 18.6$$

$$\text{Slope} = -1.90 \times 10^4 \text{ K} = \frac{-\Delta H}{R}$$

$$\Delta H = \text{-Slope} \times R = 1.90 \times 10^4 \text{ K} \times 8.3145 \text{ J/mol•K}$$

$$\Delta H = 1.58 \times 10^5 \text{ J/mol} = 158 \text{ kJ/mol}$$

For Mg, the equation of the line is:

$$\ln P = -1.70 \times 10^4 (1/T) + 18.7$$

$$\text{Slope} = -1.70 \times 10^4 \text{ K} = \frac{-\Delta H}{R}$$

$$\Delta H = \text{-Slope} \times R = 1.70 \times 10^4 \text{ K} \times 8.3145 \text{ J/mol•K}$$

$$\Delta H = 1.41 \times 10^5 \text{ J/mol} = 141 \text{ kJ/mol}$$

The bonding is stronger in Li.

66. Again we graph ln P vs 1/T

T(K)	$10^3/T$	P	ln P
273	3.66	14.4	2.67
283	3.53	26.6	3.28
293	3.41	47.9	3.87
303	3.30	81.3	4.40
313	3.19	133	4.89
323	3.10	208	5.34
353	2.83	670	6.51

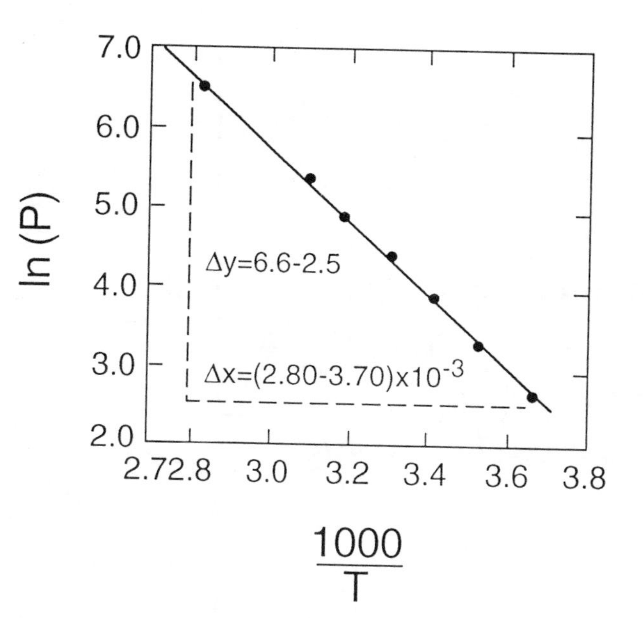

The slope of the line is -4600 K

$$-4600 \text{ K} = \frac{-\Delta H}{R} = \frac{-\Delta H}{8.3145 \text{ J/K•mol}}$$

$$\Delta H = 38 \text{ kJ/mol}$$

$$\ln\left(\frac{P_1}{P_2}\right) = \frac{\Delta H_{vap}}{R}\left(\frac{1}{T_2} - \frac{1}{T_1}\right)$$

At the normal boiling point, the vapor pressure equals 1 atm or 760. torr. At 353 K, the vapor pressure is 670. torr.

$$\ln\left(\frac{670.}{760.}\right) = \frac{38{,}000 \text{ kJ/mol}}{8.3145 \text{ J/K•mol}}\left(\frac{1}{T_2} - 2.83 \times 10^{-3}\right)$$

$$-0.126 = 4.6 \times 10^3 (1/T_2 - 2.83 \times 10^{-3})$$

$$-2.7 \times 10^{-5} + 2.83 \times 10^{-3} = 1/T_2 = 2.80 \times 10^{-3}$$

$$T_2 = 357 \text{ K}$$

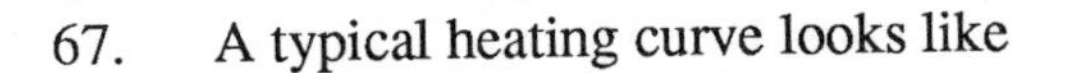

67. A typical heating curve looks like

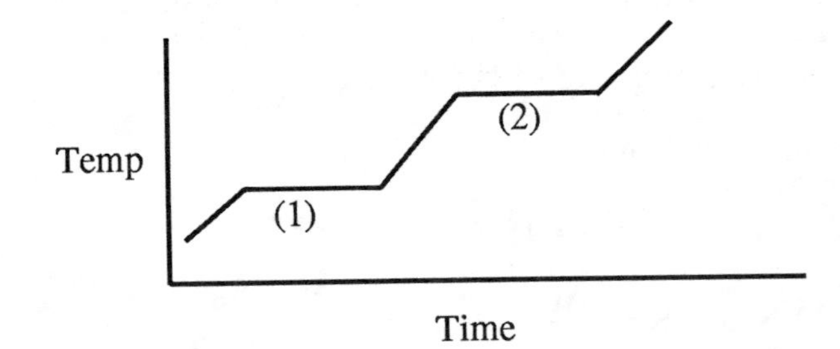

The plateau (1) gives a temperature at which the solid and liquid are in equilibrium. This temperature along with the pressure gives a point on the phase diagram. The plateau (2) will give a temperature on the liquid-vapor line.

68.

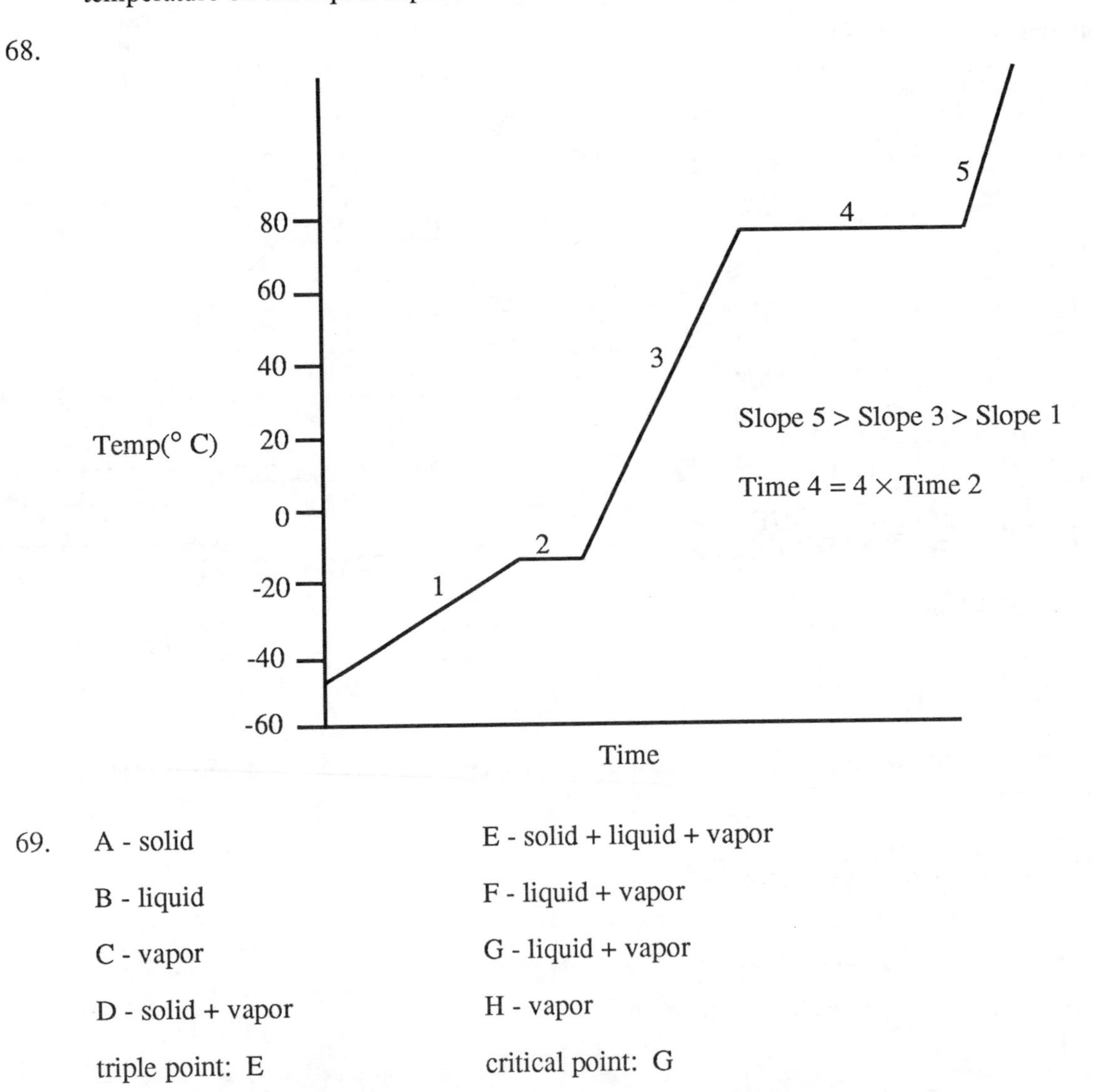

69. A - solid

B - liquid

C - vapor

D - solid + vapor

E - solid + liquid + vapor

F - liquid + vapor

G - liquid + vapor

H - vapor

triple point: E

critical point: G

normal freezing point: temperature at which solid - liquid line is at 1 atm.

normal boiling point: temperature at which liquid - vapor line is at 1 atm.

70. a. 2

b. triple point at 96°C: rhombic, monoclinic, vapor
triple point at 119°C: monoclinic, liquid, vapor

c. rhombic

d. yes

ADDITIONAL EXERCISES

71. CH_3CO_2H H-bonding

CH_2ClCO_2H H-bonding + larger electronegative atom replacing H (greater dipole)

$CH_3CO_2CH_3$ polar

We predict $CH_3CO_2CH_3$ to have the weakest interparticle forces and CH_2ClCO_2H to have the strongest. The boiling points are consistent with this view.

72.

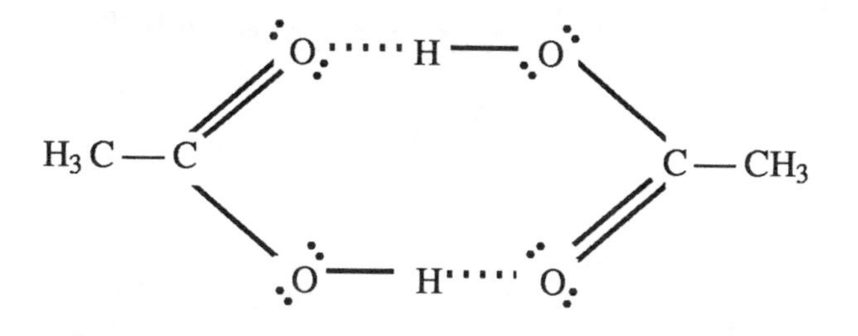

73. $n\lambda = 2d \sin\theta$, $1.54\ \text{Å} = 1.54 \times 10^{-10}\ \text{m} = 154\ \text{pm}$

$$d = \frac{n\lambda}{2 \sin\theta} = \frac{1 \times 154\ \text{pm}}{2 \times \sin 14.22°} = 313\ \text{pm} = 3.13 \times 10^{-10}\ \text{m} = 3.13\ \text{Å}$$

74. A cubic closest packed structure has a face centered cubic unit cell. A face centered cubic unit cell will contain 4 Co atoms. The length of the face diagonal of the unit cell is $4 \times 125 = l\sqrt{2}$. See solution to Exercise 10.43 for diagram.

$l = 354\ \text{pm} = 3.54 \times 10^{-8}\ \text{cm}$

$$D = \frac{\text{mass}}{\text{volume}} = \frac{4 \times 58.93\ \text{amu} \times \dfrac{1\ \text{g}}{6.022 \times 10^{23}\ \text{amu}}}{(3.54 \times 10^{-8}\ \text{cm})^3}$$

$$D = \frac{3.91 \times 10^{-22}\ \text{g}}{4.44 \times 10^{-23}\ \text{cm}^3} = 8.81\ \text{g/cm}^3 \text{ for } \beta \text{ form.}$$

There appears to be a slight difference in density between the two forms.

75. If TiO_2 conducts electricity as a liquid it would be ionic.

$$\Delta EN = (3.5 - 1.5) = 2.0$$

From this difference in electronegativities we might predict TiO_2 to be ionic. It is actually a covalent network solid. The 1.5 is the value of the EN for Ti^{2+}. Ti^{4+} should have a higher electronegativity. See Question 8.3.

76. 1. We can draw a reasonable Lewis structure.

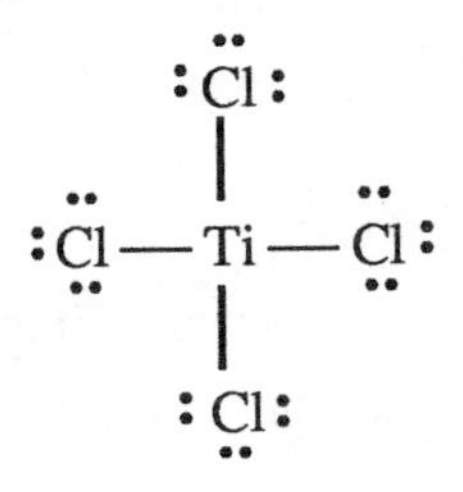

2. If we start with Ti^{4+} ions and Cl^- anions, the Ti^{4+} will have a strong attraction for the electrons on Cl^-.

3. Consider the cases:

$$H^+ + Cl^- \rightarrow H - Cl$$

empty 1s orbital — covalent bond

$$Ti^{4+} + 4\,Cl^- \rightarrow TiCl_4$$

empty 4s, 3d, 4p orbitals — covalent bonds

They really aren't very different in how the covalent bonds form. A pair of electrons on a chloride ion form a bond using empty orbitals on the cation.

77. The boiling point of HF is higher than might be expected because of especially strong hydrogen bonding in HF. ΔH_{vap} should be high also, but is lower than expected because many of the H-bonds reform in the gas phase forming dimers, H—F···H—F.

78. The reciprocal plot is most nearly linear. We would expect this since the lattice energy is given by:

$$LE = \frac{Aq_1 q_2}{r}$$

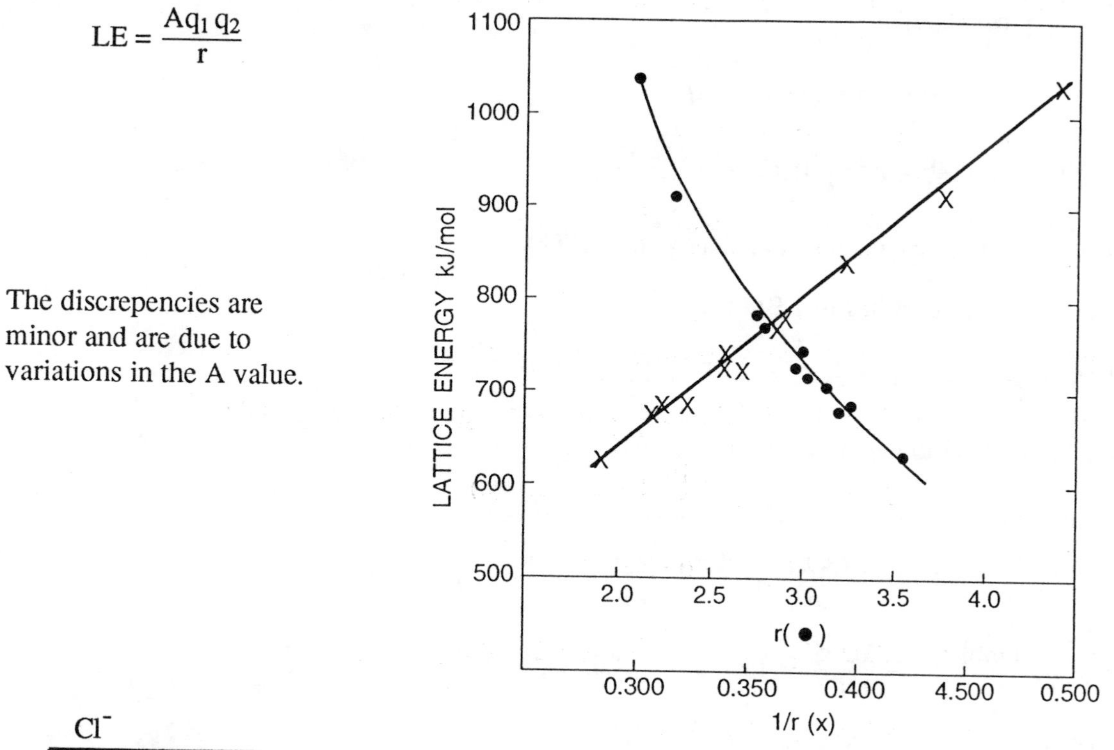

The discrepencies are minor and are due to variations in the A value.

79.

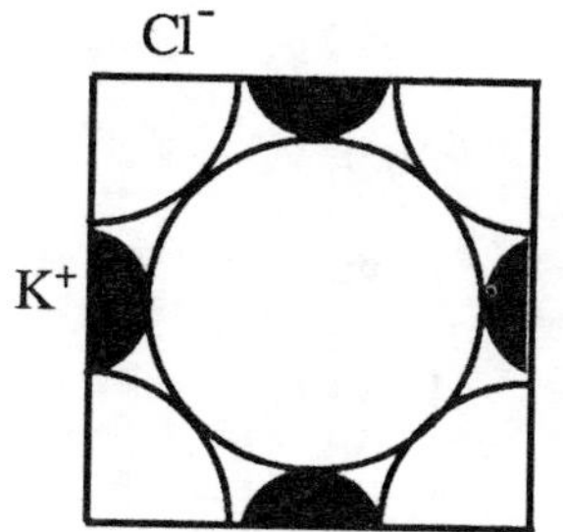

Assuming K^+ and Cl^- just touch along the edge.

$l = 2(314) = 628 \text{ pm} = 6.28 \times 10^{-8} \text{ cm}$

The unit cell contains 4 K^+ and 4 Cl^-

$$D = \frac{4(39.10) \text{ amu} + 4(35.45) \text{ amu}}{(6.28 \times 10^{-8} \text{ cm})^3} \times \frac{1 \text{ g}}{6.022 \times 10^{23} \text{ amu}}$$

$$D = \frac{2.00 \text{ g}}{\text{cm}^3}$$

80. $$\ln\left(\frac{P_1}{P_2}\right) = \frac{\Delta H_{vap}}{R}\left(\frac{1}{T_2} - \frac{1}{T_1}\right)$$

$P_1 = 760. \text{ torr}, T_1 = 630. \text{ K}$

$P_2 = ?$ $\quad T_2 = 298 \text{ K}$

$$\ln 760. - \ln P_2 = \frac{59.1 \times 10^3 \text{ J/mol}}{8.3145 \text{ J/K•mol}}\left(\frac{1}{298 \text{ K}} - \frac{1}{630. \text{ K}}\right)$$

$6.633 - \ln P_2 = 12.6$

$\ln P_2 = -6.0$

$P_2 = e^{-6.0} = 2.5 \times 10^{-3}$ torr

81. $1.00 \text{ lb} \times 454 \frac{\text{g}}{\text{lb}} = 454 \text{ g } H_2O$

A change of 1.00°F is equal to a change of 5/9°C.

The amount of heat in J in 1 Btu is:

$4.18 \frac{\text{J}}{\text{g °C}} \times 454 \text{ g} \times \frac{5}{9} \text{°C} = 1.05 \times 10^3 \text{ J or } 1.05 \text{ kJ}$

It takes 40.7 kJ to vaporize 1 mol H_2O (ΔH_{vap}).

Combining these:

$$1.00 \times 10^4 \frac{\text{Btu}}{\text{hr}} \times \frac{1.05 \text{ kJ}}{\text{Btu}} \times \frac{1 \text{ mol } H_2O}{40.7 \text{ kJ}} = 258 \text{ mol/hr}$$

or: $\frac{258 \text{ mol}}{\text{hr}} \times \frac{18.02 \text{ g}}{\text{mol}} = 4650 \text{ g/hr} = 4.65 \text{ kg/hr}$

82.

T(°C)	T(K)	1/T	P	ln P
-6.0	267.2	3.743×10^{-3}	20.0	2.996
5.0	278.2	3.595×10^{-3}	40.0	3.689
12.1	285.3	3.505×10^{-3}	50.0	4.094
21.2	294.4	3.397×10^{-3}	100.0	4.605
49.9	323.1	3.095×10^{-3}	400.0	5.991

From graph:

$\ln P = -4620/T + 20.29$

$\text{Slope} = \frac{-\Delta H_{vap}}{R} = -4.62 \times 10^3 \text{ K}^{-1}$

$\Delta H_{vap} = 4.62 \times 10^3 \text{ K}^{-1} \left(\frac{8.3145 \text{ J}}{\text{mol K}}\right)$

$= 38{,}400 \text{ J/mol} = 38.4 \text{ kJ/mol}$

At normal boiling point, $P = 760.$ torr

$\ln 760. = -4620/T_b + 20.29$

$T = \frac{-4620}{6.63 - 20.29} = 338 \text{ K} = 65\text{°C}$

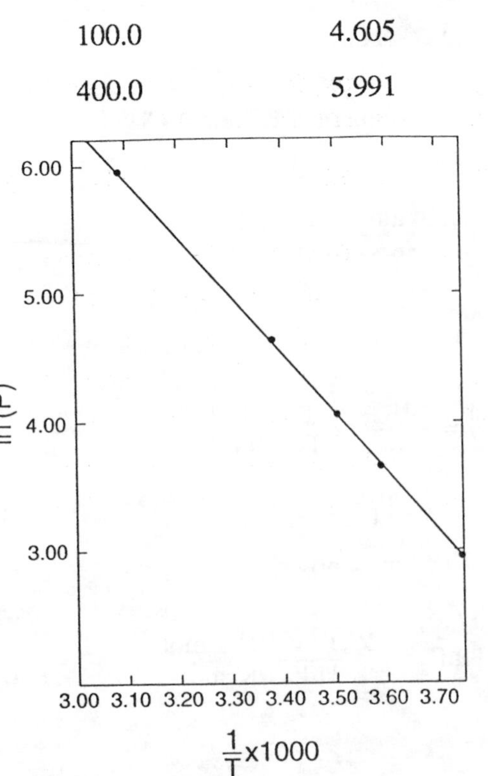

83. a. two

b. higher pressure triple point: graphite, diamond, and liquid

lower pressure triple point: graphite, liquid, and vapor

c. It is converted to diamond.

d. diamond

CHALLENGE PROBLEMS

84. A single hydrogen bond in H_2O has a strength of 21 kJ/mol. Each H_2O molecule forms two H-bonds. Thus, it should take 42 kJ/mol of energy to break all of the H-bonds in water. For the phase transitions:

$$\text{solid} \xrightarrow{6.0\text{ kJ}} \text{liquid} \xrightarrow{40.7\text{ kJ}} \text{vapor}$$

it takes 46.7 kJ/mol (ΔH_{sub}). This would be the amount of energy to disrupt all of the interparticle forces in water. Thus about, $(42 \div 46.7) \times 100 = 90\%$ of the attraction in water can be attributed to H-bonding.

85. Both molecules are capable of H-bonding. However, in oil of wintergreen the hydrogen bonding is intramolecular.

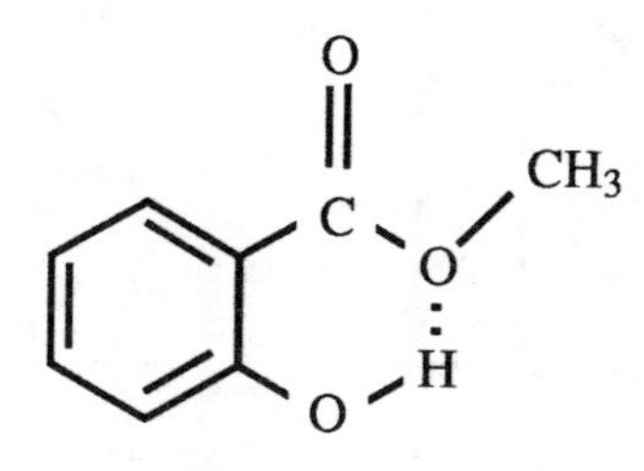

In methyl-4-hydroxybenzoate the H-bonding is intermolecular, resulting in greater intermolecular forces and a higher melting point.

86. The interparticle attractions are slightly stronger in D_2O than in H_2O.

87. NaCl, $MgCl_2$, NaF, MgF_2, AlF_3 all have very high melting points indicative of strong interparticle forces. They are all ionic solids. $SiCl_4$, SiF_4, Cl_2, F_2, PF_5 and SF_6 are non-polar covalent molecules. Only LDF are present. PCl_3 and SCl_2 are polar molecules. LDF and dipole forces are present. In these 8 molecular substances the interparticle forces are weak and the melting points low. $AlCl_3$ doesn't seem to fit in as well. From the melting point, there are much stronger forces present than in the non-metal halides, but they aren't as strong as we would expect for an ionic solid. $AlCl_3$ illustrates a gradual transition from ionic to covalent bonding; from an ionic solid to discrete molecules.

88. One B atom and one N atom together have the same number of electrons as two C-atoms. The description of physical properties sound a lot like the properties of graphite and diamond. The two forms of BN have structures similar to graphite and diamond.

89. $n\lambda = 2d \sin \theta$; 1 Å = 10^{-8} cm = 10^{-10} m = 100 pm

$$d = \frac{n\lambda}{2\sin \theta} = \frac{1 \times 71.2 \text{ pm}}{2 \sin 5.564} = 367.2 \text{ pm} = 3.672 \times 10^{-8} \text{ cm}$$

$$13.28 \frac{g}{cm^3} = \frac{n \times 178.5 \text{ amu}}{(3.672 \times 10^{-8} \text{ cm})^3} \times \frac{1 \text{ g}}{6.022 \times 10^{23} \text{ amu}}$$

where n = the number of Hf atoms per unit cell

n = 2.2 This is most consistent with body centered cubes.

See Exercise 10.45 for geometry.

$$(4r)^2 = (367.2)^2 + (367.2 \sqrt{2})^2$$

r = 159 pm = 1.59 Å

90. For the unit cell:

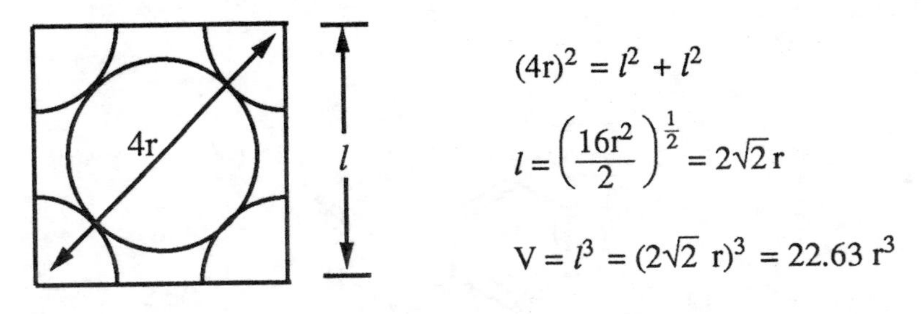

$$(4r)^2 = l^2 + l^2$$

$$l = \left(\frac{16r^2}{2}\right)^{\frac{1}{2}} = 2\sqrt{2}\,r$$

$$V = l^3 = (2\sqrt{2}\ r)^3 = 22.63\ r^3$$

There are four atoms in a face centered cubic unit cell (Ex. 10.43) each with a volume of $4/3\ \pi r^3$.

$$V_{atoms} = 4 \times \frac{4}{3} \pi r^3 = 16.76\ r^3$$

So, $\frac{V_{atom}}{V_{cube}} = \frac{16.76}{22.63} = 0.7406$ or 74.06% of the volume of each unit cell is occupied by atoms.

91. There are atoms at each of the 8 corners (shared by eight unit cells) plus one atom inside the unit cell: 2/3 of one atom plus 1/6 from each of two others. Thus, there are two atoms in the unit cell.

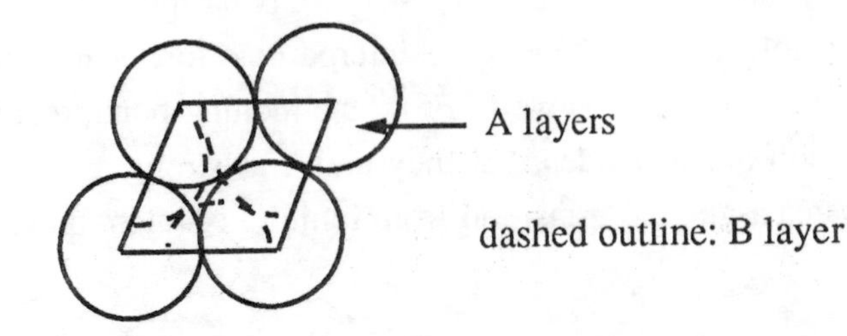

92. There are 2 tetrahedral holes per packing atom.

Let f = fraction of tetrahedral holes filled

K_2O: $\frac{2}{1} = \frac{2f}{1}$; f = 1, all tetrahedral holes are filled.

CuI: $\frac{1}{1} = \frac{2f}{1}$; $f = \frac{1}{2}$, $\frac{1}{2}$ of the tetrahedral holes are filled.

ZrI_4: $\frac{1}{4} = \frac{2f}{1}$; $f = \frac{1}{8}$, $\frac{1}{8}$ of the tetrahedral holes are filled.

93. First, we need to get the empirical formula of spinel. Out of 100 g there are:

$37.9 \text{ g Al} \times \frac{1 \text{ mol Al}}{26.98 \text{ g Al}} = 1.40 \ (2)$

$17.1 \text{ g Mg} \times \frac{1 \text{ mol Mg}}{24.31 \text{ g Mg}} = 0.703 \ (1)$ Empirical formula: Al_2MgO_4

$45.0 \text{ g O} \times \frac{1 \text{ mol O}}{16.00 \text{ g O}} = 2.81 \ (4)$

Each unit cell will contain an integral value (n) of empirical formula units. Each unit has a mass of:

$$(24.31) + 2(26.98) + 4(16.00) = 142.27 \text{ g/mol}$$

$$3.57 \text{ g/cm}^3 = \frac{n(142.27 \text{ g/mol})}{(6.022 \times 10^{23} \text{ /mol})(8.09 \times 10^{-8} \text{ cm})^3}$$

n = 8.00

Each unit cell has: 16 Al; 8 Mg; 32 O

94. Face centered cube = 4 atoms/unit cell

$$10.5 \text{ g/cm}^3 = \frac{4 \text{ (A)}}{(6.022 \times 10^{23} \text{ mol}^{-1})(4.09 \times 10^{-8} \text{ cm})^3}$$

A = 108 g/mol

The closest metal in atomic mass is silver (A = 107.9 g/mol).

95. a)

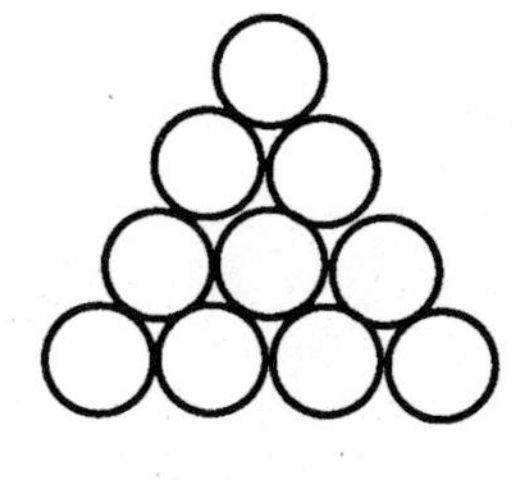

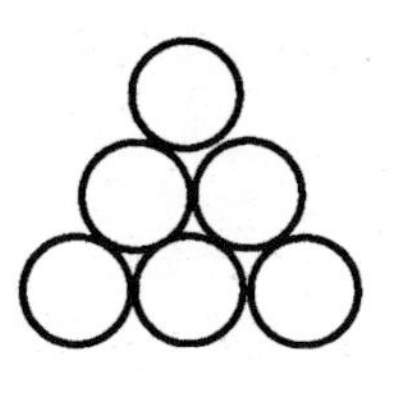

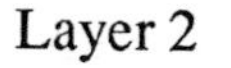

Layer 1 Layer 2 Layer 3 Layer 4

A total of 20 cannonballs will be needed.

b) Cubic closest packed, abc pattern of packing layers.

c) tetrahedron

96.

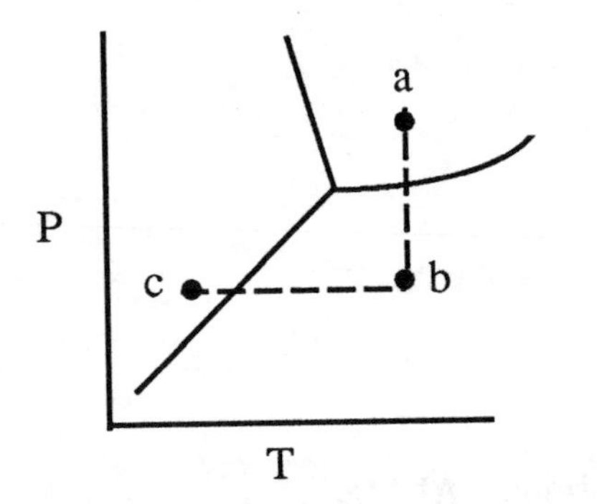

As P is lowered, we go from a to b on the phase diagram. The water boils. The evaporation of the water is endothermic and the water is cooled. b ⟶ c, leaving some ice. If the pump is left on, the ice will sublime until none is left. This is the basis of freeze drying.

97. Heat released:

$$0.250 \text{ g Na} \times \frac{1 \text{ mol}}{22.99 \text{ g}} \times \frac{368 \text{ kJ}}{2 \text{ mol}} = 2.00 \text{ kJ}$$

To melt 50.0 g of ice requires:

$$50.0 \text{ g ice} \times \frac{1 \text{ mol } H_2O}{18.02 \text{ g}} \times 6.02 \text{ kJ/mol} = 16.7 \text{ kJ}$$

The reaction doesn't release enough heat to melt all of the ice. The temperature will remain at 0°C.

CHAPTER ELEVEN: PROPERTIES OF SOLUTIONS

SOLUTION REVIEW

1. $125 \text{ g sucrose} \times \frac{1 \text{ mol}}{342.3 \text{ g}} = 0.365 \text{ mol}$

$M = \frac{0.365 \text{ mol}}{1.00 \text{ L}} = 0.365 \text{ mol/L}$

2. $0.250 \text{ L} \times \frac{0.100 \text{ mol}}{\text{L}} \times \frac{134.0 \text{ g}}{\text{mol}} = 3.35 \text{ g } Na_2C_2O_4$

3. $25.00 \times 10^{-3} \text{ L} \times \frac{0.308 \text{ mol}}{\text{L}} = 7.70 \times 10^{-3} \text{ mol}$

$\frac{7.70 \times 10^{-3} \text{ mol}}{0.500 \text{ L}} = 1.54 \times 10^{-2} \text{ mol/L}$

$NiCl_2(s) \rightarrow Ni^{2+}(aq) + 2\ Cl^-(aq)$

$M_{Ni^{2+}} = 1.54 \times 10^{-2} \text{ mol/L}; \ M_{Cl^-} = 3.08 \times 10^{-2} \text{ mol/L}$

4. $158.5 \times 10^{-3} \text{ g Cu} \times \frac{1 \text{ mol}}{63.55 \text{ g}} = 2.494 \times 10^{-3} \text{ mol Cu}$

$M_{Cu^{2+}} = \frac{2.494 \times 10^{-3} \text{ mol}}{1.00 \text{ L}} = 2.49 \times 10^{-3} \text{ mol/L}$

5. $\frac{2.8 \times 10^{-3} \text{ g}}{1.0 \times 10^{-3} \text{ L}} \times \frac{1 \text{ mol}}{112.4 \text{ g}} = 2.5 \times 10^{-2} \text{ mol/L}$

6. a. $Ca(NO_3)_2(s) \rightarrow Ca^{2+}(aq) + 2\ NO_3^-(aq)$

$M_{Ca^{2+}} = 1.06 \times 10^{-3} \text{ mol/L}; \ M_{NO_3^-} = 2.12 \times 10^{-3} \text{ mol/L}$

b. $1 \times 10^{-3} \text{ L} \times \frac{1.06 \times 10^{-3} \text{ mol } Ca^{2+}}{\text{L}} \times \frac{40.08 \text{ g } Ca^{2+}}{\text{mol}} = 4 \times 10^{-5} \text{ g } Ca^{2+} \text{ ions}$

c. $1.0 \times 10^{-6} \text{ L} \times \frac{2.12 \times 10^{-3} \text{ mol } NO_3^-}{\text{L}} \times \frac{6.02 \times 10^{23}\ NO_3^- \text{ ions}}{\text{mol } NO_3^-} = 1.3 \times 10^{15}\ NO_3^- \text{ ions}$

7. $1.00 \text{ L} \times \frac{0.040 \text{ mol HCl}}{\text{L}} = 0.040 \text{ mol HCl}$

$0.040 \text{ mol HCl} \times \frac{1 \text{ L}}{0.25 \text{ mol HCl}} = 0.16 \text{ L} = 160 \text{ mL}$

8. a. $HNO_3(l) \rightarrow H^+(aq) + NO_3^-(aq)$

b. $Na_2SO_4(s) \rightarrow 2\,Na^+(aq) + SO_4^{2-}(aq)$

c. $AlCl_3(s) \rightarrow Al^{3+}(aq) + 3\,Cl^-(aq)$

d. $SrBr_2(s) \rightarrow Sr^{2+}(aq) + 2\,Br^-(aq)$

e. $KClO_4(s) \rightarrow K^+(aq) + ClO_4^-(aq)$

f. $NH_4Br(s) \rightarrow NH_4^+(aq) + Br^-(aq)$

g. $NH_4NO_3(s) \rightarrow NH_4^+(aq) + NO_3^-(aq)$

h. $CuSO_4(s) \rightarrow Cu^{2+}(aq) + SO_4^{2-}(aq)$

i. $NaOH(s) \rightarrow Na^+(aq) + OH^-(aq)$

QUESTIONS

9. $\text{Molarity} = \dfrac{\text{moles solute}}{\text{L solution}}$ $\text{Molality} = \dfrac{\text{moles solute}}{\text{kg solvent}}$

Since volume is temperature dependent and mass isn't, then molarity is temperature dependent and molality is temperature independent. In determining ΔT_f and ΔT_b, we are interested in how some temperature depends on composition. Thus, we don't want our expression of composition to also be dependent on temperature.

10. The nature of the interparticle forces. (See introduction to this chapter for detailed answer.)

11. hydrophobic: water hating

hydrophilic: water loving

12. As the temperature increases the gas molecules will have a greater average kinetic energy. A greater fraction of the gas molecules in solution will have kinetic energy greater than the attractive forces between the gas molecules and the solvent molecules. More gas molecules escape to the vapor phase and the solubility of the gas decreases.

13. If solute-solvent attraction > solvent-solvent and solute-solute, there is a negative deviation from Raoult's Law. If solute-solvent < solvent-solvent and solute-solute attraction, then there is a positive deviation from Raoult's Law.

14. A positive deviation from Raoult's Law means the vapor pressure of the solution is greater than if the solution were ideal. At the boiling point, the vapor pressure equals atmospheric pressure. For a solution with positive deviations, it will take a lower temperature to achieve a vapor pressure of one atmosphere. So the boiling point is lower than if the solution were ideal.

15. No, the solution is not ideal. For an ideal solution, this strength of interparticle forces in the solution are the same as in the pure solute and pure solvent. This results in $\Delta H_{soln} = 0$ for an ideal solution. ΔH_{soln} for methanol/water is not zero.

16. Ion pairing can occur, resulting in fewer particles than expected. Ion pairing will increase as the concentration of electrolyte increases.

17. The osmotic pressure inside the fruit cells (and bacteria) is less than outside the cell. Water will leave the cells which will dehydrate bacteria that might be present causing them to die.

18. A strong electrolyte dissociates completely into ions in solution. A weak electrolyte dissociates only partially into ions in solution. Colligative properties depend on the total number of particles in solution. By measuring a property such as freezing point depression, boiling point elevation, or osmotic pressure, we can calculate the van't Hoff factor (i) to see if an electrolyte is strong or weak.

EXERCISES

Concentration Units:

19. $$\frac{3.0 \text{ g } H_2O_2}{100 \text{ g soln}} \times \frac{1.0 \text{ g soln}}{\text{cm}^3 \text{ soln}} \times \frac{1000 \text{ cm}^3}{\text{L}} \times \frac{1 \text{ mol } H_2O_2}{34.02 \text{ g } H_2O_2} = 0.88 \text{ mol/L}$$

$$\frac{3.0 \text{ g } H_2O_2}{97.0 \text{ g } H_2O} \times \frac{1000 \text{ g}}{\text{kg}} \times \frac{1 \text{ mol } H_2O_2}{34.02 \text{ g } H_2O_2} = 0.91 \text{ mol/kg}$$

$$3.0 \text{ g } H_2O_2 \times \frac{1 \text{ mol}}{34.02 \text{ g}} = 8.8 \times 10^{-2} \text{ mol } H_2O_2$$

$$97.0 \text{ g } H_2O \times \frac{1 \text{ mol}}{18.02 \text{ g}} = 5.38 \text{ mol } H_2O, \qquad \chi = \frac{8.8 \times 10^{-2}}{5.38 + 0.088} = 1.6 \times 10^{-2}$$

20. $$1.0 \text{ L} \times \frac{1000 \text{ mL}}{\text{L}} \times \frac{1.0 \text{ g}}{\text{mL}} = 1.0 \times 10^3 \text{ g solution}$$

$$\% \text{ NaCl} = \frac{25 \text{ g NaCl}}{1.0 \times 10^3 \text{ g solution}} \times 100 = 2.5\%$$

$$\frac{25 \text{ g NaCl}}{\text{L}} \times \frac{1 \text{ mol NaCl}}{58.44 \text{ g NaCl}} = 0.43 \; M$$

1.0×10^3 g solution: 25 g NaCl, 975 g $H_2O \approx 980$ g H_2O

$$\text{molality} = \frac{0.43 \text{ mol NaCl}}{0.98 \text{ kg solvent}} = 0.44 \text{ molal}$$

$$0.43 \text{ mol NaCl; } 980 \text{ g} \times \frac{1 \text{ mol}}{18.02 \text{ g}} = 54 \text{ mol } H_2O, \qquad \chi = \frac{0.43}{0.43 + 54} = 7.9 \times 10^{-3}$$

21. Hydrochloric acid:

$$\frac{38 \text{ g HCl}}{100. \text{ g soln}} \times \frac{1.19 \text{ g soln}}{\text{cm}^3 \text{ soln}} \times \frac{1000 \text{ cm}^3}{\text{L}} \times \frac{1 \text{ mol HCl}}{36.46 \text{ g}} = 12 \text{ mol/L}$$

$$\frac{38 \text{ g HCl}}{62 \text{ g solvent}} \times \frac{1000 \text{ g}}{\text{kg}} \times \frac{1 \text{ mol HCl}}{36.46 \text{ g}} = 17 \text{ mol/kg}$$

$$38 \text{ g HCl} \times \frac{1 \text{ mol}}{36.46 \text{ g}} = 1.0 \text{ mol HCl}$$

$$62 \text{ g } H_2O \times \frac{1 \text{ mol}}{18.02 \text{ g}} = 3.4 \text{ mol } H_2O$$

$$\chi = \frac{1.0}{3.4 + 1.0} = 0.23$$

Nitric acid:

$$\frac{70.\text{ g HNO}_3}{100.\text{ g soln}} \times \frac{1.42\text{ g soln}}{\text{cm}^3\text{ soln}} \times \frac{1000\text{ cm}^3}{\text{L}} \times \frac{1\text{ mol HNO}_3}{63.02\text{ g}} = 16\text{ mol/L}$$

$$\frac{70.\text{ g HNO}_3}{30.\text{ g solvent}} \times \frac{1000\text{ g}}{\text{kg}} \times \frac{1\text{ mol HNO}_3}{63.02\text{ g}} = 37\text{ mol/kg}$$

$$70.\text{ g HNO}_3 \times \frac{1\text{ mol}}{63.02\text{ g}} = 1.1\text{ mol HNO}_3$$

$$30.\text{ g H}_2\text{O} \times \frac{1\text{ mol}}{18.02\text{ g}} = 1.7\text{ mol H}_2\text{O}$$

$$\chi = \frac{1.1}{1.7 + 1.1} = 0.39$$

Sulfuric acid:

$$\frac{95\text{ g H}_2\text{SO}_4}{100.\text{ g soln}} \times \frac{1.84\text{ g soln}}{\text{cm}^3\text{ soln}} \times \frac{1000\text{ cm}^3}{\text{L}} \times \frac{1\text{ mol H}_2\text{SO}_4}{98.09\text{ g H}_2\text{SO}_4} = 18\text{ mol/L}$$

$$\frac{95\text{ g H}_2\text{SO}_4}{5\text{ g H}_2\text{O}} \times \frac{1000\text{ g}}{\text{kg}} \times \frac{1\text{ mol}}{98.09\text{ g}} = 194\text{ mol/kg} \approx 200\text{ mol/kg}$$

$$95\text{ g H}_2\text{SO}_4 \times \frac{1\text{ mol}}{98.09\text{ g}} = 0.97\text{ mol H}_2\text{SO}_4$$

$$5\text{ g H}_2\text{O} \times \frac{1\text{ mol}}{18.02\text{ g}} = 0.3\text{ mol H}_2\text{O}$$

$$\chi = \frac{0.97}{0.97 + 0.3} = 0.76$$

Acetic acid:

$$\frac{99\text{ g CH}_3\text{CO}_2\text{H}}{100.\text{ g soln}} \times \frac{1.05\text{ g soln}}{\text{cm}^3\text{ soln}} \times \frac{1000\text{ cm}^3}{\text{L}} \times \frac{1\text{ mol}}{60.05\text{ g}} = 17\text{ mol/L}$$

$$\frac{99\text{ g CH}_3\text{CO}_2\text{H}}{1\text{ g H}_2\text{O}} \times \frac{1000\text{ g}}{\text{kg}} \times \frac{1\text{ mol}}{60.05\text{ g}} = 1600\text{ mol/kg} \approx 2000\text{ mol/kg}$$

$$99\text{ g HOAc} \times \frac{1\text{ mol}}{60.05\text{ g}} = 1.6\text{ mol HOAc}$$

$$1\text{ g H}_2\text{O} \times \frac{1\text{ mol}}{18.02\text{ g}} = 0.06\text{ mol H}_2\text{O}$$

$$\chi = \frac{1.6}{1.6 + 0.06} = 0.96$$

Ammonia:

$$\frac{28\text{ g NH}_3}{100.\text{ g soln}} \times \frac{0.90\text{ g}}{\text{cm}^3} \times \frac{1000\text{ cm}^3}{\text{L}} \times \frac{1\text{ mol}}{17.03\text{ g}} = 15\text{ mol/L}$$

$$\frac{28\text{ g NH}_3}{72\text{ g H}_2\text{O}} \times \frac{1000\text{ g}}{\text{kg}} \times \frac{1\text{ mol}}{17.03\text{ g}} = 23\text{ mol/kg}$$

$$28 \text{ g } NH_3 \times \frac{1 \text{ mol}}{17.03 \text{ g}} = 1.6 \text{ mol } NH_3$$

$$72 \text{ g } H_2O \times \frac{1 \text{ mol}}{18.02 \text{ g}} = 4.0 \text{ mol } H_2O$$

$$\chi = \frac{1.6}{4.0 + 1.6} = 0.29$$

22. $$50 \text{ g NaOH} \times \frac{1 \text{ mol}}{40.00 \text{ g}} = 1.3 \text{ mol NaOH} \quad \text{(carry extra significant figure)}$$

$$50 \text{ g } H_2O \times \frac{1 \text{ mol}}{18.02 \text{ g}} = 2.8 \text{ mol } H_2O$$

$$\chi_{NaOH} = \frac{1.3}{1.3 + 2.8} = 0.3$$

$$m = \frac{1.3 \text{ mol NaOH}}{50 \text{ g } H_2O} \times \frac{1000 \text{ kg}}{\text{kg}} = 26 \text{ mol/kg} \approx 30 \text{ mol/kg}$$

23. $$50.0 \text{ mL toluene} \times \frac{0.867 \text{ g}}{\text{mL}} = 43.4 \text{ g toluene}$$

$$125 \text{ mL benzene} \times \frac{0.874 \text{ g}}{\text{mL}} = 109 \text{ g benzene}$$

$$\% \text{ toluene} = \frac{43.4}{43.4 + 109} \times 100 = 28.5\%$$

$$\frac{43.4 \text{ g toluene}}{175 \text{ mL soln}} \times \frac{1000 \text{ mL}}{\text{L}} \times \frac{1 \text{ mol toluene}}{92.13 \text{ g toluene}} = 2.69 \text{ mol/L}$$

$$\frac{43.4 \text{ g toluene}}{109 \text{ g benzene}} \times \frac{1000 \text{ g}}{\text{kg}} \times \frac{1 \text{ mol toluene}}{92.13 \text{ g toluene}} = 4.32 \text{ mol/kg}$$

$$43.4 \text{ g toluene} \times \frac{1 \text{ mol}}{92.13 \text{ g}} = 0.471 \text{ mol toluene}$$

$$109 \text{ g benzene} \times \frac{1 \text{ mol benzene}}{78.11 \text{ g benzene}} = 1.40 \text{ mol benzene}$$

$$\chi = \frac{0.471}{0.471 + 1.40} = 0.252$$

24. $$\frac{1 \text{ mol acetone}}{1.0 \text{ kg ethanol}} = 1.0 \text{ molal}$$

$$1.0 \times 10^3 \text{ g } C_2H_5OH \times \frac{1 \text{ mol}}{46.07 \text{ g}} = 22 \text{ mol}$$

$$\chi = \frac{1.0}{1.0 + 22} = 0.043$$

$$1.0 \text{ mol } CH_3COCH_3 \times \frac{58.08 \text{ g } CH_3COCH_3}{\text{mol } CH_3COCH_3} \times \frac{1 \text{ mL}}{0.788 \text{ g}} = 74 \text{ mL}$$

$$1.0 \times 10^3 \text{ g ethanol} \times \frac{1 \text{ mL}}{0.789 \text{ g}} = 1270 \text{ mL} \quad \text{(carry extra significant figure)}$$

$$\text{Total V} = 1270 + 74 = 1341 \text{ mL} \approx 1.3 \text{ L}, \quad \text{molarity} = \frac{1.0 \text{ mol}}{1.3 \text{ L}} = 0.77\ M$$

25. If we have 1.00 L of solution we have:

$$1.37 \text{ mol citric acid} \times \frac{192.1 \text{ g}}{\text{mol}} = 263 \text{ g citric acid}$$

$$1.00 \times 10^3 \text{ mL solution} \times \frac{1.10 \text{ g}}{\text{mL}} = 1.10 \times 10^3 \text{ g solution}$$

$$\% = \frac{263}{1.10 \times 10^3} \times 100 = 23.9\%$$

In 1.00 L of solution we have 263 g citric acid and $(1.10 \times 10^3 - 263) = 840$ g of the solvent, H_2O.

$$\frac{1.37 \text{ mol citric acid}}{0.84 \text{ g}} = 1.6 \text{ mol/kg}$$

$$840 \text{ g } H_2O \times \frac{1 \text{ mol}}{18.02 \text{ g}} = 47 \text{ mol } H_2O$$

$$\chi = \frac{1.37}{47 + 1.37} = 0.028$$

26. $$\frac{50.0 \text{ g CsCl}}{100.0 \text{ g solution}} \times 100 = 50.0\%$$

$$100.0 \text{ g solution} \times \frac{1 \text{ mL}}{1.58 \text{ g}} = 63.3 \text{ mL}$$

$$M = \frac{50.0 \text{ g}}{63.3 \text{ mL}} \times \frac{1000 \text{ mL}}{\text{L}} \times \frac{1 \text{ mol}}{168.4 \text{ g}} = 4.69 \text{ mol/L}$$

$$\frac{50.0 \text{ g CsCl}}{50.0 \text{ g solvent}} \times \frac{1000 \text{ g}}{\text{kg}} \times \frac{1 \text{ mol CsCl}}{168.4 \text{ g}} = 5.94 \text{ mol/kg}$$

$$50.0 \text{ g CsCl} \times \frac{1 \text{ mol}}{168.4 \text{ g}} = 0.297 \text{ mol CsCl}$$

$$50.0 \text{ g } H_2O \times \frac{1 \text{ mol}}{18.02 \text{ g}} = 2.77 \text{ mol } H_2O$$

$$\chi = \frac{0.297}{0.297 + 2.77} = 9.68 \times 10^{-2}$$

27. If there are 100.0 mL of wine, there are

$$12.5 \text{ mL } C_2H_5OH \times \frac{0.79 \text{ g}}{\text{mL}} = 9.9 \text{ g } C_2H_5OH$$

$$87.5 \text{ mL } H_2O \times \frac{1.0 \text{ g}}{\text{mL}} = 87.5 \text{ g } H_2O$$

$$\% \text{ ethanol} = \frac{9.9}{87.5 + 9.9} \times 100 = 10.\% \text{ by mass}$$

$$m = \frac{9.9 \text{ g } C_2H_5OH}{0.0875 \text{ kg } H_2O} \times \frac{1 \text{ mol}}{46.07 \text{ g}} = 2.5 \text{ molal}$$

28. $0.40 \text{ mol } CH_3COCH_3 \times \frac{58.08 \text{ g}}{\text{mol}} = 23 \text{ g acetone}$

$0.60 \text{ mol } C_2H_5OH \times \frac{46.07 \text{ g}}{\text{mol}} = 28 \text{ g ethanol}$

$\% \text{ acetone} = \frac{23 \text{ g}}{23 + 28 \text{ g}} \times 100 = 45\%$

Energetics of Solutions and Solubility

29. Water is a polar molecule capable of hydrogen bonding. Polar molecules, molecules capable of hydrogen bonding, and ions can be hydrated.

a. CH_3CH_2OH

b. $CHCl_3$: $CHCl_3$ is polar, CCl_4 is non-polar.

c. CH_3CO_2H

d. Na_2S (Most sulfides are insoluble, Na_2S is an exception.)

e. $AlCl_3$: Al_2O_3 is an insoluble covalent network solid.

f. CO_2: CO_2 is non-polar and not very soluble, but SiO_2 is a covalent network solid and virtually insoluble in water.

30. a. NH_3, capable of H-bonding

b. CH_3CN, polar

c. CH_3CO_2H, capable of H-bonding

d. $MgCl_2$, ionic with lower lattice energy than MgO

e. HCl, AgCl is insoluble.

f. NH_3

31.

$$KCl(s) \rightarrow K^+(g) + Cl^-(g) \qquad \Delta H = -(-715 \text{ kJ/mol})$$
$$K^+(g) + Cl^-(g) \rightarrow K^+(aq) + Cl^-(aq) \qquad \Delta H = -684 \text{ kJ/mol}$$

$$KCl(s) \rightarrow K^+(aq) + Cl^-(aq) \qquad \Delta H_{soln} = 31 \text{ kJ/mol}$$

32.

$$CsI(s) \rightarrow Cs^+(g) + I^-(g) \qquad 604 \text{ kJ}$$
$$Cs^+(g) + I^-(g) \rightarrow Cs^+(aq) + I^-(aq) \qquad \Delta H_{hyd}$$

$$CsI(s) \rightarrow Cs^+(aq) + I^-(aq) \qquad 33 \text{ kJ}$$

$33 \text{ kJ} = 604 + \Delta H_{hyd}$; $\Delta H_{hyd} = -571 \text{ kJ}$

$$CsOH(s) \rightarrow Cs^+(g) + OH^-(g) \qquad 724 \text{ kJ}$$
$$Cs^+(g) + OH^-(g) \rightarrow Cs^+(aq) + OH^-(aq) \qquad \Delta H_{hyd}$$

$$CsOH(s) \rightarrow Cs^+(aq) + OH^-(aq) \qquad -72 \text{ kJ}$$

$-72 \text{ kJ} = 724 \text{ kJ} + \Delta H_{hyd}$; $\Delta H_{hyd} = -796 \text{ kJ}$

The enthalpy of hydration of CsOH is more exothermic than for CsI. Any differences must be because of differences in hydrations between OH^- and I^-. Thus, hydroxide ion is more strongly hydrated.

33. a. Mg^{2+}, smaller, higher charge b. Be^{2+}, smaller

c. Fe^{3+}, smaller size, higher charge

34. a. Li^+, smaller b. F^-, smaller c. Cl^-, smaller

35. a. Water, $Cu(NO_3)_2$ is an ionic solid. b. CCl_4, CS_2 is a non-polar molecule.

c. Water d. CCl_4, the long non-polar chain favors a non-polar solvent

e. Water f. CCl_4

36. a. water b. water c. hexane d. water

37. As the length of the hydrocarbon chain increases, the solubility decreases. The —OH end of the alcohols can hydrogen bond with water. The hydrocarbon chain, however, cannot form H-bonds. As the chain gets longer, a greater portion of the molecule cannot interact with the water molecules; i.e. the effect of the —OH group decreases as the alcohol gets larger.

38. The micelles form so the ionic end, $—OSO_3^-$, is exposed to the water on the outside, and the non-polar hydrocarbon chain is hidden from the water by pointing toward the inside of the micelle.

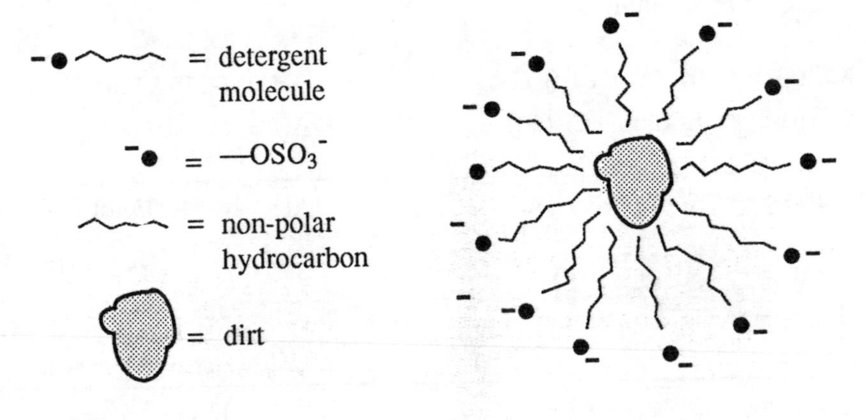

39. $CO_2 + OH^- \rightarrow HCO_3^-$

No, the reaction of CO_2 with OH^- greatly increases the solubility of CO_2 by forming the soluble bicarbonate ion.

40. $P_{gas} = kC$; $120 \text{ torr} \times \frac{1 \text{ atm}}{760 \text{ torr}} = \frac{780 \text{ atm}}{M} \times C$

$C = 2.0 \times 10^{-4}$ mol/L

Vapor Pressure of Solutions

41. $P = \chi P^\circ$

In solution:

$$50.0\ g\ C_6H_{12}O_6 \times \frac{1\ mol\ C_6H_{12}O_6}{180.16\ g\ C_6H_{12}O_6} = 0.278\ mol\ glucose$$

$$600.0\ g\ H_2O \times \frac{1\ mol}{18.02\ g} = 33.30\ mol\ H_2O$$

$$\chi_{H_2O} = \frac{33.30}{33.58} = 0.9917$$

$$P = 0.9917 \times 23.8 = 23.6\ torr$$

42. $P = \chi P^\circ$; 710.0 torr = χ(760.0 torr); $\chi = 0.9342$

43. $$25.8\ g\ CH_4N_2O \times \frac{1\ mol}{60.06\ g} = 0.430\ mol$$

$$275\ g\ H_2O \times \frac{1\ mol}{18.02\ g} = 15.3\ mol$$

$$\chi_{H_2O} = \frac{15.3}{15.3 + 0.430} = 0.973$$

$$P_{H_2O} = \chi_{H_2O}P^\circ_{H_2O} = 0.973\ (23.8\ torr) = 23.2\ torr\ at\ 25^\circ C$$

$$P_{H_2O} = 0.973\ (71.9\ torr) = 70.0\ torr\ at\ 45^\circ C$$

44. 19.6 torr = χ_{H_2O} (23.8 torr), $\chi_{H_2O} = 0.824$

$\chi_{urea} = 1 - 0.824 = 0.176$

at 45°C. $P_{H_2O} = 0.824\ (71.9\ torr) = 59.2\ torr$

45. a. $$25\ mL\ C_5H_{12} \times \frac{0.63\ g}{mL} \times \frac{1\ mol}{72.15\ g} = 0.22\ mol\ C_5H_{12}$$

$$45\ mL\ C_6H_{14} \times \frac{0.66\ g}{mL} \times \frac{1\ mol}{86.17\ g} = 0.34\ mol\ C_6H_{14}$$

In liquid: $\chi_{pentane} = \frac{0.22}{0.56} = 0.39$ and $\chi_{hexane} = 0.61$

$P_{pen} = 0.39(511\ torr) = 2.0 \times 10^2\ torr$ and $P_{hex} = 0.61(150\ torr) = 92\ torr$

$P_{total} = 2.0 \times 10^2 + 92 = 292 \approx 290\ torr$

b. In the vapor phase:

$$\chi_{pen} = \frac{P_{pen}}{P_{total}} = \frac{2.0 \times 10^2\ torr}{290\ torr} = 0.69$$

46. $\chi_{pen} + \chi_{hex} = 1$

$350 \text{ torr} = \chi_{pen}(511 \text{ torr}) + \chi_{hex}(150. \text{ torr})$

$350 = 511\ \chi_{pen} + (1 - \chi_{pen})150.$

$350 = 150. + 361\ \chi_{pen}$

$\dfrac{2.0 \times 10^2}{361} = \chi_{pen} = 0.55$

$\chi_{hex} = 1 - 0.55 = 0.45$

Composition of solution: $\chi_{pen} = 0.55,\ \chi_{hex} = 0.45$

$P_{pen} = 0.55(511) = 280$ torr and $P_{hex} = 0.45(150.) = 68$ torr

In vapor: $\chi_{pen} = \dfrac{P_{pen}}{P_{tot}} = \dfrac{280}{350} = 0.80$

$\chi_{hex} = 1 - 0.80 = 0.20$

47. $0.15 = \dfrac{P_{pen}}{P_{pen} + P_{hex}}$ (in vapor) and $P_{pen} + P_{hex} = P_{total}$

Thus, $P_{pen} = 0.15\ P_{tot}$

In the liquid:

$P_{pen} + P_{hex} = P_{tot} = \chi_{pen}(511) + \chi_{hex}(150.)$

$P_{tot} = \chi_{pen}(511) + (1 - \chi_{pen})\ (150.)$

$P_{tot} = 150. + 361\ \chi_{pen}$

$P_{pen} = \chi_{pen}(511) = 0.15\ P_{tot}$

$0.15\ P_{tot} = 0.15(150.) + 0.15(361)\ \chi_{pen}$
$-0.15\ P_{tot} = -511\ \chi_{pen}$

$0 = 22.5 + 54.2\ \chi_{pen} - 511\ \chi_{pen}$ (carry extra significant figure)

$22.5 = 457\ \chi_{pen}$

$\chi_{pen} = 0.0492 \approx 0.049$ in the liquid

48. $P_{pen} = \chi_{pen}(511 \text{ torr}) = P_{hex} = \chi_{hex}(150. \text{ torr})$

Since, $\chi_{pen} + \chi_{hex} = 1$, then: $\chi_{pen}(511) = (1 - \chi_{pen})\ (150.)$

$661\ \chi_{pen} = 150.,\ \chi_{pen} = 0.227$ and $\chi_{hex} = 0.773$

In the vapor phase, $P_{pen} = P_{hex}$. So, $n_{pen} = n_{hex}$ and $\chi_{pen} = \chi_{hex} = 0.50$

49. Compared to H_2O, d (methanol/water) will have a higher vapor pressure since methanol is more volatile than water. Both b (glucose/water) and c (NaCl/water) will have a lower vapor pressure than water. NaCl dissolves to give Na^+ ions and Cl^- ions; glucose is a nonelectrolyte. Since there are more particles, the vapor pressure of (c) will be the lowest.

50. d (methanol/water), methanol is more volatile than water.

Colligative Properties

51. $$m = \frac{4.9 \text{ g sucrose}}{175 \text{ g solvent}} \times \frac{1000 \text{ g}}{\text{kg}} \times \frac{1 \text{ mol } C_{12}H_{22}O_{11}}{342.3 \text{ g } C_{12}H_{22}O_{11}} = 0.082 \text{ molal}$$

$$\Delta T_b = K_b m = \frac{0.51°\text{C}}{\text{molal}} \times 0.082 \text{ molal} = 0.042°\text{C}$$

$T_b = 100.042°C$ at 1.0 atm barometric pressure.

$$\Delta T_f = K_f m = \frac{1.86°\text{C}}{\text{molal}} \times 0.082 \text{ molal} = 0.15°\text{C}$$

$T_f = -0.15°C$

$\Pi = MRT$

In dilute aqueous solutions the density of the solution is about 1.0 g/mL, thus we can assume M = molality = 0.082.

$$\Pi = \frac{0.082 \text{ mol}}{\text{L}} \times \frac{0.08206 \text{ L atm}}{\text{mol K}} \times 298 \text{ K} = 2.0 \text{ atm}$$

52. $$\frac{200.0 \text{ g } C_3H_8O_2}{400.0 \text{ g } H_2O} \times \frac{1000 \text{ g}}{\text{kg}} \times \frac{1 \text{ mol } C_3H_8O_2}{76.09 \text{ g } C_3H_8O_2} = 6.571 \text{ molal}$$

$$\Delta T_f = K_f m = \frac{1.86°\text{C}}{\text{molal}} \times 6.571 \text{ molal} = 12.2°\text{C}$$

$T_f = -12.2°C$

$$\Delta T_b = K_b m = \frac{0.51°\text{C}}{\text{molal}} \times 6.571 \text{ molal} = 3.4°\text{C}$$

$T_b = 103.4°C$ assuming 1.0 atm barometric pressure.

53. a. Assume molarity and molality are equal.

$$M = m = \frac{1.0 \text{ g}}{\text{L}} \times \frac{1 \text{ mol}}{9.0 \times 10^4 \text{ g}} = 1.1 \times 10^{-5} M$$

$$\text{At 298 K, } \Pi = \frac{1.1 \times 10^{-5} \text{ mol}}{\text{L}} \times \frac{0.08206 \text{ L atm}}{\text{mol K}} \times 298 \text{ K} \times \frac{760 \text{ mm Hg}}{\text{atm}}$$

$\Pi = 0.20$ mm Hg

$$\Delta T_f = K_f m = \frac{1.86°\text{C}}{\text{molal}} \times 1.1 \times 10^{-5} \text{ molal} = 2.0 \times 10^{-5} \text{ °C}$$

b. Osmotic pressure is better for determining the molecular weight of large molecules. A temperature change of 10^{-5} °C is very difficult to measure. A change in height of a column of mercury by 0.2 mm is not as hard to measure precisely.

54. $\Pi = MRT$

$$M = \frac{\Pi}{RT} = \frac{0.56 \text{ torr}}{\frac{0.08206 \text{ L atm}}{\text{mol K}} \times 298 \text{ K}} \times \frac{1 \text{ atm}}{760 \text{ torr}}$$

$M = 3.0 \times 10^{-5}\ M$

$$3.0 \times 10^{-5}\ M = \frac{20.0 \times 10^{-3} \text{ g}}{25.0 \times 10^{-3} \text{ L}} \times \frac{1}{\text{MW}}$$

MW = 27,000 g/mol

55. $\Delta T_f = K_f m$; $2.63 = (40.)\ m$

$m = 6.6 \times 10^{-2}$ mol/kg and $m = \frac{\text{moles reserpine}}{\text{kg solvent}}$

$$6.6 \times 10^{-2} = \frac{1.00 \text{ g reserpine}}{25.0 \text{ g solvent}} \times \frac{1}{\text{MW}} \times \frac{1000 \text{ g solvent}}{\text{kg solvent}}$$

$$\text{MW} = \frac{1000}{25.0 \times 6.6 \times 10^{-2}} = 606 \text{ g/mol} \approx 610 \text{ g/mol (2 S. F. for } K_f)$$

56. $$\Pi = MRT = \frac{0.1 \text{ mol}}{\text{L}} \times \frac{0.08206 \text{ L atm}}{\text{mol K}} \times 298 \text{ K} = 2.45 \text{ atm} \approx 2 \text{ atm}$$

$$\Pi = 2 \text{ atm} \times \frac{760 \text{ mm Hg}}{\text{atm}} \approx 2000 \text{ mm} \approx 2 \text{ m}$$

The osmotic pressure would support a mercury column of ≈ 2 m. The height of the fluid column will be higher because Hg is more dense than the fluid in the tree. If we assume the fluid in the tree is mostly H_2O, then the fluid has a density of 1.0 g/cm^3. The density of Hg is 13.6 g/cm^3.

height of fluid ≈ 2 m × 13.6 ≈ 30 m

Properties of Electrolyte Solutions

57. $NaCl(s) \longrightarrow Na^+(aq) + Cl^-(aq)$

So the total concentration of particles (colligative molarity) is 2 (0.5) = 1.0.

$$\Pi = MRT = \frac{1.0 \text{ mol}}{\text{L}} \times \frac{0.08206 \text{ L atm}}{\text{mol K}} \times 298 \text{ K} = 24 \text{ atm}$$

A pressure greater than 24 atm must be applied.

58. a. Water would go from right to left. Level in right arm would go down and the level in left arm would go up.

b. The levels would be equal. The concentration of NaCl would be equal in both chambers.

59. $MgCl_2$ > NaCl > HOCl > glucose

60. a. pure water b. $CaCl_2$ solution c. $CaCl_2$ solution

d. pure water e. $CaCl_2$ solution

61. $Ca(NO_3)_2(s) \longrightarrow Ca^{2+}(aq) + 2\ NO_3^-(aq)$; i = 3

$$\Delta T_f = 3 \times \frac{1.86°C}{molal} \times 0.5\ molal = 2.8°C \approx 3°C,\ T_f = -3°C$$

The measured freezing point should be higher. Because of ion association the depression of the freezing point will be less than expected, resulting in a higher freezing point.

62. $\Delta T_f = iK_f m$, For $Co(NH_2CH_2CH_2NH_2)_2Cl_3$

MW = 58.93 + 4(14.01) + 4(12.01) + 16(1.008) + 3(35.45) = 285.49 g/mol

$$m = \frac{1.00\ g}{0.0250\ kg} \times \frac{1\ mol}{285.49\ g} = 0.140\ molal$$

$$i = \frac{T_f}{K_f m} = \frac{0.52}{(1.86)(0.140)} = 2.0$$

For $Co(NH_2CH_2CH_2NH_2)_3Cl_3$ MW = 345.59

$$m = \frac{1.00\ g}{0.0250\ kg} \times \frac{1\ mol}{345.59\ g} = 0.116\ mol/kg$$

$$i = \frac{T_f}{K_f m} = \frac{0.87}{1.86 \times 0.116} = 4.0$$

63. $\Delta T_f = iK_f m$

$$i = \frac{\Delta T_f}{K_f m} = \frac{0.440}{1.86 \times 0.091} = 2.6 \text{ for } 0.091 \text{ molal } CaCl_2$$

$$i = \frac{1.330}{1.86 \times 0.279} = 2.56 \text{ for } 0.279 \text{ molal } CaCl_2$$

$$i = \frac{2.345}{1.86 \times 0.475} = 2.65 \text{ for } 0.475 \text{ molal } CaCl_2$$

64. For $CaCl_2$, i = 2.6 (From Exercise 11.63)

$$\%\ CaCl_2 \text{ ionized} = \frac{2.6 - 1.0}{3.0 - 1.0} \times 100 = 80.\%$$

For CsCl: $i = \frac{\Delta T_f}{K_f m} = \frac{0.302}{1.86 \times 0.091} = 1.8$

$\% \text{ CsCl ionized} = \frac{1.8 - 1.0}{2.0 - 1.0} \times 100 = 80\%$

Since both are about equally ionized, we can conclude that the extent of ion association is about the same in each solution.

ADDITIONAL EXERCISES

65.

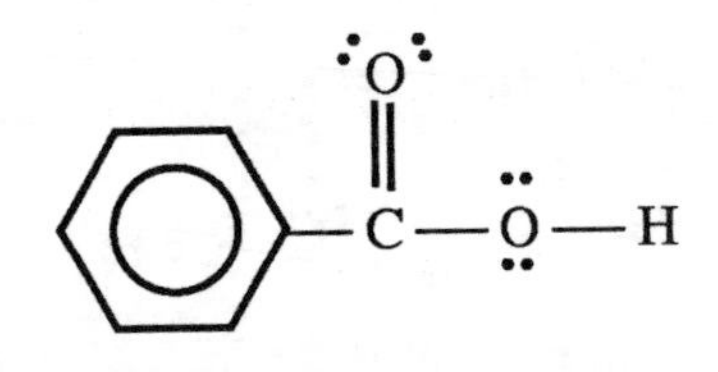

Benzoic acid is capable of hydrogen bonding. However, it is more soluble in nonpolar benzene than in water. In benzene, a non-polar hydrogen bonded dimer forms.

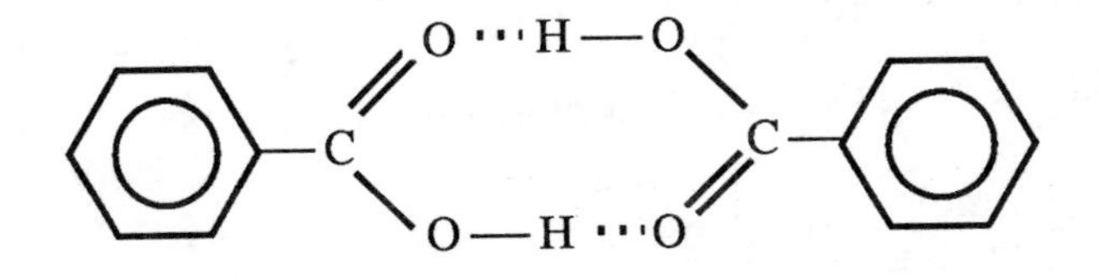

The dimer is nonpolar and thus more soluble in benzene than in water.

66. Benzoic acid would be more soluble in a basic solution because of the reaction:

$$C_6H_5CO_2H + OH^- \longrightarrow C_6H_5CO_2^- + H_2O$$

67. $\frac{40.0 \text{ g EG}}{60.0 \text{ g } H_2O} \times \frac{1000 \text{ g}}{\text{kg}} \times \frac{1 \text{ mol EG}}{62.07 \text{ g}} = 10.7 \text{ mol/kg}$

$\frac{40.0 \text{ g EG}}{100.0 \text{ g solution}} \times \frac{1.05 \text{ g}}{\text{cm}^3} \times \frac{1000 \text{ cm}^3}{\text{L}} \times \frac{1 \text{ mol}}{62.07 \text{ g}} = 6.77 \text{ mol/L}$

$40.0 \text{ g EG} \times \frac{1 \text{ mol}}{62.07 \text{ g}} = 0.644 \text{ mol EG}$

$60.0 \text{ g } H_2O \times \frac{1 \text{ mol}}{18.02 \text{ g}} = 3.33 \text{ mol } H_2O$

$\chi = \frac{0.644}{3.33 + 0.644} = 0.162$

68. a. If we use 100. mL (100. g) H_2O we need

$$\frac{2 \text{ mol KCl}}{\text{kg}} \times 0.100 \text{ kg} \times \frac{74.55 \text{ g}}{\text{mol}} = 14.9 \text{ g} \approx 15 \text{ g KCl}$$

Dissolve 15 g KCl in 100. mL H_2O. This will give us slightly more than 100 mL, but this will be the easiest way to make the solution. Since we don't know the density, we can't calculate the molarity and use a volumetric flask.

b. If we took 15 g NaOH and 85 g H_2O, the volume would be less than 100 mL. To make sure we have enough solution, use 100. mL H_2O (100 g)

$$15 = \frac{x}{100. + x}(100)$$

$$1500 + 15x = 100x$$

$$x = 17.6 \text{ g} \approx 18 \text{ g}$$

Dissolve 18 g NaOH in 100. mL H_2O.

c. In a fashion similar to 11.68b, let's use 100. mL CH_3OH

$$100.\ \text{mL} \times \frac{0.79\ \text{g}}{\text{mL}} = 79\ \text{g CH}_3\text{OH}$$

$$25 = \frac{x}{79 + x}(100)$$

$$25(79) + 25x = 100x$$

$$x = 26.3 \text{ g} \approx 26 \text{ g}$$

Dissolve 26 g NaOH in 100. mL CH_3OH

d. Dissolve 30. ml C_2H_5OH in 70. mL H_2O

69. Both $Al(OH)_3$ and NaOH are ionic. Since the lattice energy is proportiuonal to charge, the lattice energy of aluminum hydroxide is greater than that of sodium hydroxide. The attraction of water molecules for Al^{3+} and OH^- cannot overcome the larger lattice energy and $Al(OH)_3$ is insoluble.

70. The main interparticle forces are:

hexane: LDF; Chloroform, $CHCl_3$: dipole-dipole, LDF;

methanol: H-bonding; H_2O: H-bonding (two places)

There is a gradual change in the nature of the interparticle forces (weaker to stronger). Each preceding solvent is miscible in its predecessor because there is not a great change in the strengths of interparticle forces.

71. a. CF_3CF_3 and $CF_3CF_2CF_3$. Solutions will be ideal when only dispersion forces are involved and both substances are nearly the same size. This is true for the two fluorocarbons. Hydrogen bonding will be important in water/acetone solutions.

b. C_7H_{16} and C_6H_{14}. Both non-polar, similar size. Dipole forces are present in both CHF_3 and CH_3OCH_3. It is unlikely that the forces in the solution will be the same as in the two pure substances.

c. CCl_4 and CF_4 since both are non-polar substances. There will be a hydrogen bonding in aqueous solutions of phosphoric acid.

72. $50.0 \text{ g } CH_3COCH_3 \times \frac{1 \text{ mol}}{58.08 \text{ g}} = 0.861 \text{ mol acetone}$

$50.0 \text{ g } CH_3OH \times \frac{1 \text{ mol}}{32.04 \text{ g}} = 1.56 \text{ mol methanol}$

$\chi_{acetone} = \frac{0.861}{0.861 + 1.56} = 0.356$ (in solution)

$\chi_{methanol} = 1 - \chi_{acetone} = 0.644$ (in solution)

$P = P_{methanol} + P_{acetone} = 0.644(143) \text{ torr} + 0.356(271) \text{ torr}$

$P = 92.1 \text{ torr} + 96.5 \text{ torr} = 188.6 \text{ torr}$

Since partial pressures are proportional to the number of moles, then in the vapor phase:

$\chi_{acetone} = \frac{P_{acetone}}{P_{total}} = \frac{96.5}{188.6} = 0.512$

$\chi_{acetone} = 0.512$; $\chi_{methanol} = 0.488$ (vapor)

It is probably not a good assumption that this solution is ideal because of methanol-acetone hydrogen bonding.

73. $P_{total} = P_{methanol} + P_{propanol}$

$= 0.50(303) + 0.50(44.6) = 150 + 22 = 172 \text{ mm Hg} \approx 170 \text{ mm Hg}$

In the vapor: $\chi_{methanol} = \frac{P_{methanol}}{P_{total}} = \frac{150}{170} = 0.88$

$\chi_{methanol} = 0.88$ and $\chi_{propanol} = 0.12$

74. Cellophane has several —OH groups that can hydrogen bond to water molecules. Water should be "soluble" in cellophane.

75. $14.22 \text{ mg } CO_2 \times \frac{12.01 \text{ mg C}}{44.01 \text{ mg } CO_2} = 3.881 \text{ mg C}$

$\% \text{ C} = \frac{3.881}{4.80} \times 100 = 80.9\% \text{ C}$

$1.66 \text{ mg } H_2O \times \frac{2.016 \text{ mg H}}{18.02 \text{ mg } H_2O} = 0.186 \text{ mg H}$

$\% \text{ H} = \frac{0.186}{4.80} \times 100 = 3.88\% \text{ H}$

$\% \text{ O} = 100 - (80.9 + 3.88) = 15.2\%$

Out of 100 g:

$80.9 \text{ g C} \times \frac{1 \text{ mol}}{12.01 \text{ g}} = 6.74$ $\qquad \frac{6.74}{0.950} = 7.09 \approx 7$

$3.88\text{ g H} \times \dfrac{1\text{ mol}}{1.008\text{ g}} = 3.85$ $\qquad \dfrac{3.85}{0.950} = 4.05 \approx 4$

$15.2\text{ g O} \times \dfrac{1\text{ mol}}{16.00\text{ g}} = 0.950$ $\qquad \dfrac{0.950}{0.950} = 1$

Therefore, the empirical formula is C_7H_4O

$\Delta T_f = K_f m$

$$m = \frac{22.3°C}{40.°C/\text{molal}} = 0.56\text{ molal}$$

$$0.56\text{ molal} = \frac{1.32\text{ g anthraquinone}}{0.0114\text{ kg camphor}} \times \frac{1}{MW},\ MW = \frac{1.32}{0.0114 \times 0.56} = 210\text{ g/mol}$$

C_7H_4O: 7(12) + 4(1) + 16 ≈ 104 g/mol, so, molecular formula is $C_{14}H_8O_2$

76. a. $\dfrac{5.0\text{ g } C_6H_{12}O_6}{0.025\text{ kg}} \times \dfrac{1\text{ mol}}{180.2\text{ g}} = 1.1$ molal, $\Delta T_f = 1.86 \times 1.1 = 2.0°C$ and thus, $T_f = -2.0°C$

b. $m = \dfrac{5.0\text{ g NaCl}}{0.025\text{ kg}} \times \dfrac{1\text{ mol}}{58.44\text{ g}} = 3.4$ molal

NaCl is an electrolyte with i = 2.

$\Delta T_f = iK_f m = 2 \times 1.86 \times 3.4 = 13°C$ and thus, $T_f = -13°C$

c. aluminum nitrate, $Al(NO_3)_3$, i = 4

$$m = \frac{2.0\text{ g}}{0.015\text{ kg}} \times \frac{1\text{ mol}}{213.0\text{ g}} = 0.63\text{ mol/kg}$$

$\Delta T_f = iK_f m = 4 \times 1.86 \times 0.63 = 4.7°C$ and $T_f = -4.7°C$

d. $m = \dfrac{1.0\text{ g } C_6H_5CHO}{0.010\text{ kg}} \times \dfrac{1\text{ mol}}{106.1\text{ g}} = 0.94\text{ mol/kg}$

$\Delta T_f = K_f m = 5.12 \times 0.94 = 4.8°C$ and $T_f = 5.5 - 4.8 = 0.7°C$

77. $m = \dfrac{40.0\text{ g } C_2H_6O_2}{60.0\text{ g } H_2O} \times \dfrac{1000\text{ g}}{\text{kg}} \times \dfrac{1\text{ mol}}{62.07\text{ g}} = 10.7\text{ mol/kg}$

$\Delta T_f = 1.86 \times 10.7 = 19.9°C$ and $T_f = -19.9°C$

$\Delta T_b = 0.51 \times 10.7 = 5.5°C$ and $T_b = 105.5°C$

78. $\Pi = MRT;\ M = \dfrac{\Pi}{RT} = \dfrac{15}{(0.08206)(298)} = 0.61\ M$

$\dfrac{0.61\text{ mol}}{L} \times \dfrac{342.3\text{ g}}{\text{mol}} = 209$ g/L (carry extra significant figure)

Dissolve 209 g sucrose in water and dilute to 1.0 L in a volumetric flask. To get 0.61 ± 0.01 mol/L, we need 209 ± 3 g sucrose.

79. $\Delta T_f = 3.00°C = \frac{1.86°C}{\text{molal}} \times m$; $m = 1.61$ mol/kg

$$\frac{1.61 \text{ mol urea}}{\text{kg } H_2O} \times 0.150 \text{ kg } H_2O \times \frac{60.06 \text{ g urea}}{\text{mol urea}} = 14.5 \text{ g}$$

80. $\Pi = MRT$; $18.6 \text{ torr} \times \frac{1 \text{ atm}}{760 \text{ torr}} = M \times \frac{0.08206 \text{ L atm}}{\text{mol K}} \times 298 \text{ K}$

$$M = 1.00 \times 10^{-3} \text{ mol/L and } \frac{1.00 \times 10^{-3} \text{ mol}}{\text{L}} = \frac{0.15 \text{ g}}{2.0 \times 10^{-3} \text{ L}} \times \frac{1}{\text{MW}}$$

MW = 7.5×10^4 g/mol

81. Out of 100 g, there are

$$31.57 \text{ g C} \times \frac{1 \text{ mol}}{12.01 \text{ g}} = 2.629 \qquad \frac{2.629}{2.629} = 1.000$$

$$5.30 \text{ g H} \times \frac{1 \text{ mol}}{1.008 \text{ g}} = 5.26 \qquad \frac{5.26}{2.629} = 2.00$$

$$63.13 \text{ g O} \times \frac{1 \text{ mol}}{16.00 \text{ g}} = 3.946 \qquad \frac{3.946}{2.629} = 1.501$$

Empirical formula: $C_2H_4O_3$

$$\Delta T_f = K_f m; \; m = \frac{5.20°C}{1.86°C/\text{molal}} = 2.80 \text{ molal}$$

$\frac{2.80 \text{ mol}}{\text{kg}} = \frac{10.56 \text{ g}}{0.0250 \text{ kg } H_2O} \times \frac{1}{\text{MW}}$, MW = 151 g/mol (expt) and mass of $C_2H_4O_3$ = 76.05 g/mol

Molecular formula = $C_4H_8O_6$ and molar mass = 152.10 g/mol

Note: We can use the experimental molar mass to get the molecular formula. Knowing this, we can calculate the molar mass precisely from the formula.

CHALLENGE PROBLEMS

82. a. The average values for each ion are:

300. mg Na^+, 15.7 mg K^+, 5.45 mg Ca^{2+},

388 mg Cl^-, and 246 mg lactate, $C_3H_5O_3^-$

Note: Since we can accurately weigh to ± 0.1 mg on an analytical balance, we'll carry extra significant figures and calculate results to ± 0.1 mg.

The only source of lactate is $NaC_3H_5O_3$

$$246 \text{ mg lactate} \times \frac{112.06 \text{ mg } NaC_3H_5O_3}{89.07 \text{ mg } C_3H_5O_3^-} = 309.5 \text{ mg sodium lactate}$$

The only source of Ca^{2+} is $CaCl_2{\cdot}2H_2O$

$$5.45 \text{ mg } Ca^{2+} \times \frac{147.0 \text{ mg } CaCl_2{\cdot}2H_2O}{40.08 \text{ mg } Ca^{2+}} = 19.99 \text{ or } 20.0 \text{ mg } CaCl{\cdot}2H_2O$$

The only source of K^+ is KCl

$$15.7 \text{ mg } K^+ \times \frac{74.55 \text{ mg KCl}}{39.10 \text{ mg } K^+} = 29.9 \text{ mg KCl}$$

From what we have used already, let's calculate the mass of Na^+ and Cl^- added.

309.5 mg sodium lactate; 246 mg lactate, 63.5 mg Na^+

Thus, we need to add an additional 236.5 mg Na^+ to get the desired 300 mg.

$$236.5 \text{ mg } Na^+ \times \frac{58.44 \text{ mg NaCl}}{22.99 \text{ mg } Na^+} = 601.2 \text{ mg NaCl}$$

Let's check the mass of Cl^- added:

$$20.0 \text{ mg } CaCl_2{\cdot}2H_2O \times \frac{70.90 \text{ mg } Cl^-}{147.0 \text{ mg } CaCl_2{\cdot}2H_2O} = 9.6 \text{ mg } Cl^-$$

$$29.9 \text{ mg KCl} - 15.7 \text{ mg } K^+ = 14.2 \text{ mg } Cl^-$$

$$601.2 \text{ mg NaCl} - 236.5 \text{ mg } Na^+ = \underline{364.7 \text{ mg } Cl^-}$$

$$388.5 \text{ mg } Cl^-$$

Which is what we want (the average amount of Cl^-).

An analytical balance can weigh to the nearest 0.1 mg. We would use 309.5 mg sodium lactate, 20.0 mg $CaCl_2{\cdot}2H_2O$, 29.9 mg KCl, and 601.2 mg NaCl.

b. To get the range of osmotic pressure, we need to calculate the molar concentration of each ion at its minimum and maximum values. At minimum concentrations, we have:

$$\frac{285 \text{ mg } Na^+}{100 \text{ mL}} \times \frac{1 \text{ mmol}}{22.99 \text{ mg}} = 0.124\ M$$

$$\frac{14.1 \text{ mg } K^+}{100 \text{ mL}} \times \frac{1 \text{ mmol}}{39.10 \text{ mg}} = 0.00361\ M$$

$$\frac{4.9 \text{ mg } Ca^{2+}}{100 \text{ mL}} \times \frac{1 \text{ mmol}}{40.08 \text{ mg}} = 0.0012\ M$$

$$\frac{368 \text{ mg } Cl^-}{100 \text{ mL}} \times \frac{1 \text{ mmol}}{35.45 \text{ mg}} = 0.104\ M$$

$$\frac{231 \text{ mg lactate}}{100 \text{ mL}} \times \frac{1 \text{ mmol}}{89.07 \text{ mg}} = 0.0259\ M$$

total = 0.124 + 0.00361 + 0.0012 + 0.104 + 0.0259 = 0.259 M

$$\Pi = MRT = \frac{0.259\text{ mol}}{\text{L}} \times \frac{0.08206\text{ L atm}}{\text{mol K}} \times 310.\text{ K} = 6.59\text{ atm}$$

Similarly at maximum concentrations, the concentration of each ion is:

Na^+: 0.137 M | Cl^-: 0.115 M

K^+: 0.00442 M | $C_3H_5O_3^-$: 0.0293 M

Ca^{2+}: 0.0015 M

The total concentration of all ions is the sum, 0.287 M.

$$\Pi = \frac{0.287\text{ mol}}{\text{L}} \times \frac{0.08206\text{ L atm}}{\text{mol K}} \times 310.\text{ K} = 7.30\text{ atm}$$

Osmotic pressure ranges from 6.59 atm to 7.30 atm.

83. $\Delta T_f = 5.5 - 2.8 = 2.7°C$

$\Delta T_f = K_f m$

$$m = \frac{\Delta T_f}{K_f} = \frac{2.7°C}{5.12°C/\text{molal}} = 0.527\text{ molal} \approx 0.53\text{ molal}$$

Let x = mass of naphthalene (molar mass: 128 g/mol)

Then $1.60 - x$ = mass of anthracene (molar mass: 178 g/mol)

$\frac{x}{128}$ = moles naphthalene and $\frac{1.60 - x}{178}$ = moles anthracene

$$\frac{\frac{x}{128} + \frac{1.60 - x}{178}}{0.0200\text{ kg solvent}} = 0.527 \text{ (carry extra significant figure)}$$

$$\frac{178x + 1.60(128) - 128x}{(128)(178)} = 1.05 \times 10^{-2}$$

$50.x + 204.8 = 239.2$; $50.x = 239.2 - 204.8$; $50.x = 34.4$

$x = 0.688$ g naphthalene. So mixture is:

$\frac{0.688\text{ g}}{1.60\text{ g}} \times 100 = 43\%$ naphthalene by mass and 57% anthracene by mass

84. $\Delta T_f = K_f m$; $m = \frac{5.4}{1.86} = 2.9$ molal

$$\frac{2.9\text{ mol}}{\text{kg}} = \frac{\text{n}}{0.050\text{ kg}}$$

n = 0.145 mol of ions in solution (carry extra significant figure)

n = 2(x mol of $NaNO_3$) + 3(y mol $Mg(NO_3)_2$)

$6.5 \text{ g} = x \text{ mol NaNO}_3 \left(\frac{85.0 \text{ g}}{\text{mol}}\right) + y \text{ mol Mg(NO}_3)_2 \left(\frac{148 \text{ g}}{\text{mol}}\right)$

$2x + 3y = 0.145$

$85x + 148y = 6.5$

$170x + 255y = 12.33$
$170x + 296y = 13.0$

$41y = 0.67$

$y = 0.0163 \text{ mol} = 2.4 \text{ g Mg(NO}_3)_2$

Mass of $NaNO_3$ = 6.5 g - 2.4 g = 4.1 g $NaNO_3$ (Note: carry extra significant figure)

85. a. $\Delta T_f = K_f m;\ m = \frac{1.32}{5.12} = 0.258 \text{ mol/kg}$

$0.258 \text{ mol/kg} = \frac{1.22}{0.01560} \times \frac{1}{MW}$

MW = 303 g/mol

Uncertainty in temp. = $\frac{0.04}{1.32} \times 100 = 3\%$

A 3% uncertainty of 303 g/mol = 9 g/mol

So, MW = 303 ± 9 g/mol

b. No, codeine could not be eliminated since it is in the possible range.

c. We would really like the uncertainty to be ± 1 g/mol. We need a T_f to be about 10 times what it was in this problem. Two possibilities:

1. make solution ten times as concentrated (possible solubility problem)

2. use camphor (K_f = 40) as the solvent.

Even then it may be difficult to distinguish between the two. Using a mass spectrometer (see Ch. 3 in textbook) would be a much better technique.

86. a. $im = \frac{10.0°C}{1.86°C/\text{molal}} = 5.38 \text{ molal}$

$CaCl_2$ is ionic: $CaCl_2 \rightarrow Ca^{2+} + 2\,Cl^-$, i = 3

5.38 mol/kg is the total molality of all ions. So we need

$\frac{5.38}{3} = 1.79 \text{ mol CaCl}_2/\text{kg H}_2\text{O}$

$2.00 \text{ kg H}_2\text{O} \times \frac{1.79 \text{ mol CaCl}_2}{\text{kg H}_2\text{O}} \times \frac{111.0 \text{ g CaCl}_2}{\text{mol CaCl}_2} = 397 \text{ g CaCl}_2$

Dissolve 397 g $CaCl_2$ in 2.00 kg (2.00 L) of H_2O. This will make slightly more than 2 L, but this will be the easiest way since we don't know the density of the solution. We will have at least 2 L to use for whatever purpose.

b. $\frac{5.38}{2.6} = 2.1$ mol $CaCl_2$ (Solution is not ideal, i ≈ 2.6 from problem 11.63)

We need: $2.00 \text{ kg } H_2O \times \frac{2.1 \text{ mol } CaCl_2}{\text{kg } H_2O} \times \frac{111.0 \text{ g } CaCl_2}{\text{mol } CaCl_2} = 470 \text{ g } CaCl_2$

Dissolve 470 g $CaCl_2$ in 2.00 L (2.00 kg) H_2O.

CHAPTER TWELVE: CHEMICAL KINETICS

QUESTIONS

1. The rate of a chemical reaction varies with time. Consider a reaction:

$$A \longrightarrow \text{Products} \qquad \text{rate} = \frac{-\Delta[A]}{\Delta t}$$

If we graph [A] vs. t, it would roughly look like

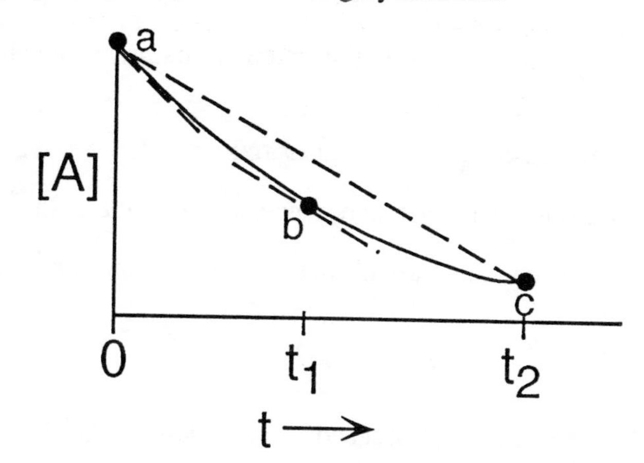

An instantaneous rate is the slope of the tangent line to the graph of [A] vs. t. We can determine the instantaneous rate at any point.

The initial rate is the instantaneous rate very near t = 0 (point a). The slope of the tangent at point b is the instantaneous rate at time, t_1.

The average rate is measured over a period of time. For example, the slope of the line connecting points a and c is the average rate of the reaction over the entire length of time, 0 to t_2.

2. a. An elementary step (reaction) is one in which the rate law can be written from the molecularity, i.e. from the balanced reaction.

   b. The mechanism of a reaction is the series of elementary reactions that occur to give the overall reaction. The sum of all the steps in the mechanism gives the balanced chemical reaction.

   c. The rate determining step is the slowest elementary reaction in any given mechanism.

3. a. The greater the frequency of collisions, the greater the opportunities for molecules to react, and, hence, the greater the rate.

   b. Chemical reactions involve the making and breaking of chemical bonds. The kinetic energy of the collision can be used to break bonds.

   c. For a reaction to occur, it is the reactive portion of each molecule that must be involved in a collision. Thus, only some collisions have the right orientation.

4. In a unimolecular reaction a single reactant molecule decomposes to products. In a biomolecular reaction, two molecules collide to give products.

5. The probability of the simultaneous collision of three molecules with the correct energy and orientation is exceedingly small.

6. a. A homogeneous catalyst is in the same phase as the reactants.

   b. A heterogeneous catalyst is in a different phase than the reactants. The catalyst is usually a solid, although a catalyst in a liquid phase can act as a heterogeneous catalyst for some gas phase reactions.

7. A catalyst increases the rate of a reaction by providing an alternate pathway (mechanism) with a lower activation energy.

8. Yes, the catalyst takes part in the mechanism, and will appear in the rate law.

9. No, the catalyzed reaction has a different mechanism and hence, a different rate law.

10. Some energy must be added to get the reaction started, that is, to overcome the activation energy barrier. Chemically what happens is:

$$\text{Energy} + H_2 \longrightarrow 2\,H$$

The hydrogen atoms initiate a chain reaction that proceeds very rapidly. Collision of H_2 and O_2 molecules at room temperature do not have sufficient kinetic energy to form hydrogen atoms and initiate the reaction.

EXERCISES

Reaction Rates

11. $$\text{Rate} = \frac{-\Delta[S_2O_3^{2-}]}{\Delta t} = \frac{0.0080\ \text{mol}}{\text{L s}}$$

I_2 is consumed at half this rate.

$$\frac{-\Delta[I_2]}{\Delta t} = 0.0040\ \text{mol/L•s}$$

$$\frac{\Delta[S_4O_6^{2-}]}{\Delta t} = 0.0040\ \text{mol/L•s} \qquad \frac{\Delta[I^-]}{\Delta t} = 0.0080\ \text{mol/L•s}$$

12. $$\frac{\Delta[H_2]}{\Delta t} = 3\,\frac{\Delta[N_2]}{\Delta t} \quad \text{and} \quad \frac{\Delta[NH_3]}{\Delta t} = -2\,\frac{\Delta[N_2]}{\Delta t}$$

So: $$\frac{1}{3}\,\frac{\Delta[H_2]}{\Delta t} = -\frac{1}{2}\,\frac{\Delta[NH_3]}{\Delta t}$$

$$\frac{\Delta[NH_3]}{\Delta t} = -\frac{2}{3}\,\frac{\Delta[H_2]}{\Delta t}$$

Ammonia is produced at a rate equal to 2/3 of the rate of consumption of hydrogen.

13. a. mol/L•s

b. Rate = k; k has units of mol/L•s

c. Rate = k[A]

$$\frac{\text{mol}}{\text{L s}} = k\left(\frac{\text{mol}}{\text{L}}\right)$$

k must have units s^{-1}

d. Rate = $k[A]^2$

$$\frac{\text{mol}}{\text{L s}} = k\left(\frac{\text{mol}}{\text{L}}\right)^2$$

k must have units L/mol•s

e. L^2/mol^2•s

14. a. molecules/cm^3•s

b. molecules/cm^3•s

c. s^{-1}

d. cm^3/molecules•s

e. cm^6/molecules^2•s

15. $$\frac{1.24 \times 10^{-12}\ \text{cm}^3}{\text{molecules s}} \times \frac{1\ \text{L}}{1000\ \text{cm}^3} \times \frac{6.022 \times 10^{23}\ \text{molecules}}{\text{mol}}$$

$k = 7.47 \times 10^8$ L/mol•s

16. $$\frac{4.4 \times 10^{-6}\ \text{L}}{\text{mol s}} \times \frac{1\ \text{mol}}{6.022 \times 10^{23}\ \text{molecules}} \times \frac{1000\ \text{cm}^3}{\text{L}} = 7.3 \times 10^{-27}\ \text{cm}^3/\text{molecules•s}$$

Rate Laws from Experimental Data: Initial Rates Method

17. a. In the first two experiments, [NO] is held constant and $[Cl_2]$ is doubled. The rates also double. Thus, the reaction is first order with respect to Cl_2.

Mathematically: Rate = $k[NO]^x\ [Cl_2]^y$

$$\frac{0.35}{0.18} = \frac{k(0.10)^x\ (0.20)^y}{k(0.10)^x\ (0.10)^y} = \frac{(0.20)^y}{(0.10)^y}$$

$1.9 = 2.0^y \qquad y \approx 1$

We get the dependence on NO from the second and third experiments.

$$\frac{1.45}{0.35} = \frac{k(0.20)^x\ (0.20)}{k(0.10)^x\ (0.20)} = \frac{(0.20)^x}{(0.10)^x}$$

$4.1 = 2^x$; $\log 4.1 = x \log 2$; $0.62 = x(0.30)$

$x = 2.1 \approx 2$

So rate = $k[NO]^2\ [Cl_2]$

b. $$\frac{0.18\ \text{mol}}{\text{L min}} = k\left(\frac{0.10\ \text{mol}}{\text{L}}\right)^2\left(\frac{0.10\ \text{mol}}{\text{L}}\right)$$

$k = 180\ L^2/mol^2$•min (1st exp.)

$k = 175\ L^2/mol^2$•min (2nd exp.)

$k = 181\ L^2/mol^2$•min (3rd exp.)

To two significant figures, $k_{mean} = 1.8 \times 10^2\ L^2/mol^2$•min

18. a. Rate = $k[I^-]^x [S_2O_8^{2-}]^y$

$$\frac{12.50 \times 10^{-6}}{6.250 \times 10^{-6}} = \frac{k(0.080)^x (0.040)^y}{k(0.040)^x (0.040)^y}$$

$2.000 = 2.0^x$, $x = 1$

$$\frac{12.50 \times 10^{-6}}{5.560 \times 10^{-6}} = \frac{k(0.080)(0.040)^y}{k(0.080)(0.020)^y}; \quad 2.248 = 2.0^y$$

$\log 2.248 = y \log 2.0$; $0.3518 = 0.30y$; $y = 1.2 \approx 1$

Rate = $k[I^-][S_2O_8^{2-}]$

b. For first experiment:

$$\frac{12.50 \times 10^{-6}\ \text{mol}}{\text{L s}} = k\left(\frac{0.080\ \text{mol}}{\text{L}}\right)\left(\frac{0.040\ \text{mol}}{\text{L}}\right) \quad \text{and } k = 3.9 \times 10^{-3}\ \text{L/mol•s}$$

The other values are:

Initial Rate (mol/L•s)	k (L/mol•s)
12.5×10^{-6}	3.9×10^{-3}
6.25×10^{-6}	3.9×10^{-3}
5.56×10^{-6}	3.5×10^{-3}
4.35×10^{-6}	3.4×10^{-3}
6.41×10^{-6}	3.6×10^{-3}

$$\text{Mean} = \frac{(3.9 + 3.9 + 3.5 + 3.4 + 3.6) \times 10^{-3}}{5}$$

$k_{mean} = 3.7 \times 10^{-3}$ L/mol•s $\pm$ 0.3 L/mol•s

19. a. Rate = $K[I^-]^x [OCl^-]^y$

$$\frac{7.91 \times 10^{-2}}{3.95 \times 10^{-2}} = \frac{k(0.12)^x (0.18)^y}{k(0.060)^x (0.18)^y} = 2.0^x$$

$2.00 = 2.0^x$, $x = 1$

$$\frac{3.95 \times 10^{-2}}{9.88 \times 10^{-3}} = \frac{k(0.060)(0.18)^y}{k(0.030)(0.090)^y}, \quad x = 1$$

$4.00 = 2.0 \times 2.0^y$; $2.0 = 2.0^y$; $y = 1$

Rate = $k[I^-][OCl^-]$

b. From the first experiment:

$$\frac{7.91 \times 10^{-2}\ \text{mol}}{\text{L s}} = k\left(\frac{0.12\ \text{mol}}{\text{L}}\right)\left(\frac{0.18\ \text{mol}}{\text{L}}\right)$$

$k = 3.7$ L/mol•s

All four experiments give the same value of k to two significant figures.

20. a. Rate = $k[NOCl]^n$

$$\frac{2.66 \times 10^4}{6.64 \times 10^3} = \frac{k(2.0 \times 10^{16})^n}{k(1.0 \times 10^{16})^n}$$

$4.01 = 2.0^n$; $n = 2$; Rate = $k[NOCl]^2$

b. $$\frac{5.98 \times 10^4\ \text{molecules}}{\text{cm}^3\ \text{s}} = k\left(\frac{3.0 \times 10^{16}\ \text{molecules}}{\text{cm}^3}\right)^2$$

$k = 6.64 \times 10^{-29}$ cm^3/molecules•s = 6.6×10^{-29} cm^3/molecules•s

The other three experiments give (6.65, 6.64, and 6.62) $\times 10^{-29}$ cm^3/molecules•s, respectively.

The mean value is 6.6×10^{-29} cm^3/molecules•s.

c. $$\frac{6.6 \times 10^{-29}\ \text{cm}^3}{\text{molecules s}} \times \frac{1\ \text{L}}{1000\ \text{cm}^3} \times \frac{6.02 \times 10^{23}\ \text{molecules}}{\text{mol}} = \frac{4.0 \times 10^{-8}\ \text{L}}{\text{mol s}}$$

21. Rate = $k[ClO_2]^x[OH^-]^y$

From the first two experiments:

$2.30 \times 10^{-1} = k(0.100)^x(0.100)^y$ and $5.75 \times 10^{-2} = k(0.0500)^x(0.100)^y$

Dividing the two rate laws:

$$4.00 = \frac{(0.100)^x}{(0.0500)^x} = 2.00^x;\ x = 2$$

Comparing the second and third experiments:

$2.30 \times 10^{-1} = k(0.100)(0.100)^y$ and $1.15 \times 10^{-1} = k(0.100)(0.0500)^y$

Dividing: $$2.00 = \frac{(0.100)^y}{(0.0500)^y} = 2.00^y;\ y = 1$$

The rate law is:

Rate = $k[ClO_2]^2[OH^-]$

2.30×10^{-1} mol/L•s = $k(0.100 \text{ mol/L})^2(0.100 \text{ mol/L})$

$k = 2.30 \times 10^2 \text{ L}^2/\text{mol}^2\text{•s}$, $k_{mean} = 2.30 \times 10^2 \text{ L}^2/\text{mol}^2\text{•s}$

22. Rate = $k[N_2O_5]^x$

The rate laws for the first two experiments are:

$2.26 \times 10^{-3} = k(0.190)^x$ and $8.90 \times 10^{-4} = k(0.075)^x$

Dividing: $2.54 = \frac{(0.190)^x}{(0.075)^x} = (2.5)^x$; $x = 1$ Rate = $k[N_2O_5]$

$k = \frac{\text{Rate}}{[N_2O_5]} = \frac{8.90 \times 10^{-4} \text{ mol/L•s}}{0.075 \text{ mol/L}} = 1.2 \times 10^{-2} \text{ s}^{-1}$, $k_{mean} = 1.2 \times 10^{-2} \text{ s}^{-1}$

Integrated Rate Laws from Experimental Data

23. The first guess to make is that the reaction is first order. For a first order reaction a graph of $\ln[C_4H_6]$ vs. t should yield a straight line.

Time (s)	195	604	1246	2180	6210
$[C_4H_6]$ (mol/L)	1.6×10^{-2}	1.5×10^{-2}	1.3×10^{-2}	1.1×10^{-2}	0.68×10^{-2}
$\ln[C_4H_6]$	-4.14	-4.20	-4.34	-4.51	-4.99
$1/[C_4H_6]$	62.5	66.7	76.9	90.9	147

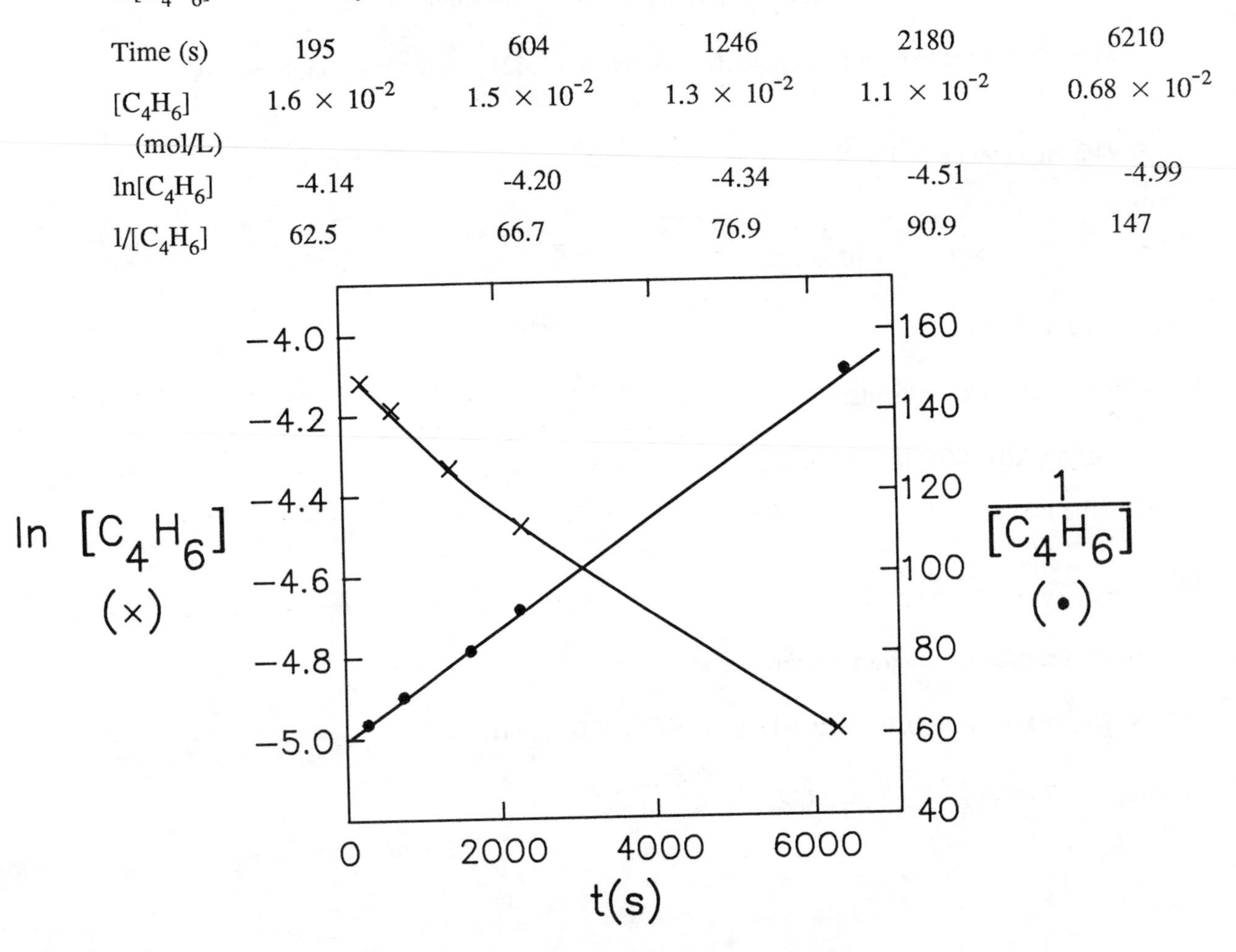

The natural log plot is not linear. The next guess is that the reaction is second order. For a second order reaction a plot of $1/[C_4H_6]$ vs. t should yield a straight line. Since we get a straight line for this graph, we conclude the reaction is second order in butadiene. Rate = $k[C_4H_6]^2$

For a second order reaction: $\frac{1}{[A]_t} = \frac{1}{[A]_o} + kt$

Thus the slope of the straight line equals the value of the rate constant. Using the points on the line at 1000 and 6000 s

$$k = \text{Slope} = \frac{144 \text{ L/mol} - 73 \text{ L/mol}}{6000 \text{ s} - 1000 \text{ s}} = 1.4 \times 10^{-2} \text{ L/mol}\bullet\text{s}$$

This value was obtained from reading the graph and "eyeballing" a straight line with a ruler.

24. This problem differs in two ways from the previous problem:
 1. a product is measured instead of a reactant and
 2. only the volume of a gas is given and not the concentration.

We can find the initial concentration of $C_6H_5N_2Cl$ from the amount of N_2 evolved after infinite time when all $C_6H_5N_2Cl$ is decomposed. It is a measure of the amount of benzene diazonium chloride originally present (assuming the reaction has gone to completion). Since PV = nRT:

$$n = \frac{PV}{RT} = \frac{1.00 \text{ atm} \times 58.3 \times 10^{-3} \text{ L}}{\frac{0.08206 \text{ L atm}}{\text{mol K}} \times 323 \text{ K}} = 2.20 \times 10^{-3} \text{ mol}$$

Since each mole of $C_6H_5N_2Cl$ that decomposes produces one mole of N_2, then the initial concentration (t = 0) of $C_6H_5N_2Cl$ was

$$\frac{2.20 \times 10^{-3} \text{ mol}}{40.0 \times 10^{-3} \text{ L}} = 0.0550 \text{ mol/L}$$

We can similarly calculate the moles of N_2 evolved at each point of the experiment, subtract that from 2.20×10^{-3} mol to get the moles of $C_6H_5N_2Cl$ remaining, and calculate $[C_6H_5N_2Cl]$ at each time. We would then use these results to make the appropriate graph to check the order of the reaction. Since the rate constant is the slope of a straight line, we would favor this approach to get a value for the rate constant.

There is a simpler way to check for the order of the reaction that saves doing a lot of math.

The quantity $(V_\infty - V_t)$

where V_∞ = 58.3 mL N_2 evolved

V_t = mL of N_2 evolved at time, t

will be proportional to the moles of $C_6H_5N_2Cl$ remaining; $(V_\infty - V_t)$ will also be proportional to the concentration of $C_6H_5N_2Cl$. Thus, we can get the same information by using $(V_\infty - V_t)$ as our measure of $[C_6H_5N_2Cl]$. If the reaction is first order a graph of $\ln(V_\infty - V_t)$ vs. t would be linear. The data for such a graph are:

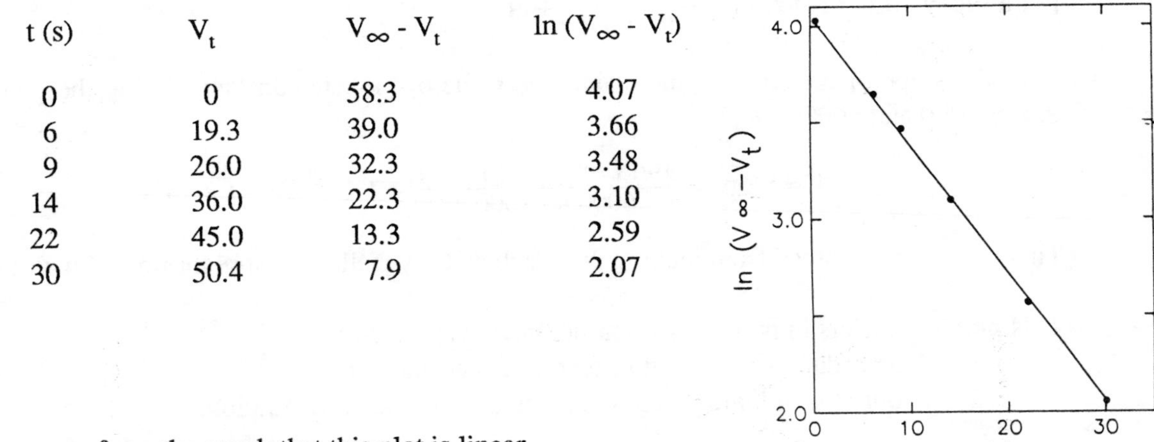

t (s)	V_t	$V_\infty - V_t$	$\ln(V_\infty - V_t)$
0	0	58.3	4.07
6	19.3	39.0	3.66
9	26.0	32.3	3.48
14	36.0	22.3	3.10
22	45.0	13.3	2.59
30	50.4	7.9	2.07

We can see from the graph that this plot is linear, so the reaction is first order.

Rate = $k[C_6H_5N_2Cl]$ with $k = 6.9 \times 10^{-2}\ s^{-1}$

25.

Time (s)	$[H_2O_2]$ (mol/L)	$\ln[H_2O_2]$
0	1.0	0
120	0.91	-0.094
300	0.78	-0.248
600	0.59	-0.528
1200	0.37	-0.994
1800	0.22	-1.51
2400	0.13	-2.04
3000	0.082	-2.50
3600	0.050	-3.00

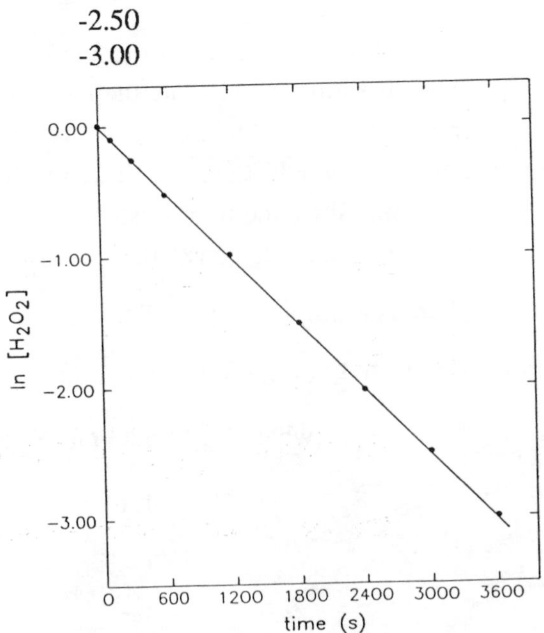

The plot $\ln[H_2O_2]$ vs t is linear.
Thus, the reaction is first order.

$\ln[H_2O_2] = -kt + \ln[H_2O_2]_o$.

Rate = $k[H_2O_2]$

$$\text{Slope} = -k = \frac{0 - (-3.00)}{0 - 3600} = -8.3 \times 10^{-4}\ s^{-1}$$

$k = 8.3 \times 10^{-4}\ s^{-1}$

26.

Time (s)	$[NO_2]$ mol/L	ln $[NO_2]$	$1/[NO_2]$
0	0.500	-0.693	2.00
1.20×10^3	0.444	-0.812	2.25
3.00×10^3	0.381	-0.965	2.62
4.50×10^3	0.340	-1.079	2.94
9.00×10^3	0.250	-1.386	4.00
1.80×10^4	0.174	-1.749	5.75

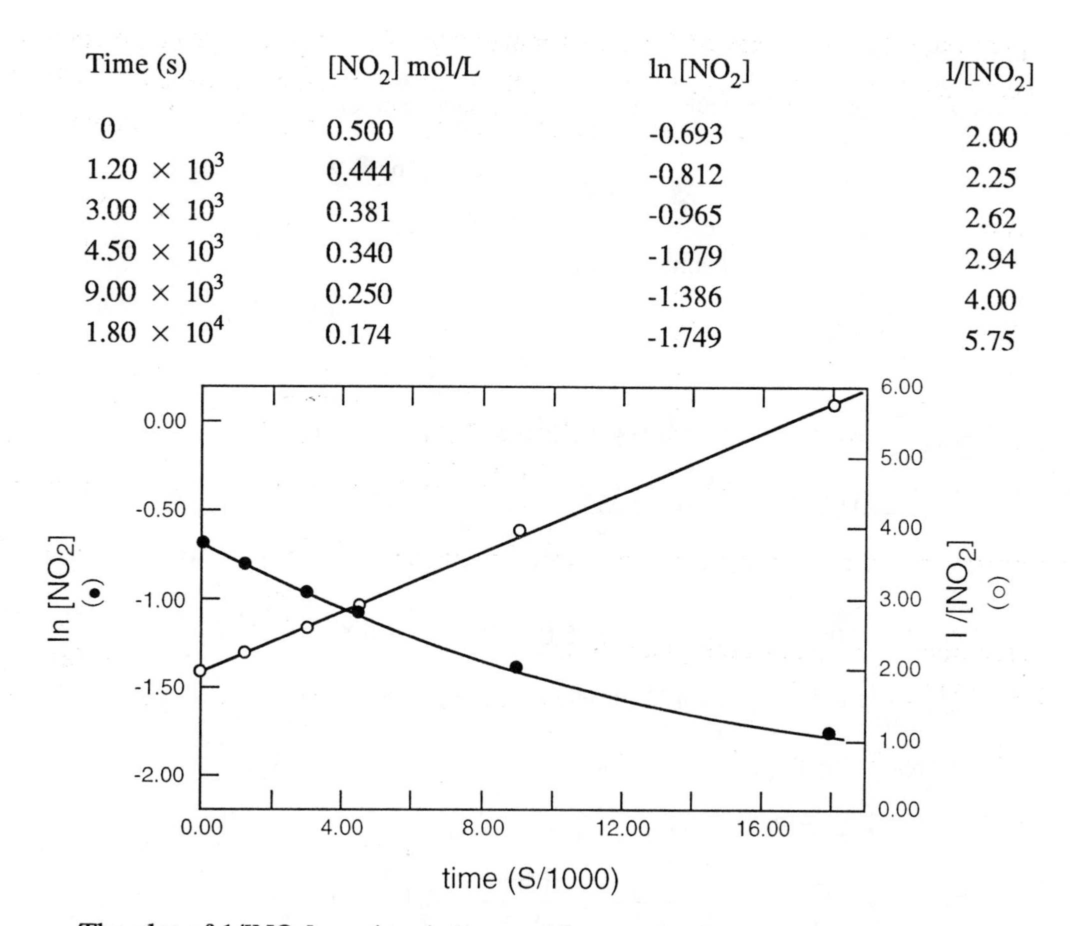

The plot of $1/[NO_2]$ vs. time is linear. The reaction is second order in NO_2. The integrated and differential rate laws are:

$$\frac{1}{[NO_2]} = kt + \frac{1}{[NO_2]_o} \quad \text{and Rate} = k[NO_2]^2$$

The slope of the plot $1/[NO_2]$ vs. t gives the value of k.

Slope = 2.10×10^{-4} L/mol•s = k

Half-Life

27. a. $k = \frac{\ln 2}{t_{1/2}} = \frac{0.69315}{t_{1/2}}$

For ^{239}Pu, $k = \frac{0.69315}{24{,}360 \text{ yr}} = 2.845 \times 10^{-5} \text{ yr}^{-1}$

$$\frac{2.845 \times 10^{-5}}{\text{yr}} \times \frac{1 \text{ yr}}{365 \text{ d}} \times \frac{1 \text{ d}}{24 \text{ hr}} \times \frac{1 \text{ hr}}{60 \text{ min}} \times \frac{1 \text{ min}}{60 \text{ s}} = 9.021 \times 10^{-13} \text{ s}^{-1}$$

For ^{241}Pu, $k = \frac{0.693}{13 \text{ yr}} = 0.0533 \text{ yr}^{-1} = 1.7 \times 10^{-9} \text{ s}^{-1}$

b. The rate constant is larger for the decay of ^{241}Pu, hence, ^{241}Pu decays more rapidly.

c. For radioactive decay:

Rate = kN where N is the number of radioactive nuclei.

$$\text{Rate} = 1.7 \times 10^{-9}\ s^{-1} \times \left(5.0\ g \times \frac{1\ mol}{241\ g} \times \frac{6.02 \times 10^{23}\ atoms}{mol}\right)$$

Rate = 2.1×10^{13} disintegrations/s

$$\ln \frac{[A]_t}{[A]_o} = -kt$$

$$\ln\left(\frac{N}{5.0}\right) = -0.0533\ yr^{-1}\ (1.0\ yr) \quad \text{(carry extra sig. fig.)}$$

$$\ln N = -0.0533 + \ln 5.0 = -0.0533 + 1.609 = 1.556$$

$N = e^{1.556} = 4.7$ g left after 1 yr

$$\ln N = -0.0533\ yr^{-1}\ (10.\ yr) + \ln 5.0 = -0.533 + 1.609 = 1.076$$

$N = e^{1.076} = 2.9$ g ^{241}Pu left after 10. years

$$\ln N = -0.0533\ yr^{-1}\ (100.\ yr) + \ln 5.0 = -5.33 + 1.609 = -3.72$$

$N = e^{-3.72} = 0.024$ g left after 100. yr

28. a. $k = \frac{0.6931}{t_{1/2}} = \frac{0.6931}{12.8\ day} \times \frac{1 day}{24\ hour} \times \frac{1\ hour}{3600\ s} = 6.27 \times 10^{-7}\ s^{-1}$

b. $\text{Rate} = kN = 6.27 \times 10^{-7}\ s^{-1}\left(28.0 \times 10^{-3}\ g \times \frac{6.022 \times 10^{23}\ atoms}{64.0\ g}\right)$

Rate = 1.65×10^{14} decays/s

c. 3% of the copper - 64 must be left.

$$\ln(0.03) = -\left(\frac{0.6931}{12.8\ day}\right)(t)$$

t = 64.8 days ≈ 60 days

29. $\ln \frac{[A]_t}{[A]_o} = -kt$

If $[A]_o = 100.$, then after 65 s, 45% of A has decayed or $[A]_{65} = 55$

$$\ln\left(\frac{55}{100.}\right) = -k(65\ s)$$

$k = 9.2 \times 10^{-3}\ s^{-1}$; $t_{1/2} = \frac{0.693}{k} = 75$ s

30. $\ln \frac{[A]_t}{[A]_o} = -kt$; $k = \frac{0.6931}{t_{1/2}} = \frac{0.6931}{14.3\ d} = 4.85 \times 10^{-2}\ d^{-1}$

If $A_o = 100.0$, after 95.0% completion $A_t = 5.0$

$\ln(0.050) = -4.85 \times 10^{-2}\ d^{-1} \times t$; $t = 62$ days

31. $t_{1/2} = \ln \frac{1}{k[A]_o}$ or $k = \frac{1}{t_{1/2}[A]_o}$

$$k = \frac{1}{143\ s\ (0.060\ mol/L)} = 0.12\ L/mol \cdot s$$

32. $\frac{1}{[A]} = kt + \frac{1}{[A]_o}$

$$\frac{1}{0.050\ mol/L} = (0.40\ L/mol \cdot min)\ t + \frac{1}{0.10\ mol/L}$$

$$t = \frac{20.-10.}{0.40}\ min = 25\ min.$$

33. a. $\ln\left(\frac{[A]_t}{[A]_o}\right) = -kt$

If reaction is 38.5% complete, then 38.5% of the original concentration is consumed, leaving 61.5%.

$[A] = 61.5\%$ of $[A]_o$ or $[A] = 0.615[A]_o$

$\ln(0.615) = -k(480\ s)$; $-0.486 = -k(480\ s)$

$k = 1.0 \times 10^{-3}\ s^{-1}$

b. $t_{1/2} = 0.693/k = 0.693/1.0 \times 10^{-3}\ s^{-1} = 693\ s = 690\ s$

c. 25% complete: $[A] = 0.75[A]_o$

$\ln(0.75) = -1.0 \times 10^{-3}(t)$

$t = 288\ s \approx 2.9 \times 10^2\ s$

75% complete: $[A] = 0.25[A]_o$

$\ln(0.25) = -1.0 \times 10^{-3}(t)$

$t = 1.39 \times 10^3\ s = 1.4 \times 10^3\ s$

or we know it takes 2 $t_{1/2}$ for reaction to be 75% complete:

$t = 2(693\ s) = 1386\ s \approx 1.4 \times 10^3\ s$

95% complete: $[A] = 0.05[A]_o$, $\ln(0.05) = -1.0 \times 10^{-3}(t)$ $t = 3 \times 10^3\ s$

34. $\ln\left(\frac{[A]_t}{[A]_o}\right) = -kt$

50.% complete in 3.5 hr, $t_{1/2} = 3.5$ hr

$k = \ln \frac{0.693}{t_{1/2}} = \frac{0.693}{3.5\text{ hr}} = 0.20\text{ hr}^{-1}$

When 88% complete, $[A] = 0.12[A]_o$

$\ln(0.12) = -0.20\text{ hr}^{-1}(t)$, $t = 10.6\text{ hr} \approx 11\text{ hr}$

Reaction Mechanisms

35. a. Rate = $k[CH_3NC]$ b. Rate = $k[O_3][NO]$

36. a. Rate = $k[O_3]$ b. Rate = $k[O_3][O]$

37. We know from Exercise 12.25 that Rate = $k[H_2O_2]$.

If the first step, $H_2O_2 \rightarrow 2\ OH$, is the slow step, then the mechanism gives the correct rate law.

38. Rate = $k[NO_2]^2$

Temperature Dependence of Rate Constants and Collision Theory

39. $k = A \exp(-E_a/RT)$

$\ln k = \frac{-E_a}{RT} + \ln A$; so $\ln k_1 = \frac{-E_a}{RT_1} + \ln A$ and $\ln k_2 = \frac{-E_a}{RT_2} + \ln A$

Subtracting the two equations, $\ln k_1 - \ln k_2 = \frac{-E_a}{R}\left(\frac{1}{T_1} - \frac{1}{T_2}\right) = \ln\left(\frac{k_1}{k_2}\right)$

$\ln 2.0 \times 10^3 - \ln k_2 = \frac{-15.0 \times 10^3\text{ J/mol}}{8.3145\text{ J/K·mol}}\left(\frac{1}{298\text{ K}} - \frac{1}{348\text{ K}}\right)$

$7.60 - \ln k_2 = -0.870$

$\ln k_2 = 8.47$

$k_2 = e^{8.47} = 4.8 \times 10^3\text{ s}^{-1}$

40. $k_1 = A \exp(-E_a/RT_1)$, $\ln k_1 = -E_a/RT_1 + \ln A$

$k_2 = A \exp(-E_a/RT_2)$, $\ln k_2 = -E_a/RT_2 + \ln A$

$\ln k_1 - \ln k_2 = -E_a/RT_1 - (-E_a/RT_2)$

$\ln\left(\frac{k_1}{k_2}\right) = \frac{E_a}{R}\left(\frac{1}{T_2} - \frac{1}{T_1}\right)$, $\ln\left(\frac{4.6 \times 10^{-2}}{8.1 \times 10^{-2}}\right) = \frac{E_a}{8.3145}\left(\frac{1}{293} - \frac{1}{273}\right)$

$$-0.566 = \frac{E_a}{8.3145}\left(-2.5 \times 10^{-4}\right)$$

$E_a = 1.9 \times 10^4$ J/mol = 19 kJ/mol

41. a. $\ln\left(\frac{k_1}{k_2}\right) = \frac{E_a}{R}\left(\frac{1}{T_2} - \frac{1}{T_1}\right)$, $\ln\left(\frac{2.45 \times 10^{-4}}{0.950}\right) = \frac{E_a}{8.3145}\left(\frac{1}{781} - \frac{1}{575}\right)$

$$-8.263 = \frac{E_a}{8.3145}\left(-4.6 \times 10^{-4}\right)$$

$E_a = 1.5 \times 10^5$ J/mol = 150 kJ/mol

$\ln k_1 = -E_a/RT + \ln A$

$$\ln (2.45 \times 10^{-4}) = \frac{-1.5 \times 10^5 \text{ J/mol}}{8.3145 \text{ J/mol•K}}\left(\frac{1}{575 \text{ K}}\right) + \ln A$$

$-8.314 = -31.38 + \ln A$

$\ln A = 23.07$

$A = e^{23.07} = 1.0 \times 10^{10}$ L/mol•s

b. $k = A \exp (-E_a/RT) = 1.0 \times 10^{10} \exp\left(\frac{-1.5 \times 10^5}{8.3145 \times 648}\right) = 8.1 \times 10^{-3}$ L/mol•s

42. If we double the rate, k doubles and

$$\frac{k_{35}}{k_{25}} = 2.00 = \frac{A \exp [-E_a/R(308)]}{A \exp [-E_a/R(298)]}$$

$$\ln 2.00 = \frac{-E_a}{308 R} + \frac{E_a}{298 R} = E_a\left(\frac{1}{298 R} - \frac{1}{308 R}\right)$$

$0.693 = E_a(4.04 \times 10^{-4} - 3.91 \times 10^{-4})$

$$\frac{0.693}{0.13 \times 10^{-4}} = E_a = 5.3 \times 10^4 \text{ J/mol} = 53 \text{ kJ/mol}$$

43. From the Arrhenius equation, $k = A \exp (-E_a/RT)$

or in logarithmic form, $\ln k = \frac{-E_a}{RT} + \ln A$

Hence, a graph of ln k vs. 1/T should yield a straight line with a slope equal to $-E_a/R$.

T (K)	1/T	k (L/mol•s)	ln k
195	5.13×10^{-3}	1.08×10^9	20.80
230	4.35×10^{-3}	2.95×10^9	21.82
260	3.85×10^{-3}	5.42×10^9	22.41
298	3.36×10^{-3}	12.0×10^9	23.21
369	2.71×10^{-3}	35.5×10^9	24.29

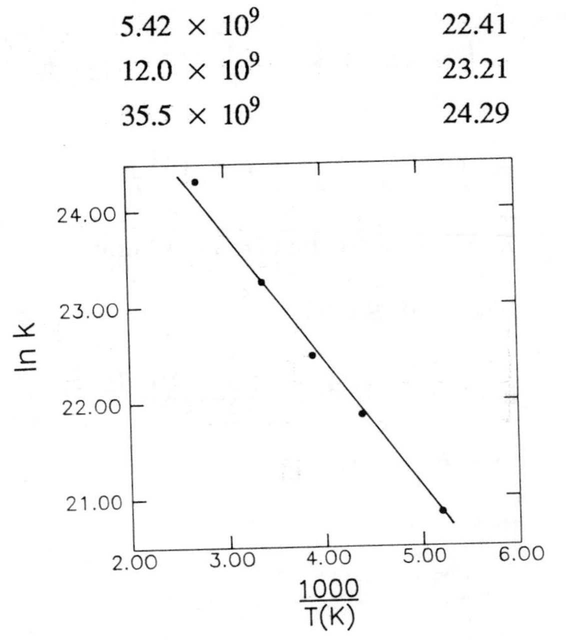

From the "eyeball" line on the graph:

$$\text{Slope} = \frac{20.95 - 23.65}{5.00 \times 10^{-3} - 3.00 \times 10^{-3}}$$

$$= \frac{-2.70}{2.00 \times 10^{-3}} = -1.35 \times 10^3 \text{ K} = \frac{-E_a}{R}$$

$$E_a = 1.35 \times 10^3 \text{ K} \times \frac{8.3145 \text{ J}}{\text{K mol}}$$

$$= 1.12 \times 10^4 \text{ J/mol or } 11.2 \text{ kJ/mol}$$

From the best straight line (by computer):

Slope = -1.43×10^3 K and E_a = 11.9 kJ/mol

44. Again we graph ln k vs. 1/T

T (K)	1/T	k (s^{-1})	ln k
338	2.96×10^{-3}	4.9×10^{-3}	-5.32
318	3.14×10^{-3}	5.0×10^{-4}	-7.60
298	3.36×10^{-3}	3.5×10^{-5}	-10.26

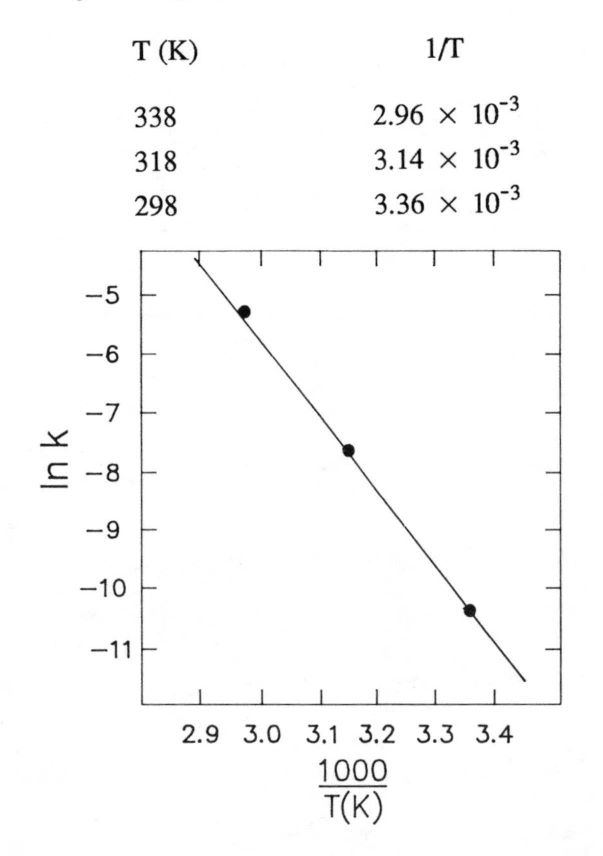

$$\text{Slope} = \frac{-10.76 - (-5.85)}{3.40 \times 10^{-3} - 3.00 \times 10^{-3}} = -1.2 \times 10^{4}\ \text{K}$$

$$\text{Slope} = \frac{-E_a}{R}$$

$$E_a = -\text{Slope} \times R = 1.2 \times 10^{4}\ \text{K} \times \frac{8.3145\ \text{J}}{\text{K mol}} = 1.0 \times 10^{5}\ \text{J/mol or } 1.0 \times 10^{2}\ \text{kJ/mol}$$

45.

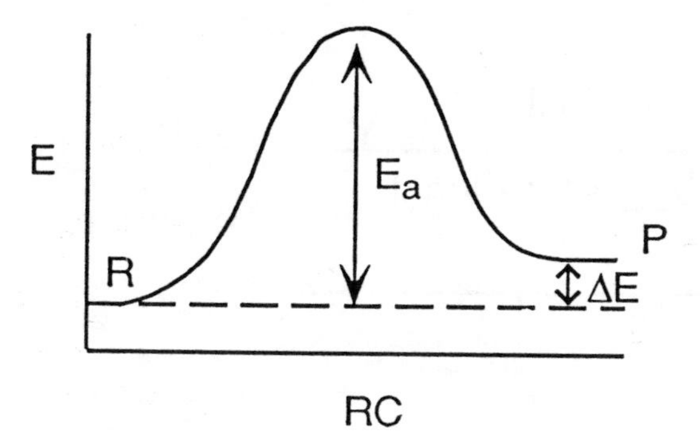

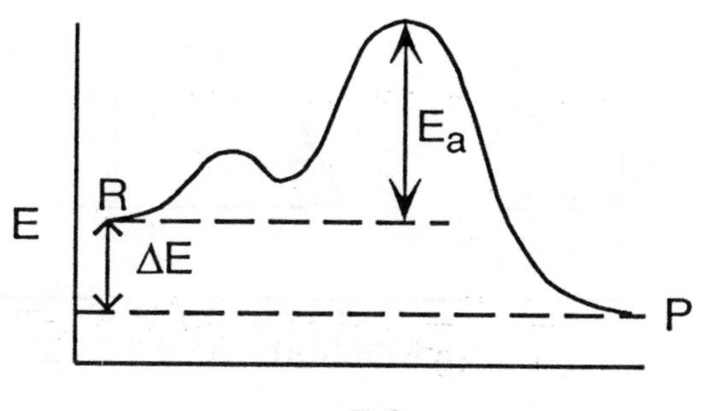

46.

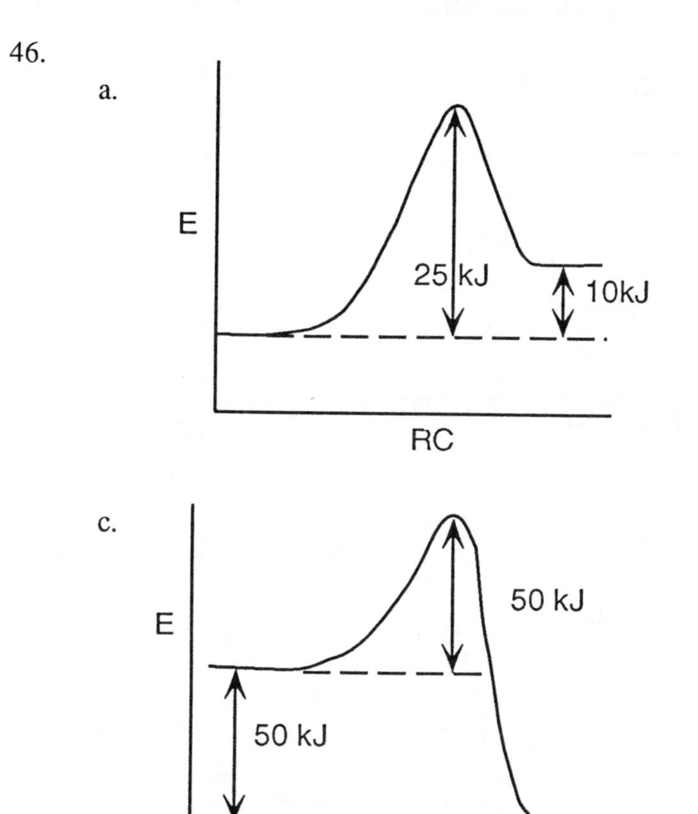

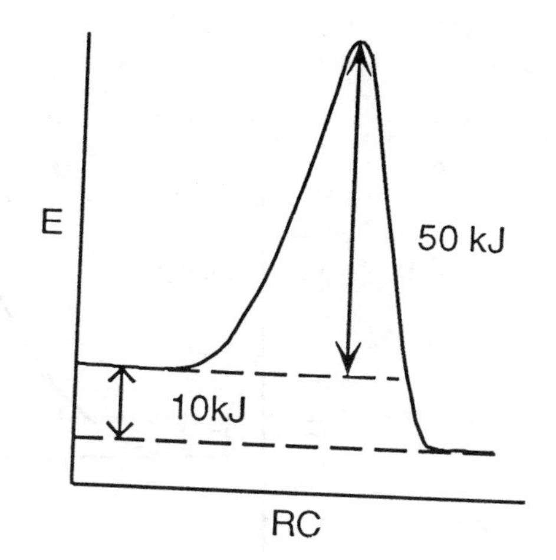

The reaction in (a) will have the greater rate since it has the lowest activation energy.

47.

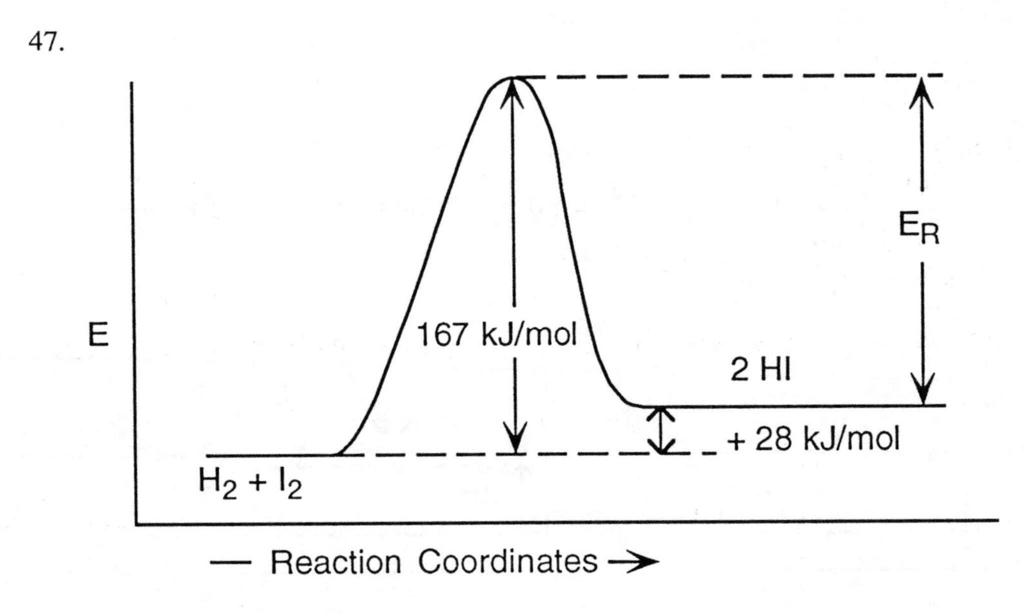

The activation energy for the reverse reaction is E_R in the diagram.

$$E_R = 167 - 28 = 139 \text{ kJ/mol}$$

48.

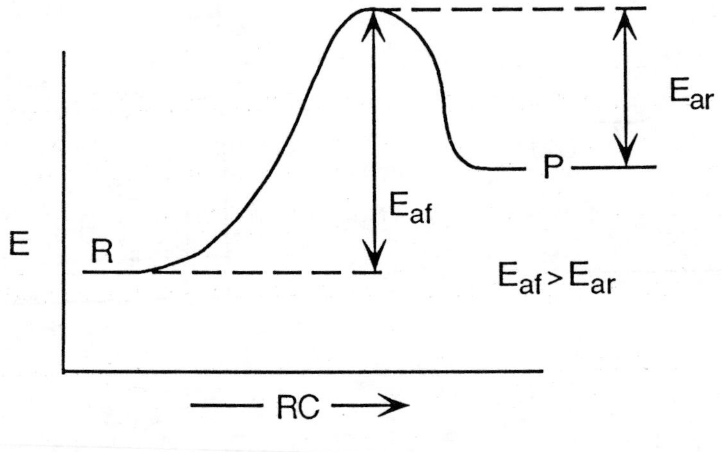

E

R

P

E_{af}

E_{ar}

$E_{af} > E_{ar}$

— RC →

The products are at a higher energy than the reactants.

Catalysis

49. a. NO is the catalyst.

b. NO_2 is an intermediate

c. $k = A \exp(-E_a/RT)$

$$\frac{k_{cat}}{k_{un}} = \frac{\exp[-E_a(cat)/RT]}{\exp[-E_a(un)/RT]}$$

$$\frac{k_{cat}}{k_{un}} = \exp\frac{E_a(un) - E_a(cat)}{RT}$$

$$\frac{k_{cat}}{k_{un}} = \exp\frac{2100 \text{ J/mol}}{8.3145 \text{ J/mol•K} \times 298 \text{ K}} = e^{0.848} = 2.3$$

The catalyzed reaction is 2.3 times faster than the uncatalyzed reaction at 25°C.

50. $O_3 + Cl \rightarrow O_2 + ClO$ (slow)

$ClO + O \rightarrow O_2 + Cl$ (fast)

$O_3 + O \rightarrow 2\ O_2$

Since the chlorine atom catalyzed reaction has the lower activation energy, the Cl catalyzed rate is faster. Hence, Cl is a more effective catalyst.

At 25°C:

$$\frac{k_{Cl}}{k_{NO}} = \exp\left(\frac{-E_a(Cl)}{RT} + \frac{E_a(NO)}{RT}\right)$$

$$= \exp\left(\frac{9800 \text{ J/mol}}{8.3145 \text{ J/mol•K} \times 298 \text{ J/mol}}\right) = e^{3.96} = 52$$

At 25°C, the Cl catalyzed reaction is roughly 52 times faster than the NO catalyzed reaction.

51. The surface of the catalyst should be:

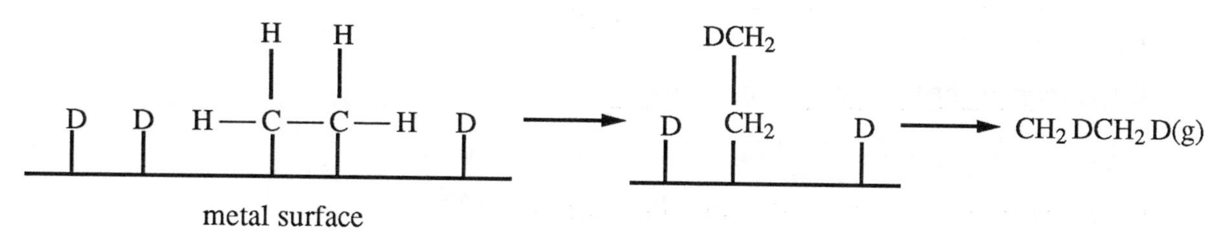

Thus, CH_2DCH_2D should be the product.

52. If the mechanism was possible then we would get:

$$C_2H_4 + D_2 \rightarrow CH_2DCH_2D$$

If we got this product, then we could conclude that this was a possible mechanism. If we got some other product, e.g. CH_3CHD_2, we could conclude our mechanism was wrong. The result is not conclusive since we might be able to conceive of other mechanisms that would give the same products as our proposed one.

ADDITIONAL EXERCISES

53. Rate = $k[Cl]^{1/2}[CHCl_3]$

$$\frac{mol}{L\,s} = k\left(\frac{mol}{L}\right)^{1/2}\left(\frac{mol}{L}\right)$$

k must have units of $L^{1/2}/mol^{-1/2}\cdot s$

54. a. $25\ min \times \frac{60\ s}{min} \times \frac{1.20 \times 10^{-4}\ mol}{L\,s} = \frac{0.18\ mol}{L}$

0.18 mol/L HI has been consumed.

[HI] = 0.25 - 0.18 = 0.07 mol/L

b. $t \times \frac{1.20 \times 10^{-4}\ mol}{L\,s} = \frac{0.250\ mol}{L}$

$t = 2.08 \times 10^3\ s$

c. $\frac{1.20 \times 10^{-4}\ mol\ HI}{L\,s} \times \frac{1\ mol\ H_2}{2\ mol\ HI} = 6.00 \times 10^{-5}\ mol/L\cdot s$

Both H_2 and I_2 are formed at the rate of 6.00×10^{-5} mol/L•s

55. Rate = $k[H_2SeO_3]^x[H^+]^y[I^-]^z$

Comparing the first and second experiments:

$$\frac{3.33 \times 10^{-7}}{1.66 \times 10^{-7}} = \frac{k(2.0 \times 10^{-4})^x\ (2.0 \times 10^{-2})^y\ (2.0 \times 10^{-2})^z}{k(1.0 \times 10^{-4})^x\ (2.0 \times 10^{-2})^y\ (2.0 \times 10^{-2})^z} = 2.0^x$$

$2.01 = 2.0^x,\ x = 1$

Comparing the first and fourth experiments:

$$\frac{6.66 \times 10^{-7}}{1.66 \times 10^{-7}} = \frac{k(1.0 \times 10^{-4})\ (4.0 \times 10^{-2})^y\ (2.0 \times 10^{-2})^z}{k(1.0 \times 10^{-4})\ (2.0 \times 10^{-2})^y\ (2.0 \times 10^{-2})^z}$$

$4.01 = 2.0^y,\ y = 2$

Comparing the first and sixth experiments:

$$\frac{13.2 \times 10^{-7}}{1.66 \times 10^{-7}} = \frac{k(1.0 \times 10^{-4})\ (2.0 \times 10^{-2})^2\ (4.0 \times 10^{-2})^z}{k(1.0 \times 10^{-4})\ (2.0 \times 10^{-2})^2\ (2.0 \times 10^{-2})^z}$$

$7.95 = 2.0^z$

$$z = \frac{\log 7.95}{\log 2} = 2.99 \approx 3$$

Rate: $k[H_2SeO_3][H^+]^2[I^-]^3$

Experiment #1:

$$\frac{1.66 \times 10^{-7}\ \text{mol}}{\text{L s}} = k\left(\frac{1.0 \times 10^{-4}\ \text{mol}}{\text{L}}\right)\left(\frac{2.0 \times 10^{-2}\ \text{mol}}{\text{L}}\right)^2\left(\frac{2.0 \times 10^{-2}\ \text{mol}}{\text{L}}\right)^3$$

$k = 5.19 \times 10^5\ L^5/mol^5{\bullet}s \approx 5.2 \times 10^5\ L^5/mol^5{\bullet}s$

For all experiments:

Exp. #	$k(L^5/\ mol^5{\bullet}s)$
1	5.19×10^5
2	5.20×10^5
3	5.20×10^5
4	5.20×10^5
5	5.25×10^5
6	5.16×10^5
7	5.25×10^5

$k_{mean} = 5.2 \times 10^5\ L^5/mol^5{\bullet}s$

56. a. First, we guess the reaction to be first order with respect to O.

Hence, a graph of ln [O] vs. t should be linear

t (s)	[O] atoms/cm^3	ln [O]
0	5.0×10^9	22.33
$10. \times 10^{-3}$	1.9×10^9	21.37
$20. \times 10^{-3}$	6.8×10^8	20.34
$30. \times 10^{-3}$	2.5×10^8	19.34

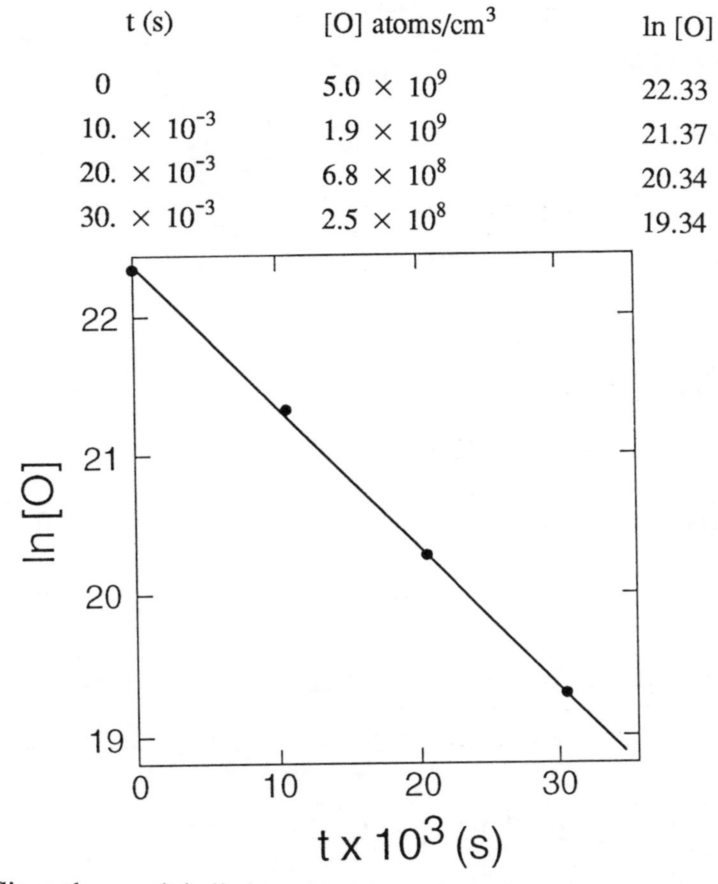

Since the graph is linear, we can conclude the reaction is first order with respect to O.

b. Rate = $k[NO_2]$ [O]

With an excess of NO_2, the rate law becomes: Rate = $k'[O]$

For a first order reaction:

$$\ln [A]_t = -kt + \ln [A]_o$$

We can get k´ from the slope of the graph:

$$k' = -\text{Slope} = -\frac{19.34 - 22.33}{(30 \times 10^{-3} - 0)\ s},\quad k' = 1.0 \times 10^2\ s^{-1}$$

However, $k' = k[NO_2]$

$1.0 \times 10^2\ s^{-1} = k(1.0 \times 10^{13}\ \text{molecules/cm}^3)$

$k = 1.0 \times 10^{-11}\ \text{cm}^3/\text{molecules}\bullet\text{s}$

57. $SO_2Cl_2(g) \rightarrow SO_2(g) + Cl_2(g)$

Let P_O = initial partial pressure of SO_2Cl_2

If $x = P_{SO_2}$ at some time, then

$x = P_{SO_2} = P_{Cl_2}$

$P_{SO_2Cl_2} = P_O - x$

$P_{tot} = P_{SO_2Cl_2} + P_{SO_2} + P_{Cl_2} = P_O - x + x + x$

$P_{tot} = P_O + x$

So:

Time(hour)	0.00	1.00	2.00	4.00	8.00	16.00
P(atm)	4.93	5.60	6.34	7.33	8.56	9.52
$P_{SO_2Cl_2}$ (atm)	4.93	4.26	3.52	2.53	1.30	0.34
$\ln P_{SO_2Cl_2}$	1.60	1.45	1.26	0.928	0.262	-1.08

$P_{SO_2Cl_2}\ \alpha\ [SO_2Cl_2]$ $P = \frac{n}{V}RT = [SO_2Cl_2]RT$

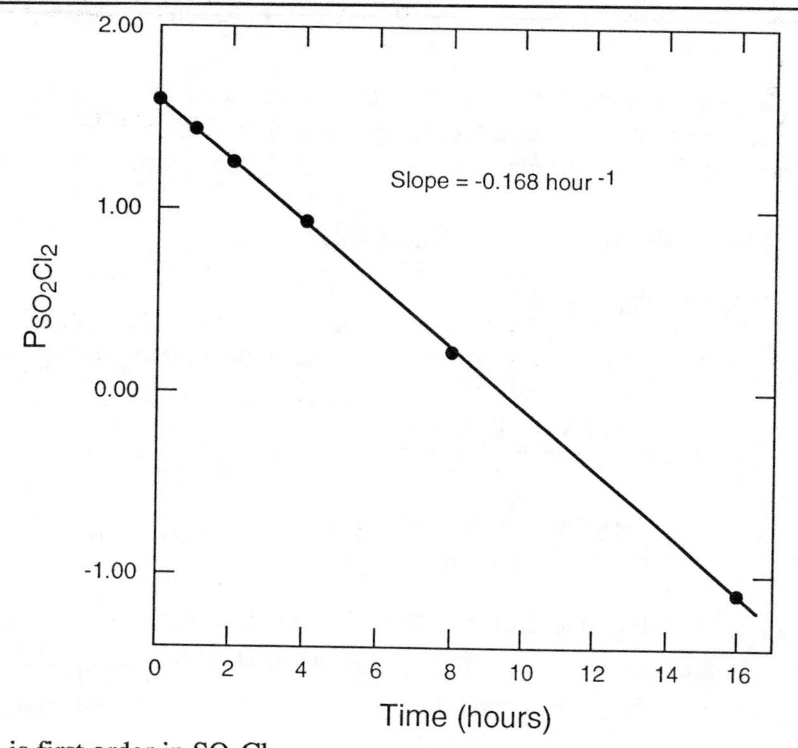

a. The reaction is first order in SO_2Cl_2

Slope of ln(P) vs. t is $-0.168\ \text{hour}^{-1} = -k$

$k = 0.168\ \text{hour}^{-1} = 4.67 \times 10^{-5}\ s^{-1}$

b. $t_{1/2} = \dfrac{0.69315}{k} = \dfrac{0.69315}{0.168} = 4.13$ hour

c. The total pressure increases at the same rate as the partial pressure of SO_2Cl_2 decreases.

$$\frac{\Delta P_{SO_2Cl_2}}{\Delta t} = \frac{-\Delta P_{tot}}{\Delta t} = -k\, P_{SO_2Cl_2}$$

$$\ln\left(\frac{P_{tot}}{P_O}\right) = kt;\ \ln\left(\frac{P_{tot}}{4.93}\right) = 0.168\ h^{-1}(0.500\ h)$$

$\ln P_{tot} = 0.0840 + 1.595 = 1.679;\ P_{tot} = e^{1.679} = 5.36$ atm

After 12.0 hours: From graph, $\ln P_{SO_2Cl_2} = -0.403$

$P_{SO_2Cl_2} = e^{-0.403} = 0.668$ atm

$P_{SO_2} = 4.93 - 0.668 = 4.26;$ $P_{Cl_2} = 4.26;$ $P_{tot} = 9.19$ atm

d. $\ln\left(\dfrac{P_{SO_2Cl_2}}{P_O}\right) = -0.168(20.0\ \text{hr}) = -3.36$

$\dfrac{P_{SO_2Cl_2}}{P_O} = e^{-3.36} = 3.47 \times 10^{-2},$ Fraction left = 0.0347 (3.47%)

58. The rate depends on the number of reactant molecules adsorbed on the surface of the catalyst. This quantity is proportional to the concentration of reactant. However, when all of the surface sites are occupied, the rate becomes independent of the concentration of reactant.

59. a. W, lower activation energy than Os catalyst.

b. $k_w = A_w \exp(-E_a(W)/RT)$

$K_{uncat} = A_{uncat} \exp(-E_a(uncat)/RT)$. Assume $A_w = A_{uncat}$

$$\frac{k_w}{k_{uncat}} = \exp\left(\frac{-E_a(W)}{RT} + \frac{E_a(uncat)}{RT}\right)$$

$$\frac{k_w}{k_{uncat}} = \exp\left(\frac{-163{,}000 \text{ J/mol} + 335{,}000 \text{ J/mol}}{(8.3145 \text{ J/mol•K})(298 \text{ K})}\right) = 1.41 \times 10^{30}$$

c. H_2 decreases the rate of the reaction. For the decomposition to occur, NH_3 molecules must be adsorbed on the surface of the catalyst. If H_2 that is produced from the reaction is also adsorbed on the catalyst, then there are fewer sites for NH_3 molecules to be adsorbed.

60. At high [S], the enzyme is completely saturated with substrate. Essentially, we have E•S and no free E and are observing the rate of decomposition of E•S.

CHALLENGE PROBLEMS

61. Rate = $k[I^-]^x [OCl^-]^y [OH^-]^z$

Comparing the first and second experiments:

$$\frac{18.7 \times 10^{-3}}{9.4 \times 10^{-3}} = \frac{k(0.0026)^x (0.012)^y (0.10)^z}{k(0.0013)^x (0.012)^y (0.10)^z}$$

$2.0 = 2.0^x$, $x = 1$

Comparing the first and third experiments:

$$\frac{9.4 \times 10^{-3}}{4.7 \times 10^{-3}} = \frac{k(0.0013)(0.012)^y (0.10)^z}{k(0.0013)(0.006)^y (0.10)^z}$$

$2.0 = 2.0^y$, $y = 2$

Comparing the first and sixth experiments:

$$\frac{4.7 \times 10^{-3}}{9.4 \times 10^{-3}} = \frac{k(0.0013)(0.012)(0.20)^z}{k(0.0013)(0.012)(0.10)^z}$$

$1/2 = 2.0^z$, $z = -1$

$$\text{Rate} = \frac{k[I^-][OCl^-]}{[OH^-]}$$

For the first experiment:

$$\frac{9.4 \times 10^{-3}\ \text{mol}}{\text{L s}} = k\ \frac{(0.0013\ \text{mol/L})(0.012\ \text{mol/L})}{(0.10\ \text{mol/L})}$$

$k = 60.3\ s^{-1} = 60.\ s^{-1}$

For all experiments:

Exp. #	1	2	3	4	5	6	7
k (s^{-1})	60.3	59.9	60.3	59.8	59.9	60.3	59.8

$k_{mean} = 6.0 \times 10^{1}\ s^{-1}$

62. a.

Heating Time	Untreated		Deacidifying		Antiox.	
(days)	s	ln s	s	ln s	s	ln s
0.00	100.0	4.605	100.1	4.606	114.6	4.741
1.00	67.9	4.22	60.8	4.11	65.2	4.18
2.00	38.9	3.66	26.8	3.29	28.1	3.34
3.00	16.1	2.78	-------	-------	11.3	2.42
6.00	6.8	1.92	-------	-------	-------	-------

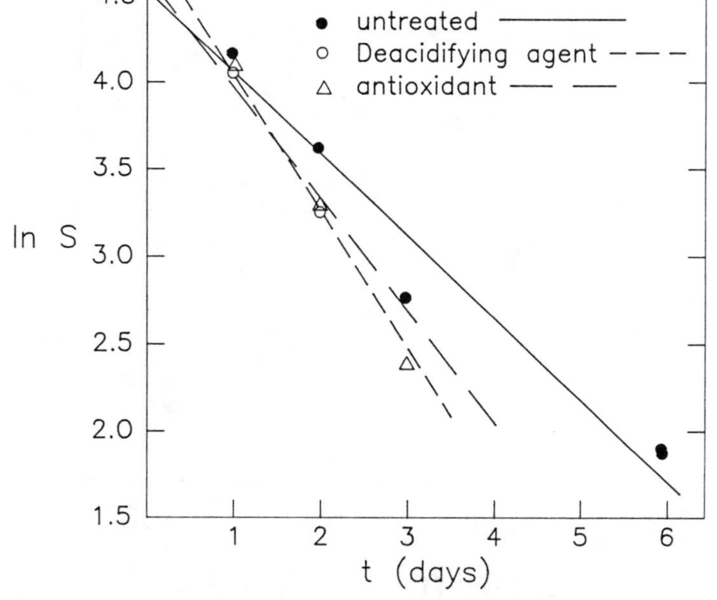

Using a calculator to fit the data by least squares:

Untreated: $\ln s = -0.465\ t + 4.55$, $k = 0.465\ \text{day}^{-1}$

Deacidifying agent: $\ln s = -0.659\ t + 4.66$, $k = 0.659\ \text{day}^{-1}$

Antioxidant: $\ln s = -0.779\ t + 4.84$, $k = 0.779\ \text{day}^{-1}$

b. No, the silk degrades more rapidly with the additives.

c. $t_{1/2} = 0.69315/k$

Untreated: $t_{1/2} = 1.49$ day

Deacidifying agent: $t_{1/2} = 1.05$ day

Antioxidant: $t_{1/2} = 0.890$ day

63. a. We check the first order dependence by graphing ln C vs. t for each set of data.

Dependence on NO

time (ms)	[NO] (molecules/cm^3)	ln [NO]
0	6.0×10^8	20.21
100.	5.0×10^8	20.03
500.	2.4×10^8	19.30
700.	1.7×10^8	18.95
1000.	9.9×10^7	18.41

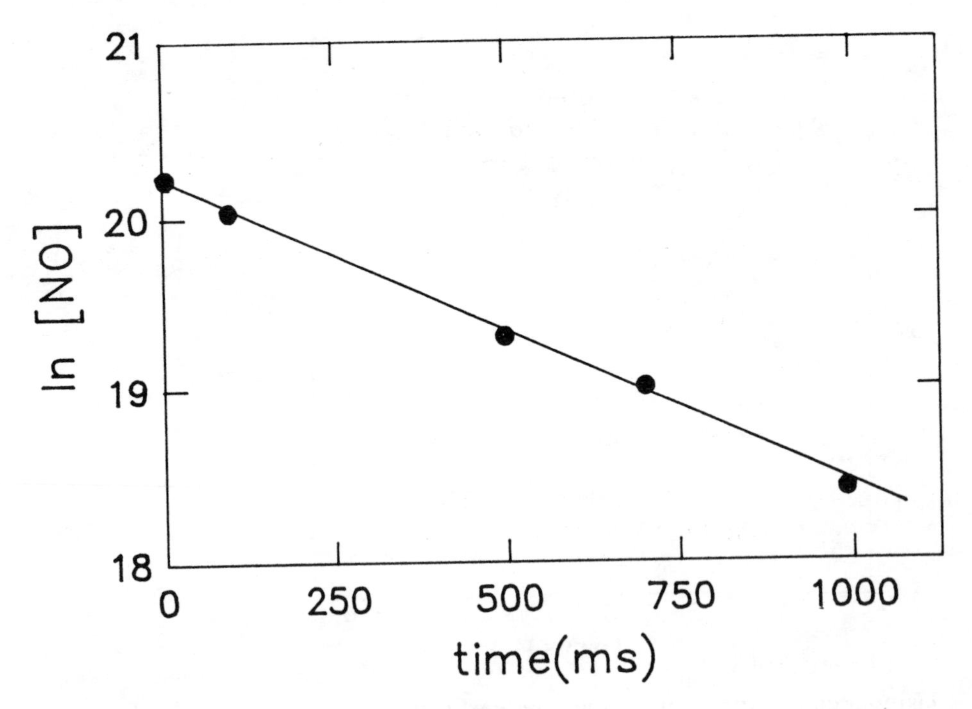

Since ln [NO] vs. t is linear, the reaction is first order with respect to nitric oxide.

We follow the same procedure for ozone. The data are:

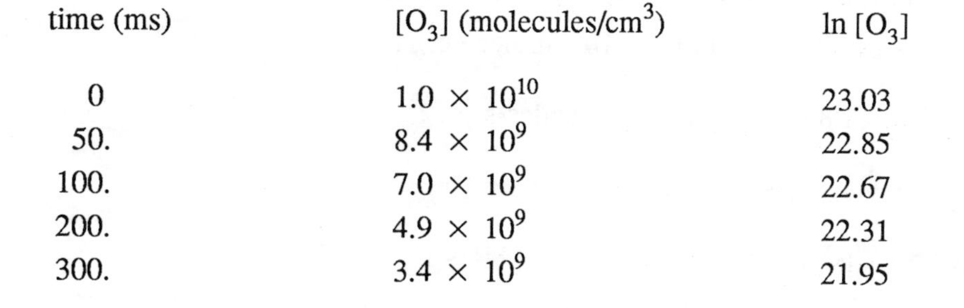

time (ms)	$[O_3]$ (molecules/cm^3)	ln $[O_3]$
0	1.0×10^{10}	23.03
50.	8.4×10^{9}	22.85
100.	7.0×10^{9}	22.67
200.	4.9×10^{9}	22.31
300.	3.4×10^{9}	21.95

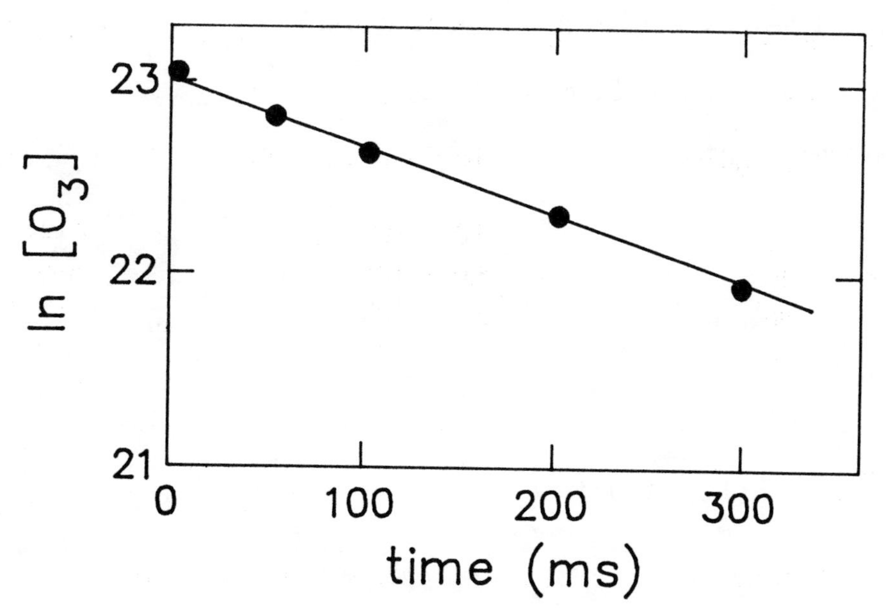

The plot of ln$[O_3]$ vs. t is linear. Hence, the reaction is first order with respect to ozone.

b. Rate = k[NO]$[O_3]$

c. Rate = k´[NO]

k´ = -(Slope from graph of ln [NO] vs. t)

$$k' = \text{-Slope} = -\frac{18.41 - 20.21}{(1000. - 0) \times 10^{-3}\ \text{s}} = 1.8\ \text{s}^{-1}$$

For ozone,

Rate = k´´$[O_3]$

and k´´ = -(Slope from ln$[O_3]$ vs. t)

$$k'' = \text{-Slope} = -\frac{(21.95 - 23.03)}{(300. - 0) \times 10^{-3}\ \text{s}}$$

$k'' = 3.6\ \text{s}^{-1}$

d. Rate = $k[NO][O_3] = k'[NO]$, so $k' = k[O_3]$

$$k' = 1.8\ s^{-1} = k(1.0 \times 10^{14}\ \text{molecules/cm}^3)$$

$$k = 1.8 \times 10^{-14}\ \text{cm}^3\text{/molecules}\bullet\text{s}$$

We can check this from the ozone data:

Rate = $k''[O_3] = k[NO][O_3]$, so $k'' = k[NO]$

$$k'' = 3.6\ s^{-1} = k(2.0 \times 10^{14}\ \text{molecules/cm}^3)$$

$$k = 1.8 \times 10^{-14}\ \text{cm}^3\text{/molecules}\bullet\text{s}$$

Both values agree.

CHAPTER THIRTEEN: CHEMICAL EQUILIBRIUM

QUESTIONS

1. a. The rates of the forward and reverse reactions are equal.

   b. There is no net change in the composition.

2. False. For example consider two cases.

   a. $k_f = 10^6\ s^{-1}$ $\quad k_r = 10^8\ s^{-1}$ $\quad$ fast reaction

   $K = 10^{-2}$ $\quad$ small K

   b. $k_f = 10^{-5}\ s^{-1}$ $\quad k_r = 10^{-9}\ s^{-1}$ $\quad$ slow reaction

   $K = 10^4$ $\quad$ large K

   The equilibrium constant is the ratio of two rate constants. This ratio does not necessarily get larger as the reactions get faster. The two examples illustrate this. If one starts with a mixture not at equilibrium, equilibrium will be reached faster if the rates of the forward and reverse reactions are fast.

3. The equilibrium constant is a number that tells us the relative concentrations (pressures) of reactants and products at equilibrium.

   An equilibrium position is a set of concentrations that satisfy the equilibrium constant expression. More than one equilibrium position can satisfy the same equilibrium constant expression.

4. For the reaction; $a\,A + b\,B \rightleftharpoons c\,C + d\,D$,

   the equilibrium constant expression is: $K_{eq} = \dfrac{[C]^c\,[D]^d}{[A]^a\,[B]^b}$

   The reaction quotient has the same form: $Q = \dfrac{[C]^c\,[D]^d}{[A]^a\,[B]^b}$

   The difference is that in the expressions for K_{eq} we use equilibrium concentrations, i.e., the [A], [B], [C], and [D] that are in equilibrium with each other. Any set of concentrations can be plugged into the reaction quotient expression. Typically, we compare a value of Q to K_{eq} to see how far we are from equilibrium.

5. When we change the pressure by adding an unreactive gas, we do not change the partial pressure of any of the substances in equilibrium with each other. In this case the equilibrium will not shift. If we change the pressure by changing the volume, we will change the partial pressures of all the substances in equilibrium by the same factor. If there are unequal numbers of gaseous particles on the two sides of the equation, the equilibrium will shift.

6. A change in volume will change the partial pressure of all reactants and products by the same factor. The shift in equilibrium depends on the number of gaseous particles on each side. An increase in volume will shift the equilibrium to the side with the greater number of particles in the gas phase. A decrease in volume will favor the side with the fewer number of gas phase particles. If there are the same number of gas phase particles on each side of the reaction, then a change in volume will not shift the equilibrium.

EXERCISES

Characteristics of Chemical Equilibrium

7. No, equilibrium is a dynamic process. Both reactions:

$$H_2O + CO \longrightarrow H_2 + CO_2$$

and

$$H_2 + CO_2 \longrightarrow H_2O + CO$$

are occurring. Thus, ^{14}C atoms will be distributed between CO and CO_2.

8. No, it doesn't matter how the equilibrium position is reached. Both mixtures will give the same equilibrium position.

The Equilibrium Constant

9. a. $K = \frac{[NO_2][O_2]}{[NO][O_3]}$ b. $K = \frac{[O_2][O]}{[O_3]}$

   c. $K = \frac{[ClO][O_2]}{[Cl][O_3]}$ d. $K = \frac{[O_2]^3}{[O_3]^2}$

10. a. $K_p = \frac{P_{NO_2} P_{O_2}}{P_{NO} P_{O_3}}$ b. $K_p = \frac{P_{O_2} P_O}{P_{O_3}}$

    c. $K_p = \frac{P_{ClO} P_{O_2}}{P_{Cl} P_{O_3}}$ d. $K_p = \frac{P_{O_2}^3}{P_{O_3}^2}$

    $K_p = K_c$ for a reaction in which the number of gaseous reactant particles equals the number of gaseous product particles. Thus, for the reactions in a and c, $K_p = K_c$.

11. $K_p = K(RT)^{\Delta n}$, $\Delta n = 2$

$$K_p = \frac{2.6 \times 10^{-5}\ mol^2}{L^2}\left(\frac{0.08206\ L\ atm}{mol\ K} \times 400.\ K\right)^2$$

$$K_p = 2.8 \times 10^{-2}\ atm^2$$

12. $K_p = K(RT)^{\Delta n}$, $K = \frac{K_p}{(RT)^{\Delta n}}$ with $\Delta n = -1$

$$K = K_p(RT) = \frac{0.25}{atm} \times \frac{0.08206\ L\ atm}{mol\ K} \times 1100\ K = 23\ L/mol$$

13. a. $K = \frac{1}{[O_2]^5}$ b. $K = [N_2O][H_2O]^2$

    c. $K = \frac{1}{[CO_2]}$ d. $K = \frac{[SO_2]^8}{[O_2]^8}$

14. a. $K_p = \frac{1}{(P_{O_2})^{3/2}}$

b. $K_p = \frac{1}{P_{CO_2}}$

c. $K_p = \frac{P_{CO} \times P_{H_2}}{P_{H_2O}}$

d. $K_p = \frac{P_{O_2}^3}{P_{H_2O}^2}$

15. $K = 278 = \frac{[SO_3]^2}{[SO_2]^2[O_2]}$ for $2\ SO_2 + O_2 \rightleftharpoons 2\ SO_3$

a. $SO_2(g) + 1/2\ O_2(g) \rightleftharpoons SO_3(g)$

$$K_{eq} = \frac{[SO_3]}{[SO_2][O_2]^{1/2}} = K^{1/2} = 16.7$$

b. $2\ SO_3(g) \rightleftharpoons 2\ SO_2(g) + O_2(g)$

$$K_{eq} = \frac{[SO_2]^2[O_2]}{[SO_3]^2} = \frac{1}{K} = 3.60 \times 10^{-3}$$

c. $SO_3(g) \rightleftharpoons SO_2(g) + 1/2\ O_2(g)$

$$K_{eq} = \frac{[SO_2][O_2]^{1/2}}{[SO_3]} = \left(\frac{1}{K}\right)^{1/2} = 6.00 \times 10^{-2}$$

d. $4\ SO_2(g) + 2\ O_2(g) \rightleftharpoons 4\ SO_3(g)$

$$K_{eq} = \frac{[SO_3]^4}{[SO_2]^4[O_2]^2} = K^2 = 7.73 \times 10^4$$

16. $H_2(g) + Br_2(g) \rightleftharpoons 2\ HBr(g)$

$$K_p = \frac{P_{HBr}^2}{(P_{H_2})(P_{Br_2})} = 3.5 \times 10^4$$

a. $HBr \rightleftharpoons 1/2\ H_2 + 1/2\ Br_2$

$$K_p' = \frac{(P_{H_2})^{1/2}(P_{Br_2})^{1/2}}{P_{HBr}} = \left(\frac{1}{K_p}\right)^{1/2} = 5.3 \times 10^{-3}$$

b. $2\ HBr \rightleftharpoons H_2 + Br_2$

$$K_p'' = \frac{(P_{H_2})(P_{Br_2})}{P_{HBr}^2} = \frac{1}{K_p} = 2.9 \times 10^{-5}$$

c. $1/2\ H_2 + 1/2\ Br_2 \rightleftharpoons HBr$

$$K_p''' = \frac{P_{HBr}}{(P_{H_2})^{1/2}(P_{Br_2})^{1/2}} = (K_p)^{1/2} = 190$$

17. $CO_2(g) + H_2(g) \rightleftharpoons CO(g) + H_2O(g)$

$$K = \frac{[CO]\,[H_2O]}{[CO_2]\,[H_2]} = \frac{(5.9\ mol/L)(12.0\ mol/L)}{(18.0\ mol/L)(20.0\ mol/L)} = 0.20$$

18. $H_2 + I_2 \rightleftharpoons 2\ HI$

$$K = \frac{[HI]^2}{[H_2]\,[I_2]} = \frac{\left(\frac{3.5\ mol}{3.0\ L}\right)^2}{\left(\frac{4.1\ mol}{3.0\ L}\right)\left(\frac{0.3\ mol}{3.0\ L}\right)} = 10$$

19. $C(s) + CO_2(g) \rightleftharpoons 2\ CO(g)$

$$K_p = \frac{P_{CO}^2}{P_{CO_2}} = \frac{(2.6\ atm)^2}{2.9\ atm} = 2.3\ atm$$

20. $NH_4Cl(s) \rightleftharpoons NH_3(g) + HCl(g)$

$K_p = P_{NH_3} \times P_{HCl}$

At equilibrium, $P_{total} = P_{NH_3} + P_{HCl}$ and $P_{NH_3} = P_{HCl}$

$P_{total} = 4.4\ atm = 2P_{NH_3}$, Thus, $P_{NH_3} = P_{HCl} = 2.2\ atm$

$K_p = (2.2\ atm)(2.2\ atm) = 4.8\ atm^2$

21.

	$PCl_5(g)$	$\rightleftharpoons$	$PCl_3(g)$	+	$Cl_2(g)$
Initial	0.50 atm		0		0
	x atm PCl_5 reacts to reach equilibrium				
Change	$-x$	$\rightarrow$	$+x$		$+x$
Equil.	$0.50 - x$		x		x

$P_{total} = P_{PCl_5} + P_{PCl_3} + P_{Cl_2} = 0.50 - x + x + x = 0.50 + x$

$0.84\ atm = 0.50 + x,\ x = 0.34\ atm$

$P_{PCl_5} = 0.16\ atm,\quad P_{PCl_3} = P_{Cl_2} = 0.34\ atm$

$$K_p = \frac{P_{PCl_3} \times P_{Cl_2}}{P_{PCl_5}} = \frac{(0.34)(0.34)}{(0.16)} = 0.72\ atm$$

$$K = \frac{K_p}{(RT)^{\Delta n}} = \frac{0.72}{(0.08206)(523)} = 0.017\ mol/L$$

22.

	$SO_2(g)$	+	$NO_2(g)$	$\rightleftharpoons$	$SO_3(g)$	+	$NO(g)$
Initial	$\frac{2.0\text{ mol}}{V}$		$\frac{2.0\text{ mol}}{V}$		0		0
Change	$\frac{-1.3\text{ mol}}{V}$		$\frac{-1.3\text{ mol}}{V}$		$\frac{+1.3\text{ mol}}{V}$		$\frac{+1.3\text{ mol}}{V}$
Equil.	$\frac{0.7\text{ mol}}{V}$		$\frac{0.7\text{ mol}}{V}$		$\frac{1.3\text{ mol}}{V}$		$\frac{1.3\text{ mol}}{V}$

$$K = \frac{[SO_3][NO]}{[SO_2][NO_2]} = \frac{(1.3)(1.3)}{(0.7)(0.7)} = 3.4 \approx 3$$

Equilibrium Calculations

23. $H_2O + Cl_2O(g) \rightleftharpoons 2\ HOCl(g)$ $\qquad K_p = K = 0.0900 = \frac{[HOCl]^2}{[H_2O][Cl_2O]}$

a. $Q = \frac{(21.0)^2}{(200.)(49.8)} = 4.43 \times 10^{-2} < K$

Reaction will proceed to the right to reach equilibrium.

$H_2O(g) + Cl_2O(g) \longrightarrow 2\ HOCl(g)$

b. $Q = \frac{(20.0)^2}{(296)(15.0)} = 0.0901 \approx K$ (at equilibrium)

c. $$Q = \frac{\left(\frac{0.084\text{ mol}}{2.0\text{ L}}\right)^2}{\left(\frac{0.98\text{ mol}}{2.0\text{ L}}\right)\left(\frac{0.080\text{ mol}}{2.0\text{ L}}\right)} = \frac{(0.084)^2}{(0.98)(0.080)} = 0.090 = K \text{ (at equilibrium)}$$

d. $$Q = \frac{\left(\frac{0.25\text{ mol}}{3.0\text{ L}}\right)^2}{\left(\frac{0.56\text{ mol}}{3.0\text{ L}}\right)\left(\frac{0.0010\text{ mol}}{3.0\text{ L}}\right)} = \frac{(0.25)^2}{(0.56)(0.0010)} = 110 > K$$

Reaction will proceed to the left to reach equilibrium.

$2\ HOCl \longrightarrow H_2O + Cl_2O$ or $H_2O + Cl_2O \longleftarrow 2\ HOCl$

24. $N_2O_4(g) \rightleftharpoons 2\ NO_2(g)$ $\qquad K_p = \frac{P_{NO_2}^2}{P_{N_2O_4}} = 0.133$

a. $Q = \frac{(0.144)^2}{(0.156)} = 0.133 = K_p$, at equilibrium

b. $Q = \frac{(0.175)^2}{(0.102)} = 0.300$

c. $Q = \frac{(0.056)^2}{(0.048)} = 0.065$

d. $Q = \frac{(0.064)^2}{(0.0308)} = 0.13 = K_p$, at equilibrium

The conditions in (a) and (d) correspond to equilibrium positions.

25. $CaCO_3(s) \rightleftharpoons CaO(s) + CO_2(g)$

$$K_p = P_{CO_2} = 1.04$$

We only need to calculate the partial pressure of CO_2. At this temperature all CO_2 will be in the gas phase.

a. $PV = nRT$

$$P = \frac{nRT}{V} = \frac{\frac{58.4\text{ g}}{44.01\text{ g/mol}}\left(\frac{0.08206\text{ L atm}}{\text{mol K}}\right)1173\text{ K}}{(50.0\text{ L})} = 2.55 > K_p$$

Reaction will proceed to the left, the mass of CaO will decrease.

b. $P = \frac{(23.76)\ (0.08206)(1173)}{(44.01)(50.0)}\text{ atm} = 1.04\text{ atm} = K_p$

At equilibrium; mass of CaO will not change.

c. Mass of CO_2 same as in part b.

$P = 1.04\text{ atm} = K_p$

At equilibrium; mass of CaO will not change.

d. $P = \frac{(4.82)\ (0.08206)(1173)}{(44.01)(50.0)}\text{ atm} = 0.211 < K_p$

Reaction will proceed to the right; the mass of CaO will increase.

26. $CH_3CO_2H + C_2H_5OH \rightleftharpoons CH_3CO_2C_2H_5 + H_2O$

$$K = \frac{[CH_3CO_2C_2H_5][H_2O]}{[CH_3CO_2H][C_2H_5OH]} = 2.2$$

a. $Q = \frac{(0.22)\ (0.010)}{(0.010)(0.010)} = 220$, $Q > K$, reaction shifts left, $[H_2O]$ decreases

b. $Q = \frac{(0.22)\ (0.0020)}{(0.0020)(0.10)} = 2.2 = K$, at equilibrium, $[H_2O]$ stays the same

c. $Q = \frac{(0.88)\ (0.12)}{(0.044)(6.0)} = 0.40$, $Q < K$, reaction shifts right, $[H_2O]$ increases

d. $Q = \frac{(4.4)\ (4.4)}{(0.88)(10.0)} = 2.2 = K$, at equilibrium, $[H_2O]$ stays the same

The conditions in (b) and (d) correspond to equilibrium positions.

e. $2.2 = \frac{[H_2O](2.0)}{(0.10)\ (5.0)}$, $[H_2O] = 0.55$ mol/L

f. Water is a product of the reaction, but it is not the solvent. Thus, the concentration of water must be included in the equilibrium expression.

27.

	$N_2(g)$ +	$O_2(g)$	$\rightleftharpoons$	$2\,NO(g)$	$K = K_p = 0.050$
Initial	0.80 atm	0.20 atm		0	
	x atm N_2 reacts to reach equilibrium				
Change	$-x$	$-x$	$\rightarrow$	$+2x$	
Equil.	$0.80 - x$	$0.20 - x$		$2x$	

$$K = 0.050 = \frac{P_{NO}^2}{P_{N_2}P_{O_2}} = \frac{(2x)^2}{(0.80 - x)\,(0.20 - x)}$$

$(0.050)\,(0.80 - x)\,(0.20 - x) = 4x^2$

$8.0 \times 10^{-3} - 0.050x + 0.050\,x^2 = 4x^2$

$3.95x^2 + 0.050x - 8.0 \times 10^{-3} = 0$

For a quadratic equation in the form: $a\,x^2 + b\,x + c = 0$

the solutions are $x = \dfrac{-b \pm \sqrt{b^2 - 4ac}}{2a}$

For this equation:

$$x = \frac{-0.050 \pm \sqrt{2.5 \times 10^{-3} + 0.1264}}{7.9}$$

$$x = \frac{-0.050 \pm \sqrt{0.1289}}{7.9}$$

Note: Rounding off intermediate answers in solving the quadratic formula and in the approximations method described below, leads to excessive round off error. In these problems we will discontinue the usual practice in this Solutions Guide and carry extra significant figures and round at the end.

$x = \dfrac{-0.050 \pm 0.359}{7.9}$, $x = 0.039$ or $x = -0.051$

A negative answer makes no physical sense, so $x = 0.039$

$P_{NO} = 2x = 0.078$ atm

There is an easier way to solve this problem in terms of algebra. There are a lot of manipulations necessary to get a solution using the quadratic formula. If we make an early mistake, it takes a lot of work to get the final answer and find out it is wrong. As a case in point it took me (KCB) about 20 minutes to find the mistake I made in initially solving this problem. In an early step I copied an exponent wrong. We can use our chemical sense to simplify the algebra. For this problem, we are trying to solve the equation:

$$\frac{4x^2}{(0.80 - x)\,(0.20 - x)} = 0.050$$

However, the value of the equilibrium constant, 0.050, is not very large; not much N_2 and O_2 will react to reach equilibrium. In this case, if we can say:

$$0.80 - x \approx 0.80$$

$$0.20 - x \approx 0.20$$

then the equation simplifies to:

$$\frac{4x^2}{(0.80)(0.20)} = 0.050 \text{ and } x = 0.045$$

When we check how good our assumptions are, we get:

$0.80 - x = 0.80 - 0.045 = 0.755$

$0.20 - x = 0.20 - 0.045 = 0.155$

These assumptions aren't very good (22.5% error in one of the assumptions). All is not lost! We can use this result as a further approximation. That is:

$$\frac{4x^2}{(0.80 - x)(0.20 - x)} \approx \frac{4x^2}{(0.80 - 0.045)(0.20 - 0.045)} = 0.050$$

Solving for x we get: $x = 0.038$. We repeat the process using 0.038 as a further guess for x.

$$\frac{4x^2}{(0.80 - 0.038)\ (0.20 - 0.038)} = 0.050,\ x = 0.039 \qquad \text{Repeat using 0.039.}$$

$$\frac{4x^2}{(0.80 - 0.039)\ (0.20 - 0.039)} = 0.050,\ x = 0.039$$

We just got the same answer in consecutive iterations; we have converged on the true answer. It is the same answer to two significant figures that we solved for using the quadratic formula. This method, called the method of successive approximations (See Appendix 1.4 in the text), at first appears to be more laborious. However, any one iteration is less time consuming than using the quadratic formula; it is easier to make an arithmetic mistake using the quadratic formula; and we can discover mistakes in a shorter period of time using successive approximations. Even more compelling, we're using our chemical sense to make approximations and, after all, this is a chemistry text. Try the method; once you get the hang of it you'll prefer successive approximation to the quadratic formula.

28. a.

	$2\ NOCl(g)$	$\rightleftharpoons$	$2\ NO(g)$	$+\ Cl_2(g)$	$K = 1.6 \times 10^{-5}$
Initial	$\frac{2.0 \text{ mol}}{2.0 \text{ L}} = 1.0\ M$		0	0	
	$2x$ mol/L NOCl reacts to reach equilibrium				
Change	$-2x$	$\rightarrow$	$+2x$	$+x$	
Equil.	$1.0 - 2x$		$2x$	x	

$$K = 1.6 \times 10^{-5} = \frac{(2x)^2(x)}{(1.0 - 2x)^2}$$

If $1.0 - 2x \approx 1.0$ (from the size of K, we know that not much reaction will occur), then

$$1.6 \times 10^{-5} = 4x^3, \quad x = 1.6 \times 10^{-2}$$

$$1.0 - 2x = 1.0 - 2(0.016) = 0.97 = 1.0 \text{ (to proper sig. fig.)}$$

Our error is about 3%, i.e. $2x$ is 3.2% of 1.0 M. If the error we introduce by making simplifying assumptions is less than 5%, we will go no further, the assumption is valid.

$$[NO] = 2x = 0.032\ M$$

$$[Cl_2] = x = 0.016\ M$$

$$[NOCl] = 1.0 - 2x = 0.97\ M \approx 1.0\ M$$

Note that the equation is a cubic equation. I have no idea how to solve a cubic equation exactly. If we aren't satisfied with the first answer, we can use successive approximations.

$$\frac{4x^3}{0.97} = 1.6 \times 10^{-5}, \qquad x = 1.57 \times 10^{-2} = 1.6 \times 10^{-2}$$

The 5% rule of thumb is pretty useful. The first approximation, to the proper number of significant figures, gave the same answer as the second try.

b.

	$2\ NOCl(g)$	$\rightleftharpoons$	$2\ NO(g)$	+ $Cl_2(g)$	$K = 1.6 \times 10^{-5}$
Before	0		2.0 M	1.0 M	
	1. 0 mol/L Cl_2 reacts completely		(K is small, reactants dominate)		
Change	+2.0	←	-2.0	-1.0	React completely
After	2.0		0	0	
	$2x$ mol/L NOCl reacts to reach equilibrium				
Change	$-2x$	→	$+2x$	$+x$	
Equil.	$2.0 - 2x$		$2x$	x	

$$1.6 \times 10^{-5} = \frac{(2x)^2(x)}{(2.0 - 2x)^2} \approx \frac{4x^3}{2.0^2} \qquad \text{(Assuming } 2.0 - 2x \approx 2.0\text{)}$$

$x^3 = 1.6 \times 10^{-5}$ and $x = 2.5 \times 10^{-2}$, Assumption good (2.5% error).

$[NOCl] = 2.0 - 0.050 = 1.95\ M = 2.0\ M$

$[NO] = 0.050\ M$, $[Cl_2] = 0.025\ M$

Note: If we do not break this problem into two parts, (a stoichiometric part and an equilibrium part), we are faced with solving a cubic equation.

	$2\ NOCl$	$\rightleftharpoons$	$2\ NO$	+ Cl_2
Initial	0		2.0 M	1.0 M
Change	$+2y$	←	$-2y$	$-y$
Equil.	$2y$		$2.0 - 2y$	$1.0 - y$

$$1.6 \times 10^{-5} = \frac{(2.0 - 2y)^2\ (1.0 - y)}{(2y)^2}$$

If we assume that y is small to simplify this problem then:

$$1.6 \times 10^{-5} = \frac{2.0^2}{4y^2}$$

We get $y = 250$ **THIS IS IMPOSSIBLE!**

To solve this equation we cannot make any simplifying assumptions; we have to find a way to solve a cubic equation. Or, we can use some chemical common sense and solve the problem an easier way.

c.

	$2\,NOCl(g)$	$\rightleftharpoons$	$2\,NO(g)$	+ $Cl_2(g)$
Initial	1.0 *M*		1.0 *M*	0
	$2x$ mol/L NOCl reacts to reach equilibrium			
Change	$-2x$	→	$+2x$	$+x$
Equil.	$1.0 - 2x$		$1.0 + 2x$	x

$$1.6 \times 10^{-5} = \frac{(1.0 + 2x)^2\,(x)}{(1.0 - 2x)^2}$$

If $1.0 + 2x \approx 1.0$ and $1.0 - 2x \approx 1.0$, then $x = 1.6 \times 10^{-5}$ Assumptions are good, so:

$[Cl_2] = 1.6 \times 10^{-5}\,M$ and $[NOCl] = [NO] = 1.0\,M$

d.

	$2\,NOCl$	$\rightleftharpoons$	$2\,NO$	+ Cl_2	
Before	0		3.0 *M*	1.0 *M*	
	1. 0 mol/L Cl_2 reacts (limiting reagent)				
Change	+2.0	←	-2.0	-1.0	React completely
After	2.0		1.0	0	New Initial
	$2x$ mol/L NOCl reacts to reach equilibrium				
Change	$-2x$	→	$+2x$	$+x$	
Equil.	$2.0 - 2x$		$1.0 + 2x$	x	

$$1.6 \times 10^{-5} = \frac{(1.0 + 2x)^2\,(x)}{(2.0 - 2x)^2} \approx \frac{x}{4.0} \quad \text{(assuming } 2.0 - 2x \approx 2.0)$$

$x = 6.4 \times 10^{-5}$ Assumptions good.

$[Cl_2] = 6.4 \times 10^{-5}\,M$, $[NOCl] = 2.0\,M$, and $[NO] = 1.0\,M$

e.

	$2\,NOCl$	$\rightleftharpoons$	$2\,NO$	+ Cl_2	
Before	2.0 *M*		2.0 *M*	1.0 *M*	
	1. 0 mol/L Cl_2 reacts				
Change	+2.0	←	-2.0	-1.0	React completely
After	4.0		0	0	New Initial

	$2\,NOCl$	$\rightleftharpoons$	$2\,NO$	+	Cl_2
New Initial	4.0 *M*		0		0
	$2x$ mol/L NOCl reacts to reach equilibrium				
Change	$-2x$	→	$+2x$		$+x$
Equil.	$4.0 - 2x$		$2x$		x

$$1.6 \times 10^{-5} = \frac{(2x)^2(x)}{(4.0 - 2x)^2} \approx \frac{4x^3}{16} \qquad \text{(assuming } 4.0 - 2x \approx 4.0\text{)}$$

$x = 4.0 \times 10^{-2}$, Assumption good (2% error).

$[Cl_2] = 0.040\ M$

$[NO] = 0.080\ M$

$[NOCl] = 4.0 - 2(0.040) = 3.92\ M \approx 3.9\ M$

f.

	$2\,NOCl$	$\rightleftharpoons$	$2\,NO$	+	Cl_2	
Before	1.0 *M*		1.0 *M*		1.0 *M*	
	1. 0 mol/L NO reacts (limiting reagent)					
Change	+1.0	←	-1.0		-0.5	React completely
After (New Initial)	2.0		0		0.5	
	$2x$ mol/L NOCl reacts to reach equilibrium					
Change	$-2x$	→	$+2x$		$+x$	
Equil.	$2.0 - 2x$		$2x$		$0.5 + x$	

$$\frac{(2x)^2(0.5 + x)}{(2.0 - 2x)^2} \approx \frac{4x^2(0.5)}{(2.0)^2} = 1.6 \times 10^{-5},\ x = 6 \times 10^{-3}, \qquad \text{Assumptions good.}$$

$[NO] = 2x = 1 \times 10^{-2}\ M$

$[Cl_2] = 0.5 + 0.006 = 0.5\ M$

$[NOCl] = 2.0 - 2(0.006) = 1.99\ M \approx 2.0\ M$

29.

	$2\,SO_2(g)$	+	$O_2(g)$	$\rightleftharpoons$	$2\,SO_3(g)$	$K_p = 0.25$
Initial	0.50 atm		0.50 atm		0	
	$2x$ atm SO_2 reacts to reach equilibrium					
Change	$-2x$		$-x$	→	$+2x$	
Equil.	$0.50 - 2x$		$0.50 - x$		$2x$	

$$\frac{P^2_{SO_3}}{P^2_{SO_2} P_{O_2}} = \frac{(2x)^2}{(0.50 - 2x)^2(0.50 - x)} \approx \frac{4x^2}{(0.50)^3} = 0.25$$

$x = 0.088$ Assumption is not good (x is 20% of 0.5).

Using method of successive approximations (Appendix 1.4):

$$\frac{4x^2}{[0.50 - 2(0.088)]^2[0.50 - (0.088)]} = \frac{4x^2}{(0.324)^2(0.412)} = 0.25; \quad x = 0.052$$

$$\frac{4x^2}{[0.50 - 2(0.052)]^2[0.50 - (0.052)]} = \frac{4x^2}{(0.396)^2(0.448)} = 0.25; \quad x = 0.066$$

$$\frac{4x^2}{(0.368)^2(0.434)} = 0.25; \quad x = 0.061$$

$$\frac{4x^2}{(0.378)^2(0.439)} = 0.25; \quad x = 0.063$$

The next trials converge at 0.062.

$P_{SO_2} = 0.50 - 2x = 0.376 = 0.38$ atm

$P_{SO_3} = 2x = 0.124 = 0.12$ atm, $P_{O_2} = 0.50 - x = 0.438 = 0.44$ atm

30. a.

	H_2	+	F_2	$\rightleftharpoons$	2 HF	$K = 1.0 \times 10^2$
Before	2.0 M		2.0 M		0	
	2. 0 mol/L H_2 reacts (K is large, products dominate)					
Change	-2.0		-2.0	→	+4.0	React completely
After	0		0		4.0	New Initial
	$2x$ mol/L HF reacts to reach equilibrium					
Change	$+x$		$+x$	←	$-2x$	
Equil.	x		x		$4.0 - 2x$	

$$K = 1.0 \times 10^2 = \frac{[HF]^2}{[H_2][F_2]} = \frac{(4.0 - 2x)^2}{x^2}$$

The equation is a perfect square.

$$10. = \frac{4.0 - 2x}{x}, \quad 10.(x) = 4.0 - 2x, \quad 12x = 4.0$$

$x = [H_2] = [F_2] = 0.33\ M$, $[HF] = 4.0 - 2x = 3.3\ M$

Alternate Solution:

	H_2	+	F_2	$\rightleftharpoons$	2 HF
Initial	2.0		2.0		0
	y mol/L H_2 reacts to reach equilibrium				
Change	$-y$		$-y$	→	$+2y$
Equil.	$2.0 - y$		$2.0 - y$		$2y$

$$\frac{4y^2}{(2.0 - y)^2} = 1.0 \times 10^2,$$ This is a perfect square.

$$\frac{2y}{2.0 - y} = 10., \ 2y = 20. - 10.(y), \ 12y = 20., \ y = 1.67$$ (carry extra significant figure)

$2y = [HF] = 3.3\ M$, $2.0 - y = [H_2] = [F_2] = 0.3\ M$

b.

	H_2 +	F_2 $\leftrightharpoons$	2 HF
Initial	0.33 M	0.33 M	3.3 M
	Add 0.50 mol H_2 to the 1.0 L flask		
New Initial	0.83	0.33	3.3
Change	$-x$ +	$-x$ $\rightarrow$	$+2x$
Equil.	$0.83 - x$	$0.33 - x$	$3.3 + 2x$

$$\frac{(3.3 + 2x)^2}{(0.83 - x)(0.33 - x)} = 1.0 \times 10^2$$

$10.9 + 6.6x + 4x^2 = 27.4 - 116x + 100.x^2$

$96x^2 - 122.6x + 16.5 = 0$

$$x = \frac{122.6 \pm [(-122.6)^2 - 4(96)(16.5)]^{1/2}}{2(96)} = \frac{122.6 \pm 93.2}{192}$$

$x = 0.15$ or $x = 1.12$, only $x = 0.15$ makes sense

$[H_2] = 0.83 - x = 0.68\ M$, $[F_2] = 0.33 - x = 0.18\ M$, and $[HF] = 3.3 + 2x = 3.6\ M$

31. a.

	$N_2O_4(g)$ $\leftrightharpoons$	$2\ NO_2(g)$	$K_p = 0.25$
Initial	0	0.050 atm	
	$2x$ atm NO_2 reacts to reach equilibrium		
Change	$+x$ $\leftarrow$	$-2x$	
Equil.	x	$0.050 - 2x$	

$$K_p = 0.25 = \frac{(0.050 - 2x)^2}{x}, \quad 0.25x = 2.5 \times 10^{-3} - 0.20x + 4x^2$$

$4x^2 - 0.45x + 2.5 \times 10^{-3} = 0$

$$x = \frac{+0.45 \pm [(-0.45)^2 - 4(4)(2.5 \times 10^{-3})]^{1/2}}{2(4)}$$

$$x = \frac{+0.45 - 0.403}{8} = 5.9 \times 10^{-3} \text{ (other root, 0.11 makes no sense)}$$

$P_{N_2O_4} = 5.9 \times 10^{-3}$ atm

$P_{NO_2} = 0.050 - 2(0.0059) = 3.8 \times 10^{-2}$ atm

b.

	$N_2O_4(g)$ $\leftrightharpoons$	$2\ NO_2(g)$	$K_p = 0.25$
Initial	0.040 atm	0	
	x atm N_2O_4 reacts to reach equilibrium		
Change	$-x$ $\rightarrow$	$+2x$	
Equil.	$0.040 - x$	$2x$	

$$\frac{(2x)^2}{0.040 - x} = 0.25$$

$$4x^2 = 0.010 - 0.25x$$

$$4x^2 + 0.25x - 0.010 = 0$$

$$x = \frac{-0.25 \pm [(-0.25)^2 - 4(4)(-0.010)]^{1/2}}{8}$$

Only way to get positive answer is:

$$x = \frac{-0.25 + 0.47}{8} = 0.028$$

$$P_{NO_2} = 2x = 0.056 \text{ atm}$$

$$P_{N_2O_4} = 0.040 - x = 0.012 \text{ atm}$$

c. $N_2O_4 \rightleftharpoons 2\ NO_2$ $\quad K_p = 0.25$

$Q = 1.0$, so some NO_2 must go to N_2O_4 ($Q > K_p$).

	N_2O_4	$\rightleftharpoons$	$2\ NO_2$
Initial	1.0 atm		1.0 atm
	$2x$ atm NO_2 reacts to reach equilibrium		
Change	$+x$	$\leftarrow$	$-2x$
Equil.	$1.0 + x$		$1.0 - 2x$

$$\frac{(1.0 - 2x)^2}{1.0 + x} = 0.25$$

$$1.0 - 4.0x + 4x^2 = 0.25 + 0.25x$$

$$4x^2 - 4.25x + 0.75 = 0$$

$$x = \frac{4.25 \pm [(-4.25)^2 - 4(4)(0.75)]^{1/2}}{8}$$

$$x = \frac{4.25 \pm 2.46}{8} \qquad x = 0.22 \text{ or } 0.84$$

$x = 0.84$, $1 - 2x < 0$ Not possible

$x = 0.22$ is the chemically correct solution.

$$P_{N_2O_4} = 1.0 + x = 1.2 \text{ atm}$$

$$P_{NO_2} = 1.0 - 2x = 0.56 \text{ atm} \approx 0.6 \text{ atm}$$

32.

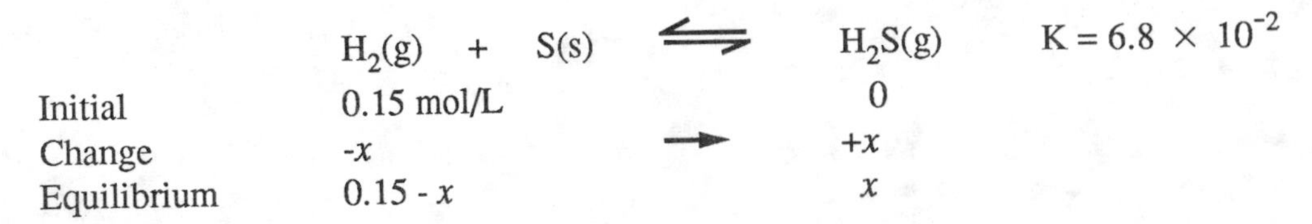

	$H_2(g)$ + $S(s)$	$\rightleftharpoons$	$H_2S(g)$	$K = 6.8 \times 10^{-2}$
Initial	0.15 mol/L		0	
Change	$-x$	$\rightarrow$	$+x$	
Equilibrium	$0.15 - x$		x	

$$K = 6.8 \times 10^{-2} = \frac{[H_2S]}{[H_2]} = \frac{x}{0.15 - x}$$

$0.0102 - 0.068x = x$

$$x = \frac{0.0102}{1.068} = 9.6 \times 10^{-3} \text{ mol/L}$$

$$P_{H_2S} = \frac{nRT}{V} = \frac{9.6 \times 10^{-3} \text{ mol} \left(\frac{0.08206 \text{ L atm}}{\text{mol K}}\right) 363 \text{ K}}{1.0 \text{ L}}$$

$P_{H_2S} = 0.29$ atm

LeChatelier's Principle

33. a. right b. right c. no effect

d. Reaction is exothermic: $\Delta H° = (-393.5) - (-110.5 - 242) = -41$ kJ

$CO(g) + H_2O(g) \rightarrow H_2(g) + CO_2(g) + \text{Heat}$

Increase T; add heat; the equilibrium shifts to the left.

34. $2\ SO_3(g) \rightleftharpoons 2\ SO_2(g) + O_2(g)$

a. increase.

b. increase $K_p = \frac{P^2_{SO_2} \times P_{O_2}}{P^2_{SO_3}}$

Half volume, double each partial pressure before equilibrium shifts.

$$Q = \frac{(2\ P_{SO_2})^2 \times 2\ P_{O_2}}{(2\ P_{SO_3})^2} = 2\ K_p,\ Q > K_p$$

Equilibrium shifts to the left, increasing the number of moles of sulfur trioxide.

c. No effect. The partial pressures of sulfur trioxide, sulfur dioxide, and oxygen are unchanged.

d. Increase. $\text{Heat} + 2\ SO_3 \rightleftharpoons 2\ SO_2 + O_2$, Decrease T, remove heat, shift to left.

e. no effect f. Decrease

35. a. right b. left c. right d. no effect e. no effect

f. An increase in temperature will shift the equilibrium to the right, since the reaction is endothermic, $\Delta H° = 2(25.9) = 51.8$ kJ.

36. a. shift to left

b. Since the reaction is endothermic ($\Delta H° = 207$ kJ), an increase in temperature will shift the equilibrium to the right.

c. no effect d. shift to right e. no effect

37. We can do two things in choosing a solvent. First we must avoid water. Any extra water we add from the solvent tends to push the equilibrium to the left. This eliminates water and 95% ethanol as solvent choices. Of the remaining two solvents, acetonitrile will not take part in the reaction, whereas ethanol is a reactant. If we use ethanol as the solvent it will drive the equilibrium to the right, thereby reducing the concentrations of the objectionable butyric acid to a minimum. Thus, the best solvent is 100% ethanol.

38. $CoCl_2(s) + 6\,H_2O(g) \rightleftharpoons CoCl_2{\cdot}6\,H_2O$

If rain is imminent, there would be a lot of water vapor in the air. Reaction would shift to the right and would take on the color of $CoCl_2{\cdot}6\,H_2O$, pink.

Kinetics and Equilibrium

39. The catalyst increased the rates of both reactions. Equilibrium was reached more rapidly with the catalyst. The uncatalyzed reaction is too slow to be practical.

A catalyst increases the rate of a reaction by providing an alternate pathway with a lower activation energy. Thus, in the new pathway, the activation energy will be lowered for both the forward and reverse reactions.

40. The value of k_f, k_r, the forward rate, and the reverse rate will all be increased by adding a catalyst. The ratio k_f/k_r (which equals the equilibrium constant) is unchanged.

41. $$K = \frac{k_f}{k_r} = \frac{[Br^-][SO_4^{2-}]^3}{[BrO_3^-][SO_3^{2-}]^3}$$

Rate forward reaction = Rate reverse reaction

$$k_f\,[BrO_3^-]\,[SO_3^{2-}]\,[H^+] = k_r\,[Br^-]^a\,[SO_4^{2-}]^b\,[H^+]^c\,[SO_3^{2-}]^d\,[BrO_3^-]^e$$

$$\frac{k_f}{k_r} = \frac{[Br^-]^a\,[SO_4^{2-}]^b\,[H^+]^c\,[SO_3^{2-}]^d\,[BrO_3^-]^e}{[H^+]\,[BrO_3^-]\,[SO_3^{2-}]} = K = \frac{[Br^-]\,[SO_4^{2-}]^3}{[BrO_3^-]\,[SO_3^{2-}]^3}$$

For the two concentration expressions to be the same: $a = 1, b = 3, c = 1, d = -2, e = 0$

$$Rate_r = k_r\,\frac{[Br^-][SO_4^{2-}]^3\,[H^+]}{[SO_3^{2-}]^2}$$

42. $CO(g) + Cl_2(g) \rightleftharpoons COCl_2(g)$

$$K = \frac{[COCl_2]}{[CO]\,[Cl_2]} = \frac{k_f}{k_r}$$

$Rate_f = Rate_r$

$k_f\,[CO]\,[Cl_2]^{3/2} = k_r\,[COCl_2]^a\,[Cl_2]^b$

To satisfy the Law of Mass Action, then:

$Rate_r = k_r\,[COCl_2]\,[Cl_2]^{1/2}$

ADDITIONAL EXERCISES

43. $2\,O_3(g) \rightleftharpoons 3\,O_2(g) \quad K_p = \frac{P^3_{O_2}}{P^2_{O_3}}$

a. $O_3(g) \rightleftharpoons 3/2\,O_2(g)$

$$K'_p = \frac{P^{3/2}_{O_2}}{P_{O_3}} = K_p^{1/2}$$

b. $3\,O_2(g) \rightleftharpoons 2\,O_3(g)$

$$K''_p = \frac{P^2_{O_3}}{P^3_{O_2}} = \frac{1}{K_p}$$

44. The units for both reactions are:

$$\frac{\text{molecules/cm}^3}{(\text{molecules/cm}^3)\,(\text{molecules/cm}^3)} = \frac{\text{cm}^3}{\text{molecules}}$$

a. $$K = \frac{1.26 \times 10^{-11}\ \text{cm}^3}{\text{molecules}} \times \frac{1\ \text{L}}{1000\ \text{cm}^3} \times \frac{6.022 \times 10^{23}\ \text{molecules}}{\text{mol}}$$

$K = 7.59 \times 10^9$ L/mol

$K_p = K(RT)^{\Delta n}; \qquad \Delta n = -1$

$$K_p = \frac{7.59 \times 10^9\ \text{L/mol}}{\left(\frac{0.08206\ \text{L atm}}{\text{mol K}}\right) \times 300.\ \text{K}} = 3.08 \times 10^8\ \text{atm}^{-1}$$, (assuming 3 S.F. in temp.)

b. $$K = \frac{2.09 \times 10^{-12}\ \text{cm}^3}{\text{molecules}} \times \frac{1\ \text{L}}{1000\ \text{cm}^3} \times \frac{6.022 \times 10^{23}\ \text{molecules}}{\text{mol}}$$

$K = 1.26 \times 10^9$ L/mol

$K_p = K(RT)^{\Delta n}; \qquad \Delta n = -1$

$$K_p = \frac{1.26 \times 10^9\ \text{L/mol}}{\left(\frac{0.08206\ \text{L atm}}{\text{mol K}}\right) \times 300.\ \text{K}} = 5.12 \times 10^7\ \text{atm}^{-1}$$

45. a.

$$Na_2O(s) \rightleftharpoons 2\ Na(l) + 1/2\ O_2(g) \qquad K_1$$
$$2\ Na(l) + O_2(g) \rightleftharpoons Na_2O_2(s) \qquad 1/K_3$$

$$Na_2O(s) + 1/2\ O_2(g) \rightleftharpoons Na_2O_2(s) \qquad K_{eq} = (K_1)(1/K_3)$$

$$K_{eq} = \frac{2 \times 10^{-25}}{5 \times 10^{-29}} = 4 \times 10^3$$

b.

$$NaO(g) \rightleftharpoons Na(l) + 1/2\ O_2(g) \qquad K_2$$
$$Na_2O(s) \rightleftharpoons 2\ Na(l) + 1/2\ O_2(g) \qquad K_1$$
$$2\ Na(l) + O_2(g) \rightleftharpoons Na_2O_2(s) \qquad 1/K_3$$

$$NaO(g) + Na_2O(s) \rightleftharpoons Na_2O_2(s) + Na(l)$$

$$K_{eq} = \frac{K_1 K_2}{K_3} = 8 \times 10^{-2}$$

c.

$$2\ NaO(g) \rightleftharpoons 2\ Na(l) + O_2(g) \qquad K_2^2$$
$$2\ Na(l) + O_2(g) \rightleftharpoons Na_2O_2(s) \qquad 1/K_3$$

$$2\ NaO(g) \rightleftharpoons Na_2O_2(s)$$

$$K_{eq} = \frac{K_2^2}{K_3} = 8 \times 10^{18}$$

46.

$$O + NO \rightleftharpoons NO_2 \qquad K_{eq} = 1/6.8 \times 10^{-49} = 1.5 \times 10^{48}$$
$$NO_2 + O_2 \rightleftharpoons NO + O_3 \qquad K_{eq} = 1/5.8 \times 10^{-34} = 1.7 \times 10^{33}$$

$$O_2(g) + O(g) \rightleftharpoons O_3(g) \qquad K = (1.5 \times 10^{48})(1.7 \times 10^{33}) = 2.6 \times 10^{81}$$

47.

	$Fe^{3+}(aq)$ +	$SCN^-(aq)$	$\rightleftharpoons$	$FeSCN^{2+}(aq)$	$K = 1.1 \times 10^3$
Before	0.10	2.0		0	
	0.10 mol/L Fe^{3+} reacts completely (large K, products dominate)				
Change	-0.10	-0.10	$\rightarrow$	+0.10	Reaction goes to completion.
After	0	1.9		0.10	New Initial
	x mol/L $FeSCN^{2+}$ reacts				
Change	$+x$	$+x$	$\leftarrow$	$-x$	Go back to equilibrium
Equil.	x	$1.9 + x$		$0.10 - x$	

$$K = 1.1 \times 10^3 = \frac{[FeSCN^{2+}]}{[Fe^{3+}][SCN^-]} = \frac{0.10 - x}{(x)(1.9 + x)} \approx \frac{0.10}{1.9x}$$

$x = 4.8 \times 10^{-5}$; $0.10 - x = 0.10 - 0.000048 = 0.099952 = 0.10$

Good Assumption!

$x = [Fe^{3+}] = 4.8 \times 10^{-5}\ M$ $[FeSCN^{2+}] = 0.10\ M$

$[SCN^-] = 1.9\ M$

48.

	$2\ CO(g)$	+	$O_2(g)$	$\rightleftharpoons$	$2\ CO_2$	
Initial	$\frac{2.0\ mol}{5.0\ L} = 0.40$		0.40		0	
React to completion	-0.40		-0.20		+0.40	(K is large)
New Initial	0		0.20		0.40	
Equilibrium	$2x$		$0.20 + x$		$0.40 - 2x$	($2x\ M\ CO_2$ reacts)

$$K = 5.0 \times 10^3 = \frac{[CO_2]^2}{[CO]^2\ [O_2]} = \frac{(0.40 - 2x)^2}{4x^2\ (0.20 + x)}$$

If x is small, $0.20 + x \approx 0.20$ and $0.40 - 2x \approx 0.40$

$$5.0 \times 10^3 \approx \frac{(0.40)^2}{4x^2\,(0.20)},\ x = 6.3 \times 10^{-3}\ mol/L$$

$0.20 + x = 0.206$, we are only 3% off. Assumption is good.

$[CO] = 2x = 1.3 \times 10^{-2}$ mol/L

$[O_2] = 0.20 + x = 0.206 \approx 0.21$ mol/L

$[CO_2] = 0.40 - 2x = 0.39$ mol/L

49. When SO_2 is added, the equilibrium will shift to the left, producing more SO_2Cl_2 and some energy. The energy increases the temperature of the reaction mixture.

50.

$$C{\equiv}O\ (g) + 2\ H{-}H(g) \rightleftharpoons H{-}\underset{\substack{|\\H}}{\overset{\substack{H\\|}}{C}}{-}O{-}H(g)$$

Bonds broken:

$C{\equiv}O$	1072 kJ/mol
$2\ H{-}H$	432 kJ/mol

Bonds made:

$3\ C{-}H$	413 kJ/mol
$C{-}O$	358 kJ/mol
$O{-}H$	467 kJ/mol

$\Delta H = 1072 + 2(432) - [3(413) + 358 + 467]$

$\Delta H = 1936 \text{ kJ} - 2064 \text{ kJ} = -128 \text{ kJ}$ Exothermic

We would try to keep the temperature as low as possible. In practice this process is run at about 250°C to get the best trade off between rate of reaction and equilibrium position.

51. $H_2O(g) + Cl_2O(g) \rightleftharpoons 2\ HOCl(g)$ $K = 0.090$

a. The initial concentrations of H_2O and Cl_2O are:

$$\frac{1.0 \text{ g } H_2O}{1.0 \text{ L}} \times \frac{1 \text{ mol}}{18.02 \text{ g}} = 5.5 \times 10^{-2} \text{ mol/L}$$

$$\frac{2.0 \text{ g } Cl_2O}{1.0 \text{ L}} \times \frac{1 \text{ mol}}{86.90 \text{ g}} = 2.3 \times 10^{-2} \text{ mol/L}$$

	$H_2O(g)$ +	$Cl_2O(g)$	$\rightleftharpoons$	$2\ HOCl(g)$
Initial	5.5×10^{-2}	2.3×10^{-2}		0
	x mol/L H_2O reacts			
Change	$-x$	$-x$	$\rightarrow$	$+2x$
Equil.	$5.5 \times 10^{-2} - x$	$2.3 \times 10^{-2} - x$		$2x$

$$K = 0.090 = \frac{(2x)^2}{(5.5 \times 10^{-2} - x)\ (2.3 \times 10^{-2} - x)}$$

Assume x is small, so:

$$\frac{4x^2}{(0.055)\ (0.023)} = 0.090, \ x = 0.0053$$

Assumption is not good. Refining our guess:

$$\frac{4x^2}{(0.055 - 0.0053))\ (0.023 - 0.0053)} = 0.090, \ x = 0.0041$$

$$\frac{4x^2}{(0.055 - 0.0041))\ (0.023 - 0.0041)} = 0.090, \ x = 0.0047$$

$$\frac{4x^2}{(0.055 - 0.0047))\ (0.023 - 0.0047)} = 0.090, \ x = 0.0046$$

$$\frac{4x^2}{(0.055 - 0.0046))\ (0.023 - 0.0046)} = 0.090, \ x = 0.0046$$

We have two identical successive answers, so we have converged on the answer.

$x = 4.6 \times 10^{-3}$

$[HOCl] = 2x = 9.2 \times 10^{-3}$ mol/L

$[H_2O] = 0.055 - x = 0.050$ mol/L

$[Cl_2O] = 0.023 - x = 0.018$ mol/L

b.

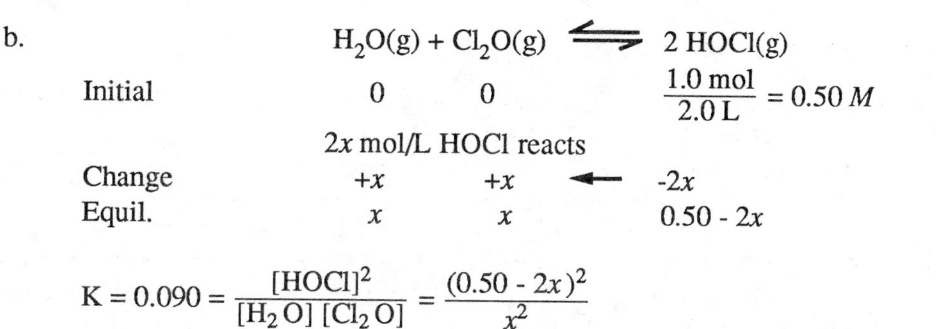

	$H_2O(g)$ +	$Cl_2O(g)$	$\rightleftharpoons$	2 HOCl(g)
Initial	0	0		$\frac{1.0 \text{ mol}}{2.0 \text{ L}} = 0.50\ M$
	$2x$ mol/L HOCl reacts			
Change	$+x$	$+x$	←	$-2x$
Equil.	x	x		$0.50 - 2x$

$$K = 0.090 = \frac{[HOCl]^2}{[H_2O][Cl_2O]} = \frac{(0.50 - 2x)^2}{x^2}$$

We can solve this exactly very quickly. The expression is a perfect square, so we can take the square root of each side:

$$0.30 = \frac{0.50 - 2x}{x}$$

$$0.30x = 0.50 - 2x$$

$$2.30x = 0.50$$

$x = 0.217$ (carry extra significant figure)

$x = [H_2O] = [Cl_2O] = 0.217 = 0.22\ M$

$[HOCl] = 0.50 - 2x = 0.50 - 0.434 = 0.07\ M$

52.

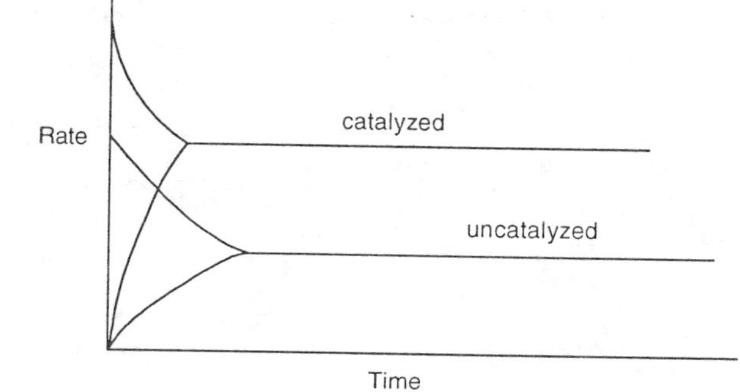

The rates would become equal in less time at a higher value for the catalyzed reaction, however, the ratio k_f/k_r would be unchanged.

CHALLENGE PROBLEMS

53. a.

	2 NO(g)	+	$Br_2(g)$	$\rightleftharpoons$	2 NOBr(g)
Initial	98.4 torr		41.3 torr		0
Change	$-2x$		$-x$	→	$2x$
Equilibrium	$98.4 - 2x$		$41.3 - x$		$2x$

$P_{total} = (98.4 - 2x) + (41.3 - x) + 2x = 139.7 - x$

$P_{total} = 110.5 = 139.7 - x$

$x = 29.2$ torr

$P_{NO} = 98.4 - 2(29.2) = 40.0$ torr $= 0.0526$ atm

$P_{Br_2} = 41.3 - 29.2 = 12.1$ torr $= 0.0159$ atm

$P_{NOBr} = 2(29.2) = 58.4$ torr $= 0.0768$ atm

$$K_p = \frac{P^2_{NOBr}}{P^2_{NO} P_{Br_2}} = \frac{(0.0768 \text{ atm})^2}{(0.0526 \text{ atm})^2 (0.0159 \text{ atm})} = 134 \text{ atm}^{-1}$$

b.

	$2\ NO(g)$	+	$Br_2(g)$	$\rightleftharpoons$	$2\ NOBr(g)$
Initial	0.30		0.30		0
Change	$-2x$		$-x$	$\rightarrow$	$+2x$
Equilibrium	$0.30 - 2x$		$0.30 - x$		$2x$

This would yield a cubic equation. There is no way we can simplify the task. K_p is pretty large, so we will solve this problem in two steps; assume reaction goes to completion then solve the back equilibrium problem.

	$2\ NO$	+	Br_2	$\rightleftharpoons$	$2\ NOBr$
Initial	0.30		0.30		0
Change (React Completely)	-0.30		-0.15	$\rightarrow$	+0.30
New Initial	0		0.15		0.30
Change	$+2y$		$+y$	$\leftarrow$	$-2y$
Equilibrium	$2y$		$0.15 + y$		$0.30 - 2y$

$$\frac{(0.30 - 2y)^2}{(2y)^2\ (0.15 + y)} = 134$$

$$\frac{(0.30 - 2y)^2}{(0.15 + y)} = 134 \times 4y^2 = 536y^2$$

If y is small: $\frac{(0.30)^2}{0.15} = 536y^2$ and $y = 0.034$

But $0.15 + y = 0.184$; assumption is not good (23% error).

Use 0.034 as approximation for y.

$$\frac{(0.30 - 0.068)^2}{0.15 + 0.034} = 536y^2;\ y = 0.023$$

$$\frac{(0.30 - 0.046)^2}{0.173} = 536y^2;\ y = 0.026$$

$$\frac{(0.30 - 0.052)^2}{0.176} = 536y^2;\ y = 0.026$$

So: $P_{NO} = 2y = 0.052$ atm

$P_{Br_2} = 0.15 + y = 0.176$ atm ≈ 0.18 atm

$P_{NOBr} = 0.30 - 2y = 0.25$ atm

54. a. $\Delta H° = 2\ \Delta H_f°(NH_3) = 2\ mol\ (-46\ kJ/mol) = -92\ kJ/mol$ The reaction is exothermic.

Note that K decreases as T increases. This is true for all exothermic reactions.

b. The value of the equilibrium constant will be most favorable at low temperatures. The rate will be faster as the temperature is increased. Intermediate temperatures are chosen as the best compromise between thermodynamics and kinetics. Since rates are proportional to partial pressures, high pressures will increase the rate and also drive the equilibrium to the right.

55. a. $N_2(g) + O_2(g) \rightleftharpoons 2\ NO(g)$

$$K_p = 1 \times 10^{-31} = \frac{P_{NO}^2}{P_{N_2}P_{O_2}} = \frac{P_{NO}^2}{(0.8)(0.2)}$$

$P_{NO} = 1 \times 10^{-16}$ atm

PV = nRT

$$n = \frac{PV}{RT} = \frac{(1 \times 10^{-16}\ atm)\ (1.0 \times 10^{-3}\ L)}{\left(\frac{0.08206\ L\ atm}{mol\ K}\right)(298\ K)} = 4 \times 10^{-21}\ mol$$

$$\frac{4 \times 10^{-21}\ mol}{cm^3} \times \frac{6.02 \times 10^{23}\ molecules}{mol} = \frac{2 \times 10^3\ molecules}{cm^3}$$

b. There is more NO in the atmosphere than we would expect from the value of K. The answer must lie in the rates of the reaction. At 25°C the rates of both reactions:

$$N_2 + O_2 \rightarrow 2\ NO \text{ and } 2\ NO \rightarrow N_2 + O_2$$

are essentially zero. Very strong bonds must be broken; the activation energy is very high. Nitric oxide is produced in high energy or high temperature environments. In nature some NO is produced by lightning. The primary manmade source is from automobiles. The production of NO is endothermic ($\Delta H_f° = +\ 90$ kJ/mol). At high temperature, K will increase and the rates of the reaction increases; so NO is produced. Once the NO gets into a more normal environment in the atmosphere, it doesn't go back to N_2 and O_2 because of the slow rate.

56. a.

	$I^-(aq)$	+	$I_2(aq)$	$\rightleftharpoons$	$I_3^-(aq)$	
Before	0.20		0.10		0	
	0.10 mol/L I^- reacts					
Change	-0.10		-0.10	$\rightarrow$	+0.10	React completely. From the value of K = 710, we know equilibrium lies far to the right.

	$I^-(aq)$	+	$I_2(aq)$	$\rightleftharpoons$	$I_3^-(aq)$	
New Initial	0.10		0		0.10	New Initial conditions to use in solving equilibrium problem.
	x mol/L I_3^- reacts					
Change	$+x$		$+x$	$\leftarrow$	$-x$	
Equil.	$0.10 + x$		x		$0.10 - x$	

$$\frac{[I_3^-]}{[I_2][I^-]} = 710 = \frac{0.10 - x}{(x)(0.10 + x)} \approx \frac{1}{x}$$

$x = 1.4 \times 10^{-3}$, $0.10 - x = 0.0986 \approx 0.10$ Assumptions are good.

$x = [I_2] = 1.4 \times 10^{-3}$ $\quad \dfrac{[I_3^-]}{[I_2]} = \dfrac{0.10}{1.4 \times 10^{-3}} = 71$

b.

	$I^-(aq)$	+	$I_2(aq)$	$\rightleftharpoons$	$I_3^-(aq)$	
Before	1.5		0.10		0	
	0.10 mol/L I_2 reacts completely					
Change	-0.10		-0.10	$\rightarrow$	+0.10	React completely
After	1.4		0		0.10	New initial conditions
	x mol/L I_3^- reacts					
Change	$+x$		$+x$	$\leftarrow$	$-x$	
Equil.	$1.4 + x$		x		$0.10 - x$	

$$\frac{0.10 - x}{(1.4 + x)(x)} = 710 = \frac{0.10}{1.4x} \quad x = 1.0 \times 10^{-4} = [I_2]$$ Assumptions are good.

$$\frac{[I_3^-]}{[I_2]} = \frac{0.10}{1.0 \times 10^{-4}} = 1.0 \times 10^3$$

c.

	$I^-(aq)$	+	$I_2(aq)$	$\rightleftharpoons$	$I_3^-(aq)$	
Before	5.0		0.10		0	
	0.10 mol/L I_2 reacts completely					
Change	-0.10		-0.10	$\rightarrow$	+0.10	React completely
After	4.9		0		0.10	New initial conditions
	x mol/L I_3^- reacts					
Change	$+x$		$+x$	$\leftarrow$	$-x$	
Equil.	$4.9 + x$		x		$0.10 - x$	

$$\frac{0.10 - x}{(4.9 + x)(x)} = 710 \approx \frac{0.10}{4.9x} \quad x = [I_2] = 2.9 \times 10^{-5}$$

$\dfrac{[I_3^-]}{[I_2]} = 3.4 \times 10^3$ Assumptions good.

57. $CCl_4(g) \rightleftharpoons C(s) + 2\,Cl_2(g)$ $K_p = 0.76$ atm

	$CCl_4(g)$		$2\,Cl_2(g)$
Initial	P_O		0
Change	-p	→	2p
Equil.	P_O -p		2p

$P_{total} = P_O - p + 2p = P_O + p = 1.2$ atm

$\frac{(2p)^2}{P_O - p} = 0.76$, $4p^2 = 0.76\,P_O - 0.76\,p$, $P_O = \frac{4p^2 + 0.76\,p}{0.76}$

$\frac{4p^2}{0.76} + p + p = 1.2$ atm, $5.3p^2 + 2p - 1.2 = 0$

$p = \frac{-2 \pm (4 + 25)^{1/2}}{2(5.3)} = 0.32$ atm and $P_O + 0.32 = 1.2$. Thus, $P_O = 0.9$ atm

58. $N_2(g) + 3\,H_2(g) \rightleftharpoons 2\,NH_3(g)$

$$\frac{(P_{NH_3})^2}{(P_{N_2})\,(P_{H_2})^3} = 6.5 \times 10^{-3}\ \text{atm}^{-2}$$

1.0 atm	N_2	+	$3\,H_2$	$2\,NH_3$
Initial	0.25		0.75	0
Equil.	0.25 - x		0.75 - $3x$	$2x$

$$\frac{(2x)^2}{(0.75 - 3x)^3\,(0.25 - x)} = 6.5 \times 10^{-3}$$

$x = 1.2 \times 10^{-2}$, $P_{NH_3} = 2x = 0.024$ atm

10 atm	N_2	+	$3\,H_2$	⇌	$2\,NH_3$
Initial	2.5		7.5		0
Equil.	2.5 - x		7.5 - $3x$		$2x$

$$\frac{(2x)^2}{(2.5 - x)\,(7.5 - 3x)^3} = 6.5 \times 10^{-3}$$

$x = 0.69$ atm, $P_{NH_3} = 1.4$ atm

100 atm $\frac{4x^2}{(25 - x)\,(75 - 3x)^3} = 6.5 \times 10^{-3}$

By successive approximations: $x = 16$ atm, $P_{NH_3} = 32$ atm

1000 atm

	N_2	+	$3\ H_2$	$\rightleftharpoons$	$2\ NH_3$
Initial	250		750		0
	250 atm N_2 reacts completely				
New Initial	0		0		5.0×10^2
Equil.	x		$3x$		$5.0 \times 10^2 - 2x$

$$\frac{(500 - 2x)^2}{(3x)^3 x} = 6.5 \times 10^{-3}$$

Assume x is small.

$$\frac{(500)^2}{(3x)^3 x} = 6.5 \times 10^{-3},\ x = 34.5$$ Assumption bad (14% error).

Solve by successive approximations using $x = 34.5$:

$$\frac{(500 - 2x)^2}{6.5 \times 10^{-3}} = 27x^4$$

$$\frac{(429)^2}{6.5 \times 10^{-3}} = 27x^4,\ x = 32$$

$$\frac{(436)^2}{6.5 \times 10^{-3}} = 27x^4,\ x = 32$$

$P_{NH_3} = 5.0 \times 10^2 - 2x = 440$ atm

The results are plotted as log P_{NH_3} vs. log P_{tot}. The plot will asymptotically approach the line given by log P_{NH_3} = log P_{tot} at the upper limit.

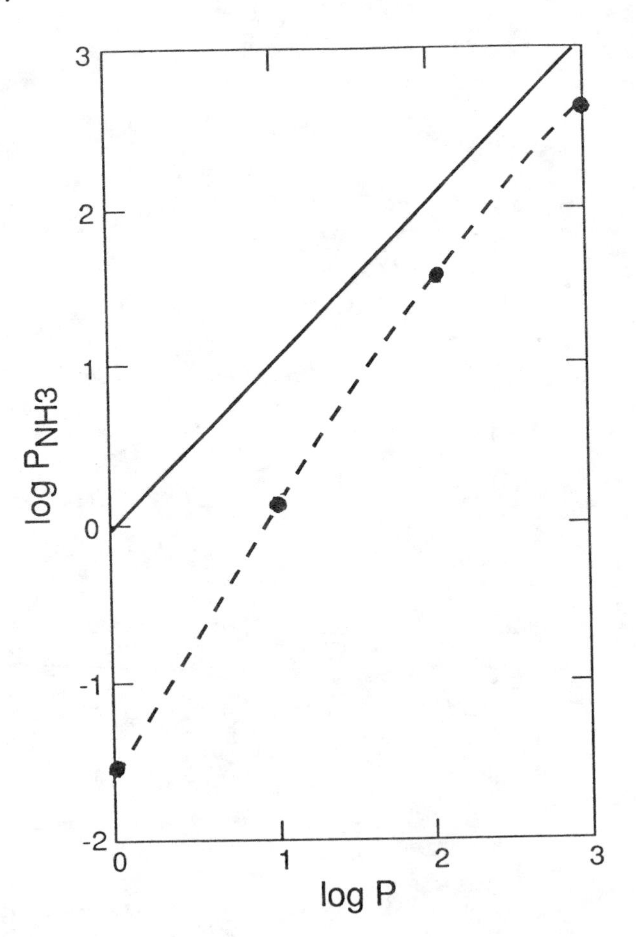

CHAPTER FOURTEEN: ACIDS AND BASES

QUESTIONS

1. a. A strong acid is 100% dissociated in water.
   b. A strong base is 100% dissociated in water.
   c. A weak acid is much less than 100% dissociated in water.
   d. A weak base is one that only a small percentage of the molecules react with water to form OH^-.

2. If we added an acid HX that was a stronger acid than H_3O^+, then the reaction:

$$HX + H_2O \rightarrow H_3O^+ + X^-$$

would go to 100% completion because H_2O is a much stronger base than X^-. Thus, strong acids are "leveled" in water. They all generate H_3O^+. If a base stronger than OH^- were added to water, the reaction:

$$\text{Base} + H_2O \rightarrow (\text{H Base})^+ + OH^-$$

would go to 100% completion because H_2O would be a much stronger acid than $(\text{H Base})^+$.

3. $H_2O \rightleftharpoons H^+ + OH^-$ $\quad K_w = [H^+][OH^-] = 1.0 \times 10^{-14}$

Neutral solution $[H^+] = [OH^-]$; $[H^+] = 1.0 \times 10^{-7}$ *M*; pH = 7.00

4. 10.78 (4 S. F.), 6.78 (3 S. F.), 0.78 (2 S. F.)

If these were pH values, however, all three contain only two significant figures. Thus, the $[H^+]$ can be expressed to two significant figures in each case.

$$pH = 10.78,\ [H^+] = 10^{-10.78} = 10^{0.22} \times 10^{-11} = 1.7 \times 10^{-11}\ \text{mol/L}$$

$$pH = 6.78,\ [H^+] = 10^{-6.78} = 10^{0.22} \times 10^{-7} = 1.7 \times 10^{-7}\ \text{mol/L}$$

$$pH = 0.78,\ [H^+] = 10^{-0.78} = 10^{0.22} \times 10^{-1} = 1.7 \times 10^{-1}\ \text{mol/L}$$

A pH value is a logarithm. The numbers to the left of the decimal point only help to identify the power of ten when we express $[H^+]$ in scientific notation.

5. a. Arrhenius acid: generate H^+ in water
   b. Brönsted-Lowry acid: proton donor
   c. Lewis acid: electron pair acceptor

6. The Lewis definition is most general. It can apply to all Arrhenius and Brönsted-Lowry acids, since H^+ has an empty orbital and can form bonds by accepting a pair of electrons. There are reactions that can only be considered as Lewis acid/base reactions. For example,

$$BF_3(g) + NH_3(g) \longrightarrow F_3B\text{—}NH_3(s)$$

If NH_3 is something we usually consider a base, it seems logical that it should be a base in this reaction. It is using the Lewis definition.

7. Nitrogen

8. a. weaker bond = stronger acid
   b. greater electronegativity = stronger acid
   c. more oxygen atoms = stronger acid.

9. a. weaker bond = weaker base (stronger bond = stronger base)
   b. greater electronegativity = weaker base
   c. more oxygen atoms = weaker base

10. A Lewis acid must have an empty orbital to accept an electron pair and a Lewis base must have an unshared pair of electrons.

EXERCISES

Nature of Acids and Bases

11. In deciding whether a substance is an acid, base, strong, or weak, we should keep in mind three things:

    1. There are only a few important strong acids and bases.

    2. All other acids and bases are weak.

    3. Structural features to look for to identify a substance as an acid or a base: —CO_2H groups (acid) or the presence of N (base) in organic compounds.

    a. weak acid   b. weak acid   c. weak base

    d. strong base   e. weak base   f. weak acid

12. a. weak acid   b. weak base   c. strong acid

    d. weak acid   e. strong acid   f. strong acid

13. HNO_3: strong acid   HOCl: $K_a = 3.5 \times 10^{-8}$ NH_4^+: $K_a = 5.6 \times 10^{-10}$

    H_3O^+ is the strongest acid that can exist in water.

$$H_3O^+ > HNO_3 > HOCl > NH_4^+$$

All are acidic in water, so all of these acids are stronger acids than H_2O.

$$H_3O^+ > HNO_3 > HOCl > NH_4^+ > H_2O$$

14. NH_3, OCl^-, and NO_3^- are the conjugate bases of the acids in the previous exercise. OH^- is the strongest base that can exist in H_2O. The strength of bases will be:

$$OH^- > NH_3 > OCl^- >> NO_3^-$$

NH_3 and OCl^- are basic in water. Thus, they are better bases than H_2O. HNO_3 is a strong acid. Water is a much better base than the nitrate ion (the conjugate base of a strong acid).

$$OH^- > NH_3 > OCl^- > H_2O >> NO_3^-$$

15. a. HI is a strong acid and $K_a = K_w = 10^{-14}$ for H_2O.

HI is a stronger acid than H_2O.

b. H_2O, $K_a = 10^{-14}$; $HClO_2$, $K_a = 1.2 \times 10^{-2}$

$HClO_2$ is a stronger acid than H_2O.

c. HF, $K_a = 7.2 \times 10^{-4}$; HCN, $K_a = 6.2 \times 10^{-10}$

HF is a stronger acid than HCN.

16. a. H_2O b. ClO_2^- c. CN^-

17.

	Acid	Base	Conjugate Base of Acid	Conjugate Acid of Base
a.	H_2O	H_2O	OH^-	H_3O^+
b.	CH_3COCH_3	CH_3O^-	$CH_3COCH_2^-$	CH_3OH
c.	H_2S	NH_3	HS^-	NH_4^+
d.	H_2SO_4	H_2O	HSO_4^-	H_3O^+
e.	H_3O^+	OH^-	H_2O	H_2O
f.	H_2O	$H_2PO_4^-$	OH^-	H_3PO_4
g.	HCN	H_2O	CN^-	H_3O^+

18.

	Acid	Base	Conjugate Base of Acid	Conjugate Acid of Base
a.	$H_2PO_4^-$	H_2O	HPO_4^{2-}	H_3O^+
b.	$H_2PO_4^-$	$H_2PO_4^-$	HPO_4^{2-}	H_3PO_4
c.	$Fe(H_2O)_6^{3+}$	H_2O	$Fe(H_2O)_5(OH)^{2+}$	H_3O^+

	Acid	Base	Conjugate Base of Acid	Conjugate Acid of Base
d.	HCN	CO_3^{2-}	CN^-	HCO_3^-
e.	$CO_2 + H_2O$*	H_2O	HCO_3^-	H_3O^+
f.	H_2O	NH_3	OH^-	NH_4^+
g.	H_2O	C_5H_5N	OH^-	$C_5H_5NH^+$

*$CO_2 + H_2O$ = carbonic acid, H_2CO_3

19. a. $H_3PO_4 + H_2O \rightleftharpoons H_3O^+ + H_2PO_4^-$ $\quad K_{a_1} = \frac{[H_3O^+][H_2PO_4^-]}{[H_3PO_4]}$

or $H_3PO_4 \rightleftharpoons H^+ + H_2PO_4^-$ $\quad K_{a_1} = \frac{[H^+][H_2PO_4^-]}{[H_3PO_4]}$

Only the first reaction will be written for the rest, but both are equivalent.

b. $H_2PO_4^- + H_2O \rightleftharpoons H_3O^+ + HPO_4^{2-}$ $\quad K_{a_2} = \frac{[H_3O^+][HPO_4^{2-}]}{[H_2PO_4^-]}$

c. $HPO_4^{2-} + H_2O \rightleftharpoons H_3O^+ + PO_4^{3-}$ $\quad K_{a_3} = \frac{[H_3O^+][PO_4^{3-}]}{[HPO_4^{2-}]}$

d. $HNO_2 + H_2O \rightleftharpoons H_3O^+ + NO_2^-$ $\quad K_a = \frac{[H_3O^+][NO_2^-]}{[HNO_2]}$

e. $Ti(H_2O)_6^{4+} + H_2O \rightleftharpoons Ti(H_2O)_5(OH)^{3+} + H_3O^+$ $\quad K_a = \frac{[H_3O^+][Ti(H_2O)_5(OH)^{3+}]}{[Ti(H_2O)_6^{4+}]}$

20. a. $HCN + H_2O \rightleftharpoons H_3O^+ + CN^-$ $\quad K_a = \frac{[H_3O^+][CN^-]}{[HCN]}$

b. $CH_3CO_2H + H_2O \rightleftharpoons CH_3CO_2^- + H_3O^+$ $\quad K_a = \frac{[H_3O^+][CH_3CO_2^-]}{[CH_3CO_2H]}$

c. $C_6H_5OH + H_2O \rightleftharpoons C_6H_5O^- + H_3O^+$ $\quad K_a = \frac{[H_3O^+][C_6H_5O^-]}{[C_6H_5OH]}$

d. $C_6H_5CO_2H + H_2O \rightleftharpoons H_3O^+ + C_6H_5CO_2^-$ $\quad K_a = \frac{[H_3O^+][C_6H_5CO_2^-]}{[C_6H_5CO_2H]}$

e. $H_2NCH_2CO_2H + H_2O \rightleftharpoons H_2N{-}CH_2{-}CO_2^- + H_3O^+$

$$K_a = \frac{[H_3O^+][H_2NCH_2CO_2^-]}{[H_2NCH_2CO_2H]}$$

Note: Glycine does not exist in this form in water. Glycine dissolved in H_2O exists as $^+H_3N—CH_2—CO_2^-$, in acid solution as $^+H_3NCH_2CO_2H$, and in base as $H_2NCH_2CO_2^-$.

21. a. $PO_4^{3-} + H_2O \rightleftharpoons HPO_4^{2-} + OH^-$ $\quad K_b = \frac{[OH^-][HPO_4^{2-}]}{[PO_4^{3-}]}$

b. $HPO_4^{2-} + H_2O \rightleftharpoons H_2PO_4^- + OH^-$ $\quad K_b = \frac{[OH^-][H_2PO_4^-]}{[HPO_4^{2-}]}$

c. $H_2PO_4^- + H_2O \rightleftharpoons H_3PO_4 + OH^-$ $\quad K_b = \frac{[OH^-][H_3PO_4]}{[H_2PO_4^-]}$

d. $NH_3 + H_2O \rightleftharpoons NH_4^+ + OH^-$ $\quad K_b = \frac{[NH_4^+][OH^-]}{[NH_3]}$

e. $CN^- + H_2O \rightleftharpoons OH^- + HCN$ $\quad K_b = \frac{[OH^-][HCN]}{[CN^-]}$

22. a. $C_5H_5N + H_2O \rightleftharpoons C_5H_5NH^+ + OH^-$ $\quad K_b = \frac{[OH^-][C_5H_5NH^+]}{[C_5H_5N]}$

b. $H_2NCH_2CO_2H + H_2O \rightleftharpoons {}^+H_3NCH_2CO_2H + OH^-$ $\quad K_b = \frac{[^+H_3NCH_2CO_2H][OH^-]}{[H_2NCH_2CO_2H]}$

c. $CH_3CH_2NH_2 + H_2O \rightleftharpoons CH_3CH_2NH_3^+ + OH^-$ $\quad K_b = \frac{[OH^-][CH_3CH_2NH_3^+]}{[CH_3CH_2NH_2]}$

d. $C_6H_5NH_2 + H_2O \rightleftharpoons C_6H_5NH_3^+ + OH^-$ $\quad K_b = \frac{[OH^-][C_6H_5NH_3^+]}{[C_6H_5NH_2]}$

e. $(CH_3)_2NH + H_2O \rightleftharpoons (CH_3)_2NH_2^+ + OH^-$ $\quad K_b = \frac{[OH^-][(CH_3)_2NH_2^+]}{[(CH_3)_2NH]}$

23. $H^- + H_2O \rightarrow H_2 + OH^-$; $OCH_3^- + H_2O \rightarrow CH_3OH + OH^-$

24. a. $HClO_4(aq) \rightarrow H^+(aq) + ClO_4^-(aq)$

b. $CH_3CH_2CO_2H(aq) \rightleftharpoons H^+(aq) + CH_3CH_2CO_2^-(aq)$

c. $CH_3CO_2H(aq) + NaOH(aq) \rightarrow H_2O(l) + NaCH_3CO_2(aq)$

d. $NH_4^+(aq) \rightleftharpoons NH_3(aq) + H^+(aq)$

Autoionization of Water and the pH Scale

25. a. $pH = -\log(1.4 \times 10^{-3}) = 2.85$ b. $pH = -\log(2.5 \times 10^{-10}) = 9.60$

c. $pH = -\log(6.1) = -0.79$ Note: If $[H^+] > 1$ *M*, pH is negative.

d. $[OH^-] = 3.5 \times 10^{-2}$ *M*; pOH = 1.46

pH + pOH = 14; pH = 12.54

$$\text{or } [H^+] = \frac{1.0 \times 10^{-14}}{3.5 \times 10^{-2}} = 2.9 \times 10^{-13}\ M$$

$pH = -\log(2.9 \times 10^{-13}); = 12.54$

e. $[OH^-] = 8 \times 10^{-11}$ *M*; pOH = 10.1; pH = 3.9

f. $[OH^-] = 5.0$ *M*; pOH = -0.70; pH = 14.70

Note: if $[OH^-] > 1.0$ *M*, then pOH is negative and pH > 14.

g. pOH = 10.5, pH = 14.0 - 10.5 = 3.5

h. pOH = 2.3, pH = 14.0 - 2.3 = 11.7

26. a. $pH = -\log(1.0) = 0.00$

b. $pH = -\log(3.0 \times 10^{-2}) = 1.52$

c. $pH = -\log(4.6 \times 10^{-8}) = 7.34$

d. $pOH = -\log(6.2 \times 10^{-5}) = 4.21$ pH = 14.00 - 4.21 = 9.79

e. $pOH = -\log(6.2 \times 10^{-9}) = 8.21$ pH = 14.00 - 8.21 = 5.79

f. pH = 14.0 - 2.6 = 11.4

g. pH = 14.0 - 9.1 = 4.9

27. a. pH = 7.41; $[H^+] = 10^{-7.41} = 3.9 \times 10^{-8}$ *M*

pOH = 14 - 7.41 = 6.59; $[OH^-] = 10^{-6.59} = 2.6 \times 10^{-7}$ *M*

$$\text{or } [OH^-] = \frac{K_w}{[H^+]} = \frac{1.0 \times 10^{-14}}{3.9 \times 10^{-8}} = 2.6 \times 10^{-7}\ M$$

b. pH = 15.3; $[H^+] = 10^{-15.3} = 5 \times 10^{-16}$ *M*

pOH = 14 - pH = -1.3; $[OH^-] = 10^{1.3} = 20$ *M*

c. $H = -1.0$

$[H^+] = 10\ M$

$pOH = 14.0 - pH = 15.0$

$[OH^-] = 1 \times 10^{-15}\ M$

d. $pH = 3.2$

$[H^+] = 10^{-3.2} = 6 \times 10^{-4}\ M$

$pOH = 14.0 - pH = 10.8$

$[OH^-] = 10^{-10.8} = 2 \times 10^{-11}\ M$

e. $pOH = 5.0$; $[OH^-] = 1 \times 10^{-5}\ M$

$pH = 9.0$; $[H^+] = 1 \times 10^{-9}\ M$

f. $pOH = 9.6$

$[OH^-] = 10^{-9.6} = 2.5 \times 10^{-10}\ M = 3 \times 10^{-10}\ M$ (1 significant figure)

$pH = 4.4$; $[H^+] = 10^{-4.4} = 4 \times 10^{-5}\ M$

28. a. $pH = 2.66$, $[H^+] = 10^{-2.66} = 2.2 \times 10^{-3}\ M$

$pOH = 14.00 - 2.66 = 11.34$, $[OH^-] = 10^{-11.34} = 4.6 \times 10^{-12}\ M$

b. $pH = 8.66$, $[H^+] = 10^{-8.66} = 2.2 \times 10^{-9}\ M$

$pOH = 14.00 - 8.66 = 5.34$, $[OH^-] = 10^{-5.34} = 4.6 \times 10^{-6}\ M$

c. $pH = -0.34$, $[H^+] = 10^{0.34} = 2.2\ M$

$pOH = 14.00 + 0.34 = 14.34$, $[OH^-] = 10^{-14.34} = 4.6 \times 10^{-15}\ M$

d. $pOH = 2.48$, $[OH^-] = 10^{-2.48} = 3.3 \times 10^{-3}\ M$

$pH = 14.00 - 2.48 = 11.52$, $[H^+] = 10^{-11.52} = 3.0 \times 10^{-12}\ M$

e. $pOH = 8.48$, $[OH^-] = 10^{-8.48} = 3.3 \times 10^{-9}\ M$

$pH = 14.00 - 8.48 = 5.52$, $[H^+] = 10^{-5.52} = 3.0 \times 10^{-6}\ M$

29. a. Since the value of the equilibrium constant increases as the temperature increases, the reaction is endothermic.

b. $H_2O \rightleftharpoons H^+ + OH^-$ $K_w = 5.47 \times 10^{-14} = [H^+][OH^-]$

In pure water $[H^+] = [OH^-]$:

$5.47 \times 10^{-14} = [H^+]^2$; $[H^+] = 2.34 \times 10^{-7}\ M$

$pH = -\log [H^+] = -\log (2.34 \times 10^{-7}) = 6.631$

c. A neutral solution of water at 50°C has:

$$[H^+] = [OH^-]; \; [H^+] = 2.34 \times 10^{-7} \, M; \; pH = 6.631$$

Obviously, the condition that $[H^+] = [OH^-]$ is the most general definition of a neutral solution.

d.

Temp (°C)	Temp(K)	1/T	K_w	ln K_w
0	273	3.66×10^{-3}	1.14×10^{-15}	-34.408
25	298	3.36×10^{-3}	1.00×10^{-14}	-32.236
35	308	3.25×10^{-3}	2.09×10^{-14}	-31.499
40	313	3.19×10^{-3}	2.92×10^{-14}	-31.165
50	323	3.10×10^{-3}	5.47×10^{-14}	-30.537

From the graph: 37°C = 310. K; $1/T = 3.23 \times 10^{-3}$

$$\ln K_w = -31.38; \; K_w = 2.35 \times 10^{-14}$$

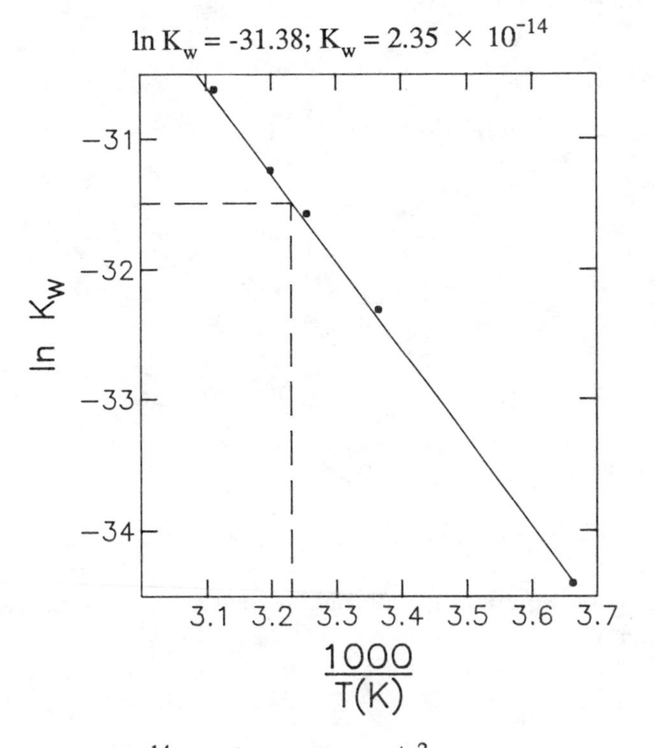

e. At 37°C, $2.35 \times 10^{-14} = [H^+][OH^-] = [H^+]^2$

$[H^+] = 1.53 \times 10^{-7}$; pH = 6.815

30. The equation of the line in Exercise 29 is:

$$\ln K_w = -6.91 \times 10^3 \left(\frac{1}{T}\right) - 9.09$$

$T = 374°C + 273 = 647\ K$

$$\ln K_w = -6.91 \times 10^3 \left(\frac{1}{647}\right) - 9.09 = -19.8$$

$$K_w = e^{-19.8} = 2.52 \times 10^{-9}$$

Note: Neutral water at this temperature has pH = pOH = 4.3. This is one reason supercritical water is so reactive. In fact, the effect of pressure also increases K_w, so that supercritical water has even higher concentrations of H^+ and OH^- than estimated here.

Solutions of Acids

31. a. $H^+(aq)$, $Cl^-(aq)$, and H_2O (HCl is a strong acid.) $[H^+] = 0.250$, pH = 0.602

b. $H^+(aq)$, $Br^-(aq)$, and H_2O (HBr is a strong acid.) pH = 0.602

c. H_2O, $H^+(aq)$, $ClO_4^-(aq)$ ($HClO_4$ is a strong acid.) pH = 0.602

d. H_2O, $H^+(aq)$, $NO_3^-(aq)$ (HNO_3 is a strong acid.) pH = 0.602

32. a. $HNO_2(aq)$ and H_2O are the major species. HNO_2 is a stronger acid than H_2O. We will consider it the major source of H^+. We will ignore H^+ contribution from water.

	HNO_2	$\rightleftharpoons$	H^+	+	NO_2^-
Initial	0.250 *M*		$0(10^{-7})$		0
	x mol/L HNO_2 dissociates to reach equilibrium				
Change	$-x$	$\rightarrow$	$+x$		$+x$
Equil	$0.250 - x$		x		x

$$K_a = \frac{[H^+][NO_2^-]}{[HNO_2]} = 4.0 \times 10^{-4} = \frac{x^2}{0.250 - x} \approx \frac{x^2}{0.250} \quad \text{(assuming } x << 0.250\text{)}$$

$$x = 0.010\ M;\ 0.250 - x = 0.240,\ 4\%\ \text{error, i.e.}\ \frac{x}{0.250} = 0.04$$

Assumptions are good by the 5% rule. $x = 0.010\ M = [H^+]$; pH = 2.00

b. $CH_3CO_2H(aq)$ and H_2O are the major species. We will consider CH_3CO_2H as the major source of H^+.

	CH_3CO_2H	$\rightleftharpoons$	H^+	+	$CH_3CO_2^-$
Initial	0.250 *M*		$0(10^{-7})$		0
	x mol/L CH_3CO_2H dissociates to reach equilibrium				
Change	$-x$	$\rightarrow$	$+x$		$+x$
Equil	$0.250 - x$		x		x

$$K_a = \frac{[H^+][CH_3CO_2^-]}{[CH_3CO_2H]} = 1.8 \times 10^{-5} = \frac{x^2}{0.250 - x} \approx \frac{x^2}{0.250}$$

$x = 2.1 \times 10^{-3}$ *M*; $0.250 - x = 0.248$, Assumptions good (~ 1% error).

$[H^+] = x = 2.1 \times 10^{-3}$ *M*; pH = 2.68

c. $NH_4^+(aq)$, $Cl^-(aq)$ and H_2O are the major species. The major equilibrium is the dissociation of NH_4^+ ion.

	NH_4^+	⇌	H^+	+	NH_3
Initial	0.250 *M*		$0(10^{-7})$		0
	x mol/L NH_4^+ dissociates to reach equilibrium				
Change	$-x$	→	$+x$		$+x$
Equil	$0.250 - x$		x		x

$$K_a = \frac{K_w}{K_b} = \frac{1.0 \times 10^{-14}}{1.8 \times 10^{-5}} = 5.6 \times 10^{-10}$$

$$5.6 \times 10^{-10} = \frac{[H^+][NH_3]}{[NH_4^+]} = \frac{x^2}{0.250 - x} \approx \frac{x^2}{0.250}$$

$x = [H^+] = 1.2 \times 10^{-5}$; Assumptions good.

$pH = -\log(1.2 \times 10^{-5}) = 4.92$

d. HCN(aq) and H_2O are the major species. HCN is the major source of H^+.

	HCN	⇌	H^+	+	CN^-
Initial	0.250 *M*		$0(10^{-7})$		0
	x mol/L HCN dissociates to reach equilibrium				
Change	$-x$	→	$+x$		$+x$
Equil	$0.250 - x$		x		x

$$K_a = 6.2 \times 10^{-10} = \frac{[H^+][CN^-]}{[HCN]} = \frac{x^2}{0.250 - x} \approx \frac{x^2}{0.250}$$

$x = [H^+] = 1.2 \times 10^{-5}$ Assumptions good.

$pH = -\log(1.2 \times 10^{-5}) = 4.92$

33. a. $HCl \rightarrow H^+ + Cl^-$ Strong acids dissociate completely.

$$\frac{0.1 \text{ mol HCl}}{\text{L}} \times \frac{1 \text{ mol } H^+}{\text{mol HCl}} = 0.1\ M = [H^+]$$

$pH = -\log(0.1) = 1.0$

b. $[H^+] = 0.1\ M$ $pH = 1.0$

c. $[H^+] = 0.1\ M$ $pH = 1.0$

34. a. $[H^+] = 3.0 \times 10^{-5}\ M$ $pH = 4.52$

b. $[H^+] = 2.0 \times 10^{-2}\ M$ $pH = 1.70$

c. $[H^+] = 4.0\ M$ $pH = -0.60$

35. $$50.0\text{ mL con. HCl soln} \times \frac{1.19\text{ g}}{\text{mL}} \times \frac{38\text{ g HCl}}{100.\text{ g con. HCl soln}} \times \frac{1\text{ mol HCl}}{36.46\text{ g}} = 0.62\text{ mol HCl}$$

$$20.0\text{ mL con. HNO}_3\text{ soln} \times \frac{1.42\text{ g}}{\text{mL}} \times \frac{70.\text{ g HNO}_3}{100.\text{ g soln}} \times \frac{1\text{ mol HNO}_3}{63.02\text{ g HNO}_3} = 0.32\text{ mol HNO}_3$$

$$HCl + H_2O \rightarrow H_3O^+ + Cl^- \text{ and } HNO_3 + H_2O \rightarrow H_3O^+ + NO_3^-$$

So we will have 0.62 + 0.32 = 0.94 mol of H_3O^+ in the final solution.

$$[H^+] = \frac{0.94\text{ mol}}{0.500\text{ L}} = 1.9\ M$$

$$[OH^-] = \frac{1.0 \times 10^{-14}}{1.9} = 5.3 \times 10^{-15}\ M\ ;\ pH = -0.28$$

36. $$90.0 \times 10^{-3}\text{ L} \times 5.00\ \frac{\text{mol}}{\text{L}} = 0.450\text{ mol H}^+\text{ from HCl}$$

$$30.0 \times 10^{-3}\text{ L} \times 8.00\ \frac{\text{mol}}{\text{L}} = 0.240\text{ mol H}^+\text{ from HNO}_3$$

$$[H^+] = \frac{0.450\text{ mol} + 0.240\text{ mol}}{1.00\text{ L}} = 0.690\ M$$

$pH = 0.161$, $pOH = 13.839$, $[OH^-] = 10^{-13.839} = 1.45 \times 10^{-14}\ M$

37. 0.050 mol/L CH_3CO_2H

Major species: CH_3CO_2H, H_2O

We will consider the acetic acid, $K_a = 1.8 \times 10^{-5}$, as a more important source of H^+ than water, i.e., we will ignore the H^+ contribution from water. For our purposes, this assumption will almost always hold true.

We will use the abbreviations HOAC for CH_3CO_2H and OAc^- for $CH_3CO_2^-$.

	HOAc	$\rightleftharpoons$	H^+	+	OAc^-
Initial	0.050 mol/L		$0(10^{-7})$		0
	x mol/L HOAc dissociates to reach equilibrium				
Change	$-x$	$\rightarrow$	$+x$		$+x$
Equil	$0.050 - x$		x		x

$$K_a = 1.8 \times 10^{-5} = \frac{[H^+][OAc^-]}{[HOAc]} = \frac{x^2}{0.050 - x} \approx \frac{x^2}{0.050} \quad \text{(assuming } x << 0.050\text{)}$$

$x = [H^+] = 9.5 \times 10^{-4}\ M$ Assumptions good (~ 2% error).

pH = 3.02

Note: Assumption is good if % error in assumption is < 5%. Both assumptions we have made are good by the 5% rule, i.e., we can ignore H^+ contribution from water and $x << 0.050$. Always check your assumptions. If the assumptions fail, we must solve exactly using either the quadratic equation or the method of successive approximations (see Appendix 1.4 of text).

For 0.10 M HOAc, we follow the same procedure. The final equation is:

$$1.8 \times 10^{-5} = \frac{x^2}{0.10 - x} \approx \frac{x^2}{0.10},\ x = [H^+] = 1.34 \times 10^{-3}\ M \approx 1.3 \times 10^{-3}\ M$$

Assumptions good. pH = 2.89

For 0.40 M HOAc:

$$1.8 \times 10^{-5} = \frac{x^2}{0.40 - x} \approx \frac{x^2}{0.40},\ x = [H^+] = 2.7 \times 10^{-3}\ M$$

Assumptions good. pH = 2.57

38. For propanoic acid, major species are $CH_3CH_2CO_2H$ and H_2O. We will consider propanoic acid as the major source of H^+, i.e., ignore contribution from H_2O. We will use the abbreviation: HOPr for $CH_3CH_2CO_2H$ and OPr^- for $CH_3CH_2CO_2^-$.

	HOPr	$\rightleftharpoons$	H^+	+	OPr^-
Initial	0.050 mol/L		$0(10^{-7})$		0
	x mol/L HOPr dissociates to reach equilibrium				
Change	$-x$	$\rightarrow$	$+x$		$+x$
Equil	$0.050 - x$		x		x

$$K_a = \frac{[H^+][OPr^-]}{[HOPr]} = 1.3 \times 10^{-5} = \frac{x^2}{0.050 - x} \approx \frac{x^2}{0.050} \quad \text{(assuming } x << 0.050\text{)}$$

$x = [H^+] = 8.1 \times 10^{-4}\ M$ Assumptions good (1.6% error).

Note: Assumption is good if % error in assumption is < 5%.

pH = 3.09

For 0.10 M HOPr, we follow the same procedure. The final equation is:

$$1.3 \times 10^{-5} = \frac{x^2}{0.10 - x} \approx \frac{x^2}{0.10},\ x = [H^+] = 1.1 \times 10^{-3}\ M,\ pH = 2.96 \quad \text{Assumptions good.}$$

For 0.40 M HOPr:

$$1.3 \times 10^{-5} = \frac{x^2}{0.40 - x} \approx \frac{x^2}{0.40}, \ x = [H^+] = 2.3 \times 10^{-3}\ M, \text{pH} = 2.64 \ \text{Assumptions good.}$$

39. Major species: $HC_6H_2Cl_3O$ and H_2O. We will consider trichlorophenol as the major source of H^+, i.e., ignore H^+ contribution from H_2O.

	$HC_6H_2Cl_3O$	$\rightleftharpoons$	H^+	+	$C_6H_2Cl_3O^-$
Initial	0.05 M		$0(10^{-7})$		0
	x mol/L trichlorophenol dissociates to reach equilibrium				
Change	$-x$	$\longrightarrow$	$+x$		$+x$
Equil	$0.05 - x$		x		x

$$K_a = 1 \times 10^{-6} = \frac{[H^+][C_6H_2Cl_3O^-]}{[HC_6H_2Cl_3O]} = \frac{x^2}{0.05 - x} \approx \frac{x^2}{0.05} \quad \text{(assuming } x << 0.05\text{)}$$

$x = [H^+] = 2 \times 10^{-4}\ M$ Assumptions good (0.4% error).

pH = 3.7

$[C_6H_2Cl_3O^-] = [H^+] = 2 \times 10^{-4}\ M$, $[OH^-] = 5 \times 10^{-11}\ M$

$[HC_6H_2Cl_3O] = 0.05 - x = 0.05 - 0.0002 = 0.04998 = 0.05\ M$

40. In all four parts of this problem the major species are the weak acid and water. In all we will consider the weak acid to be the major source of H^+, i.e., ignore H^+ contribution from H_2O.

a.

	$HC_2H_3O_2$ + H_2O	$\rightleftharpoons$	H_3O^+	+	$C_3H_3O_2^-$
Initial	0.20 M		$0(10^{-7})$		0
	x mol/L $HC_2H_3O_2$ dissociates to reach equilibrium				
Change	$-x$	$\longrightarrow$	$+x$		$+x$
Equil	$0.20 - x$		x		x

$$K_a = 1.8 \times 10^{-5} = \frac{[H^+][C_2H_3O_2^-]}{[HC_2H_3O_2]} = \frac{x^2}{0.20 - x} \approx \frac{x^2}{0.20}$$

$x = [H^+] = 1.9 \times 10^{-3}\ M$

We have made two assumptions which we must check.

1. $0.20 - x \approx 0.20$

   $0.20 - x = 0.20 - 0.002 = 0.198 = 0.20$ Good assumption (1% error).
   If the percent error in assumption is < 5%, assumption is valid.

2. Acetic acid is the major source of H^+, i.e., we can ignore $10^{-7}\ M$ H^+ already present in neutral H_2O.

   $[H^+]$ from $HC_2H_3O_2 = 2 \times 10^{-3} >> 10^{-7}$ This assumption is valid.

In future problems we will always begin the problem solving process by making these assumptions and we will always check them. However, we may not explicitly state that the assumptions are valid. We will always state when the assumptions are not valid and we have to use other techniques to solve the problem.

Remember, anytime we make an assumption, we must check its validity before the solution to the problem is complete.

$[H^+] = [C_2H_3O_2^-] = 1.9 \times 10^{-3}\ M$, $[OH^-] = 5.3 \times 10^{-12}\ M$

$[HC_2H_3O_2] = 0.20 - x = 0.198 \approx 0.20\ M$

$pH = -\log(1.9 \times 10^{-3}) = 2.72$

b.

	HNO_2	$\rightleftharpoons$	H^+	+	NO_2^-	$K_a = 4.0 \times 10^{-4}$
Initial	1.5 M		$0(10^{-7})$		0	
	x mol/L HNO_2 dissociates to reach equilibrium					
Change	$-x$	$\rightarrow$	$+x$		$+x$	
Equil	$1.5 - x$		x		x	

$$K_a = 4.0 \times 10^{-4} = \frac{[H^+][NO_2^-]}{[HNO_2]} = \frac{x^2}{1.5 - x} \approx \frac{x^2}{1.5}$$

$x = [H^+] = 2.4 \times 10^{-2}\ M$

Assumptions good: $10^{-7} << 2.4 \times 10^{-2} << 1.5$

$[H^+] = [NO_2^-] = 2.4 \times 10^{-2}\ M$, $[OH^-] = 4.2 \times 10^{-13}$

$[HNO_2] = 1.5 - x = 1.48 \approx 1.5\ M$; $pH = 1.62$

c.

	HF	$\rightleftharpoons$	H^+	+	F^-	$K_a = 7.2 \times 10^{-4}$
Initial	0.020 M		$0(10^{-7})$		0	
	x mol/L HF dissociates to reach equilibrium					
Change	$-x$	$\rightarrow$	$+x$		$+x$	
Equil	$0.020 - x$		x		x	

$$K_a = 7.2 \times 10^{-4} = \frac{[H^+][F^-]}{[HF]} = \frac{x^2}{0.020 - x} \approx \frac{x^2}{0.020} \quad \text{(assuming } x << 0.020\text{)}$$

$x = [H^+] = 3.8 \times 10^{-3} >> 10^{-7}$

But $0.020 - x = 0.020 - 0.004 = 0.016$ (x is 20% of 0.020)

The assumption $x << 0.020$ is not good (x is more than 5% of 0.020). We must continue. We can solve the equation by using the quadratic formula or by the method of successive approximations (see Appendix 1.4 of text). We let 0.016 M be a new approximation for [HF]. That is, try $x = 0.004$; so $0.020 - 0.004 = 0.016$, then solve for a new value of x.

$$\frac{x^2}{0.020 - x} \approx \frac{x^2}{0.016} = 7.2 \times 10^{-4},\ x = 3.4 \times 10^{-3}$$

We use this new value of x to further refine our estimate of [HF], i.e. $0.020 - x = 0.020 - 0.0034 = 0.0166$ (carry extra significant figure).

$$\frac{x^2}{0.020 - x} \approx \frac{x^2}{0.0166} = 7.2 \times 10^{-4},\ x = 3.5 \times 10^{-3}$$

We repeat, until we get a self-consistent answer. In this case it will be:

$$x = 3.5 \times 10^{-3}$$

So: $[H^+] = [F^-] = x = 3.5 \times 10^{-3}\ M$; $[OH^-] = 2.9 \times 10^{-12}\ M$

$[HF] = 0.020 - x = 0.020 - 0.0035 = 0.017\ M$; pH = 2.46

d.

$$CH_3CH(OH)CO_2H \rightleftharpoons CH_3CH(OH)CO_2^- + H^+$$

HLac $\qquad$ Lac^-

	HLac	$\rightleftharpoons$	Lac^-	+	H^+	$K_a = 1.4 \times 10^{-4}$
Initial	0.83 M		0		$0(10^{-7})$	
	x mol/L HLac dissociates to reach equilibrium					
Change	$-x$	$\rightarrow$	$+x$		$+x$	
Equil	$0.83 - x$		x		x	

$$K_a = 1.4 \times 10^{-4} = \frac{[H^+][Lac^-]}{[HLac]} = \frac{x^2}{0.83 - x} \approx \frac{x^2}{0.83}$$

$x = [H^+] = 1.1 \times 10^{-2} >> 10^{-7}$

$0.83 - 0.01 = 0.82$ Assumptions good (1.3% error).

So: $[H^+] = [Lac^-] = 1.1 \times 10^{-2}$, $[OH^-] = 9.1 \times 10^{-13}$

$[HLac] = 0.83 - 0.01 = 0.82\ M$; pH = 1.96

41.

	$B(OH)_3$ + H_2O	$\rightleftharpoons$	$B(OH)_4^-$	+	H^+
Initial	0.50 M		0		$0(10^{-7})$
	x mol/L $B(OH)_3$ reacts to reach equilibrium				
Change	$-x$	$\rightarrow$	$+x$		$+x$
Equil	$0.50 - x$		x		x

$$K_a = 5.8 \times 10^{-10} = \frac{[H^+][B(OH)_4^-]}{[B(OH)_3]} = \frac{x^2}{0.50 - x}$$

If $0.50 - x \approx 0.50$, then $5.8 \times 10^{-10} \approx \frac{x^2}{0.50}$

$x = [H^+] = 1.7 \times 10^{-5}\ M$; pH = 4.77 Assumptions good.

42.

	HCO_2H + H_2O	$\rightleftharpoons$	H_3O^+	+ HCO_2^-	$K_a = 1.8 \times 10^{-4}$
Initial	0.025 *M*		$0(10^{-7})$	0	
	x mol/L HCO_2H dissociates to reach equilibrium				
Change	$-x$	$\rightarrow$	$+x$	$+x$	
Equil	$0.025 - x$		x	x	

$K_a = 1.8 \times 10^{-4} = \frac{[H^+][HCO_2^-]}{[HCO_2H]} = \frac{x^2}{0.025 - x} \approx \frac{x^2}{0.025}$, $x = [H^+] = 2.1 \times 10^{-3}$

$0.025 - x = 0.025 - 0.0021 = 0.023$, Assumption $x << 0.025$ is not good ($\approx 8\%$ error).

Solve exactly using method of successive approximations.

$\frac{x^2}{0.025 - x} \approx \frac{x^2}{0.023} = 1.8 \times 10^{-4}$

$x = 2.0 \times 10^{-3}$ which we will get consistently.

$x = [H^+] = 2.0 \times 10^{-3}$; pH = 2.70

43. $0.56\ g\ C_6H_5CO_2H(HBz) \times \frac{1\ mol\ HBz}{122.1\ g} = 4.6 \times 10^{-3}$ mol

Initial concentration of benzoic acid is $\frac{4.6 \times 10^{-3}\ mol}{L}$

	HBz	$\rightleftharpoons$	H^+	+ Bz^-
Initial	$4.6 \times 10^{-3}\ M$		$0(10^{-7})$	0
	x mol/L HBz dissociates to reach equilibrium			
Change	$-x$	$\rightarrow$	$+x$	$+x$
Equil	$4.6 \times 10^{-3} - x$		x	x

$K_a = 6.4 \times 10^{-5} = \frac{[H^+][Bz^-]}{[HBz]} = \frac{x^2}{4.6 \times 10^{-3} - x} \approx \frac{x^2}{4.6 \times 10^{-3}}$

$x = [H^+] = 5.4 \times 10^{-4} >> 10^{-7}$

$0.0046 - x = 0.0046 - 0.00054 = 0.0041$

Assumption is not good (0.00054 is greater than 5% of 0.0046).

When the assumption(s) fail, we must solve exactly using the quadratic formula or the method of successive approximations (see Appendix 1.4 of text).

Using successive approximations:

$$\frac{x^2}{0.0041} = 6.4 \times 10^{-5},\ x = 5.1 \times 10^{-4}$$

$$\frac{x^2}{(0.0046 - 0.00051)} = \frac{x^2}{0.00041} = 6.4 \times 10^{-5},\ x = 5.1 \times 10^{-4}$$

So $x = [H^+] = [Bz^-] = 5.1 \times 10^{-4}\ M$

$[HBz] = 4.6 \times 10^{-3} - x = 4.1 \times 10^{-3}\ M$

pH = 3.29; pOH = 10.71; $[OH^-] = 10^{-10.71} = 1.9 \times 10^{-11}\ M$

44. $20.0\ \text{mL glacial acetic acid} \times \frac{1.05\ \text{g}}{\text{mL}} \times \frac{1\ \text{mol}}{60.05\ \text{g}} = 0.350\ \text{mol}$

Initial concentration of $HC_2H_3O_2$ is $\frac{0.350\ \text{mol}}{0.250\ \text{L}} = 1.40\ M$

A common abbreviation used by chemists for acetic acid, $HC_2H_3O_2$, is HOAc. The abbreviation for $C_2H_3O_2^-$ is OAc^-.

	HOAc	⇌	H^+ +	OAc^-	$K_a = 1.8 \times 10^{-5}$
Initial	1.40 M		$0(10^{-7})$	0	
	x mol/L $HC_2H_3O_2$ dissociates to reach equilibrium				
Change	$-x$	→	$+x$	$+x$	
Equil	$1.40 - x$		x	x	

$$K_a = 1.8 \times 10^{-5} = \frac{[H^+][OAc^-]}{[HOAc]} = \frac{x^2}{1.40 - x} \approx \frac{x^2}{1.40}$$

$x = [H^+] = 5.0 \times 10^{-3}\ M$; pH = 2.30 Assumptions good.

45. Major species: HIO_3, H_2O

Major source of H^+: HIO_3

	HIO_3	⇌	H^+ +	IO_3^-
Initial	0.50 M		$0(10^{-7})$	0
	x mol/L HIO_3 dissociates to reach equilibrium			
Change	$-x$	→	$+x$	$+x$
Equil	$0.50 - x$		x	x

$$K_a = 0.17 = \frac{[H^+][IO_3^-]}{[HIO_3]} = \frac{x^2}{0.50 - x}$$

If $0.50 - x \approx 0.50$, then:

$x = 0.29$ Assumption is bad (60% error). Use 0.29 as an estimate of x and use the method of successive approximations to solve (Appendix 1.4 of text).

$$0.17 = \frac{x^2}{0.50 - x} \approx \frac{x^2}{0.50 - 0.29},\ x = 0.20$$

$$0.17 = \frac{x^2}{0.50 - 0.20},\ x = 0.23;\ 0.17 = \frac{x^2}{0.50 - 0.23},\ x = 0.21$$

$$0.17 = \frac{x^2}{0.50 - 0.21},\ x = 0.22;\ 0.17 = \frac{x^2}{0.50 - 0.22},\ x = 0.22$$

$x = [H^+] = 0.22$ mol/L; pH = -log(0.22) = 0.66

46. For H_2SO_4, the first dissociation occurs to completion. Hydrogen sulfate ion, HSO_4^-, is a weak acid with $K_a = 1.2 \times 10^{-2}$. Major species are H^+, HSO_4^-, and H_2O. We will treat the equilibrium:

	HSO_4^-	$\rightleftharpoons$	H^+	+	SO_4^{2-}
Initial	0.0010 *M*		0.0010 *M*		0
	x mol/L HSO_4^- dissociates to reach equilibrium				
Change	$-x$	→	$+x$		$+x$
Equil	$0.0010 - x$		$0.0010 + x$		x

$$K_a = 0.012 = \frac{(0.0010 + x)(x)}{(0.0010 - x)} \approx x$$

$x = 0.012$ Assumption is not good. We will use the quadratic formula to solve exactly.

$$1.2 \times 10^{-5} - 0.012x = x^2 + 0.0010\,x$$

$$x^2 + 0.013x - 1.2 \times 10^{-5} = 0$$

$$x = \frac{-0.013 \pm (1.69 \times 10^{-4} + 4.80 \times 10^{-5})^{1/2}}{2} = \frac{-0.013 \pm 0.0147}{2}$$

$x = 8.5 \times 10^{-4}$ Note: We will always carry extra significant figures when using the quadratic formula.

$[H^+] = 0.0010 + x = 0.0010 + 0.00085 = 0.0019\ M$; pH = 2.72

47. HCl is a strong acid. It will give 0.10 *M* H^+, 0.10 M Cl^-. Major species are: H^+, Cl^-, HOCl and H_2O.

	HOCl	$\rightleftharpoons$	H^+	+	OCl^-
Initial	0.10 *M*		0.10 *M*		0
	x mol/L HOCl dissociates to reach equilibrium				
Change	$-x$	→	$+x$		$+x$
Equil	$0.10 - x$		$0.10 + x$		x

$$K_a = 3.5 \times 10^{-8} = \frac{[H^+][OCl^-]}{[HOCl]} = \frac{(0.10 + x)(x)}{0.10 - x} \approx x$$

Assumption is good. We are really assuming that HCl is the only important source of H^+. The contribution to $[H^+]$ from the HOCl is negligible. Therefore,

$[H^+] = 0.10\ M$; pH = 1.00

48. HNO_3 is a strong acid, giving an initial concentration of H^+ equal to 0.050 *M*. Consider the equilibrium:

	$HC_2H_3O_2$	$\rightleftharpoons$	H^+	+	$C_2H_3O_2^-$	$K_a = 1.8 \times 10^{-5}$
Initial	0.50 *M*		0.050 *M*		0	
	x mol/L $HC_2H_3O_2$ dissociates to reach equilibrium					
Change	$-x$	$\rightarrow$	$+x$		$+x$	
Equil	$0.50 - x$		$0.050 + x$		x	

$$K_a = 1.8 \times 10^{-5} = \frac{[H^+][OAc^-]}{[HOAc]} = \frac{(0.050 + x)(x)}{(0.50 - x)} \approx \frac{0.050\,(x)}{0.50}$$

$x = 1.8 \times 10^{-4}$; $0.050 + x \approx 0.050$; Assumption is good. $[H^+] = 0.050$ and pH = 1.30.

49.

	$HClO_2$	$\rightleftharpoons$	H^+	+	ClO_2^-	$K_a = 1.2 \times 10^{-2}$
Initial	0.22 *M*		$0(10^{-7})$		0	
	x mol/L $HClO_2$ dissociates to reach equilibrium					
Change	$-x$	$\rightarrow$	$+x$		$+x$	
Equil	$0.22 - x$		x		x	

$$K_a = 1.2 \times 10^{-2} = \frac{[H^+][ClO^-]}{[HClO_2]} = \frac{x^2}{(0.22 - x)} \approx \frac{x^2}{0.22}$$

$x = 5.1 \times 10^{-2}$

The assumption that *x* is small is not good (23% error). Using the method of successive approximations (see Appendix 1.4 of text):

$$\frac{x^2}{0.169} = 1.2 \times 10^{-2},\ x = 4.5 \times 10^{-2}$$

$$\frac{x^2}{0.175} = 1.2 \times 10^{-2},\ x = 4.6 \times 10^{-2}$$ We can stop here (4.6×10^{-2} will repeat).

$[H^+] = [ClO_2^-] = x = 4.6 \times 10^{-2}\ M$; % ionized $= \dfrac{4.6 \times 10^{-2}}{0.22} \times 100 = 21\%$

50. $HF \rightleftharpoons H^+ + F^-$

$[H^+]_e = [F^-]_e = 0.081\ [HF]_o = (0.081)(0.100) = 8.1 \times 10^{-3}\ M$

$[HF]_e = 0.100 - 8.1 \times 10^{-3} = 0.092\ M$

$$K_a = \frac{[H^+]_e\,[F^-]_e}{[HF]_e} = \frac{(8.1 \times 10^{-3})^2}{(9.2 \times 10^{-2})} = 7.1 \times 10^{-4}$$

51. a.

	HOAc	$\rightleftharpoons$	H^+	+	OAc^-
Initial	0.50 M		$0(10^{-7})$		0
	x mol/L HOAc ($HC_2H_3O_2$) dissociates to reach equilibrium				
Change	$-x$	$\rightarrow$	$+x$		$+x$
Equil	$0.50 - x$		x		x

$$K_a = 1.8 \times 10^{-5} = \frac{[H^+][OAc^-]}{[HOAc]} = \frac{x^2}{(0.50 - x)} \approx \frac{x^2}{0.50}$$

$x = [H^+] = [OAc^-] = 3.0 \times 10^{-3}\ M$ Assumptions good.

$$\%\ \text{ionized} = \frac{[OAc^-]}{0.50\ M} \times 100 = 0.60\%$$

b. The set-up for (b) and (c) is similar to (a) except the final equation is slightly different.

$$K_a = 1.8 \times 10^{-5} = \frac{x^2}{(0.050 - x)} \approx \frac{x^2}{0.050}$$

$x = [OAc^-] = 9.5 \times 10^{-4}$ Assumptions good.

$$\%\ \text{ionized} = \frac{9.5 \times 10^{-4}}{0.050} \times 100 = 1.9\%$$

c. $$K_a = 1.8 \times 10^{-5} = \frac{x^2}{(0.0050 - x)} \approx \frac{x^2}{0.0050}$$

$x = [OAc^-] = 3.0 \times 10^{-4}$

$0.0050 - x = 0.0047$ Assumption that x is negligible is borderline (6% error).

Using successive approximations:

$1.8 \times 10^{-5} = \dfrac{x^2}{0.0047}$, $x = 2.9 \times 10^{-4}$, next trial gives 2.9×10^{-4}

$$\%\ \text{ionized} = \frac{2.9 \times 10^{-4}}{5.0 \times 10^{-3}} \times 100 = 5.8\%$$

Note: As the solution is more dilute, the percent ionization increases. This is what we would predict using Le Chatelier's Principle.

52. a.

	HOCl	$\rightleftharpoons$	H^+	+	OCl^-	$K_a = 3.5 \times 10^{-8}$
Initial	0.100 M		$0(10^{-7})$		0	
	x mol/L HOCl dissociates to reach equilibrium					
Change	$-x$	$\rightarrow$	$+x$		$+x$	
Equil	$0.100 - x$		x		x	

$$K_a = 3.5 \times 10^{-8} = \frac{[H^+][OCl^-]}{[HOCl]} = \frac{x^2}{0.100 - x} \approx \frac{x^2}{0.100}$$

$x = [OCl^-] = 5.9 \times 10^{-5}\ M$; % ionized $= \frac{5.9 \times 10^{-5}}{0.100} \times 100 = 0.059\%$ Assumptions good.

b.

	HCN	$\rightleftharpoons$	H^+	+	CN^-	$K_a = 6.2 \times 10^{-10}$
Initial	0.100 M		$0 (10^{-7})$		0	
	x mol/L HCN dissociates to reach equilibrium					
Change	$-x$	$\rightarrow$	$+x$		$+x$	
Equil	$0.100 - x$		x		x	

$$K_a = 6.2 \times 10^{-10} = \frac{[H^+][CN^-]}{[HCN]} = \frac{x^2}{(0.100 - x)} \approx \frac{x^2}{0.100}$$

$x = [CN^-] = 7.9 \times 10^{-6}\ M$ Assumptions good.

$$\% \text{ ionized} = \frac{7.9 \times 10^{-6}}{0.100} \times 100 = 7.9 \times 10^{-3}\ \%$$

c. HCl is a strong acid; it is 100% ionized in solution.

Note: For the same initial concentration, the percent ionization increases as the strength of the acid increases (as K_a increases).

53. $HOBr \rightleftharpoons H^+ + OBr^-$

From normal setup: $[H^+] = [OBr^-] = x$

$[HOBr] = 0.063 - x = 0.063 - [H^+]$

$pH = 4.95$, $[H^+] = 10^{-4.95} = 1.1 \times 10^{-5}$

$$K_a = \frac{[H^+][OBr^-]}{[HOBr]} = \frac{(1.1 \times 10^{-5})^2}{(0.063 - 1.1 \times 10^{-5})} = 1.9 \times 10^{-9}$$

54. $CCl_3CO_2H \rightleftharpoons H^+ + CCl_3CO_2^-$

From normal setup:

$[H^+] = [CCl_3CO_2^-]$ and $[CCl_3CO_2H] = 0.050 - [H^+]$

$pH = 1.4$, $[H^+] = 10^{-1.4} = 4 \times 10^{-2}$

$$K_a = \frac{[H^+][CCl_3CO_2^-]}{[CCl_3CO_2H]} = \frac{(4 \times 10^{-2})^2}{0.050 - 0.04}$$

$K_a = 1.6 \times 10^{-1} = 2 \times 10^{-1}$ (pH has only 1 significant figure)

55. a.

	H_3AsO_4	$\rightleftharpoons$	H^+	+	$H_2AsO_4^-$	$K_{a_1} = 5 \times 10^{-3}$
Initial	0.10 *M*		$0(10^{-7})$		0	
	x mol/L H_3AsO_4 dissociates to reach equilibrium					
Change	$-x$	$\rightarrow$	$+x$		$+x$	
Equil	$0.10 - x$		x		x	

$$K_{a_1} = 5 \times 10^{-3} = \frac{[H^+][H_2AsO_4^-]}{[H_3AsO_4]} = \frac{x^2}{(0.10 - x)} \approx \frac{x^2}{0.10}$$

$x = 2 \times 10^{-2}$, Assumption bad (x is 20% of 0.10). Using successive approximations:

$$\frac{x^2}{0.08} = 5 \times 10^{-3},\ x = 2 \times 10^{-2}$$

$x = [H^+] = 2 \times 10^{-2}\ M$; pH = 1.7

b.

	H_2CO_3	$\rightleftharpoons$	H^+	+	HCO_3^-	$K_{a_1} = 4.3 \times 10^{-7}$
Initial	0.10 *M*		$0(10^{-7})$		0	
	x mol/L H_2CO_3 dissociates					
Change	$-x$	$\rightarrow$	$+x$		$+x$	
Equil	$0.10 - x$		x		x	

$$K_a = 4.3 \times 10^{-7} = \frac{[H^+][HCO_3^-]}{[H_2CO_3]} = \frac{x^2}{(0.10 - x)} \approx \frac{x^2}{0.10}$$

$x = [H^+] = 2.1 \times 10^{-4}$ Assumptions good. pH = 3.68

56. The reactions are:

$H_3PO_4 \rightleftharpoons H^+ + H_2PO_4^-$ $\quad K_{a_1} = 7.5 \times 10^{-3}$

$H_2PO_4^- \rightleftharpoons H^+ + HPO_4^{2-}$ $\quad K_{a_2} = 6.2 \times 10^{-8}$

$HPO_4^{2-} \rightleftharpoons H^+ + PO_4^{3-}$ $\quad K_{a_3} = 4.8 \times 10^{-13}$

We will deal with the reactions in order of importance, beginning with the largest K_a.

$$K_{a_1} = 7.5 \times 10^{-3} = \frac{[H^+][H_2PO_4^-]}{[H_3PO_4]}$$

	H_3PO_4	$\rightleftharpoons$	H^+	+	$H_2PO_4^-$
Initial	0.100 *M*		$0(10^{-7})$		0
	x mol/L H_3PO_4 dissociates to reach equilibrium				
Change	$-x$	$\rightarrow$	$+x$		$+x$
Equil	$0.100 - x$		x		x

$7.5 \times 10^{-3} = \frac{x^2}{0.100 - x}$, $x = 2.4 \times 10^{-2}\ M$ (by using successive approx. or quadratic formula)

$[H^+] = [H_2PO_4^-] = 2.4 \times 10^{-2}\ M$

$[H_3PO_4] = 0.100 - 0.024 = 0.076\ M$

For $K_{a_2} = \frac{[H^+][HPO_4^{2-}]}{[H_2PO_4^-]}$, we use the previously calculated concentration of H^+ and $H_2PO_4^-$ to calculate the concentration of HPO_4^{2-}.

$6.2 \times 10^{-8} = \frac{(2.4 \times 10^{-2})[HPO_4^{2-}]}{(2.4 \times 10^{-2})}$; $[HPO_4^{2-}] = 6.2 \times 10^{-8}\ M$

We repeat the process using K_{a_3} to get $[PO_4^{3-}]$.

$K_{a_3} = 4.8 \times 10^{-13} = \frac{[H^+][PO_4^{3-}]}{[HPO_4^{2-}]} = \frac{(2.4 \times 10^{-2})[PO_4^{3-}]}{(6.2 \times 10^{-8})}$

$[PO_4^{3-}] = 1.2 \times 10^{-18}\ M$

So in 0.100 M analytical concentration of H_3PO_4:

$[H_3PO_4] = 7.6 \times 10^{-2}\ M$; $[H^+] = [H_2PO_4^-] = 2.4 \times 10^{-2}\ M$

$[HPO_4^{2-}] = 6.2 \times 10^{-8}\ M$; $[PO_4^{3-}] = 1.2 \times 10^{-18}\ M$

$[OH^-] = 1.0 \times 10^{-14}/[H^+] = 4.2 \times 10^{-13}$

Solutions of Bases

57. NO_3^-: $K_b \approx 0$ since HNO_3 is a strong acid. H_2O: $K_b = 10^{-14}$

NH_3: $K_b = 1.8 \times 10^{-5}$; CH_3NH_2: $K_b = 4.38 \times 10^{-4}$

$CH_3NH_2 > NH_3 > H_2O >> NO_3^-$

58. Excluding water these are the conjugate acids of the bases in the previous exercise: the stronger the base, the weaker the conjugate acid.

$HNO_3 > NH_4^+ > CH_3NH_3^+ > H_2O$

59. a. NH_3 b. NH_3 c. OH^- d. CH_3NH_2

60. a. HNO_3 b. NH_4^+ c. NH_4^+

61. $2.48 \text{ g TlOH} \times \frac{1 \text{ mol TlOH}}{221.4 \text{ g}} = 1.12 \times 10^{-2} \text{ mol}$

TlOH is a strong base, so $[OH^-] = \frac{1.12 \times 10^{-2} \text{ mol}}{L}$

pOH = 1.951; pH = 12.049

62. a. $[OH^-] = 0.25$ *M*; pOH = 0.60; pH = 13.40

b. $Ba(OH)_2 \rightarrow Ba^{2+} + 2\ OH^-$

$[OH^-] = 2(0.00040) = 8.0 \times 10^{-4}$; pOH = 3.10; pH = 10.90

c. $\frac{25 \text{ g KOH}}{L} \times \frac{1 \text{ mol}}{56.11 \text{ g}} = 0.45 \text{ mol/L}$

KOH is a strong base, so $[OH^-] = 0.45$ *M*; pOH = 0.35; pH = 13.65

d. $\frac{150.0 \text{ g NaOH}}{L} \times \frac{1 \text{ mol}}{40.00 \text{ g}} = 3.750\ M$

NaOH is a strong base. $[OH^-] = 3.750$ *M*

pOH = -0.5740 and pH = 14.5740

Although we are justified in calculating the answer to four decimal places, in reality the pH can only be measured to ± 0.01 pH units.

63. a. Major species: K^+, OH^-, H_2O

$[OH^-] = 0.150$, pOH = -log(0.150) = 0.824

pH = 14.000 - pOH = 13.176

b. Major species: Cs^+, OH^-, H_2O

$[OH^-] = 0.150$, pOH = 0.824, pH = 13.176

c. Major species: NH_3, H_2O

We will consider NH_3 as the major source of OH^-, i.e. ignore OH^- contribution from H_2O.

	$NH_3 + H_2O$	$\rightleftharpoons$	NH_4^+	+	OH^-
Initial	0.150 *M*		0		$0(10^{-7})$
	x mol/L NH_3 reacts with H_2O to reach equilibrium				
Change	$-x$	$\rightarrow$	$+x$		$+x$
Equil	$0.150 - x$		x		x

$$K_b = \frac{[NH_4^+][OH^-]}{[NH_3]} = 1.8 \times 10^{-5} = \frac{x^2}{0.150 - x} \approx \frac{x^2}{0.150}$$

$x = 1.6 \times 10^{-3}$; Assumptions good ($1 \times 10^{-7} << x << 0.150$).

$x = [OH^-] = 1.6 \times 10^{-3}$, pOH = 2.80, pH = 11.20

d. Major species: pyridine (C_5H_5N), H_2O

We will consider pyridine as the major source of OH^-.

	$C_5H_5N + H_2O$	$\rightleftharpoons$	$C_5H_5NH^+$	+	OH^-
Initial	0.150 *M*		0		$0(10^{-7})$
	x mol/L pyridine reacts with water				
Change	$-x$	$\rightarrow$	$+x$		$+x$
Equil	$0.150 - x$		x		x

$$K_b = \frac{[C_5H_5NH^+][OH^-]}{[C_5H_5N]} = 1.7 \times 10^{-9} = \frac{x^2}{0.150 - x} \approx \frac{x^2}{0.150}$$

$x = 1.6 \times 10^{-5}$; Assumptions good by 5% rule.

$x = [OH^-] = 1.6 \times 10^{-5}$, pOH = 4.80; pH = 9.20

e. Major species: CH_3NH_2, H_2O

The major source of OH^- is CH_3NH_2.

	$CH_3NH_2 + H_2O$	$\rightleftharpoons$	$CH_3NH_3^+$	+	OH^-
Initial	0.150 *M*		0		$0(10^{-7})$
	x mol/L CH_3NH_2 reacts with water				
Change	$-x$	$\rightarrow$	$+x$		$+x$
Equil	$0.150 - x$		x		x

$$K_b = 4.38 \times 10^{-4} = \frac{[CH_3NH_3^+][OH^-]}{[CH_3NH_2]} = \frac{x^2}{0.150 - x} \approx \frac{x^2}{0.150}$$

$x = 8.11 \times 10^{-3}$; Assumption is borderline (5.4% error).

Using successive approximations:

$$4.38 \times 10^{-4} = \frac{x^2}{0.150 - 0.00811},\ x = 7.88 \times 10^{-3}$$

$$4.38 \times 10^{-4} = \frac{x^2}{0.150 - 0.00788},\ x = 7.89 \times 10^{-3} \text{ (consistent answer)}$$

$x = [OH^-] = 7.89 \times 10^{-3}$, pOH = 2.103

pH = 11.897 (We could only measure 11.90 ± 0.01 at best.)

64. a. Major species: Na^+, K^+, OH^-, H_2O

$[OH^-] = 0.050 + 0.050 = 0.100\ M$

pH = 13.000

b. Major species: Na^+, OH^-, NH_3, H_2O

Consider the equilibrium:

	NH_3 + H_2O	⇌	NH_4^+ +	OH^-
Initial	0.50 M		0	0.050 M
	x mol/L NH_3 reacts with H_2O to reach equilibrium			
Change	$-x$	→	$+x$	$+x$
Equil	$0.50 - x$		x	$0.050 + x$

$$K_b = 1.8 \times 10^{-5} = \frac{[NH_4^+][OH^-]}{[NH_3]} = \frac{(x)(0.050 + x)}{(0.50 - x)} \approx \frac{x\,(0.050)}{(0.50)}$$

$x = 1.8 \times 10^{-4}$; Our assumption that x is small is good (0.36% error).

Thus, $[OH^-] = 0.050$, pH = 12.70

c. Major species: Ba^{2+}, OH^-, Na^+, H_2O

$[OH^-] = 2(0.0010) + 0.020 = 0.022\ M$

pH = 12.34

65.

	H_2NNH_2 + H_2O	⇌	$H_2NNH_3^+$ +	OH^-	$K_b = 3.0 \times 10^{-6}$
Initial	2.0 M		0	$0(10^{-7})$	
	x mol/L H_2NNH_2 reacts with H_2O to reach equilibrium				
Change	$-x$	→	$+x$	$+x$	
Equil	$2.0 - x$		x	x	

$$K_b = 3.0 \times 10^{-6} = \frac{[H_2NNH_3^+][OH^-]}{[H_2NNH_2]} = \frac{x^2}{2.0 - x} \approx \frac{x^2}{2.0}$$

$x = [OH^-] = 2.4 \times 10^{-3}\ M$; pOH = 2.62; pH = 11.38 Assumptions good by 5% rule.

$[H_2NNH_3^+] = [OH^-] = 2.4 \times 10^{-3}\ M$, $[H_2NNH_2] = 2.0 - x = 2.0\ M$

$[H^+] = 10^{-11.38} = 4.2 \times 10^{-12}$

66. $HONH_2 + H_2O \rightleftharpoons HONH_3^+ + OH^-$ $K_b = 1.1 \times 10^{-8}$

Initial: $HONH_2$ = 1.0 mol/L; $HONH_3^+$ = 0; OH^- = $0(10^{-7})$

x mol/L hydroxylamine reacts with H_2O to reach equilibrium

Change: $-x$ → $+x$, $+x$

Equil: $1.0 - x$, x, x

$$K_b = 1.1 \times 10^{-8} = \frac{[HONH_3^+][OH^-]}{[HONH_2]} = \frac{x^2}{1.0 - x} \approx \frac{x^2}{1.0}$$

$x = [OH^-] = 1.049 \times 10^{-4} \approx 1.0 \times 10^{-4}$ *M*; pOH = 4.00; pH = 10.00 Assumptions good.

$[HONH_3^+] = [OH^-] = 1.0 \times 10^{-4}$ *M*

$[HONH_2] = 1.0 - x \approx 1.0$ *M*, $[H^+] = 1.0 \times 10^{-10}$

67. a. $C_2H_5NH_2 + H_2O \rightleftharpoons C_2H_5NH_3^+ + OH^-$ $K_b = 5.6 \times 10^{-4}$

Initial: $C_2H_5NH_2$ = 0.20 *M*; $C_2H_5NH_3^+$ = 0; OH^- = $0(10^{-7})$

x mol/L $C_2H_5NH_2$ reacts with H_2O to reach equilibrium

Change: $-x$ → $+x$, $+x$

Equil: $0.20 - x$, x, x

$$K_b = \frac{[C_2H_5NH_3^+][OH^-]}{[C_2H_5NH_2]} = \frac{x^2}{0.20 - x} \approx \frac{x^2}{0.20}$$

$x = 1.1 \times 10^{-2}$; 0.200 - 0.011 = 0.189 Assumption borderline (~ 5% error).

Using successive approximations:

$$\frac{x^2}{0.189} = 5.6 \times 10^{-4},\ x = 1.0 \times 10^{-2}\ M \quad \text{(consistent answer)}$$

$x = [OH^-] = 1.0 \times 10^{-2}$ *M*

$$[H^+] = \frac{K_w}{[OH^-]} = \frac{1.0 \times 10^{-14}}{1.0 \times 10^{-2}} = 1.0 \times 10^{-12}\ M;\ pH = 12.00$$

b. $Et_2NH + H_2O \rightleftharpoons Et_2NH_2^+ + OH^-$ Et = $-C_2H_5$ $K_b = 1.3 \times 10^{-3}$

Initial: Et_2NH = 0.20 *M*; $Et_2NH_2^+$ = 0; OH^- = $0(10^{-7})$

x mol/L $(C_2H_5)_2NH$ reacts with H_2O

Change: $-x$ → $+x$, $+x$

Equil: $0.20 - x$, x, x

$$K_b = 1.3 \times 10^{-3} = \frac{[Et_2NH_2^+][OH^-]}{[Et_2NH]} = \frac{x^2}{0.20 - x} \approx \frac{x^2}{0.20}$$

$x = 1.6 \times 10^{-2}$ Assumption bad (x is 8% of 0.20).

Using successive approximations:

$\frac{x^2}{0.184} = 1.3 \times 10^{-3}$; $x = 1.55 \times 10^{-2}$ (carry extra significant figure)

$\frac{x^2}{0.185} = 1.3 \times 10^{-3}$; $x = 1.55 \times 10^{-2}$

$[OH^-] = x = 1.55 \times 10^{-2}\ M = 1.6 \times 10^{-2}\ M$; $[H^+] = 6.45 \times 10^{-13}\ M = 6.5 \times 10^{-13}\ M$; pH = 12.19

c. $Et_3N + H_2O \rightleftharpoons Et_3NH^+ + OH^-$ $K_b = 4.0 \times 10^{-4}$

Equil $0.20 - x$ x x

$4.0 \times 10^{-4} = \frac{x^2}{0.20 - x} \approx \frac{x^2}{0.20}$; $x = 8.9 \times 10^{-3}$; Assumptions good

$[OH^-] = 8.9 \times 10^{-3}\ M$; $[H^+] = 1.1 \times 10^{-12}\ M$; pH = 11.96

68. a. $C_6H_5NH_2 + H_2O \rightleftharpoons C_6H_5NH_3^+ + OH^-$ $K_b = 3.8 \times 10^{-10}$

Equil $0.20 - x$ x x

$3.8 \times 10^{-10} = \frac{x^2}{0.20 - x} \approx \frac{x^2}{0.20}$

$x = [OH^-] = 8.7 \times 10^{-6}\ M$ Assumptions good.

$[H^+] = 1.1 \times 10^{-9}\ M$; pH = 8.96

b. $C_5H_5N + H_2O \rightleftharpoons C_5H_5NH^+ + OH^-$ $K_b = 1.7 \times 10^{-9}$

Equil $0.20 - x$ x x

$K_b = 1.7 \times 10^{-9} = \frac{x^2}{0.20 - x} \approx \frac{x^2}{0.20}$; $x = 1.8 \times 10^{-5}$ Assumptions good.

$[OH^-] = 1.8 \times 10^{-5}\ M$; $[H^+] = 5.6 \times 10^{-10}\ M$; pH = 9.25

c. $HONH_2 + H_2O \rightleftharpoons HONH_3^+ + OH^-$ $K_b = 1.1 \times 10^{-8}$

Equil $0.20 - x$ x x

$K_b = 1.1 \times 10^{-8} = \frac{x^2}{0.20 - x} \approx \frac{x^2}{0.20}$

$x = [OH^-] = 4.7 \times 10^{-5}\ M$ Assumptions good.

$[H^+] = 2.1 \times 10^{-10}\ M$; pH = 9.68

69. a. $NH_3 + H_2O \rightleftharpoons NH_4^+ + OH^-$ $K_b = 1.8 \times 10^{-5}$

Equil $0.10 - x$ x x

$$1.8 \times 10^{-5} = \frac{x^2}{0.10 - x} \approx \frac{x^2}{0.10}$$

$x = [NH_4^+] = 1.3 \times 10^{-3}$; % ionized = $\frac{1.3 \times 10^{-3}}{0.10} \times 100 = 1.3\%$ Assumptions good.

b. $NH_3 + H_2O \rightleftharpoons NH_4^+ + OH^-$

Equil $0.010 - x$ x x

$$1.8 \times 10^{-5} = \frac{x^2}{0.010 - x} \approx \frac{x^2}{0.010}$$

$x = [NH_4^+] = 4.2 \times 10^{-4}$ Assumptions good.

$$\% \text{ ionized} = \frac{4.2 \times 10^{-4}}{0.010} \times 100 = 4.2\%$$

Note: For the same base, the percent ionized increases as the initial concentration decreases.

70. a. $HONH_2 + H_2O \rightleftharpoons HONH_3^+ + OH^-$ $K_b = 1.1 \times 10^{-8}$

Initial $0.10\ M$ 0 $0(10^{-7})$

x mol/L $HONH_2$ reacts with water to reach equilibrium

Change $-x$ → $+x$ $+x$

Equil $0.10 - x$ x x

$1.1 \times 10^{-8} = \frac{x^2}{0.10 - x} \approx \frac{x^2}{0.10}$; $x = [OH^-] = 3.3 \times 10^{-5}\ M$ Assumptions good.

$$\% \text{ ionized} = \frac{3.3 \times 10^{-5}}{0.10} \times 100 = 0.033\% \text{ ionized}$$

b. $CH_3NH_2 + H_2O \rightleftharpoons CH_3NH_3^+ + OH^-$ $K_b = 4.38 \times 10^{-4}$

Equil $0.10 - x$ x x

$4.38 \times 10^{-4} = \frac{x^2}{0.10 - x} \approx \frac{x^2}{0.10}$; $x = 6.6 \times 10^{-3}$; Assumption fails the 5% rule.

Using successive approximations:

$$\frac{x^2}{0.093} = 4.38 \times 10^{-4};\ x = 6.4 \times 10^{-3}$$

$$\frac{x^2}{0.094} = 4.38 \times 10^{-4};\ x = 6.4 \times 10^{-3}\ M = [CH_3NH_3^+]$$

$$\% \text{ ionized} = \frac{6.4 \times 10^{-3}}{0.10} \times 100 = 6.4\%$$

Note: As K_b increases, the percent ionization increases for the same concentration.

71. $$\frac{5.0 \text{ mg}}{10.0 \text{ mL}} \times \frac{1 \text{ mmol}}{299.4 \text{ mg}} = 1.7 \times 10^{-3} \frac{\text{mmol}}{\text{mL}} = 1.7 \times 10^{-3}\ M$$

	Cod	+	H_2O	$\rightleftharpoons$	$CodH^+$	+	OH^-	$K_b = 10^{-6.05} = 8.9 \times 10^{-7}$
Initial	$1.7 \times 10^{-3}\ M$				0		$0 (10^{-7})$	
	x mol/L codeine reacts with H_2O to reach equilibrium							
Change	$-x$			$\rightarrow$	$+x$		$+x$	
Equil	$1.7 \times 10^{-3} - x$				x		x	

$$8.9 \times 10^{-7} = \frac{x^2}{1.7 \times 10^{-3} - x} \approx \frac{x^2}{1.7 \times 10^{-3}};\ x = 3.9 \times 10^{-5}$$ Assumptions good.

$[OH^-] = 3.9 \times 10^{-5}$; $[H^+] = 2.6 \times 10^{-10}$; pH = 9.59

72. $$\frac{1.0 \text{ g}}{1.9 \text{ L}} \times \frac{1 \text{ mol quinine}}{324.4 \text{ g}} = 1.6 \times 10^{-3}\ M \text{ in quinine}$$

	Q	+	H_2O	$\rightleftharpoons$	QH^+	+	OH^-	$K_{b_1} = 10^{-5.1} = 8 \times 10^{-6}$
Initial	$1.6 \times 10^{-3}\ M$				0		$0 (10^{-7})$	
	x mol/L quinine reacts with H_2O to reach equilibrium							
Change	$-x$			$\rightarrow$	$+x$		$+x$	
Equil	$1.6 \times 10^{-3} - x$				x		x	

$$K_b = 8 \times 10^{-6} = \frac{[QH^+][OH^-]}{[Q]} = \frac{x^2}{1.6 \times 10^{-3} - x} \approx \frac{x^2}{1.6 \times 10^{-3}}$$

$x = 1 \times 10^{-4}$ Assumption fails the 5% rule (~7% error).

Using successive approximations:

$$\frac{x^2}{1.5 \times 10^{-3}} = 8 \times 10^{-6};\ x = 1 \times 10^{-4}$$

$x = [OH^-] = 1 \times 10^{-4}\ M$ (only 1 significant figure allowed by pK value)

pOH = 4.0; pH = 10.0

73. PT = p-toluidine

$$PT + H_2O \rightleftharpoons PTH^+ + OH^-;\ K_b = \frac{[PTH^+][OH^-]}{[PT]}$$

From normal setup:

$x = [PTH^+] = [OH^-]$; $[PT] = 0.016 - [OH^-]$

$pH = 8.60$; $pOH = 5.40$; $[OH^-] = 4.0 \times 10^{-6}$

$$K_b = \frac{(4.0 \times 10^{-6})^2}{(0.016 - 4.0 \times 10^{-6})} = 1.0 \times 10^{-9}$$

74. pyr = pyrrolidine, C_4H_8NH

$$pyr + H_2O \rightleftharpoons pyrH^+ + OH^-;\ K_b = \frac{[OH^-][pyrH^+]}{[pyr]}$$

From normal setup:

$x = [OH^-] = [pyrH^+]$; $[pyr] = 1.00 \times 10^{-3} - [OH^-]$

$pH = 10.82$; $pOH = 3.18$; $[OH^-] = 10^{-3.18} = 6.6 \times 10^{-4}$

$$K_b = \frac{(6.6 \times 10^{-4})^2}{(1.00 \times 10^{-3} - 6.6 \times 10^{-4})} = 1.3 \times 10^{-3}$$

Acid/Base Properties of Salts

75. a. $NaNO_3 \rightarrow Na^+ + NO_3^-$ neutral (Na^+ and NO_3^- have no acidic/basic properties)

b. $NaNO_2 \rightarrow Na^+ + NO_2^-$ basic

NO_2^- is a weak base; it is the conjugate base of the weak acid, HNO_2.

$$NO_2^- + H_2O \rightleftharpoons HNO_2 + OH^-$$

c. $NH_4NO_3 \rightarrow NH_4^+ + NO_3^-$ acidic

NH_4^+, weak acid: $NH_4^+ \rightleftharpoons H^+ + NH_3$

d. $NH_4NO_2 \rightarrow NH_4^+ + NO_3^-$

NH_4^+ is a weak acid and NO_2^- is a weak base.

$$NH_4^+ \rightleftharpoons NH_3 + H^+ \quad K_a = \frac{K_w}{K_b} = \frac{1.0 \times 10^{-14}}{1.8 \times 10^{-5}} = 5.6 \times 10^{-10}$$

$$NO_2^- + H_2O \rightleftharpoons HNO_2 + OH^- \quad K_b = \frac{K_w}{K_a} = \frac{1.0 \times 10^{-14}}{4.0 \times 10^{-4}} = 2.5 \times 10^{-11}$$

NH_4^+ is stronger as an acid in water than NO_2^- is as a base, $K_a(NH_4^+) > K_b(NO_2^-)$. Therefore, the solution is acidic.

e. $Na_2CO_3 \longrightarrow 2\,Na^+ + CO_3^{2-}$, basic

Na^+ has no effect. CO_3^{2-} is a weak base. $CO_3^{2-} + H_2O \rightleftharpoons HCO_3^- + OH^-$

f. $NaF \longrightarrow Na^+ + F^-$; F^- weak base; $F^- + H_2O \rightleftharpoons HF + OH^-$ solution is basic.

76. a. $KCl \longrightarrow K^+ + Cl^-$ Neither K^+ nor Cl^- will change the pH of water. neutral

b. $NH_4C_2H_3O_2 \rightleftharpoons \underset{\text{weak acid}}{NH_4^+} + \underset{\text{weak base}}{C_2H_3O_2^-}$

$$NH_4^+ \rightleftharpoons NH_3 + H^+ \qquad K_a = \frac{K_w}{K_b} = \frac{1.0 \times 10^{-14}}{1.8 \times 10^{-5}} = 5.6 \times 10^{-10}$$

$$OAc^- + H_2O \rightleftharpoons HOAc + OH^- \qquad K_b = \frac{K_w}{K_a} = \frac{1.0 \times 10^{-14}}{1.8 \times 10^{-5}} = 5.6 \times 10^{-10}$$

The acid and base are equal in strength; the solution is neutral.

c. $CH_3NH_3Br \longrightarrow CH_3NH_3^+ + Br^-$, acidic

$CH_3NH_3^+$ is conjugate acid of CH_3NH_2. $CH_3NH_3^+ \rightleftharpoons H^+ + CH_3NH_3$

Br^- has no effect (conjugate base of a strong acid).

d. $NaHCO_3 \longrightarrow Na^+ + HCO_3^-$

HCO_3^- can be either an acid or a base.

$$HCO_3^- \rightleftharpoons H^+ + CO_3^{2-} \qquad K_{a_2} = 5.6 \times 10^{-11}$$

$$HCO_3^- + H_2O \longrightarrow H_2CO_3 + OH^- \qquad K_b = \frac{K_w}{K_{a_1}} = 2.3 \times 10^{-8}$$

HCO_3^- is a stronger base than an acid; the solution is basic.

e. $NH_4F \longrightarrow NH_4^+ + F^-$

NH_4^+, weak acid, $K_a = K_w/K_b\,(NH_3) = 5.6 \times 10^{-10}$, $NH_4^+ \rightleftharpoons H^+ + NH_3$

F^-, weak base, $K_b = K_w/K_a(HF) = 1.4 \times 10^{-11}$, $F^- + H_2O \rightleftharpoons HF + OH^-$

NH_4^+ is stronger as an acid than F^- is as a base. Solution is acidic.

77. KOH: strong base; KBr: neutral

KCN: CN^- is weak base, $K_b = 1.0 \times 10^{-14}/6.2 \times 10^{-10} = 1.6 \times 10^{-5}$

NH_4Br: NH_4^+ is weak acid, $K_a = 5.6 \times 10^{-10}$

NH_4CN: slightly basic, CN^- is a stronger base compared to NH_4^+ as an acid.

HCN: weak acid, $K_a = 6.2 \times 10^{-10}$

acidic → basic: HCN, NH_4Br, KBr, NH_4CN, KCN, KOH

78. HNO_2, $K_a = 4.0 \times 10^{-4}$; NO_2^-, $K_b = 2.5 \times 10^{-11}$

NH_4^+, $K_a = 10^{-14}/K_b = 5.6 \times 10^{-10}$; HNO_3: Strong acid

Most acidic → most basic: $HNO_3 > HNO_2 > NH_4NO_3 > NH_4NO_2 > H_2O > KNO_2$

79. a. $CH_3NH_3Cl \rightarrow CH_3NH_3^+ + Cl^-$ Methylammonium ion is a weak acid.

$CH_3NH_3^+ \rightleftharpoons CH_3NH_2 + H^+$ Ignore Cl^-, conjugate base of a strong acid.

$$K_a = \frac{[CH_3NH_2][H^+]}{[CH_3NH_3^+]} = \frac{[CH_3NH_2][H^+][OH^-]}{[CH_3NH_3^+][OH^-]} = \frac{K_w}{K_b} = \frac{1.00 \times 10^{-14}}{4.38 \times 10^{-4}} = 2.28 \times 10^{-11}$$

	$CH_3NH_3^+$	$\rightleftharpoons$	CH_3NH_2	+	H^+
Initial	0.10 M		0		$0(10^{-7})$
	x mol/L $CH_3NH_3^+$ dissociates to reach equilibrium				
Change	$-x$	→	$+x$		$+x$
Equil	$0.10 - x$		x		x

$$2.28 \times 10^{-11} = \frac{x^2}{0.10 - x} \approx \frac{x^2}{0.10}$$

$x = [H^+] = 1.5 \times 10^{-6}\ M$; pH = 5.82 Assumptions good.

b. $NaCN \rightarrow Na^+ + CN^-$ Cyanide ion is a weak base. Ignore Na^+.

$CN^- + H_2O \rightleftharpoons HCN + OH^-$ $K_b = \frac{K_w}{K_a} = \frac{1.0 \times 10^{-14}}{6.2 \times 10^{-10}} = 1.6 \times 10^{-5}$

	$CN^- + H_2O$	$\rightleftharpoons$	HCN	OH^-
Initial	0.050 M		0	$0(10^{-7})$
	x mol/L CN^- reacts with H_2O			
Change	$-x$	→	$+x$	$+x$
Equil	$0.050 - x$		x	x

$$K_b = 1.6 \times 10^{-5} = \frac{[HCN][OH^-]}{[CN^-]} = \frac{x^2}{0.050 - x} \approx \frac{x^2}{0.050}$$

$x = [OH^-] = 8.9 \times 10^{-4}$ *M*; pOH = 3.05; pH = 10.95 Assumptions good.

c. $CO_3^{2-} + H_2O \rightarrow HCO_3^- + OH^-$ $K_b = \frac{K_w}{K_a} = \frac{1.0 \times 10^{-14}}{5.6 \times 10^{-11}} = 1.8 \times 10^{-4}$

	CO_3^{2-}		HCO_3^-	OH^-
Initial	0.20 *M*		0	$0(10^{-7})$
	x mol/L CO_3^{2-} reacts with H_2O			
Change	$-x$	→	$+x$	$+x$
Equil	$0.20 - x$		x	x

$$K_b = 1.8 \times 10^{-4} = \frac{[HCO_3^-][OH^-]}{[CO_3^{2-}]} = \frac{x^2}{0.20 - x} \approx \frac{x^2}{0.20}$$

$x = 6.0 \times 10^{-3} = [OH^-]$; pOH = 2.22; pH = 11.78 Assumptions good.

80. a. $NaNO_2 \rightarrow Na^+ + NO_2^-$ NO_2^- is a weak base. Ignore Na^+.

$NO_2^- + H_2O \rightarrow HNO_2 + OH^-$ $K_b = \frac{K_w}{K_a} = \frac{1.0 \times 10^{-14}}{4.0 \times 10^{-4}} = 2.5 \times 10^{-11}$

	NO_2^-		HNO_2	OH^-
Initial	0.12 *M*		0	$0(10^{-7})$
	x mol/L NO_2^- reacts with H_2O to reach equilibrium			
Change	$-x$	→	$+x$	$+x$
Equil	$0.12 - x$		x	x

$$K_b = 2.5 \times 10^{-11} = \frac{[OH^-][HNO_2]}{[NO_2^-]} = \frac{x^2}{0.12 - x} \approx \frac{x^2}{0.12}$$

$x = [OH^-] = 1.7 \times 10^{-6}$; pOH = 5.77; pH = 8.23 Assumptions good.

b. $NaOCl \rightarrow Na^+ + OCl^-$, OCl^- is a weak base. Ignore Na^+.

$OCl^- + H_2O \rightleftharpoons HOCl + OH^-$ $K_b = \frac{K_w}{K_a} = \frac{1.0 \times 10^{-14}}{3.5 \times 10^{-8}} = 2.9 \times 10^{-7}$

	OCl^-		$HOCl$	OH^-
Initial	0.45 *M*		0	$0(10^{-7})$
	x mol/L OCl^- reacts with H_2O			
Change	$-x$	→	$+x$	$+x$
Equil	$0.45 - x$		x	x

$$K_b = 2.9 \times 10^{-7} = \frac{[HOCl][OH^-]}{[OCl^-]} = \frac{x^2}{0.45 - x} \approx \frac{x^2}{0.45}$$

$x = [OH^-] = 3.6 \times 10^{-4}$; pOH = 3.44; pH = 10.56 Assumptions good.

c. $NH_4I \rightarrow NH_4^+ + I^-$ NH_4^+ is a weak acid. Ignore I^-, conjugate base of strong acid.

	$NH_4^+ \rightleftharpoons$	NH_3 +	H^+
Initial	0.40 M	0	$0(10^{-7})$
	x mol/L NH_4^+ dissociates		
Change	$-x$ $\rightarrow$	$+x$	$+x$
Equil	$0.40 - x$	x	x

$K_b = K_w/1.8 \times 10^{-5} = 5.6 \times 10^{-10}$

$$5.6 \times 10^{-10} = \frac{x^2}{0.40 - x} \approx \frac{x^2}{0.40}, \; x = [H^+] = 1.5 \times 10^{-5}$$ Assumptions good.

pH = 4.82

81. $NaN_3 \rightarrow Na^+ + N_3^-$; Azide, N_3^- is a weak base.

	$N_3^- + H_2O \rightleftharpoons$	HN_3 +	OH^-
Initial	0.010 M	0	$0(10^{-7})$
	x mol/L N_3^- reacts with H_2O to reach equilibrium		
Change	$-x$ $\rightarrow$	$+x$	$+x$
Equil	$0.010 - x$	x	x

$$K_b = \frac{K_w}{K_a} = \frac{1.0 \times 10^{-14}}{1.9 \times 10^{-5}} = 5.3 \times 10^{-10}$$

$$K_b = \frac{[HN_3][OH^-]}{[N_3^-]} = 5.3 \times 10^{-10} = \frac{x^2}{0.010 - x} \approx \frac{x^2}{0.010}$$

$$x = [OH^-] = 2.3 \times 10^{-6}; \; [H^+] = \frac{1.0 \times 10^{-14}}{2.3 \times 10^{-6}} = 4.3 \times 10^{-9}$$ Assumptions good.

$[HN_3] = [OH^-] = 2.3 \times 10^{-6}\,M$; $[Na^+] = 0.010\,M$

$[N_3^-] = 0.010 - 2.3 \times 10^{-6} \approx 0.010\,M$

82. $EtNH_3Cl \rightarrow EtNH_3^+ + Cl^-$

$$\left[\begin{array}{c} \quad\quad\quad\quad\quad H \\ \quad\quad\quad\quad\quad | \\ CH_3CH_2 - N - H \\ \quad\quad\quad\quad\quad | \\ \quad\quad\quad\quad\quad H \end{array} \right]^+$$

Major species: $EtNH_3^+$ (weak acid), Cl^- (ignore), H_2O

	$EtNH_3^+ \rightleftharpoons$	$EtNH_2$	+ H^+
Initial	0.25 *M*	0	$0(10^{-7})$
	x mol/L $EtNH_3^+$ dissociates		
Change	$-x$ →	$+x$	$+x$
Equil	$0.25 - x$	x	x

$K_a = K_w/5.6 \times 10^{-4} = 1.8 \times 10^{-11}$

$$K_a = 1.8 \times 10^{-11} = \frac{[EtNH_2][H^+]}{[EtNH_3^+]} = \frac{x^2}{0.25 - x} \approx \frac{x^2}{0.25}$$

$x = [H^+] = 2.1 \times 10^{-6}$ *M*, pH = 5.68 Assumptions good.

$[EtNH_2] = [H^+] = 2.1 \times 10^{-6}$ *M*, $[EtNH_3^+] = 0.25$ *M*, $[Cl^-] = 0.25$ *M*

$[OH^-] = K_w/2.1 \times 10^{-6} = 4.8 \times 10^{-9}$

83. The stronger an acid, the weaker its conjugate base, since $K_aK_b = K_w$. Acetic acid is a stronger acid than hypochlorous acid. It's conjugate base, $C_2H_3O_2^-$ is a weaker base than the conjugate base of HOCl, OCl^-. Thus, the hypochlorite ion, OCl^-, is a stronger base than the acetate ion, $C_2H_3O_2^-$.

84. Methylamine is a stronger base than ammonia. Thus, NH_4^+ is a stronger acid than $CH_3NH_3^+$.

Relationships Between Structure and Strengths of Acids and Bases

85. a. $HBrO < HBrO_2 < HBrO_3$

The greater the number of O-atoms in an oxy-acid, the stronger the acid.

b. $HAsO_4^{2-} < H_2AsO_4^- < H_3AsO_4$

It gets progressively more difficult to remove successive H^+ from a polyprotic acid because the H^+ being removed is attracted to a more negatively charged anion.

86. a. $HClO_2 > HBrO_2 > HIO_2$

b. $H_3PO_4 > H_3AsO_4$

In both cases H is bonded to O. All have the same number of oxygens. The only difference is the electronegativity of the central atom, and the trend is the same as if H was bonded to the central atom.

87. a. $BrO_3^- < BrO_2^- < BrO^-$

Acids are stronger as more oxygens are present. Thus, the conjugate base of the acid is weaker as more oxygen atoms are present.

b. $H_2PO_4^- < HPO_4^{2-} < PO_4^{3-}$

For the conjugate bases of a polyprotic acid, the more highly negatively charged ions are more strongly attracted to H^+.

88. a. CaO, basic; $CaO + H_2O \rightarrow Ca^{2+}(aq) + 2\ OH^-(aq)$

b. Na_2O, basic; $Na_2O + H_2O \rightarrow 2\ Na^+(aq) + 2\ OH^-(aq)$

c. SO_2, acidic: $SO_2 + H_2O \rightarrow H_2SO_3(aq)$; $H_2SO_3(aq) \rightleftharpoons H^+(aq) + HSO_3^-(aq)$

d. Cl_2O, acidic; $Cl_2O + H_2O \rightarrow 2\ HOCl(aq)$; $HOCl(aq) \rightleftharpoons H^+(aq) + OCl^-(aq)$

e. P_4O_{10}, acidic; $P_4O_{10} + 6\ H_2O \rightarrow 4\ H_3PO_4(aq)$; $H_3PO_4(aq) \rightleftharpoons H^+(aq) + H_2PO_4^-(aq)$

f. NO_2, acidic; $2NO_2 + H_2O \rightarrow HNO_3(aq) + HNO_2(aq)$, HNO_3 strong acid, HNO_2 weak acid.

Lewis Acids and Bases

89. a. $B(OH)_3$, acid; H_2O, base b. Ag^+, acid; NH_3, base c. BF_3, acid; NH_3, base

90. a. I_2, acid; I^-, base b. $Zn(OH)_2$, acid; OH^-, base c. Fe^{3+}, acid; SCN^-, base

91. $Al(OH)_3(s) + 3\ H^+(aq) \rightarrow Al^{3+}(aq) + 3\ H_2O(l)$

$Al(OH)_3(s) + OH^-(aq) \rightarrow Al(OH)_4^-(aq)$

92. $Zn(OH)_2 + 2\ H^+(aq) \rightarrow 2\ H_2O(l) + Zn^{2+}(aq)$

$Zn(OH)_2(s) + 2\ OH^-(aq) \rightarrow Zn(OH)_4^{2-}(aq)$

93. Fe^{3+} should be the stronger Lewis acid. It is smaller with a greater positive charge and will be more strongly attracted to lone pairs of electrons.

94. Ti^{4+}, small, high positive charge. Ti^{4+} will be a good Lewis acid. Cl^- has a lone pair of electrons to donate. It is a Lewis base.

$$Ti^{4+} + 4\ [:\ddot{\underset{..}{Cl}}:]^- \rightarrow TiCl_4$$

:Cl:
:Cl—Ti—Cl:
:Cl:

ADDITIONAL EXERCISES

95. Both are strong acids.

$0.050 \text{ mol/L} \times 50.0 \text{ mL} = 2.5 \text{ mmol HCl}$

$0.10 \text{ mol/L} \times 150.0 \text{ mL} = 15 \text{ mmol } HNO_3$

$$[H^+] = \frac{2.5 \text{ mmol} + 15 \text{ mmol}}{200.0 \text{ mL}} = 0.088\ M,\ [OH^-] = 1.1 \times 10^{-13}$$

$$[Cl^-] = \frac{2.5 \text{ mmol}}{200.0 \text{ mL}} = 0.0125 \text{ mol/L} \approx 0.013\ M$$

$$[NO_3^-] = \frac{15 \text{ mmol}}{200.0 \text{ mL}} = 0.075\ M$$

96. If we just take the negative log of 10^{-12} we get pH = 12. This can't be. We can't add a tiny amount of acid to a neutral solution and get a strongly basic (pH = 12) solution. There are two sources of H^+: water and the HCl. So far we've considered the acid we add to water as the major source of H^+. In this case that is not true. The major source of H^+ is water (10^{-7} *M*) and not the HCl (10^{-12} *M*).

So the $[H^+] \approx 10^{-7} + 10^{-12} \approx 10^{-7}\ M$ and pH = 7.00

This makes sense. If we add a minute amount of acid to a neutral solution we would still expect to have a roughly neutral solution.

97. $NaHSO_4 \rightarrow Na^+ + HSO_4^-$ HSO_4^- is a weak acid with $K_a = 0.012$

H_2SO_4 is a strong acid, so the reaction $HSO_4^- + H_2O \rightarrow H_2SO_4 + OH^-$ does not occur to any appreciable extent.

The solution is acidic and the reaction is:

$$HSO_4^- + H_2O \rightleftharpoons H_3O^+ + SO_4^{2-}$$

CO_3^{2-} is a base, so the reaction:

$$CO_3^{2-} + HSO_4^- \rightleftharpoons HCO_3^- + SO_4^{2-} \text{ can occur.}$$

98. $K_aK_b = K_w$; $-\log K_aK_b = -\log K_w$

$-\log K_a - \log K_b = -\log K_w$; $pK_a + pK_b = pK_w = 14.00$ (at 25 °C)

99.

	HBz	$\rightleftharpoons$	H^+	+	Bz^-
Initial	C		~0		0
	x mol/L HBz dissociates to reach equilibrium				
Change	$-x$	$\rightarrow$	$+x$		$+x$
Equil	$C - x$		x		x

$K_a = \frac{[H^+][Bz^-]}{[HBz]} = 6.4 \times 10^{-5} = \frac{x^2}{C - x}$, and $x = [H^+]$

$6.4 \times 10^{-5} = \frac{[H^+]^2}{C - [H^+]}$

pH = 2.8; $[H^+] = 10^{-2.8} = 1.6 \times 10^{-3}$ (carry 1 extra sig. fig.)

$C - 1.6 \times 10^{-3} = \frac{(1.6 \times 10^{-3})^2}{6.4 \times 10^{-5}} = 4.0 \times 10^{-2}$

$C = 4.0 \times 10^{-2} + 0.16 \times 10^{-2} = 4.16 \times 10^{-2}\ M$

The molar solubility is 4.16×10^{-2} mol/L $\approx 4 \times 10^{-2}$ mol/L

$\frac{4 \times 10^{-2}\ \text{mol}}{\text{L}} \times \frac{122.1\ \text{g}}{\text{mol}} \times 0.1\ \text{L} = \frac{0.5\ \text{g}}{100\ \text{mL}}$

100. The relevant reactions are:

$H_2CO_3 \rightleftharpoons H^+ + HCO_3^-$ $\quad K_{a_1} = 4.3 \times 10^{-7}$

$HCO_3^- \rightleftharpoons H^+ + CO_3^{2-}$ $\quad K_{a_2} = 5.6 \times 10^{-11}$

Initially, we deal only with the first reaction (since $K_{a_1} >> K_{a_2}$) and then let those results control values for the concentrations in the second reaction.

	H_2CO_3	$\rightleftharpoons$	H^+	+	HCO_3^-	$K_{a_1} = 4.3 \times 10^{-7}$
Initial	0.010 M		~0		0	
	x mol/L H_2CO_3 dissociates to reach equilibrium					
Change	$-x$	$\rightarrow$	$+x$		$+x$	
Equil	$0.010 - x$		x		x	

$K_{a_1} = 4.3 \times 10^{-7} = \frac{[H^+][HCO_3^-]}{[H_2CO_3]} = \frac{x^2}{0.010 - x} \approx \frac{x^2}{0.010}$

$x = 6.6 \times 10^{-5}\ M = [H^+] = [HCO_3^-]$ Assumptions good.

	HCO_3^-	$\rightleftharpoons$	H^+	+	CO_3^{2-}	$K_{a_2} = 5.6 \times 10^{-11}$
Initial	$6.6 \times 10^{-5}\ M$		$6.6 \times 10^{-5}\ M$		0	
	y mol/L HCO_3^- dissociates to reach equilibrium					
Change	$-y$	$\rightarrow$	$+y$		$+y$	
Equil	$6.6 \times 10^{-5} - y$		$6.6 \times 10^{-5} + y$		y	

If y is small, then $[H^+] = [HCO_3^-]$:

$$K_{a_2} = 5.6 \times 10^{-11} = \frac{[H^+][CO_3^{2-}]}{[HCO_3^-]} \approx y$$

$[CO_3^{2-}] = 5.6 \times 10^{-11}\ M$ Assumption good.

The amount of H^+ from the second dissociation is $5.6 \times 10^{-11}\ M$ or:

$$\frac{5.6 \times 10^{-11}}{6.6 \times 10^{-5}} \times 100 = 8.5 \times 10^{-5}\ \%$$

The result justifies our treating the equilibria separately. If the second dissociation contributed a significant amount of H^+ we would have to treat both equilibria simultaneously. The reaction that occurs when acid is added to a solution of HCO_3^- is:

$$HCO_3^- + H^+ \longrightarrow (H_2CO_3) \longrightarrow H_2O + CO_2$$

The bubbles are $CO_2(g)$ and this implies that discrete H_2CO_3 molecules are not stable. We should write $H_2O + CO_2(aq)$ or just $CO_2(aq)$ for what we call carbonic acid. It is for convenience, however, that we write $H_2CO_3(aq)$.

101. The N-atom is protonated in each case.

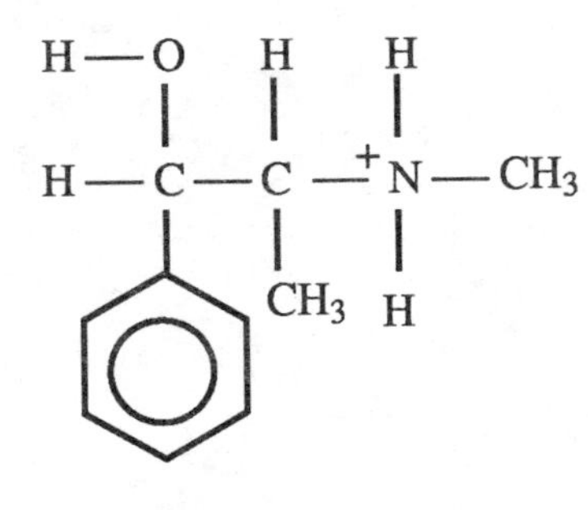

conjugate acid of ephedrine

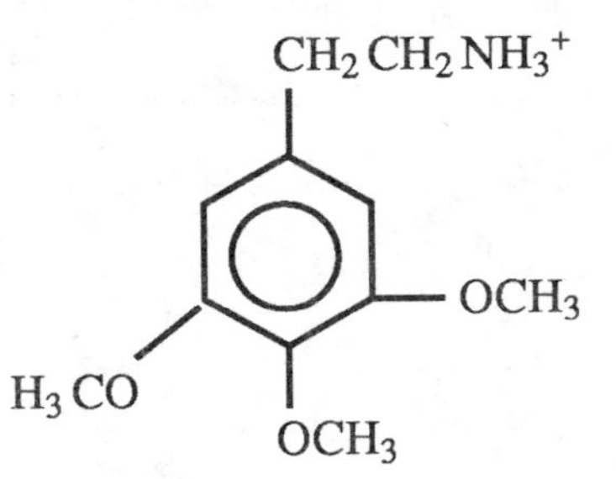

conjugate acid of mescaline

102. a. $NH_3 + H_3O^+ \rightleftharpoons NH_4^+ + H_2O$

$$K_{eq} = \frac{[NH_4^+]}{[NH_3][H^+]} = \frac{1}{K_a \text{ of } NH_4^+} = \frac{K_b}{K_w} = \frac{1.8 \times 10^{-5}}{1.0 \times 10^{-14}} = 1.8 \times 10^9$$

b. $NO_2^- + H_3O^+ \rightleftharpoons H_2O + HNO_2$

$$K_{eq} = \frac{[HNO_2]}{[NO_2^-][H^+]} = \frac{1}{K_a} = \frac{1}{4.0 \times 10^{-4}} = 2.5 \times 10^3$$

c. $NH_4^+ + CH_3CO_2^- \rightleftharpoons NH_3 + CH_3CO_2H$

$$K_{eq} = \frac{[NH_3][CH_3CO_2H]}{[NH_4^+][CH_3CO_2^-]} \times \frac{[H^+]}{[H^+]}$$

$$K_{eq} = \frac{K_a \text{ of } NH_4^+}{K_a \text{ of HOAc}} = \frac{K_w}{(K_b \text{ of } NH_3)(K_a \text{ of HOAc})} = \frac{1.0 \times 10^{-14}}{(1.8 \times 10^{-5})(1.8 \times 10^{-5})}$$

$$= 3.1 \times 10^{-5}$$

d. $H_3O^+ + OH^- \rightleftharpoons 2\,H_2O$; $K_{eq} = \frac{1}{K_w} = 1.0 \times 10^{14}$

e. $NH_4^+ + OH^- \rightleftharpoons NH_3 + H_2O$; $K_{eq} = \frac{1}{K_b} = 5.6 \times 10^4$

f. $HNO_2 + OH^- \rightleftharpoons H_2O + NO_2^-$

$$K_{eq} = \frac{[NO_2^-]}{[HNO_2][OH^-]} \times \frac{[H^+]}{[H^+]} = \frac{K_a}{K_w} = \frac{4.0 \times 10^{-4}}{1.0 \times 10^{-14}} = \frac{1}{K_b} = 4.0 \times 10^{10}$$

103. a. In the lungs, there is a lot of O_2; the equilibrium favors $Hb(O_2)_4$. In the cells there is a deficiency of O_2; the equilibrium favors HbH_4^{4+}.

b. CO_2 is a weak acid, $CO_2 + H_2O \rightleftharpoons HCO_3^- + H^+$. Removing CO_2 essentially decreases H^+. $Hb(O_2)_4$ is favored and O_2 is not released by hemoglobin in the cells. Breathing into a paper bag increases $[CO_2]$ in the blood.

c. CO_2 builds up in the blood and it becomes too acidic, driving the equilibrium to the left. Hemoglobin can't bind O_2 as strongly in the lungs. Bicarbonate ion acts as a base in water and neutralizes the excess acidity.

CHALLENGE PROBLEMS

104.

$HX(g) \rightarrow H(g) + X(g)$	ΔH = H—X bond energy
$H(g) \rightarrow H^+(g) + e^-$	ΔH = ionization energy of hydrogen atom
$X(g) + e^- \rightarrow X^-(g)$	ΔH = electron affinity of X
$HX(g) \rightarrow H^+(g) + X^-(g)$	

a.

$HF \rightarrow H + F$	ΔH = 565 kJ
$H \rightarrow H^+ + e^-$	ΔH = 1312 kJ
$F + e^- \rightarrow F^-$	ΔH = -328 kJ
$HF(g) \rightarrow H^+(g) + F^-(g)$	ΔH = 1549 kJ

For HCl: $\Delta H = 427\ kJ + 1312\ kJ - 349\ kJ = 1390.\ kJ$

For HBr: $\Delta H = 363 + 1312 - 325 = 1350.\ kJ$

For HI: $\Delta H = 295 + 1312 - 295 = 1312\ kJ$

b. The ionization of H is common to all acids. An acid will be a stronger acid if the bond to the hydrogen atom is weak or if the electron affinity of X is more favorable. In this example we calculated the enthalpy to dissociate the hydrogen halide, thus ΔH correlates with the acid strengths of HF, HCl, HBr, and HI.

105. Weak bonds to H and the presence of atoms with favorable electron affinities.

106. a. Bond energies decrease as acid strength increases.

H—O, 467 kJ/mol; H—S, 363 kJ/mol; H—Se, 276 kJ/mol

$H_2O < H_2S < H_2Se$

b. Electronegative elements increase acid strengths. For more complicated molecules like these, we look for atoms with high electronegativity. Electronegativity is defined as the ability of an atom to attract electrons to itself in a molecule.

$CH_3CO_2H < FCH_2CO_2H < F_2CHCO_2H < CF_3CO_2H$

c. O has a large electronegativity and CH_3- weakens acids.

$CH_3NH_3^+ < NH_4^+ < HONH_3^+$

107. Base strength is enhanced if bonds to hydrogen are strong. The presence of atoms with high electronegativities can decrease base strength.

a. $SeH^- < SH^- < OH^-$ Same order as bond energies.

b. Bond energies N—H (391 kJ/mol), P—H (322 kJ/mol)

stronger bond, stronger base $PH_3 < NH_3$

c. $HONH_2 < NH_3$ Presence of O decreases base strength.

108.

$HX(aq) \rightarrow HX(g)$ $-\Delta H_{hyd}(HX)$

$HX(g) \rightarrow H(g) + X(g)$ D_{HX}

$H(g) \rightarrow H^+(g) + e^-$ IE

$X(g) + e^- \rightarrow X^-(g)$ EA

$H^+(g) \rightarrow H^+(aq)$ $\Delta H_{hyd}(H^+)$

$X^-(g) \rightarrow X^-(aq)$ $\Delta H_{hyd}(X^-)$

$HX(aq) \rightleftharpoons H^+(aq) + X^-(aq)$

If the hydration energy of $X^-(aq)$ is very favorable (large and negative), then acid strength is increased.

109. a. An acid will generate NH_4^+ in liquid ammonia and a base will generate NH_2^-. NH_4^+ corresponds to H^+ and NH_2^- corresponds to OH^-.

b. $[NH_4^+] = [NH_2^-]$

c. $2\ Na(s) + 2\ NH_3(l) \longrightarrow 2\ Na^+ + 2\ NH_2^- + H_2(g)$

d. Ammonia is more basic than water. The acidity of substance is enhanced in liquid ammonia. For example, acetic acid is a strong acid in ammonia.

110. $H_3Cit + HCO_3^- \rightleftharpoons H_2Cit^- + H_2CO_3\quad (H_2O + CO_2)$

$$K_{eq} = \frac{[H_2Cit^-][H_2CO_3]}{[H_3Cit][HCO_3^-]} \times \frac{[H^+]}{[H^+]} = \frac{K_{a_1}\text{ (citric acid)}}{K_{a_1}\text{ (carbonic acid)}}$$

$$K_{eq} = \frac{8.4 \times 10^{-4}}{4.3 \times 10^{-7}} = 1950 = 2.0 \times 10^3$$

$H_3Cit + 3\ HCO_3^- \rightleftharpoons Cit^{3-} + 3\ H_2CO_3\ (3\ H_2O + 3\ CO_2)$

$$K_{eq} = \frac{[Cit^{3-}][H_2CO_3]^3}{[H_3Cit][HCO_3^-]^3} \times \frac{[H^+]^3}{[H^+]^3} = \frac{K_{a_1}K_{a_2}K_{a_3}\text{ (citric acid)}}{K_{a_1}^3\text{ (carbonic acid)}}$$

$$K_{eq} = \frac{(8.4 \times 10^{-4})(1.8 \times 10^{-5})(4.0 \times 10^{-6})}{(4.3 \times 10^{-7})^3} = 7.6 \times 10^5$$

111. a. $HCO_3^- + HCO_3^- \rightleftharpoons H_2CO_3 + CO_3^{2-}$

$$K_{eq} = \frac{[H_2CO_3][CO_3^{2-}]}{[HCO_3^-][HCO_3^-]} \times \frac{[H^+]}{[H^+]} = \frac{K_{a_2}}{K_{a_1}} = \frac{5.6 \times 10^{-11}}{4.3 \times 10^{-7}} = 1.3 \times 10^{-4}$$

b. $[H_2CO_3] = [CO_3^{2-}]$ if the reaction in (a) is considered to be the only reaction.

c. $H_2CO_3 \rightleftharpoons 2\ H^+ + CO_3^{2-}$; $\quad K_{eq} = \frac{[H^+]^2[CO_3^{2-}]}{[H_2CO_3]} = K_{a_1}K_{a_2}$

Since, $[H_2CO_3] = [CO_3^{2-}]$ from part b, $[H^+]^2 = K_{a_1}K_{a_2}$

$[H^+] = (K_{a_1}K_{a_2})^{1/2}$; $\quad pH = \frac{pK_{a_1} + pK_{a_2}}{2}$

d. $[H^+] = [(4.3 \times 10^{-7})(5.6 \times 10^{-11})]^{1/2}$; $[H^+] = 4.9 \times 10^{-9}$; $pH = 8.31$

112.

	H_2O	$\rightleftharpoons$	H^+	+	OH^-
Initial			$1.0 \times 10^{-7} M$		$1.0 \times 10^{-7} + 1.0 \times 10^{-7} M$
	x mol/L H^+ reacts with OH^- to reach equilibrium				
Change		$\leftarrow$	$-x$		$-x$
Equil			$1.0 \times 10^{-7} - x$		$2.0 \times 10^{-7} - x$

$(1.0 \times 10^{-7} - x)(2.0 \times 10^{-7} - x) = 1.0 \times 10^{-14}$

If x is small:

$[OH^-] = 2.0 \times 10^{-7}$; pOH = 6.70; pH = 7.30

Close but not the correct answer.

Solving the quadratic: $x^2 - 3.0 \times 10^{-7}x + 1.0 \times 10^{-14} = 0$

$$x = \frac{3.0 \times 10^{-7} \pm [9.0 \times 10^{-14} - 4.0 \times 10^{-14}]^{1/2}}{2}$$

$$x = \frac{3.0 \times 10^{-7} \pm 2.24 \times 10^{-7}}{2} = 2.6 \times 10^{-7} \text{ or } 3.8 \times 10^{-8}$$

Only 3.8×10^{-8} makes sense.

$[H^+] = 1.0 \times 10^{-7} - 0.38 \times 10^{-7} = 0.62 \times 10^{-7} = 6.2 \times 10^{-8} \approx 6 \times 10^{-8}$ mol/L

pH = 7.2

113.

	HBrO	$\rightleftharpoons$	H^+	+	BrO^-	$K_a = 2 \times 10^{-9}$
Initial	$1.0 \times 10^{-6} M$		$0(10^{-7})$		0	As usual, assume H^+ contribution from autoionization of water is negligible.
	x mol/L HBrO dissociates					
Change	$-x$	$\rightarrow$	$+x$		$+x$	
Equil	$1.0 \times 10^{-6} - x$		x		x	

$$\frac{x^2}{1.0 \times 10^{-6} - x} \approx \frac{x^2}{1.0 \times 10^{-6}} = 2 \times 10^{-9}$$

$x = [H^+] = 4 \times 10^{-8}$; pH = 7.4 Assumption that we can ignore H^+ contribution from H_2O is bad.

This answer is impossible. We can't add a small amount of a weak acid to a neutral solution and get a basic solution. In the correct solution, we would have to take into account the autoionization of water and solve exactly. This is beyond the scope of this text book.

114. $HIO_3 \rightleftharpoons H^+ + IO_3^-$ pH = 1.20, $[H^+] = 0.063\ M$

$[H^+] = 0.063\ M$

$[HIO_3] + [IO_3^-] = 0.10\ M$

$$K_a = \frac{[H^+][IO_3^-]}{[HIO_3]} = 0.17 = 0.063\ \frac{[IO_3^-]}{[HIO_3]}$$

$[IO_3^-] = 2.7\ [HIO_3]$

$[HIO_3] + 2.7\ [HIO_3] = 0.10$

$[HIO_3] = 0.027\ M$

$[IO_3^-] = 0.073\ M$

115. 25.0 mL × 2.00 mmol/mL = 50.0 mmol H^+ from the strong acid HCl. The strong acid is in excess and will neutralize PO_4^{3-} to H_3PO_4 and neutralize OH^- to H_2O.

$$5.0\ \text{mmol}\ PO_4^{3-} \times \frac{3\ \text{mmol}\ H^+}{\text{mmol}\ PO_4^{3-}} + 5.0\ \text{mmol}\ OH^- \times \frac{1\ \text{mmol}\ H^+}{\text{mmol}\ OH^-}$$

$= 20.$ mmol H^+ needed to neutralize bases.

After the neutralization reactions, solution contains 30. mmol H^+ and 5.0 mmol acetic acid. The contribution of H^+ from the weak acid, acetic acid, will be negligible. Ignoring H^+ from acetic acid:

$$[H^+] = \frac{30.\ \text{mmol}}{250.0\ \text{mL}} = 0.12\ \text{mol/L},\ pH = 0.92$$

CHAPTER FIFTEEN: APPLICATIONS OF AQUEOUS EQUILIBRIA

QUESTIONS

1. A common ion is an ion that appears in an equilibrium reaction but came from a source other than that reaction. Addition of a common ion (H^+ or NO_2^-) to the reaction, $HNO_2 \rightleftharpoons H^+ + NO_2^-$, will drive the equilibrium to the left.

2. A buffered solution must contain both a weak acid and a weak base. Buffer solutions are useful for controlling the pH of a solution.

3. The capacity of a buffer is a measure of how much strong acid or base the buffer can neutralize. All the buffers listed have the same pH. The 1.0 *M* buffer has the greatest capacity; the 0.01 *M* buffer the least capacity.

4. No, as long as there is both a weak acid and a weak base present, the solution will be buffered. If the concentrations are the same, the buffer will have the same capacity towards both added H^+ and OH^-.

5. Between the starting point of the titration and the equivalence point, we are dealing with a buffer solution. Thus,

$$pH = pK_a + \log \frac{[Base]}{[Acid]}$$

Halfway to the equivalence point: [Base] = [Acid] so $pH = pK_a + \log 1 = pK_a$

6. Yes, at any point this can be done. In Chapter 7 we calculated the K_a from a solution of only the weak acid. In the buffer region, we can calculate the ratio of the basic to acidic form of the weak acid and use the Henderson-Hasselbalch equation:

$$pH = pK_a + \log \frac{[Base]}{[Acid]}$$

7. No, since there are three colored forms there must be two proton transfer reactions. Thus, there must be at least two acidic protons in the acid (orange) form of thymol blue.

8. Equivalence point: moles acid = moles base. End point: indicator changes color. We want the indicator to tell us when we have reached the equivalence point. We can detect the end point and assume it is the equivalence point for doing stoichiometric calculations. They don't have to be as close as 0.01 pH units, since at the equivalence point the pH is changing very rapidly with added titrant. The range over which an indicator changes color only needs to include the pH of the equivalence point.

9. The two forms of an indicator are different colors. To see only one color that form must be in ten fold excess over the other. To go from

$$\frac{[HIn]}{[In^-]} = 10 \text{ to } \frac{[HIn]}{[In^-]} = 0.1 \text{ requires a change of 2 pH units.}$$

10. If the number of ions in the two salts are the same the K_{sp}'s can be compared i.e., 1:1 electrolytes can be compared to each other; 2:1 electrolytes can be compared to each other, etc.

EXERCISES

Buffers

11. When strong acid or base is added to an acetic acid/sodium acetate mixture, the strong acid/base is neutralized. The reaction goes to completion. The strong acid/base is replaced with a weak acid/base.

$$H^+ + CH_3CO_2^- \rightarrow CH_3CO_2H;\ OH^- + CH_3CO_2H \rightarrow CH_3CO_2^- + H_2O$$

12. $CO_3^{2-} + H^+ \rightarrow HCO_3^-$; $HCO_3^- + OH^- \rightarrow H_2O + CO_3^{2-}$

13. For $HOAc \rightleftharpoons H^+ + OAc^-$ $K_a = 1.8 \times 10^{-5}$ $pK_a = 4.74$

a.

	HOAc	$\rightleftharpoons$	H^+	+	OAc^-
Initial	0.10 *M*		~0		0.25 *M*
	x mol/L HOAc dissociates to reach equilibrium				
Change	$-x$	$\rightarrow$	$+x$		$+x$
Equil	$0.10 - x$		x		$0.25 + x$

$$1.8 \times 10^{-5} = \frac{x(0.25 + x)}{(0.10 - x)} \approx \frac{x(0.25)}{0.10} \quad \text{(assuming } 0.25 + x \approx 0.25 \text{ and } 0.10 - x \approx 0.10\text{)}$$

$x = [H^+] = 7.2 \times 10^{-6}$, pH = 5.14 Assumptions good by the 5% rule.

Alternatively, we can use the Henderson-Hasselbalch equation.

$$pH = pK_a + \log \frac{[Base]}{[Acid]}$$

$$pH = 4.74 + \log \frac{(0.25)}{(0.10)} = 4.74 + 0.40 = 5.14$$

The Henderson-Hasselbalch equation will be valid when assumptions of the type, $0.10 - x \approx 0.10$, that we just made are valid. From a practical standpoint, this will almost always be true for useful buffer solutions. Note: The Henderson-Hasselbalch equation can only be used to solve for the pH of buffer solutions.

b. $pH = 4.74 + \log \frac{(0.10)}{(0.25)} = 4.34$

c. $pH = 4.74 + \log \frac{(0.20)}{(0.080)} = 4.74 + 0.40 = 5.14$

d. $pH = 4.74 + \log \frac{(0.080)}{(0.20)} = 4.34$

14. a. $HNO_2 \rightleftharpoons H^+ + NO_2^-$ $\quad K_a = 4.0 \times 10^{-4}$

This is a buffer solution. Using the Henderson-Hasselbalch equation:

$$pH = pK_a + \log \frac{[Base]}{[Acid]} = 3.40 + \log \frac{(0.15)}{(0.10)}$$

$$pH = 3.40 + 0.18 = 3.58$$

b. $25.0 \text{ g } CH_3CO_2H \times \frac{1 \text{ mol}}{60.05 \text{ g}} = 0.416 \text{ mol HOAc}$

Remember we write acetic acid several different ways.
acetic acid: HOAc, CH_3CO_2H, $HC_2H_3O_2$; acetate ion: OAc^-, $CH_3CO_2^-$, $C_2H_3O_2^-$

$$[HOAc] = \frac{0.416 \text{ mol}}{0.500 \text{ L}} = 0.832\ M$$

$$40.0 \text{ g } CH_3CO_2Na \times \frac{1 \text{ mol}}{82.03 \text{ g}} = 0.488 \text{ mol } OAc^-$$

$$[OAc^-] = \frac{0.488 \text{ mol}}{0.500 \text{ L}} = 0.976\ M$$

$$pH = pK_a + \log \frac{[Base]}{[Acid]}$$

$K_a = 1.8 \times 10^{-5}$; $pK_a = 4.74$

$$pH = 4.74 + \log \frac{(0.976)}{(0.832)} = 4.74 + 0.0693 = 4.81$$

c. $50.0 \text{ mL} \times \frac{1.0 \text{ mmol}}{\text{mL}} = 50. \text{ mmol HOCl}$

$$30.0 \text{ mL} \times \frac{0.80 \text{ mmol NaOH}}{\text{mL}} = 24 \text{ mmol } OH^-$$

The strong base reacts with the best acid available, HOCl. The reaction goes to completion. Whenever strong base or strong acid react, the reaction is always assumed to go to completion.

	HOCl	+	OH^-	→	OCl^-	+	H_2O
Before	50. mmol		24 mmol		0		
	24 mmol OH^- reacts completely						
Change	-24		-24	→	+24		
After	26 mmol		0		24 mmol		

After reaction, the solution contains 26 mmol HOCl and 24 mmol OCl^- in 250 mL of solution. This is a buffer solution. Using the Henderson-Hasselbalch equation:

$pH = pK_a + \log \frac{[Base]}{[Acid]}$ and $K_a = 3.5 \times 10^{-8}$, $pk_a = 7.46$

$$pH = 7.46 + \log \frac{\frac{24 \text{ mmol OCl}^-}{250 \text{ mL}}}{\frac{26 \text{ mmol HOCl}}{250 \text{ mL}}} = 7.46 - 0.035 = 7.43$$

d. OH^- reacts completely with best acid present, NH_4^+.

$$100.0 \text{ g } NH_4Cl \times \frac{1 \text{ mol}}{53.49 \text{ g}} = 1.870 \text{ mol } NH_4^+, \ [NH_4^+] = 1.870\ M$$

$$65.0 \text{ g NaOH} \times \frac{1 \text{ mol}}{40.00 \text{ g}} = 1.63 \text{ mol } OH^-, \ [NaOH] = 1.63\ M$$

	NH_4^+	+	OH^-	→	NH_3 + H_2O
Before	1.87 mol/L		1.63 mol/L		0
	1.63 mol/L OH^- reacts completely				
Change	-1.63		-1.63	→	+1.63
After	0.24 mol/L		0		1.63 mol/L

After reaction, a buffer solution results.

If we use $NH_4^+ \rightleftharpoons H^+ + NH_3$

$$pH = pK_a + \log \frac{[NH_3]}{[NH_4^+]}$$

$$K_a = \frac{K_w}{K_b} = \frac{1.0 \times 10^{-14}}{1.8 \times 10^{-5}} = 5.6 \times 10^{-10}$$

$$pH = 9.25 + \log \frac{1.63}{0.24} = 9.25 + 0.83 = 10.08$$

e. $$26.4 \text{ g } CH_3CO_2Na \times \frac{1 \text{ mol}}{82.03 \text{ g}} = 0.322 \text{ mol NaOAc}$$

$$50.0 \text{ mL HCl} \times \frac{1 \text{ L}}{1000 \text{ mL}} \times \frac{6.00 \text{ mol HCl}}{\text{L}} = 0.300 \text{ mol HCl}$$

H^+ reacts completely with best base present, OAc^-. (Strong acids always react to completion.)

	H^+	+	OAc^-	→	HOAc
Before	0.300 mol		0.322 mol		0
	0.300 mol H^+ reacts completely				
Change	-0.300		-0.300	→	+0.300
After	0		0.022 mol		0.300 mol

After reaction, a buffer solution results. Using the Henderson-Hasselbalch equation:

$$pH = pK_a + \log \frac{[OAc^-]}{[HOAc]} = 4.74 + \log \frac{(0.022 \text{ mol}/0.500 \text{ L})}{(0.300 \text{ mol}/0.500 \text{ L})}$$

$$pH = 4.74 - 1.13 = 3.61$$

15. a. $CH_3CH_2CO_2H = HOPr$

$CH_3CH_2CO_2^- = OPr^-$

Weak acid problem:

	$HOPr$	$\rightleftharpoons$	H^+	+	OPr^-	$K_a = 1.3 \times 10^{-5}$
Initial	0.10 M		~0		0	
	x mol/L HOPr dissociates to reach equilibrium					
Change	$-x$	$\rightarrow$	$+x$		$+x$	
Equil	$0.10 - x$		x		x	

$$K_a = 1.3 \times 10^{-5} = \frac{[H^+][OPr^-]}{[HOPr]} = \frac{x^2}{0.10 - x} \approx \frac{x^2}{0.10}$$

$x = [H^+] = 1.1 \times 10^{-3}\ M$; pH = 2.94 Assumptions good.

b. Weak base problem:

	$OPr^- + H_2O$	$\rightleftharpoons$	$HOPr$	+	OH^-	$K_b = \frac{K_w}{K_a} = 7.7 \times 10^{-10}$
Initial	0.10 M		0		~0	
	x mol/L OPr^- reacts with H_2O to reach equilibrium					
Change	$-x$	$\rightarrow$	$+x$		$+x$	
Equil	$0.10 - x$		x		x	

$$K_b = 7.7 \times 10^{-10} = \frac{[HOPr][OH^-]}{[OPr^-]} = \frac{x^2}{0.10 - x} \approx \frac{x^2}{0.10}$$

$x = [OH^-] = 8.8 \times 10^{-6}\ M$; pOH = 5.06; pH = 8.94 Assumptions good.

c. pure H_2O, $[H^+] = [OH^-] = 10^{-7}$; pH = 7.00

d. Buffer problem:

	$HOPr$	$\rightleftharpoons$	H^+	+	OPr^-	$K_a = 1.3 \times 10^{-5}$
Initial	0.10 M		~0		0.10 M	
	x mol/L HOPr dissociates to reach equilibrium					
Change	$-x$	$\rightarrow$	$+x$		$+x$	
Equil	$0.10 - x$		x		$0.10 + x$	

$$1.3 \times 10^{-5} = \frac{(0.10 + x)(x)}{(0.10 - x)} \approx \frac{(0.10)(x)}{0.10} = x$$

$[H^+] = 1.3 \times 10^{-5}$; pH = 4.89 Assumptions good.

Or using the Henderson-Hasselbalch equation:

$$pH = pK_a + \log \frac{[Base]}{[Acid]}$$

$$pH = pK_a + \log \frac{(0.10)}{(0.10)} = pK_a = 4.89$$

16. a. Weak base problem:

	NH_3 + H_2O	⇌	NH_4^+	+ OH^-
Initial	0.10 mol/L		0	~0
	x mol/L NH_3 reacts with H_2O to reach equilibrium			
Change	$-x$	→	$+x$	$+x$
Equil	$0.10 - x$		x	x

$$K_b = 1.8 \times 10^{-5} = \frac{x^2}{(0.10 - x)} \approx \frac{x^2}{0.10}$$

$x = [OH^-] = 1.3 \times 10^{-3}$ mol/L, pH = 11.11 Assumptions good.

b. Weak acid problem:

	NH_4^+	⇌	NH_3 +	H^+
Initial	0.10 M		0	~0
	x mol/L NH_4^+ dissociates to reach equilibrium			
Change	$-x$	→	$+x$	$+x$
Equil	$0.10 - x$		x	x

$$K_a = \frac{K_w}{K_b} = 5.6 \times 10^{-10} = \frac{[NH_3][H^+]}{[NH_4^+]} = \frac{x^2}{0.10 - x} \approx \frac{x^2}{0.10}$$

$x = [H^+] = 7.5 \times 10^{-6}$ mol/L, pH = 5.12 Assumptions good.

c. Pure H_2O, pH = 7.00

d. Buffer problem:

$$pH = pK_a + \log \frac{[Base]}{[Acid]} = 9.25 + \log \frac{[NH_3]}{[NH_4^+]} = 9.25 + \log \frac{(0.10)}{(0.10)} = 9.25$$

17. a.

	HOPr	⇌	H^+ +	OPr^-
Initial	0.10 M		0.020 M	0
	x mol/L HOPr dissociates to reach equilibrium			
Change	$-x$	→	$+x$	$+x$
Equil	$0.10 - x$		$0.020 + x$	x

$[H^+] = 0.020 + x \approx 0.020$; pH = 1.70 Assumptions good.

Note: H^+ contribution from the weak acid, HOPr, was negligible. The pH of the solution can be determined by only considering the amount of strong acid present.

b. H^+ reacts completely with the best base present, OPr^-.

	OPr^- +	H^+	$\rightarrow$	HOPr	
Before	0.10 *M*	0.020 *M*		0	
Change	-0.020	-0.020	$\rightarrow$	+0.020	Reacts completely
After	0.08 *M*	0		0.020 *M*	

After reaction, a weak acid, HOPr, and its conjugate base, OPr^-, are present. This is a buffer solution. Using the Henderson-Hasselbalch equation:

$$pH = pK_a + \log\frac{[Base]}{[Acid]} = 4.89 + \log\frac{(0.08)}{(0.020)} = 5.5$$ Assumptions good.

c. $[H^+] = 0.020$; pH = 1.70

d. Added H^+ reacts completely with best base present, OPr^-.

	OPr^- +	H^+	$\rightarrow$	HOPr	
Before	0.10 *M*	0.020 *M*		0.10 *M*	
Change	-0.020	-0.020	$\rightarrow$	+0.020	Reacts completely
After	0.08	0		0.12	

A buffer solution results (weak acid and conjugate base). Using Henderson-Hasselbalch equation:

$$pH = pK_a + \log\frac{[Base]}{[Acid]} = 4.89 + \log\frac{(0.08)}{(0.12)} = 4.7$$

18. a. Added H^+ reacts completely with NH_3 (best base present) to form NH_4^+.

	NH_3 +	H^+	$\rightarrow$	NH_4^+	
Before	0.10 *M*	0.020 *M*		0	
Change	-0.020	-0.020	$\rightarrow$	+0.020	Reacts completely
After	0.08	0		0.020	

After this reaction, a buffer solution exists, i.e., a weak acid (NH_4^+) and its conjugate base (NH_3) present at the same time. Using the Henderson-Hasselbalch equation to solve for the pH:

$$pH = 9.25 + \log\frac{(0.08)}{(0.020)} = 9.25 + 0.6 = 9.85 \approx 9.9$$

b.

	NH_4^+ $\rightleftharpoons$	NH_3 +	H^+
Initial	0.10 *M*	0	0.020 *M*

Consider only H^+ from HCl, $[H^+] = 0.020$, pH = 1.70

Note: H^+ contribution from NH_4^+ (a weak acid) will be negligible.

c. $[H^+] = 0.020$, pH = 1.70

d. Major species: H_2O, Cl^-, NH_3, NH_4^+, H^+

0.10 0.10 0.020

H^+ will react completely with NH_3, the best base present.

	NH_3 +	H^+	→	NH_4^+	
Before	0.10 *M*	0.020 *M*		0.10 *M*	
Change	-0.020	-0.020	→	+0.020	Reacts completely
After	0.08	0		0.12	

A buffer solution results after reaction. Using Henderson-Hasselbalch equation:

$$pH = 9.25 + \log\frac{[NH_3]}{[NH_4^+]} = 9.25 + \log\frac{(0.08)}{(0.12)} = 9.25 - 0.2 = 9.1$$

19. a. HOPr, OH^-

0.10 *M* 0.020 *M*

OH^- will react completely with the best acid present, HOPr.

	HOPr +	OH^-	→	OPr^- +	H_2O	
Before	0.10 *M*	0.020 *M*		0		
Change	-0.020	-0.020	→	+0.020		Reacts completely
After	0.08	0		0.020		

A buffer solution results after the reaction. Using Henderson-Hasselbalch equation:

$$pH = pK_a + \log\frac{[Base]}{[Acid]} = 4.89 + \log\frac{(0.020)}{(0.08)} = 4.3$$

b.

	OPr^- + H_2O	⇌	HOPr +	OH^-
Initial	0.10 *M*		0	0.020 *M*

x mol/L OPr^- reacts with H_2O to reach equilibrium

Change	$-x$	→	$+x$	$+x$
Equil	$0.10 - x$		x	$0.020 + x$

$[OH^-] = 0.020 + x \approx 0.020$; pOH = 1.70; pH = 12.30 Assumption good.

Note: OH^- contribution from the weak base, OPr^-, was negligible. pH can be determined by only considering the amount of strong base present.

c. $[OH^-] = 0.020$; pOH = 1.70; pH = 12.30

d. HOPr, OPr^- OH^-

0.10 *M* 0.10 *M* 0.020 *M*

OH^- will react completely with HOPr, the best acid present.

	HOPr	+	OH^-	→	OPr^-	+	H_2O	
Before	0.10 *M*		0.020 *M*		0.10 *M*			
Change	-0.020		-0.020	→	+0.020			Reacts completely
After	0.08		0		0.12			

Using the Henderson-Hasselbalch equation to solve for the pH of the resulting buffer solution:

$$pH = pK_a + \log \frac{[Base]}{[Acid]} = 4.89 + \log \frac{(0.12)}{(0.08)} = 5.1$$

20. a.

	NH_3	+	H_2O	⇌	NH_4^+	+	OH^-
Initial	0.10 *M*				0		0.020 *M*

OH^- contribution from the weak base, NH_3, will be negligible. Consider only added strong base as source of OH^-.

$[OH^-] = 0.020$, $pOH = 1.70$, $pH = 12.30$

b. Added strong base will react to completion with the best acid present, NH_4^+.

	OH^-	+	NH_4^+	→	NH_3	+	H_2O	
Before	0.020 *M*		0.10 *M*		0			
Change	-0.020		-0.020	→	+0.020			Reacts completely
After	0		0.08		0.020			

The resulting solution is a buffer (weak acid and conjugate base). Using Henderson-Hasselbalch equation:

$$pH = 9.25 + \log\left(\frac{0.020}{0.08}\right) = 9.25 - 0.6 = 8.7$$

c. $[OH^-] = 0.020$, $pOH = 1.70$, $pH = 12.30$

d. Major species: H_2O, Na^+, NH_3, NH_4^+, OH^-

0.10 0.10 0.020

Again added strong base reacts completely with best acid present, NH_4^+.

	NH_4^+	+	OH^-	→	NH_3	+	H_2O	
Before	0.10 *M*		0.020 *M*		0.10 *M*			
Change	-0.020		-0.020	→	+0.020			Reacts completely
After	0.08		0		0.12			

A buffer solution results. Using Henderson-Hasselbalch equation:

$$pH = 9.25 + \log\frac{[NH_3]}{[NH_4^+]} = 9.25 + \log\left(\frac{0.12}{0.08}\right) = 9.25 + 0.18 = 9.4$$

21. Consider all of the results to Exercises 15, 17, and 19.

Solution	Initial pH	after added acid	after added base
a	2.94	1.70	4.3
b	8.94	5.5	12.30
c	7.00	1.70	12.30
d	4.89	4.7	5.1

The solution in (d) is a buffer; it contains both weak acid (HOPr) and a weak base (OPr^-). It resists changes in pH when strong acid or base is added.

22. Consider all of the results to Exercises 16, 18, and 20.

Solution	Initial pH	after added acid	after added base
a	11.11	9.9	12.30
b	5.12	1.70	8.7
c	7.00	1.70	12.30
d	9.25	9.1	9.4

The solution in (d) is a buffer; it shows the greatest resistance to a change in pH when strong acid or base is added.

23. The reaction, $H^+ + NH_3 \rightarrow NH_4^+$, goes to completion. After this reaction occurs there must be both NH_3 and NH_4^+ in solution for it to be a buffer. After the reaction the solutions contain:

a. 0.05 *M* NH_4^+ and 0.05 *M* H^+

b. 0.05 *M* NH_4^+

c. 0.05 *M* NH_4^+ and 0.05 *M* H^+

d. 0.05 *M* NH_4^+ and 0.05 *M* NH_3

Thus, only the combination in (d) results in a buffer.

24. a. No, strong acid and neutral salt.

b. Yes, HNO_2 - weak acid, NO_2^- - weak base

c. Yes, 250 mL × 0.10 mmol/mL = 25 mmol CH_3CO_2H

500 mg KOH × 1 mmol/56.11 mg = 9 mmol OH^-

$OH^- + CH_3CO_2H \rightarrow CH_3CO_2^- + H_2O$ Reacts completely

After the reaction goes to completion, both a weak acid and a weak base are present: 16 mmol CH_3CO_2H and 9 mmol $CH_3CO_2^-$ ion. This is a buffer solution.

d. No, both CO_3^{2-} and PO_4^{3-} ions are weak bases. No acid is present.

25. $NH_4^+ \rightleftharpoons H^+ + NH_3$ $\quad K_a = \frac{K_w}{K_b} = 5.6 \times 10^{-10}$ $\quad pK_a = 9.25$

a. $9.00 = 9.25 + \log \frac{[NH_3]}{[NH_4^+]}$

$\log \frac{[NH_3]}{[NH_4^+]} = -0.25$

$\frac{[NH_3]}{[NH_4^+]} = 10^{-0.25} = 0.56$

b. $8.80 = 9.25 + \log \frac{[NH_3]}{[NH_4^+]}$

$\frac{[NH_3]}{[NH_4^+]} = 10^{-0.45} = 0.35$

c. $10.00 = 9.25 + \log \frac{[NH_3]}{[NH_4^+]}$

$\frac{[NH_3]}{[NH_4^+]} = 10^{0.75} = 5.6$

d. $9.60 = 9.25 + \log \frac{[NH_3]}{[NH_4^+]}$

$\frac{[NH_3]}{[NH_4^+]} = 10^{0.35} = 2.2$

26. $pH = pK_a + \log \frac{[Base]}{[Acid]}$; $7.41 = 6.37 + \log \frac{[HCO_3^-]}{[H_2CO_3]}$

$\frac{[HCO_3^-]}{[H_2CO_3]} = 10^{1.04} = 11,$ $\quad \frac{[H_2CO_3]}{[HCO_3^-]} = 0.91$

27. When OH^- is added, it converts HOAc into OAc^-:

$$HOAc + OH^- \rightarrow OAc^- + H_2O \quad \text{Reacts completely}$$

The moles of OAc^- produced equals the moles of OH^- added. Since the volume is 1.0 L (assuming no volume change), then:

$[OAc^-] + [HOAc] = 2.0\ M$ and $[OAc^-]$ produced = $[OH^-]$ added

a. $pH = pK_a + \log \frac{[OAc^-]}{[HOAc]}$

For $pH = pK_a$, $\log \frac{[OAc^-]}{[HOAc]} = 0$

Therefore, $\frac{[OAc^-]}{[HOAc]} = 1.0$ and $[OAc^-] = [HOAc]$

Since $[OAc^-] + [HOAc] = 2.0$, then $[OAc^-] = [HOAc] = 1.0\ M$

To produce 1.0 L of a 1.0 *M* OAc^- solution we need to add 1.0 mol of NaOH to 1.0 L of the HOAc solution.

$1.0 \text{ mol of NaOH} \times \frac{40.00 \text{ g}}{\text{mol NaOH}} = 40. \text{ g NaOH}$

b. $4.00 = 4.74 + \log \frac{[OAc^-]}{[HOAc]}$; $\frac{[OAc^-]}{[HOAc]} = 10^{-0.74} = 0.18$

$[OAc^-] = 0.18\ [HOAc]$ or $[HOAc] = 5.6\ [OAc^-]$

$[OAc^-] + [HOAc] = 2.0\ M$; $[OAc^-] + 5.6\ [OAc^-] = 2.0\ M$

$[OAc^-] = \frac{2.0\ M}{6.6} = 0.30\ M$

To produce 0.30 M OAc^- solution we need to add 0.30 mol of NaOH to 1.0 L of the HOAc solution.

$0.30\ \text{mol} \times \frac{40.00\ g}{mol} = 12\ g\ NaOH$

c. $5.00 = 4.74 + \log \frac{[OAc^-]}{[HOAc]}$; $\frac{[OAc^-]}{[HOAc]} = 10^{0.26} = 1.8$

$1.8\ [HOAc] = [OAc^-]$ or $[HOAc] = 0.56\ [OAc^-]$; $[HOAc] + [OAc^-] = 2.0\ M$

$1.56\ [OAc^-] = 2.0\ M$; $[OAc^-] = 1.3\ M$

We need to add 1.3 mol of NaOH or $1.3\ \text{mol} \times \frac{40.00\ g}{mol} = 51\ g\ NaOH$

28. sodium acetate, $Na_2C_2H_3O_2$, NaOAc; 1.0 mol in 1.0 L

$pH = pK_a + \log \frac{[OAc^-]}{[HOAc]}$

$H^+ + OAc^- \longrightarrow HOAc$ Reacts completely

Added H^+ will convert OAc^- into HOAc. Let x = mol H^+ (HCl) added. A buffer solution results after H^+ reacts with OAc^-. Therefore (assuming V = 1.0 L):

$pH = pK_a + \log \frac{1.0\ \text{mol} - x}{x}$

a. Since $pH = pK_a$, then $\frac{1.0 - x}{x} = 1.0$, $x = 0.50$ mol HCl

b. $4.20 = 4.74 + \log \left(\frac{1.0 - x}{x}\right)$

$\frac{1.0 - x}{x} = 10^{-0.54} = 0.29$

$1.29x = 1.0$, $x = 0.78$ mol HCl

c. $5.00 = 4.74 + \log\left(\frac{1.0 - x}{x}\right)$

$\frac{1.0 - x}{x} = 10^{0.26} = 1.8, \quad 2.8x = 1.0, \quad x = 0.36$ mol HCl

29. $75.0 \text{ g } CH_3CO_2Na \times \frac{1 \text{ mol}}{82.03 \text{ g}} = 0.914$ mol NaOAc; $[OAc^-] = \frac{0.914 \text{ mol}}{0.5000 \text{ L}} = 1.83\ M$

$pH = pK_a + \log\frac{[OAc^-]}{[HOAc]} = 4.74 + \log\left(\frac{1.83}{0.64}\right) = 5.20$

30. $50.0 \text{ g } NH_4Cl \times \frac{1 \text{ mol } NH_4Cl}{53.49 \text{ g } NH_4Cl} = 0.935$ mol NH_4Cl added to 1.0 L

$pH = pK_a + \log\frac{[NH_3]}{[NH_4^+]} = 9.25 + \log\left(\frac{0.75}{0.935}\right) = 9.25 - 0.096 = 9.15$

31. $[H^+]$ added $= \frac{0.010 \text{ mol}}{0.25 \text{ L}} = \frac{0.040 \text{ mol}}{\text{L}}$ The added H^+ converts NH_3 to NH_4^+.

a.

	NH_4^+	$\rightleftharpoons$	NH_3 +	H^+	
Before	0.15 *M*		0.050 *M*	0.040 *M*	
Change	+0.040	←	-0.040	-0.040	React completely
After	0.19		0.010	0	

A buffer solution still exists. Using Henderson-Hasselbalch equation:

$pH = pK_a + \log\frac{[NH_3]}{[NH_4^+]} = 9.25 + \log\left(\frac{0.010}{0.19}\right) = 7.97$

b.

	NH_4^+	$\rightleftharpoons$	NH_3 +	H^+	
Before	1.5 *M*		0.50 *M*	0.040 *M*	
Change	+0.040	←	-0.040	-0.040	React completely
After	1.54		0.46	0	(carry extra sig. fig.)

To calculate the pH, plug in the new buffer concentrations into the Henderson-Hasselbalch equation.

$pH = pK_a + \log\frac{[NH_3]}{[NH_4^+]} = 9.25 + \log\left(\frac{0.46}{1.54}\right) = 8.73$

The two buffers differ in their capacity. Solution (b) has the greatest capacity, i.e., largest concentrations. Buffers with greater capacities will be able to absorb more acid or base.

32. a. Major species: H_2O, Na^+, OH^-, HOPr, OPr^-

			OH^-	HOPr	OPr^-
			0.15 *M*	0.050 *M*	0.080 *M*

OH^- will react completely with HOPr. HOPr is limiting reagent.

	$HOPr$ +	OH^-	→	OPr^- + H_2O	
Before	0.050 *M*	0.15 *M*		0.080 *M*	
Change	-0.050	-0.050	→	+0.050	React completely
After	0	0.10		0.130	

OH^- is in excess. OH^- contribution from the weak base, OPr^-, will be negligible. Consider only the excess strong base to determine pH.

$[OH^-] = 0.10$, pOH = 1.00, pH = 13.00

Note: Original pH of buffer is: $pH = 4.89 + \log\left(\frac{0.080}{0.050}\right) = 5.09$

b. OH^- will react completely with HOPr. OH^- is limiting reagent.

	$HOPr$ +	OH^-	→	OPr^- + H_2O	
Before	0.50 *M*	0.15 *M*		0.80 *M*	
Change	-0.15	-0.15	→	+0.15	React completely
After	0.35	0		0.95	

A buffer solution results.

$$pH = 4.89 + \log\left(\frac{0.95}{0.35}\right) = 4.89 + 0.43 = 5.32$$

c. Although a & b both started out as buffers with the same pH, the solution in "a" is no longer a buffer after 0.15 mol NaOH were added. We added more strong base than there was weak acid present in the buffer.

Acid-Base Titrations

33.

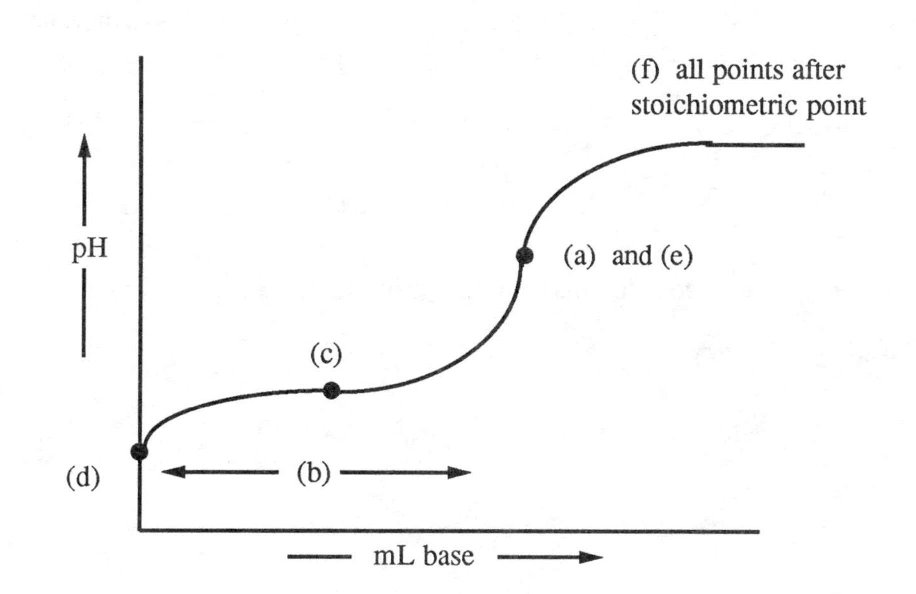

34.

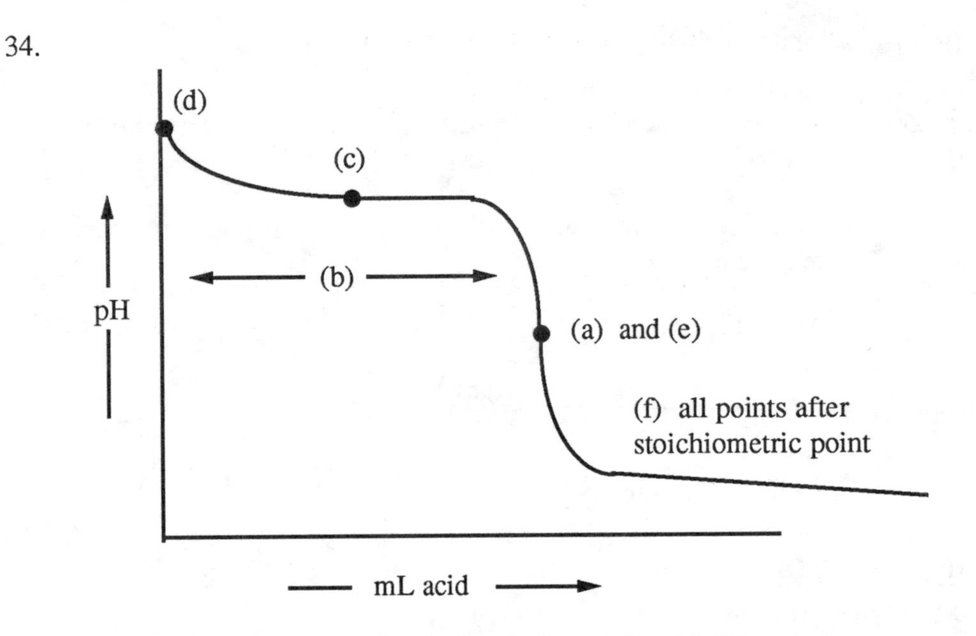

35. At the beginning of the titration, only the weak acid, HLac, present:

	HLac	$\rightleftharpoons$	H^+	+	Lac^-	$K_a = 1.4 \times 10^{-4}$	HLac: $HC_3H_5O_3$
Initial	0.100 *M*		~0		0		Lac^-: $C_3H_5O_3^-$
	x mol/L HLac dissociates too reach equilibrium						
Change	$-x$	$\rightarrow$	$+x$		$+x$		
Equil	$0.100 - x$		x		x		

$$1.4 \times 10^{-4} = \frac{x^2}{0.100 - x} \approx \frac{x^2}{0.100}$$

$x = [H^+] = 3.7 \times 10^{-3}$ *M*; pH = 2.43   Assumptions good.

Up to the stoichiometric point, we calculate the pH using the Henderson-Hasselbalch equation. This is the buffer region. For example, 4.0 mL of NaOH added:

$$\text{initial mmol HLac} = 25.0 \text{ mL} \times \frac{0.100 \text{ mmol}}{\text{mL}} = 2.50 \text{ mmol}$$

$$\text{mmol OH}^- \text{ added} = 4.0 \text{ mL} \times \frac{0.100 \text{ mmol}}{\text{mL}} = 0.40 \text{ mmol OH}^-$$

The base converts 0.40 mmoles HLac to 0.40 mmoles Lac^- according to the equation:

$$HLac + OH^- \rightarrow Lac^- + H_2O$$

mmol HLac remaining = 2.50 - 0.40 = 2.10,   mmol Lac^- produced = 0.40

We have a buffer solution. Using the Henderson-Hasselbalch equation:

$$pH = pK_a + \log \frac{[Lac^-]}{[HLac]} = 3.86 + \log \frac{(0.40)}{(2.10)}$$ (total volume cancels, so we can use the ratio of moles)

$pH = 3.14$

Other points in the buffer region are calculated in a similar fashion. Do a stoichiometry problem first, followed by a buffer problem. The buffer region includes all points up to 24.9 mL OH^- added.

At the stoichiometric point, we have added enough OH^- to convert all of the HLac (2.50 mmol) into its conjugate base, Lac^-. All that is present is a weak base. To determine the pH, we perform a weak base calculation.

$$[Lac^-] = \frac{2.50 \text{ mmol}}{50.0 \text{ mL}} = 0.0500\ M$$

	Lac^- + H_2O	$\rightleftharpoons$	HLac	+ OH^-
Initial	$0.0500\ M$		0	~0
	x mol/L Lac^- reacts with H_2O to reach equilibrium			
Change	$-x$	$\rightarrow$	$+x$	$+x$
Equil	$0.0500 - x$		x	x

$$K_b = \frac{1.00 \times 10^{-14}}{1.4 \times 10^{-4}} = 7.1 \times 10^{-11}$$

$$\frac{x^2}{0.0500 - x} \approx \frac{x^2}{0.0500} = 7.1 \times 10^{-11}$$

$x = [OH^-] = 1.9 \times 10^{-6}\ M$; $pOH = 5.72$; $pH = 8.28$ Assumptions good.

Past the stoichiometric point, we have added more than 2.50 mmol of NaOH. The pH will be determined by the excess OH^- ion present.

At 25.1 mL: OH^- added $= 25.1 \text{ mL} \times \frac{0.100 \text{ mmol}}{\text{mL}} = 2.51$ mmol

excess $OH^- = 2.51 - 2.50 = 0.01$ mmol

$$[OH^-] = \frac{0.01 \text{ mmol}}{50.1 \text{ mL}} = 2 \times 10^{-4}\ M;\ pOH = 3.7;\ pH = 10.3$$

All results are listed in Table 15.1 at the end of the solution to Exercise 15.38.

36. At the beginning of the titration, a weak acid problem:

	HOPr	$\rightleftharpoons$	H^+	+ OPr^-
Initial	$0.100\ M$		~0	0
	x mol/L propanoic acid dissociates to reach equilibrium			
Change	$-x$	$\rightarrow$	$+x$	$+x$
Equil	$0.100 - x$		x	x

$$K_a = \frac{[H^+][OPr^-]}{[HOPr]} = 1.3 \times 10^{-5} = \frac{x^2}{0.100 - x} \approx \frac{x^2}{0.100}$$

$x = [H^+] = 1.1 \times 10^{-3}$, pH = 2.96 Assumptions good.

Buffer region, e.g. at 24.0 mL OH^- added:

$$\text{initial mmol HOPr} = 25.0 \text{ mL} \times \frac{0.100 \text{ mmol}}{\text{mL}} = 2.50 \text{ mmol}$$

$$\text{mmol } OH^- \text{ added} = 24.0 \text{ mL} \times \frac{0.100 \text{ mmol}}{\text{mL}} = 2.40 \text{ mmol}$$

The base converts HOPr to OPr^-.

	HOPr	+	OH^-	$\longrightarrow$	OPr^- + H_2O	
Before	2.50 mmol		2.40 mmol		0	
Change	-2.40		-2.40	$\longrightarrow$	+2.40	React completely
After	0.10 mmol		0		2.40 mmol	

A buffer solution results.

$$pH = pK_a + \log \frac{[\text{Base}]}{[\text{Acid}]} = 4.89 + \log \frac{[OPr^-]}{[HOPr]}$$

$$pH = 4.89 + \log\left(\frac{2.40}{0.10}\right) = 4.89 + 1.38 = 6.27$$ (Volume cancels, so we can use moles.)

All points in the buffer region, 4.0 mL to 24.9 mL are calculated this way. See Table 15.1 at the end of Exercise 15.38 for all the results.

At the stoichiometric point, only a weak base (OPr^-) present:

	OPr^- + H_2O	$\rightleftharpoons$	OH^-	+ HOPr
Initial	$\frac{2.50 \text{ mmol}}{50.0 \text{ mL}} = 0.0500\ M$		~0	0
	x mol/L OPr^- reacts with H_2O to reach equilibrium			
Change	$-x$	$\longrightarrow$	$+x$	$+x$
Equil	$0.0500 - x$		x	x

$$K_b = \frac{[OH^-][HOPr]}{[OPr^-]} = \frac{K_w}{K_a} = 7.7 \times 10^{-10} = \frac{x^2}{0.0500 - x} \approx \frac{x^2}{0.0500}$$

$x = 6.2 \times 10^{-6}\ M = [OH^-]$, pOH = 5.21, pH = 8.79 Assumptions good.

Beyond the stoichiometric point the pH is determined by the excess added base. The results are identical to those same points in Exercise 35. (See Table 15.1)

For example at 26 mL NaOH added:

$$[OH^-] = \frac{2.6 \text{ mmol} - 2.50 \text{ mmol}}{51.0 \text{ mL}} = 2 \times 10^{-3} M$$

pOH = 2.7, pH = 11.3

37. At beginning of titration, only a weak base, NH_3, present:

	NH_3 + H_2O	⇌	NH_4^+	+	OH^-	$K_b = 1.8 \times 10^{-5}$
Equil.	$0.100 - x$		x		x	

$$\frac{x^2}{0.100 - x} \approx \frac{x^2}{0.100} = 1.8 \times 10^{-5}$$

$x = [OH^-] = 1.3 \times 10^{-3}$; pOH = 2.89; pH = 11.11 Assumptions good.

In buffer region (4 - 24.9 mL):

Use $pH = pK_a + \log \frac{[Base]}{[Acid]}$ $K_a = \frac{1.0 \times 10^{-14}}{1.8 \times 10^{-5}} = 5.6 \times 10^{-10}$ $pK_a = 9.25$

$$pH = 9.25 + \log \frac{[NH_3]}{[NH_4^+]}$$

For example, after 8.0 mL HCl added:

$$\text{initial mmol } NH_3 = 25.0 \text{ mL} \times \frac{0.100 \text{ mmol}}{\text{mL}} = 2.50 \text{ mmol}$$

$$\text{mmol } H^+ \text{ added} = 8.0 \text{ mL} \times \frac{0.100 \text{ mmol}}{\text{mL}} = 0.80 \text{ mmol}$$

Added H^+ converts: $NH_3 + H^+ \rightarrow NH_4^+$

mmol NH_3 remaining = 2.50 - 0.80 = 1.70 mmol, mmol NH_4^+ produced = 0.80 mmol

$pH = 9.25 + \log \frac{1.70}{0.80} = 9.58$ (mole ratios can be used since total volume cancels)

At stoichiometric point, enough HCl has been added to convert all of the weak base (NH_3) into its conjugate acid, NH_4^+. Perform a weak acid calculation, $[NH_4^+]_o = \frac{2.50 \text{ mmol}}{50.0 \text{ mL}} = 0.0500\, M$:

	NH_4^+	⇌	H^+	+	NH_3	$K_a = 5.6 \times 10^{-10}$
Equil	$0.0500 - x$		x		x	

$$5.6 \times 10^{-10} = \frac{x^2}{0.0500 - x} \approx \frac{x^2}{0.0500}$$

$x = [H^+] = 5.3 \times 10^{-6} M$; pH = 5.28 Assumptions good.

Beyond the stoichiometric point, the pH is determined by the excess H^+.

At 28 mL: H^+ added = 28 mL $\times \dfrac{0.100 \text{ mmol}}{\text{mL}}$ = 2.8 mmol

Excess H^+ = 2.8 mmol - 2.5 mmol = 0.3 mmol

$[H^+] = \dfrac{0.3 \text{ mmol}}{53 \text{ mL}} = 6 \times 10^{-3}\ M$; pH = 2.2

The results are in Table 15.1.

38. Initially, a weak base problem:

	py	+	H_2O	$\rightleftharpoons$	Hpy^+	+	OH^-	py is pyridine
Equil	$0.100 - x$				x		x	

$$K_b = \frac{[Hpy^+][OH^-]}{[py]} = \frac{x^2}{0.100 - x} \approx \frac{x^2}{0.100} = 1.7 \times 10^{-9}$$

$x = [OH^-] = 1.3 \times 10^{-5}$; pOH = 4.89; pH = 9.11 Assumptions good.

Buffer region: $K_a = \dfrac{K_w}{K_b} = \dfrac{1.0 \times 10^{-14}}{1.7 \times 10^{-9}} = 5.9 \times 10^{-6}$, $pK_a = 5.23$

Added acid converts: $py + H^+ \rightarrow Hpy^+$. Determine moles of py and Hpy^+ after reaction and use the Henderson-Hasselbalch equation to solve for the pH.

$$pH = 5.23 + \log \frac{[py]}{[Hpy^+]}$$

At stoichiometric point, a weak acid problem:

	Hpy^+	$\rightleftharpoons$	py	+	H^+	$K_a = 5.9 \times 10^{-6}$
Equil	$0.0500 - x$		x		x	

$$5.9 \times 10^{-6} = \frac{x^2}{0.0500 - x} \approx \frac{x^2}{0.0500}$$

$x = [H^+] = 5.4 \times 10^{-4}$; pH = 3.27 Assumptions good.

Beyond the equivalence point, the pH is the same as in Exercise 15.37.

All results are summarized in Table 15.1.

Table 15.1: Summary of results in Exercises 15.35 - 15.38
(Graph below)

mL added titrant	pH Exercise 15.35	Exercise 15.36	Exercise 15.37	Exercise 15.38
0.0	2.43	2.96	11.11	9.11
4.0	3.14	4.17	9.97	5.95
8.0	3.53	4.56	9.58	5.56
12.5	3.86	4.89	9.25	5.23
20.0	4.46	5.49	8.65	4.63
24.0	5.24	6.27	7.87	3.85
24.5	5.6	6.6	7.6	3.5
24.9	6.3	7.3	6.9	----
25.0	8.28	8.79	5.28	3.27
25.1	10.3	10.3	3.7	----
26.	11.3	11.3	2.7	2.7
28.	11.8	11.8	2.2	2.2
30.	12.0	12.0	2.0	2.0

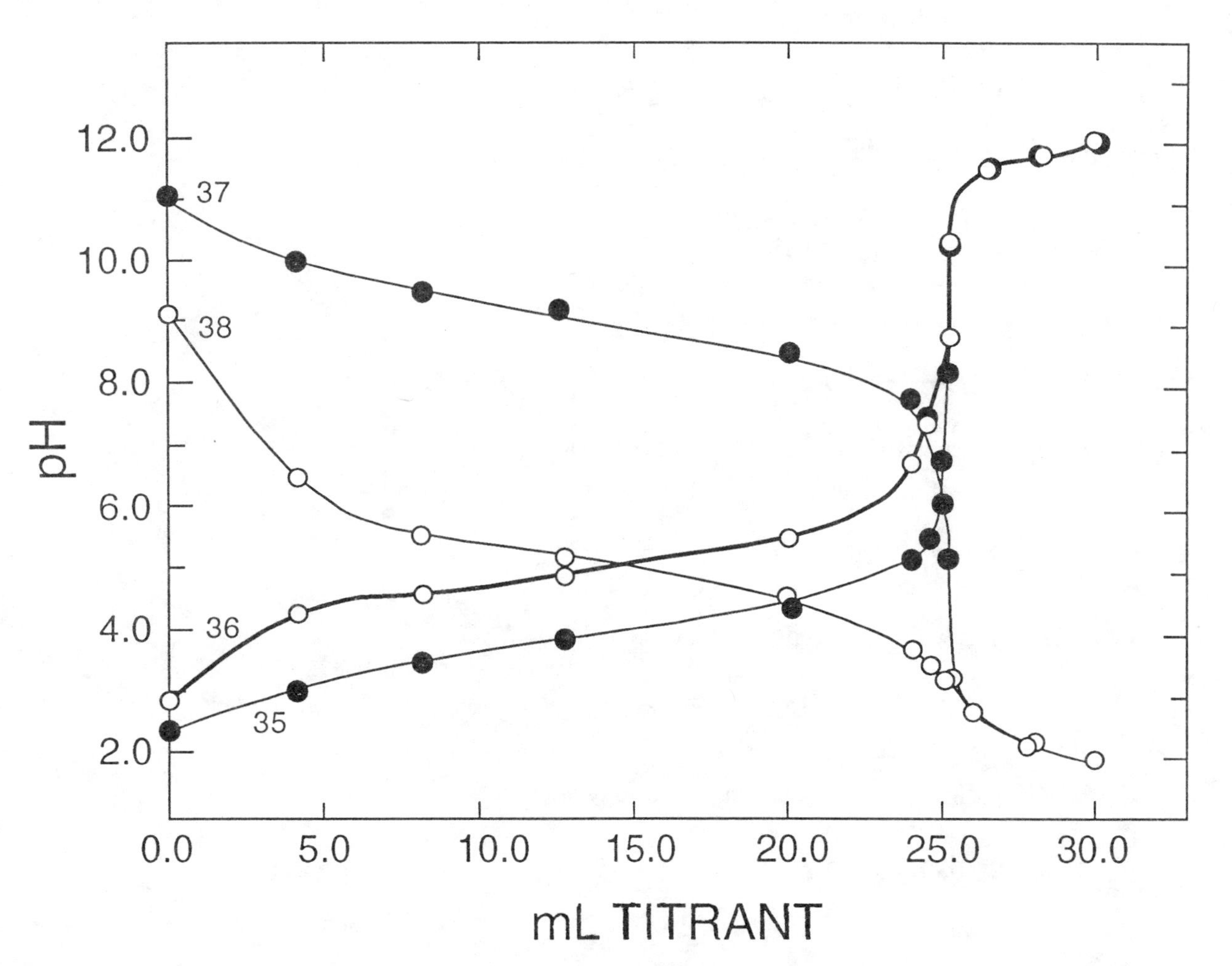

39. a. $0.104 \text{ g NaC}_2\text{H}_3\text{O}_2 \times \frac{1 \text{ mol NaC}_2\text{H}_3\text{O}_2}{82.03 \text{ g}} = 1.27 \times 10^{-3} \text{ mol} = 1.27 \text{ mmol NaC}_2\text{H}_3\text{O}_2$

$1.27 \text{ mmol NaC}_2\text{H}_3\text{O}_2 \times \frac{1 \text{ mmol HCl}}{\text{mmol NaC}_2\text{H}_3\text{O}_2} \times \frac{1 \text{ mL}}{0.0996 \text{ mmol HCl}} = 12.7 \text{ mL HCl}$

Add H^+ reacts with the acetate ion to completion:

$$H^+ + C_2H_3O_2^- \rightarrow HC_2H_3O_2$$

The titration converts the acetate ion into acetic acid. At the equivalence point, there is a solution of only acetic acid with a concentration of:

$$\frac{1.27 \text{ mmol}}{25 \text{ mL} + 12.7 \text{ mL}} = 3.4 \times 10^{-2}\ M$$

The acetic acid dissociates, making the equivalence point acidic:

	HOAc	$\rightleftharpoons$	H^+	+	OAc^-
Initial	0.034 M		~0		0
	x mol/L HOAc dissociates to reach equilibrium				
Change	$-x$	$\rightarrow$	$+x$		$+x$
Equil	$0.034 - x$		x		x

$$K_a = 1.8 \times 10^{-5} = \frac{[H^+][OAc^-]}{[HOAc]} = \frac{x^2}{0.034 - x} \approx \frac{x^2}{0.034}$$

$x = [H^+] = 7.8 \times 10^{-4}\ M$; pH = 3.11 Assumptions good.

b. $50.00 \text{ mL} \times \frac{0.0426 \text{ mmol HOCl}}{\text{mL}} = 2.13 \text{ mmol HOCl}$

$2.13 \text{ mmol HOCl} \times \frac{1 \text{ mmol NaOH}}{\text{mmol HOCl}} \times \frac{1 \text{ mL}}{0.1028 \text{ mol NaOH}} = 20.7 \text{ mL of NaOH soln.}$

Added OH^- reacts with HOCl to completion:

$$OH^- + HOCl \rightarrow OCl^-$$

The titration converts HOCl into OCl^- ion. At the equivalence point, there are 2.13 mmol of OCl^- in 70.7 mL of solution (50.0 mL + 20.7 mL). The concentration of OCl^- is:

$$\frac{2.13 \text{ mmol}}{70.7 \text{ mL}} = 3.01 \times 10^{-2} \text{ mol/L}$$

The OCl^- reacts with H_2O, to produce a basic solution.

	OCl^-	+	H_2O	$\rightleftharpoons$	HOCl	+	OH^-
Initial	0.0301 M				0		~0
	x mol/L OCl^- reacts with H_2O to reach equilibrium						
Change	$-x$			$\rightarrow$	$+x$		$+x$
Equil	$0.0301 - x$				x		x

$$K_b = \frac{K_w}{K_a} = \frac{1.0 \times 10^{-14}}{3.5 \times 10^{-8}} = 2.9 \times 10^{-7} = \frac{[HOCl][OH^-]}{[OCl^-]} = \frac{x^2}{0.0301 - x}$$

$$2.9 \times 10^{-7} \approx \frac{x^2}{0.0301};\ x = [OH^-] = 9.3 \times 10^{-5};\ pOH = 4.03;\ pH = 9.97$$

Assumptions good.

40. a. HBr is a strong acid; NaOH is a strong base. At the equivalence point we will have an aqueous solution of NaBr. Neither Na^+ nor Br^- reacts with water; pH = 7.00.

b. Assuming quinine is monobasic with $pK_b = 5.1$;

$K_b = 10^{-5.1} = 7.9 \times 10^{-6}$ (carry one extra significant figure)

Solubility of quinine 1.00 g/1900.0 mL

Quinine (Q): $C_{20}H_{24}N_2O_2$: 324.4 g/mol

$$\frac{1.00\ g}{1.900\ L} \times \frac{1\ mol}{324.4\ g} = 1.62 \times 10^{-3}\ mol/L$$

$$\frac{1.62 \times 10^{-3}\ mmol\ Q}{mL} \times 100.0\ mL = 1.62 \times 10^{-1}\ mmol\ Q$$

$$H^+ + Q \longrightarrow HQ^+$$

$$1.62 \times 10^{-1}\ mmol\ H^+ \times \frac{1\ mL}{0.0200\ mmol} = 8.1\ mL\ of\ HCl\ needed\ to\ titrate\ Q$$

At stoichiometric point, H^+ will convert all of the quinine, Q, into the conjugate acid of quinine, HQ^+.

	HQ^+	$\rightleftharpoons$	H^+	+	Q
Initial	$\frac{0.162\ mmol}{108.1\ mL} = 1.50 \times 10^{-3}\ M$		~0		0
	x mol/L HQ^+ dissociates to reach equilibrium				
Change	$-x$	$\longrightarrow$	$+x$		$+x$
Equil	$1.50 \times 10^{-3} - x$		x		x

$$K_a = \frac{K_w}{K_b} = 1.3 \times 10^{-9} = \frac{x^2}{1.50 \times 10^{-3} - x} \approx \frac{x^2}{1.50 \times 10^{-3}}$$

$x = [H^+] = 1.4 \times 10^{-6}$, pH = 5.86 = 5.9 Assumptions good.

41. The pH will be less than about 0.5.

42. pH > 5 for bromcresol green to be blue.

pH < 8 for thymol blue to be yellow.

pH is between 5 and 8.

43. a. yellow b. orange c. blue d. bluish green

44. a. yellow c. green (Both yellow and blue forms are present.)

b. yellow d. colorless

45.

Exercise	pH at stoich. pt.	indicator
15.35	8.28	phenolphthalein
15.37	5.28	bromcresol green
15.39a	3.11	erythrosin B or 2,4-dinitrophenol
15.39b	9.97	thymolphthalein

titration in 15.39a not feasible, pH break too small

46.

Exercise	pH at stoich. pt.	indicator
15.36	8.79	phenolphthalein
15.38	3.27	erythrosin B or 2,4-dinitrophenol
15.40a	7.00	bromthymol blue
15.40b	5.9	bromcresol purple

titration in 15.38 not feasible, pH break too small

Solubility Equilibria

47. a. $AgC_2H_3O_2(s) \rightleftharpoons Ag^+(aq) + C_2H_3O_2^-(aq)$; $K_{sp} = [Ag^+]\,[C_2H_3O_2^-]$

b. $MnS(s) \rightleftharpoons Mn^{2+}(aq) + S^{2-}(aq)$; $K_{sp} = [Mn^{2+}]\,[S^{2-}]$

c. $Al(OH)_3(s) \rightleftharpoons Al^{3+}(aq) + 3\,OH^-(aq)$; $K_{sp} = [Al^{3+}]\,[OH^-]^3$

d. $Ca_3(PO_4)_2(s) \rightleftharpoons 3\,Ca^{2+}(aq) + 2\,PO_4^{3-}(aq)$; $K_{sp} = [Ca^{2+}]^3\,[PO_4^{3-}]^2$

48. a. $Ca_5(PO_4)_3(OH)\,(s) \rightleftharpoons 5\,Ca^{2+}(aq) + 3\,PO_4^{3-}(aq) + OH^-(aq)$

$K_{sp} = [Ca^{2+}]^5\,[PO_4^{3-}]^3\,[OH^-]$

b. $PbI_2(s) \rightleftharpoons Pb^{2+}(aq) + 2\,I^-(aq)$; $K_{sp} = [Pb^{2+}]\,[I^-]^2$

c. $Cu_2S(s) \rightleftharpoons 2\,Cu^+(aq) + S^{2-}(aq)$; $K_{sp} = [Cu^+]^2\,[S^{2-}]$

d. $Ni(OH)_2(s) \rightleftharpoons Ni^{2+}(aq) + 2\,OH^-(aq)$; $K_{sp} = [Ni^{2+}]\,[OH^-]^2$

49. a. $CaC_2O_4(s) \rightleftharpoons Ca^{2+}(aq) + C_2O_4^{2-}(aq)$

$[Ca^{2+}] = [C_2O_4^{2-}] = 4.8 \times 10^{-5}$ mol/L

$K_{sp} = [Ca^{2+}][C_2O_4^{2-}] = (4.8 \times 10^{-5})^2 = 2.3 \times 10^{-9}$

b. $PbBr_2(s) \rightleftharpoons Pb^{2+}(aq) + 2\,Br^-(aq)$

s = solubility in mol/L → s 2s s = 2.14×10^{-2} *M*

$K_{sp} = [Pb^{2+}][Br^-]^2 = (2.14 \times 10^{-2})(4.28 \times 10^{-2})^2 = 3.92 \times 10^{-5}$

c. $BiI_3(s) \rightleftharpoons Bi^{3+}(aq) + 3\,I^-(aq)$

s = solubility in mol/L → s 3s s = 1.32×10^{-5} *M*

$[I^-] = 3s = 3.96 \times 10^{-5}$

$K_{sp} = [Bi^{3+}][I^-]^3 = (1.32 \times 10^{-5})(3.96 \times 10^{-5})^3 = 8.20 \times 10^{-19}$

50. a. $FeC_2O_4(s) \rightleftharpoons Fe^{2+}(aq) + C_2O_4^{2-}(aq)$

s = solubility in mol/L → s s

$$s = \frac{65.9 \times 10^{-3}\ g}{L} \times \frac{1\ mol}{143.9\ g} = 4.58 \times 10^{-4}\ mol/L$$

$[Fe^{2+}] = [C_2O_4^{2-}] = 4.58 \times 10^{-4}$ *M*

$K_{sp} = [Fe^{2+}][C_2O_4^{2-}] = (4.58 \times 10^{-4})^2 = 2.10 \times 10^{-7}$

b. $Cu(IO_4)_2(s) \rightleftharpoons Cu^{2+}(aq) + 2\,IO_4^-(aq)$

s = solubility in mol/L → s 2s

$$s = \frac{0.146\ g\ Cu(IO_4)_2}{0.100\ L} \times \frac{1\ mol}{445.3\ g} = 3.28 \times 10^{-3}\ mol/L$$

$[Cu^{2+}] = 3.28 \times 10^{-3}$ *M*; $[IO_4^-] = 6.56 \times 10^{-3}$ *M*

$K_{sp} = [Cu^{2+}][IO_4^-]^2 = (3.28 \times 10^{-3})(6.56 \times 10^{-3})^2$

$K_{sp} = 1.41 \times 10^{-7}$

c. $Li_2CO_3(s) \rightleftharpoons 2\,Li^+(aq) + CO_3^{2-}(aq)$

s = solubility in mol/L $\rightarrow$ 2s s

$$s = \frac{5.48\text{ g}}{\text{L}} \times \frac{1\text{ mol}}{73.89\text{ g}} = 7.42 \times 10^{-2}\text{ mol/L}$$

$[Li^+] = 1.48 \times 10^{-1}\,M$; $[CO_3^{2-}] = 7.42 \times 10^{-2}\,M$

$K_{sp} = [Li^+]^2\,[CO_3^{2-}] = (1.48 \times 10^{-1})^2\,(7.42 \times 10^{-2}) = 1.63 \times 10^{-3}$

51. s = solubility in mol/L of solid

a. $Al(OH)_3(s) \rightleftharpoons Al^{3+}(aq) + 3\,OH^-(aq)$

	$Al(OH)_3(s)$		$Al^{3+}(aq)$	$3\,OH^-(aq)$
Initial			0	10^{-7} from water
	s mol/L $Al(OH)_3(s)$ dissolves to reach equilibrium			
Change	-s	$\rightarrow$	+s	+3s
Equil			s	$10^{-7} + 3s$

$K_{sp} = 2 \times 10^{-32} = [Al^{3+}]\,[OH^-]^3 = (s)\,(10^{-7} + 3s)^3 = s(10^{-7})^3$ Assuming $10^{-7} + 3s \approx 10^{-7}$

$$s = \frac{2 \times 10^{-32}}{10^{-21}} = 2 \times 10^{-11}\text{ mol/L}$$ Assumption good.

$$\frac{2 \times 10^{-11}\text{ mol}}{\text{L}} \times \frac{78.00\text{ g}}{\text{mol}} = 2 \times 10^{-9}\text{ g/L}$$

b. $Be(IO_4)_2 \rightleftharpoons Be^{2+} + 2\,IO_4^-$

	$Be(IO_4)_2$		Be^{2+}	$2\,IO_4^-$
Initial			0	0
	s mol/L dissolves to reach equilibrium			
Change	-s	$\rightarrow$	+s	+2s
Equil			s	2s

$K_{sp} = 1.57 \times 10^{-9} = [Be^{2+}]\,[IO_4^-]^2 = (s)\,(2s)^2 = 4s^3$, $s = \left(\frac{1.57 \times 10^{-9}}{4}\right)^{\frac{1}{3}}$

$s = 7.32 \times 10^{-4}$ mol/L; $\frac{7.32 \times 10^{-4}\text{ mol}}{\text{L}} \times \frac{390.8\text{ g}}{\text{mol}} = 0.286\text{ g/L}$

c. $CaSO_4 \rightleftharpoons Ca^{2+} + SO_4^{2-}$

	$CaSO_4$		Ca^{2+}	SO_4^{2-}
Initial			0	0
	s mol/L dissolves to reach equilibrium			
Change	-s	$\rightarrow$	+s	+s
Equil			s	s

$K_{sp} = 6.1 \times 10^{-5} = [Ca^{2+}][SO_4^{2-}] = s^2,\ s = (6.1 \times 10^{-5})^{\frac{1}{2}}$

$s = 7.8 \times 10^{-3}$ mol/L; $\frac{7.8 \times 10^{-3}\ \text{mol}}{\text{L}} \times \frac{136.2\ \text{g}}{\text{mol}} = 1.1$ g/L

d.

	$CaCO_3$	$\rightleftharpoons$	Ca^{2+} +	CO_3^{2-}
Initial			0	0
	s mol/L dissolves to reach equilibrium			
Change	-s	→	+s	+s
Equil			s	s

$K_{sp} = 8.7 \times 10^{-9} = [Ca^{2+}][CO_3^{2-}] = s^2,\ s = (8.7 \times 10^{-9})^{\frac{1}{2}}$

$s = 9.3 \times 10^{-5}$ mol/L; $\frac{9.3 \times 10^{-5}\ \text{mol}}{\text{L}} \times \frac{100.1\ \text{g}}{\text{mol}} = 9.3 \times 10^{-3}$ g/L

Note: This ignores any reaction between CO_3^{2-} and H_2O to form HCO_3^-.

e.

	$Mg(NH_4)PO_4$	$\rightleftharpoons$	Mg^{2+} +	NH_4^+ +	PO_4^{3-}
Initial			0	0	0
	s mol/L dissolves to reach equilibrium				
Change	-s	→	+s	+s	+s
Equil			s	s	s

$K_{sp} = 3 \times 10^{-13} = [Mg^{2+}][NH_4^+][PO_4^{3-}] = s^3,\ s = (3 \times 10^{-13})^{\frac{1}{3}}$

$s = 7 \times 10^{-5}$ mol/L; $\frac{7 \times 10^{-5}\ \text{mol}}{\text{L}} \times \frac{137.3\ \text{g}}{\text{mol}} = 1 \times 10^{-2}$ g/L

Note: We are ignoring any reaction between NH_4^+ and PO_4^{3-}.

52. s = solubility in mol/L of solid

a.

	$Hg_2Cl_2(s)$	$\rightleftharpoons$	$Hg_2^{2+}(aq)$ +	$2\ Cl^-(aq)$
Initial			0	0
	s mol/L dissolves to reach equilibrium			
Change	-s	→	+s	+2s
Equil			s	2s

$K_{sp} = 1.1 \times 10^{-18} = [Hg_2^{2+}][Cl^-]^2 = (s)(2s)^2 = 4s^3,\ s = 6.5 \times 10^{-7}$ mol/L

$\frac{6.5 \times 10^{-7}\ \text{mol}}{\text{L}} \times \frac{472.1\ \text{g}}{\text{mol}} = \frac{3.1 \times 10^{-4}\ \text{g}}{\text{L}}$

b.

$SrSO_4$	$\rightleftharpoons$	Sr^{2+} +	SO_4^{2-}
s mol/L dissolves	→	s	s

$K_{sp} = 3.2 \times 10^{-7} = [Sr^{2+}][SO_4^{2-}] = s^2$

$s = 5.7 \times 10^{-4}$ mol/L

$$\frac{5.7 \times 10^{-4} \text{ mol}}{\text{L}} \times \frac{183.7 \text{ g}}{\text{mol}} = \frac{0.10 \text{ g}}{\text{L}}$$

c. $Ag_2CO_3 \rightleftharpoons 2\,Ag^+ + CO_3^{2-}$ (ignore $CO_3^{2-} + H_2O \rightleftharpoons HCO_3^- + OH^-$)

s mol/L dissolves $\rightarrow$ 2s s

$K_{sp} = 8.1 \times 10^{-12} = [Ag^+]^2[CO_3^{2-}] = (2s)^2(s) = 4s^3$

$s = 1.3 \times 10^{-4}$ mol/L; $\dfrac{1.3 \times 10^{-4} \text{ mol}}{\text{L}} \times \dfrac{275.8 \text{ g}}{\text{mol}} = \dfrac{3.6 \times 10^{-2} \text{ g}}{\text{L}}$

d. $Ag_2Cl_2O_7 \rightleftharpoons 2\,Ag^+ + Cl_2O_7^{2-}$

s mol/L dissolves $\rightarrow$ 2s s

$K_{sp} = 2 \times 10^{-7} = [Ag^+]^2[Cl_2O_7^{2-}] = (2s)^2(s) = 4s^3$

$s = 3.7 \times 10^{-3}$ mol/L $\approx 4 \times 10^{-3}$ mol/L

$$\frac{4 \times 10^{-3} \text{ mol}}{\text{L}} \times \frac{398.7 \text{ g}}{\text{mol}} = 1.6 \text{ g/L} \approx 2 \text{ g/L}$$

e. $Ag_2CrO_4 \rightleftharpoons 2\,Ag^+ + CrO_4^{2-}$ Ignoring weak base properties of CrO_4^{2-}

s mol/L dissolves $\rightarrow$ 2s s

$K_{sp} = 9.0 \times 10^{-12} = [Ag^+]^2[CrO_4^{2-}] = (2s)^2(s) = 4s^3$

$s = 1.3 \times 10^{-4}$ mol/L

$$\frac{1.3 \times 10^{-4} \text{ mol}}{\text{L}} \times \frac{331.8 \text{ g}}{\text{mol}} = 4.3 \times 10^{-2} \text{ g/L}$$

53. If the anion in the salt can act as a base in water, the solubility will increase as the solution becomes more acidic.

This is true for $Al(OH)_3$, $CaCO_3$, $Mg(NH_4)PO_4$, Ag_2CO_3, and Ag_2CrO_4.

$Al(OH)_3$, $Al(OH)_3 + 3\,H^+ \rightarrow Al^{3+} + 3\,H_2O$

$CaCO_3 + H^+ \rightarrow Ca^{2+} + HCO_3^- \xrightarrow{H^+} Ca^{2+} + H_2O + CO_2(g)$

$$Mg(NH_4)PO_4 + H^+ \rightarrow Mg^{2+} + NH_4^+ + HPO_4^{2-}$$

$$HPO_4^{2-} \xrightarrow{H^+} H_2PO_4^- \xrightarrow{H^+} H_3PO_4$$

$$Ag_2CO_3 + H^+ \rightarrow 2\,Ag^+ + HCO_3^- \xrightarrow{H^+} 2\,Ag^+ + H_2O + CO_2(g)$$

$$2Ag_2CrO_4 + 2\,H^+ \rightarrow 4\,Ag^+ + Cr_2O_7^{2-} + H_2O$$

$CaSO_4$ and $SrSO_4$ will be slightly more soluble in acid. Since,

$$MSO_4 + H^+ \rightarrow M^{2+} + HSO_4^-$$

occurs. However, K_b for SO_4^{2-} is only about 10^{-12}, and the effect of acid on these salts is not as great as on the others.

54. a. AgF b. PbO c. $Sr(NO_2)_2$ d. $Ni(CN)_2$

All the above have anions that are weak bases. The anions of the other choices are conjugate bases of strong acids. They have no basic properties in water.

55. s = solubility in mol/L

a.

	$Fe(OH)_3(s)$	$\rightleftharpoons$	$Fe^{3+}(aq)$	+	$3\,OH^-(aq)$
Initial			0		$1 \times 10^{-7}\,M$
	s mol/L dissolves to reach equilibrium				
Change	-s	$\rightarrow$	+s		+3s
Equil			s		$1 \times 10^{-7} + 3s$

$$K_{sp} = 4 \times 10^{-38} = [Fe^{3+}]\,[OH^-]^3 = (s)\,(1 \times 10^{-7} + 3s)^3 \approx s(1 \times 10^{-7})^3$$

$s = 4 \times 10^{-17}$ mol/L Assumption good ($s << 1 \times 10^{-7}$).

b. $Fe(OH)_3 \rightleftharpoons Fe^{3+} + 3\,OH^-$ pH = 5.0, pOH = 9.0, $[OH^-] = 1 \times 10^{-9}$

	$Fe(OH)_3$	$\rightleftharpoons$	Fe^{3+}	+	$3\,OH^-$	
Initial			0		$1 \times 10^{-9}\,M$	(buffered)
	s mol/L dissolves to reach equilibrium					
Change	-s	$\rightarrow$	+s		-----	(assume no pH change in buffer)
Equil			s		1×10^{-9}	

$$K_{sp} = 4 \times 10^{-38} = [Fe^{3+}]\,[OH^-]^3 = (s)\,(1 \times 10^{-9})^3$$

$s = 4 \times 10^{-11}$ mol/L

c. pH = 11.0; pOH = 3.0, $[OH^-] = 1 \times 10^{-3}$ *M*

	$Fe(OH)_3$	$\rightleftharpoons$	Fe^{3+}	+	$3\ OH^-$	
Initial			0		0.001 *M*	(buffered)
	s mol/L dissolves to reach equilibrium					
Change	-s	$\rightarrow$	+s		-----	(assume no pH change)
Equil			s		0.001	

$K_{sp} = 4 \times 10^{-38} = [Fe^{3+}]\,[OH^-]^3 = (s)\,(0.001)^3$

$s = 4 \times 10^{-29}$ mol/L

56. s = solubility in mol/L

a.

$PbI_2(s)$	$\rightleftharpoons$	$Pb^{2+}(aq)$	+	$2\ I(aq)^-$
s mol/L dissolves	$\rightarrow$	s		2s

$K_{sp} = 1.4 \times 10^{-8} = [Pb^{2+}]\,[I^-]^2 = 4s^3$; $s = 1.5 \times 10^{-3}$ mol/L

b.

	PbI_2	$\rightleftharpoons$	Pb^{2+}	+	$2\ I^-$
Initial			0.10 *M*		0
	s mol/L PbI_2 dissolves to reach equilibrium				
Change	-s	$\rightarrow$	+s		+2s
Equil			0.10 + s		2s

$1.4 \times 10^{-8} = (0.10 + s)\,(2s)^2 \approx (0.10)\,(2s)^2 = 0.40\ s^2$

$s = 1.9 \times 10^{-4}$ mol/L Assumption good.

c.

	PbI_2	$\rightleftharpoons$	Pb^{2+}	+	$2\ I^-$
Initial			0		0.010 *M*
	s mol/L PbI_2 dissolves to reach equilibrium				
Change	-s	$\rightarrow$	+s		+2s
Equil			s		0.010 + 2s

$1.4 \times 10^{-8} = (s)\,(0.010 + 2s)^2 \approx (s)\,(0.010)^2$; $s = 1.4 \times 10^{-4}$ mol/L Assumption good.

57. After mixing $[Pb^{2+}]_o = [Cl^-]_o = 0.010\ M$

$[Pb^{2+}]_o\,[Cl^-]_o^{\,2} = (0.010)\,(0.010)^2 = 1.0 \times 10^{-6} < K_{sp}\ (1.6 \times 10^{-5})$

No precipitate will form. Thus,

$[Pb^{2+}] = [Cl^-] = 0.010\ M$, $[NO_3^-] = 2(0.010) = 0.020\ M$

$[Na^+] = [Cl^-] = 0.010\ M$

58. $$[Ba^{2+}]_o = \frac{75.0\text{ mL} \times \frac{0.020\text{ mmol}}{\text{mL}}}{200.\text{ mL}} = 7.5 \times 10^{-3}\ M$$

$$[SO_4^{2-}]_o = \frac{125\text{ mL} \times \frac{0.040\text{ mmol}}{\text{mL}}}{200.\text{ mL}} = 2.5 \times 10^{-2}\ M$$

$[Ba^{2+}]_o\,[SO_4^{2-}]_o = (7.5 \times 10^{-3})(2.5 \times 10^{-2}) = 1.9 \times 10^{-4} > K_{sp}\ (1.5 \times 10^{-9})$

A precipitate of $BaSO_4(s)$ will form.

	$BaSO_4(s)$	$\rightleftharpoons$	$Ba^{2+}(aq)$	+	$SO_4^{2-}(aq)$	
Before			0.0075 M		0.025 M	
	0.0075 mol/L Ba^{2+} reacts with SO_4^{2-}					
Change		←	-0.0075		-0.0075	React completely
After			0		0.0175	New I (carry extra S.F.)
	s mol/L $BaSO_4$ dissolves to reach equilibrium					
Change	-s		+s		+s	
Equil			s		0.0175 + s	

$K_{sp} = 1.5 \times 10^{-9} = [Ba^{2+}]\,[SO_4^{2-}] = (s)\,(0.0175 + s)$

$1.5 \times 10^{-9} \approx (0.0175)s$

$s = 8.6 \times 10^{-8}$; $[Ba^{2+}] = 8.6 \times 10^{-8}\ M$; $[SO_4^{2-}] = 0.018\ M$ Assumption good.

Complex Ion Equilibria

59. a.

$$Co^{2+} + NH_3 \rightleftharpoons CoNH_3^{2+} \qquad K_1$$
$$CoNH_3^{2+} + NH_3 \rightleftharpoons Co(NH_3)_2^{2+} \qquad K_2$$
$$Co(NH_3)_2^{2+} + NH_3 \rightleftharpoons Co(NH_3)_3^{2+} \qquad K_3$$
$$Co(NH_3)_3^{2+} + NH_3 \rightleftharpoons Co(NH_3)_4^{2+} \qquad K_4$$
$$Co(NH_3)_4^{2+} + NH_3 \rightleftharpoons Co(NH_3)_5^{2+} \qquad K_5$$
$$Co(NH_3)_5^{2+} + NH_3 \rightleftharpoons Co(NH_3)_6^{2+} \qquad K_6$$

$$Co^{2+} + 6\,NH_3 \rightleftharpoons Co(NH_3)_6^{2+} \qquad K_f = K_1K_2K_3K_4K_5K_6$$

b.

$$Fe^{3+} + SCN^- \rightleftharpoons FeSCN^{2+} \qquad K_1$$
$$FeSCN^{2+} + SCN^- \rightleftharpoons Fe(SCN)_2^{+} \qquad K_2$$
$$Fe(SCN)_2^{+} + SCN^- \rightleftharpoons Fe(SCN)_3 \qquad K_3$$
$$Fe(SCN)_3 + SCN^- \rightleftharpoons Fe(SCN)_4^{-} \qquad K_4$$
$$Fe(SCN)_4^{-} + SCN^- \rightleftharpoons Fe(SCN)_5^{2-} \qquad K_5$$
$$Fe(SCN)_5^{2-} + SCN^- \rightleftharpoons Fe(SCN)_6^{3-} \qquad K_6$$

$$Fe^{3+} + 6\,SCN^- \rightleftharpoons Fe(SCN)_6^{3-} \qquad K_f = K_1K_2K_3K_4K_5K_6$$

c. $Ag^+ + NH_3 \rightleftharpoons AgNH_3^+$ K_1

$AgNH_3^+ + NH_3 \rightleftharpoons Ag(NH_3)_2^+$ K_2

$Ag^+ + 2\ NH_3 \rightleftharpoons Ag(NH_3)_2^+$ $K_f = K_1K_2$

60. a. $Ni^{2+} + CN^- \rightleftharpoons NiCN^+$ K_1

$NiCN^+ + CN^- \rightleftharpoons Ni(CN)_2$ K_2

$Ni(CN)_2 + CN^- \rightleftharpoons Ni(CN)_3^-$ K_3

$Ni(CN)_3^- + CN^- \rightleftharpoons Ni(CN)_4^{2-}$ K_4

$Ni^{2+} + 4\ CN^- \rightleftharpoons Ni(CN)_4^{2-}$ $K_f = K_1K_2K_3K_4$

b. $Fe^{3+} + Cl^- \rightleftharpoons FeCl^{2+}$ K_1

$FeCl^{2+} + Cl^- \rightleftharpoons FeCl_2^+$ K_2

$FeCl_2^+ + Cl^- \rightleftharpoons FeCl_3$ K_3

$FeCl_3 + Cl^- \rightleftharpoons FeCl_4^-$ K_4

$Fe^{3+} + 4\ Cl^- \rightleftharpoons FeCl_4^-$ $K_f = K_1K_2K_3K_4$

c. $Mn^{2+} + C_2O_4^{2-} \rightleftharpoons MnC_2O_4$ K_1

$MnC_2O_4 + C_2O_4^{2-} \rightleftharpoons Mn(C_2O_4)_2^{2-}$ K_2

$Mn^{2+} + 2\ C_2O_4^{2-} \rightleftharpoons Mn(C_2O_4)_2^{2-}$ $K_f = K_1K_2$

61. An ammonia solution is basic. Initially the reaction that occurs is:

$$Cu^{2+}(aq) + 2\ OH^-(aq) \rightarrow Cu(OH)_2(s) \text{ (white ppt.)}$$

As the concentration of NH_3 increases the complex ion, $Cu(NH_3)_4^{2+}$, forms:

$$Cu(OH)_2(s) + 4\ NH_3(aq) \rightleftharpoons Cu(NH_3)_4^{2+}(aq) + 2\ OH^-(aq)$$

62. $Ag^+(aq) + Cl^-(aq) \rightleftharpoons AgCl(s)$; white ppt.

$AgCl(s) + 2\ NH_3(aq) \rightleftharpoons Ag(NH_3)_2^+(aq) + Cl^-(aq)$

$Ag(NH_3)_2^+(aq) + Br^-(aq) \rightleftharpoons AgBr(s) + 2\ NH_3(aq)$; pale yellow ppt.

$AgBr(s) + 2\ S_2O_3^{2-}(aq) \rightleftharpoons Ag(S_2O_3)_2^{3-}(aq) + Br^-(aq)$

$Ag(S_2O_3)_2^{3-}(aq) + I^-(aq) \rightleftharpoons AgI(s) + 2\ S_2O_3^{2-}(aq)$; yellow ppt.

K_{sp} (AgCl) > K_{sp} (AgBr) > K_{sp} (AgI)

K_f $(Ag(S_2O_3)_2^{3-})$ > K_f $(Ag(NH_3)_2^+)$

63. $\dfrac{65 \text{ g KI}}{0.500 \text{ L}} \times \dfrac{1 \text{ mol}}{166.0 \text{ g}} = \dfrac{0.78 \text{ mol}}{\text{L}}$

	Hg^{2+}	+ $4 I^-$	$\rightleftharpoons$	HgI_4^{2-}	$K_f = 1.0 \times 10^{30}$
Before	0.010 *M*	0.78 *M*		0	
Change	-0.010	-0.040	→	+0.010	React completely (K large)
After (New Initial)	0	0.74		0.010	

x mol/L HgI_4^{2-} dissociates to reach equilibrium

Change	$+x$	$+4x$	←	$-x$
Equil	x	$0.74 + 4x$		$0.010 - x$

$$K_f = 1.0 \times 10^{30} = \frac{[HgI_4^{2-}]}{[Hg^{2+}][I^-]^4} = \frac{(0.010 - x)}{(x)(0.74 + 4x)^4}$$

$1.0 \times 10^{30} \approx \dfrac{(0.010)}{(x)(0.74)^4}$; $x = [Hg^{2+}] = 3.3 \times 10^{-32}$ mol/L   Assumptions good.

Note: 3.3×10^{-32} mol/L corresponds to one Hg^{2+} ion per 5×10^7 L. It is very reasonable to approach the equilibrium in two steps. The reaction really does go to completion.

64.

	Ni^{2+}	+ $6 NH_3$	$\rightleftharpoons$	$Ni(NH_3)_6^{2+}$	$K_f = 5.5 \times 10^8$
Initial	0	3.0 *M*		0.10 mol/0.50 L = 0.20 *M*	

x mol/L $Ni(NH_3)_6^{2+}$ dissociates to reach equilibrium

Change	$+x$	$+6x$	←	$-x$
Equil	x	$3.0 + 6x$		$0.20 - x$

$$K_f = 5.5 \times 10^8 = \frac{[Ni(NH_3)_6^{2+}]}{[Ni^{2+}][NH_3]^6} = \frac{(0.20 - x)}{(x)(3.0 + 6x)^6};\ 5.5 \times 10^8 \approx \frac{0.20}{(x)(3.0)^6}$$

$x = [Ni^{2+}] = 5.0 \times 10^{-13}$ *M*; $[Ni(NH_3)_6^{2+}] = 0.20$ *M*   Assumptions good.

ADDITIONAL EXERCISES

65. $NH_3 + H_2O \rightleftharpoons NH_4^+ + OH^-$ $K_b = \dfrac{[NH_4^+][OH^-]}{[NH_3]}$

$$-\log K_b = -\log [OH^-] - \log \frac{[NH_4^+]}{[NH_3]}$$

$$-\log [OH^-] = -\log K_b + \log \frac{[NH_4^+]}{[NH_3]}$$

$$pOH = pK_b + \log \frac{[NH_4^+]}{[NH_3]} \text{ or } pOH = pK_b + \log \frac{[Acid]}{[Base]}$$

66. a. $\frac{28.6 \text{ g borax}}{\text{L}} \times \frac{1 \text{ mol}}{381.4 \text{ g}} = \frac{0.0750 \text{ mol borax}}{\text{L}}$

$\frac{0.0750 \text{ mol borax}}{\text{L}} \times \frac{2 \text{ mol B(OH)}_3}{\text{mol borax}} = 0.150\ M$

The concentration of $B(OH)_4^-$ is also 0.150 M.

$B(OH)_3 + H_2O \rightleftharpoons B(OH)_4^- + H^+$ $\quad K_a = 5.8 \times 10^{-10}$

$pH = pK_a + \log \frac{[\text{Base}]}{[\text{Acid}]} = pK_a + \log 1.00 = pK_a = 9.24$

b. $100.\ \text{mL} \times \frac{0.10 \text{ mmol}}{\text{mL}} = 10.$ mmol NaOH added or 0.010 mol NaOH

For $B(OH)_3$: $\frac{0.150 \text{ mol B(OH)}_3}{\text{L}} \times 1.0 \text{ L} = 0.15 \text{ mol B(OH)}_3$

Added OH^- converts $B(OH)_3$ to $B(OH)_4^-$: $B(OH)_3 + OH^- \rightarrow B(OH)_4^-$

mol $B(OH)_3$ = 0.15 - 0.010 = 0.14 mol; mol $B(OH)_4^-$ = 0.15 + 0.010 = 0.16 mol

$[B(OH)_3] = \frac{0.14 \text{ mol}}{1.1 \text{ L}}$; Similarly $[B(OH)_4^-] = \frac{0.16 \text{ mol}}{1.1 \text{ L}}$

$pH = pK_a + \log \frac{[\text{Base}]}{[\text{Acid}]} = 9.24 + \log \frac{[B(OH)_4^-]}{[B(OH)_3]}$

$pH = 9.24 + \log \frac{(0.16)}{(0.14)} = 9.30$

67. a. The optimum pH for a buffer is when pH = pK_a. At this pH a buffer will have equal capacity for both added acid and base.

$K_b = 1.19 \times 10^{-6}$; $K_a = K_w/K_b = 8.40 \times 10^{-9}$; $pH = pK_a = 8.076 = 8.08$

b. $pH = pK_a + \log \frac{[\text{TRIS}]}{[\text{TRIS-H}^+]}$

$7.0 = 8.08 + \log \frac{[\text{TRIS}]}{[\text{TRIS-H}^+]}$

$\frac{[\text{TRIS}]}{[\text{TRIS-H}^+]} = 10^{-1.08} = 0.083 \approx 0.08$ at pH = 7.0

$9.0 = 8.08 + \log \frac{[\text{TRIS}]}{[\text{TRIS-H}^+]}$

$\frac{[\text{TRIS}]}{[\text{TRIS-H}^+]} = 10^{0.92} = 8.3 \approx 8$ at pH = 9.0

c. $$\frac{50.0\text{ g TRIS}}{2.0\text{ L}} \times \frac{1\text{ mol}}{121.1\text{ g}} = 0.206\ M = 0.21\ M = [\text{TRIS}]$$

$$\frac{65.0\text{ g TRIS-HCl}}{2.0\text{ L}} \times \frac{1\text{ mol}}{157.6\text{ g}} = 0.206\ M = 0.21\ M = [\text{TRIS-HCl}] = [\text{TRIS-H}^+]$$

$$\text{pH} = \text{p}K_a + \log\frac{[\text{TRIS}]}{[\text{TRIS-H}^+]} = 8.08 + \log\frac{(0.21)}{(0.21)} = 8.08$$

Amount of H^+ added from HCl is:

$$0.50 \times 10^{-3}\text{ L} \times \frac{12\text{ mol}}{\text{L}} = 6.0 \times 10^{-3}\text{ mol}$$

The H^+ from HCl will convert TRIS to TRIS-H^+. The reaction is:

	TRIS	+ H^+	→	TRIS-H^+	
Before	0.21 M	$\frac{6.0 \times 10^{-3}}{0.2005} = 0.030\ M$		0.21 M	
Change	-0.030	-0.030	→	+0.030	Reacts completely
After	0.18	0		0.24	

Now use the Henderson-Hasselbalch equation to solve the buffer problem.

$$\text{pH} = 8.08 + \log\frac{0.18}{0.24} = 7.96$$

68. $$\text{pH} = \text{p}K_a + \log\frac{[(CH_3)_2AsO_2^-]}{[(CH_3)_2AsO_2H]};\quad 6.60 = 6.19 + \log\frac{[\text{Cac}^-]}{[\text{HCac}]}$$

$$\frac{[\text{Cac}^-]}{[\text{HCac}]} = 10^{0.41} = 2.6;\quad [\text{Cac}^-] = 2.6\,[\text{HCac}]$$

$[\text{Cac}^-] + [\text{HCac}] = 0.25;\quad 3.6\,[\text{HCac}] = 0.25$

$[\text{HCac}] = 0.069$ and $[\text{Cac}^-] = 0.18$

$$0.500\text{ L} \times \frac{0.069\text{ mol }(CH_3)_2AsO_2H}{\text{L}} \times \frac{138.0\text{ g}}{\text{mol}} = 4.8\text{ g cacodylic acid}$$

$$0.500\text{ L} \times \frac{0.18\text{ mol }(CH_3)_2AsO_2^-}{\text{L}} \times \frac{160.0\text{ g }(CH_3)_2AsO_2Na}{\text{mol}}$$

= 14.4 g sodium cacodylate = 14 g

69.

cacodylic acid	$\text{p}K_a$ = 6.19
TRIS-HCl	8.08
benzoic acid	4.19
acetic acid	4.74
HF	3.14
NH_4Cl	9.25

a. potassium fluoride + HCl

b. benzoic acid + NaOH

c. sodium acetate + acetic acid

d. $(CH_3)_2AsO_2Na$ + HCl:

This is the best choice to produce a conjugate acid/base pair. Actually the best choice is an equimolar mixture ammonium chloride and sodium acetate. NH_4^+ is a weak acid ($K_a = 5.6 \times 10^{-10}$). $C_2H_3O_2^-$ is a weak base ($K_b = 5.6 \times 10^{-10}$). A mixture of the two will give a buffer at pH = 7 since the weak acid and weak base are the same strengths. $NH_4C_2H_3O_2$ is commercially available and its solutions are used as pH = 7 buffers.

e. ammonium chloride + NaOH

70. a. $pH = pK_a = 4.19$

b. $[Bz^-]$ will increase to 0.12 *M*, [HBz] will decrease to 0.08 *M*. Using Henderson-Hasselbalch equation:

$$pH = pK_a + \log \frac{[Bz^-]}{[HBz]}; \quad pH = 4.19 + \log \frac{(0.12)}{(0.08)} = 4.4$$

c.

	Bz^- +	H_2O ⇌	HBz +	OH^-
Initial	0.12 *M*		0.08 *M*	~0
	x mol/L Bz^- reacts with H_2O to reach equilibrium			
Change	$-x$	→	$+x$	$+x$
Equil	$0.12 - x$		$0.08 + x$	x

$$K_b = \frac{K_w}{K_a} = \frac{1.0 \times 10^{-14}}{6.4 \times 10^{-5}} = \frac{(0.08 + x)(x)}{(0.12 - x)} \approx \frac{(0.08)(x)}{0.12} \quad \text{(carry 2 S.F. until end)}$$

$x = [OH^-] = 2.3 \times 10^{-10}$; pOH = 9.6; pH = 4.4 Assumptions good.

d. We get the same answer as we should. Both equilibria involve the two major species: benzoic acid and benzoate anion. Both the equilibria must hold true. K_b is related to K_a by K_w. $[OH^-]$ is also related to $[H^+]$ by K_w.

71. a. $CH_3CO_2H + OH^- \rightleftharpoons CH_3CO_2^- + H_2O$

$$K_{eq} = \frac{[OAc^-]}{[HOAc][OH^-]} \times \frac{[H^+]}{[H^+]} = \frac{K_a}{K_w} = \frac{1.8 \times 10^{-5}}{1.0 \times 10^{-14}} = 1.8 \times 10^9$$

b. $CH_3CO_2^- + H^+ \rightleftharpoons CH_3CO_2H$

$$K_{eq} = \frac{[HOAc]}{[H^+][OAc^-]} = \frac{1}{K_a} = 5.6 \times 10^4$$

c. $CH_3CO_2H + NH_3 \rightleftharpoons CH_3CO_2^- + NH_4^+$

$$K_{eq} = \frac{[OAc^-][NH_4^+]}{[HOAc][NH_3]} \times \frac{[H^+]}{[H^+]} = \frac{K_a\ (HOAc)}{K_a\ (NH_4^+)}$$

$$= \frac{K_a(HOAc)\ K_b(NH_3)}{K_w} = 3.2 \times 10^4$$

d. $HCl(aq) + NaOH\ (aq) \rightarrow NaCl(aq) + H_2O$

Net ionic equation is: $H^+ + OH^- \rightleftharpoons H_2O$

$$K_{eq} = \frac{1}{K_w} = 1.0 \times 10^{14}$$

Although all four reactions have large equilibrium constants, only (a), (b), and (d) can be useful in a titration. The solution in (c) results in a buffer; it still contains both a weak acid and a weak base. For a titration to be useful there must be a large change in pH for a small amount of added titrant near the equivalence point. This condition is precisely what a buffer is used to prevent.

72. NaOH added = $50.0\ mL \times \frac{0.500\ mmol}{mL} = 25.0$ mmol NaOH

NaOH left unreacted = $31.92\ mL \times \frac{0.289\ mmol}{mL} = 9.22$ mmol NaOH

NaOH reacted with aspirin = 25.0 - 9.22 = 15.8 mmol

$$15.8\ mmol\ NaOH \times \frac{1\ mmol\ aspirin}{2\ mmol\ NaOH} \times \frac{180.2\ mg}{mmol} = 1420\ mg = 1.42\ g$$

$$Purity = \frac{1.42\ g}{1.427\ g} \times 100 = 99.5\%$$

A strong base is titrated by a strong acid. The equivalence point is at pH = 7. Bromthymol blue would be the best indicator.

73. At equivalence point, $C_6H_4(CO_2)_2^{2-} = P^{2-}$ is the major species. It is a weak base in water.

	P^{2-} + H_2O	$\rightleftharpoons$	HP^-	+ OH^-	
Initial	$\frac{0.5\ g}{0.1\ L} \times \frac{1\ mol}{204.2\ g} = 0.025\ M$		0	~0	(carry extra sig. fig.)
	x mol/L P^{2-} reacts with H_2O to reach equilibrium				
Change	$-x$	$\rightarrow$	$+x$	$+x$	
Equil	$0.025 - x$		x	x	

$$K_b = \frac{[HP^-][OH^-]}{[P^{2-}]} = \frac{K_w}{K_{a_2}} = \frac{1.0 \times 10^{-14}}{10^{-5.51}} = 3.2 \times 10^{-9} = \frac{x^2}{0.025 - x} \approx \frac{x^2}{0.025}$$

$x = [OH^-] = 8.9 \times 10^{-6}$; pH = 8.9 Assumptions good.

Phenolphthalein would be the best indicator for this titration.

Note: Although HP^- is also a weak base, it is a much weaker base than P^{2-}. We can ignore the OH^- contribution from HP^-.

74. Color will change over a range: $pH = pK_a \pm 1 = 5.3 \pm 1$

Useful pH range of methyl red would be about 4.3 (red) - 6.3 (yellow). In titrating a weak acid with base, the color would change from red to yellow starting at pH ~ 4.3. In titrating a weak base with acid, the color change would be from yellow to red starting at pH ~ 6.3. Methyl red would be suitable for the titrations in Exercise 37 and 40b, since the equivalence points occur at a pH between 4.5-6.

75. $Ca_5(PO_4)_3(OH)(s) \rightleftharpoons 5\,Ca^{2+} + 3\,PO_4^{3-} + OH^-$

s = solubility (mol/L) → 5s 3s $s + 1.0 \times 10^{-7} \approx s$

$K_{sp} = 6.8 \times 10^{-37} = [Ca^{2+}]^5\,[PO_4^{3-}]^3\,[OH^-] = (5s)^5(3s)^3(s)$

$6.8 \times 10^{-37} = (3125)(27)s^9$; $s = 2.7 \times 10^{-5}$ mol/L

The solubility of hydroxyapatite will increase as a solution gets more acidic, since both phosphate and hydroxide can react with H^+.

$Ca_5(PO_4)_3(F) \rightleftharpoons 5\,Ca^{2+} + 3\,PO_4^{3-} + F^-$

s = solubility (mol/L) → 5s 3s s

$1 \times 10^{-60} = (5s)^5(3s)^3(s) = (3125)(27)s^9$; $s = 6 \times 10^{-8}$ mol/L

The hydroxyapatite in the tooth enamel is converted to the less soluble fluorapatite by fluoride treated water. The less soluble fluorapatite will then be more difficult to remove making teeth less susceptible to decay.

76. $100.\text{ mL} \times \dfrac{0.20\text{ mmol } K_2C_2O_4}{\text{mL}} = 20.\text{ mmol } K_2C_2O_4$

$150.\text{ mL} \times \dfrac{0.25\text{ mmol } BaBr_2}{\text{mL}} = 37.5\text{ mmol } BaBr_2$ (carry an extra sig. fig.)

	$Ba^{2+}(aq)$	+	$C_2O_4^{2-}(aq)$	→	$BaC_2O_4(s)$	
Before	37.5 mmol		20. mmol		0	
Change	-20.		-20.	→	+20.	React completely
After	17.5		0		20.	

$[Ba^{2+}] = \dfrac{17.5\text{ mmol}}{250.\text{ mL}} = 7.0 \times 10^{-2}$ mol/L

$[K^+] = \dfrac{40.\text{ mmol}}{250.\text{ mL}} = 0.16$ mol/L; $[Br^-] = \dfrac{75.\text{ mmol}}{250.\text{ mL}} = 0.30$ mol/L

	$BaC_2O_4(s)$	$\rightleftharpoons$	$Ba^{2+}(aq)$	+	$C_2O_4^{2-}(aq)$
Initial			0.070		0
	s mol/L BaC_2O_4 dissolves to reach equilibrium				
Change	-s	$\rightarrow$	+s		+s
Equil			0.070 + s		s

$K_{sp} = 2.3 \times 10^{-8} = [Ba^{2+}][C_2O_4^{2-}] = (0.070 + s)(s) \approx 0.070\ s$

$s = [C_2O_4^{2-}] = 3.3 \times 10^{-7}$ mol/L, $[Ba^{2+}] = 0.070\ M$ Assumption good.

77. $Fe(OH)_3 \rightleftharpoons Fe^{3+} + 3\ OH^-$ $K_{sp} = [Fe^{3+}][OH^-]^3 = 4 \times 10^{-38}$

pH = 7.41; pOH = 6.59; $[OH^-] = 2.6 \times 10^{-7}$

$$[Fe^{3+}] = \frac{4 \times 10^{-38}}{(2.6 \times 10^{-7})^3} = 2 \times 10^{-18}\ M$$

The lowest level of iron in serum:

$$\frac{60 \times 10^{-6}\ g}{0.1\ L} \times \frac{1\ mol}{55.85\ g} = 1 \times 10^{-5}\ M$$

The actual concentration of iron is much greater than expected when considering only the solubility of $Fe(OH)_3$. Thus, there must be complexing agents present to increase the solubility.

78. For a titration of a strong base with a strong acid, we calculate the amount of excess OH^- or H^+ and then calculate the pH.

0 mL: $[H^+] = 0.100$ from HNO_3, pH = 1.000

4.0 mL: initial $H^+ = 25.0\ mL \times \frac{0.100\ mmol}{mL} = 2.50\ mmol\ H^+$

OH^- added $= 4.0\ mL \times \frac{0.100\ mmol}{mL} = 0.40\ mmol\ OH^-$

0.40 mmol OH^- reacts completely with 0.40 mmol H^+: $OH^- + H^+ \rightarrow H_2O$

$[H^+] = \frac{2.50 - 0.40}{29.0} = 7.24 \times 10^{-2}\ M$, pH = 1.140

8.0 mL: $[H^+] = \frac{2.50 - 0.80}{33.0} = 5.15 \times 10^{-2}\ M$, pH = 1.288

12.5 mL: $[H^+] = \frac{2.50 - 1.25}{37.5} = 3.33 \times 10^{-2}\ M$, pH = 1.478

20.0 mL: $[H^+] = \frac{2.50 - 2.00}{45.0} = 1.1 \times 10^{-2}\ M$, pH = 1.96

24.0 mL: $[H^+] = \dfrac{2.50 - 2.40}{49.0} = 2.0 \times 10^{-3}\ M$, pH = 2.70

24.5 mL: $[H^+] = \dfrac{2.50 - 2.45}{49.5} = 1 \times 10^{-3}\ M$, pH = 3.0

24.9 mL: $[H^+] = \dfrac{2.50 - 2.49}{49.9} = 2 \times 10^{-4}\ M$, pH = 3.7

25.0 mL: neutral solution, pH = 7.00; stoichiometric point

25.1 mL: base in excess, $[OH^-] = \dfrac{2.51 - 2.50}{50.1} = 2 \times 10^{-4}$, pOH = 3.7, pH = 10.3

26 mL: $[OH^-] = \dfrac{2.6 - 2.50}{51.0} = 2 \times 10^{-3}$, pOH = 2.7, pH = 11.3

28 mL: $[OH^-] = \dfrac{2.8 - 2.50}{53.0} = 6 \times 10^{-3}$, pOH = 2.2, pH = 11.8

30 mL: $[OH^-] = \dfrac{3.0 - 2.50}{55.0} = 9 \times 10^{-3}$, pOH = 2.0 pH = 12.0

pH
12.0
10.0
8.0
6.0
4.0
2.0
8.78
0.0 5.0 10.0 15.0 20.0 25.0 30.0
mL TITRANT

79.

	Cr^{3+}	+ H_2EDTA^{2-}	$\rightleftharpoons$	$CrEDTA^-$	+ $2\ H^+$	
Before	0.0010 *M*	0.050 *M*		0	1×10^{-6} *M*	(buffer pH = 6.0)
Change	-0.0010	-0.0010	→	+0.0010	buffered (no change)	(React completely)
After (New Initial)	0	0.049		0.0010	1×10^{-6} (buffer)	

x mol/L $CrEDTA^-$ dissociates to reach equilibrium

Change	$+x$	$+x$	←	$-x$	-----
Equil	x	$0.049 + x$		$0.0010 - x$	1×10^{-6} (buffer)

$$K_f = 1.0 \times 10^{23} = \frac{[CrEDTA^-][H^+]^2}{[Cr^{3+}][H_2EDTA^{2-}]} = \frac{(0.0010 - x)(1 \times 10^{-6})^2}{(x)(0.049 + x)}$$

$$1.0 \times 10^{23} \approx \frac{(0.0010)(1 \times 10^{-12})}{(x)(0.049)}$$

$x = [Cr^{3+}] = 2 \times 10^{-37}\ M$ Assumptions good.

CHALLENGE PROBLEMS

80. At 4 mL added:

$$\left| \frac{\Delta pH}{\Delta mL} \right| = \left| \frac{2.43 - 3.14}{0 - 4.0} \right| = 0.18$$

Using the results from Table 15.1. As can be seen from the graph, the advantage of this approach is that it is much easier to accurately determine the location of the equivalence point.

mL	pH	I ΔpH/ΔmL
0	2.43	0
4	3.14	0.18
8	3.53	0.098
12.5	3.86	0.073
20	4.46	0.080
24	5.24	0.20
24.5	5.6	0.7
24.9	6.3	2
25.0	8.28	20
25.1	10.3	20
26	11.3	1
28	11.8	0.3
30	12.0	0.1

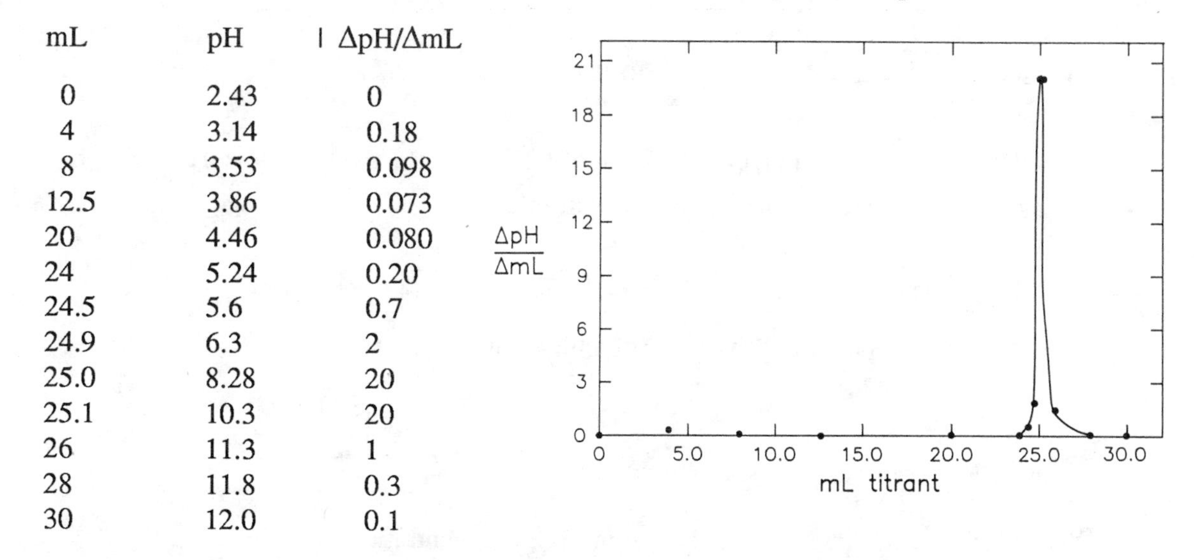

81. Phenolphthalein will change color at pH ~ 9. Phenolphthalein will mark the second end point. Therefore, at the phenolphthalein end point we will have titrated both protons on malonic acid.

$$H_2Mal + 2\ OH^- \longrightarrow 2\ H_2O + Mal^{2-}$$

$$31.50\ mL \times \frac{0.0984\ mmol\ NaOH}{mL} \times \frac{1\ mmol\ H_2Mal}{2\ mmol\ NaOH} = 1.55\ mmol\ H_2Mal$$

$$[H_2Mal] = \frac{1.55\ mmol}{25.00\ mL} = 0.0620\ M$$

82. The results from each indicator tells us something about the pH. The conclusions are summarized below:

Results from	pH
bromphenol blue	> 4.0
bromcresol purple	< 6.0
bromcresol green*	pH ~ pK_a ~ 4.8 ~ 5
alizaren	< 6.5

* For bromcresol green, the resultant color is green. This is a combination of the extremes (yellow and blue). This occurs when pH ~ pK_a of indicator.

The pH of the solution is about 5 (from bromcresol green result).

$HX \rightleftharpoons H^+ + X^-$

For normal setup:

$[H^+] = [X^-]$, $[HX] = 1.0 - [H^+] \approx 1.0\,M$

$$K_a = \frac{[H^+][X^-]}{[HX]} \approx \frac{(10^{-5})(10^{-5})}{1.0} \approx 10^{-10}$$

83. It will be more soluble in base. Addition of OH^- will effectively remove H^+:
$H^+ + OH^- \rightarrow H_2O$. This drives the equilibrium:

$$H_4SiO_4 \rightleftharpoons H_3SiO_4^- + H^+$$

to the right. This in turn, drives the solubility reaction:

$$SiO_2(s) + 2\,H_2O \rightleftharpoons H_4SiO_4$$

to the right, increasing the solubility of the silica.

84. The solubility of SiO_2 will equal the sum of the concentration of H_4SiO_4 and $H_3SiO_4^-$.

$SiO_2(s) + 2\,H_2O \rightleftharpoons H_4SiO_4(aq)$

$K = 2 \times 10^{-3} = [H_4SiO_4]$

$H_4SiO_4 \rightleftharpoons H_3SiO_4^- + H^+$

$$K_a = 10^{-9.46} = 3.5 \times 10^{-10} = \frac{[H_3SiO_4^-][H^+]}{[H_4SiO_4]}$$

$$3.5 \times 10^{-10} = \frac{[H_3SiO_4^-](1 \times 10^{-7})}{2 \times 10^{-3}}$$

$[H_3SiO_4^-] = 7 \times 10^{-6}\,M$

Solubility = $2 \times 10^{-3} + 7 \times 10^{-6} \approx 2 \times 10^{-3}$ mol/L at pH = 7.0

At pH = 10.0

$$3.5 \times 10^{-10} = \frac{[H_3SiO_4^-][H^+]}{[H_4SiO_4]} = \frac{[H_3SiO_4^-](1 \times 10^{-10})}{2 \times 10^{-3}}$$

$[H_3SiO_4^-] = 7 \times 10^{-3}\ M$

$s = 7 \times 10^{-3} + 2 \times 10^{-3} = 9 \times 10^{-3}$ mol/L at pH = 10.0

85. $K_{sp} = 6.4 \times 10^{-9} = [Mg^{2+}][F^-]^2$

$6.4 \times 10^{-9} = (0.00375 - y)(0.0625 - 2y)^2$

This is a cubic equation. No simplifying assumptions can be made since *y* is relatively large. There is a formula for solving a cubic equation (I've heard), but I've never seen it. We could use a computer or a programmable calculator to solve the equation by numerical methods. By the time we've done all that, we could have solved the problem several times over using the approximations based on our "chemical common sense."

86. a.

$$Pb(OH)_2 \rightleftharpoons Pb^{2+} + 2\,OH^-$$

s mol/L dissolves $\rightarrow$ s 2s (ignore OH^- from H_2O)

$K_{sp} = 1.2 \times 10^{-15} = [Pb^{2+}][OH^-]^2 = 4s^3$

$s = [Pb^{2+}] = 6.7 \times 10^{-6}\ M$ Assumption to ignore water is good.

b.

	$Pb(OH)_2$	$\rightleftharpoons$	Pb^{2+}	+ $2\,OH^-$	pH = 13.0, pOH = 1.0,
Initial			0	0.1 M	$[OH^-] = 0.1\ M$
	s mol/L $Pb(OH)_2$ dissolves to reach equilibrium				
Change	-s	$\rightarrow$	+s	-----	
Equil			s	0.1	(buffered solution)

$1.2 \times 10^{-15} = (s)(0.1)^2$

$s = [Pb^{2+}] = 1.2 \times 10^{-13}\ M = 1 \times 10^{-13}\ M$

c.

	Pb^{2+}	+ $EDTA^{4-}$	$\rightleftharpoons$	$PbEDTA^{2-}$	$K_f = 1.1 \times 10^{18}$
Before	0.010 *M*	0.050 *M*		0	
	0.010 mol/L Pb^{2+} reacts				
Change	-0.010	-0.010	$\rightarrow$	+0.010	React completely (large K)
After	0	0.040		0.010	New Initial
	x mol/L $PbEDTA^{2-}$ dissociates to reach equilibrium				
Change	+*x*	+*x*	$\leftarrow$	-*x*	
Equil	*x*	0.040 + *x*		0.010 - *x*	

$$1.1 \times 10^{18} = \frac{(0.010 - x)}{(x)\,(0.040 + x)} \approx \frac{(0.010)}{x(0.040)}$$

$x = [Pb^{2+}] = 2.3 \times 10^{-19}\ M$ Assumptions good.

We calculate the solubility quotient, Q:

$$[Pb^{2+}]\,[OH^-]^2 = (2.3 \times 10^{-19})\,(0.1)^2 = 2.3 \times 10^{-21}$$

$$2.3 \times 10^{-21} < K_{sp}$$

No precipitate of $Pb(OH)_2(s)$ will form.

87.

$$Cu(OH)_2 \rightleftharpoons Cu^{2+} + 2\,OH^- \qquad K_{sp} = 1.6 \times 10^{-19}$$

$$Cu^{2+} + 4\,NH_3 \rightleftharpoons Cu(NH_3)_4^{\,2+} \qquad K_f = 1.0 \times 10^{13}$$

$$Cu(OH)_2 + 4\,NH_3 \rightleftharpoons Cu(NH_3)_4^{\,2+} + 2\,OH^- \qquad K = K_{sp}K_f = 1.6 \times 10^{-6}$$

88.

	$Cu(OH)_2(s)$	+ $4\,NH_3(aq)$	$\rightleftharpoons$	$Cu(NH_3)_4^{\,2+}(aq)$	+ $2\,OH^-(aq)$
Initial		5.0 *M*		0	0.010 *M*
	s mol/L $Cu(OH)_2$ dissolves to reach equilibrium				
Change	-s	-4s	$\rightarrow$	+s	+2s
Equil		5.0 - 4s		s	0.010 + 2s

$$1.6 \times 10^{-6} = \frac{[Cu(NH_3)_4^{\,2+}]\,[OH^-]^2}{[NH_3]^4} = \frac{s(0.010 + 2s)^2}{(5.0 - 4s)^4}$$

If s is small: $1.6 \times 10^{-6} = \dfrac{s(0.010)^2}{(5.0)^4}$

s = 10. mol/L Assumptions are not good. We will have to solve the problem by successive approximations.

$$s_{calc} = \frac{1.6 \times 10^{-6}(5.0 - 4s_{guess})^4}{(0.010 + 2s_{guess})^2},$$ We will start with $s_{guess} = 0.10$ mol/L.

The results from six trials are :

s_{guess}: 0.10, 0.050, 0.060, 0.055, 0.058, 0.056

s_{calc}: 0.016, 0.070, 0.049, 0.058, 0.052, 0.056

Thus, the solubility is 0.056 mol/L.

89. $1.0 \text{ mL} \times \frac{1.0 \text{ mmol}}{\text{mL}} = 1.0 \text{ mmol } Cd^{2+}$ added to the ammonia solution.

Thus, $[Cd^{2+}]_o = 1.0 \times 10^{-3}$ mol/L. We will treat each equilibrium separately, considering the one with the largest K first.

	Cd^{2+}	+	$4\ NH_3$	$\rightleftharpoons$	$Cd(NH_3)_4^{2+}$	
Before	$1.0 \times 10^{-3}\ M$		$5.0\ M$		0	
Change	-1.0×10^{-3}		-4.0×10^{-3}	$\rightarrow$	$+1.0 \times 10^{-3}$	React completely
After (New Initial)	0		$4.996 \approx 5.0$		1.0×10^{-3}	
	x mol/L $Cd(NH_3)_4^{2+}$ dissociates to reach equilibrium					
Change	$+x$		$+4x$	$\leftarrow$	$-x$	
Equil	x		$5.0 + 4x$		$0.0010 - x$	

$$K_f = 1.0 \times 10^7 = \frac{(0.0010 - x)}{(x)\,(5.0 + 4x)^4} \approx \frac{(0.0010)}{(x)\,(5.0)^4}$$

$x = [Cd^{2+}] = 1.6 \times 10^{-13}\ M$. Assumptions good. This is the maximum $[Cd^{2+}]$ possible. If a precipitate forms, $[Cd^{2+}]$ will be less than this value. Calculate the pH of a 5.0 M NH_3 solution.

	NH_3	+	H_2O	$\rightleftharpoons$	NH_4^+	+	OH^-	$K_b = 1.8 \times 10^{-5}$
Initial	$5.0\ M$				0		~0	
	y mol/L NH_3 reacts with H_2O to reach equilibrium							
Change	$-y$			$\rightarrow$	$+y$		$+y$	
Equil	$5.0 - y$				y		y	

$$K_b = 1.8 \times 10^{-5} = \frac{[NH_4^+]\,[OH^-]}{[NH_3]} = \frac{y^2}{5.0 - y} \approx \frac{y^2}{5.0}$$

$y = [OH^-] = 9.5 \times 10^{-3}\ M$ Assumptions good.

We now calculate the value of the solubility quotient.

$[Cd^{2+}]\,[OH^-]^2 = (1.6 \times 10^{-13})\,(9.5 \times 10^{-3})^2 = 1.4 \times 10^{-7}$

$1.4 \times 10^{-17} < K_{sp}\ (5.9 \times 10^{-15})$ Therefore, no precipitate forms.

CHAPTER SIXTEEN: SPONTANEITY, ENTROPY AND FREE ENERGY

QUESTIONS

1. A spontaneous process is one that occurs without any outside intervention.

2. a. Entropy is a measure of disorder.

   b. The quantity $T\Delta S$ has units of energy.

3. a. Entropy increases; there is a greater volume accessible to the randomly moving gas molecules. Positional disorder will increase.

   b. The positional entropy doesn't change. There is no change in volume and thus, no change in the numbers of position of the molecules. The total entropy increases because the increase in temperature increases the energy disorder.

   c. Entropy decreases since volume decreases.

4. a. The system is the portion of the universe in which we are interested.

   b. The surroundings are everything else in the universe.

5. Living organisms need an external energy source to carry out these processes. Green plants use the energy from sunlight to produce glucose from carbon dioxide and water by photosynthesis. The energy released from the metabolism of glucose (food) helps drive the synthesis of proteins. A living cell is not a closed system, i.e. a system that has no external energy source.

6. No, using tabulated ΔG_f° values we have specified a temperature of 25°C. Further, if gases or solutions are involved, we have specified partial pressures of 1 atm and solute concentration of 1 molar. At other temperatures and compositions the reaction may not be spontaneous. A negative ΔG° means the reaction is spontaneous under standard conditions.

7. The more widely something is dispersed, the greater the disorder. We must do work to overcome this disorder. In terms of the 2nd law it would be more advantageous to prevent contamination of the environment rather than clean it up later.

8. As any process occurs, ΔS_{univ} will increase; ΔS_{univ} cannot decrease. Time also goes in one direction, just as ΔS_{univ} goes in one direction.

EXERCISES

Spontaneity and Entropy

9. a and c. Note: "a" requires external energy because it is an endothermic process. It is spontaneous, however, since ΔS is favorable and provides the driving force.

10. a and b

11. We draw all of the possible arrangements of the two particles in the three levels.

2kJ	___	___	x	___	x	xx
1 kJ	___	x	___	xx	x	___
0 kJ	xx	x	x	___	___	___
Total E =	0kJ	1kJ	2kJ	2kJ	3kJ	4kJ

The most likely total energy is 2kJ.

12.

2kJ	___	___	AB	___	___	B	A	B	A
1kJ	___	AB	___	B	A	___	___	A	B
0kJ	AB	___	___	A	B	A	B	___	___
E_T =	0kJ	2kJ	4kJ	1kJ	1kJ	2kJ	2kJ	3kJ	3kJ

The most likely total energy is 2kJ (three possibilities).

13.

2kJ	___	___	x	___	x	___
1kJ	___	x	___	xx	x	xxx
0kJ	xxx	xx	xx	x	x	___
E_T (kJ) =	0	1	2	2	3	3

2kJ	xx	x	xx	xxx
1kJ	___	xx	x	___
0kJ	x	___	___	___
E_T (kJ) =	4	4	5	6

A total energy of 2, 3, or 4 kJ is equally probable.

14. Consider the states in Exercise 16.13. We will start with those and see if there are more ways to get that state if the particles are different.

E = 0

___
___ only 1 way
xxx

E = 1

___
x 3 ways (this can be either particle A, B, or C at E = 1 kJ).
xx

E = 2

x (A, B, or C)
___ 3 ways
xx

___
xx 3 ways
x

Total (E = 2): 6 ways

E = 3	C	B	C	A	B	A		___		
	B	C	A	C	A	B	6 ways	xxx	1 way	
	A	A	B	B	C	C		___		Total (E = 3): 7 ways

E = 4	xx		x			
	___	3 ways	xx	3 ways		Total (E = 4): 6 ways
	x	(A, B, or C)	___			
E = 5	xx			E = 6	xxx	
	x	3 ways (A, B, or C)			___	1 way
	___				___	

There are more ways to get E = 3kJ (7 ways) so 3kJ is the most likely total energy.

15. Of the three phases (solid, liquid, and gas), solids are most ordered and gases are most disordered. Thus, a, b, and c involve an increase in entropy.

16. a and c

17. a. N_2O: It is a more complex molecule, larger positional disorder.

    b. H_2 at 100°C and 0.5 atm: Higher temperature and lower pressure means greater volume and hence, greater positional entropy.

    c. N_2 at STP has the greater volume. d. H_2O (l)

18. a. He at 10°C. Higher T and greater V.

    b. He at 1 atm. Greater volume of gas.

    c. He at STP. Higher T and greater V.

    d. He(g) at 5 K. S > 0. For He(s) at 0 K, S = 0.

Entropy and the Second Law of Thermodynamics: Free Energy

19. a. $C_{12}H_{22}O_{11}$ larger molecule, more complex, larger positional disorder.

    b. H_2O (0°C) higher temp., S = 0 at 0 K.

    c. $H_2S(g)$ A gas has greater disorder than a liquid.

20. a. He (10 K) S = 0 at 0 K b. N_2O; more complicated molecule.

    c. HCl; more electrons in HCl, larger positional disorder.

21. a. Decrease in disorder; ΔS (-) b. Increase in disorder; ΔS (+)

c. Decrease in disorder; ΔS (-) The number of moles of gas decreases.

22. a. Decrease in disorder; ΔS (-) The number of moles of gas decreases.

b. Decrease in disorder; ΔS(-). The number of moles of gas decreases.

c. Increase in disorder; ΔS (+). Dissolution generally increases positional disorder.

23. a. $H_2(g) + 1/2\ O_2(g) \longrightarrow H_2O(g)$

$\Delta S° = S°(H_2O(g)) - [S°(H_2) + 1/2\ S°(O_2)]$

$\Delta S° = 1$ mol (189 J/K•mol) - [1 mol (131 J/K•mol) + 1/2 mol (205 J/K•mol)]

$\Delta S° = 189$ J/K - 234 J/K = -45 J/K

b. $3\ O_2(g) \longrightarrow 2\ O_3(g)$

$\Delta S° = 2$ mol(239 J/mol•K) - 3 mol (205 J/mol•K) = -137 J/K

c. $N_2(g) + O_2(g) \longrightarrow 2\ NO(g)$

$\Delta S° = 2(211) - (192 + 205) = +25$ J/K

24. a. $H_2(g) + 1/2\ O_2(g) \longrightarrow H_2O(l)$

$\Delta S° = S°(H_2O(l)) - [S°(H_2(g)) + 1/2\ S°(O_2)(g)]$

= 1 mol (70 J/K•mol) - [1 mol (131 J/K•mol) + 1/2 mol (205 J/K•mol)]

= 70 J/K - 234 J/K = -164 J/K

b. $N_2(g) + 3\ H_2(g) \longrightarrow 2\ NH_3(g)$

$\Delta S° = 2\ (193) - [1(192) + 3(131)] = -199$ J/K

c. $HCl(g) \longrightarrow H^+(aq) + Cl^-(aq)$

$\Delta S° = 1$ mol $H^+(0)$ + 1 mol Cl^- (57 J/K•mol) - 1 mol HCl (187 J/K•mol) = -130. J/K

25. $CS_2(g) + 3\ O_2(g) \longrightarrow CO_2(g) + 2\ SO_2(g)$

$\Delta S° = S°(CO_2) + 2\ S°(SO_2) - [3\ S°(O_2) + S°(CS_2)]$

-143 J/K = 214 J/K + 2(248 J/K) - 3(205 J/K) - (1 mol) S°

S° = 238 J/K•mol

26. $2\ Al(s) + 3\ Br_2(l) \longrightarrow 2\ AlBr_2(s)$

-144 J/K = 2 mol[S°($AlBr_3$)] - [2(28 J/K) + 3(152 J/K)]

S° = 184 J/K•mol

27. a. $\Delta G = \Delta H - T\Delta S$; $\Delta G = 25 \times 10^3$ J - (300 K) (5 J/K)

Note: We must be consistent on units. Typically enthalpy values are in kJ and entropy in J/K. We must use the same energy units for both if we are using the equation $\Delta G = \Delta H - T\Delta S$.

$\Delta G = +$ 23,500 J $\approx$ 24,000 K Not spontaneous

b. ΔG = 25,000 J - (300 K) (100 J/K) = -5000 J Spontaneous

c. Without calculating ΔG, we know this reaction will be spontaneous at all temperatures. ΔH is negative and ΔS is positive ($-T\Delta S < 0$). ΔG will always be less than zero.

d. ΔG = -10,000 J - (200 K) (-40 J/K) = -2000 J Spontaneous

28. a. ΔG = -10,000 J - (300 K) (-40 J/K) = +2000 J Not spontaneous

b. ΔG = -10,000 J - (500 K) (-40 J/K) = +10,000 J Not spontaneous

c. Since ΔH is positive and ΔS is negative ($-T\Delta S$ is positive) the value of ΔG will be greater than zero regardless of the temperature. Thus, this change is not spontaneous at any temperature.

d. Not spontaneous, See (c).

29. At the boiling point $\Delta G = 0$, so $\Delta H = T\Delta S$

$$\Delta S = \frac{\Delta H}{T} = \frac{31.4 \text{ kJ/mol}}{(273.2 + 61.7) \text{ K}} = 9.38 \times 10^{-2} \text{ kJ/mol•K}$$

or 93.8 J/K•mol

30. $\Delta G = 0$ so $\Delta H = T\Delta S$

$$T = \frac{\Delta H}{\Delta S} = \frac{58.51 \times 10^3 \text{ J/mol}}{92.92 \text{ J/K•mol}} = 629.7 \text{ K}$$

Free Energy and Chemical Reactions

31. a. $CH_4(g) + 2\,O_2(g) \rightarrow CO_2(g) + 2\,H_2O\,(g)$

	$CH_4(g)$	$2\,O_2(g)$	$CO_2(g)$	$2\,H_2O\,(g)$	
ΔH_f°	-76 kJ/mol	0	-393.5	-242	
ΔG_f°	-51 kJ/mol	0	-394	-229	Data from Appendix 4
S°	186 J/K•mol	205	214	189	

ΔH° = 2 mol (-242 kJ/mol) + 1 mol (-393.5 kJ/mol) - [1 mol (-75 kJ/mol)] = -803 kJ

ΔS° = 2 mol (189 J/K•mol) + 1 mol (214 J/K•mol)

- [1 mol (186 J/K•mol) + 2 mol (205 J/K•mol)] = -4 J/K

There are two ways to get $\Delta G°$:

$$\Delta G° = \Delta H° - T\Delta S° = -803 \times 10^3 \text{ J} - 298 \text{ K} (-4 \text{ J/K})$$

$$= -8.018 \times 10^5 \text{ J} = -802 \text{ kJ}$$

or from $\Delta G_f°$, we get

$$\Delta G° = 2 \text{ mol} (-229 \text{ kJ mol}) + 1 \text{ mol} (-394 \text{ kJ/mol}) - [1 \text{ mol} (-51 \text{ kJ/mol})]$$

$\Delta G° = -801$ kJ (Answers are the same within roundoff error)

b. $6\ CO_2(g) + 6\ H_2O(l) \rightarrow C_6H_{12}O_6(s) + 6\ O_2(g)$

	$6\ CO_2(g)$	$6\ H_2O(l)$	$C_6H_{12}O_6(s)$	$6\ O_2(g)$
$\Delta H_f°$	-393.5 kJ/mol	-286	-1275	0
S°	214 J/K•mol	70	212	205

$\Delta H° = -1275 - [6(-286) + 6(-393.5)] = 2802$ kJ

$\Delta S° = 6(205) + 212 - [6(214) + 6(70)] = -262$ J/K

$\Delta G° = 2802 \text{ kJ} - 298 \text{ K} (-0.262 \text{ kJ/K}) = 2880.$ kJ

c. $P_4O_{10}(s) + 6\ H_2O(l) \rightarrow 4\ H_3PO_4(s)$

	$P_4O_{10}(s)$	$6\ H_2O(l)$	$4\ H_3PO_4(s)$
$\Delta H_f°$ (kJ/mol)	-2984	-286	-1279
S° (J/K•mol)	229	70	110

$\Delta H° = 4 \text{ mol} (-1279 \text{ kJ/mol}) - [1 \text{ mol} (-2984 \text{ kJ/mol}) + 6 \text{ mol} (-286 \text{ kJ/mol})]$

$\Delta H° = -416$ kJ

$\Delta S° = 4(110) - [229 + 6(70)] = -209$ J/K

$\Delta G° = \Delta H° - T\Delta S° = -416 \text{ kJ} - 298 \text{ K}(-0.209 \text{ kJ/K}) = -354$ kJ

d. $HCl(g) + NH_3(g) \rightarrow NH_4Cl(s)$

	$HCl(g)$	$NH_3(g)$	$NH_4Cl(s)$
$\Delta H_f°$ (kJ/mol)	-92	-46	-314
S° (J/K•mol)	187	193	96

$\Delta H° = -314 - [-92 - 46] = -176$ kJ

$\Delta S° = 96 - [187 + 193] = -284$ J/K

$\Delta G° = \Delta H° - T\Delta S° = -176$ kJ $- (298$ K$)(-0.284$ kJ/K$) = -91$ kJ

32. a.

$$CH_2{-}CH_2(g)\ (\text{ring with O}) + HCN(g) \longrightarrow CH_2{=}CHCN(g) + H_2O(l)$$

	CH_2CH_2O	$HCN(g)$	$CH_2{=}CHCN(g)$	$H_2O(l)$
$\Delta H_f°$ (kJ/mol)	-53	135.1	195.4	-286
S° (J/K•mol)	242	202	274	70

$\Delta H° = 195.4 - 286 - (-53 + 135.1) = -173$ kJ

$\Delta S° = 274 + 70 - 242 - 202 = -100.$ J/K

$\Delta G° = \Delta H° - T\Delta S° = -173$ kJ $- 298$ K $(-0.100$ kJ/K$) = -143$ kJ

b. $HC{\equiv}CH(g) + HCN(g) \longrightarrow CH_2{=}CHCN(g)$

For $C_2H_2(g)$ $\Delta H_f° = 227$ kJ/mol and S° = 201 J/K mol

$\Delta H° = 195.4 - [135.1 + 227] = -167$ kJ

$\Delta S° = 274 - [202 + 201] = -129$ J/K

We will use 70°C for T, or T = 343 K (carry extra significant figure)

$\Delta G° = \Delta H° - T\Delta S° = -167$ kJ $- 343$ K $(-0.129$ kJ/K$)$

$\Delta G° = -167 + 44 = -123$ kJ

c. $4\,CH_2{=}CHCH_3(g) + 6\,NO(g) \longrightarrow 4\,CH_2{=}CHCN(g) + 6\,H_2O(g) + N_2(g)$

	$CH_2{=}CHCH_3(g)$	$NO(g)$	$CH_2{=}CHCN(g)$	$H_2O(g)$	$N_2(g)$
$\Delta H_f°$ (kJ/mol)	20.9	90	195.4	-242	0
S° (J/K•mol)	266.9	211	274	189	192

$\Delta H° = 6(-242) + 4(195.4) - [4(20.9) + 6(90)] = -1294$ kJ

$\Delta S° = 192 + 6(189) + 4(274) - [6(211) + 4(266.9)] = 88$ J/K

We will use 700°C = 973 K for T. (carry extra significant figures)

$\Delta G° = \Delta H° - T\Delta S°$

$\Delta G° = -1294\text{ kJ} - 973\text{ K }(0.088\text{ kJ/K}) = -1294\text{ kJ} - 86\text{ kJ} = -1380\text{ kJ}$

33. $SF_4(g) + F_2(g) \longrightarrow SF_6(g)$

$-374\text{ kJ} = -1105\text{ kJ} - \Delta G_f°(SF_4)$

$\Delta G_f° = -731$ kJ/mol

34. $2\ Al(OH)_3(s) \longrightarrow Al_2O_3(s) + 3\ H_2O(g)$

$7\text{ kJ} = 3(-299\text{ kJ}) - 1582\text{ kJ} - (2\text{ mol})\Delta G_f°\ [Al(OH)_3]$

$\Delta G_f° = -1138$ kJ/mol

35. $P_4\ (s,\alpha) \longrightarrow P_4\ (s,\beta)$

a. At T < -76.9°C, this reaction is spontaneous. Thus, the sign of ΔG is (-). At 76.9°C, ΔG = 0 and above 76.9 °C, the sign of ΔG is (+). This is consistent with ΔH (-) and ΔS (-).

b. Since the sign of ΔS is negative, then the β form has the more ordered structure.

36. Endothermic: ΔH (+); Spontaneous: ΔG (-)

Enthalpy is not favorable, so ΔS must provide the driving force for the change. Thus, ΔS is positive. There is an increase in disorder, so the original enzyme has the more ordered structure.

37. $H_2(g) \longrightarrow 2\ H(g)$

a. A bond is broken; ΔH (+); Increase in disorder; ΔS (+)

b. $\Delta G = \Delta H - T\Delta S$

For the reaction to be spontaneous, the entropy term must be dominate. The reaction will be spontaneous at higher temperatures.

38. $2\ NH_3(g) + 3\ O_2(g) + 2\ CH_4(g) \longrightarrow 2\ HCN(g) + 6\ H_2O(g)$

	$2\ NH_3(g)$	$3\ O_2(g)$	$2\ CH_4(g)$	$2\ HCN(g)$	$6\ H_2O(g)$
$\Delta H_f°$ (kJ/mol)	-46	0	-75	135.1	-242
S° (J/K•mol)	193	205	186	202	189

$\Delta H° = 6(-242) + 2(135.1) - [2(-75) + 2(-46)] = -940.\text{ kJ}$

$\Delta S° = 2(202) + 6(189) - [2(193) + 3(205) + 2(186)] = +165\text{ J/K}$

From the signs of ΔH and ΔS, this reaction is spontaneous at all temperatures. It will cost money to heat the reaction mixture. Since there is no thermodynamic reason to do this, then the purpose of the elevated temperature must be to increase the rate of the reaction.

39. $NO(g) + O_3(g) \rightleftharpoons NO_2(g) + O_2(g)$

$\Delta G° = \Sigma\, \Delta G_f^\circ$ (Products - $\Sigma\, \Delta G_f^\circ$ (Reactants)

= 1 mol (52 kJ/mol) - [1 mol (87 kJ/mol) + 1 mol (163 kJ/mol)]

$\Delta G° = -198$ kJ and $\Delta G° = -RT \ln K$

$$K = \exp \frac{-\Delta G°}{RT} = \exp\left(\frac{+1.98 \times 10^5 \text{ J}}{8.3145 \text{ J/K•mol } (298 \text{ K})}\right) = e^{79.912} = 5.07 \times 10^{34}$$

40. $2\, H_2S(g) + SO_2(g) \rightleftharpoons 3\, S(s) + 2\, H_2O(g)$

	$2\, H_2S(g)$	$SO_2(g)$	$3\, S(s)$	$2\, H_2O(g)$
ΔG_f° (kJ/mol)	-34	-300	0	-229

$\Delta G° = 2$ mol (-229 kJ/mol) - [2 mol (-34 kJ/mol) + 1 mol (-300 kJ/mol)] = -90. kJ

$$K = \exp\left(\frac{-\Delta G°}{RT}\right) = \exp\left(\frac{90. \times 10^3}{(8.3145)\,(298)}\right) = e^{36.32} = 5.9 \times 10^{15}$$

Since there is a decrease in the number of moles of gaseous species, $\Delta S°$ is negative. Since, $\Delta G°$ is negative, then $\Delta H°$ must be negative. The reaction will be spontaneous at low temperatures (the $\Delta H°$ term is dominate at low temperature).

41. a. $\Delta G° = -RT \ln K$

$\Delta G° = -(8.3145$ J/K•mol) (298 K) $\ln(1.00 \times 10^{-14}) = 7.99 \times 10^4$ J/mol = 79.9 kJ/mol

b. $\Delta G° = -RT \ln K = -(8.3145$ J/K•mol) (313 K) $\ln(2.92 \times 10^{-14})$

$= 8.11 \times 10^4$ J/mol = 81.1 kJ/mol

42. a. Necessary Data

	ΔH_f° (kJ/mol)	S° (J/K•mol))
$NH_3(g)$	-46	193
$O_2(g)$	0	205
$NO(g)$	90	211
$H_2O(g)$	-242	189
$NO_2(g)$	34	240
$HNO_3(l)$	-174	156
$H_2O(l)$	-286	70

$4\ NH_3(g) + 5\ O_2(g) \longrightarrow 4\ NO(g) + 6\ H_2O(g)$

$\Delta H° = 6(-242) + 4(90) - [4(-46)] = -908$ kJ

$\Delta S° = 4(211) + 6(189) - [4(193) + 5(205)] = 181$ J/K

$\Delta G° = -908$ kJ $- 298$ K $(0.181$ kJ/K$) = -962$ kJ

$\Delta G° = -RT \ln K$

$$\ln K = \frac{-\Delta G°}{RT} = \left(\frac{+962 \times 10^3 \text{ J}}{(8.3145 \text{ J/K•mol})\ (298 \text{ K})}\right) = 388$$

$\ln K = 2.303 \log K$; $\log K = 168$; $K = 10^{168}$

$2\ NO(g) + O_2(g) \longrightarrow 2\ NO_2(g)$

$\Delta H° = 2(34) -[2(90)] = -112$ kJ

$\Delta S° = 2(240) - [2(211) + (205)] = -147$ J/K

$\Delta G° = -112$ kJ $- (298$ K$)\ (-0.147$ kJ/K$) = -68$ kJ

$$K = \exp \frac{-\Delta G°}{RT} = \exp\left(\frac{+68{,}000 \text{ J}}{(8.3145 \text{ J/K•mol})\ (298 \text{ K})}\right)$$

$K = e^{27.44} = 8.3 \times 10^{11}$

$3\ NO_2(g) + H_2O\ (l) \longrightarrow 2\ HNO_3(l) + NO(g)$

$\Delta H° = 2(-174) + (90) - [3(34) + (-286)] = -74$ kJ

$\Delta S° = 2(156) + (211) - [3(240) + (70)] = -267$ J/K

$\Delta G° = -74$ kJ $- (298$ K$)\ (-0.267$ kJ/K$) = +6$ kJ

$$K = \exp \frac{-\Delta G°}{RT} = \exp\left(\frac{-6000 \text{ J}}{(8.3145 \text{ J/K•mol})\ (298 \text{ K})}\right) = e^{-2.4} = 9 \times 10^{-2}$$

b. $\Delta G° = -RT \ln K$, $825°C = (825 + 273)K = 1098$ K

$\Delta G° = -908$ kJ $- (1098$ K$)\ (0.181$ kJ/K$) = -1107$ kJ

$\Delta G°$ is temperature dependent. We assume $\Delta H°$ and $\Delta S°$ do not depend on T.

$$K = \exp \frac{-\Delta G°}{RT} = \exp\left(\frac{1.107 \times 10^6 \text{ J}}{(8.3145 \text{ J/K•mol})\ (1098 \text{ K})}\right) = e^{121.258} = 4.589 \times 10^{52}$$

c. There is no thermodynamic reason for the elevated temperature since $\Delta H°$ is negative and $\Delta S°$ is positive. Thus, the purpose of high temperature must be to increase the rate of the reaction.

Free Energy and Pressure

43. From Exercise 16.39, we get $\Delta G° = -198$ kJ

$$\Delta G = \Delta G° + RT \ln \frac{P_{NO_2} P_{O_2}}{P_{NO} P_{O_3}}$$

$$\Delta G = -198 \text{ kJ} + \frac{8.3145 \text{ J/K•mol}}{1000 \text{ J/kJ}} (298 \text{ K}) \ln \frac{(1.00 \times 10^{-7})(1.00 \times 10^{-3})}{(1.00 \times 10^{-6})(2.00 \times 10^{-6})}$$

$$= -198 \text{ kJ} + 9.69 \text{ kJ} = -188 \text{ kJ}$$

44. $$\Delta G = \Delta G° + RT \ln \frac{P^2_{H_2O}}{P^2_{H_2S} P_{SO_2}}$$

$$\Delta G = -90. \text{ kJ} + \frac{(8.3145)(298)}{1000} \text{ kJ} \ln \frac{(0.030)^2}{(1.0 \times 10^{-4})^2 (0.010)}$$

$$\Delta G = -90. + 39.7 = -50. \text{ kJ}$$

45. $N_2(g) + 3 H_2(g) \rightleftharpoons 2 NH_3(g)$

$\Delta H° = 2 \Delta H_f° (NH_3) = 2(-46) = -92$ kJ

$\Delta G° = 2 \Delta G_f° (NH_3) = 2(-17) = -34$ kJ

$\Delta S° = 2(193) - [192 + 3(131)]$

$\Delta S° = -199$ J/K

$$K = \exp \frac{-\Delta G°}{RT} = \exp \left(\frac{34{,}000 \text{ J}}{(8.3145 \text{ J/K•mol})(298 \text{ K})} \right)$$

$K = e^{13.72} = 9.1 \times 10^5$

a. $$\Delta G = \Delta G° + RT \ln \frac{P^2_{NH_3}}{P_{N_2} P^3_{H_2}}$$

$$\Delta G = -34 \text{ kJ} + \frac{(8.3145 \text{ J/K•mol})(298 \text{ K})}{1000 \text{ J/kJ}} \ln \frac{(50.)^2}{(200.)(200.)^3}$$

$$\Delta G = -34 \text{ kJ} - 33 \text{ kJ} = -67 \text{ kJ}$$

b. $$\Delta G = -34 \text{ kJ} + \frac{(8.3145 \text{ J/K•mol})(298 \text{ K})}{1000 \text{ J/kJ}} \ln \frac{(200.)^2}{(200.)(600.)^3}$$

$$\Delta G = -34 \text{ kJ} - 34.4 \text{ kJ} = -68 \text{ kJ}$$

c. $\Delta G°_{100} = \Delta H° - T\Delta S°$, $\Delta G°$ depends on T. Assume $\Delta H°$ and $\Delta S°$ are T independent.

$\Delta G°_{100} = -92 \text{ kJ} - (100. \text{ K})(-0.199 \text{ kJ/K})$

$\Delta G^\circ_{100} = -72$ kJ

$\Delta G = \Delta G^\circ + RT \ln Q$

$$\Delta G = -72 \text{ kJ} + \frac{(8.3145 \text{ J/K}\cdot\text{mol})(100.\text{ K})}{1000 \text{ J/kJ}} \ln \frac{(10.)^2}{(50.)(200.)^3} = -72 - 13 = -85 \text{ kJ}$$

d. $\Delta G^\circ_{700} = -92 \text{ kJ} - (700.\text{ K})(-0.199 \text{ kJ/K}) = 47 \text{ kJ}$

$$\Delta G = 47 \text{ kJ} + \frac{(8.3145 \text{ J/K}\cdot\text{mol})(700.\text{ K})}{1000 \text{ J/kJ}} \ln \frac{(10.)^2}{(50.)(200.)^3}$$

$\Delta G = 47 \text{ kJ} - 88 \text{ kJ} = -41 \text{ kJ}$

46. a. $\Delta G^\circ = -RT \ln K = -(8.3145 \text{ J/K}\cdot\text{mol})(298 \text{ K}) \ln 0.090$

$\Delta G^\circ = 6.0 \times 10^3 \text{ J/mol} = 6.0 \text{ kJ/mol}$

b. $H—O—H + Cl—O—Cl \longrightarrow 2\ H—O—Cl$

On each side of the reaction there are 2 H — O bonds and 2 O — Cl bonds. Both sides have the same number and type of bonds. Thus, $\Delta H^\circ \approx 0$

c. $\Delta G^\circ = \Delta H^\circ - T\Delta S^\circ$

$$\Delta S^\circ = \frac{\Delta H^\circ - \Delta G^\circ}{T} = \frac{0 - 6.0 \times 10^3 \text{ J}}{298 \text{ K}} = -20.\text{ J/K}$$

d. $\Delta H^\circ = 0 = 2\,\Delta H_f^\circ(\text{HOCl}) - [1 \text{ mol}(80.3 \text{ kJ/mol}) + 1 \text{ mol}(-242 \text{ kJ/mol})]$

$0 = 2\,\Delta H_f^\circ + 162;\ \ \Delta H_f^\circ = -81 \text{ kJ/mol}$

$-20.\text{ J/K} = 2\,S^\circ(\text{HOCl}) - [1 \text{ mol}(266.1 \text{ J/K}\cdot\text{mol}) + 1 \text{ mol}(189 \text{ J/K}\cdot\text{mol})]$

$-20.\text{ J/K} = 2\,S^\circ - 455 \text{ J/K};\ \ S^\circ = 218 \text{ J/K}\cdot\text{mol}$

e. $\Delta G^\circ_{500} = 0 - (500 \text{ K})(-20.\text{ J/K}) = +10{,}000 \text{ J}$ Assume ΔH° and ΔS° are T independent.

$\Delta G^\circ = -RT \ln K$

$$K = \exp\left(\frac{-\Delta G^\circ}{RT}\right) = \exp\left(\frac{-10{,}000}{(8.3145)(500)}\right) = e^{-2.4} = 0.09$$

f. $$\Delta G = \Delta G^\circ + RT \ln \frac{P^2_{HOCl}}{P_{H_2O}P_{Cl_2O}}$$

We should express all P's in atm. However, we perform the pressure conversion the same number of times in the numerator and denominator, so the factors of 760 torr/atm will all cancel. Thus, we can use the pressures in units of torr.

$$\Delta G = 6.0 \text{ kJ} + \frac{(8.3145 \text{ J/K}\cdot\text{mol})(298 \text{ K})}{1000 \text{ J/kJ}} \ln\left(\frac{(0.10)^2}{(18)(2.0)}\right) = 6.0 - 20. = -14 \text{ kJ}$$

47. a.

$$C \equiv O(g) + 2\,H—H(g) \longrightarrow H—\underset{H}{\overset{H}{\underset{|}{\overset{|}{C}}}}—O—H(g)$$

$\Delta S° \approx 100(-2) = -200$ J/K

Bonds broken:

C≡O	1072 kJ/mol
2 H—H	432 kJ/mol

Bonds made:

3 C—H	413 kJ/mol
1 C—O	358 kJ/mol
1 O—H	467 kJ/mol

$\Delta H° = 1072 + 2(432) - [3(413) + 358 + 467] = -128$ kJ

$\Delta G° = \Delta H° - T\Delta S° = -128$ kJ - (298 K) (-0.2 kJ/K) = -68 kJ ≈ - 70 kJ

b. $\Delta H° = -201 - [-110.5] = -91$ kJ

$\Delta S° = 240 - [198 + 2(131)] = -220.$ J/K

$\Delta G° = -91$ kJ - (298 K) (-0.220 kJ/K) = -25 kJ

The values are reasonably close for $\Delta H°$ and $\Delta S°$ (within 30%). $\Delta G°$ is the right order of magnitude. Even a factor of 3 difference (as in $\Delta G°$) translates into less than 1 power of 10 in the equilibrium constant. The approximations give good ball park estimates.

c. Since ΔS is unfavorable and ΔH is favorable, we would run the reaction at relatively low temperatures.

$\Delta G = 0$; $\Delta H - T\Delta S = 0$

$$T = \frac{\Delta H}{\Delta S} = \frac{91 \times 10^3\ \text{J}}{220\ \text{J/K}} = 410\ \text{K}$$

Reaction would be spontaneous at $T < 410$ K. Note: We may need high temperatures to get the reaction to occur at a reasonable rate.

48. a.

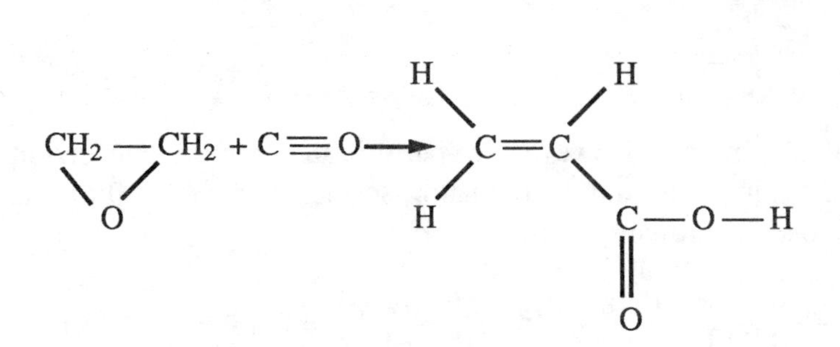

Bonds broken:		Bonds made:	
1 C—O	358 kJ/mol	1 C=C	614 kJ/mol
1 C≡O	1072 kJ/mol	1 C=O	799 kJ/mol
1 C—H	413 kJ/mol	1 O—H	467 kJ/mol

$\Delta H° = 358 + 1072 + 413 - (614 + 799 + 467) = -37$ kJ

$\Delta S° \approx 100\ \Delta n \approx -200$ J/K = -0.2 kJ/K

$\Delta G° = \Delta H° - T\Delta S° = -37$ kJ - (298 K) (-0.2 kJ/K) = + 23 kJ = 20 kJ

b. Reaction is spontaneous at low temperatures, since both $\Delta H°$ and $\Delta S°$ are negative.

$$T < \frac{37{,}000}{200} = 185 \text{ K} \approx 200 \text{ K}$$

At T < 200 K, the reaction is spontaneous. This may be too low, however, for a reasonable rate.

ADDITIONAL EXERCISES

49. There are more ways to roll a seven. We can consider all of the possible throws by constructing a table.

one die	1	2	3	4	5	6	
1	2	3	4	5	6	7	
2	3	4	5	6	7	8	
3	4	5	6	7	8	9	sum of the two dice
4	5	6	7	8	9	10	
5	6	7	8	9	10	11	
6	7	8	9	10	11	12	

There are six ways to get a seven, more than any other number. The seven is not favored by energy; rather it is favored by probability. To change the probability we would have to expend energy (do work).

50. $\Delta S_{univ} = \Delta S_{sys} + \Delta S_{surr}$; $\Delta S_{surr} = \dfrac{-\Delta H_{sys}}{T}$

$$\Delta S_{univ} = \Delta S_{sys} - \frac{\Delta H_{sys}}{T}$$

For a spontaneous process: $\Delta S_{univ} > 0$

So: $\Delta S_{sys} - \frac{\Delta H_{sys}}{T} > 0$

or $\Delta H_{sys} - T\Delta S_{sys} < 0$ for a spontaneous process.

Since $\Delta G_{sys} = \Delta H_{sys} - T\Delta S_{sys}$

then $\Delta G_{sys} < 0$ for a spontaneous process.

51. $\Delta G = 0$ so $\Delta H = T\Delta S$

$$\Delta S = \frac{\Delta H}{T} = \frac{35.23 \times 10^3 \text{ J/mol}}{3650 \text{ K}} = 9.65 \text{ J/K}\cdot\text{mol}$$

52. S (rhombic) $\longrightarrow$ S (monoclinic)

$\Delta H° = 0.30$ kJ/mol, $\Delta S° = 32.55 - 31.88 = 0.67$ J/K

For $\Delta G° = 0 = \Delta H° - T\Delta S°$; $\Delta H° = T\Delta S°$

$$T = \frac{\Delta H°}{\Delta S°} = \frac{3.0 \times 10^2 \text{ J/mol}}{0.67 \text{ J/K}\cdot\text{mol}} = 450 \text{ K}$$

53. It appears that the sum of the two processes is no net change. This is not so, ΔS_{univ} has increased even though it looks as if we have gone through a cyclic process.

54. $C_2H_4(g) + H_2O(g) \longrightarrow CH_3CH_2OH(l)$

$\Delta H° = -278 - (52 - 242) = -88$ kJ

$\Delta S° = 161 - (219 + 189) = -247$ J/K

When $\Delta G° = 0$, $\Delta H° = T\Delta S°$

$$T = \frac{\Delta H°}{\Delta S°} = \frac{-88 \times 10^3 \text{ J}}{-247 \text{ J/K}} = 360 \text{ K}$$

The reaction is spontaneous at temperatures below 360 K.

$C_2H_6(g) + H_2O(g) \longrightarrow CH_3CH_2OH\ (l) + H_2(g)$

$\Delta H° = -278 - (-84.7 - 242) = +49$ kJ

$\Delta S° = 131 + 161 - (229.5 + 189) = -127$ J/K

This reaction can never be spontaneous. Thus, the reaction

$$C_2H_4(g) + H_2O(g) \longrightarrow C_2H_5OH\ (l)$$

would be preferred. Note: This is a standard method for producing alcohols.

55. $CH_4(g) + CO_2(g) \longrightarrow CH_3CO_2H\ (l)$

$\Delta H° = -484 - [-75 + (-393.5)] = -16$ kJ

$\Delta S° = 160 - [186 + 214] = -240.$ J/K

$\Delta G° = \Delta H° - T\Delta S° = -16$ kJ $- (298$ K$)(-0.240$ kJ/K$) = +56$ kJ

This reaction is spontaneous only at a temperature below $T = \frac{\Delta H°}{\Delta S°} = 67$ K.

This is not practical. Substances will be in condensed phases and rates will be very slow.

$CH_3OH(g) + CO(g) \longrightarrow CH_3CO_2H(l)$

$\Delta H° = -484 - [-110.5 + (-201)] = -173$ kJ

$\Delta S° = 160 - [198 + 240] = -278$ J/K

$\Delta G° = -173$ kJ $- (298$ K$)(-0.278$ kJ/K$) = -90.$ kJ

This reaction is spontaneous at a temperature below $T = \frac{\Delta H°}{\Delta S°} = 622$ K.

Thus, the reaction of CH_3OH and CO will be preferred. It is spontaneous at high enough temperatures that the rates of reaction should be reasonable.

56. From our answer to Exercise 16.41 we get $\Delta G° = 79.9$ kJ

$$\Delta G = \Delta G° + RT \ln [H^+][OH^-]$$

$$\Delta G = 79.9 + \frac{(8.3145 \text{ J/K•mol})(298 \text{ K})}{1000 \text{ J/kJ}} \ln [H^+][OH^-]$$

a. $\Delta G = 79.9 - 79.9 = 0$, At equilibrium

b. $\Delta G = 0$, At equilibrium

c. $\Delta G = -34$ kJ, Shift to right

d. $\Delta G = 45.7$ kJ, Shift to left

e. $\Delta G = 79.9$ kJ, Shift to left

LeChatelier's Principle gives the same results.

57. K^+ (blood) $\rightleftharpoons$ K^+ (muscle) $\quad \Delta G° = 0$, so $\Delta G = RT \ln \left(\frac{[K^+]_m}{[K^+]_b}\right)$

$\Delta G = \frac{8.3145 \text{ J}}{\text{K mol}}(310.\text{ K}) \ln \left(\frac{0.15}{0.0050}\right)$; $\Delta G = 8.8 \times 10^3$ J/mol $= 8.8$ kJ/mol $=$ work

$$\frac{8.8 \text{ kJ}}{\text{mol } K^+} \times \frac{1 \text{ mol ATP}}{30.5 \text{ kJ}} = 0.29 \text{ mol ATP}$$

Other ions will have to be transported in order to maintain electroneutrality. Either anions must be transported into the cells, or cations (Na^+) in the cell must be transported to the blood. The latter is what happens: $[Na^+]$ in blood is greater than $[Na^+]$ in cells as a result of this pumping.

58. If we go through a cycle, it may look like nothing has changed when we get back to the starting point. This is not true; there has been a change. The change that occurred is that the entropy of the universe has increased.

59. As ΔS_{univ} continually increases, there will be less energy available for work. When no work can be done, the world ends.

60. The introduction of mistakes is an effect of entropy. The purpose of redundant information is to provide a control to check the "correctness" of the transmitted information.

See: *Grammatical Man: Information, Entropy, and Language* by Jeremy Campbell, Simon and Schuster (Touchstone Book), 1982.

CHALLENGE PROBLEMS

61. Arrangement I:

$S = k \ln W$
$W = 1$
$S = k \ln 1 = 0$

Arrangement II.

$W = 4$
$S = k \ln 4 = 1.38 \times 10^{-23}$ J/K ln 4
$S = 1.91 \times 10^{-23}$ J/K

Arrangement III:

$W = 6$
$S = k \ln 6 = 2.47 \times 10^{-23}$ J/K

62. $\Delta G° = \Delta H° - T\Delta S° = -RT \ln K$

Dividing by -RT, we get: $\ln K = \frac{-\Delta H°}{RT} + \frac{\Delta S°}{R}$

If we graph ln K vs 1/T we get a straight line. For an endothermic process the slope is negative.

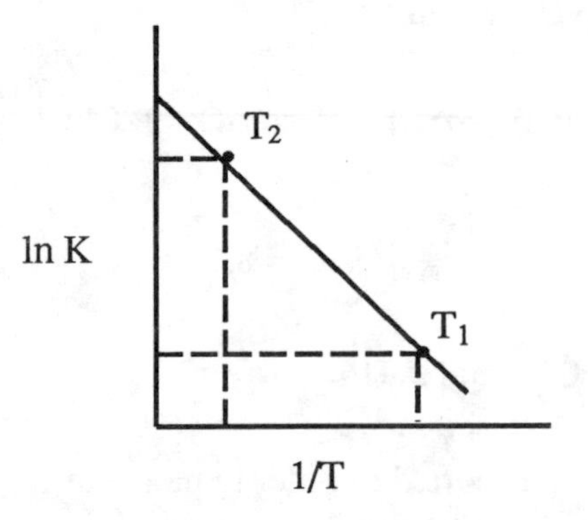

As we increase the temperature from T_1 to T_2 ($1/T_2 < 1/T_1$) we can see that ln K and hence, K increases. A larger value of K means some more reactants can be converted to products.

63. We shall graph ln K vs 1/T

Temp (°C)	T (K)	1000/T	K	ln K
0	273	3.66	1.14×10^{-15}	-34.41
25	298	3.36	1.00×10^{-14}	-32.24
35	308	3.25	2.09×10^{-14}	-31.50
40	313	3.19	2.92×10^{-14}	-31.17
50	323	3.10	5.47×10^{-14}	-30.54

The graph was drawn previously for Exercise 14.29. The equation of the straight line is:

$$\ln K = -6.91 \times 10^3 \left(\frac{1}{T}\right) - 9.09$$

$$\text{Slope} = -6.91 \times 10^3 = \frac{-\Delta H°}{R}; \qquad \Delta H° = 5.74 \times 10^4\ J = 57.4\ kJ$$

$$\text{Intercept} = -9.09 = \frac{\Delta S°}{R}; \qquad \Delta S° = -75.6\ J/K$$

64. a. $k_f = A \exp\left(\frac{-E_a}{RT}\right)$ and $k_r = A \exp\left(\frac{-(E_a - \Delta G°)}{RT}\right)$

The activation energy of the reverse reaction is $E_a + (-\Delta G°) = E_a - \Delta G°$. ($\Delta G°$ is a negative number as drawn in the diagram.)

$$\frac{k_f}{k_r} = \frac{A \exp\left(\frac{-E_a}{RT}\right)}{A \exp\left(\frac{-(E_a - \Delta G°)}{RT}\right)}$$

If the A factors are equal:

$$\frac{k_f}{k_r} = \left(\frac{-E_a}{RT} + \frac{(E_a - \Delta G°)}{RT}\right) = \exp\left(\frac{-\Delta G°}{RT}\right)$$

$$\Delta G° = -RT \ln K, \quad \text{so } K = \exp\left(\frac{-\Delta G°}{RT}\right) \qquad \text{Thus, } K = \frac{k_f}{k_r}$$

b. The catalyst will change the activation energy of both the forward and reverse reaction but not change $\Delta G°$. As a state function, $\Delta G°$ depends only on the initial and final conditions, not on the path.

65. a. $\Delta G° = -RT \ln K$

$K = \exp(-\Delta G°/RT)$

$$K = \exp\left(\frac{30{,}500\ J/mol}{8.3145\ J/K{\bullet}mol \times 298\ K}\right) = 2.22 \times 10^5$$

b. $C_6H_{12}O_6(s) + 6\ O_2(g) \longrightarrow 6\ CO_2(g) + 6\ H_2O(l)$

$\Delta G° = 6\ mol\ (-394\ kJ/mol) + 6\ mol\ (-237\ kJ/mol) - 1\ mol\ (-911\ kJ/mol)$

$\Delta G° = -2875\ kJ$

$$\frac{2875\ kJ}{mol\ glucose} \times \frac{1\ mol\ ATP}{30.5\ kJ} = 94.3\ mol\ ATP$$

This is an overstatement. The assumption that all of the free energy goes into this reaction is false. Actually only 38 moles of ATP are produced by metabolism of one mole of glucose.

66. $\Delta G°\ (CO) = -RT \ln K_{CO}$ and $\Delta G°(O_2) = -RT \ln K_{O_2}$

$\Delta G°\ (CO) - \Delta G°(O_2) = -RT \ln K_{CO} + RT \ln K_{O_2} = -RT \ln K_{CO}/K_{O_2}$

$\Delta G°\ (CO) - \Delta G°(O_2) = -(8.3145\ J/K{\bullet}mol)\ (298\ K) \ln 210 = -13{,}200\ J/mol = -13\ kJ/mol$

67. The hydration of the dissociated ions results in a more ordered "structure" in solution. The smallest ion (F^-) has the largest charge density and should show the most ordering.

68. Because of hydrogen bonding there is greater "structure" (or order) in liquid water than in most other liquids. Thus, there is a greater increase in disorder when water evaporates or a greater increase in entropy.

69. At equilibrium:

$$P_{H_2} = \frac{nRT}{V} = \frac{\left(\dfrac{1.10 \times 10^{13}\ molecules}{6.022 \times 10^{23}\ molecules/mol}\right)\left(\dfrac{0.08206\ L\ atm}{mol\ K}\right)(298\ K)}{1.0\ L}$$

$P_{H_2} = 4.5 \times 10^{-10}\ atm$

Essentially all of the H_2 and Br_2 have reacted, therefore $P_{HBr} = 2.0\ atm$.

Since we began with equal moles of H_2 and Br_2, we will have equal moles of H_2 and Br_2 at equilibrium. Therefore, $P_{H_2} = P_{Br_2} = 4.5 \times 10^{-10}\ atm$

$$K = \frac{P^2_{HBr}}{P_{H_2}P_{Br_2}} = \frac{(2.0)^2}{(4.5 \times 10^{-10})^2} = 2.0 \times 10^{19}$$

$\Delta G° = -RT \ln K = -(8.3145\ J/K{\bullet}mol)\ (298\ K) \ln (2.0 \times 10^{19})$

$\Delta G° = -1.1 \times 10^5\ J/mol = -110\ kJ/mol$

$\Delta G° = \Delta H° - T\Delta S°$

$$\Delta S° = \frac{\Delta H° - \Delta G°}{T} = \frac{-103{,}800\ J/mol - (-110{,}000\ J/mol)}{298\ K}$$

$\Delta S° = 21\ J/K{\bullet}mol = 20\ J/K{\bullet}mol$

CHAPTER SEVENTEEN: ELECTROCHEMISTRY

REVIEW OF OXIDATION - REDUCTION REACTIONS

1. Oxidation: increase in oxidation number
   loss of electrons

   Reduction: decrease in oxidation number
   gain of electrons

2. a. H (+1), 0 (-2), N (+5) b. Cl (-1), Cu (+2)

   c. zero d. H (+1), 0 (-1)

   e. Mg (+2), 0 (-2), S (+6) f. zero

   g. Pb (+2), 0 (-2), S (+6) h. O (-2), Pb (+4)

   i. Na (+1), 0 (-2), C (+3) j. O (-2), C (+4)

   k. $(NH_4)_2Ce(SO_4)_3$ contains NH_4^+ ions and SO_4^{2-} ions. Thus, cerium exists as the Ce^{4+} ion.
   H (+1), N (-3), Ce (+4), S (+6), O (-2)

   l. O (-2), Cr (+3)

3.

	Redox?	Ox. Agent	Red. Agent	Substance Oxidized	Substance Reduced
a.	Yes	H_2O	CH_4	CH_4 (C)	H_2O (H)
b.	Yes	$AgNO_3$	Cu	Cu	$AgNO_3$ (Ag)
c.	No	-----	-----	-----	-----
d.	Yes	HCl	Zn	Zn	HCl (H)
e.	No	-----	-----	-----	-----
f.	Yes	HCl	Fe	Fe	HCl (H)

4. a. $Cr \rightarrow Cr^{3+} + 3\ e^-$

$$NO_3^- \rightarrow NO$$
$$4\ H^+ + NO_3^- \rightarrow NO + 2\ H_2O$$
$$3\ e^- + 4\ H^+ + NO_3^- \rightarrow NO + 2\ H_2O$$

$$Cr \rightarrow Cr^{3+} + 3\ e^-$$
$$3\ e^- + 4\ H^+ + NO_3^- \rightarrow NO + 2\ H_2O$$

$$4\ H^+(aq) + NO_3^-(aq) + Cr(s) \rightarrow Cr^{3+}(aq) + NO(g) + 2\ H_2O(l)$$

b. $(Al \rightarrow Al^{3+} + 3\,e^-) \times 5$

$$MnO_4^- \rightarrow Mn^{2+}$$
$$8\,H^+ + MnO_4^- \rightarrow Mn^{2+} + 4\,H_2O$$
$$(5\,e^- + 8\,H^+ + MnO_4^- \rightarrow Mn^{2+} + 4\,H_2O) \times 3$$

$$5\,Al \rightarrow 5\,Al^{3+} + 15\,e^-$$
$$15\,e^- + 24\,H^+ + 3\,MnO_4^- \rightarrow 3\,Mn^{2+} + 12\,H_2O$$

$$24\,H^+(aq) + 3\,MnO_4^-(aq) + 5\,Al(s) \rightarrow 5\,Al^{3+}(aq) + 3\,Mn^{2+}(aq) + 12\,H_2O(l)$$

c. $(Ce^{4+} + e^- \rightarrow Ce^{3+}) \times 6$

$$CH_3OH \rightarrow CO_2$$
$$H_2O + CH_3OH \rightarrow CO_2 + 6\,H^+$$
$$H_2O + CH_3OH \rightarrow CO_2 + 6\,H^+ + 6\,e^-$$

$$6\,Ce^{4+} + 6\,e^- \rightarrow 6\,Ce^{3+}$$
$$H_2O + CH_3OH \rightarrow CO_2 + 6\,H^+ + 6\,e^-$$

$$H_2O(l) + CH_3OH(aq) + 6\,Ce^{4+}(aq) \rightarrow 6\,Ce^{3+}(aq) + CO_2(g) + 6\,H^+(aq)$$

d.

$$SO_3^{2-} \rightarrow SO_4^{2-}$$
$$H_2O + SO_3^{2-} \rightarrow SO_4^{2-} + 2\,H^+$$
$$(H_2O + SO_3^{2-} \rightarrow SO_4^{2-} + 2\,H^+ + 2\,e^-) \times 5$$

$$(5\,e^- + 8\,H^+ + MnO_4^- \rightarrow Mn^{2+} + 4\,H_2O) \times 2$$
See 17.4 b

$$5\,H_2O + 5\,SO_3^{2-} \rightarrow 5\,SO_4^{2-} + 10\,H^+ + 10\,e^-$$
$$10\,e^- + 16\,H^+ + 2\,MnO_4^- \rightarrow 2\,Mn^{2+} + 8\,H_2O$$

$$6\,H^+(aq) + 5\,SO_3^{2-}(aq) + 2\,MnO_4^-(aq) \rightarrow 5\,SO_4^{2-}(aq) + 2\,Mn^{2+}(aq) + 3\,H_2O(l)$$

e.

$$PO_3^{3-} \rightarrow PO_4^{3-}$$
$$H_2O + PO_3^{3-} \rightarrow PO_4^{3-} + 2\,H^+ + 2\,e^-$$
$$2\,OH^- + H_2O + PO_3^{3-} \rightarrow PO_4^{3-} + 2\,H^+ + 2\,OH^- + 2\,e^-$$
$$(2\,OH^- + PO_3^{3-} \rightarrow PO_4^{3-} + H_2O + 2\,e^-) \times 3$$

$$MnO_4^- \rightarrow MnO_2$$
$$3\,e^- + 4\,H^+ + MnO_4^- \rightarrow MnO_2 + 2\,H_2O$$
$$3\,e^- + 4\,H^+ + 4\,OH^- + MnO_4 \rightarrow MnO_2 + 2\,H_2O + 4\,OH^-$$
$$(3\,e^- + 2\,H_2O + MnO_4^- \rightarrow MnO_2 + 4\,OH^-) \times 2$$

$$6\,OH^- + 3\,PO_3^{3-} \rightarrow 3\,PO_4^{3-} + 3\,H_2O + 6\,e^-$$
$$6\,e^- + 4\,H_2O + 2\,MnO_4^- \rightarrow 2\,MnO_2 + 8\,OH^-$$

$$H_2O\,(l) + 3\,PO_3^{3-}(aq) + 2\,MnO_4^-(aq) \rightarrow 3\,PO_4^{3-}(aq) + 2\,MnO_2(s) + 2\,OH^-(aq)$$

f.

$$Mg \rightarrow Mg(OH)_2$$
$$2\,OH^- + Mg \rightarrow Mg(OH)_2 + 2\,e^-$$

$$OCl^- \rightarrow Cl^-$$
$$2\,e^- + 2\,H^+ + OCl^- \rightarrow Cl^- + H_2O$$
$$2\,e^- + 2\,OH^- + 2\,H^+ + OCl^- \rightarrow Cl^- + H_2O + 2\,OH^-$$
$$2\,e^- + H_2O + OCl^- \rightarrow Cl^- + 2\,OH^-$$

$$2\,OH^- + Mg \rightarrow Mg(OH)_2 + 2\,e^-$$
$$2\,e^- + H_2O + OCl^- \rightarrow Cl^- + 2\,OH^-$$

$$H_2O(l) + Mg(s) + OCl^-(aq) \rightarrow Mg(OH)_2(s) + Cl^-(aq)$$

g.

$$Ce^{4+} \rightarrow Ce(OH)_3$$
$$(e^- + 3\,OH^- + Ce^{4+} \rightarrow Ce(OH)_3) \times 40$$

$$Ni(CN)_4^{2-} \rightarrow Ni(OH)_2 + CO_3^{2-} + NO_3^-$$
$$2\,OH^- + Ni(CN)_4^{2-} \rightarrow Ni(OH)_2 + 4\,CO_3^{2-} + 4\,NO_3^-$$
$$24\,H_2O + 2\,OH^- + Ni(CN)_4^{2-} \rightarrow Ni(OH)_2 + 4\,CO_3^{2-} + 4\,NO_3^- + 48\,H^+ + 40\,e^-$$
$$24\,H_2O + 48\,OH^- + 2\,OH^- + Ni(CN)_4^{2-} \rightarrow Ni(OH)_2 + 4\,CO_3^{2-} + 4\,NO_3^{2-} + 48\,H^+ + 48\,OH^- + 40\,e^-$$
$$50\,OH^- + Ni(CN)_4^{2-} \rightarrow Ni(OH)_2 + 4\,CO_3^{2-} + 4\,NO_3^- + 24\,H_2O + 40\,e^-$$

$$40\,e^- + 120\,OH^- + 40\,Ce^{4+} \rightarrow 40\,Ce(OH)_3$$
$$50\,OH^- + Ni(CN)_4^{2-} \rightarrow Ni(OH)_2 + 4\,CO_3^{2-} + 4\,NO_3^- + 24\,H_2O + 40\,e^-$$

$$170\,OH^-(aq) + 40\,Ce^{4+}(aq) + Ni(CN)_4^{2-}(aq) \rightarrow Ni(OH)_2(s) + 40\,Ce(OH)_3(s) + 4\,CO_3^{2-}(aq) + 4\,NO_3^-(aq) + 24\,H_2O(l)$$

h.

$$H_2CO \rightarrow HCO_3^-$$
$$2\,H_2O + H_2CO \rightarrow HCO_3^- + 5\,H^+ + 4\,e^-$$
$$5\,OH^- + 2\,H_2O + H_2CO \rightarrow HCO_3^- + 5\,H^+ + 5\,OH^- + 4\,e^-$$
$$5\,OH^- + H_2CO \rightarrow HCO_3^- + 3\,H_2O + 4\,e^-$$

$$Ag(NH_3)_2^+ \rightarrow Ag + 2\,NH_3$$
$$(e^- + Ag(NH_3)_2^+ \rightarrow Ag + 2\,NH_3) \times 4$$

$$5\,OH^- + H_2CO \rightarrow HCO_3^- + 3\,H_2O + 4\,e^-$$
$$4\,e^- + 4\,Ag(NH_3)_2^+ \rightarrow 4\,Ag + 8\,NH_3$$

$$5\,OH^-(aq) + H_2CO(aq) + 4\,Ag(NH_3)_2^+(aq) \rightarrow 4\,Ag(s) + 8\,NH_3(aq) + HCO_3^-(aq) + 3\,H_2O(l)$$

QUESTIONS:

5. In a galvanic cell a spontaneous reaction occurs, producing an electric current. In an electrolytic cell electricity is used to force a reaction to occur that is not spontaneous.

6. The salt bridge completes the electrical circuit and allows counter ions to flow into the two cell compartments to maintain electrical neutrality. It also separates the cathode and anode compartments such that electrical energy can be extracted. If the cathode and anode compartments are not separated the cell would rapidly go to equilibrium without any useful work being done.

7. a. cathode: electrode at which reduction occurs.

   b. anode: electrode at which oxidation occurs.

   c. oxidation half-reaction: half reaction in which e^- are products.

   d. reduction half-reaction: half reaction in which e^- are reactants.

8. a. See text for the example of the electrolysis of copper. Purifying metals by electrolysis is possible because of the selectivity of an electrode reaction. The anode is made up of the impure metal. The metal of interest and all more easily oxidized metals are oxidized at the anode. The metal of interest is the only metal plated at the cathode. The metal ions of metals that could plate out at the cathode in preference to the metal we are purifying will not be in solution, since these metals were not oxidized at the anode.

   b. A more easily oxidized metal is placed in electrical contact with the metals we are trying to protect. It is oxidized in preference to the protected metal. The protected metal becomes the cathode; thus, cathodic protection.

9. 1. protective coating 2. alloying 3. cathodic protection

10. As a battery discharges, E_{cell} decreases, eventually reaching zero. A charged battery is not at equilibrium. At equilibrium $E_{cell} = 0$ and $\Delta G = 0$. We can get no work out of an equilibrium system. A battery is useful to us, because it can do work as it approaches equilibrium.

EXERCISES:

Galvanic Cells

11.

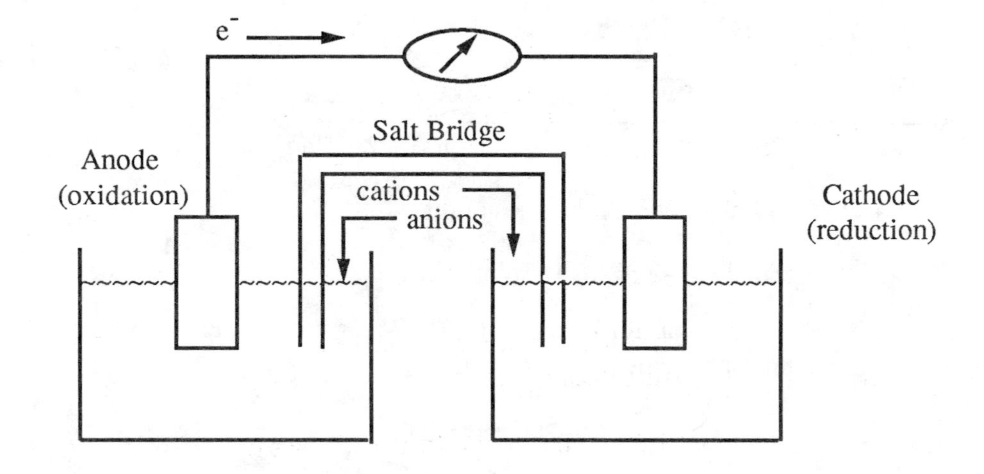

The diagram for all cells will look like this. The contents of each half cell will be identified for each reaction.

a.

$$7\ H_2O + 2\ Cr^{3+} \rightarrow Cr_2O_7^{2-} + 14\ H^+ + 6\ e^-$$
$$(Cl_2 + 2\ e^- \rightarrow 2\ Cl^-) \times 3$$

$$7\ H_2O\ (l) + 2\ Cr^{3+}(aq) + 3\ Cl_2(g) \rightarrow Cr_2O_7^{2-}(aq) + 6\ Cl^-(aq) + 14\ H^+(aq)$$

Cathode: Pt electrode; $Cl_2(g)$ bubbled into solution, Cl^- in solution

Anode: Pt electrode; Cr^{3+}, H^+, and $Cr_2O_7^{2-}$ in solution

We need a nonreactive metal to use as the electrode in each case, since all of the reactants and products are in solution. Pt is the most common choice. Other possibilities are gold and graphite.

b.

$$Cu^{2+} + 2\ e^- \rightarrow Cu$$ Cathode: Cu electrode; Cu^{2+} in solution

$$Mg \rightarrow Mg^{2+} + 2\ e^-$$ Anode: Mg electrode; Mg^{2+} in solution

$$Cu^{2+}(aq) + Mg(s) \rightarrow Cu(s) + Mg^{2+}(aq)$$

12. a.

$$IO_3^- + Fe^{2+} \rightarrow Fe^{3+} + I_2$$

$$5\ e^- + 6\ H^+ + IO_3^- \rightarrow 1/2\ I_2 + 3\ H_2O$$
$$(Fe^{2+} \rightarrow Fe^{3+} + e^-) \times 5$$

$$6\ H^+ + IO_3^- + 5\ Fe^{2+} \rightarrow 5\ Fe^{3+} + 1/2\ I_2 + 3\ H_2O$$

or $12\ H^+(aq) + 2\ IO_3^-(aq) + 10\ Fe^{2+}(aq) \rightarrow 10\ Fe^{3+}(aq) + I_2(aq) + 6\ H_2O(l)$

Cathode: Pt electrode; IO_3^-, I_2 and H^+ in solution

Anode: Pt electrode; Fe^{2+} and Fe^{3+} in solution

b.

$$(Ag^+ + e^- \rightarrow Ag) \times 2$$
$$Zn \rightarrow Zn^{2+} + 2\,e^-$$

$$Zn(s) + 2\,Ag^+(aq) \rightarrow 2\,Ag(s) + Zn^{2+}(aq)$$

Cathode: Ag electrode; Ag^+ in solution

Anode: Zn electrode; Zn^{2+} in solution

13. The diagram will look like that for Exercise 17.11. We get the reaction by remembering that the reduction occurs at the cathode in a galvanic cell.

a.

$$Cl_2 + 2\,e^- \rightarrow 2\,Cl^-$$
$$2\,Br^- \rightarrow Br_2 + 2\,e^-$$

$$Cl_2(g) + 2\,Br^-(aq) \rightarrow Br_2(aq) + 2\,Cl^-(aq)$$

Cathode: Pt electrode; $Cl_2(g)$ bubbled in, Cl^- in solution

Anode: Pt electrode; Br_2 and Br^- in solution

b.

$$(2\,e^- + 2\,H^+ + IO_4^- \rightarrow IO_3^- + H_2O) \times 5$$
$$(4\,H_2O + Mn^{2+} \rightarrow MnO_4^- + 8\,H^+ + 5\,e^-) \times 2$$

$$10\,H^+ + 5\,IO_4^- + 8\,H_2O + 2\,Mn^{2+} \rightarrow 5\,IO_3^- + 5\,H_2O + 2\,MnO_4^- + 16\,H^+$$

This simplifies to:

$$3\,H_2O\,(l) + 5\,IO_4^-\,(aq) + 2\,Mn^{2+}(aq) \rightarrow 5\,IO_3^-(aq) + 2\,MnO_4^-(aq) + 6\,H^+(aq)$$

Cathode: Pt electrode; IO_4^-, IO_3^- and H^+ in solution

Anode: Pt electrode; Mn^{2+}, MnO_4^- and H^+ in solution

14. a.

$$(Ni^{2+} + 2\,e^- \rightarrow Ni) \times 3$$
$$(Al \rightarrow Al^{3+} + 3\,e^-) \times 2$$

$$3\,Ni^{2+}(aq) + 2\,Al(s) \rightarrow 2\,Al^{3+}(aq) + 3\,Ni(s)$$

Cathode: Ni electrode; Ni^{2+} in solution

Anode: Al electrode; Al^{3+} in solution

b.

$$Co^{3+} + e^- \rightarrow Co^{2+}$$
$$Fe^{2+} \rightarrow Fe^{3+} + e^-$$

$$Co^{3+}(aq) + Fe^{2+}(aq) \rightarrow Fe^{3+}(aq) + Co^{2+}(aq)$$

Cathode: Pt electrode; Co^{3+} and Co^{2+} in solution

Anode: Pt electrode; Fe^{2+} and Fe^{3+} in solution

15. 11a. Pt | Cr^{3+} (1 M), $Cr_2O_7^{2-}$(1 M), H^+ (1 M) || Cl_2 (1 atm), Cl^- (1 M) | Pt

11b. Mg | Mg^{2+} (1 M) || Cu^{2+} (1 M) | Cu

13a. Pt | Br^- (1 M), Br_2 (1 M) || Cl_2 (1 atm), Cl^- (1 M) | Pt

13b. Pt | Mn^{2+} (1 M), MnO_4^- (1 M), H^+ (1 M) || H^+ (1 M), IO_4^- (1 M), IO_3^- (1 M) | Pt

16. 12a. Pt | Fe^{2+} (1 M), Fe^{3+} (1 M) || H^+ (1 M), IO_3^- (1 M), I_2 (1 M) | Pt

12b. Zn | Zn^{2+} (1 M) || Ag^+ (1 M) | Ag

14a. Al | Al^{3+} (1 M) || Ni^{2+} (1 M) | Ni

14b. Pt | Fe^{2+} (1 M), Fe^{3+} (1 M) || Co^{3+} (1 M), Co^{2+} (1 M) | Pt

Cell Potential, Standard Reduction Potentials, and Free Energy

17. 11a.

$$3\ Cl_2 + 6\ e^- \rightarrow 6\ Cl^- \qquad E° = 1.36\ V$$
$$2\ Cr^{3+} + 7\ H_2O \rightarrow 14\ H^+ + Cr_2O_7^{2-} + 6\ e^- \qquad -E° = -1.33\ V$$

$$2\ Cr^{3+} + 7\ H_2O + 3\ Cl_2 \rightarrow 6\ Cl^- + 14\ H^+ + Cr_2O_7^{2-} \qquad E°_{cell} = 0.03\ V$$

11b.

$$Cu^{2+} + 2\ e^- \rightarrow Cu \qquad E° = 0.34\ V$$
$$Mg \rightarrow Mg^{2+} + 2\ e^- \qquad -E° = -\ (-2.37\ V)$$

$$Cu^{2+} + Mg \rightarrow Mg^{2+} + Cu \qquad E°_{cell} = 2.71\ V$$

13a.

$$Cl_2 + 2\ e^- \rightarrow 2\ Cl^- \qquad E° = 1.36\ V$$
$$2\ Br^- \rightarrow Br_2 + 2\ e^- \qquad -E° = -1.09\ V$$

$$Cl_2 + 2\ Br^- \rightarrow Br_2 + 2\ Cl^- \qquad E°_{cell} = 0.27\ V$$

13b.

$$10\ e^- + 10\ H^+ + 5\ IO_4^- \rightarrow 5\ IO_3^- + 5\ H_2O \qquad E° = 1.60\ V$$
$$8\ H_2O + 2\ Mn^{2+} \rightarrow 2\ MnO_4^- + 16\ H^+ + 10\ e^- \qquad -E° = -1.51\ V$$

$$3\ H_2O + 5\ IO_4^- + 2\ Mn^{2+} \rightarrow 5\ IO_3^- + 2\ MnO_4^- + 6\ H^+ \qquad E°_{cell} = 0.09\ V$$

18. 12a.

$$5\,e^- + 6\,H^+ + IO_3^- \rightarrow 1/2\,I_2 + 3\,H_2O \qquad E^\circ = 1.20\text{ V}$$
$$5\,Fe^{2+} \rightarrow 5\,Fe^{3+} + 5\,e^- \qquad -E^\circ = -0.77\text{ V}$$

$$6\,H^+ + IO_3^- + 5\,Fe^{2+} \rightarrow 5\,Fe^{3+} + 1/2\,I_2 + 3\,H_2O \qquad E^\circ_{cell} = 0.43\text{ V}$$

12b.

$$2\,Ag^+ + 2\,e^- \rightarrow 2\,Ag \qquad E^\circ = 0.80\text{ V}$$
$$Zn \rightarrow Zn^{2+} + 2\,e^- \qquad -E^\circ = -(-0.76\text{ V})$$

$$2\,Ag^+ + Zn \rightarrow 2\,Ag + Zn^{2+} \qquad E^\circ_{cell} = 1.56\text{ V}$$

14a.

$$3\,Ni^{2+} + 6\,e^- \rightarrow 3\,Ni \qquad E^\circ = -0.23\text{ V}$$
$$2\,Al \rightarrow 2\,Al^{3+} + 6\,e^- \qquad -E^\circ = -(-1.66\text{ V})$$

$$3\,Ni^{2+} + 2\,Al \rightarrow 3\,Ni + 2\,Al^{3+} \qquad E^\circ_{cell} = 1.43\text{ V}$$

14b.

$$Co^{3+} + e^- \rightarrow Co^{2+} \qquad E^\circ = 1.95\text{ V}$$
$$Fe^{2+} \rightarrow Fe^{3+} + e^- \qquad -E^\circ = -0.77\text{ V}$$

$$Co^{3+} + Fe^{2+} \rightarrow Co^{2+} + Fe^{3+} \qquad E^\circ_{cell} = 1.18\text{ V}$$

19. a.

$$2\,Ag^+ + 2\,e^- \rightarrow 2\,Ag \qquad E^\circ = 0.80\text{ V}$$
$$Cu \rightarrow Cu^{2+} + 2\,e^- \qquad -E^\circ = -0.34\text{ V}$$

$$2\,Ag^+ + Cu \rightarrow Cu^{2+} + 2\,Ag \qquad E^\circ_{cell} = 0.46\text{ V}$$ Spontaneous

b.

$$Zn^{2+} + 2\,e^- \rightarrow Zn \qquad E^\circ = -0.76\text{ V}$$
$$Ni \rightarrow Ni^{2+} + 2\,e^- \qquad -E^\circ = -(-0.23\text{ V})$$

$$Zn^{2+} + Ni \rightarrow Zn + Ni^{2+} \qquad E^\circ_{cell} = -0.53\text{ V}$$ Not spontaneous

20. a.

$$(5\,e^- + 8\,H^+ + MnO_4^- \rightarrow Mn^{2+} + 4\,H_2O) \times 2 \qquad E^\circ = 1.51\text{ V}$$
$$(2\,I^- \rightarrow I_2 + 2\,e^-) \times 5 \qquad -E^\circ = -0.54\text{ V}$$

$$16\,H^+ + 2\,MnO_4^- + 10\,I^- \rightarrow 5\,I_2 + 2\,Mn^{2+} + 8\,H_2O \qquad E^\circ_{cell} = 0.97\text{ V}$$
Spontaneous

b.

$$(5\,e^- + 8\,H^+ + MnO_4^- \rightarrow Mn^{2+} + 4\,H_2O) \times 2 \qquad E^\circ = 1.51\text{ V}$$
$$(2\,F^- \rightarrow F_2 + 2\,e^-) \times 5 \qquad -E^\circ = -2.87\text{ V}$$

$$16\,H^+ + 2\,MnO_4^- + 10\,F^- \rightarrow 5\,F_2 + 2\,Mn^{2+} + 8\,H_2O \qquad E^\circ_{cell} = -1.36\text{ V}$$

Not spontaneous

21. a. $2\ H^+ + 2\ e^- \rightarrow H_2$ $E° = 0.0$ V

$Cu \rightarrow Cu^{2+} + 2\ e^-$ $-E° = -0.34$ V

$E°_{cell} = -0.34$ V; No, $H^+(aq)$ cannot oxidize Cu to Cu^{2+} at standard conditions.

b. $2\ H^+ + 2\ e^- \rightarrow H_2$ $E° = 0.0$ V

$Mg \rightarrow Mg^{2+} + 2\ e^-$ $-E° = -(-2.37$ V)

$E°_{cell} = 2.37$ V; Yes

c. $Fe^{3+} + e^- \rightarrow Fe^{2+}$ $E° = 0.77$ V

$2\ I^- \rightarrow I_2 + 2\ e^-$ $-E° = -0.54$ V

$E°_{cell} = 0.23$ V; Yes

d. $Fe^{3+} + e^- \rightarrow Fe^{2+}$ $E° = 0.77$ V

$Fe^{3+} + 3\ e^- \rightarrow Fe$ $E° = -0.036$ V

$Br_2 + 2\ e^- \rightarrow 2\ Br^-$ $E° = 1.09$ V

No, for $2\ Fe^{3+} + 2\ Br^- \rightarrow Br_2 + 2\ Fe^{2+}$ $E°_{cell} = -0.32$ V

The reaction is not spontaneous ($E°_{cell} < 0$). Fe^{3+} cannot oxidize Br^- at standard conditions.

22. a. $H_2 \rightarrow 2\ H^+ + 2\ e^-$ $-E° = 0.0$ V

$Ag^+ + e^- \rightarrow Ag$ $E° = 0.80$ V

$E°_{cell} = 0.80$ V; Yes, H_2 will reduce Ag^+ at standard conditions ($E°_{cell} > 0$).

b. No, $E°_{cell} = -0.23$ V

c. $Fe^{2+} \rightarrow Fe^{3+} + e^-$ $-E° = -(0.77$ V)

$VO_2^+ + 2\ H^+ + e^- \rightarrow VO^{2+} + H_2O$ $E° = 1.00$ V

$E°_{cell} = 0.23$ V; Yes

d. $Fe^{2+} \rightarrow Fe^{3+} + e^-$ $-E° = -0.77$ V

$Cr^{3+} + e^- \rightarrow Cr^{2+}$ $E° = -0.50$ V

$E°_{cell} = -1.27$ V; No

e. $Fe^{2+} \rightarrow Fe^{3+} + e^-$ $-E° = -0.77$

$Cr^{3+} + 3\ e^- \rightarrow Cr$ $E° = -0.73$

$E°_{cell} = -1.50$ V; No

f. $Fe^{2+} \rightarrow Fe^{3+} + e^-$ $-E° = -0.77$

$Sn^{2+} + 2\ e^- \rightarrow Sn$ $E° = -0.14$

$E°_{cell} = -0.91$; No

23. The general form of a reduction half reaction is $Ox + ne^- \rightarrow Red$. Oxidizing agents (Ox) are on the left hand side. We look for the largest E° values to correspond to the best oxidizing agents.

	MnO_4^-	> Cl_2	> $Cr_2O_7^{2-}$	> Fe^{3+}	> Fe^{2+}	> Mg^{2+}
E°	1.51	1.36	1.33	0.77	-0.44	-2.37

24. Again if we write a half reaction as: $Ox + ne^- \rightarrow Red$, reducing agents (Red) are on the right side of the reaction. The more negative E°, the better the reducing agent.

	Li	> Zn	> H_2	> Fe^{2+}	> Cr^{3+}	> F^-
E°	-3.05	-0.76	0.0	0.77	1.33	2.87

25. $Br_2 + 2\ e^- \rightarrow 2\ Br^-$ $E° = 1.09$ V

$2\ H^+ + 2\ e^- \rightarrow H_2$ $E° = 0.00$ V

$Cd^{2+} + 2\ e^- \rightarrow Cd$ $E° = -0.40$ V

$La^{3+} + 3\ e^- \rightarrow La$ $E° = -2.37$ V

$Ca^{2+} + 2\ e^- \rightarrow Ca$ $E° = -2.76$ V

a. Oxidizing agents are on the left side of the half reaction. Br_2 is the best oxidizing agent.

b. Reducing agents are on the right side of the half reaction. Ca is the best reducing agent.

c. $MnO_4^- + 8\ H^+ + 5\ e^- \rightarrow Mn^{2+} + 4\ H_2O$

$E° = 1.51$ V Permanganate can oxidize Br^-, H_2, Ca, and Cd.

d. $Zn^{2+} + 2\ e^- \rightarrow Zn$, $E° = -0.76$ V So for $Zn \rightarrow Zn^{2+} + 2\ e^-$, $-E° = 0.76$ V.
Thus, zinc can reduce Br_2 and H^+.

26. $Ce^{4+} + e^- \rightarrow Ce^{3+}$ $E° = 1.70$ V

$Fe^{3+} + e^- \rightarrow Fe^{2+}$ $E° = 0.77$ V

$Fe^{3+} + 3\,e^- \rightarrow Fe$ $E° = -0.036$ V

$Sn^{2+} + 2\,e^- \rightarrow Sn$ $E° = -0.14$ V

$Ni^{2+} + 2\,e^- \rightarrow Ni$ $E° = -0.23$ V

$Fe^{2+} + 2\,e^- \rightarrow Fe$ $E° = -0.44$ V

$Mg^{2+} + 2\,e^- \rightarrow Mg$ $E° = -2.37$ V

a. Ce^{4+} is the strongest oxidizing agent. (largest E°)

b. Mg is the strongest reducing agent. (largest -E°)

c. Yes, Ce^{4+} will oxidize the Fe to Fe^{3+}. ($E°_{cell} > 0$)

d. Fe, Sn and Mg can be oxidized by H^+. ($E°_{cell} > 0$)

e. Ce^{4+} and Fe^{3+} can be reduced by H_2(g). ($E°_{cell} > 0$)

f. Sn and Mg will reduce Fe^{3+} to Fe^{2+}. While at standard condition Sn and Mg should also reduce Fe^{3+} to Fe. Concentrations might be adjusted to prevent this from happening. It is easier to adjust concentrations using Sn since it has the smallest $E°_{cell}$, and it would be the preferred reducing agent.

27. a. $E°_{red} > 0.80$ V oxidize Hg; $E°_{red} < 0.91$ V not oxidize Hg_2^{2+}

No half reaction in this table fits this requirement. However, by changing concentrations, Ag^+ may be able to work.

b. $E°_{red} > 1.09$ V, oxidize Br^-

$E°_{red} < 1.36$ V, not oxidize Cl^-

$Cr_2O_7^{2-}$, O_2, MnO_2, and IO_3^- are all possible.

c. $Ni^{2+} + 2\,e^- \rightarrow Ni$ $E° = -0.23$ V

$Mn^{2+} + 2\,e^- \rightarrow Mn$ $E° = -1.18$ V

Any oxidizing agent with $-0.23\text{ V} > E°_{red} > -1.18$ V will work. $PbSO_4$, Cd^{2+}, Fe^{2+}, Cr^{3+}, Zn^{2+} and H_2O will be able to do this.

28. a. $E^\circ_{red} < 0.77$ V reduce Fe^{3+} to Fe^{2+}

$E^\circ_{red} > -0.44$ V not reduce Fe^{2+} to Fe

H_2O_2, MnO_4^{2-}, I^-, Cu, OH^-, Hg (in 1 *M* Cl^-), Ag (in 1 *M* Cl^-), H_2SO_3, H_2, Pb and Cd are all capable of this.

b. $Ag^+ + e^- \rightarrow Ag$ $E^\circ = 0.80$ V

$O_2 + 2H^+ + 2e^- \rightarrow H_2O_2$ $E^\circ = 0.68$ V

Fe^{2+} will reduce Ag^+ to Ag but not O_2 to H_2O_2.

29. Reduce I_2 to I^-, $E^\circ_{red} < 0.54$ V

Not reduce Cu^{2+} to Cu, $E^\circ_{red} > 0.34$ V

Yes, $0.34 < E^\circ_{red} < 0.54$

30. a. $Sn^{4+} + 2e^- \rightarrow Sn^{2+}$ $E^\circ = 0.15$ V

$Sn^{2+} + 2e^- \rightarrow Sn$ $E^\circ = -0.14$ V

Need $0.15\ V > E^\circ_{red} > -0.14$ V; H_2, Fe, and Pb are all possible.

b. Look for oxidizing agents in the same range. H^+, Fe^{3+}, and Pb^{2+} are all possible.

31. a.

$$(ClO_2^- \rightarrow ClO_2 + e^-) \times 2 \qquad -E^\circ = -0.954\ V$$

$$Cl_2 + 2e^- \rightarrow 2Cl^- \qquad E^\circ = 1.36\ V$$

$$2ClO_2^- + Cl_2 \rightarrow 2ClO_2 + 2Cl^- \qquad E^\circ_{cell} = 0.41\ V$$

$\Delta G^\circ = -nFE^\circ$

$\Delta G^\circ = -(2 \text{ mol } e^-)(96{,}485 \text{ C/mol } e^-)(0.41 \text{ J/C}) = -7.91 \times 10^4 \text{ J} = -79 \text{ kJ}$

$\Delta G^\circ = -RT \ln K$; so $K = \exp(-\Delta G^\circ/RT)$

$K = \exp[(7.9 \times 10^4 \text{ J}) / (8.3145 \text{ J/K•mol})(298 \text{ K}) = 7.0 \times 10^{13}$

or $E^\circ = \dfrac{0.0592}{n} \log K^\circ$; $\log K = \dfrac{nE^\circ}{0.0592} = \dfrac{2(0.41)}{0.0592} = 13.85$

$K = 10^{13.85} = 7.1 \times 10^{13}$

b.

$$(H_2O + ClO_2 \longrightarrow ClO_3^- + 2\,H^+ + e^-) \times 5$$
$$5\,e^- + 4\,H^+ + ClO_2 \longrightarrow Cl^- + 2\,H_2O$$

$$5\,H_2O + 5\,ClO_2 + 4\,H^+ + ClO_2 \longrightarrow 5\,ClO_3^- + 10\,H^+ + Cl^- + 2\,H_2O$$

$$3\,H_2O\,(l) + 6\,ClO_2(g) \longrightarrow 5\,ClO_3^-(aq) + Cl^-(aq) + 6\,H^+(aq)$$

32. a.

$$Mn + NO_3^- \longrightarrow Mn^{2+} + NO$$

$$(Mn \longrightarrow Mn^{2+} + 2\,e^-) \times 3 \qquad -E° = 1.18\text{ V}$$
$$(4\,H^+ + NO_3^- + 3\,e^- \longrightarrow NO + 2\,H_2O) \times 2 \qquad E° = 0.96\text{ V}$$

$$3\,Mn + 8\,H^+ + 2\,NO_3^- \longrightarrow 2\,NO + 4\,H_2O + 3\,Mn^{2+} \qquad E°_{cell} = 2.14\text{ V}$$

$$IO_4^- + Mn^{2+} \longrightarrow IO_3^- + MnO_4^-$$

$$5 \times (2\,e^- + 2\,H^+ + IO_4^- \longrightarrow IO_3^- + H_2O) \qquad E° = 1.60\text{ V}$$
$$2 \times (Mn^{2+} + 4\,H_2O \longrightarrow MnO_4^- + 8\,H^+ + 5\,e^-) \qquad -E° = -1.51\text{ V}$$

$$5\,IO_4^- + 2\,Mn^{2+} + 3\,H_2O \longrightarrow 5\,IO_3^- + 2\,MnO_4^- + 6\,H^+ \qquad E°_{cell} = 0.09\text{ V}$$

b. Nitric acid oxidation:

$E° = 2.14$ V, $\Delta G° = -nFE°$

$\Delta G° = -(6\text{ mol e}^-)(96{,}485\text{ C/mol e}^-)(2.14\text{ J/C}) = -1.24 \times 10^6\text{ J} = -1240\text{ kJ}$

$$E° = \frac{0.0592}{n}\log K°\,; \quad \log K = \frac{nE°}{0.0592} = \frac{6(2.14)}{0.0592} = 217; \quad K = 10^{217}$$

Periodate oxidation:

$E° = 0.09$ V

$\Delta G° = -(10\text{ mol e}^-)(96{,}485\text{ C/mol e}^-)(0.09\text{ J/C}) = -87\text{ kJ} = -90\text{ kJ}$

$$\log K = \frac{10(0.09)}{0.0592} = 15.20$$

$$K = 10^{15.20} = 2 \times 10^{15}$$

33. 11a. $E° = +0.03$ V, $\Delta G° = -nFE°$

$\Delta G° = -(6\text{ mol e}^-)(96{,}485\text{ C/mol e}^-)(0.03\text{ J/C}) = -1.7 \times 10^4\text{ J} = -20\text{ kJ}$

$$\log K = \frac{nE°}{0.0592} = \frac{6(0.03)}{0.0592} = 3.04; \; K = 10^{3.04} = 1 \times 10^3$$

11b. $E° = 2.71$ V

$\Delta G° = -(2 \text{ mol } e^-)(96{,}485 \text{ C/mol } e^-)(2.71 \text{ J/C}) = -5.23 \times 10^5 \text{ J} = -523 \text{ kJ}$

$\log K = \frac{2(2.71)}{0.0592} = 91.55; K = 3.55 \times 10^{91}$

13a. $E° = 0.27$ V

$\Delta G° = -(2 \text{ mol } e^-)(96{,}485 \text{ C/mol } e^-)(0.27 \text{ J/C}) = -5.21 \times 10^4 \text{ J} = -52 \text{ kJ}$

$\log K = \frac{2(0.27)}{0.0592} = 9.12; K = 1.3 \times 10^9$

13b. $E° = 0.09$ V

$\Delta G° = -(10 \text{ mol } e^-)(96{,}485 \text{ C/mol } e^-)(0.09 \text{ J/C}) = -8.7 \times 10^4 \text{ J} = -90 \text{ kJ}$

$\log K = \frac{10(0.09)}{0.0592} = 15.20; K = 2 \times 10^{15}$

34. 12a. $E° = 0.43$ V, $\Delta G° = -nFE°$

$\Delta G° = -(5 \text{ mol } e^-)(96{,}485 \text{ C/mol } e^-)(0.43 \text{ J/C}) = -2.07 \times 10^5 \text{ J} = -210 \text{ kJ}$

$\log K = \frac{nE°}{0.0592} = \frac{5(0.43)}{0.0592} = 36.32; K = 2.1 \times 10^{36}$

This is for reaction written as:

$$6\ H^+ + IO_3^- + 5\ Fe^{2+} \longrightarrow 5\ Fe^{3+} + 1/2\ I_2 + 3\ H_2O$$

12b. $E° = 1.56$ V

$\Delta G° = -(2 \text{ mol } e^-)(96{,}485 \text{ C/mol } e^-)(1.56 \text{ J/C}) = -3.01 \times 10^5 \text{ J} = -301 \text{ kJ}$

$\log K = \frac{2(1.56)}{0.0592} = 52.70; K = 5.01 \times 10^{52}$

14a. $E° = 1.43$ V

$\Delta G° = -(6 \text{ mol } e^-)(96{,}485 \text{ C/mol } e^-)(1.43 \text{ J/C}) = -8.28 \times 10^5 \text{ J} = -828 \text{ kJ}$

$\log K = \frac{6(1.43)}{0.0592} = 144.93 \approx 145; K = 10^{145}$

14b. $E° = 1.18$ V

$\Delta G° = -(1 \text{ mol } e^-)(96{,}485 \text{ C/mol } e^-)(1.18 \text{ J/C}) = -1.14 \times 10^5 \text{ J} = -114 \text{ kJ}$

$\log K = \frac{(1)(1.18)}{0.0592} = 19.93; K = 8.51 \times 10^{19}$

35. $\Delta G° = -nFE° = -(1 \text{ mol } e^-)(96{,}485 \text{ C/mol } e^-)(0.80 \text{ V})$

$\Delta G° = -77{,}200 \text{ J} = -77 \text{ kJ}$

$-77 \text{ kJ} = \Delta G_f°(Ag) - \Delta G_f°(Ag^+)$, $\Delta G_f°$ for e^- equals zero.

$-77 \text{ kJ} = 0 - \Delta G_f°(Ag^+)$; $\Delta G_f° = 77 \text{ kJ/mol}$

36. $Fe^{2+} + 2\,e^- \longrightarrow Fe$ $E° = -0.44 \text{ V}$

$\Delta G° = -nFE° = -(2 \text{ mol } e^-)(96{,}485 \text{ C/mol } e^-)(-0.44 \text{ V})$

$\Delta G° = 84{,}900 \text{ J} = 85 \text{ kJ}$

$85 \text{ kJ} = 0 - \Delta G_f°(Fe^{2+})$

$\Delta G_f°[Fe^{2+}(aq)] = -85 \text{ kJ}$

We can get $\Delta G_f°(Fe^{3+})$ two ways.

$Fe^{3+} + e^- \longrightarrow Fe^{2+}$ $E° = 0.77 \text{ V}$

$\Delta G° = -(1 \text{ mol e})(96{,}485 \text{ C/mol } e^-)(0.77 \text{ V}) = 74{,}300 \text{ J} = -74 \text{ kJ}$

	$\Delta G°$
$Fe^{2+} \longrightarrow Fe^{3+} + e^-$	+74 kJ
$Fe \longrightarrow Fe^{2+} + 2\,e^-$	-85 kJ
$Fe \longrightarrow Fe^{3+} + 3\,e^-$	-11 kJ

$-11 \text{ kJ} = \Delta G_f°(Fe^{3+}) - 0$; $\Delta G_f°[Fe^{3+}(aq)] = -11 \text{ kJ/mol}$

or $Fe^{3+} + 3\,e^- \longrightarrow Fe$ $E° = -0.036 \text{ V}$

$\Delta G° = -(3 \text{ mol } e^-)(96{,}485 \text{ C/mol } e^-)(-0.036 \text{ V})$

$\Delta G° = 10{,}400 \text{ J} \approx 10. \text{ kJ}$

$10. \text{ kJ} = 0 - \Delta G_f°$ Thus, $\Delta G_f° = -10. \text{ kJ/mol}$ The difference is round off error.

37. $2\,H_2O + 2\,e^- \longrightarrow H_2 + 2\,OH^-$

$\Delta G° = 2(-157) - 2(-237) = +160. \text{ kJ}$

$\Delta G° = -nFE°$

$$E° = \frac{-\Delta G°}{nF} = \frac{-1.60 \times 10^5 \text{ J}}{(2 \text{ mol } e^-)(96{,}485 \text{ C/mol } e^-)} = -0.829 \text{ V}$$

The two values agree. (-0.83 V in Table 17.1)

38. $CH_3OH(l) + 3/2\ O_2(g) \rightarrow CO_2(g) + 2\ H_2O(l)$

$\Delta G° = 2(-237) + (-394) - [-166] = -702$ kJ

Balance the half reaction: $CH_3OH \rightarrow CO_2$

$H_2O + CH_3OH \rightarrow CO_2 + 6\ H^+ + 6\ e^-$

$O_2 + 4\ H^+ + 4\ e^- \rightarrow 2\ H_2O$

For 3/2 O_2, 6 moles of electrons will be transferred. So for this reaction, n = 6.

$\Delta G° = -nFE°$

$$E° = \frac{-\Delta G°}{nF} = \frac{702{,}000\ J}{(6\ mol\ e^-)(96{,}485\ C/mol\ e^-)} = 1.21\ V$$

39. $PbSO_4(s) + 2\ e^- \rightarrow Pb + SO_4^{2-}$ $\quad E° = -0.35$ V

$Pb \rightarrow Pb^{2+} + 2\ e^-$ $\quad -E° = -(-0.13$ V)

$PbSO_4(s) \rightarrow Pb^{2+}(aq) + SO_4^{2-}(aq)$ $\quad E°_{cell} = -0.22$ V

$K = K_{sp}$ and $E°_{cell} = \frac{0.0592}{n} \log K_{sp}$

$$\log K_{sp} = \frac{nE°}{0.0592} = \frac{2(-0.22)}{0.0592} = -7.43$$

$K_{sp} = 10^{-7.43} = 3.7 \times 10^{-8}$

40. $CuI + e^- \rightarrow Cu + I^-$ $\quad E° = ?$

$Cu \rightarrow Cu^+ + e^-$ $\quad -E° = -0.52$ V

$CuI(s) \rightarrow Cu^+(aq) + I^-(aq)$ $\quad E°_{cell} = E° - 0.52$ V

$K_{sp} = 1.1 \times 10^{-12}$

$$E°_{cell} = \frac{0.0592}{n} \log K_{sp} = \frac{0.0592}{1} \log 1.1 \times 10^{-12}$$

$E°_{cell} = -0.71$ V; $\quad -0.71 = E° - 0.52$; $\quad E° = -0.19$ V

The Nernst Equation

41. For the half cell on the left, $E_l = E° = 0.80$ V since $[Ag^+] = 1.0$ *M*. Let's calculate E for the half cell on the right, and thus figure out what reaction occurs. The half cell with the smallest reduction potential is the anode.

a. $E_r = E° = 0.80$ V

Since $E_l = E_r$, then $E_{cell} = 0$, No reaction occurs. Concentration cells only produce a voltage when the ion concentrations are not equal.

b. For $Ag^+ + e^- \longrightarrow Ag$

$$E_r = E° - \frac{0.0592}{1} \log \frac{1}{[Ag^+]}$$

$E_r = 0.80 \text{ V} + 0.0592 \log [Ag^+]$

$E_r = 0.80 \text{ V} + 0.0592 \log 2.0 = 0.80 + 0.018 = 0.82$ V

The half cell with 2 *M* Ag^+ is the cathode. Electrons flow from anode to cathode, to the right in the diagram.

$E_{cell} = E_r - E_l = 0.82 - 0.80 = 0.02$ V

Note: The driving force for the reaction is to equalize the concentration of Ag^+ in the two half cells. The half cell with the largest Ag^+ concentrations will always be the cathode.

c. $E_r = 0.80 + 0.0592 \log 0.10 = 0.80 - 0.059 = 0.74$ V

Half cell with 0.10 *M* Ag^+ is the anode. Electrons move to the cathode (to the left in the diagram).

$E_{cell} = E_l - E_r = 0.80 - 0.74 = 0.06$ V

d. $E_r = 0.80 + 0.0592 \log (4.0 \times 10^{-5}) = 0.80 - 0.26 = 0.54$ V

$E_{cell} = E_l - E_r = 0.80 - 0.54 = 0.26$ V

Half cell with 4.0×10^{-5} *M* Ag^+ is the anode. Electrons move from anode to cathode (to the left in the diagram).

e. If the concentrations are the same in each half cell, $E_l = E_r$ and no reaction occurs. $E_{cell} = 0$.

42. a. $E_{cell} = 0$; Concentration cells only produce a voltage when the ion concentrations are different. Since $[Zn^{2+}] = 1.0$ *M* in both compartments, then no cell potential is produced.

b. For the half cell on the left, $E_l = E° = -0.76$ V since $[Zn^{2+}] = 1.0\ M$.. For the half cell on the right:

$$E_r = E° - \frac{0.0592}{n} \log \frac{1}{[Zn^{2+}]} = -0.76 - \frac{0.0592}{2} \log \frac{1}{2.0} = -0.76 \text{ V} + 0.009 \text{ V}$$

$E_r = -0.75$ V

The right hand half cell will be the cathode since it has the largest (most positive) reduction potential. Electrons will move from left to right in the diagram.

$E_{cell} = E_r - E_l = -0.75 \text{ V} - (-0.76 \text{ V}) = +0.01$ V

c. $$E_r = -0.76 - \frac{0.0592}{2} \log \frac{1}{0.10} = -0.79 \text{ V}$$

The half cell on the left will be the cathode. Electrons will move from right to left in the diagram.

$E_{cell} = E_l - E_r = -0.76 \text{ V} - (-0.79 \text{ V}) = +0.03$ V

d. $$E_r = -0.76 - \frac{0.0592}{n} \log \frac{1}{4.0 \times 10^{-5}} = -0.76 - 0.13 = -0.89 \text{ V}$$

The half cell on the left will be the cathode. Electrons will move from right to left in the diagram.

$E_{cell} = -0.76 \text{ V} - (-0.89) = +0.13$ V

e. $E_{cell} = 0$; since $[Zn^{2+}]$ are equal, then $E_l = E_r$ and $E_{cell} = 0$

43. $5e^- + 8H^+ + MnO_4^- \rightarrow Mn^{2+} + 4H_2O$ $\quad E° = 1.51$ V

$(Fe^{2+} \rightarrow Fe^{3+} + e^-) \times 5$ $\quad -E° = -0.77$ V

$8H^+ + MnO_4^- + 5Fe^{2+} \rightarrow 5Fe^{3+} + Mn^{2+} + H_2O$ $\quad E°_{cell} = 0.74$ V

$$E = E° - \frac{0.0592}{n} \log Q$$

$$E = 0.74 \text{ V} - \frac{0.0592}{5} \log \frac{[Fe^{3+}]^5 [Mn^{2+}]}{[Fe^{2+}]^5 [MnO_4^-] [H^+]^8}$$

$$E = 0.74 - \frac{0.0592}{5} \log \frac{(1 \times 10^{-6})^5 (1 \times 10^{-6})}{(1 \times 10^{-3})^5 (1 \times 10^{-2}) (1 \times 10^{-4})^8}$$

$$E = 0.74 - \frac{0.0592}{5} \log 1 \times 10^{13} = 0.74 - 0.15 = 0.59 \text{ V} = 0.6 \text{ V}$$

Yes, E > 0 so reaction will occur as written.

44. a. In base, $Co(OH)_2$ will precipitate. $K_{sp} = 2.5 \times 10^{-16} = [Co^{2+}][OH^-]^2$.

We need E° for: $Co(OH)_2 + 2e^- \rightarrow Co + 2OH^-$.

We look at the effect of the solubility on:

$$Co^{2+} + 2e^- \rightarrow Co, \quad E = -0.277 - \frac{0.0592}{2}\log\frac{1}{[Co^{2+}]}$$

Using the K_{sp} expression, substitute for $[Co^{2+}]$.

$$E = -0.277 - \frac{0.0592}{2}\log\frac{[OH^-]^2}{K_{sp}}$$

If $[OH^-] = 1.0\,M$, then the E we calculate is E° for:

$$Co(OH)_2 + 2e^- \rightarrow Co + 2OH^-.$$

$$E = -0.277 - \frac{0.0592}{2}\log\frac{(1.0)^2}{2.5 \times 10^{-16}} = -0.74\ V$$

$(2OH^- + Co \rightarrow Co(OH)_2 + 2e^-) \times 2$	$-E° = 0.74$ V
$O_2 + 2H_2O + 4e^- \rightarrow 4OH^-$	$E° = 0.40$ V
$2Co + O_2 + 2H_2O \rightarrow 2Co(OH)_2$	$E° = 1.14$ V

Yes, the corrosion reaction is spontaneous.

b. $O_2 + 4H^+ + 4e^- \rightarrow 2H_2O$ $E° = 1.23$ V

This half reaction is even more positive than the reduction of O_2 in base. Yes, corrosion will occur in 1.0 *M* H^+.

c. Using the reaction: $2Co + O_2 + 4H^+ \rightarrow 2H_2O + 2Co^{2+}$, $E° = 1.23\ V - (-0.277) = 1.51$ V

$$E = 1.51\ V - \frac{0.0592}{4}\log\frac{[Co^{2+}]^2}{P_{O_2}[H^+]^4},\ [H^+] = 1.0 \times 10^{-7}$$

$$E = 1.51\ V - \frac{0.0592}{4}\log(1.0 \times 10^{28}) - \frac{0.0592}{4}\log\frac{[Co^{2+}]^2}{P_{O_2}}$$

$$E = 1.51\ V - 0.41\ V - \frac{0.0592}{4}\log\frac{[Co^{2+}]^2}{P_{O_2}}$$

Typically, $[Co^{2+}]$ won't be very large and $P_{O_2} \approx 0.2$ atm.

E will still be greater than zero and the corrosion reaction is spontaneous at pH = 7.

45. $Cu^{2+} + H_2 \longrightarrow 2\,H^+ + Cu$ $\quad E° = 0.34\ V - 0.0\ V = 0.34\ V$

a. $E = E° - \frac{0.0592}{2} \log \frac{1}{[Cu^{2+}]}$ since $P_{H_2} = 1$ atm and $[H^+] = 1$ mol/L

$E = E° + \frac{0.0592}{2} \log [Cu^{2+}]$

$E = 0.34 + \frac{0.0592}{2} \log (2.5 \times 10^{-4}) = 0.23\ V$

b. $E = 0.34 + \frac{0.0592}{2} \log [Cu^{2+}]$

$Cu(OH)_2 \rightleftharpoons Cu^{2+} + 2\,OH^-$ $\quad K_{sp} = 1.6 \times 10^{-19}$

s mol/L dissolves $\longrightarrow$ s $\quad 0.10 + 2\,s$

$1.6 \times 10^{-19} = (s)\,(0.10 + 2s)^2 \approx s\,(0.10)^2$

$s = [Cu^{2+}] = 1.6 \times 10^{-17}$ Assumption good.

$E = E° + \frac{0.0592}{2} \log [Cu^{2+}] = 0.34 + \frac{0.0592}{2} \log (1.6 \times 10^{-17})$

E = 0.34 - 0.50 = -0.16 V, not spontaneous, reverse reaction occurs. $E_{cell} = 0.16$ V, Cu electrode becomes the anode.

c. $0.195 = 0.34 + \frac{0.0592}{2} \log [Cu^{2+}]$, $\log [Cu^{2+}] = -4.899$

$[Cu^{2+}] = 10^{-4.899} = 1.3 \times 10^{-5}\ M$

Note: If cell is reversed as in (b), $0.195 = -0.34 - \frac{0.0592}{2} \log [Cu^{2+}]$ and $[Cu^{2+}] = 8.4 \times 10^{-19}$

d. $E = E° + \frac{0.0592}{2} \log [Cu^{2+}]$

Graph E vs. $\log [Cu^{2+}]$. The slope will be 0.0296 V or 29.6 mV.

46. a. $2\,Ag^+ + Cu \longrightarrow Cu^{2+} + 2\,Ag$

$E°_{cell} = 0.80 - 0.34 = 0.46$ V; Ag electrode is the cathode.

Since $[Ag^+] = 1.0$, then $E = 0.46\ V - \frac{0.0592}{2} \log [Cu^{2+}]$

$E = 0.46 - \frac{0.0592}{2} \log 1.3 \times 10^{-5} = 0.46 + 0.14 = 0.60\ V$

b. $Cu^{2+} + 4\,NH_3 \rightleftharpoons Cu(NH_3)_4^{2+}$

$K_f = \frac{[Cu(NH_3)_4^{2+}]}{[Cu^{2+}]\,[NH_3]^4} = 1.0 \times 10^{13} = \frac{0.010}{[Cu^{2+}]\,(5.0)^4}$

$[Cu^{2+}] = 1.6 \times 10^{-18}$

From 46a, $E = 0.46\ V - \frac{0.0592}{2} \log [Cu^{2+}]$

$E = 0.46 - \frac{0.0592}{2} \log 1.6 \times 10^{-18} = 0.46 + 0.53 = 0.99\ V$

c. $0.58\ V = 0.46\ V - \frac{0.0592}{2} \log [Cu^{2+}]$

$\log [Cu^{2+}] = -4.05$, $[Cu^{2+}] = 8.9 \times 10^{-5}\ M$

47. The potential oxidizing agents are NO_3^- and H^+. Hydrogen ion cannot oxidize Pt under any condition. Nitrate cannot oxidize Pt unless there is Cl^- in the solution. Aqua regia has both Cl^- and NO_3^-: The nitrate oxidizes Pt and it is complexed by the Cl^-.

$$12\ Cl^- + 3\ Pt \rightarrow 3\ PtCl_4^{2-} + 6\ e^- \qquad -E° = -0.76$$

$$2\ NO_3^- + 8\ H^+ + 6\ e^- \rightarrow 2\ NO + 4\ H_2O \qquad E° = 0.96$$

$$12\ Cl^- + 3\ Pt + 2\ NO_3^- + 8\ H^+ \rightarrow 3\ PtCl_4^{2-} + 2\ NO + 4\ H_2O \qquad E°_{cell} = +0.20\ V$$

48. See Exercise 17.11a for the balanced half reactions for the equation:

$$3\ Cl_2 + 2\ Cr^{3+} + 7\ H_2O \rightleftharpoons 14\ H^+ + Cr_2O_7^{2-} + 6\ Cl^-$$

$E° = (1.36 + -1.33) = 0.03$

$$E = E° - \frac{0.0592}{6} \log \frac{[Cr_2O_7^{2-}]\,[H^+]^{14}\,[Cl^-]^6}{[Cr^{3+}]^2\, P_{Cl_2}^3}$$

When $K_2Cr_2O_7$ and Cl^- are added to concentrated H_2SO_4, the term $-\frac{0.0592}{6} \log Q$ is negative. The $[H^+]$ in concentrated H_2SO_4 is 18 M. $[H^+]^{14} \approx (18)^{14} = 3.7 \times 10^{17}$. The log of a large number is positive. E becomes negative, which means the reverse action becomes spontaneous.

$$14\ H^+ + Cr_2O_7^{2-} + 6\ Cl^- \rightarrow 2\ Cr^{3+} + 7\ H_2O + 3\ Cl_2$$

The pungent fumes were $Cl_2(g)$.

Electrolysis

49. a. $Al^{3+} + 3\ e^- \rightarrow Al$

$$1.0 \times 10^3\ g\ Al \times \frac{1\ mol\ Al}{26.98\ g\ Al} \times \frac{3\ mol\ e^-}{mol\ Al} \times \frac{96{,}485\ C}{mol\ e^-} \times \frac{1\ s}{100.0\ C}$$

$$= 1.07 \times 10^5\ s = 3.0 \times 10^1\ hours$$

b. $1.0 \text{ g Ni} \times \frac{1 \text{ mol}}{58.69 \text{ g}} \times \frac{2 \text{ mol e}^-}{\text{mol Ni}} \times \frac{96{,}485 \text{ C}}{\text{mol e}^-} \times \frac{1 \text{ s}}{100.0 \text{ C}} = 33 \text{ s}$

c. $5.0 \text{ mol Ag} \times \frac{1 \text{ mol e}^-}{\text{mol Ag}} \times \frac{96{,}485 \text{ C}}{\text{mol e}^-} \times \frac{1 \text{ s}}{100.0 \text{ C}} = 4.8 \times 10^3 \text{ s} = 1.3 \text{ hr}$

50. 15 A = 15 C/s

$\frac{15 \text{ C}}{\text{s}} \times \frac{60 \text{ s}}{\text{min}} \times \frac{60 \text{ min}}{\text{hr}} = 5.4 \times 10^4$ C of charge passed in 1 hour

a. $Co^{2+} + 2 e^- \rightarrow Co$

$5.4 \times 10^4 \text{ C} \times \frac{1 \text{ mol e}^-}{96{,}485 \text{ C}} \times \frac{1 \text{ mol Co}}{2 \text{ mol e}^-} \times \frac{58.93 \text{ g}}{\text{mol Co}} = 16 \text{ g Co}$

b. $5.4 \times 10^4 \text{ C} \times \frac{1 \text{ mol e}^-}{96{,}485 \text{ C}} \times \frac{1 \text{ mol Hf}}{4 \text{ mol e}^-} \times \frac{178.5 \text{ g}}{\text{mol Hf}} = 25 \text{ g Hf}$

c. $I_2 + 2 e^- \rightarrow 2 I^-$

$5.4 \times 10^4 \text{ C} \times \frac{1 \text{ mol e}^-}{96{,}485 \text{ C}} \times \frac{1 \text{ mol } I_2}{2 \text{ mol e}^-} \times \frac{253.8 \text{ g } I_2}{\text{mol } I_2} = 71 \text{ g } I_2$

d. Cr is in the +6 oxidation state in CrO_3. It will take 6 mol e^- to produce 1 mol Cr from 1 mol CrO_3.

$5.4 \times 10^4 \text{ C} \times \frac{1 \text{ mol e}^-}{96{,}485 \text{ C}} \times \frac{1 \text{ mol Cr}}{6 \text{ mol e}^-} \times \frac{52.00 \text{ g Cr}}{\text{mol Cr}} = 4.9 \text{ g Cr}$

51. a. Cathode: reduction: $K^+ + e^- \rightarrow K$ $E° = -2.92$ V

Anode: oxidation: $2F^- \rightarrow F_2 + 2 e^-$ $-E° = -2.87$ V

b. Cathode: easier to reduce H_2O than K^+

$2 H_2O + 2 e^- \rightarrow H_2 + 2 OH^-$ $E° = -0.83$ V

Anode: easier to oxidize H_2O than F^-

$2 H_2O \rightarrow 4 H^+ + O_2 + 4 e^-$ $-E° = -1.23$ V

c. Species present that can be reduced: H_2O_2, SO_4^{2-}, H^+ and H_2O

Possible cathode reactions:

$$2\,H_2O + 2\,e^- \rightarrow H_2 + 2\,OH^- \qquad E° = -0.83\ V$$

$$2\,H^+ + 2\,e^- \rightarrow H_2 \qquad E° = 0.00\ V$$

$$SO_4^{2-} + 4\,H^+ + 2\,e^- \rightarrow H_2SO_3 + H_2O \qquad E° = 0.20\ V$$

$$H_2O_2 + 2\,H^+ + 2\,e^- \rightarrow 2\,H_2O \qquad E° = 1.78\ V$$

Reduction of H_2O_2 will occur (most positive E°) at the cathode.

Possible anode reactions:

$$2\,H_2O \rightarrow O_2 + 4\,H^+ + 4\,e^- \qquad -E° = -1.23\ V$$

$$H_2O_2 \rightarrow O_2 + 2\,H^+ + 2\,e^- \qquad -E° = -0.68\ V$$

Easier to oxidize H_2O_2 than H_2O, so

$H_2O_2 \rightarrow O_2 + 2\,H^+ + 2\,e^-$ occurs at the anode.

52. a. Cathode: reduction: $Cu^{2+} + 2\,e^- \rightarrow Cu$ $\qquad E° = 0.34\ V$

Anode: oxidation: $2\,Cl^- \rightarrow Cl_2 + 2\,e^-$ $\qquad -E° = -1.36\ V$

b. Same as a. From E° values we would predict H_2O would be oxidized (-E° = -1.23 V), but because of the overvoltage, Cl^- is oxidized rather than H_2O (see text) .

c. Cathode: $Mg^{2+} + 2\,e^- \rightarrow Mg$

Anode: $2\,Cl^- \rightarrow Cl_2 + 2\,e^-$

53. $$600.\ s \times \frac{5.0\ C}{s} \times \frac{1\ mol\ e^-}{96{,}485\ C} \times \frac{1\ mol\ M^{3+}}{3\ mol\ e^-} = 1.036 \times 10^{-2}\ mol$$ (carry extra significant figures)

$$\text{Atomic mass} = \frac{1.18\ g}{1.036 \times 10^{-2}\ mol} = \frac{114\ g}{mol} \approx \frac{110\ g}{mol}$$

The element is most likely indium, In. Indium forms 3+ ions, Cd, Ag and Pd do not.

54. $$74.6\ s \times \frac{2.50\ C}{s} \times \frac{1\ mol\ e^-}{96{,}485\ C} \times \frac{1\ mol\ M^{2+}}{2\ mol\ e^-} = 9.66 \times 10^{-4}\ mol$$

$$\text{Atomic Mass} = \frac{0.1086\ g}{9.66 \times 10^{-4} mol} = \frac{112\ g}{mol}$$

The metal is Cd. Cd has the correct atomic mass and does form Cd^{2+} ions in solution.

55. F_2 is produced at the anode.

$$2\ F^- \longrightarrow F_2 + 2\ e^-$$

$$2.00\ \text{hr} \times \frac{60\ \text{min}}{\text{hr}} \times \frac{60\ \text{s}}{\text{min}} \times \frac{10.0\ \text{C}}{\text{s}} \times \frac{1\ \text{mol e}^-}{96{,}485\ \text{C}} = 0.746\ \text{mol e}^-$$

$$0.746\ \text{mol e}^- \times \frac{1\ \text{mol F}_2}{2\ \text{mol e}^-} = 0.373\ \text{mol F}_2$$

$$V = \frac{nRT}{P} = \frac{(0.373\ \text{mol})\left(\frac{0.08206\ \text{L atm}}{\text{mol K}}\right)(298\ \text{K})}{1.00\ \text{atm}} = 9.12\ \text{L}$$

K is produced at the cathode.

$$K^+ + e^- \longrightarrow K$$

$$0.746\ \text{mol e}^- \times \frac{1\ \text{mol K}}{\text{mol e}^-} \times \frac{39.10\ \text{g K}}{\text{mol K}} = 29.2\ \text{g K}$$

56.

$$(2\ H_2O \longrightarrow 4\ H^+ + O_2 + 4\ e^-) \times 1/2$$
$$2\ e^- + 2\ H_2O \longrightarrow H_2 + 2\ OH^-$$

$$H_2O \longrightarrow H_2 + 1/2\ O_2$$

So, n = 2 for reaction as it is written.

$$15.0\ \text{min} \times \frac{60\ \text{s}}{\text{min}} \times \frac{2.50\ \text{C}}{\text{s}} \times \frac{1\ \text{mol e}^-}{96{,}485\ \text{C}} \times \frac{1\ \text{mol H}_2}{2\ \text{mol e}^-} = 1.17 \times 10^{-2}\ \text{mol H}_2$$

At STP, 1 mol of an ideal gas occupies a volume of 22.41 L.

$$1.17 \times 10^{-2}\ \text{mol H}_2 \times \frac{22.41\ \text{L}}{\text{mol H}_2} = 0.262\ \text{L} = 262\ \text{mL of H}_2$$

$$1.17 \times 10^{-2}\ \text{mol H}_2 \times \frac{0.5\ \text{mol O}_2}{\text{mol H}_2} \times \frac{22.41\ \text{L mol}}{\text{O}_2} = 0.131\ \text{L} = 131\ \text{mL O}_2$$

The same answers are obtained using the ideal gas equation.

57. $Au(CN)_4^- + 3\ e^- \longrightarrow Au + 4\ CN^-$

Water is oxidized: $2\ H_2O \longrightarrow O_2 + 4\ H^+ + 4\ e^-$

Actually, this process is done in basic solution. You don't want $H^+ + CN^- \longrightarrow HCN(g)$ occurring. So oxidation of water in base is:

$$4\ OH^- \longrightarrow O_2 + 2\ H_2O + 4\ e^-$$

Overall reaction:

$$4\ Au(CN)_4^-(aq) + 12\ OH^-(aq) \longrightarrow 4\ Au(s) + 16\ CN^-(aq) + 3\ O_2(g) + 6\ H_2O(l)$$

$$1.00 \times 10^3\ g\ Au \times \frac{1\ mol\ Au}{197.0\ g} \times \frac{3\ mol\ O_2}{4\ mol\ Au} = 3.81\ mol\ O_2$$

PV = nRT

$$V = \frac{nRT}{P} = \frac{3.81\ mol\left(\frac{0.08206\ L\ atm}{mol\ K}\right)(298\ K)}{740\ torr} \times \frac{760\ torr}{atm} = 96\ L$$

58. Anode reaction: $2\ Cl^- \longrightarrow Cl_2 + 2\ e^-$

Cathode reaction: $2\ H_2O + 2\ e^- \longrightarrow H_2 + 2\ OH^-$

If 6.00 L of H_2 are produced, then 6.00 L of Cl_2 will also be produced.

59. $$\frac{1.0 \times 10^6\ g}{hr} \times \frac{1\ hr}{60\ min} \times \frac{1\ min}{60\ s} \times \frac{1\ mol\ C_6H_8N_2}{108.1\ g\ C_6H_8N_2} \times \frac{2\ mol\ e^-}{mol\ C_6H_8N_2} \times \frac{96{,}485\ C}{mol\ e^-}$$

$= 5.0 \times 10^5$ C/s or current of 5.0×10^5 A.

60. $Al^{3+} + 3\ e^- \longrightarrow Al$

$$2000\ lb\ Al \times \frac{454\ g}{lb} \times \frac{1\ mol\ Al}{26.98\ g} \times \frac{3\ mol\ e^-}{mol\ Al} \times \frac{96{,}485\ C}{mol\ e^-} = 1 \times 10^{10}\ C \text{ of electricity needed}$$

$$1 \times 10^{10}\ C = i \times 24\ hr \times \frac{60\ min}{hr} \times \frac{60\ s}{min},\ i = \text{current}$$

$i = 1 \times 10^5\ C/s = 1 \times 10^5\ A$

61. $$2.30\ min \times \frac{60\ s}{min} = 138\ s$$

$$138\ s \times \frac{2.00\ C}{s} \times \frac{1\ mol\ e^-}{96{,}485\ C} \times \frac{1\ mol\ Ag}{mol\ e^-} = 2.86 \times 10^{-3}\ mol\ Ag$$

$$\text{Conc.} = \frac{2.86 \times 10^{-3}\ mol}{0.250\ L} = 1.14 \times 10^{-2}\ mol/L$$

62. $$0.50\ L \times \frac{0.010\ mol\ Pt^{4+}}{L} = 5.0 \times 10^{-3}\ mol\ Pt^{4+} \text{ in solution}$$

$$5.0 \times 10^{-3}\ mol\ Pt^{4+} \times \frac{4\ mol\ e^-}{mol\ Pt^{4+}} \times \frac{96{,}485\ C}{mol\ e^-} \times 0.99 = 1.9 \times 10^3\ C \text{ to plate out 99\% of the Pt}$$

$1.9 \times 10^3\ C = i \times t = 4.00\ A \times t$

$t = 4.8 \times 10^2\ s$

ADDITIONAL EXERCISES

63. The half reaction for the SCE is:

$$Hg_2Cl_2(s) + 2\ e^- \rightarrow 2\ Hg + 2\ Cl^- \qquad E_{SCE} = +0.242\ V$$

a. $Cu^{2+} + 2\ e^- \rightarrow Cu$ $\qquad E° = 0.34\ V$

$E_{cell} = 0.10\ V$, SCE is anode

b. $Fe^{3+} + e^- \rightarrow Fe^{2+}$ $\qquad E° = 0.77\ V$

$E_{cell} = 0.53\ V$, SCE is anode

c. $AgCl + e^- \rightarrow Ag + Cl^-$ $\qquad E° = 0.22\ V$

$E_{cell} = 0.02\ V$, SCE is cathode

d. $Al^{3+} + 3\ e^- \rightarrow Al$ $\qquad E° = -1.66\ V$

$E_{cell} = 1.90\ V$, SCE is cathode

e. $Ni^{2+} + 2\ e^- \rightarrow Ni$ $\qquad E° = -0.23\ V$

$E_{cell} = 0.47\ V$, SCE is cathode

64. a. $E° = 0.52 - 0.16 = 0.36\ V$, spontaneous

b. $E° = -0.14\ V - (0.15\ V) = -0.29\ V$, not spontaneous

c. $E° = -0.44\ V - (0.77\ V) = -1.21\ V$, not spontaneous

d. $E° = 1.65 - 1.21 = 0.44\ V$, spontaneous

65. $2\ Cu^+ \rightarrow Cu + Cu^{2+} \qquad E°_{cell} = 0.52\ V - 0.16\ V = 0.36\ V$

$\Delta G° = -nFE° = -(1\ mol\ e^-)\ (96{,}485\ C/mol\ e^-)\ (0.36\ V)$

$\Delta G° = -34{,}700\ CV = -34.7\ kJ = -35\ kJ$

$$\log K = \frac{nE°}{0.0592} = \frac{0.36}{0.0592} = 6.08;\ K = 1.2 \times 10^6$$

66. $2\ Cu^+(aq) \rightarrow Cu^{2+}(aq) + Cu(s)$

$\Delta G° = -nFE° = -(1\ mol\ e^-)\ (96{,}485\ C/mol\ e)\ (0.36\ V) = -34{,}700\ J = -35\ kJ$

$$\log K = \frac{nE°}{0.0592} = \frac{0.36}{0.0592} = 6.08$$

$K = 1.2 \times 10^6$

$2\ HClO_2 \rightarrow ClO_3^- + H^+ + HClO$

$\Delta G° = -nFE° = -(2 \text{ mol e}^-)(96{,}485 \text{ C/mol e}^-)(0.44 \text{ V})$

$\Delta G° = -84{,}900 \text{ J} = -84.9 \text{ kJ} = -85 \text{ kJ}$

$\log K = \frac{nE°}{0.0592} = \frac{2(0.44)}{0.0592} = 14.86;\ K = 7.2 \times 10^{14}$

67. $H_2O_2 + 2\,H^+ + 2\,e^- \rightarrow 2\,H_2O$ $\quad E° = 1.78$ V

$O_2 + 2\,H^+ + 2\,e^- \rightarrow H_2O_2$ $\quad E° = 0.68$ V

$H_2O_2 + 2\,H^+ + 2\,e^- \rightarrow 2\,H_2O$ $\quad E° = 1.78$ V $\quad H_2O_2$ as an oxidizing agent

$H_2O_2 \rightarrow O_2 + 2\,H^+ + 2\,e^-$ $\quad -E° = -0.68$ V $\quad H_2O_2$ as a reducing agent

---

$2\,H_2O_2 \rightarrow 2\,H_2O + O_2$ $\quad E°_{cell} = 1.10$ V

68. a. Possible reaction: $I_2 + 2\,Cl^- \rightarrow 2\,I^- + Cl_2$

$E°_{cell} = -0.82$ V, not spontaneous, no reaction occurs

b. Possible reaction: $Cl_2 + 2\,I^- \rightarrow I_2 + 2\,Cl^-$

$E°_{cell} = 0.82$ V, spontaneous

c. Possible reaction: $2\,Ag + Cu^{2+} \rightarrow Cu + 2\,Ag^+$

$E°_{cell} = -0.46$ V, no reaction occurs

d. Possible reaction: $Pb + Cu^{2+} \rightarrow Cu + Pb^{2+}$

$E°_{cell} = +0.47$ V, spontaneous

However, $PbCl_2$ is insoluble. $Pb^{2+}(aq) + 2\,Cl^-(aq) \rightarrow PbCl_2(s)$

This will drive the reaction further to the right. The reaction that will occur is:

$$Pb(s) + Cu^{2+}(aq) + 2\,Cl^-(aq) \rightarrow PbCl_2(s) + Cu(s)$$

e. Possible reaction: $4\,Fe^{2+} + 4\,H^+ + O_2 \rightarrow 4\,Fe^{3+} + 2\,H_2O$

$E°_{cell} = +0.46$ V, spontaneous

69. 68b, d, and e are spontaneous.

b. $Cl_2 + 2\,I^- \rightarrow I_2 + 2\,Cl^-$; $E° = +0.82$ V

$\Delta G° = -nFE° = -(2 \text{ mol e}^-)(96{,}485 \text{ C/mol e}^-)(0.82 \text{ J/C})$

$\Delta G° = -1.58 \times 10^5 \text{ J} = -160 \text{ kJ}$

$\Delta G° = -RT \ln K$

$\log K = \frac{nE°}{0.0592} = \frac{2(0.82)}{0.0592} = 27.70$

$K = 10^{27.70} = 5.0 \times 10^{27}$

d. $Pb + Cu^{2+} \rightarrow Cu + Pb^{2+}$ $E° = 0.47$ V Ignore $PbCl_2(s)$ formation.

$\Delta G° = -nFE° = -(2 \text{ mol } e^-)(96{,}485 \text{ C/mol } e^-)(0.47 \text{ J/C})$

$\Delta G° = -9.07 \times 10^4 \text{ J} = -91 \text{ kJ}$

$\log K = \frac{2(0.47)}{0.0592} = 15.88$

$K = 7.6 \times 10^{15}$

e.

$$(Fe^{2+} \rightarrow Fe^{3+} + e^-) \times 4 \qquad -E° = -0.77 \text{ V}$$
$$4\,H^+ + O_2 + 4\,e^- \rightarrow 2\,H_2O \qquad E° = 1.23 \text{ V}$$

$$4\,H^+ + O_2 + 4\,Fe^{2+} \rightarrow 4\,Fe^{3+} + 2\,H_2O \qquad E°_{cell} = 0.46 \text{ V}$$

$\Delta G° = -nFE° = -(4 \text{ mol } e^-)(96{,}485 \text{ C/mol } e^-)(0.46 \text{ J/C}) = -180 \text{ kJ}$

$\log K = \frac{4(0.46)}{0.0592} = 31.08$; $K = 1.2 \times 10^{31}$

70. a.

$$Cu^+ + e^- \rightarrow Cu \qquad E° = 0.52 \text{ V}$$
$$Cu^+ \rightarrow Cu^{2+} + e^- \qquad -E° = -0.16 \text{ V}$$

$$2\,Cu^+ \rightarrow Cu + Cu^{2+} \qquad E°_{cell} = 0.36 \text{ V}$$

$\Delta G° = -nFE° = -(1 \text{ mol } e^-)(96{,}485 \text{ C/mol } e^-)(0.36 \text{ J/C})$

$\Delta G° = -3.47 \times 10^4 \text{ J} = -35 \text{ kJ}$

$\log K = \frac{1(0.36)}{0.0592} = 6.08$; $K = 1.2 \times 10^6$

b.

$$2\,e^- + 2\,H_2O \rightarrow 2\,OH^- + H_2 \qquad E° = -0.83 \text{ V}$$
$$2\,Na \rightarrow 2\,Na^+ + 2\,e^- \qquad -E° = 2.71 \text{ V}$$

$$2\,Na + 2\,H_2O \rightarrow 2\,NaOH + H_2 \qquad E°_{cell} = 1.88 \text{ V}$$

$\Delta G° = -(2 \text{ mol } e^-)(96{,}485 \text{ C/mol } e^-)(1.88 \text{ J/C}) = -3.63 \times 10^5 \text{ J} = -363 \text{ kJ}$

$\log K = \frac{2(1.88)}{0.0592} = 63.514$; $K = 3.27 \times 10^{63}$

c.

$$3\,H^+ + 3\,e^- \rightarrow 3/2\,H_2 \qquad E° = 0.0\ V$$
$$Al \rightarrow Al^{3+} + 3\,e^- \qquad -E° = 1.66\ V$$

$$3\,H^+ + Al \rightarrow Al^{3+} + 3/2\,H_2 \qquad E°_{cell} = 1.66\ V$$

$\Delta G° = -(3\text{ mol }e^-)\ (96{,}485\text{ C/mol }e^-)\ (1.66\text{ J/C}) = -4.80 \times 10^5\text{ J} = -480.\text{ kJ}$

$\log K = \dfrac{3(1.66)}{0.0592} = 84.122; \qquad K = 1.32 \times 10^{84}$

d.

$$Br_2 + 2\,e^- \rightarrow 2\,Br^- \qquad E° = 1.09\ V$$
$$2\,I^- \rightarrow I_2 + 2\,e^- \qquad -E° = -0.54\ V$$

$$Br_2 + 2\,I^- \rightarrow 2\,Br^- + I_2 \qquad E°_{cell} = 0.55\ V$$

$\Delta G° = -(2\text{ mol }e^-)\ (96{,}485\text{ C/mol }e^-)\ (0.55\text{ J/C}) = -1.1 \times 10^5\text{ J} = -110\text{ kJ}$

$\log K = \dfrac{2(0.55)}{0.0592} = 18.58; \qquad K = 3.8 \times 10^{18}$

71. $Zn^{2+} + 2\,e^- \rightarrow Zn \qquad E° = -0.76\ V$

$Fe^{2+} + 2\,e^- \rightarrow Fe \qquad E° = -0.44\ V$

It is easier to oxidize Zn than Fe, so the Zn will be oxidized protecting the iron of the *Monitor's* hull.

72.

$$Al^{3+} + 3\,e^- \rightarrow Al \qquad E° = -1.66\ V$$
$$Al + 6\,F^- \rightarrow AlF_6^{\,3-} + 3\,e^- \qquad -E° = -(-2.07\ V)$$

$$Al^{3+} + 6\,F^- \rightarrow AlF_6^{\,3-} \qquad E°_{cell} = 0.41\ V$$

$\log K_f = \dfrac{nE°}{0.0592} = \dfrac{3(0.41)}{0.0592} = 21.78$

$K_f = 10^{21.78} = 6.0 \times 10^{20}$

73.

$$2\,H^+ + 2\,e^- \rightarrow H_2 \qquad E° = 0.0000\ V$$
$$D_2 \rightarrow 2\,D^+ + 2\,e^- \qquad -E° = 0.0034\ V$$

$$2\,H^+ + D_2 \rightarrow 2\,D^+ + H_2 \qquad E°_{cell} = 0.0034\ V$$

$\Delta G° = -(2\text{ mol }e^-)\ (96{,}485\text{ C/mol }e^-)\ (0.0034\text{ J/C}) = -660\text{ J}$

$\log K = \dfrac{2(0.0034)}{0.0592} = 0.115; \; K = 1.3$

74. a. $E = E° - \frac{RT}{nF} \ln Q$ or $E = E° - \frac{0.0592}{n} \log_{10} Q$ (at 25° C)

For $Cu^{2+} + 2\,e^- \rightarrow Cu$ $E° = 0.34$ V

$$E = E° - \frac{0.0592}{n} \log 1/[Cu^{2+}] = E° + \frac{0.0592}{n} \log [Cu^{2+}]$$

$[Cu^{2+}] = 0.10\ M$

$$E = 0.34 + \frac{0.0592}{2} \log (0.10) = 0.34 - 0.030 = 0.31 \text{ V}$$

b. $E = 0.34 + \frac{0.0592}{2} \log 2.0 = 0.34 \text{ V} + 8.9 \times 10^{-3} \text{ V} = 0.35 \text{ V}$

c. $E = 0.34 + \frac{0.0592}{2} \log 1.0 \times 10^{-4} = 0.34 - 0.12 = 0.22 \text{ V}$

d. $5\,e^- + 8\,H^+ + MnO_4^- \rightarrow Mn^{2+} + 4\,H_2O$ $E° = 1.51$ V

$$E = E° - \frac{0.0592}{5} \log \frac{[Mn^{2+}]}{[MnO_4^-]\,[H^+]^8}$$

$$E = 1.51 \text{ V} - \frac{0.0592}{5} \log \left[\frac{(0.010)}{(0.10)\,(1.0 \times 10^{-3})^8} \right]$$

$$E = 1.51 - \frac{0.0592}{5}(23) = 1.51 \text{ V} - 0.27 \text{ V} = 1.24 \text{ V}$$

e. $E = 1.51 - \frac{0.0592}{5} \log \left[\frac{(0.010)}{(0.10)\,(0.10)^8} \right]$

$$E = 1.51 - \frac{0.0592}{5}(7.0) = 1.51 \text{ V} - 0.083 = 1.43 \text{ V}$$

75. a. $Hg_2Cl_2 + 2\,e^- \rightarrow 2\,Hg + 2\,Cl^-$ $E_{SCE} = 0.242$ V

$H_2 \rightarrow 2\,H^+ + 2\,e^-$ $-E° = 0.000$ V

$HgCl_2 + H_2 \rightarrow 2\,H^+ + 2\,Cl^- + 2\,Hg$ $E_{cell} = 0.242$ V

b. The hydrogen half cell is the oxidation half reaction; thus, the hydrogen electrode is the anode.

c. Ignoring concentrations from the SCE (they are constant).

$$E = 0.242 \text{ V} - \frac{0.0592}{2} \log \frac{[H^+]^2}{P_{H_2}}$$

If we keep $P_{H_2} = 1$ atm:

$$E = 0.242 \text{ V} - \frac{0.0592}{2} \log [H^+]^2$$

Since, $\log [H^+]^2 = 2 \log [H^+] = -2$ pH

Then $E = 0.242 - 0.0592 \log [H^+]$ or $E = 0.242\ V + 0.0592\ pH$

d. i) $E = 0.242 - 0.0592 \log [H^+] = 0.242 - 0.0592 \log (1.0 \times 10^{-3})$

$E = 0.42\ V$

ii) $E = 0.242 - 0.0592 \log 2.5 = 0.242 - 0.024 = 0.218\ V$

iii) $E = 0.242 - 0.0592 \log (1.0 \times 10^{-9}) = 0.242 + 0.53 = 0.77\ V$

e. $E = 0.242\ V + 0.0592\ pH$

$0.285 = 0.242 + 0.0592\ pH$; $pH = 0.726$, $[H^+] = 0.188\ M$

f. Primarily the reason is convenience. It is inconvenient to deal with the H_2 gas and particularly to keep P_{H_2} constant. Gas cylinders are bulky; and H_2 presents a fire and explosion hazard.

76. Cathode: $O_2 + 2\ H_2O + 4\ e^- \rightarrow 4\ OH^-$ $\quad E° = 0.40\ V$

Anode: $(Fe \rightarrow Fe^{2+} + 2\ e^-) \times 2$ $\quad -E° = 0.44\ V$

$O_2 + 2\ H_2O + 2\ Fe \rightarrow 2\ Fe^{2+} + 4\ OH^-$ $\quad E°_{cell} = 0.84\ V$

Note: ignore $Fe(OH)_2(s)$ formation.

$$E = 0.84\ V - \frac{0.0592}{4} \log\left(\frac{[OH^-]^4\ [Fe^{2+}]^2}{P_{O_2}}\right)$$

If $[OH^-]$ is very small, say $1 \times 10^{-14} = [OH^-]$ ($[H^+] = 1\ M$), the log term will become more positive. Thus, as the solution becomes more acidic, E_{cell} becomes more positive, and the corrosion process becomes more spontaneous.

77. $Fe^{2+} + 2\ e^- \rightarrow Fe$ $\quad E = E° = -0.44\ V$

$Ag^+ + e^- \rightarrow Ag$ $\quad E = E° - \frac{0.0592}{1} \log \frac{1}{[Ag^+]}$

$E = 0.80 + 0.0592 \log [Ag^+] = 0.80 + 0.0592 \log (0.010) = 0.68\ V$

It is still easier to plate out Ag.

78. $Au^{3+} + 3\ e^- \rightarrow Au$, $\quad E° = 1.50\ V$

$$E = E° - \frac{0.0592}{3} \log \frac{1}{[Au^{3+}]}$$

$$E = 1.50 + \frac{0.0592}{3} \log (1.0 \times 10^{-6}) = 1.38\ V$$

The concentration effects on the reduction of Fe^{3+} and Ni^{2+} will not be as great as for Au^{3+}. Au will plate out first since it has the largest reduction potential.

79. $Pt^{2+} + 2\ e^- \rightarrow Pt \qquad E° = 1.2\ V$

$Pd^{2+} + 2\ e^- \rightarrow Pd \qquad E° = 0.99\ V$

$Ni^{2+} + 2\ e^- \rightarrow Ni \qquad E° = -0.23\ V$

It looks like the electrolysis should work. The only problem might be that some Pd will begin to plate before Pt^{2+} is gone. To check this, let's calculate the potential of a Pt half cell when 99.9% of the Pt is gone.

$$E = E° - \frac{0.0592}{2} \log \frac{1}{[Pt^{2+}]} = 1.2\ V - \frac{0.0592}{2} \log\left(\frac{1}{10^{-3}}\right)$$

$E = 1.2\ V - 0.089\ V = 1.1\ V$

So no Pd^{2+} will begin to plate. Therefore, the technique is feasible.

80. The pertinent half reaction is:

$NO_3^- + 4\ H^+ + 3\ e^- \rightarrow NO + 2\ H_2O \qquad E° = 0.96\ V$

The $NaNO_3$ solution will have $[H^+] = 1.0 \times 10^{-7}\ M$ instead of 1 M.

$$E = 0.96 - \frac{0.0592}{3} \log\left(\frac{P_{NO}}{[H^+]^4\ [NO_3^-]}\right)$$

$$E = 0.96 - 0.55 - \frac{0.0592}{3} \log \frac{P_{NO}}{[NO_3^-]} = 0.41 - \frac{0.0592}{3} \log \frac{P_{NO}}{[NO_3^-]}$$

At pH = 7 it is still easier to reduce NO_3^- than water (E° = -0.83 V). So it will be possible to produce NO from the electrolysis of $NaNO_3$ solution.

81. For the Hall process it requires 15 kWh of energy.

$$15\ kWh = \frac{15000\ J}{s} \times 1\ hr \times \frac{60\ min}{hr} \times \frac{60\ s}{min} = 5.4 \times 10^7\ J \text{ or } 5.4 \times 10^4\ kJ$$

To melt Al it requires:

$$1.0 \times 10^3\ g\ Al \times \frac{1\ mol}{26.98\ g} \times \frac{10.7\ kJ}{mol\ Al} = 4.0 \times 10^2\ kJ$$

It is feasible to recycle Al because it takes less than 1% of the energy required to produce the same amount of Al by the Hall process.

82. Consider the strongest oxidizing agent combined with the strongest reducing agent.

$$F_2 + 2\,e^- \rightarrow 2\,F^- \qquad E^\circ = 2.87\text{ V}$$

$$(Li \rightarrow Li^+ + e^-) \times 2 \qquad -E^\circ = +3.05\text{ V}$$

$$F_2 + 2\,Li \rightarrow 2\,Li^+ + 2\,F^- \qquad E^\circ_{cell} = 5.92\text{ V}$$

The claim is impossible. The strongest oxidizing agent and reducing agent when combined only give E° of about 6 Vat standard conditions.

83. $$1/2\,O_2 + 2\,H^+ + 2\,e^- \rightarrow H_2O \qquad E^\circ = 1.23\text{ V}$$

$$H_2 \rightarrow 2\,H^+ + 2\,e^- \qquad -E^\circ = 0.00\text{ V}$$

$$H_2 + 1/2\,O_2 \rightarrow H_2O \qquad E^\circ_{cell} = 1.23\text{ V}$$

$\Delta G^\circ = -nFE^\circ = -(2\text{ mol }e^-)\,(96{,}485\text{ C/mol }e^-)\,(1.23\text{ V})$

$\Delta G^\circ = -237{,}000\text{ J} = -237\text{ kJ}$

$$1.00 \times 10^3\text{ g }H_2O \times \frac{1\text{ mol }H_2O}{18.02\text{ g}} \times \frac{-237\text{ kJ}}{\text{mol }H_2O} = -13{,}200\text{ kJ} = w_{max}$$

The work done can be no larger than the free energy change. The best that could happen is that all of the free energy released goes into doing work, but this does not occur in any real process.

Fuel cells are more efficient in converting chemical energy to electrical energy; they are also less massive. The major disadvantage is that they are expensive.

CHALLENGE PROBLEMS

84. $\Delta G^\circ = -nFE^\circ = \Delta H^\circ - T\Delta S^\circ$; solving for E°:

$$E^\circ = \frac{T\Delta S^\circ}{nF} - \frac{\Delta H^\circ}{nF}$$

If we graph E° vs. T we should get a straight line. The slope of the line is equal to $\Delta S^\circ/nF$ and the y-intercept is equal to $-\Delta H^\circ/nF$.

E° will have little temperature dependence for cell reactions with ΔS° close to zero.

85. a. $$(2\,H^+ + 2\,e^- \rightarrow H_2) \times 2 \qquad E^\circ = 0.0\text{ V}$$

$$2\,H_2O \rightarrow O_2 + 4\,H^+ + 4\,e^- \qquad -E^\circ = -1.23\text{ V}$$

$$2\,H_2O \rightarrow 2\,H_2 + O_2 \qquad E^\circ_{cell} = -1.23\text{ V}$$

$\Delta G^\circ = -nFE^\circ = -(4\text{ mol }e^-)\,(96{,}485\text{ C/mol }e^-)\,(-1.23\text{ V})$

$\Delta G^\circ = 4.75 \times 10^5\text{ J} = 475\text{ kJ}$

b. $\Delta H° = -2\Delta H_f° (H_2O) = -2 \text{ mol } (-286 \text{ kJ/mol}) = 572 \text{ kJ}$

$\Delta S° = 2 \text{ mol } (131 \text{ J/K•mol}) + 1 \text{ mol } (205 \text{ J/K•mol}) - 2 \text{ mol } (70 \text{ J/K•mol})$

$\Delta S° = 327 \text{ J/K}$

c. at 90°C, T = 363 K (carry extra significant figure in T)

$\Delta G° = 572 \text{ kJ} - (363 \text{ K}) (0.327 \text{ kJ/K}) = 453 \text{ kJ} = 450 \text{ kJ}$

$\Delta G° = -nFE°$

$$E° = \frac{-\Delta G°}{nF} = \frac{-4.5 \times 10^5 \text{ J}}{(4 \text{ mol } e^-) (96{,}485 \text{ C/mol } e^-)} = -1.17 \text{ V} = -1.2 \text{ V}$$

at 0°C, $\Delta G° = 572 \text{ kJ} - (273 \text{ K}) (0.327 \text{ kJ/K}) = 483 \text{ kJ}$

$$E° = \frac{-\Delta G°}{nF} = \frac{-4.83 \times 10^5 \text{ J}}{(4 \text{ mol } e^-) (96{,}485 \text{ C/mol } e^-)} = -1.25 \text{ V}$$

86. a. n = 2 for this reaction.

$$E = E° - \frac{0.0592}{2} \log \left(\frac{1}{[H^+]^2 [HSO_4^-]^2} \right)$$

$$E = 2.04 \text{ V} + \frac{0.0592}{2} \log ([H^+]^2 [HSO_4^-]^2)$$

$$E = 2.04 \text{ V} + \frac{0.0592}{2} \log [(4.5)^2 (4.5)^2]$$

$E = 2.04 \text{ V} + 0.077 = 2.12 \text{ V}$

b. We can calculate $\Delta G°$ from $\Delta G° = \Delta H° - T\Delta S°$ and then E° from $\Delta G° = -nFE°$; or we can use the equation derived in Exercise 17.84.

$$E° = \frac{T\Delta S° - \Delta H°}{nF} \qquad T = -20°C = 253 \text{ K} \qquad \text{(carry extra sig. fig.)}$$

Note: $\Delta S°$ and $\Delta H°$ must be in J since 1 J = 1 Coul × Volt

$$E° = \frac{(253 \text{ K}) (263.5 \text{ J/K}) + 315.9 \times 10^3 \text{ J}}{(2 \text{ mol } e^-) (96{,}485 \text{ C/mol } e^-)} = 1.98 \text{ V}$$

c. We must use:

$$E = E° - \frac{RT}{nF} \ln Q = E° + \frac{RT}{nF} \ln ([H^+]^2 [HSO_4^-]^2)$$

$$E = 1.98 + \frac{(8.3145 \text{ J/K•mol}) (253 \text{ K})}{(2 \text{ mol } e^-) (96{,}485 \text{ C/mol } e^-)} \ln [(4.5)^2 (4.5)^2]$$

$E = 1.98 \text{ V} + 0.066 \text{ V} = 2.05 \text{ V}$

87.
$$Ag^+ + e^- \rightarrow Ag \qquad E° = 0.80 \text{ V}$$
$$Ag + 2\,S_2O_3^{2-} \rightarrow Ag(S_2O_3)_2^{3-} + e^- \qquad -E° = -0.017 \text{ V}$$

$$Ag^+ + 2\,S_2O_3^{2-} \rightarrow Ag(S_2O_3)_2^{3-} \qquad E°_{cell} = 0.78 \text{ V}$$

$$\log K = \frac{nE°}{0.0592} = \frac{(1)(0.78)}{0.0592} = 13.18$$

$$K = 10^{13.18} = 1.5 \times 10^{13}$$

88. For $O_2 + 4\,H^+ + 4\,e^- \rightarrow 2\,H_2O$

$$E = E° - \frac{0.0592}{4} \log \frac{1}{[H^+]^4}$$

$P_{O_2} = 1$ atm so it doesn't appear in the Nernst equation.

$$[H^+][OH^-] = 1.0 \times 10^{-14};\ [H^+] = \frac{1.0 \times 10^{-14}}{[OH^-]}$$

$$E = 1.23 \text{ V} - \frac{0.0592}{4} \log \left(\frac{[OH^-]}{1.0 \times 10^{-14}}\right)^4$$

$$E = 1.23 \text{ V} - \frac{0.0592}{4} \log (1.0 \times 10^{56}) - \frac{0.0592}{4} \log [OH^-]^4$$

$$E = 1.23 \text{ V} - 0.83 - \frac{0.0592}{4} \log [OH^-]^4$$

$$E = 0.40 \text{ V} - \frac{0.0592}{4} \log [OH^-]^4$$

For $O_2 + H_2O + 4\,e^- \rightarrow 4\,OH^-$

$E = E° - \frac{0.0592}{4} \log [OH^-]^4$ since $P_{O_2} = 1$. Comparing to the equation above,

$E°_{base} = 0.40$ V

89. a. $$E = E° - \frac{0.0592}{n} \log \frac{[OH^-]^5}{[CrO_4^{2-}]}$$

pH = 7.40 and pOH = 6.60; $[OH^-] = 10^{-6.60} = 2.5 \times 10^{-7}$ *M*

$$E = -0.13 - \frac{0.0592}{3} \log \frac{(2.5 \times 10^{-7})^5}{1.0 \times 10^{-6}}$$

$$E = -0.13 - \frac{0.0592}{3} \log (9.8 \times 10^{-28}) = -0.13 + 0.53 = +0.40 \text{ V}$$

b. $$E = E° - \frac{0.0592}{n} \log \frac{[Cr^{3+}]^2}{[Cr_2O_7^{2-}][H^+]^{14}}$$

$$E = 1.33\text{ V} - \frac{0.0592}{6} \log \frac{(1.0 \times 10^{-6})^2}{(1.0 \times 10^{-6})(1.0 \times 10^{-2})^{14}}$$

$$E = 1.33 - \frac{0.0592}{6} \log (1 \times 10^{22}) = 1.33 - 0.22 = 1.11\text{ V}$$

90. a. $Ag_2CrO_4 + 2\,e^- \longrightarrow 2\,Ag + CrO_4^{2-}$ $\quad E° = 0.446$

$Hg_2Cl_2 + 2\,e^- \longrightarrow Hg + 2\,Cl^-$ $\quad E_{SCE} = 0.242$

SCE will be oxidation half reaction, $E_{cell} = 0.204$ V

$\Delta G = -2(96{,}485)(0.204)\text{J} = -3.94 \times 10^4\text{ J} = -39.4\text{ kJ}$

b. In SCE, we assume all concentrations are constant.

$$E = E° - \frac{0.0592}{2} \log [CrO_4^{2-}]$$

$$E = 0.204\text{ V} - \frac{0.0592}{2} \log [CrO_4^{2-}]$$

c. $$E = 0.204 - \frac{0.0592}{2} \log (1.0 \times 10^{-5})$$

$E = 0.204 + 0.15 = 0.35$ V

d. $$0.504 = 0.204 - \frac{0.0592}{2} \log [CrO_4^{2-}]$$

$\log [CrO_4^{2-}] = -10.135$

$[CrO_4^{2-}] = 10^{-10.135} = 7.33 \times 10^{-11}\ M$

91. a. $E_{meas} = E_{ref} + 0.05916\text{ pH}$

$0.480\text{ V} = 0.250\text{ V} + 0.05916\text{ pH}$

$$pH = \frac{0.480 - 0.250}{0.05916} = 3.888$$

Uncertainty $= \pm 1$ mV $= \pm 0.001$ V

$$pH_{max} = \frac{0.481 - 0.250}{0.05916} = 3.905; \quad 3.905 - 3.888 = 0.017$$

So if the uncertainty in potential is $\pm$ 0.001 V, the uncertainty in pH is $\pm$ 0.017 or about $\pm$ 0.02 pH units.

For this measurement, $[H^+] = 1.29 \times 10^{-4}$

For an error of + 1 mV, $[H^+] = 1.25 \times 10^{-4}$

For an error of -1 mV, $[H^+] = 1.35 \times 10^{-4}$

So the uncertainty in $[H^+]$ is $\pm 0.06 \times 10^{-4}$

b. From the previous example, we will be within ± 0.02 pH units if we measure the potential to the nearest ± 0.001 V (1 mV).

92. a. $E_{meas} = E_{ref} - 0.05916 \log [F^-]$

$0.4462 = 0.2420 - 0.05916 \log [F^-]$

$\log [F^-] = -3.4517$

$[F^-] = 3.534 \times 10^{-4}$ mol/L

b. pH = 9.00 pOH = 5.00

$[OH^-] = 1.0 \times 10^{-5}$

$0.4462 = 0.2420 - 0.05916 \log ([F^-] + 1.0 \times 10^{-4})$

$\log ([F^-] + 1.0 \times 10^{-4}) = -3.452$

$[F^-] + 1.0 \times 10^{-4} = 3.534 \times 10^{-4}$

$[F^-] = 2.5 \times 10^{-4}$

True value is 2.5×10^{-4} and by ignoring the $[OH^-]$ we would say $[F^-]$ was 3.5×10^{-4}.

$$\text{So, \% Error} = \frac{1.0 \times 10^{-4}}{2.5 \times 10^{-4}} = 40.\%$$

c. $[F^-] = 2.5 \times 10^{-4}$

$$\frac{[F^-]}{k[OH^-]} = 50. = \frac{2.5 \times 10^{-4}}{10.0\,[OH^-]}$$

$$[OH^-] = \frac{2.5 \times 10^{-4}}{10. \times 50.} = 5.0 \times 10^{-7}$$

pOH = 6.30; pH = 7.70

d. $HF \rightleftharpoons H^+ + F^-$ $K_a = 7.2 \times 10^{-4}$

$$\frac{[H^+][F^-]}{[HF]} = 7.2 \times 10^{-4}$$

If 99% is F^-, then $[F^-]/[HF] = 99$

$99[H^+] = 7.2 \times 10^{-4}$; $[H^+] = 7.3 \times 10^{-6}$; pH = 5.14

e. The buffer controls the pH so that there is little HF present and so that there is little response to OH^-. Typically a buffer of pH = 6 is used.

CHAPTER EIGHTEEN: THE REPRESENTATIVE ELEMENTS - GROUP 1A THROUGH 4A

QUESTIONS

1. The gravity of the earth is not strong enough to keep H_2 in the atmosphere.

2. 1. Ammonia production and 2. Hydrogenation of vegetable oils

3. Ionic, covalent, and metallic (or insterstitial)
The ionic and covalent hydrides are true compounds obeying the law of Definite Proportions and differ from each other in the type of bonding. The interstitial hydrides are more like solid solutions of hydrogen and a transition metal and do not obey the Laws of Definite Proportions.

4. The small size of the Li^+ cation means that there is a much greater attraction to water. The attraction to water is not so great for other alkali metal ions. Thus, lithium salts tend to absorb water.

5. Hydrogen forms many compounds in which the oxidation state is +1, as the Group 1A elements. For example H_2SO_4 and HCl compared to Na_2SO_4 and NaCl. On the other hand hydrogen forms diatomic H_2 molecules and is a nonmetal, while the Group 1A elements are metals. Hydrogen also forms hydride anion, H^-, which the Group 1A metals do not.

6. $4\ KO_2(s) + 2\ CO_2(g) \rightarrow 2\ K_2CO_3(s) + 3\ O_2(g)$

Potassium superoxide can react with exhaled CO_2 to produce O_2 which then can be breathed.

7. The metals are all easily oxidized. They must be produced in the absence of materials (H_2O, O_2) that are capable of oxidizing them.

8. The acidity decreases. Solutions of Be^{2+} are acidic, while solutions of the other M^{2+} ions are neutral.

9. Planes of carbon atoms slide easily along each other. Graphite is not volatile. The lubricant will not be lost when used in a high vacuum environment.

10. Quartz: Crystalline long range order. The structure is an ordered arrangement of 12 membered rings containing 6 - Si and 6 - O atoms.

Amorphous SiO_2: No long range order. Irregular arrangement that contains many different ring sizes. See Sec. 10.5.

11. The bonds in SnX_4 compounds have a large covalent character. SnX_4 acts as discrete molecules held together by dispersion forces. SnX_2 compounds are ionic and held in the solid state by strong ionic forces.

12. Group 3A elements have one fewer valence electron than Si or Ge. This leaves places for electrons in the semiconductor. This is called a p-type semiconductor.

13. Size decreases from left to right and increases going down. So going one element right and one element down would result in a similar size for the two elements diagonal to each other. The ionization energies will also be similar for the diagonal elements. Electron affinities are harder to predict, but the similar size and ionization energies would lead to similar properties.

14. The "inert pair effect" refers to the difficulty of removing the pair of 6s electrons from some of the elements in the sixth period of the periodic table. Thus, both Tl^+ and Tl^{3+} ions and Pb^{2+} and Pb^{4+} ions are important in the chemistry of Tl and Pb.

EXERCISES

Group 1A Elements

15. a. $\Delta H° = -110.5 - [-242 - 75] = 207$ kJ

$\Delta S° = 3(131) + 198 - [186 + 189] = 216$ J/K

b. $\Delta G° = \Delta H° - T\Delta S°$ and when $\Delta G = 0$,

$$T = \frac{\Delta H°}{\Delta S°} = \frac{207 \times 10^3 \text{ J}}{216 \text{ J/K}} = 958 \text{ K}$$

Reaction is spontaneous at $T > 958$ K (when all partial pressures = 1 atm). Pressure won't affect equilibrium if the volume of the reaction vessel is constant.

16. For $3\,Fe(s) + 4\,H_2O(g) \longrightarrow Fe_3O_4(s) + 4\,H_2(g)$

a. $\Delta H° = -1117 - [4(-242)] = -149$ kJ

$\Delta S° = 146 + 4(131) - [3(27) + 4(189)] = -167$ J/K

b. When $\Delta G° = 0$, $T = \frac{\Delta H°}{\Delta S°} = \frac{-149 \times 10^3 \text{ J}}{-167 \text{ J/K}} = 892$ K

Reaction is spontaneous at $T < 892$ K. $P_{H_2} = P_{H_2O} = 1$ atm.

For $C(s) + H_2O(g) \longrightarrow CO(g) + H_2(g)$

a. $\Delta H° = -110.5 - (-242) = +132$ kJ

$\Delta S° = 198 + 131 - [6 + 189] = +134$ J/K

$$T = \frac{\Delta H°}{\Delta S°} = \frac{132 \times 10^3 \text{ J}}{134 \text{ J/K}} = 985 \text{ K}$$

b. Spontaneous at $T > 985$ K (when all partial pressures = 1 atm).

17. sodium oxide: Na_2O, sodium superoxide: NaO_2, sodium peroxide: Na_2O_2

18. a. rubidium nitride: Rb_3N b. potassium peroxide: K_2O_2

19. a. $Li_3N(s) + 3\,HCl(aq) \longrightarrow 3\,LiCl(aq) + NH_3(aq)$

b. $Rb_2O(s) + H_2O(l) \longrightarrow 2\,RbOH(aq)$

c. $Cs_2O_2(s) + 2\,H_2O(l) \longrightarrow 2\,CsOH(aq) + H_2O_2(aq)$

d. $NaH(s) + H_2O(l) \longrightarrow NaOH(aq) + H_2(g)$

20. $K(s) + O_2(g) \rightarrow KO_2(s)$; $16\ K(s) + S_8(s) \rightarrow 8\ K_2S(s)$

$12\ K(s) + P_4(s) \rightarrow 4\ K_3P(s)$; $2\ K(s) + H_2(g) \rightarrow 2\ KH(s)$

$2\ K(s) + 2\ H_2O(l) \rightarrow H_2(g) + 2\ KOH(aq)$

21. $2\ Li(s) + 2\ C_2H_2(g) \rightarrow 2\ LiC_2H(s) + H_2(g)$. It is an oxidation-reduction reaction.

22. $NaH(s) + H_2O(l) \rightarrow Na^+(aq) + OH^-(aq) + H_2(g)$

Acid-Base: A proton is transferred from an acid, H_2O, to a base, H^-, forming the conjugate base of water, OH^-, and the conjugate acid of H^-, H_2.

Oxidation-reduction: The oxidation number of H is -1 in NaH, +1 in H_2O, and zero in H_2. Thus, an electron is transferred from the hydride ion to water.

Group 2A Elements

23. barium oxide: BaO, barium peroxide: BaO_2

24. strontium carbonate: $SrCO_3$, magnesium sulfate: $MgSO_4$

25. $Mg_3N_2(s) + 6\ H_2O(l) \rightarrow 2\ NH_3(g) + 3\ Mg^{2+}(aq) + 6\ OH^-(aq)$

$Mg_3P_2(s) + 6\ H_2O(l) \rightarrow 2\ PH_3(g) + 3\ Mg^{2+}(aq) + 6\ OH^-(aq)$

26. $2\ Ca(s) + O_2(g) \rightarrow 2\ CaO(s)$; $8\ Ca(s) + S_8(s) \rightarrow 8\ CaS(s)$

$3\ Ca(s) + N_2(g) \rightarrow Ca_3N_2(s)$; $6\ Ca(s) + P_4(s) \rightarrow 2\ Ca_3P_2(s)$

$Ca(s) + H_2(g) \rightarrow CaH_2(s)$; $Ca(s) + 2\ H_2O(l) \rightarrow H_2(g) + Ca(OH)_2(aq)$

27.

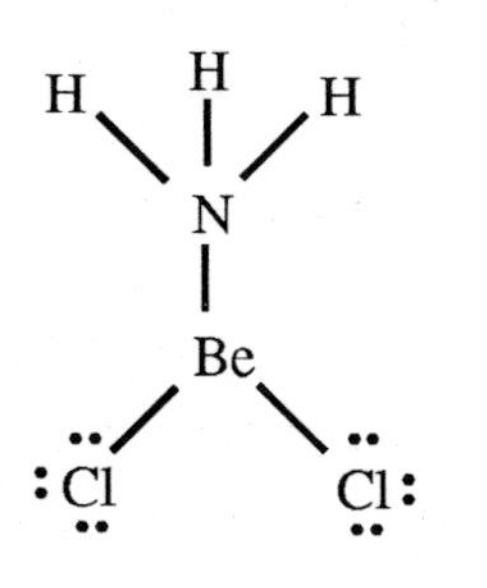

Trigonal planar

Be uses sp^2 hybrid orbitals.
N uses sp^3 hybrid orbitals.
$BeCl_2$ is a Lewis acid.

28. $BeCl_2(NH_3)_2$ would form in excess ammonia. A structure for this molecule can be drawn that obeys the octet rule for all atoms.

```
    H    :Cl:    H
    |     |      |
H — N  —  Be  —  N — H
    |     |      |
    H    :Cl:    H
```

$BeCl_2(NH_3)_2$ has 2 + 2(7) + 2(5) + 6(1) = 32 valence electrons.

29.

$$Be + H_2O \rightarrow Be(OH)_4^{2-} + H_2$$

$$Be + 4\ OH^- \rightarrow Be(OH)_4^{2-} + 2\ e^-$$

$$2\ H_2O + 2\ e^- \rightarrow H_2 + 2\ OH^-$$

$$Be(s) + 2\ H_2O(l) + 2\ OH^-(aq) \rightarrow Be(OH)_4^{2-}(aq) + H_2(g)$$

Be is the reducing agent. H_2O is the oxidizing agent.

30. $Mg(OH)_2(s) \rightleftharpoons Mg^{2+}(aq) + 2\ OH^-(aq)$ $K_{sp} = 8.9 \times 10^{-12}$

$[Mg^{2+}]\ [OH^-]^2 = 8.9 \times 10^{-12}$, pH = 8.0, pOH = 6.0, $[OH^-] = 1 \times 10^{-6}$

$[Mg^{2+}]\ (1 \times 10^{-6})^2 = 8.9 \times 10^{-12}$, $[Mg^{2+}] = 9\ M$

Considering only the solubility product equilibria we calculate that the solubility is 9 mol $Mg(OH)_2$/L. In fact, the solubility is less than this and is controlled by other factors.

Group 3A Elements

31. a. Thallium (I) hydroxide b. Indium(III) sulfide c. Gallium(III) oxide

32. a. Indium(III) phosphide b. Thallium(III) fluoride c. Gallium(III) arsenide but is commonly called gallium arsenide

33. $B_2H_6 + O_2 \rightarrow B(OH)_3$

$B_2H_6 + O_2 \rightarrow 2\ B(OH)_3$ (6 extra O atoms)

$B_2H_6 + 3\ O_2 \rightarrow 2\ B(OH)_3$

34. $B_2O_3 + Mg \rightarrow MgO + B$, balance B first.

$B_2O_3 + Mg \rightarrow MgO + 2\ B$, balance O, then Mg

$B_2O_3 + 3\ Mg \rightarrow 3\ MgO + 2\ B$

35. $In_2O_3(s) + 6\ H^+(aq) \longrightarrow 2\ In^{3+}(aq) + 3\ H_2O(l)$

$In_2O_3(s) + OH^-(aq) \longrightarrow$ No Reaction.

$Ga_2O_3(s) + 6\ H^+(aq) \longrightarrow 2\ Ga^{3+}(aq) + 3\ H_2O(l)$

$Ga_2O_3(s) + 2\ OH^-(aq) + 3\ H_2O(l) \longrightarrow 2\ Ga(OH)_4^-(aq)$

36. Al_2O_3 is an amphoteric oxide.

Base: $Al_2O_3(s) + 6\ H^+(aq) \longrightarrow 2\ Al^{3+}(aq) + 3\ H_2O(l)$

Lewis acid: $Al_2O_3(s) + 3\ H_2O(l) + 2\ OH^-(aq) \longrightarrow 2\ Al(OH)_4^-(aq)$

37. $2\ In(s) + 3\ F_2(g) \longrightarrow 2\ InF_3(s)$

$2\ In(s) + 3\ Cl_2(g) \longrightarrow 2\ InCl_3(s)$

$4\ In(s) + 3\ O_2(g) \longrightarrow 2\ In_2O_3(s)$

$2\ In(s) + 6\ HCl(aq) \longrightarrow 3\ H_2(g) + 2\ InCl_3(aq)$

or $2\ In(s) + 6\ HCl(g) \longrightarrow 3\ H_2(g) + 2\ InCl_3(s)$

38. $2\ Al(s) + 2\ NaOH(aq) + 6\ H_2O(l) \longrightarrow 2\ Al(OH)_4^-(aq) + 2\ Na^+(aq) + 3\ H_2(g)$

Group 4A Elements

39. a. linear b. sp

40. CS_2 has 4 + 2(6) = 16 valence electrons.

:S=C=S: linear

C_3S_2 has 3(4) + 2(6) = 24 valence electrons.

:S=C=C=C=S: linear

41. a. $K_2SiF_6 + K \longrightarrow KF + Si$

$K_2SiF_6(s) + 4\ K(l) \longrightarrow 6\ KF(s) + Si(s)$

b. K_2SiF_6 is an ionic compound, composed of K^+ cations and SiF_6^{2-} anions. The SiF_6^{2-} anion is held together by covalent bonds. The structure is:

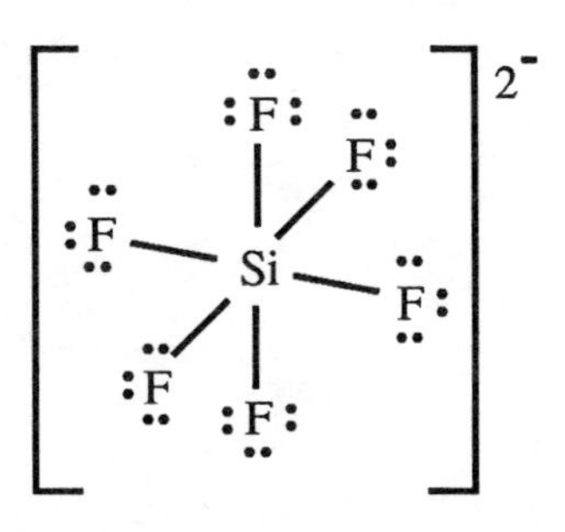

The anion is octahedral.

42. $SiO_2 + HF \rightarrow SiF_4$; H_2O must be the other product.

$SiO_2(s) + 4\ HF(aq) \rightarrow SiF_4(g) + 2\ H_2O(l)$

43. First, we must balance the equation:

$$Pb \rightarrow Pb(OH)_2$$

$$(Pb + 2\ OH^- \rightarrow Pb(OH)_2 + 2\ e^-) \times 2$$

$$H_2O + O_2 \rightarrow OH^-$$

$$2\ H_2O + O_2 \rightarrow 4\ OH^-$$

$$4\ e^- + 2\ H_2O + O_2 \rightarrow 4\ OH^-$$

$$2\ Pb + 4\ OH^- \rightarrow 2\ Pb(OH)_2 + 4\ e^- \qquad -E° = +0.57\ V$$

$$4\ e^- + 2\ H_2O + O_2 \rightarrow 4\ OH^- \qquad E° = +0.40\ V$$

$$2\ Pb + 2\ H_2O + O_2 \rightarrow 2\ Pb(OH)_2 \qquad E°_{cell} = +0.97\ V$$

Fe pipes corrode more easily than Pb pipes. However, the corrosion of Pb pipes is still spontaneous and Pb(II) is very toxic. Pb pipes were extensively used by the Romans and it has been proposed that chronic lead poisoning is one of the factors leading to the decline and fall of the Roman Empire.

44. $Pb(OH)_2 \rightleftharpoons Pb^{2+} + 2\ OH^-$ $\qquad K_{sp} = 1.2 \times 10^{-15} = [Pb^{2+}][OH^-]^2$

s = solubility in mol/L $\rightarrow$ s $\quad$ 2s $\quad$ (Ignore OH^- from H_2O)

$K_{sp} = (s)(2s)^2 = 1.2 \times 10^{-15}$; $4s^3 = 1.2 \times 10^{-15}$; $s = 6.7 \times 10^{-6}$ mol/L $\quad$ Assumption good.

45. Tin (II) fluoride

46. PbO: lead(II) oxide, $\qquad PbO_2$: lead(IV) oxide

47. The π electrons are free to move in graphite, thus giving it a greater conductivity (lower resistance). The electrons have the greatest mobility within sheets of carbon atoms. Electrons in diamond are not mobile (high resistance). The structure of diamond is uniform in all directions; thus, there is no directional dependence of the resistivity.

48. SiC would have a covalent network structure similar to diamond.

ADDITIONAL EXERCISES

49. a. $2\ Na + 2\ NH_3 \rightarrow 2\ NaNH_2 + H_2$

b. $\frac{251.4\ g}{1000.\ g + 251.4\ g} \times 100 = 20.09\%$ by mass

$\text{mol Na} = 251.4\ g \times \frac{1\ mol}{22.99\ g} = 10.94\ \text{mol Na}$

$\text{mol } NH_3 = 1000.\ g \times \frac{1\ mol}{17.03\ g} = 58.72\ \text{mol } NH_3$

$\chi_{Na} = \frac{10.94}{10.94 + 58.72} = 0.1570$

$\text{Molality} = \frac{251.4\ g\ Na}{kg} \times \frac{1\ mol\ Na}{22.99\ g\ Na} = 10.94\ mol/kg$

50. $K^+(out) \rightleftharpoons K^+(in)$

$E = E° - \frac{0.0592}{1} \log \frac{[K^+]_{in}}{[K^+]_{out}}$; $E° = 0$

$E = -0.0592 \log \frac{(0.15)}{5.0 \times 10^{-3}} = -0.087\ V$

$\Delta G = -nFE = -(1\ mol\ e^-)(96{,}485\ C/mol\ e^-)(-0.087\ J/C)$

$\Delta G = 8400\ J = 8.4\ kJ = \text{work}$

51.

	Li	Na	K	Rb	Cs	Fr
Atomic number	3	11	19	37	55	87
MP	180	98	63	39	29	(≈22)

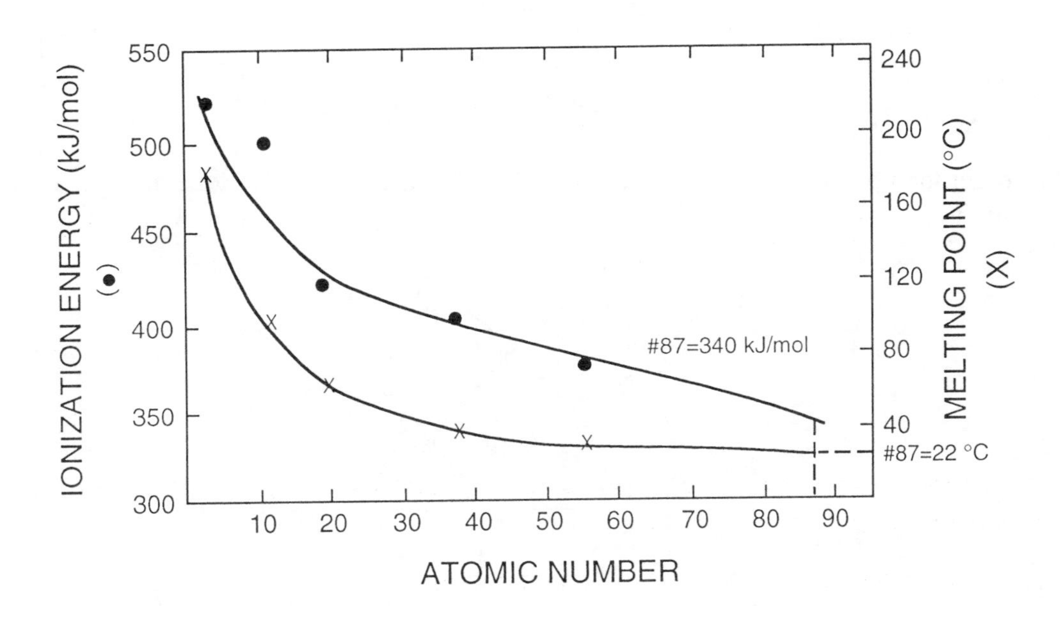

From the graph, we would estimate the melting point of Fr to be around 20°C. Thus, it would be a liquid at room temperature.

52. In the gas phase, linear diatomic molecules would exist.

$$:\ddot{\underset{..}{F}}-\text{Be}-\ddot{\underset{..}{F}}:$$

In the solid state, BeF_2 would exist as a polymeric solid with a structure:

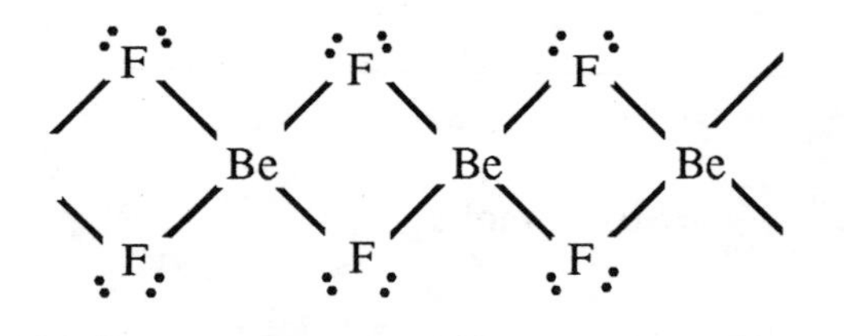

53. $1.00 \times 10^3 \text{ kg} \times \frac{1000 \text{ g}}{\text{kg}} \times \frac{1 \text{ mol Ca}}{40.08 \text{ g}} \times \frac{2 \text{ mol e}^-}{\text{mol Ca}} \times \frac{96{,}485 \text{ C}}{\text{mol e}^-} = 4.81 \times 10^9 \text{ C}$

$i = \frac{4.81 \times 10^9 \text{ C}}{8.00 \text{ hr}} \times \frac{1 \text{ hr}}{60 \text{ min}} \times \frac{1 \text{ min}}{60 \text{ s}} = \frac{1.67 \times 10^5 \text{ C}}{\text{s}} = 1.67 \times 10^5 \text{ A}$

$1.00 \times 10^3 \text{ kg Ca} \times \frac{70.90 \text{ g } Cl_2}{40.08 \text{ g Ca}} = 1.77 \times 10^3 \text{ kg of } Cl_2$

54. Be^{2+} ion is a Lewis acid. The ion in solution is $Be(H_2O)_4{}^{2+}$. The acidic solution results from the reaction:

$$Be(H_2O)_4{}^{2+} \rightleftharpoons Be(H_2O)_3(OH)^+ + H^+$$

55. A single $AlCl_3$ has the Lewis structure:

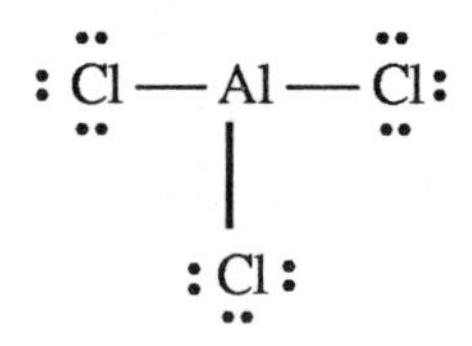

The Al has room for a pair of electrons. It can act as a Lewis acid. The only lone pairs are on Cl, so the structure is:

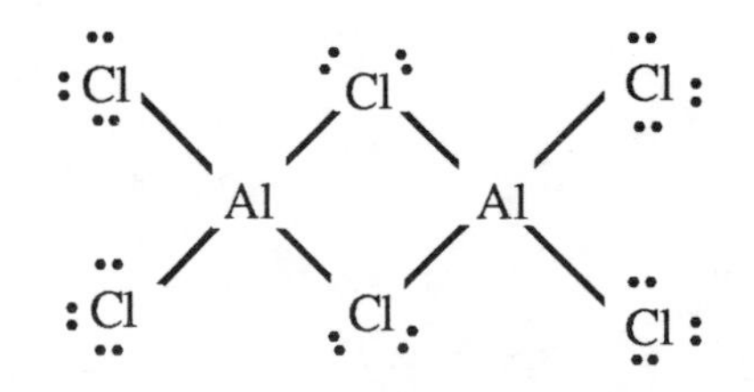

56. Element 113 would fall below Tl in the periodic table.

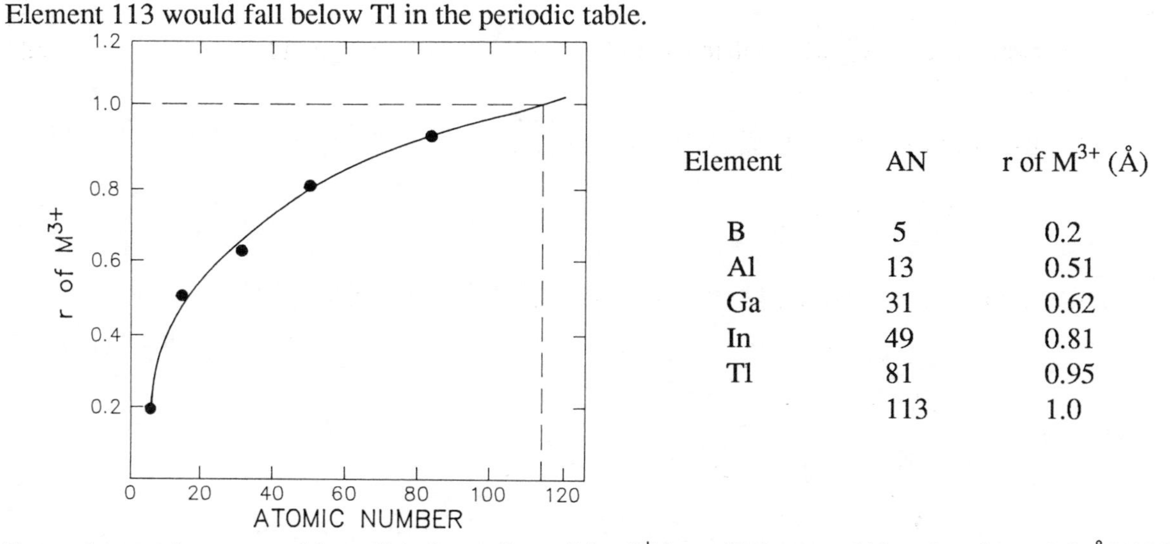

Element	AN	r of M^{3+} (Å)
B	5	0.2
Al	13	0.51
Ga	31	0.62
In	49	0.81
Tl	81	0.95
	113	1.0

From the graph, we would predict the radius of the 3^+ ion of Element 113 to be about 1.0 Å (100 pm).

From the data in Table 18.9 we would expect the ionization energy of element 113 to be close to that of Tl.

57. $LiAlH_4$ H -1 (metal hydride), Li +1

Al +3; $0 = +1 + x + 4(-1)$; $x = +3$

58.

$Tl^{3+} + 2\,e^- \longrightarrow Tl^+$	$E° = +1.25$ V
$3\,I^- \longrightarrow I_3^- + 2\,e^-$	$-E° = -0.55$ V
$Tl^{3+} + 3\,I^- \longrightarrow Tl^+ + I_3^-$	$E°_{cell} = +0.70$ V

In solution, Tl^{3+} can oxidize I^- to I_3^-. Thus, we expect TlI_3 to be Thallium(I) triiodide.

59.

Ga(I)	$[Ar]3d^{10}4s^2$	no unpaired e^-
Ga(III)	$[Ar]3d^{10}$	no unpaired e^-
Ga(II)	$[Ar]3d^{10}4s^1$	1 unpaired e^-

If the compound contained Ga(II) it would be paramagnetic. This can easily be determined by measuring the mass of a sample in the presence and absence of a magnetic field. Paramagnetic compounds will have an apparent greater mass in a magnetic field.

60. a. Out of 100 g of compound there are:

$$44.4 \text{ g Ca} \times \frac{1 \text{ mol}}{40.08 \text{ g}} = 1.11 \text{ mol Ca}$$

$$20.0 \text{ g Al} \times \frac{1 \text{ mol}}{26.98 \text{ g}} = 0.741 \text{ mol Al}$$

$$35.6 \text{ g O} \times \frac{1 \text{ mol}}{16.00 \text{ g}} = 2.23 \text{ mol O}$$

$$\frac{1.11}{0.741} = 1.5, \qquad \frac{0.741}{0.741} = 1, \qquad \frac{2.23}{0.741} = 3, \quad Ca_3Al_2O_6 = \text{empirical formula}$$

b. $Ca_9Al_6O_{18}$

c. There are covalent bonds between Al and O atoms in the $[Al_6O_{18}]^{18-}$ anion; sp^3 hybrid orbitals on aluminum overlap with sp^3 hybrid orbitals on oxygen to form the sigma bonds.

61.

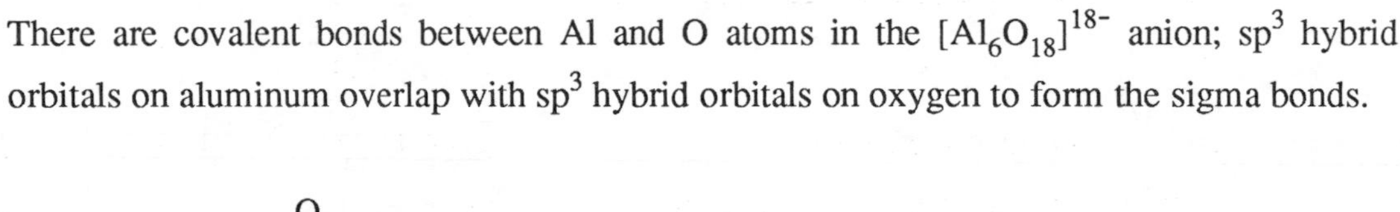

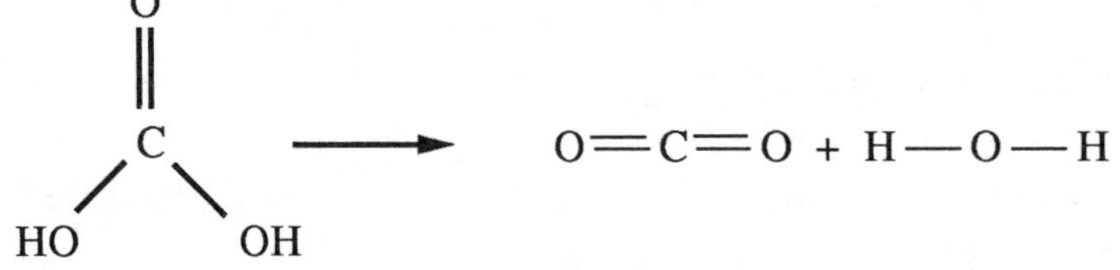

Bonds broken: 2 C—O 358 kJ/mol

Bonds made: C=O 799 kJ/mol

$\Delta H = 716 - 799 = -83$ kJ

ΔH is favorable for the decomposition of H_2CO_3 to CO_2 and H_2O. ΔS is also favorable for the decomposition as there is an increase in disorder. Hence, H_2CO_3 will spontaneously decompose to CO_2 and H_2O. Carbonic acid should be written as $CO_2(aq)$.

62. Pb_3O_4. We assign -2 to the oxidation number of O. The sum of the oxidation numbers of Pb must be +8. We get this if two are Pb(II) and one is Pb(IV). So 2/3 of the lead is Pb(II).

63. Sn and Pb can reduce H^+ to H_2

$$Sn(s) + 2\,H^+(aq) \rightarrow Sn^{2+}(aq) + H_2(g)$$

$$Pb(s) + 2\,H^+(aq) \rightarrow Pb^{2+}(aq) + H_2(g)$$

64.

$CH_3(CH_2)_6CH_2$

:Cl: — Sn

:Cl: $CH_2(CH_2)_6CH_3$

The compound is held together by covalent bonds. Bond angles are roughly tetrahedral.

CHALLENGE PROBLEMS

65. a. Na^+ can oxidize Na^- to Na.

$$Na^- \longrightarrow Na + e^- \qquad \Delta H = +52.9 \text{ kJ}$$

$$Na^+ + e^- \longrightarrow Na \qquad \Delta H = -495 \text{ kJ}$$

$$Na^+ + Na^- \longrightarrow 2\ Na \qquad \Delta H = -442 \text{ kJ}$$

The purpose of the cryptand is to encapsulate the Na^+ ion so that it does not come in contact with the Na^- ion and oxidize sodium to sodium metal.

66. The crown ether surrounds the cation.

The ion pair

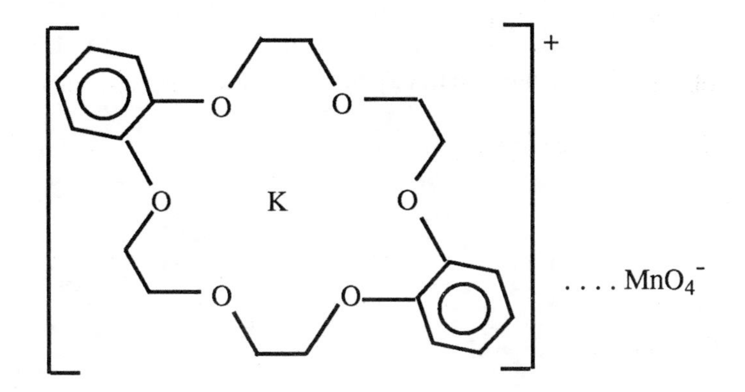

is soluble in non-polar solvents because the solvent is attracted to the crown ether.

67. White tin is stable at normal temperatures. Gray tin is stable at temperatures below 13.2°C. Thus for the phase change:

$$Sn(gray) \longrightarrow Sn(white)$$

ΔG is (-) at T > 13.2°C and ΔG is (+) at T < 13.2°C

This is only possible if ΔH is (+) and ΔS is (+). Thus, gray tin has the more ordered structure.

68. $SiCl_4(l) + 2\ H_2O(l) \longrightarrow SiO_2(s) + 4\ H^+(aq) + 4\ Cl^-(aq)$

$\Delta H° = 4(-167) + (-911) - [-687 + 2(-286)] = -320.$ kJ

$\Delta S° = 4(57) + (42) - [240 + 2(70)] = -110.$ J/K

$-T\Delta S$ at 298 K = 32.8 kJ

The reaction is spontaneous because ΔH is very favorable. There are, overall, stronger bonds in SiO_2 and HCl(aq) than in $SiCl_4$ and H_2O.

The corresponding reaction for CCl_4 is:

$CCl_4(l) + 2\ H_2O(l) \longrightarrow CO_2(g) + 4\ H^+(aq) + 4\ Cl^-(aq)$

$\Delta H° = 4(-167) + (-393.5) - [-135 + 2(-286)] = -355$ kJ

$\Delta S° = 4(57) + 214 - [216 + 2(70)] = +86$ J/K

Thermodynamics predicts that this reaction would also be spontaneous.

The answer must lie in kinetics. $SiCl_4$ reacts because an activated complex can form by a water molecule attaching to silicon in $SiCl_4$. The activated complex requires silicon to form a fifth bond. Silicon has low energy 3d orbitals available to expand the octet. Carbon will not break the octet rule, therefore, CCl_4 cannot form this activated complex. CCl_4 and H_2O require a different pathway to get to products. The different pathway has a higher activation energy and, in turn, the reaction is much slower. (See Exercise 18.69.)

69. Carbon is much smaller than Si and cannot form a fifth bond in the transition state since carbon has no low energy d orbitals available to expand the octet.

CHAPTER NINETEEN: THE REPRESENTATIVE ELEMENTS - GROUP 5A THROUGH 8A

QUESTIONS

1. The reaction $N_2(g) + 3\ H_2(g) \longrightarrow 2\ NH_3(g)$ is exothermic. Thus, K_p decreases as the temperature increases. Lower temperatures are favored for maximum yield of ammonia. However, at lower temperatures the rate is slow; without a catalyst the rate is too slow for the process to be feasible. The discovery of a catalyst increased the rate of reaction at a lower temperature favored by thermodynamics.

2. White phosphorus consists of P_4 tetrahedra that react readily with oxygen. In red phosphorus the P_4 tetrahedra are bonded to each other in chains and are less reactive. They need a source of energy to react with oxygen, such as when one strikes a match. Black phosphorus is crystalline with the P atoms more tightly bonded in the crystal and fairly unreactive towards oxygen.

3. The pollution provides nitrogen and phosphorous nutrients so the algae can grow. The algae consume oxygen, causing fish to die.

4. In upper atmosphere O_3 acts as a filter for UV radiation:

$$O_3 \xrightarrow{h\nu} O_2 + O$$

O_3 is also a powerful oxidizing agent. It irritates the lungs and eyes and at higher concentration it is toxic. The smell of a "fresh spring day" is O_3 formed during lightning discharges. Toxic materials don't necessarily smell bad. HCN smells like almonds, for example.

5. Plastic sulfur consists of long S_n chains of sulfur atoms. As plastic sulfur becomes brittle the long chains break down into S_8 rings.

6. Sulfur forms polysulfide ions, S_n^{2-}, which are soluble.

e.g., $S_8 + S^{2-} \rightleftharpoons S_9^{2-}$

Nitric acid oxidizes S^{2-} to S.

7. Fluorine is the most reactive of the halogens because it is the most electronegative and the bond in the F_2 molecule is very weak.

8. In Mendeleev's time none of the noble gases were known. Since an entire family was missing, no gaps seemed to appear in the periodic arrangement. Mendeleev had no evidence to predict the existence of such a family.

9. Helium is unreactive and doesn't combine with any other elements. It is a very light gas and would easily escape the earth's gravitational pull as the planet was formed.

10. The heavier members are not really inert. Xe and Kr have been shown to react and form compounds with other elements.

EXERCISES

Group 5A Elements

11.

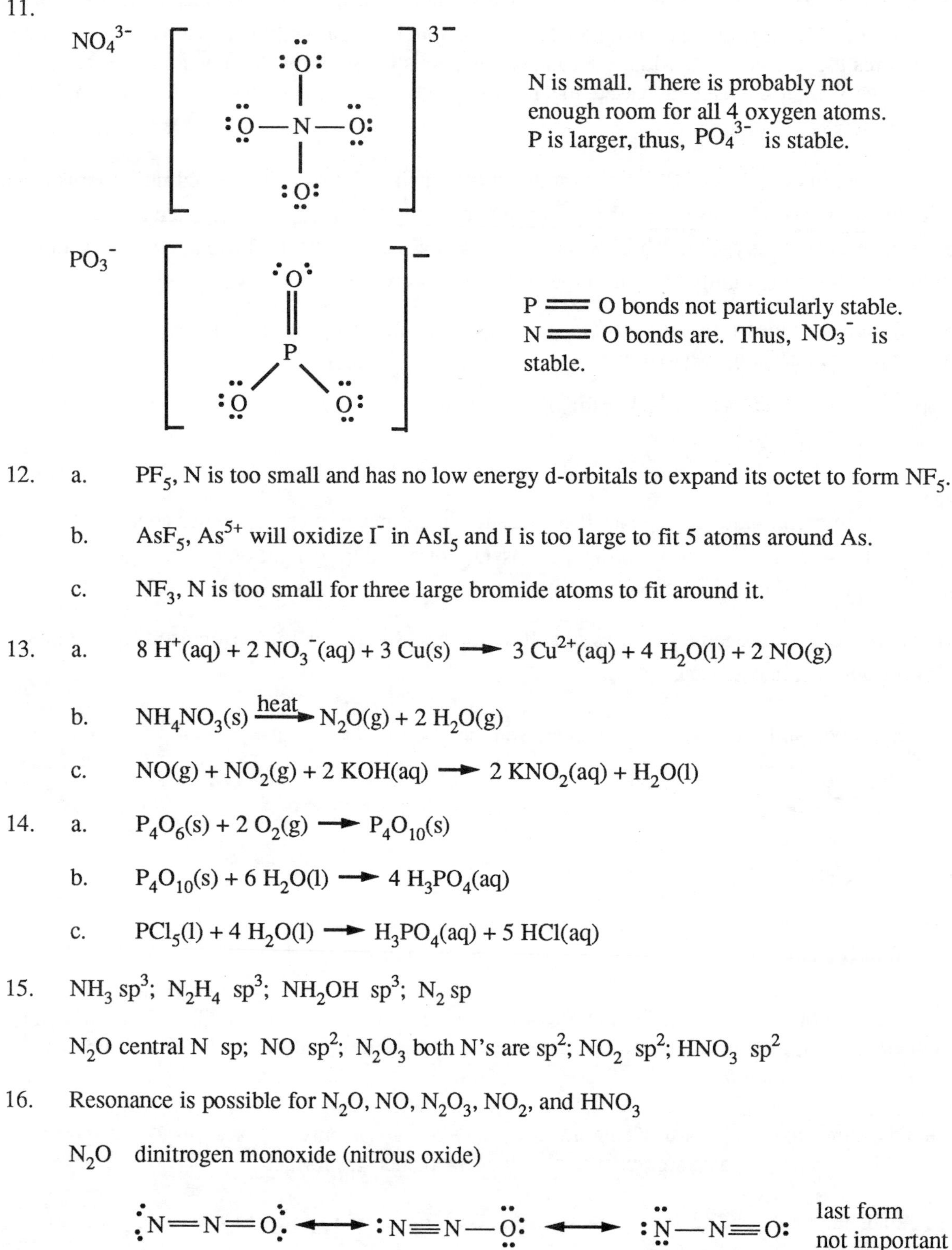

NO_4^{3-} — N is small. There is probably not enough room for all 4 oxygen atoms. P is larger, thus, PO_4^{3-} is stable.

PO_3^- — P ═ O bonds not particularly stable. N ═ O bonds are. Thus, NO_3^- is stable.

12. a. PF_5, N is too small and has no low energy d-orbitals to expand its octet to form NF_5.

b. AsF_5, As^{5+} will oxidize I^- in AsI_5 and I is too large to fit 5 atoms around As.

c. NF_3, N is too small for three large bromide atoms to fit around it.

13. a. $8\ H^+(aq) + 2\ NO_3^-(aq) + 3\ Cu(s) \rightarrow 3\ Cu^{2+}(aq) + 4\ H_2O(l) + 2\ NO(g)$

b. $NH_4NO_3(s) \xrightarrow{heat} N_2O(g) + 2\ H_2O(g)$

c. $NO(g) + NO_2(g) + 2\ KOH(aq) \rightarrow 2\ KNO_2(aq) + H_2O(l)$

14. a. $P_4O_6(s) + 2\ O_2(g) \rightarrow P_4O_{10}(s)$

b. $P_4O_{10}(s) + 6\ H_2O(l) \rightarrow 4\ H_3PO_4(aq)$

c. $PCl_5(l) + 4\ H_2O(l) \rightarrow H_3PO_4(aq) + 5\ HCl(aq)$

15. NH_3 sp^3; N_2H_4 sp^3; NH_2OH sp^3; N_2 sp

N_2O central N sp; NO sp^2; N_2O_3 both N's are sp^2; NO_2 sp^2; HNO_3 sp^2

16. Resonance is possible for N_2O, NO, N_2O_3, NO_2, and HNO_3

N_2O dinitrogen monoxide (nitrous oxide)

:N=N=O: ⟷ :N≡N—O: ⟷ :N—N≡O: last form not important

NO nitrogen monoxide (nitric oxide)

N=O ⟷ N=O ⟷ N=O

N_2O_3 dinitrogen trioxide

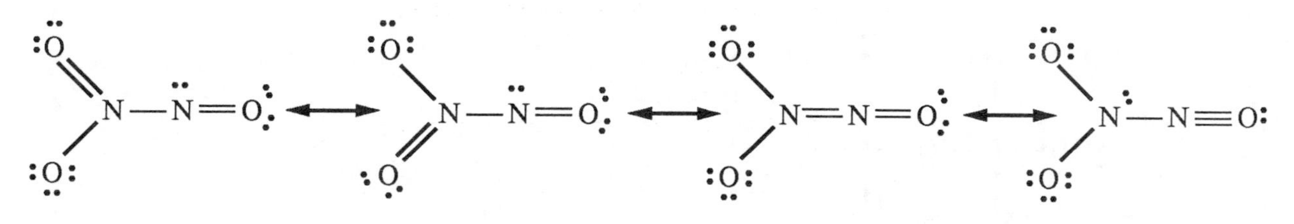

last 2 not important

NO_2 nitrogen dioxide

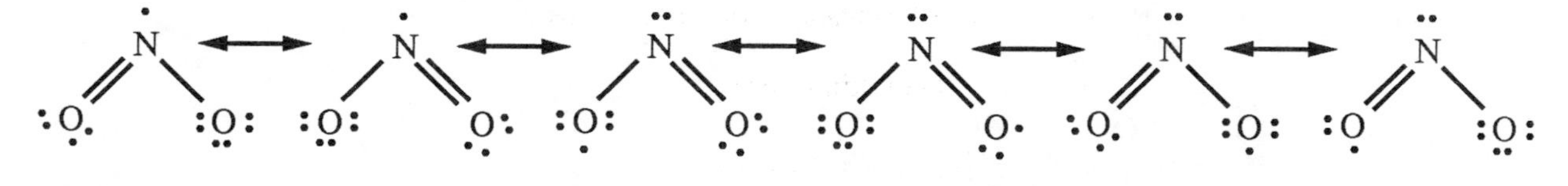

HNO_3 nitric acid

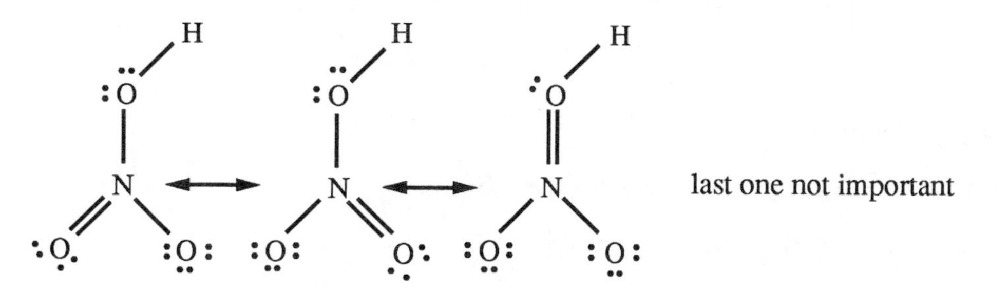

last one not important

17. $\Delta H° = (90\text{ kJ}) \times 2 = 180.\text{ kJ}$ and $\Delta G° = (87\text{ kJ}) \times 2 = 174\text{ kJ}$

$\Delta S° = 2(211\text{ J/K}) - [192 + 205] = 25\text{ J/K}$

At high temperature the reaction $N_2 + O_2 \rightarrow 2\ NO$ becomes spontaneous. In the atmosphere, even though $2\ NO \rightarrow N_2 + O_2$ is spontaneous, it doesn't occur because the rate is slow.

18. $N_2H_4 + 2\ F_2 \rightarrow 4\ HF + N_2$

Bonds broken:

1 N—N	(160 kJ/mol)
4 N—H	(391 kJ/mol)
2 F—F	(154 kJ/mol)

Bonds made:

4 H—F	(565 kJ/mol)
N ≡ N	(941 kJ/mol)

$\Delta H = 160 + 4(391) + 2(154) - [4(565) + 941]$

$\Delta H = 2032 - 3201 = -1169\text{ kJ}$

19. a. $H_3PO_4 > H_3PO_3$ More O - atoms, stronger acid

b. $H_3PO_4 > H_2PO_4^- > HPO_4^{2-}$

20. $H_4P_2O_6$:

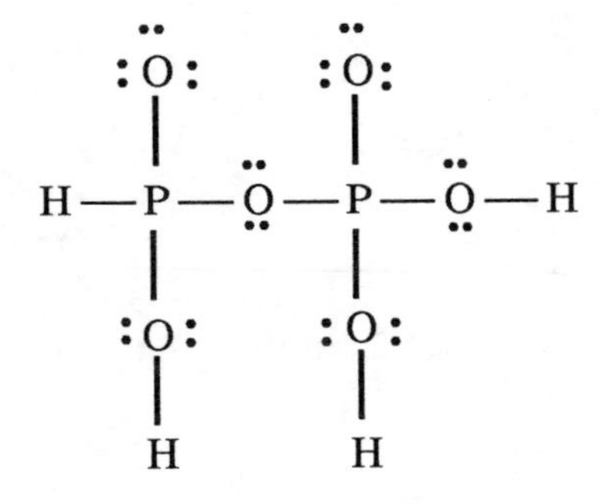

$H_4P_2O_5$:

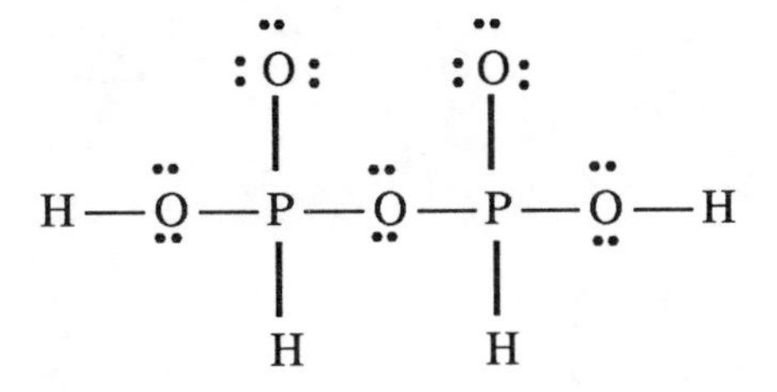

21. Production of antimony:

$$2\ Sb_2S_3(s) + 9\ O_2(g) \rightarrow 2\ Sb_2O_3(s) + 6\ SO_2(g)$$

$$2\ Sb_2O_3(s) + 3\ C(s) \rightarrow 4\ Sb(s) + 3\ CO_2(g)$$

Production of bismuth:

$$2\ Bi_2S_3(s) + 9\ O_2(g) \rightarrow 2\ Bi_2O_3(s) + 6\ SO_2(g)$$

$$2\ Bi_2O_3(s) + 3\ C(s) \rightarrow 4\ Bi(s) + 3\ CO_2(g)$$

22. $4\ As(s) + 3\ O_2(g) \rightarrow As_4O_6(s)$; $4\ As(s) + 5\ O_2(g) \rightarrow As_4O_{10}(s)$

$As_4O_6(s) + 6\ H_2O(l) \rightarrow 4\ H_3AsO_4(aq)$; $As_4O_{10}(s) + 6\ H_2O(l) \rightarrow 4\ H_3AsO_4(aq)$

Group 6A Elements

23. $O{=}O{-}O \rightarrow O{=}O + O$

Break O—O

$$\Delta H = +146\ \text{kJ/mol} \times \frac{1\ \text{mol}}{6.022 \times 10^{23}} = 2.42 \times 10^{-22}\ \text{kJ} = 2.42 \times 10^{-19}\ \text{J}$$

A photon of light must contain at least 2.42×10^{-19} J.

$$E_{photon} = \frac{hc}{\lambda},\ \lambda = \frac{hc}{E} = \frac{(6.626 \times 10^{-34}\ J\ s)(2.998 \times 10^{8}\ m/s)}{2.42 \times 10^{-19}\ J}$$

$$\lambda = 8.21 \times 10^{-7}\ m = 821\ nm$$

24. Light from violet to green will work.

25. a. oxidation - reduction reaction

b. NO, see (c)

c.

$$(S^{2-} \longrightarrow S + 2\,e^-) \times 3$$
$$(3\,e^- + 3\,H^+ + HNO_3 \longrightarrow NO + 2\,H_2O) \times 2$$

$$6\,H^+ + 2\,HNO_3 + 3\,S^{2-} \longrightarrow 3\,S + 2\,NO + 4\,H_2O$$

26. $H_2SeO_4 + 3\,SO_2(g) \longrightarrow Se(s) + 3\,SO_3(g) + H_2O(l)$

27. SF_5^- has 6 + 5(7) + 1 = 42 valence electrons.

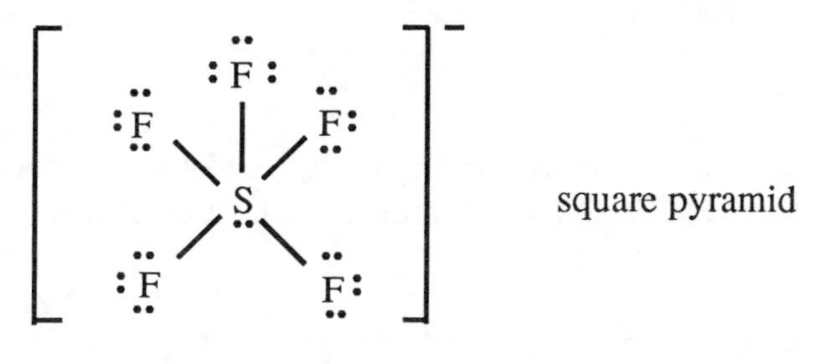

square pyramid

28. $OTeF_5^-$ has 6 + 6 + 5(7) + 1 = 48 valence electrons.

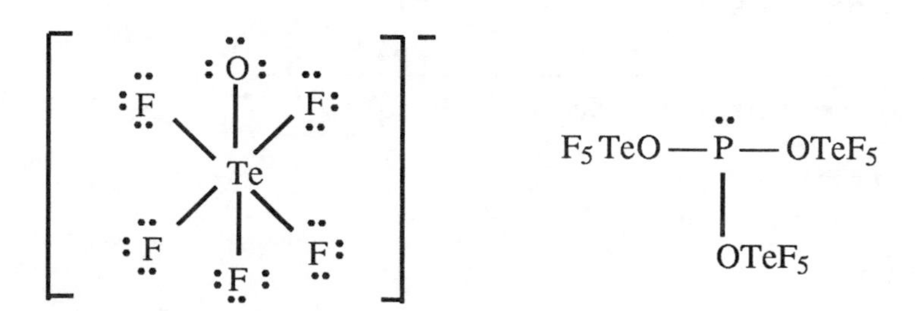

Group 7A Elements

29. a. ClF_5, 7 + 5(7) = 42 e^-

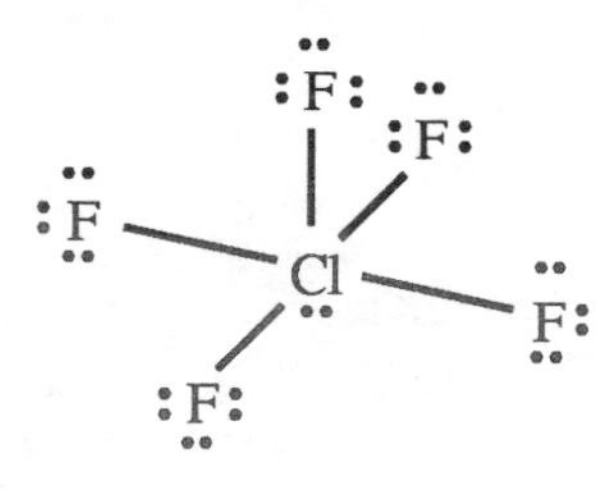

Square pyramid

b. IF_3, 7 + 3(7) = 28 e^-

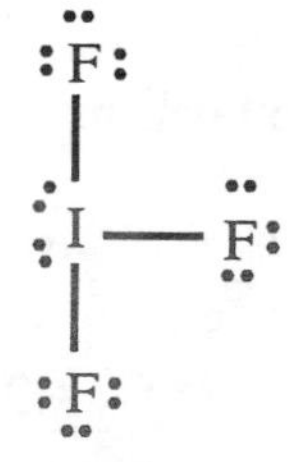

T-shaped

c. Cl_2O_7, 2(7) + 7(6) = 56 e^-

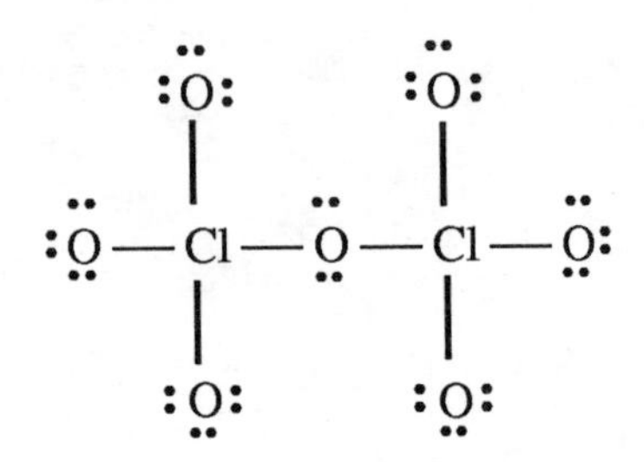

The four O atoms are tetrahedrally arranged about each Cl. The Cl—O—Cl bond angle is close to the tetrahedral bond angle.

d. $FBrO_2$, 7 + 7 + 2(6) = 26 e^-

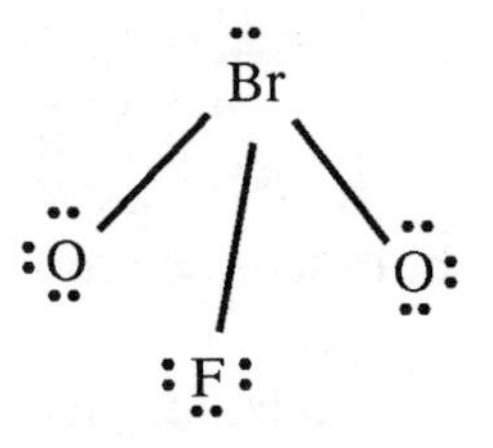

Trigonal pyramidal

30.

:F—O—O—F:

	F	O	O	F
Formal Charge	0	0	0	0
Oxid. Number	-1	+1	+1	-1

Oxidation numbers are more useful. We are forced to assign +1 as the oxidation number to oxygen. Oxygen is very electronegative and +1 is not a stable oxidation state for this element.

31.

$$BrO_3^- + XeF_2 \rightarrow BrO_4^- + Xe + HF$$

$$H_2O + BrO_3^- \rightarrow BrO_4^- + 2\,H^+ + 2\,e^-$$

$$2\,e^- + 2\,H^+ + XeF_2 \rightarrow Xe + 2\,HF$$

$$H_2O(l) + BrO_3^-(aq) + XeF_2(aq) \rightarrow BrO_4^-(aq) + Xe(g) + 2\,HF(aq)$$

32. a. $F_2 + H_2O \rightarrow HOF + HF$

$$2\,HOF \rightarrow 2\,HF + O_2$$

$$HOF + H_2O \rightarrow HF + H_2O_2 \quad \text{(acid)}$$

In dilute base HOF exists as OF^- and HF as F^-.

$$OF^- + H_2O \rightarrow F^- + O_2 \quad \text{(base)}$$

Balanced half reactions are:

$$(2\,e^- + H_2O + OF^- \rightarrow F^- + 2\,OH^-) \times 2$$

$$4\,OH^- \rightarrow O_2 + 2\,H_2O + 4\,e^-$$

Overall balanced reaction in dilute base:

$2\ OF^- \rightarrow O_2 + 2\ F^-$

b. HOF: Assign +1 to H, -1 to F. Oxidation number of oxygen is zero. Oxygen is very electronegative. A zero oxidation state would not be very stable, as it would be a powerful oxidizing agent.

33. a.

$2\ e^- + Cl_2 \rightarrow 2\ Cl^-$	$E° = +1.36\ V$
$2\ H_2O + Cl_2 \rightarrow 2\ OCl^- + 4\ H^+ + 2\ e^-$	$-E° = -1.63\ V$
$2\ H_2O + 2\ Cl_2 \rightarrow 2\ Cl^- + 2\ OCl^- + 4\ H^+$	$E° = -0.27\ V$

b. $E° < 0$, Reaction is not spontaneous.

c. Reaction is favored as solution becomes more basic.

34. A disproportionation reaction is an oxidation-reduction reaction in which one species will act as both an oxidizing and reducing agent. The species reacts with itself forming products of higher and lower oxidation states. For example $2\ Cu^+ \rightarrow Cu + Cu^{2+}$ is a disproportionation reaction.

$HClO_2$ will disproportionate:

$HClO_2 + 2\ H^+ + 2\ e^- \rightarrow HClO + H_2O$	$E° = 1.65\ V$
$HClO_2 + H_2O \rightarrow ClO_3^- + 3\ H^+ + 2\ e^-$	$-E° = -1.21\ V$
$2\ HClO_2 \rightarrow HClO + ClO_3^- + H^+$	$E°_{cell} = 0.44\ V$

Group 8A Elements: The Noble Gases

35. Xe has one more valence electron than I. Thus, the isoelectric species will have I and one extra electron substituted for Xe, giving a species with a net minus one charge.

a. IO_4^- b. IO_3^- c. IF_2^- d. IF_4^- e. IF_6^- f. IOF_3^-

36. XeO_3 has 8 + 3(6) = 26 valence electrons.

```
 ..   ..   ..
:O — Xe — O:
 ..   |    ..
      |
     :O:
      ..
```

trigonal pyramid

XeO_4 has 8 + 4(6) = 32 valence electrons.

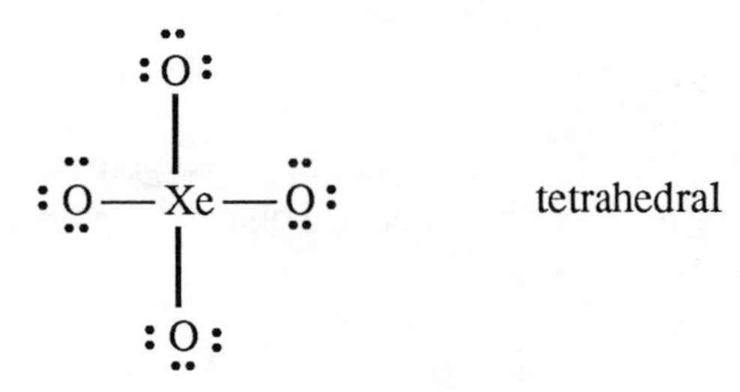

tetrahedral

$XeOF_4$ has 8 + 6 + 4(7) = 42 valence electrons.

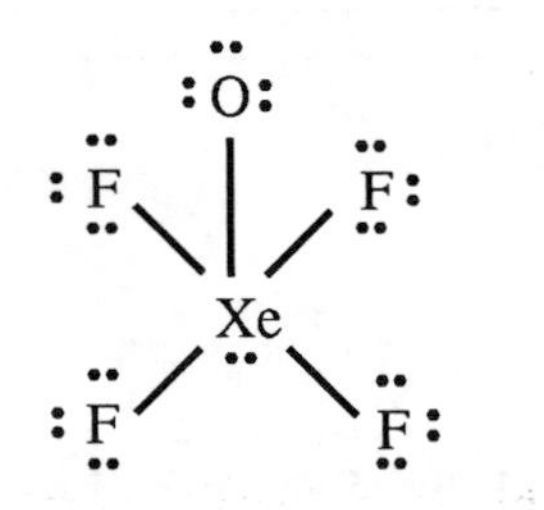

square pyramid

$XeOF_2$ has 8 + 6 + 2(7) = 28 valence electrons.

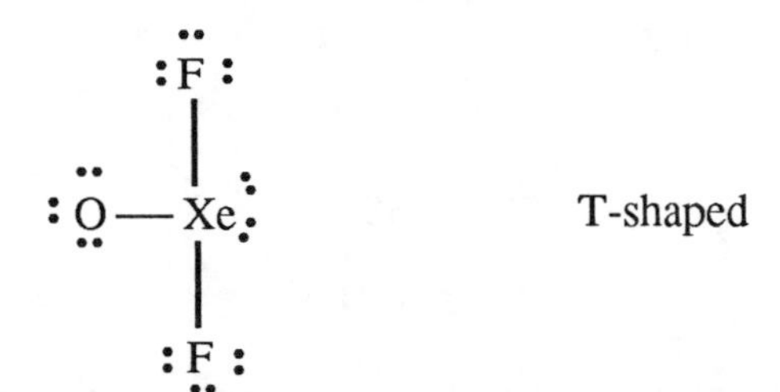

T-shaped

XeO_3F_2 has 8 + 3(6) + 2(7) = 40 valence electrons.

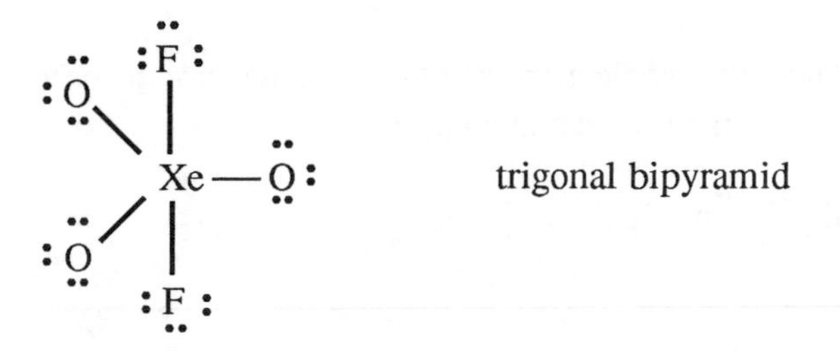

trigonal bipyramid

37. XeF_2 can react with oxygen to produce explosive xenon oxides and oxyfluorides.

38. $10.0\ \text{m} \times 5.0\ \text{m} \times 3.0\ \text{m} = 1.5 \times 10^2\ \text{m}^3$

$$1.5 \times 10^2\ \text{m}^3 \times \left(\frac{10\ \text{dm}}{\text{m}}\right)^3 \times \frac{1\ \text{L}}{\text{dm}^3} \times \frac{9 \times 10^{-6}\ \text{L Xe}}{100\ \text{L air}} = 1 \times 10^{-2}\ \text{L of Xe in the room.}$$

$$PV = nRT \qquad n = \frac{PV}{RT} = \frac{(1\ \text{atm})\,(1 \times 10^{-2}\ \text{L})}{\left(\frac{0.08206\ \text{L atm}}{\text{mol K}}\right)(298\ \text{K})} = 4 \times 10^{-4}\ \text{mol Xe}$$

$$4 \times 10^{-4} \text{ mol Xe} \times \frac{131.3 \text{ g}}{\text{mol}} = 5 \times 10^{-2} \text{ g Xe}$$

$$4 \times 10^{-4} \text{ mol Xe} \times \frac{6.022 \times 10^{23} \text{ atom}}{\text{mol}} = 2 \times 10^{20} \text{ atoms Xe}$$

A 2 L breath contains:

$$2 \text{ L air} \times \frac{9 \times 10^{-6} \text{ L Xe}}{100 \text{ L air}} = 2 \times 10^{-7} \text{ L Xe}$$

$$n = \frac{PV}{RT} = \frac{(1 \text{ atm})(2 \times 10^{-7} \text{ L})}{\left(\frac{0.08206 \text{ L atm}}{\text{mol K}}\right)(298 \text{ K})} = 8 \times 10^{-9} \text{ mol Xe}$$

$$8 \times 10^{-9} \text{ mol Xe} \times \frac{6.022 \times 10^{23} \text{ atom}}{\text{mol}} = 5 \times 10^{15} \text{ atoms of Xe}$$

ADDITIONAL EXERCISES

39. As the halogen atoms get larger, it becomes more difficult to fit three halogen atoms around the small N, and the NX_3 molecule becomes less stable.

40.

41. $3 \text{ NO(g)} \rightleftharpoons N_2O\text{(g)} + NO_2\text{(g)}$

$\Delta H° = 82 + 34 - 3(90) = -154$ kJ

$\Delta S° = 220 + 240 - 3(211) = -173$ J/K

$\Delta G° = \Delta H° - T\Delta S° = -154 - 298(-0.173) = -102$ kJ

$$\Delta G° = 0 \text{ when } T = \frac{\Delta H°}{\Delta S°} = \frac{-154{,}000 \text{ J}}{-173 \text{ J/K}} = 890. \text{ K}$$

The reaction is spontaneous at temperatures below 890. K (all partial pressures = 1 atm).

42. In solid state: NO_2^+ and NO_3^-. In gas phase: molecular N_2O_5

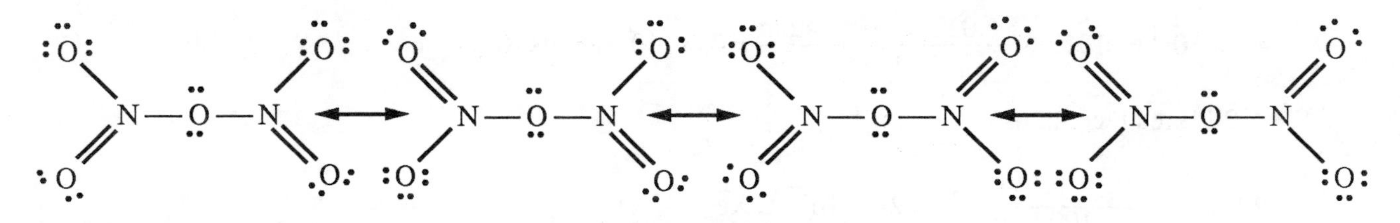

43. OCN^- has 6 + 4 + 5 + 1 = 16 valence electrons.

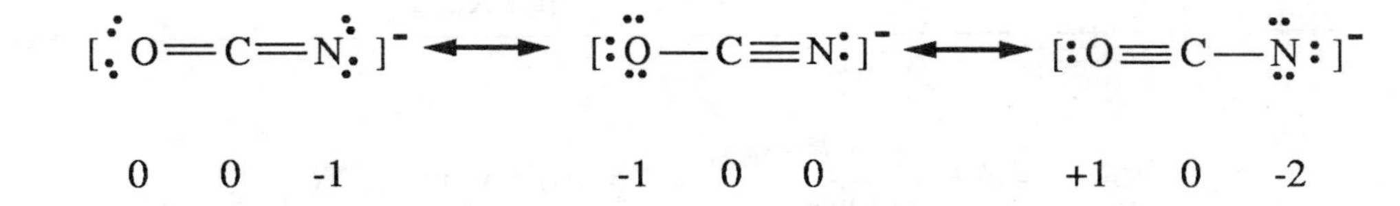

Formal Charge	0 0 -1	-1 0 0	+1 0 -2

Only the first two resonance structures should be important. The third places a positive formal charge on the most electronegative atom in the ion and a -2 charge on N.

CNO^-

$[:C{=}N{=}O:]^- \longleftrightarrow [:C{\equiv}N{-}O:]^- \longleftrightarrow [:C{-}N{\equiv}O:]^-$

Formal Charge	-2 +1 0	-1 +1 -1	-3 +1 +1

All of the resonance structures for fulminate involve greater formal charges than in cyanate, making fulminate more reactive (less stable).

44. $AsCl_4^+$, 5 + 4(7) - 1 = 32 e^- $AsCl_6^-$, 5 + 6(7) + 1 = 48 e^-

$[AsCl_4]^+$: Cl, Cl—As—Cl, Cl

$[AsCl_6]^-$: Cl, Cl, Cl, As, Cl, Cl, Cl

The reaction is a Lewis acid-base reaction. Chloride ion acts as a Lewis base as it is transferred from one $AsCl_5$ to another. As is the Lewis acid.

45.

		Bond order	# unpaired e^-
M.O.	NO	2.5	1
	NO^+	3	0
	NO^-	2	2

Lewis NO^+ $[:N\equiv O:]^+$

NO $\dot{N}=O$ ⟷ $N=O$ ⟷ $N=O$

NO^- $[:N=O:]^-$

Lewis structures are not adequate for NO and NO^-. M.O. model gives correct results for all three species. For NO, Lewis structures fail for odd electron species. For NO^-, Lewis structures fail to predict that NO^- is paramagnetic.

46. $Mg^{2+} + P_3O_{10}^{5-} \rightleftharpoons MgP_3O_{10}^{3-}$

$$\text{Initial } [Mg^{2+}] = \frac{50 \times 10^{-3}\ g}{L} \times \frac{1\ mol}{24.31\ g} = 2 \times 10^{-3}\ M$$

$$\text{Initial } [P_3O_{10}^{5-}] = \frac{40\ g\ Na_5P_3O_{10}}{L} \times \frac{1\ mol}{367.9\ g} = 0.1\ M$$

Assume reaction goes to completion, since K is large ($10^{8.6}$).

	Mg^{2+}	+	$P_3O_{10}^{5-}$	$\rightleftharpoons$	$MgP_3O_{10}^{3-}$	
Before	$2 \times 10^{-3}\ M$		$0.1\ M$		0	
Change	-2×10^{-3}		-2×10^{-3}	→	$+2 \times 10^{-3}$	React completely
After	0		0.1		2×10^{-3}	New initial
	x mol/L $MgP_3O_{10}^{3-}$ dissociates to reach equilibrium					
Change	$+x$		$+x$	←	$-x$	
Equil	x		$0.1 + x$		$2 \times 10^{-3} - x$	

$$K = 10^{8.6} = 4 \times 10^8 = \frac{[MgP_3O_{10}^{3-}]}{[Mg^{2+}][P_3O_{10}^{5-}]} = \frac{(2 \times 10^{-3} - x)}{(x)(0.1 + x)}$$

$$4 \times 10^8 \approx \frac{2 \times 10^{-3}}{(x)(0.1)},\ x = [Mg^{2+}] = 5 \times 10^{-11}\ M$$ Assumptions good.

47. a.

$$Mn^{2+} + NaBiO_3 \rightarrow MnO_4^- + BiO_3^{3-}$$

$$(4\ H_2O + Mn^{2+} \rightarrow MnO_4^- + 8\ H^+ + 5\ e^-) \times 2$$

$$(2\ e^- + NaBiO_3 \rightarrow BiO_3^{3-} + Na^+) \times 5$$

$$8\ H_2O(l) + 2\ Mn^{2+}(aq) + 5\ NaBiO_3(s) \rightarrow 2\ MnO_4^-(aq) + 16\ H^+(aq) + 5\ BiO_3^{3-}(aq) + 5\ Na^+(aq)$$

b. Bismuthate exists as a covalent network solid : $(BiO_3^-)_x$.

48. Hypochlorite can act as an oxidizing agent. For example it is capable of oxidizing I^- to I_2. If a solution containing I^- turns brown when BiOCl is added, then BiOCl is bismuth(I)hypochlorite. If not, then it is bismuthylchloride.

49. $$2.42 \text{ eV} \times \frac{96.5 \text{ kJ/mol}}{\text{eV}} \times \frac{1 \text{ mol photons}}{6.022 \times 10^{23} \text{ photons}} \times \frac{1000 \text{ J}}{\text{kJ}} = \frac{3.88 \times 10^{-19} \text{ J}}{\text{photon}}$$

$$E = \frac{hc}{\lambda}$$

$$\lambda = \frac{hc}{E} = \frac{(6.626 \times 10^{-34} \text{ J s})(2.998 \times 10^{8} \text{ m/s})}{3.88 \times 10^{-19} \text{ J}} = 5.12 \times 10^{-7} \text{ m} = 512 \text{ nm}, \quad \text{Green light}$$

50. TeF_5^- $6 + 5(7) + 1 = 42\ e^-$

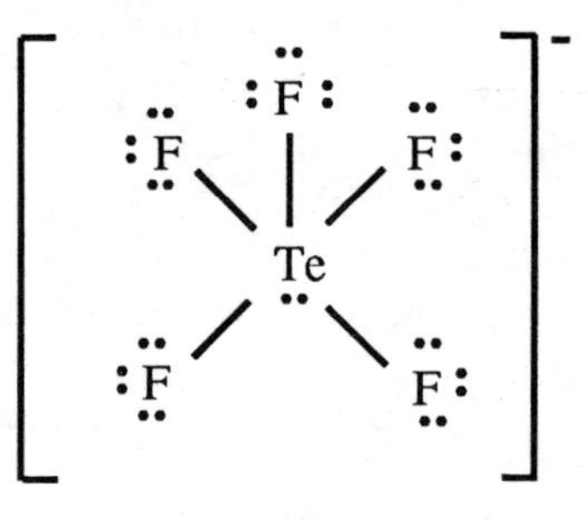

The lone pair of electrons exerts a stronger repulsion than the bonding pairs, pushing the four equatorial F's away from the lone pair.

51. a.

$$ClO_3^- + 2\,H_2O \rightarrow ClO_4^- + 2\,H^+ + 2\,e^- \qquad -E° = -1.19 \text{ V}$$
$$2\,H^+ + 2\,e^- \rightarrow H_2 \qquad E° = 0.0 \text{ V}$$

$$ClO_3^- + H_2O \rightarrow ClO_4^- + H_2 \qquad E°_{cell} = -1.19 \text{ V}$$

Therefore, a potential of 1.19 V must be applied (assuming standard conditions).

b. $3\,Al + 3\,NH_4ClO_4 \rightarrow Al_2O_3 + AlCl_3 + 3\,NO + 6\,H_2O(g)$

$\Delta H° = 3(90) + (-704) + (-1676) + 6\,(-242) - 3(-295)$

$\Delta H° = -2677$ kJ

$$7 \times 10^5 \text{ kg} \times \frac{1000 \text{ g}}{\text{kg}} \times \frac{1 \text{ mol}}{117.49 \text{ g}} \times \frac{2677 \text{ kJ}}{3 \text{ mol } NH_4ClO_4} = 5 \times 10^9 \text{ kJ}$$

CHALLENGE PROBLEMS

52. For

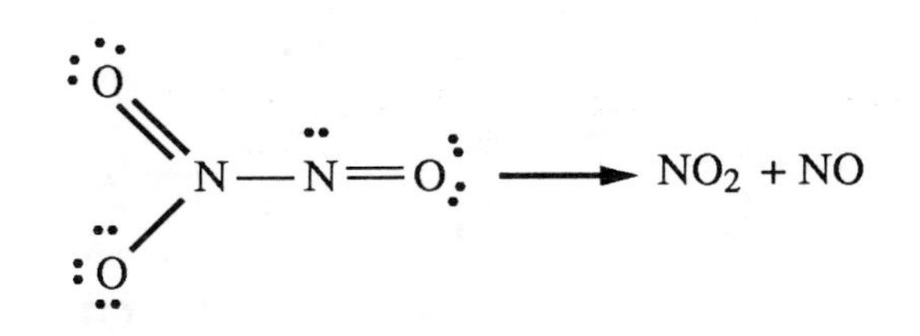

the activation energy must in some way involve the breaking of a nitrogen-nitrogen single bond. For the reaction

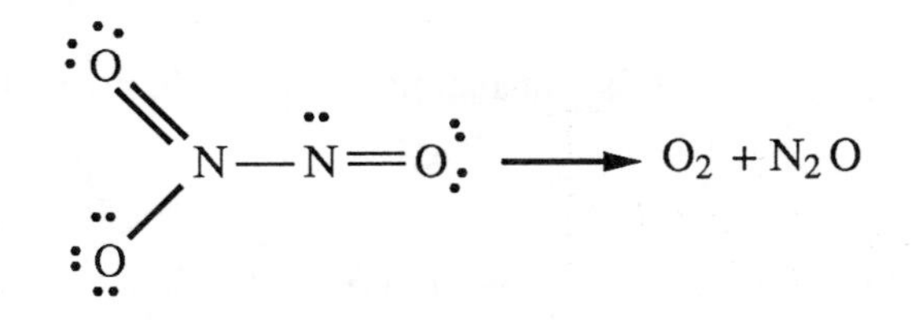

at some point nitrogen-oxygen bonds must be broken. N — N single bonds (160 kJ/mol) are weaker than N — O single bonds (201 kJ/mol). In addition, resonance structures indicate that there is more double bond character to the N — O bonds than the N — N bond. Thus, NO_2 and NO are preferred by kinetics because of the lower activation energy.

53. For $NCl_3 \rightarrow NCl_2 + Cl$, only the N — Cl bond is being broken. For O = N — Cl $\rightarrow$ NO + Cl, when the N — Cl bond is broken, the NO bond gets stronger (bond order increases from 2.0 to 2.5). This makes ΔH for the reaction smaller than just the energy necessary to break the N — Cl bond.

54. a. SbF_5 HSO_3F $H_2SO_3F^+$

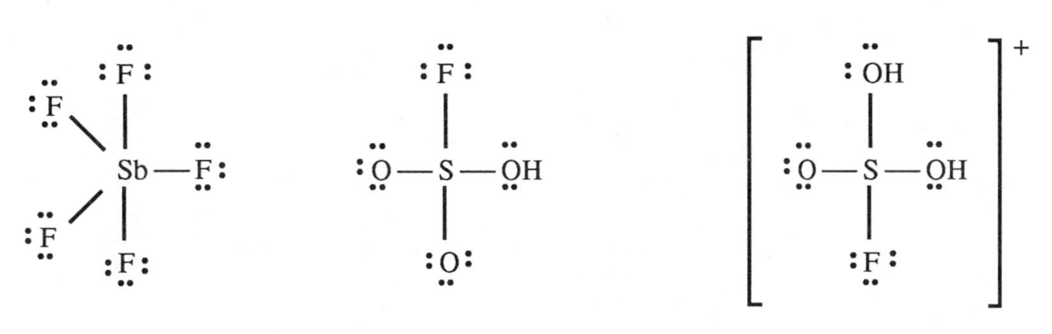

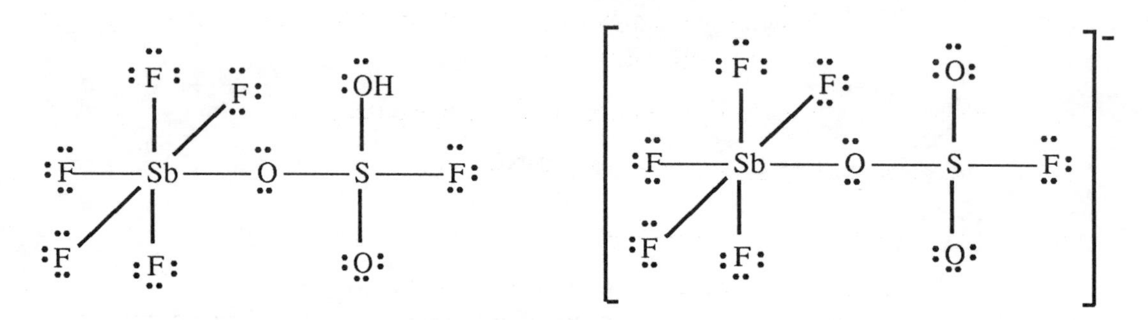

b. The active protonating species is $H_2SO_3F^+$.

55. a. As we go down the family, K_a increases. This is consistent with the bond to hydrogen getting weaker.

b. Po is below Te, so K_a should be larger. The K_a for H_2Po should be on the order of 10^{-2} or 10^{-1}.

56. a. $AgCl(s) \xrightarrow{h\nu} Ag(s) + Cl$

The reactive chlorine atom is trapped in the crystal. When the light is removed it reacts with silver atoms to reform AgCl, i.e., the reverse reaction occurs. In pure silver chloride, the Cl atoms escape, making the reverse reaction impossible.

b. Overtime chlorine is lost and the dark silver metal is permanent.

CHAPTER TWENTY: TRANSITION METALS AND COORDINATION CHEMISTRY

QUESTIONS

1. a. ligand: Species that donates a pair of electrons to form a covalent bond to a metal ion (a Lewis base).

   b. chelate: Ligand that can form more than one bond.

   c. bidentate: Ligand that can form two bonds.

   d. complex ion: Metal ion plus ligands.

2. A ligand must have at least one unshared pair of electrons.

3. Both electrons in the bond originally came from the same atom.

4. Since transition metals form bonds to species with lone pairs of electrons they are Lewis acids (electron pair acceptors).

5. a. isomers: Species with the same formulas but different properties. See text for examples of the following types of isomers.

   b. structural isomers: Isomers that have one or more bonds that are different.

   c. stereoisomers: Isomers that contain the same bonds but differ in how the atoms are arranged in space.

   d. coordination isomers: Isomers that differ in what atoms are found in the coordination sphere.

   e. linkage isomers: Isomers that differ in how one or more ligands are attached to the transition metal.

   f. geometric isomers: (cis-trans isomerism) Isomers that differ in the position of atoms with respect to a rigid ring, bond, or each other.

   g. optical isomerism: Isomers that differ by being non-superimposable mirror images of each other; that is, they are different in the same way our left and right hands differ.

6. a. Ligand that will give complex ions with the maximum number of unpaired electrons.

   b. Ligand that will give complex ions with the minimum number of unpaired electrons.

   c. Complex with minimum number of unpaired electrons.

   d. Complex with maximum number of unpaired electrons.

7. Cu^{2+}: $[Ar]3d^9$, Cu^+: $[Ar]3d^{10}$

   Cu(II) is d^9; Cu(I) is d^{10}. Color is a result of the electron transfer between split d orbitals. This cannot occur for the filled d orbitals in Cu(I).

8. No, the d-orbitals are completely filled (see question 7 above).

9. Sc^{3+} has no unpaired electrons. V^{3+} and Ti^{3+} have unpaired d-electrons present. Color of transition metal compounds results from the presence of unpaired d electrons.

10. CN^- and CO form much stronger complexes with Fe(II) than O_2. Thus, O_2 will not be transported by hemoglobin in the presence of CN^- and CO.

11. $Fe_2O_3(s) + 6\ H_2C_2O_4(aq) \longrightarrow 2\ Fe(C_2O_4)_3^{3-}(aq) + 3\ H_2O(l) + 6\ H^+(aq)$
The oxalate anion forms a soluble complex ion with the iron in rust.

12. There are a large number of valence electrons that can be removed.

13. There is a steady decrease in the atomic radii of the lanthanide elements, going from left to right. As a result of the lanthanide contraction the properties of the 4d and 5d elements in each group are very similar because of the similar size of atoms and ions (See Exercise 7.123).

14. See text for examples.

 a. Roasting: converting sulfide minerals to oxides by heating in air below their melting points.

 b. Smelting: reducing metal ions to the free metal.

 c. Flotation: separation of mineral particles in an ore from the gangue that depends on the greater wetability of the mineral particles.

 d. Leaching: the extraction of metals from ores using aqueous solutions.

 e. Gangue: the impurities (such as clay or sand) in an ore.

15. Advantages: cheap energy cost

 less pollution

 Disadvantages: chemicals used in hydrometallurgy are expensive.

16. Impurities are more soluble in the molten metal than in the solid metal. As the molten zone moves down a metal the impurities are swept along with the liquid.

EXERCISES

Transition Metals

17. a. Ni: $[Ar]4s^2 3d^8$

 b. Cd: $[Kr]5s^2 4d^{10}$

 c. Zr: $[Kr]5s^2 4d^2$

 d. Nd: $[Xe]6s^2 5d^1 4f^3$ or $[Xe]6s^2 4f^4$

18. a. Mo: $[Kr]5s^2 4d^4$ or $[Kr]5s^1 4d^5$

 b. Ru: $[Kr]5s^2 4d^6$

 c. Ir: $[Xe]6s^2 4f^{14} 5d^7$

 d. Mn: $[Ar]4s^2 3d^5$

19. a. Ni^{2+}: $[Ar]3d^8$

 b. Cd^{2+}: $[Kr]4d^{10}$

 c. Zr^{3+}: $[Kr]4d^1$; Zr^{4+}: [Kr]

 d. Nd^{3+}: $[Xe]4f^3$

20. a. Mo^{4+}: $[Kr]4d^2$

c. Ir^+: $[Xe]6s^14f^{14}5d^7$ or $[Xe]4f^{14}5d^8$
Ir^{3+}: $[Xe]4f^{14}5d^6$

b. Ru^{2+}: $[Kr]4d^6$; Ru^{3+}: $[Kr]4d^5$

d. Mn^{2+}: $[Ar]3d^5$

21. a. Co: $[Ar]4s^23d^7$
Co^{2+}: $[Ar]3d^7$
Co^{3+}: $[Ar]3d^6$

b. Pt: $[Xe]6s^14f^{14}5d^9$
Pt^{2+}: $[Xe]4f^{14}5d^8$
Pt^{4+}: $[Xe]4f^{14}5d^6$

c. Fe: $[Ar]4s^23d^6$
Fe^{2+}: $[Ar]3d^6$
Fe^{3+}: $[Ar]3d^5$

22. a. Au: $[Xe]6s^14f^{14}5d^{10}$
Au^+: $[Xe]4f^{14}5d^{10}$
Au^{3+}: $[Xe]4f^{14}5d^8$

b. Cu: $[Ar]4s^13d^{10}$
Cu^+: $[Ar]3d^{10}$
Cu^{2+}: $[Ar]3d^9$

c. V: $[Ar]4s^23d^3$
V^{2+}: $[Ar]3d^3$
V^{3+}: $[Ar]3d^2$

23. $1.00 \times 10^6 \text{ g FeTiO}_3 \times \dfrac{47.88 \text{ g Ti}}{151.7 \text{ g FeTiO}_3} = 3.16 \times 10^5$ g or 316 kg

24.

$$Fe^{2+} \rightarrow Fe^{3+} + e^- \qquad \Delta H = 2.957 \times 10^6 \text{ J}$$
$$e^- + Ti^{4+} \rightarrow Ti^{3+} \qquad \Delta H = -4.175 \times 10^6 \text{ J}$$

$$Fe^{2+} + Ti^{4+} \rightarrow Fe^{3+} + Ti^{3+} \qquad \Delta H = -1.218 \times 10^6 \text{ J}$$

From this, we would predict that Fe^{3+} and Ti^{3+} is more stable than Fe^{2+} and Ti^{4+}.

$$Fe^{3+} + e^- \rightarrow Fe^{2+} \qquad E° = 0.77 \text{ V}$$
$$Ti^{3+} + H_2O \rightarrow TiO^{2+} + 2\,H^+ + e^- \qquad -E° = -0.099 \text{ V}$$

$$Fe^{3+} + Ti^{3+} + H_2O \rightarrow Fe^{2+} + TiO^{2+} + 2\,H^+ \qquad E°_{cell} = 0.67 \text{ V}$$

From E° values, we would predict Fe(II) and Ti(IV) to be more stable. The electrochemical data are consistent with the information in the text. These data are for solutions while ionization energies are for gas phase reactions. Solution data are more representative of the conditions from which ilmenite was initially formed.

25. a. Molybdenum(IV) sulfide and Molybdenum(VI) oxide

b. MoS_2 +4 MoO_3 +6

$(NH_4)_2Mo_2O_7$ +6 $(NH_4)_6Mo_7O_{24}\cdot 4\,H_2O$ +6

c. $2\,MoS_2 + 7\,O_2 \rightarrow 2\,MoO_3 + 4\,SO_2$

$2\,NH_3 + 2\,MoO_3 + H_2O \rightarrow (NH_4)_2Mo_2O_7$

$6\,NH_3 + 7\,MoO_3 + 7\,H_2O \rightarrow (NH_4)_6Mo_7O_{24}\cdot 4\,H_2O$

26. pyrolusite, manganese(IV) oxide

rhodochrosite, manganese(II) carbonate

Coordination Compounds

27. a. pentaamminechlororuthenium(III) ion

b. hexacyanoferrate(II) ion

c. tris(ethylenediamine) manganese(II) ion

d. pentaamminenitrocobalt(III) ion

28. a. tetrachloroferrate(III) ion

b. hexaamminenickel(II) ion

c. tetrathiocyanatocobaltate(II) ion

d. hexaaquatitanium(III) ion

29. a. hexaamminecobalt(II) chloride

b. hexaaquacobalt(III) iodide

c. potassium tetrachloroplatinate(II)

d. potassium hexachloroplatinate(II)

30. a. sodium tris(oxalato)nickelate(II)

b. potassium tetrachlorocobaltate(II)

c. tetraamminecopper(II) sulfate

d. chlorobis(ethylenediamine) thiocyanatocobalt(III) chloride

31. a. $[Co(C_5H_5N)_6]Cl_3$ b. $[Cr(NH_3)_5I]I_2$

c. $[Ni(NH_2CH_2CH_2NH_2)_3]Br_2$ d. $K_2[Ni(CN)_4]$ e. $[Pt(NH_3)_4Cl_2]PtCl_4$

32. a. $FeCl_4^-$ b. $[Ru(NH_3)_5H_2O]^{3+}$

c. $[Pt(C_5H_5N)_5I]^{3+}$ d. $[Pt(NH_3)Cl_3]^-$

33. a.

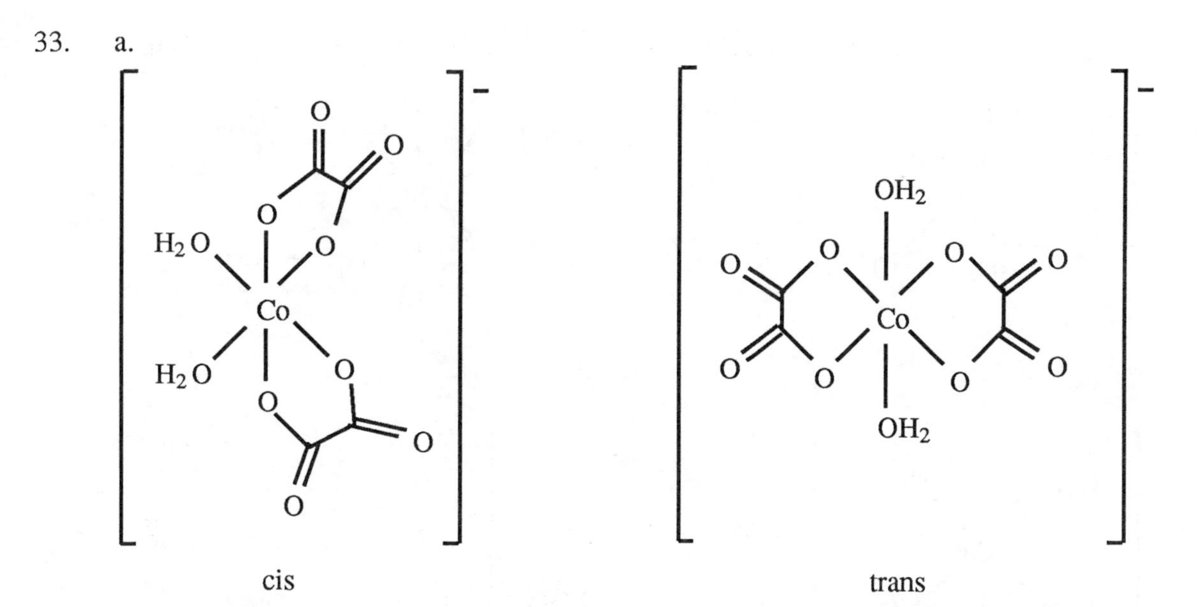

cis trans

b.

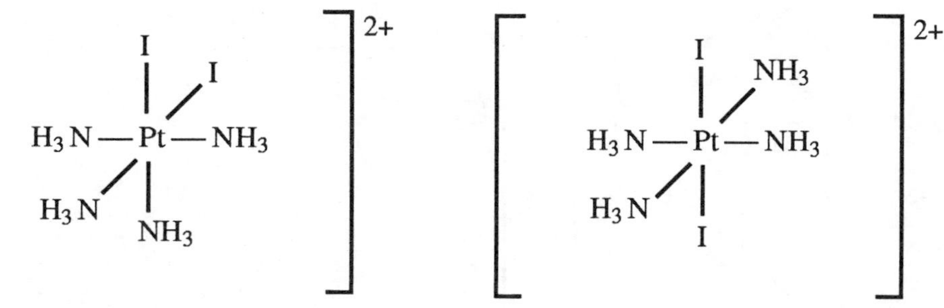
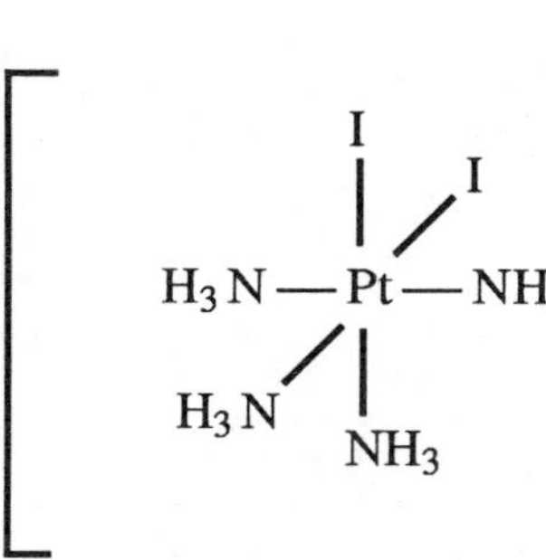
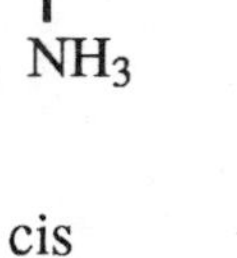

cis

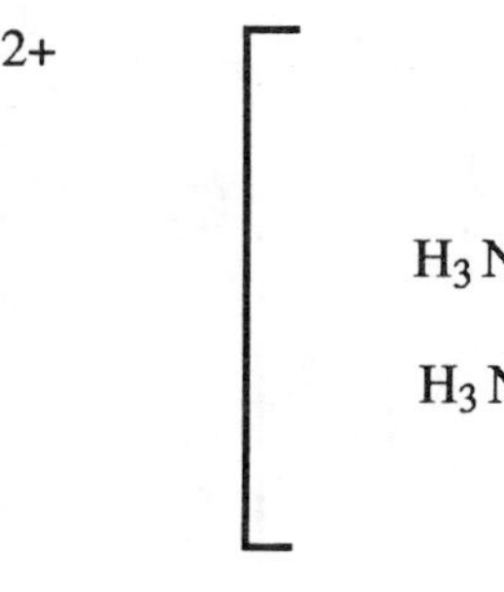

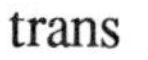
trans

c.

Cl
Cl
H_3N — Ir — Cl
H_3N
NH_3

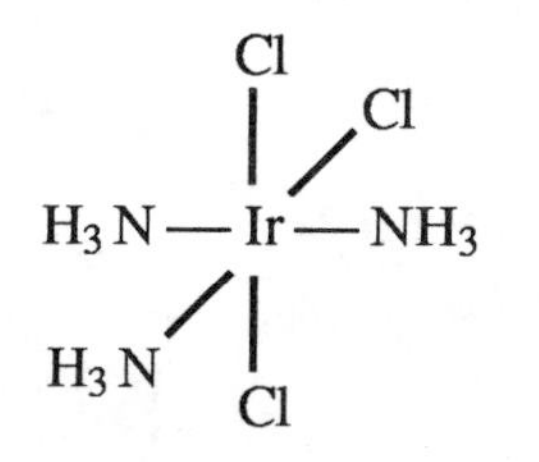

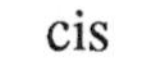
cis

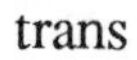
trans

d.

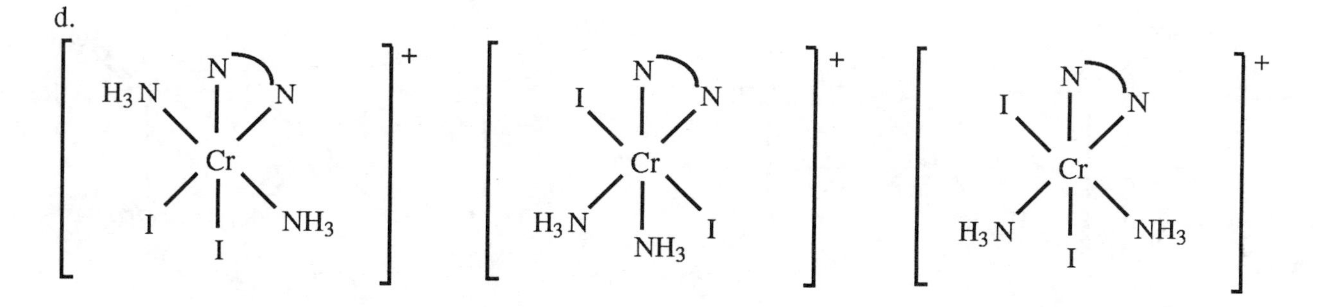

34. a.

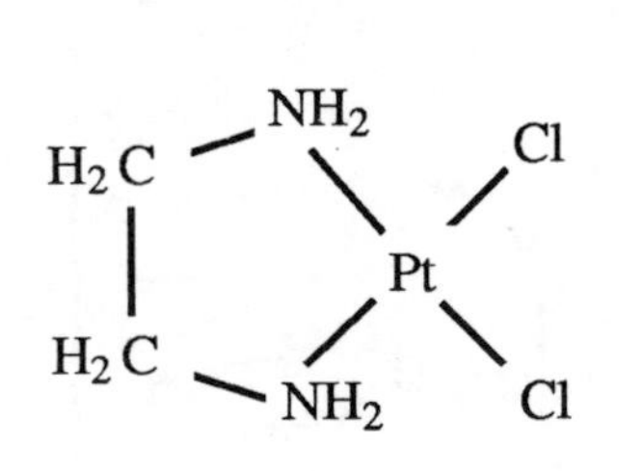

b.

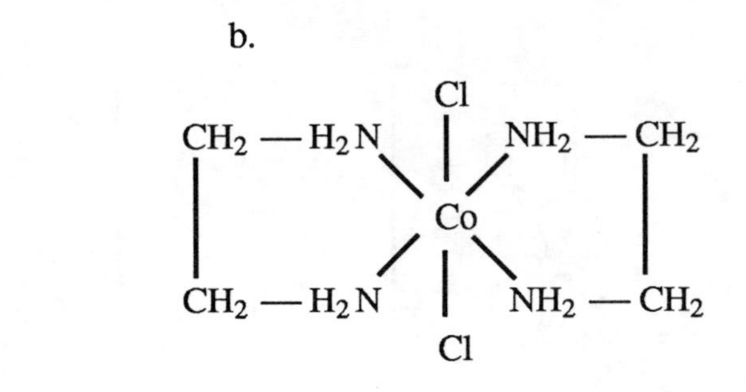

c.

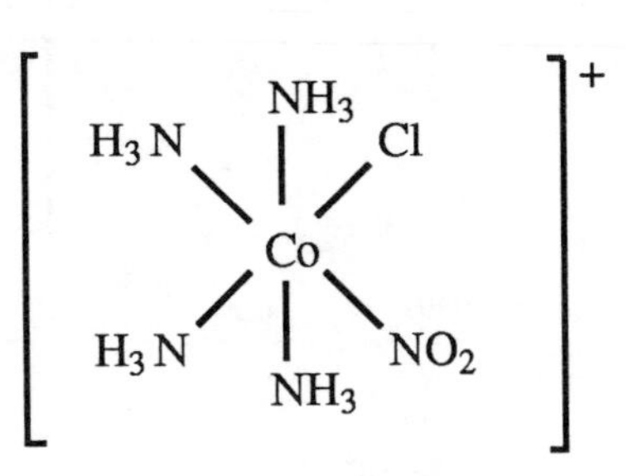

d.

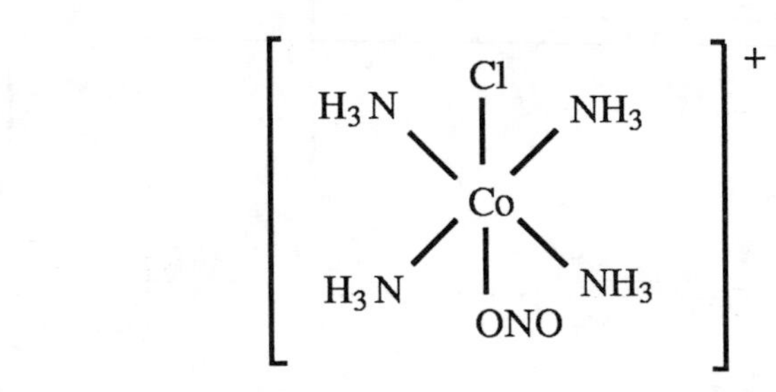

e.

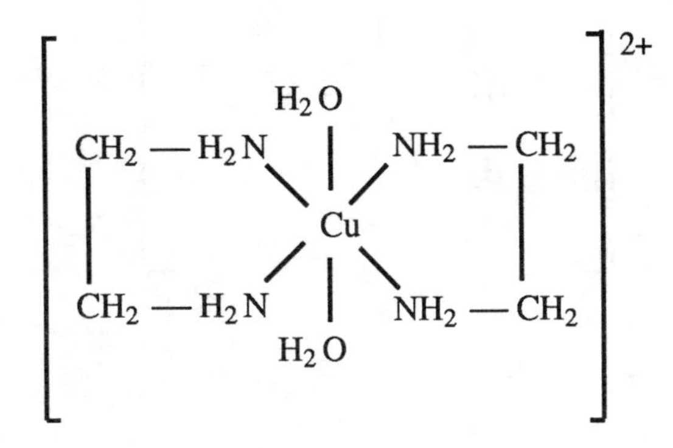

35.

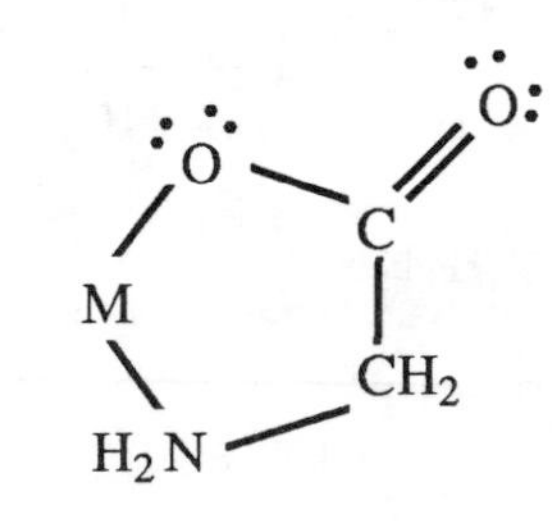

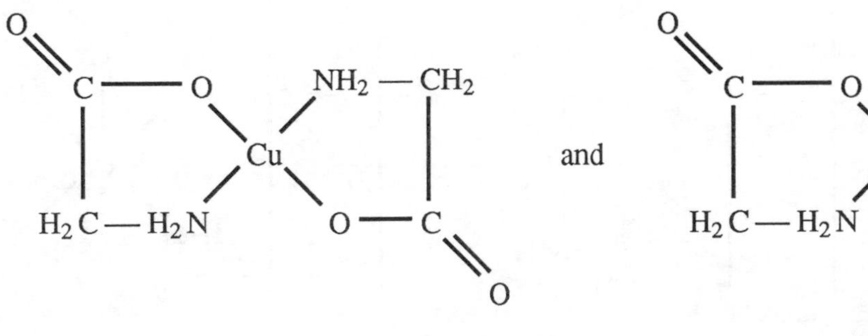

and

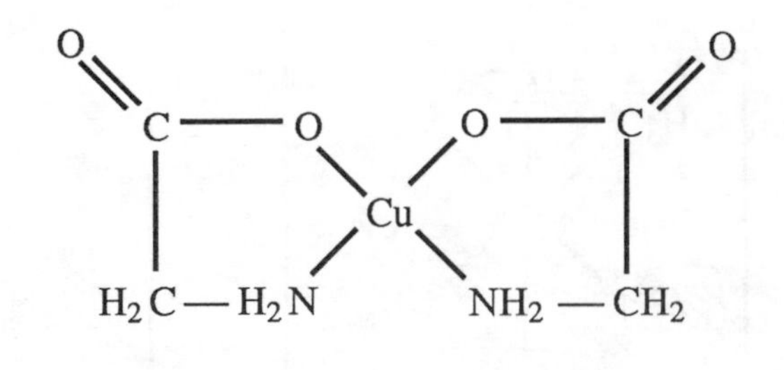

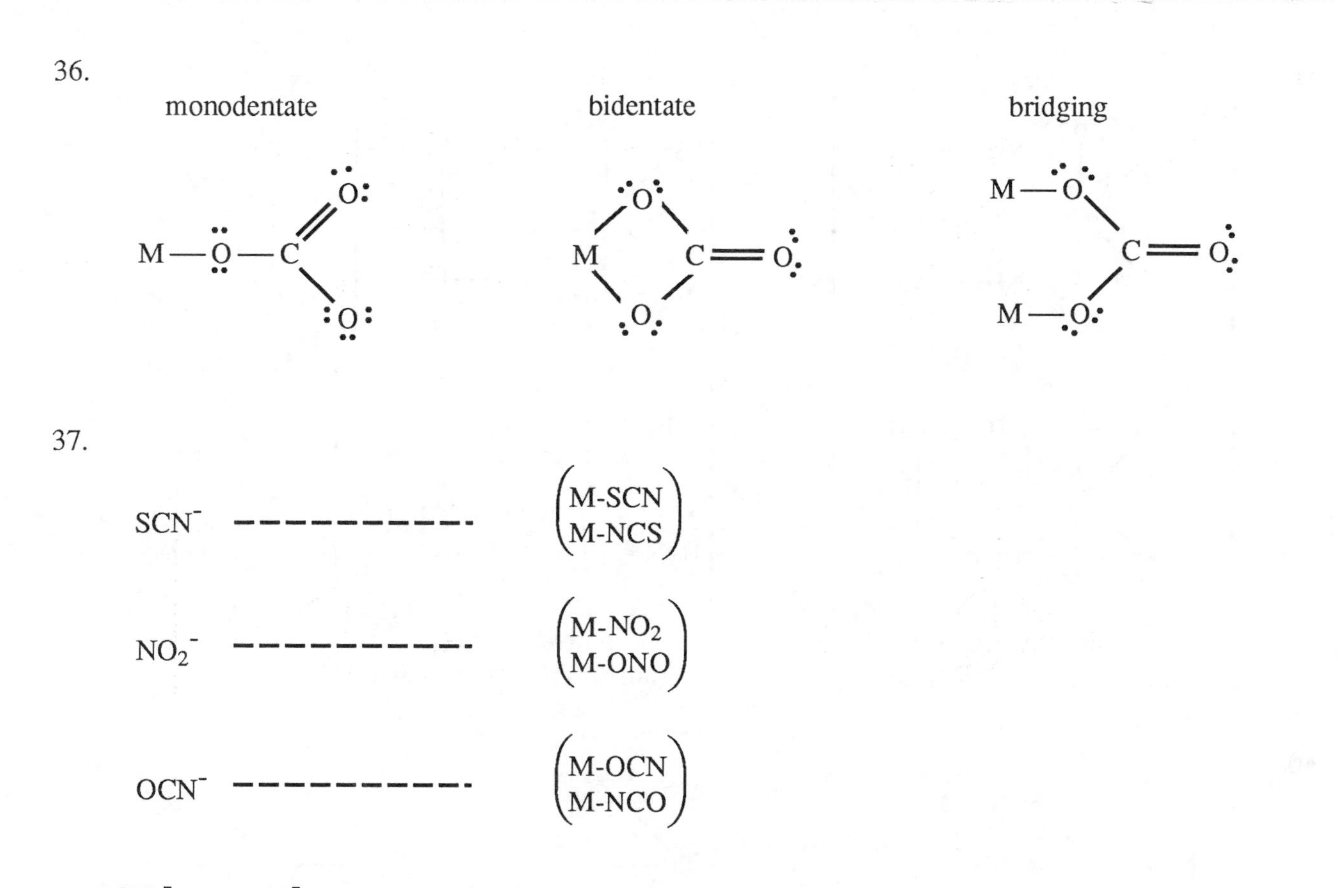

N_3^-, en, and I^- are not capable of linkage isomerism.

38.

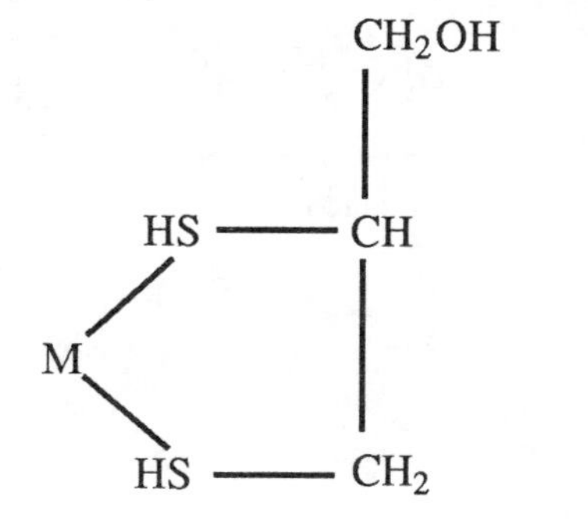

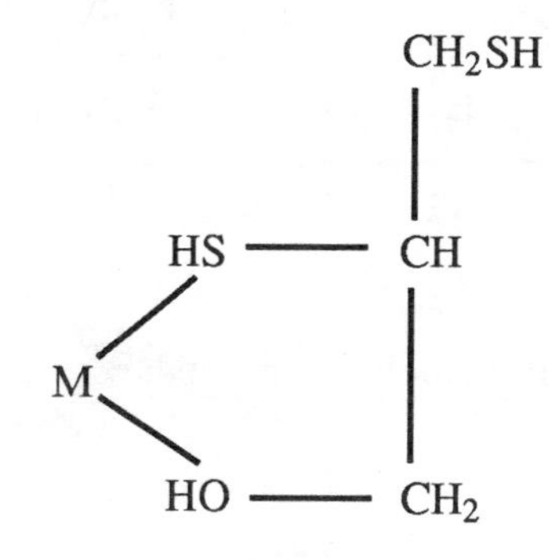

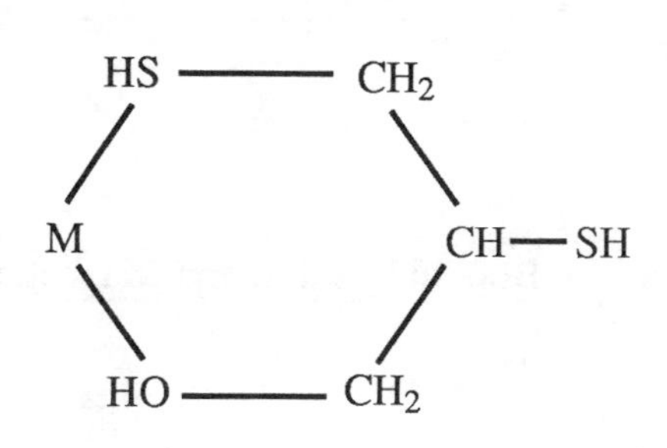

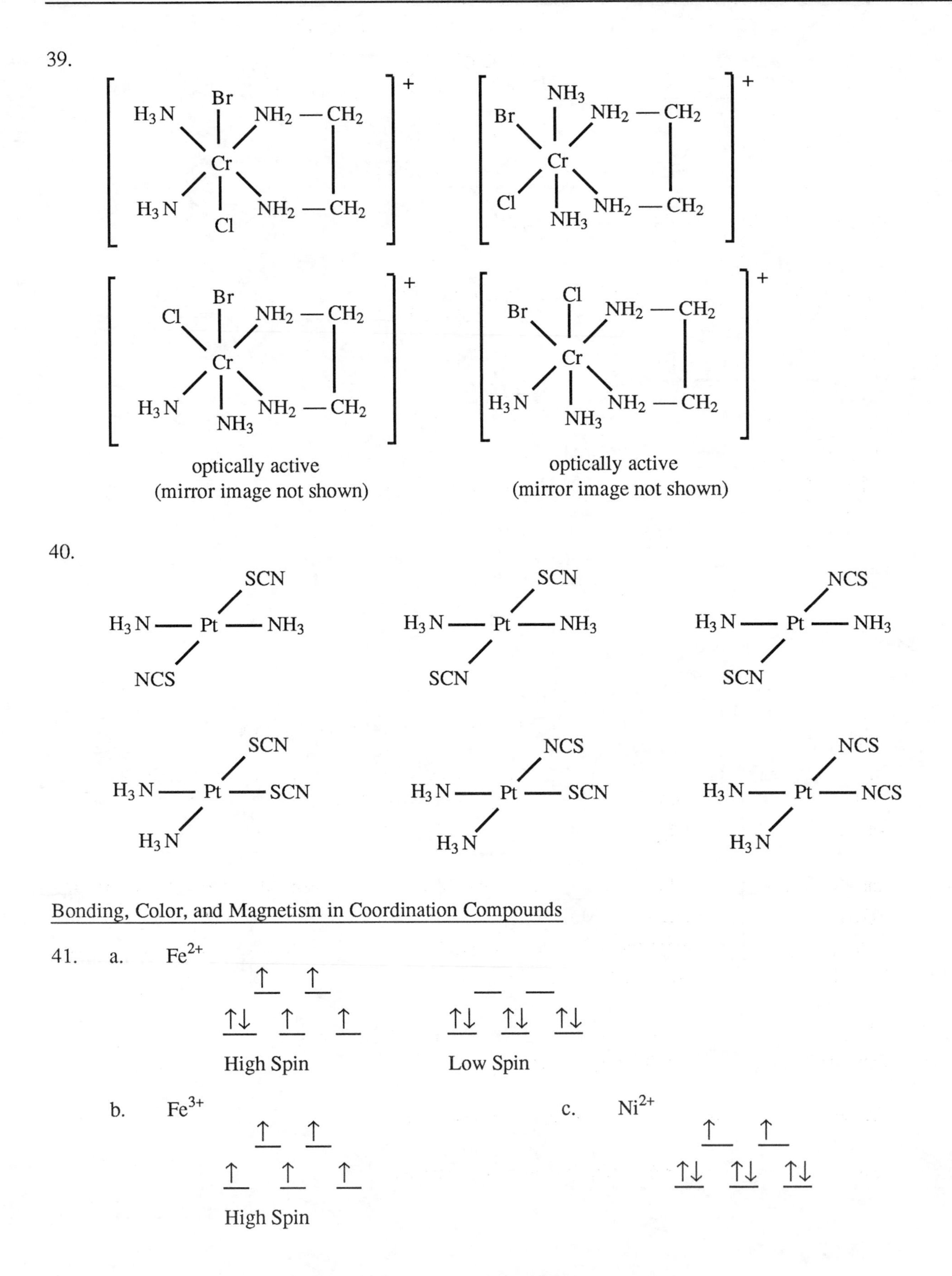
39.
optically active
(mirror image not shown)
optically active
(mirror image not shown)
40.
Bonding, Color, and Magnetism in Coordination Compounds
41. a. Fe2+
High Spin
Low Spin
b. Fe3+
High Spin
c. Ni2+

d. Zn^{2+}

↑↓ ↑↓

↑↓ ↑↓ ↑↓

e. Co^{2+}

High Spin	Low Spin
↑ ↑	↑ __
↑↓ ↑↓ ↑	↑↓ ↑↓ ↑↓

42. a. Ru^{2+} $[Kr]4d^6$, no unpaired e^- b. Fe^{3+} $[Ar]3d^5$, 1 unpaired e^-

c. Ni^{2+} $[Ar]3d^8$, 2 unpaired e^- d. V^{3+} $[Ar]3d^2$, 2 unpaired e^-

e. Co^{2+} $[Ar]3d^7$, 3 unpaired e^- (High spin because of weak field ligand and +2 charge.)

43. $NiCl_4^{2-}$ is tetrahedral and $Ni(CN)_4^{2-}$ is square planar.

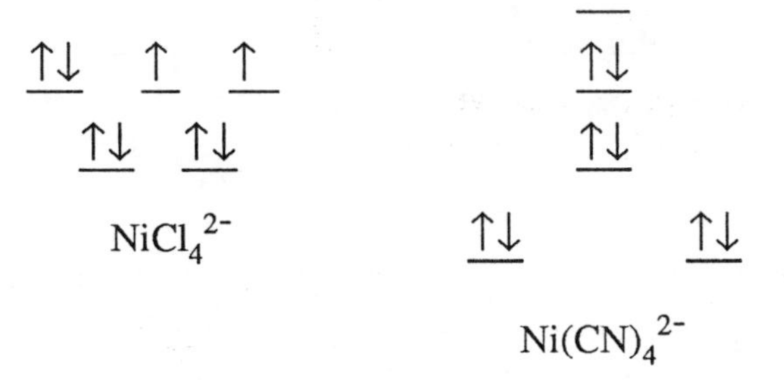

44. Co^{2+}: $[Ar]3d^7$

↑ ↑ ↑ t_{2g}

↑↓ ↑↓ e_g small Δ, Tetrahedral complexes are always high spin.

Ions with 2 or 7 d-electrons should form the most stable tetrahedral complexes, since they have the greatest number of electrons in the lower energy e_g orbitals as compared to the number of electrons in the higher energy t_{2g} orbitals.

45. Transition compounds exhibit the color complementary to that absorbed. Using Table 20.16, $Ni(H_2O)_6Cl_2$ absorbs red light and $Ni(NH_3)_6Cl_2$ absorbs yellow-green light. $Ni(NH_3)_6Cl_2$ absorbs the higher energy light, therefore, Δ is larger for $Ni(NH_3)_6Cl_2$. NH_3 is a stronger field ligand than H_2O, consistent with the spectrochemical series.

46. Fe^{3+} complexes have one unpaired electron when a strong field (low spin) case and five unpaired electrons when a weak field (high spin) case.

$Fe(CN)_6^{3-}$: Strong field; $Fe(SCN)_6^{3-}$: Weak field

Cyanide, CN^-, is a stronger field ligand than thiocyanate, SCN^-.

Metallurgy

47. a. $3\ Fe_2O_3 + CO \rightarrow 2\ Fe_3O_4 + CO_2$

$\Delta H° = -393.5 + 2(-1117) - [3(-826) + (-110.5)]$

$\Delta H° = -39$ kJ

$\Delta S° = 214 + 2(146) - [3(90) + 198]$

$\Delta S° = 38$ J/K

b. At 800°C, T = 800 + 273 = 1073 K (carry extra significant figures in T)

$\Delta G° = -39$ kJ - 1073 K(0.038 kJ/K)

= -39 kJ - 41 kJ = -80 kJ

48. The reaction $3\ Fe + C \rightleftharpoons Fe_3C$ is endothermic. At high temperatures Fe_3C forms. If the steel is cooled slowly, there is time for the equilibrium to shift back to the left producing a ductile steel. If cooling is rapid, there is not enough time for the equilibrium to shift back to the left, Fe_3C is still present in the steel and the steel is more brittle.

ADDITIONAL EXERCISES

49. a. 4-O on faces × 1/2 O/face = 2-O atoms

2-O atoms inside body, Total: 4-O atoms

8 Ti on corners × 1/8 Ti/corner + 1 Ti/body center = 2 Ti atoms

Formula of unit cell Ti_2O_4: empirical formula TiO_2

b.
$$\overset{+4\ -2}{2\ TiO_2} + \overset{0}{3\ C} + \overset{0}{4\ Cl_2} \rightarrow \overset{+4\ -1}{2\ TiCl_4} + \overset{+4\ -2}{CO_2} + \overset{+2\ -2}{2\ CO}$$

C is being oxidized. Cl_2 is oxidizing agent.

Cl is being reduced. C is reducing agent.

$$\overset{+4\ -1}{TiCl_4} + \overset{0}{O_2} \rightarrow \overset{+4\ -2}{TiO_2} + \overset{0}{2\ Cl_2}$$

Cl is being oxidized. O_2 is oxidizing agent.

O is being reduced. $TiCl_4$ is reducing agent.

50. TiF_4: ionic substance containing Ti^{4+} ions and F^- ions. $TiCl_4$, $TiBr_4$, and TiI_4: covalent substances containing discrete, tetrahedral TiX_4 molecules. As the molecule gets larger, the bp and mp increase, because the London dispersion forces increase.

51. a. $2\ CoAs_2(s) + 4\ O_2(g) \rightarrow 2\ CoO(s) + As_4O_6(s)$

As_4O_6 tetraarsenic hexoxide

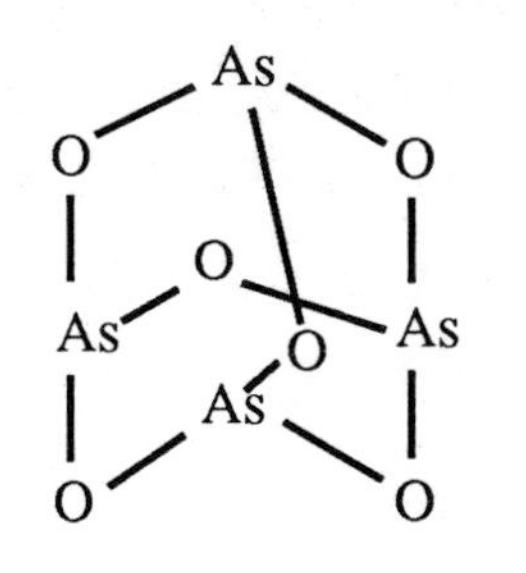

As_4O_6 has a cage structure similar to P_4O_6

b. $Co^{2+}(aq) + OCl^-(aq) \rightarrow Co(OH)_3(s) + Cl^-(aq)$

$$(Co^{2+} + 3\ OH^- \rightarrow Co(OH)_3 + e^-) \times 2$$
$$2\ e^- + H_2O + OCl^- \rightarrow Cl^- + 2\ OH^-$$

$$2\ Co^{2+}(aq) + 4\ OH^-(aq) + H_2O(l) + OCl^-(aq) \rightarrow 2\ Co(OH)_3(s) + Cl^-(aq)$$

c.

	$Co(OH)_3(s)$	$\rightleftharpoons$	$Co^{3+}(aq)$	+	$3\ OH^-(aq)$
Initial	0		0		$1.0 \times 10^{-7}\ M$
	s mol/L $Co(OH)_3$ dissolves to reach equilibrium				
Change	-s	$\rightarrow$	+s		+3s
Equil.			s		$1.0 \times 10^{-7} + 3s$

$$2.5 \times 10^{-43} = [Co^{3+}][OH^-]^3 = s(1.0 \times 10^{-7} + 3s)^3$$

$$\approx s(1.0 \times 10^{-21})$$

$s = [Co^{3+}] = 2.5 \times 10^{-22}$ mol/L Assumption good.

d. $Co(OH)_3 \rightleftharpoons Co^{3+} + 3\ OH^-$ pH = 10.00 pOH = 4.00 $[OH^-] = 1.0 \times 10^{-4}\ M$

$$2.5 \times 10^{-43} = [Co^{3+}][OH^-]^3 = [Co^{3+}]\ (1.0 \times 10^{-4})^3$$

$[Co^{3+}] = 2.5 \times 10^{-31}\ M$

52.

$(2\ OH^- + Zn \rightarrow Zn(OH)_2(s) + 2\ e^-) \times 2$ $-E° = 1.24$ V (calc. from K_{sp})

$4\ e^- + O_2 + 2\ H_2O \rightarrow 4\ OH^-$ $E° = 0.40$ V

$2\ Zn(s) + O_2(g) + 2\ H_2O(l) \rightarrow 2\ Zn(OH)_2(s)$ $E°_{cell} = 1.64$ V

$$\Delta G° = -nFE° = -(4 \text{ mol } e^-)\ (96{,}485 \text{ C/mol } e^-)(1.64 \text{ J/C})$$

$$\Delta G° = -6.33 \times 10^5 \text{ J} = -633 \text{ kJ}$$

$$\log K = \frac{nE°}{0.0592} = \frac{4(1.64)}{0.0592} = 111 \qquad K \approx 10^{111}$$

53.

$$2\ Au(CN)_2^- + 2\ e^- \longrightarrow 2\ Au + 4\ CN^- \qquad E° = -0.60\ V$$

$$Zn + 4\ CN^- \longrightarrow Zn(CN)_4^{2-} + 2\ e^- \qquad -E° = 1.26\ V$$

$$2\ Au(CN)_2^- + Zn \longrightarrow 2\ Au + Zn(CN)_4^{2-} \qquad E°_{cell} = 0.66\ V$$

$\Delta G° = -nFE° = -(2\text{ mol e}^-)(96{,}485\text{ C/mol e}^-)(0.66\text{ J/C})$

$= -1.27 \times 10^5\text{ J} = -130\text{ kJ}$

$$\log K = \frac{nE°}{0.0592} = \frac{2(0.66)}{0.0592} = 22.30$$

$K = 2.0 \times 10^{22}$

54. $Ni(CO)_4$ is composed of 4 CO molecules and Ni. Thus, nickel has an oxidation state of zero.

55. a. 2 b. 3 c. 4 d. 4

56. $BaCl_2$ gives no precipitate: SO_4^{2-} must be in the coordination sphere. A precipitate with $AgNO_3$ means that the Cl^- is not in the coordination sphere. Since there are only four ammonia molecules in the coordination sphere, the SO_4^{2-} must be acting as a bidentate ligand.

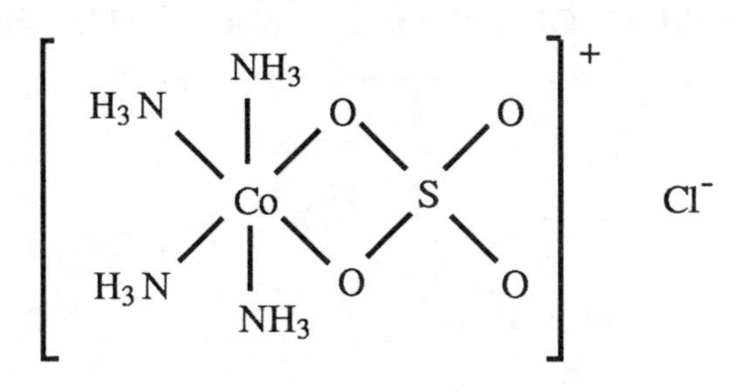

57. a.

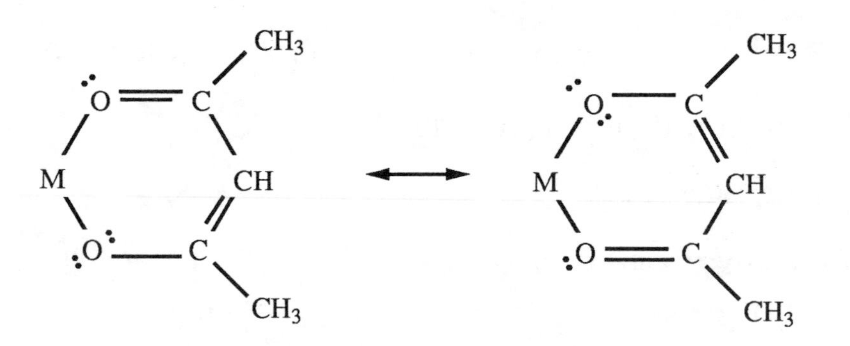

b. cis- $Cr(acac)_2(H_2O)_2$ and $Cr(acac)_3$ are optically active.

c. $$0.112\text{ g }Eu_2O_3 \times \frac{304.0\text{ g Eu}}{352.0\text{ g }Eu_2O_3} = 0.0967\text{ g Eu}$$

$$\%\ Eu = \frac{0.0967\text{ g}}{0.286\text{ g}} \times 100 = 33.8\%\ Eu$$

$\%\ O = 100 - (33.8 + 40.1 + 4.68) = 21.4\%\ O$

Out of 100 g of compound:

$33.8 \text{ g Eu} \times \frac{1 \text{ mol}}{152.0 \text{ g}} = 0.222 \text{ mol Eu}$

$40.1 \text{ g C} \times \frac{1 \text{ mol}}{12.01 \text{ g}} = 3.34 \text{ mol C}$

$4.68 \text{ g H} \times \frac{1 \text{ mol}}{1.008 \text{ g}} = 4.64 \text{ mol H}$

$21.4 \text{ g O} \times \frac{1 \text{ mol}}{16.00 \text{ g}} = 1.34 \text{ mol O}$

$\frac{3.34}{0.222} = 15.0, \frac{4.64}{0.222} = 20.9, \frac{1.34}{0.222} = 6.04$

Formula is: $EuC_{15}H_{21}O_6$

Each $acac^-$ is $C_5H_7O_2^-$; So Formula is: $Eu(acac)_3$

58.

There is no mirror plane in the complex, thus it is optically active.

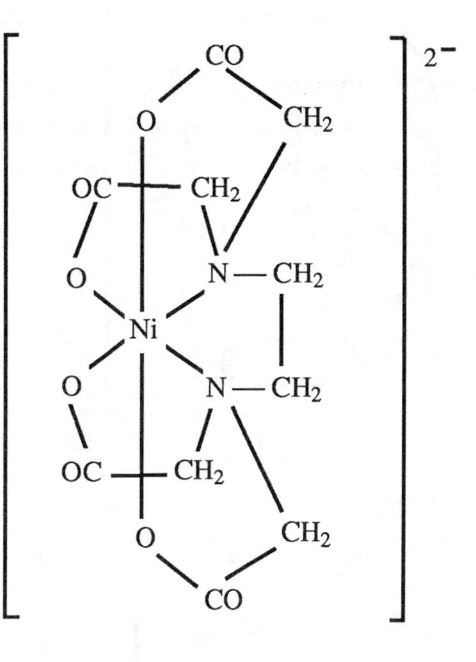

59. a. $Ru(phen)_3^{2+}$ exhibits optical isomerism.

b. Ru^{2+}: $[Kr]4d^6$, Ru^{2+} is strong field case.

— — e_g

↑↓ ↑↓ ↑↓ t_{2g}

60. Cr^{2+} complexes should be used. Cr^{2+}: $[Ar]3d^4$; High spin complexes have 4 unpaired electrons and low spin complexes have 2 unpaired electrons. Ni^{2+}: $[Ar]3d^8$; Ni^{2+} octahedral complexes will always have 2 unpaired electrons, whether high or low spin. Therefore, Ni^{2+} complexes cannot be used to distinguish weak from strong field ligands. Alternatively, the ligand field strengths can be measured using visible spectra. Either Cr^{2+} or Ni^{2+} complexes can be used for this method.

61. CN^- is a weak base.

$Ni^{2+}(aq) + 2\ OH^-(aq) \rightarrow Ni(OH)_2(s)$

$Ni(OH)_2(s) + 4\ CN^-(aq) \rightarrow Ni(CN)_4^{2-}(aq) + 2\ OH^-(aq)$

62. a. six

b.

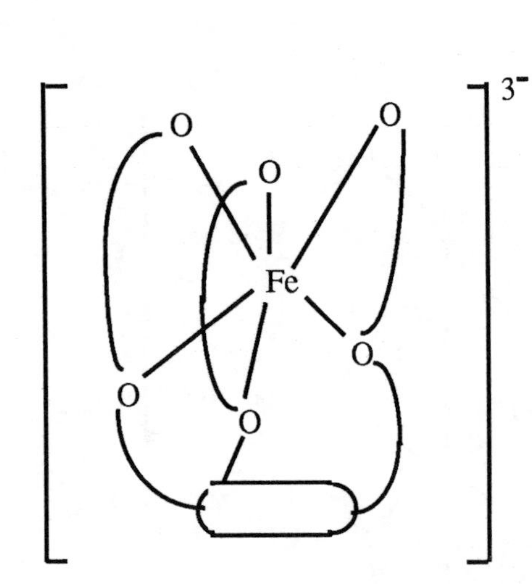

63. a.

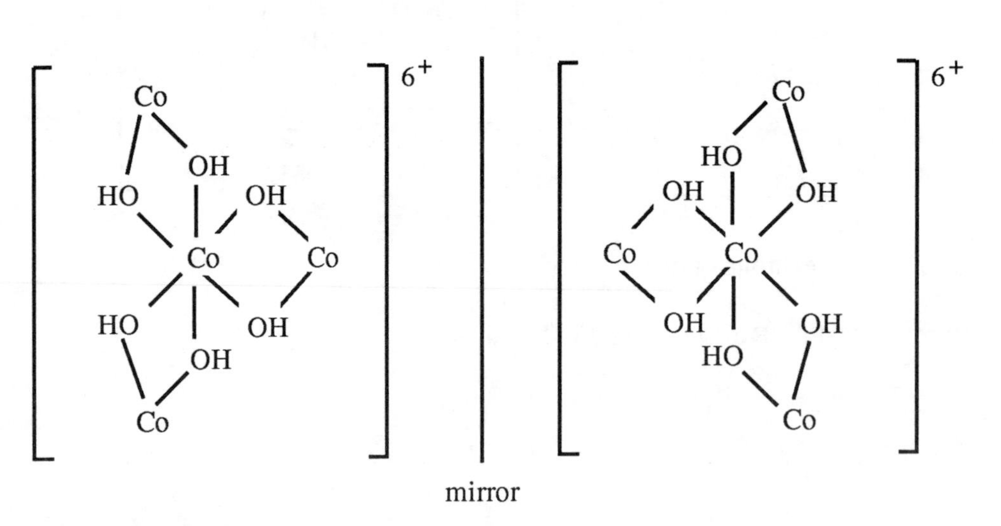

NH$_3$ molecules are not shown.

b. All are Co(III).

c. none

64. a.

$$Cu(OH)_2 \rightleftharpoons Cu^{2+} + 2\,OH^- \qquad K_{sp} = 1.6 \times 10^{-19}$$

$$Cu^{2+} + 4\,NH_3 \rightleftharpoons Cu(NH_3)_4^{2+} \qquad K_f = 1.0 \times 10^{13}$$

$$Cu(OH)_2 + 4\,NH_3 \rightleftharpoons Cu(NH_3)_4^{2+} + 2\,OH^- \qquad K_{eq} = 1.6 \times 10^{-6}$$

b.

$$Cd(NH_3)_4^{2+} \rightleftharpoons Cd^{2+} + 4\,NH_3 \qquad K = 1/K_f = 1/1.3 \times 10^7$$

$$Cd^{2+} + 4\,CN^- \rightleftharpoons Cd(CN)_4^{2-} \qquad K_f = 5.8 \times 10^{18}$$

$$Cd(NH_3)_4^{2+} + 4\,CN^- \rightleftharpoons Cd(CN)_4^{2-} + 4\,NH_3 \qquad K_{eq} = 4.5 \times 10^{11}$$

CHALLENGE PROBLEMS

65. i. $0.0203 \text{ g } CrO_3 \times \frac{52.00 \text{ g Cr}}{100.0 \text{ g } CrO_3} = 0.0106 \text{ g Cr}$

$\%Cr = \frac{0.0106}{0.105} \times 100 = 10.1\% \text{ Cr}$

ii. $32.92 \text{ mL HCl} \times \frac{0.100 \text{ mmol HCl}}{\text{mL}} \times \frac{1 \text{ mmol } NH_3}{\text{mmol HCl}} \times \frac{17.03 \text{ mg } NH_3}{\text{mmol}} = 56.1 \text{ mg } NH_3$

$\%\, NH_3 = \frac{56.1 \text{ mg}}{341 \text{ mg}} \times 100 = 16.5\% \text{ } NH_3$

iii. 73.53 + 16.5 + 10.1 = 100.1

So the compound is composed of only Cr, NH_3 and I.

Out of 100 g of compound:

$10.1 \text{ g Cr} \times \frac{1 \text{ mol}}{52.00 \text{ g}} = 0.194 \qquad \frac{0.194}{0.194} = 1$

$16.5 \text{ g } NH_3 \times \frac{1 \text{ mol}}{17.03 \text{ g}} = 0.969 \qquad \frac{0.969}{0.194} = 5$

$73.53 \text{ g I} \times \frac{1 \text{ mol}}{126.9 \text{ g}} = 0.5794 \qquad \frac{0.5794}{0.194} = 3$

$Cr(NH_3)_5I_3$ is the empirical formula. Cr(III) forms octahedral complexes.

So compound A is made of $[Cr(NH_3)_5I]^{2+}$ and two I^- ions or $[Cr(NH_3)_5I]I_2$.

iv. $\Delta T_f = iK_fm$, i = 3 ions

$m = \frac{0.601 \text{ g complex}}{10.0 \text{ g } H_2O} \times \frac{1 \text{ mol complex}}{517.9 \text{ g complex}} \times \frac{1000 \text{ g } H_2O}{\text{kg}} = 0.116 \text{ molal}$

$\Delta T_f = 3 \times 1.86°C/\text{molal} \times 0.116 \text{ molal} = 0.65°C$ This is close to the measured value.

This is consistent with the formula $[Cr(NH_3)_5I]I_2$.

66.

$$\overset{\text{II}}{(H_2O)_5Cr}—Cl—\overset{\text{III}}{Co(NH_3)_5} \longrightarrow \overset{\text{III}}{(H_2O)_5Cr}—Cl—\overset{\text{II}}{Co(NH_3)_5} \longrightarrow Cr(H_2O)_5Cl^{2+} + \text{Co(II) complex.}$$

Yes. After the oxidation, the ligands on Cr(III) won't exchange. Since Cl^- is in the coordination sphere it must have formed a bond to Cr(II) before the electron transfer occurred.

67. No, since in all three cases six bonds are formed between Ni^{2+} and nitrogen.

$\Delta S°$ for formation of the complex ion is most negative for 6 NH_3 molecules react with a metal ion (7 independent species becomes 1). For penten reacting with a metal ion, 2 independent species become 1, so $\Delta S°$ is less negative. Thus, the chelate effect occurs because the more bonds a chelating agent can form to the metal, the more favorable $\Delta S°$ is for the formation of the complex ion, and the larger the formation constant.

68.

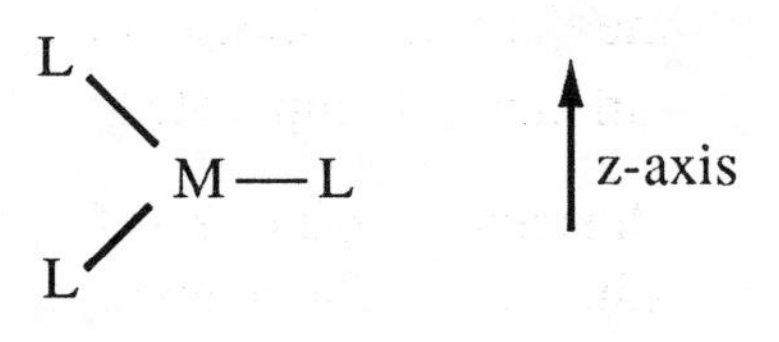

The $d_{x^2-y^2}$ and d_{xy} are in the plane of the three ligands and should be destabilized the most. The d_{z^2} has some electron density in the xy plane (the doughnut) and should be destabilized a lesser amount. The d_{xz} and d_{yz} have no electron density in the plane and should be lowest in energy.

— — $d_{x^2-y^2}$, d_{xy}

— d_{z^2}

— — d_{xz}, d_{yz}

69.

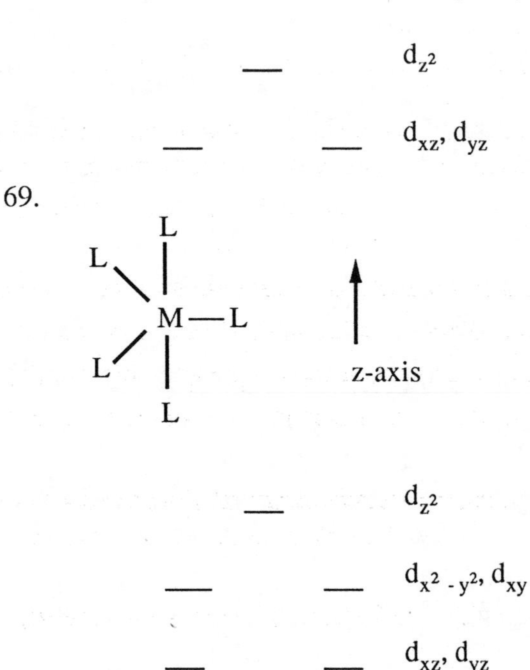

— d_{z^2}

— — $d_{x^2-y^2}$, d_{xy}

— — d_{xz}, d_{yz}

The d_{z^2} will be destabilized much more than in the trigonal planar case. (See Exercise 20.68.)

CHAPTER TWENTY-ONE: THE NUCLEUS - A CHEMIST'S VIEW

QUESTIONS

1. Fission: splitting of a heavy nucleus into two (or more) lighter nuclei.

   Fusion: combining of two light nuclei to form a heavier nucleus.

   Fusion is more likely for elements lighter than Fe; fission is more likely for elements heavier than Fe.

2. Characteristic frequencies of energies emitted in reaction suggest that discrete energy levels exist in the nucleus. Extra stability of certain numbers of nucleons and the predominance of nuclei with even numbers of nucleons suggest that nuclear structure might be described by using quantum numbers.

3. ^{14}C levels in the atmosphere are constant or the ^{14}C level at the time the plant died can be calculated. Constant ^{14}C level is a poor assumption and accounting for variation is complicated.

4. A nonradioactive substance can be put in equilibrium with a radioactive substance. The two materials can then be checked to see if all the radioactivity remains in the original material or if it has been scrambled by the equilibrium.

5. No, coal fired power plants also pose risks. A partial list of risks:

Coal	Nuclear
Air pollution	Radiation exposure to workers
Coal mine accidents	Disposal of wastes
Health risks to miners (black lung disease)	Meltdown
	Terrorists
	Public fear

6. The maximum binding energy per nucleon occurs at about Fe. Smaller nuclei become more stable by fusing to form heavier nuclei. Larger nuclei form more stable nuclei by splitting to form lighter nuclei.

7. For fusion reactions, a collision of sufficient energy must occur between two positively charged particles to initiate the reaction. This requires high temperatures. In fission, an electrically neutral neutron collides with the positively charged nucleus. This has much lower activation energy.

8. The temperatures of fusion reactions are so high that all physical containers would be destroyed. At these high temperatures most of the electrons are stripped from the atoms. A plasma of gaseous ions is formed which can be controlled by magnetic fields.

9. Magnetic fields are needed to contain the fusion reaction. Superconductors allow the production of very strong magnetic fields by being able to carry large electric currents. Stronger magnetic fields should be more capable of containing the fusion reaction.

10. Moderator: slows the neutrons

Control rods: absorb neutrons to slow or halt the fission.

EXERCISES:

Radioactive Decay and Nuclear Transformations

11. a. ${}^{3}_{1}H \rightarrow {}^{0}_{-1}e + {}^{3}_{2}He$

b. ${}^{8}_{3}Li \rightarrow {}^{8}_{4}Be + {}^{0}_{-1}e$

${}^{8}_{4}Be \rightarrow 2\,{}^{4}_{2}He$

${}^{8}_{3}Li \rightarrow 2\,{}^{4}_{2}He + {}^{0}_{-1}e$

c. ${}^{7}_{4}Be + {}^{0}_{-1}e \rightarrow {}^{7}_{3}Li$

d. ${}^{8}_{5}B \rightarrow {}^{8}_{4}Be + {}^{0}_{+1}e$

e. ${}^{32}_{15}P \rightarrow {}^{32}_{16}S + {}^{0}_{-1}e$

12. a. ${}^{60}_{27}Co \rightarrow {}^{60}_{28}Ni + {}^{0}_{-1}e$

b. ${}^{97}_{43}Tc + {}^{0}_{-1}e \rightarrow {}^{97}_{42}Mo$

c. ${}^{99}_{43}Tc \rightarrow {}^{99}_{44}Ru + {}^{0}_{-1}e$

d. ${}^{239}_{94}Pu \rightarrow {}^{235}_{92}U + {}^{4}_{2}He$

13. a. ${}^{1}_{1}H + {}^{14}_{7}N \rightarrow {}^{11}_{6}C + {}^{4}_{2}He$

b. $2\,{}^{3}_{2}He \rightarrow {}^{4}_{2}He + 2\,{}^{1}_{1}H$

c. ${}^{1}_{1}H + {}^{1}_{1}H \rightarrow {}^{2}_{1}H + {}^{0}_{+1}e$ (positron)

d. ${}^{1}_{1}H + {}^{12}_{6}C \rightarrow {}^{13}_{7}N$

14. a. ${}^{240}_{95}Am + {}^{4}_{2}He \rightarrow {}^{243}_{97}Bk + {}^{1}_{0}n$

b. ${}^{238}_{92}U + {}^{12}_{6}C \rightarrow {}^{244}_{98}Cf + 6\,{}^{1}_{0}n$

c. ${}^{249}_{98}Cf + {}^{18}_{8}O \rightarrow {}^{263}_{106}Unh + 4\,{}^{1}_{0}n$

d. ${}^{249}_{98}Cf + {}^{10}_{5}B \rightarrow {}^{257}_{103}Lr + 2\,{}^{1}_{0}n$

15. ${}^{8}B$ and ${}^{9}B$ contain too many protons or too few neutrons.

${}^{1}_{1}p \rightarrow {}^{1}_{0}n + {}^{0}_{+1}e$ (positron emission)

${}^{1}_{1}p + {}^{0}_{-1}e \rightarrow {}^{1}_{0}n$ (electron capture)

${}^{8}B$ and ${}^{9}B$ might decay by either positron emission or electron capture.

${}^{12}B$ and ${}^{13}B$ contain too many neutrons or too few protons. Beta emission converts neutrons to protons, so we expect ${}^{12}B$ and ${}^{13}B$ to be β-emitters.

16. ${}^{53}_{26}Fe$: too many protons

${}^{53}_{26}Fe \rightarrow {}^{53}_{25}Mn + {}^{0}_{+1}e$ positron emission

or

${}^{53}_{26}Fe + {}^{0}_{-1}e \rightarrow {}^{53}_{25}Mn$ electron capture

${}^{59}_{26}Fe$: too many neutrons, ${}^{59}_{26}Fe \rightarrow {}^{59}_{27}Co + {}^{0}_{-1}e$ beta emission

17. a. Complete decay is 7α and 4β.

$$^{235}_{92}U \rightarrow ^{207}_{82}Pb + 7\,^{4}_{2}He + 4\,^{0}_{-1}e$$

b. $$^{235}_{92}U \rightarrow ^{231}_{90}Th + ^{4}_{2}He \rightarrow ^{231}_{91}Pa + ^{0}_{-1}e \rightarrow ^{227}_{89}Ac + ^{4}_{2}He$$

$$\searrow$$

$$^{215}_{84}Po + ^{4}_{2}He \leftarrow ^{219}_{86}Rn + ^{4}_{2}He \leftarrow ^{223}_{88}Ra + ^{4}_{2}He \leftarrow ^{227}_{90}Th + ^{0}_{-1}e$$

$$\searrow$$

$$^{4}_{2}He + ^{211}_{82}Pb \rightarrow ^{211}_{83}Bi + ^{0}_{-1}e \rightarrow ^{207}_{81}Tl + ^{4}_{2}He \rightarrow ^{207}_{82}Pb + ^{0}_{-1}e$$

18. a. $$^{241}_{95}Am \rightarrow ^{4}_{2}He + ^{237}_{93}Np$$

b. $$^{241}_{95}Am \rightarrow 8\,^{4}_{2}He + 4\,^{0}_{-1}e + ^{209}_{83}Bi$$

The final product is $^{209}_{83}Bi$.

c. $$^{241}_{95}Am \rightarrow ^{237}_{93}Np + \alpha \rightarrow ^{233}_{91}Pa + \alpha \rightarrow ^{233}_{92}U + \beta \rightarrow ^{229}_{90}Th + \alpha \rightarrow ^{225}_{88}Ra + \alpha$$

$$\swarrow$$

$$^{213}_{84}Po + \beta \leftarrow ^{213}_{83}Bi + \alpha \leftarrow ^{217}_{85}At + \alpha \leftarrow ^{221}_{87}Fr + \alpha \leftarrow ^{225}_{89}Ac + \beta$$

$$\downarrow$$

$$^{209}_{82}Pb + \alpha \rightarrow ^{209}_{83}Bi + \beta$$

The intermediate radionuclides are:

$^{237}_{93}Np$, $^{233}_{91}Pa$, $^{233}_{92}U$, $^{229}_{90}Th$, $^{225}_{88}Ra$, $^{225}_{89}Ac$, $^{221}_{87}Fr$, $^{217}_{85}At$, $^{213}_{83}Bi$, $^{213}_{84}Po$, and $^{209}_{82}Pb$.

Kinetics of Radioactive Decay

19. For $t_{1/2}$ = 12,000 yr

$$k = \frac{\ln(2)}{t_{1/2}} = \frac{0.693}{t_{1/2}} = \frac{0.693}{12{,}000\text{ yr}} \times \frac{1\text{ yr}}{365\text{ d}} \times \frac{1\text{ d}}{24\text{ hr}} \times \frac{1\text{ hr}}{3600\text{ s}}$$

$$k = 1.8 \times 10^{-12}\text{ s}^{-1}$$

$$\text{Rate} = kN = 1.8 \times 10^{-12}\text{ s}^{-1} \times 6.02 \times 10^{23}$$

$$= 1.1 \times 10^{12}\text{ disint./s}$$

For $t_{1/2}$ = 12 hr

$$k = \frac{0.693}{t_{1/2}} = \frac{0.693}{12\text{ hr}} \times \frac{1\text{ hr}}{3600\text{ s}} = 1.6 \times 10^{-5}\text{ s}^{-1}$$

$$\text{Rate} = 1.6 \times 10^{-5}\ \text{s}^{-1} \times 6.02 \times 10^{23} = 9.6 \times 10^{18}\ \text{disint./s}$$

For $t_{1/2}$ = 12 min

$$\text{Rate} = \frac{0.693}{12\ \text{min}} \times \frac{1\ \text{min}}{60\ \text{s}} \times 6.02 \times 10^{23} = 5.8 \times 10^{20}\ \text{disint./s}$$

For $t_{1/2}$ = 12 s

$$\text{Rate} = \frac{0.693}{12\ \text{s}} \times 6.02 \times 10^{23} = 3.5 \times 10^{22}\ \text{disint./s}$$

20. a. $$175\ \text{mg Na}_3\text{PO}_4 \times \frac{32.0\ \text{mg}\ ^{32}\text{P}}{165.0\ \text{mg Na}_3{}^{32}\text{PO}_4} = 33.9\ \text{mg}\ ^{32}\text{P}$$

$$33.9 \times 10^{-3}\ \text{g} \times \frac{1\ \text{mol}}{32.0\ \text{g}} \times 6.022 \times 10^{23}\ \frac{\text{atoms}}{\text{mol}} = 6.38 \times 10^{20}\ \text{atoms}$$

$$\text{Rate} = kN = \frac{0.69315}{14.3\ \text{d}} \times \frac{1\ \text{d}}{24\ \text{hr}} \times \frac{1\ \text{hr}}{3600\ \text{s}} \times 6.38 \times 10^{20}\ \text{atoms} = 3.58 \times 10^{14}\ \text{disint./s}$$

$$\text{Rate of decay} = \frac{3.58 \times 10^{14}\ \text{disint.}}{\text{s}} \times \frac{1\ \text{Ci}}{\frac{3.7 \times 10^{10}\ \text{disint.}}{\text{s}}} \times \frac{1000\ \text{mCi}}{\text{Ci}}$$

$$= 9.7 \times 10^{6}\ \text{mCi}$$

b. $$\text{Rate of decay} = \frac{0.693}{24{,}000\ \text{yr}} \times \frac{1\ \text{yr}}{365\ \text{d}} \times \frac{1\ \text{d}}{24\ \text{hr}} \times \frac{1\ \text{hr}}{3600\ \text{s}} \times 6.02 \times 10^{23}$$

$$= \frac{5.5 \times 10^{11}\ \text{dis.}}{\text{s}}$$

$$\text{Rate} = \frac{5.5 \times 10^{11}\ \text{disint.}}{\text{s}} \times \frac{1\ \text{Ci}}{\frac{3.7 \times 10^{10}\ \text{disint.}}{\text{s}}} \times \frac{1000\ \text{mCi}}{\text{Ci}} = 1.5 \times 10^{4}\ \text{mCi}$$

21. $$175\ \text{mg Na}_3{}^{32}\text{PO}_4 \times \frac{32.0\ \text{mg}\ ^{32}\text{P}}{165.0\ \text{mg Na}_3{}^{32}\text{PO}_4} = 33.9\ \text{mg}\ ^{32}\text{P}, \quad k = \frac{\ln(2)}{t_{1/2}}$$

$$\ln\left(\frac{N}{N_o}\right) = \frac{-0.69315\ t}{t_{1/2}};\ \ln\left(\frac{m}{33.9}\right) = \frac{-0.69315\ (35.0)}{14.3}$$

$\ln(m) = -1.697 + 3.523 = 1.826$; $m = e^{1.826} = 6.21$ mg ^{32}P remains

22. $$\ln\left(\frac{N}{N_o}\right) = \frac{-0.693\ t}{t_{1/2}};\ \ln\left(\frac{5.0\mu\text{g}}{m}\right) = \frac{-0.693\ (2.0)}{4.5} = -0.31$$

$\ln 5.0 - \ln(m) = -0.31$; $1.61 + 0.31 = \ln(m)$

$m = e^{1.92} = 6.8$ μg of ^{47}Ca

$$6.8\ \mu\text{g}\ ^{47}\text{Ca} \times \frac{107.0\ \mu\text{g}\ ^{47}\text{CaCO}_3}{47.00\ \mu\text{g}\ ^{47}\text{Ca}} = 15\ \mu\text{g}\ ^{47}\text{CaCO}_3$$

23. $\ln\left(\frac{N}{N_o}\right) = -kt = \frac{-0.69315\ t}{t_{1/2}}$; If 10.0% decays, then 90.0% is left.

$\ln\left(\frac{90.0}{100.0}\right) = \frac{-0.69315\ t}{5.26\ yr}$; t = 0.800 years

24. a. $10.0\ mCi \times \frac{\frac{3.7 \times 10^7\ disintegrations}{s}}{mCi} = 3.7 \times 10^8$ disintegrations/s

Rate = kN; $\frac{3.7 \times 10^8\ disintegrations}{s} = \frac{0.69315}{2.87\ hr} \times \frac{1\ hr}{3600\ s} \times N$

$N = 5.5 \times 10^{12}$ atoms of ^{38}S

$5.5 \times 10^{12}\ atoms\ ^{38}S \times \frac{1\ mol\ ^{38}S}{6.02 \times 10^{23}\ atoms} \times \frac{1\ mol\ Na_2{}^{38}SO_4}{mol\ S}$

$= 9.1 \times 10^{-12}\ mol\ Na_2{}^{38}SO_4$

$9.1 \times 10^{-12}\ mol\ Na_2{}^{38}SO_4 \times \frac{148.0\ g\ Na_2{}^{38}SO_4}{mol\ Na_2{}^{38}SO_4} = 1.3 \times 10^{-9}\ g = 1.3\ ng$

b. 99.99% decays, 0.01% left. $\ln\left(\frac{0.01}{100}\right) = \frac{-0.693\ t}{2.87\ hr}$; t = 38.1 hours ≈ 40 hours

25. $t_{1/2} = 5370$ yr; $k = \frac{\ln(2)}{t_{1/2}}$, $\ln\left(\frac{N}{N_o}\right) = -kt$

$\ln\left(\frac{N}{N_o}\right) = \frac{-0.693\ t}{t_{1/2}} = \frac{-0.693\ (2200\ yr)}{5730\ yr} = -0.27$

$\frac{N}{N_o} = e^{-0.27} = 0.76$

26. $\ln\left(\frac{N}{N_o}\right) = \frac{-0.693\ t}{t_{1/2}}$; $\ln\left(\frac{N}{15.3}\right) = \frac{-0.693\ (15{,}000\ yr)}{5730\ yr}$

ln N = -1.81 + ln 15.3 = -1.81 + 2.728 = 0.92

N = 2.5 disintegrations per minute per g of C

If we had 10. mg C we would see:

$10.\ mg \times \frac{1\ g}{1000\ mg} \times \frac{2.5\ disintegrations}{min \bullet g} = \frac{0.025\ disintegrations}{min}$

It would take roughly 40 min to see a single disintegration. This is too long to wait and the background radiation would probably be much greater than the ^{14}C activity. Thus, ^{14}C dating is not practical for small samples.

27. $t_{1/2} = 4.5 \times 10^9$ years

Since 4.5×10^9 years is equal to the half-life, one half of the ^{238}U atoms will have been converted to ^{206}Pb. The numbers of atoms of ^{206}Pb and ^{238}U will be equal. Thus, the mass ratio is:

$$\frac{206}{238} = 0.866$$

28. a. The decay of ^{40}K is not the sole source of ^{40}Ca.

b. Decay of ^{40}K is the sole source of ^{40}Ar and no ^{40}Ar is lost over the years.

c. $\frac{0.95 \text{ g } ^{40}\text{Ar}}{1.00 \text{ g } ^{40}\text{K}}$ = current mass ratio

0.95 g of ^{40}K decayed to give ^{40}Ar. This 0.95 g of ^{40}K is only 10.7% of the total ^{40}K that decayed, or:

$0.107 (m) = 0.95$

m = 8.9 g (the total mass of ^{40}K that decayed)

Mass of ^{40}K when the rock was formed was 1.00 g + 8.9 g = 9.9 g.

$$\ln\left(\frac{1.00 \text{ g } ^{40}\text{K}}{9.9 \text{ g } ^{40}\text{K}}\right) = \frac{-0.693\ t}{1.27 \times 10^9 \text{ yr}}$$

$t = 4.2 \times 10^9$ years old

d. If some ^{40}Ar escaped then the measured ratio of $^{40}Ar/^{40}K$ is less than it should be. We would calculate the age of the rocks to be less than it actually is.

Energy Changes in Nuclear Reactions

29. $E = mc^2, m = \frac{E}{c^2} = \frac{3.9 \times 10^{23} \text{ kg m}^2/\text{s}^2}{(3.00 \times 10^8 \text{ m/s})^2} = 4.3 \times 10^6 \text{ kg}$

The sun loses 4.3×10^6 kg of mass each second.

30. $\frac{1.8 \times 10^{14} \text{ kJ}}{\text{s}} \times \frac{1000 \text{ J}}{\text{kJ}} \times \frac{3600 \text{ s}}{\text{hr}} \times \frac{24 \text{ hr}}{\text{day}} = 1.6 \times 10^{22} \text{ J}$

$E = mc^2, m = \frac{E}{c^2} = \frac{1.6 \times 10^{22} \text{ J}}{(3.00 \times 10^8 \text{ m/s})^2} = 1.8 \times 10^5$ kg of solar material provides 1 day of solar energy to the earth.

$1.6 \times 10^{22} \text{J} \times \frac{1 \text{ kJ}}{1000 \text{ J}} \times \frac{1 \text{ g}}{32 \text{ kJ}} \times \frac{1 \text{ kg}}{1000 \text{ g}} = 5.0 \times 10^{14}$ kg of coal is needed to provide the same amount of energy.

31. $12\,^1_1H + 12\,^0_1n + 12\,^0_{-1}e \longrightarrow\ ^{24}_{12}Mg$ mass of proton = 1.00728 amu.

$\Delta m = 23.9850 - [12(1.00728) + 12(1.00866) + 12(5.49 \times 10^{-4})]$ amu

$\Delta m = -0.2129$ amu

$$E = mc^2 = 0.2129 \text{ amu} \times \frac{1 \text{ g}}{6.022 \times 10^{23} \text{ amu}} \times \frac{1 \text{ kg}}{1000 \text{ g}} (2.9979 \times 10^8 \text{ m/s})^2$$

$$E = 3.177 \times 10^{-11} \text{ J}$$

$$\frac{\text{BE}}{\text{nucleon}} = \frac{3.177 \times 10^{-11} \text{ J}}{24} = \frac{1.324 \times 10^{-12} \text{ J}}{\text{nucleon}}$$

For ^{27}Mg

$$12\ {}^1_1\text{H} + 15\ {}^1_0\text{n} + 12\ {}^{\ 0}_{-1}\text{e} \longrightarrow {}^{27}_{12}\text{Mg}$$

$$\Delta m = 26.9843 - [12(1.00728) + 15(1.00866) + 12(5.49 \times 10^{-4})] \text{ amu}$$

$$\Delta m = -0.2395 \text{ amu}$$

$$E = mc^2 = 0.2395 \text{ amu} \times \frac{1 \text{ g}}{6.022 \times 10^{23} \text{ amu}} \times \frac{1 \text{ kg}}{1000 \text{ g}} (2.9979 \times 10^8 \text{ m/s})^2$$

$$E = 3.574 \times 10^{-11} \text{ J}$$

$$\frac{\text{BE}}{\text{nucleon}} = \frac{3.574 \times 10^{-11} \text{ J}}{27 \text{ nucleons}} = \frac{1.324 \times 10^{-12} \text{ J}}{\text{nucleon}}$$

32. ${}^1_1\text{H} + {}^1_0\text{n} \longrightarrow {}^2_1\text{H}$ Atomic mass - mass of electrons = mass of nucleus

$$\Delta m = 2.01410 - 0.000549 - [1.00728 + 1.00866] = -2.39 \times 10^{-3} \text{ amu}$$

$$E = 2.39 \times 10^{-3} \text{ amu} \times \frac{1 \times 10^{-3} \text{ kg}}{6.022 \times 10^{23} \text{ amu}} (2.998 \times 10^8 \text{ m/s})^2$$

$$E = 3.57 \times 10^{-13} \text{ J}$$

$$\frac{\text{BE}}{\text{nucleon}} = \frac{3.57 \times 10^{-13} \text{ J}}{2 \text{ nucleon}} = 1.79 \times 10^{-13} \text{ J/nucleon}$$

$${}^1_1\text{H} + 2\ {}^1_0\text{n} \longrightarrow {}^3_1\text{H}$$

$$\Delta m = 3.01605 - 0.000549 - [1.00728 + 2(1.00866)] = -9.10 \times 10^{-3} \text{ amu}$$

$$E = 9.10 \times 10^{-3} \text{ amu} \times \frac{1 \times 10^{-3} \text{ kg}}{6.022 \times 10^{23} \text{ amu}} \times (2.998 \times 10^8 \text{ m/s})^2$$

$$E = 1.36 \times 10^{-12} \text{ J}$$

$$\frac{\text{BE}}{\text{nucleon}} = \frac{1.36 \times 10^{-12} \text{ J}}{3 \text{ nucleons}} = 4.53 \times 10^{-13} \text{ J/nucleon}$$

33. $^1_1H + ^1_0n \longrightarrow 2\,^1_1H + ^1_0n + ^{\;1}_{-1}H$ Mass $^{\;1}_{-1}H$ = mass ^1_1H

$\Delta m = +2(1.00728) = 2.01456$ amu

$$E = 2.01456 \text{ amu} \times \frac{1 \times 10^{-3}\text{ kg}}{6.02214 \times 10^{23}\text{ amu}} \times (2.997925 \times 10^8 \text{ m/s})^2$$

$= 3.00657 \times 10^{-10}$ J of energy is absorbed per atom or 1.81060×10^{14} J/mol.

The source of energy is the kinetic energy of the proton and the neutron in the particle accelerator.

34. $\Delta m = -2(5.486 \times 10^{-4} \text{ amu}) = -10.97 \times 10^{-4}$ amu

$$E = 10.97 \times 10^{-4} \text{ amu} \times \frac{1 \times 10^{-3}\text{ kg}}{6.022 \times 10^{23}\text{ amu}} \times (2.998 \times 10^8 \text{ m/s})^2$$

$E = 1.637 \times 10^{-13}$ J

$E_{photon} = 1/2(1.637 \times 10^{-13}\text{ J}) = 8.185 \times 10^{-14}$ J

$$E_{photon} = \frac{hc}{\lambda}$$

$$\lambda = \frac{hc}{E} = \frac{6.626 \times 10^{-34}\text{ J s} \times 2.998 \times 10^8 \text{ m/s}}{8.185 \times 10^{-14}\text{ J}}$$

$\lambda = 2.427 \times 10^{-12}$ m or 2.427×10^{-3} nm

35. $^1_1H + ^1_1H \longrightarrow ^2_1H + ^{\;0}_{+1}e$

$\Delta m = (2.01410 - m_e + m_e) - 2(1.00782 - m_e)$

$\Delta m = 2.01410 - 2(1.00782) + 2(0.000549) = -4.4 \times 10^{-4}$ amu

When two mol of ^1_1H undergoes fusion, $\Delta m = -4.4 \times 10^{-4}$ g.

$E = 4.4 \times 10^{-7} \text{ kg} \times (3.00 \times 10^8 \text{ m/s})^2 = 4.0 \times 10^{10}$ J

$$\frac{4.0 \times 10^{10}\text{ J}}{2 \text{ mol } ^1_1H} \times \frac{1 \text{ mol}}{1.00782 \text{ g}} = 2.0 \times 10^{10} \text{ J/g}$$

36. $^2_1H + ^3_1H \longrightarrow ^2_1H + ^{\;0}_{+1}e$ Mass of electrons cancel.

$\Delta m = [4.00260 + 1.00866 - (2.01410 + 3.01605)] \text{ amu} = -1.889 \times 10^{-2}$ amu

For production of 1.0 mol of ^4_2He:

$\Delta m = -1.889 \times 10^{-2} \text{ g} = -1.889 \times 10^{-5}$ kg

$E = 1.889 \times 10^{-5} \text{ kg} \times (2.9979 \times 10^8 \text{ m/s})^2 = 1.698 \times 10^{12}$ J/mol

For 1 atom of ^4_2He:

$$1.698 \times 10^{12} \text{ J/mol} \times \frac{1 \text{ mol}}{6.022 \times 10^{23} \text{ atom}} = 2.820 \times 10^{-12} \text{ J/atom}$$

Detection, Uses, and Health Effects of Radiation

37. The Geiger-Müller tube has a certain response time. After the gas in the tube ionizes to produce a "count" some time must elapse for the gas to return to an electrically neutral state. The response of the tube levels because at high activities radioactive particles are entering the tube faster than the tube can respond to them.

38. Not all of the emitted radiation enter the Geiger-Müller tube. The fraction of radiation entering the tube must be constant.

39. Assuming that (1) the radionuclide is long lived enough such that no significant decay occurs during the time of the experiment, the total counts of radioactivity injected are:

$$0.10 \text{ mL} \times \frac{5.0 \times 10^3 \text{ cpm}}{\text{mL}} = 5.0 \times 10^2 \text{ cpm}$$

Assuming that (2) the total activity is uniformly distributed only in the rats blood, the blood volume is:

$$\frac{48 \text{ cpm}}{\text{mL}} \times V = 5.0 \times 10^2; V = 10.4 \text{ mL} = 10. \text{ mL}$$

40. All evolved O_2 comes from water. From just looking at the balanced reaction this result looks strange. However, if we view this as an electrochemical process, the two half reactions are:

$$24 e^- + 24 H^+ + 6 CO_2 \rightarrow C_6H_{12}O_6 + 6 H_2O$$

$$O_2 \rightarrow 2 H_2O + 4 H^+ + 4 e^-$$

The protons and electrons generated by the oxidation of O_2 to water are used in the reduction of CO_2 to glucose. The oxidation of O_2 is the half reaction that is coupled to light absorption by chlorophyll.

41. Release of Sr is probably more harmful. Xe is chemically unreactive, Sr can be easily oxidized to Sr^{2+}. Strontium is in the same family as calcium and could be absorbed and concentrated in the body in a fashion similar to Ca. This puts the radioactive Sr in the bones; red blood cells are produced in bone marrow. Xe would not be readily incorporated in the body.

The chemical properties determine where a radioactive material may be concentrated in the body or how easily it may be excreted. The length of time of exposure and what is exposed to radiation significantly affects the health hazard.

42. i) and ii) mean that Pu is not a significant threat outside the body. Our skin is sufficient to keep out the α particles. If Pu gets inside the body, it is easily oxidized to Pu^{4+} (iv) which is chemically similar to Fe^{3+} (iii). Thus, Pu^{4+} will concentrate in tissues where Fe^{3+} is found. One of these is the bone marrow where red blood cells are produced. Here α particles cause considerable damage.

ADDITIONAL EXERCISES

43. Mass of nucleus = mass of atom - mass of electrons

$$= 6.015126 - 3(0.0005486) = 6.013480 \text{ amu}$$

$$3\,^1_1\text{H} + 3\,^1_0\text{n} \longrightarrow \,^6_3\text{Li}$$

$\Delta m = 6.013480 - [3(1.00728) + 3(1.00866)] = -0.03434$ amu

For 1 mol of ^{6}Li, the mass defect is 0.03434 g.

$$E = mc^2 = 3.434 \times 10^{-2} \text{ g} \times \frac{1 \text{ kg}}{1000 \text{ g}} \times (2.9979 \times 10^8 \text{ m/s})^2$$

$E = 3.086 \times 10^{12}$ J/mol

44. $$\ln\left(\frac{N}{N_o}\right) = \frac{-0.69315\ (47.0 \text{ yr})}{28.8 \text{ yr}} = -1.131;\ \frac{N}{N_o} = e^{-1.131} = 0.323$$

32.3% of the ^{90}Sr remains (as of July 16, 1992).

45. mass of nucleus = 2.01410 - 0.000549 = 2.01355 amu

$$u_{rms} = \left(\frac{3RT}{M}\right)^{1/2} = \left(\frac{3(8.3145 \text{ J/K•mol})\ (4 \times 10^7 \text{ K})}{2.01355 \text{ g } (\frac{1 \text{ kg}}{1000 \text{ g}})}\right)^{\frac{1}{2}} = 7 \times 10^5 \text{ m/s}$$

$$E_K = \frac{1}{2}mu^2 = \frac{1}{2}\left(2.01355 \text{ amu} \times \frac{1 \times 10^{-3} \text{ kg}}{6.022 \times 10^{23} \text{ amu}}\right)(7 \times 10^5 \text{ m/s})^2$$

$E_K = 8 \times 10^{-16}$ J

46. $$20{,}000 \text{ ton TNT} \times \frac{4 \times 10^9 \text{ J}}{\text{ton TNT}} \times \frac{1 \text{ mol } ^{235}\text{U}}{2 \times 10^{13} \text{ J}} \times \frac{235 \text{ g } ^{235}\text{U}}{\text{mol } ^{235}\text{U}} = 940 \text{ g } ^{235}\text{U} \approx 900 \text{ g } ^{235}\text{U}$$

This assumes all of the ^{235}U undergoes fission.

47. a. $^{12}_6$C: It takes part in the reaction (1st step) but is regenerated in the last step. $^{12}_6$C is not consumed.

b. ^{13}N, ^{13}C, ^{14}N, ^{15}O, and ^{15}N are intermediates.

c. $$4\,^1_1\text{H} \longrightarrow \,^4_2\text{He} + 2\,^{\ 0}_{+1}\text{e}$$

$\Delta m = 4.00260 - 2\,m_e + 2\,m_e - 4(1.00782 - m_e)$

$\Delta m = 4.00260 + 4(0.000549) - 4(1.00782)$; $\Delta m = -0.02648$ amu

For 4 mol ^1_1H, $\Delta m = -0.02648$ g

$E = 2.648 \times 10^{-5}$ kg $(2.9979 \times 10^8 \text{ m/s})^2 = 2.380 \times 10^{12}$ J

$$\frac{2.380 \times 10^{12} \text{ J}}{4 \text{ mol H}} = \frac{5.950 \times 10^{11} \text{ J}}{\text{mol H}}$$

CHALLENGE PROBLEMS

48. The radiation may cause nuclei in the metal to undergo nuclear reaction. This changes the identity of the element and as the electrons rearrange themselves, the new atom may not fit in the crystal lattice. One result is the crystal becomes brittle. One needs to look for materials that do not undergo nuclear reactions when subjected to radiation (particularly neutrons) or ones that produce products of a similar atomic size as the original atoms.

49. a. For $2\ H_2O + 2\ e^- \rightarrow H_2 + 2\ OH^-$, $E° = -0.83$ V

$E^\circ_{H_2O} - E^\circ_{Zr} = +1.53$ V

Yes, the reduction of H_2O to H_2 by Zr is spontaneous, $E° > 0$ (at standard conditions).

b.

$4\ H_2O + 4\ e^- \rightarrow 2\ H_2 + 4\ OH^-$

$Zr + 4\ OH^- \rightarrow ZrO_2{\bullet}H_2O + H_2O + 4\ e^-$

$3\ H_2O(l) + Zr(s) \rightarrow 2\ H_2(g) + ZrO_2{\bullet}H_2O(s)$

c. $E° = -0.83 \text{ V} - (-2.36 \text{ V}) = +1.53$ V

$\Delta G° = -nFE° = -(4 \text{ mol e}^-) (96{,}485 \text{ C/mol e}^-) (1.53 \text{ J/C})$

$\Delta G° = -5.90 \times 10^5 \text{ J} = -590.\ \text{kJ}$

From the Nernst equation:

$E = E° - \frac{0.0592}{n} \log Q$ At equilibrium, E = 0 and Q = K

$E° = \frac{0.0592}{n} \log K$; $\log K = \frac{4(1.53)}{0.0592} = 103$

$K \approx 10^{103}$

d. $1.00 \times 10^3 \text{ kg Zr} \times \frac{1000 \text{ g}}{\text{kg}} \times \frac{1 \text{ mol Zr}}{91.22 \text{ g Zr}} \times \frac{2 \text{ mol } H_2}{\text{mol Zr}} = 2.19 \times 10^4 \text{ mol } H_2$

$2.19 \times 10^4 \text{ mol } H_2 \times \frac{2.016 \text{ g } H_2}{\text{mol } H_2} = 4.42 \times 10^4 \text{ g } H_2$

$PV = nRT$

$$V = \frac{nRT}{P} = \frac{(2.19 \times 10^4 \text{ mol}) (0.08206 \text{ L atm/mol}{\bullet}\text{K}) (1273 \text{ K})}{1 \text{ atm}} = 2 \times 10^6 \text{ L}$$

e. Probably yes, less radioactivity overall was released by venting the H_2, than what would have been released if the H_2 exploded inside the reactor (as happened at Chernobyl). Neither alternative is pleasant, but venting the radioactive hydrogen is the less unpleasant of the two alternatives.

50. $\Delta x \bullet \Delta(mv) \geq h/4\pi$; $\Delta(mv) = v\Delta m$; $\Delta x(v\Delta m) = h/4\pi$

$$\Delta m = \frac{h}{4\pi(\Delta x)(v)} = \frac{6.63 \times 10^{-34}\ \text{J s}}{4(3.14)\ (1 \times 10^{-35}\ \text{m})\ (3.0 \times 10^{7}\ \text{m/s})} = 2 \times 10^{-7}\ \text{kg}$$

Mass of electron, $m_e = 9.1 \times 10^{-31}$ kg

Mass of proton, $m_p = 1.7 \times 10^{-27}$ kg

$$\frac{\Delta m}{m_e} = \frac{2 \times 10^{-7}}{9.1 \times 10^{-31}} = 2 \times 10^{23};\quad \frac{\Delta m}{m_p} = \frac{2 \times 10^{-7}}{1.7 \times 10^{-27}} = 1 \times 10^{20}$$

The uncertainty in the superstring mass is 2×10^{23} times the electron mass and 1×10^{20} times the proton mass.

CHAPTER TWENTY-TWO: ORGANIC CHEMISTRY

QUESTIONS

1. There is only one consecutive chain of C-atoms. They are not all in a true straight line since the bond angle at each carbon is a tetrahedral angle of 109.5°.

2. Structural isomers: difference in bonding, either the kinds of bonds present or the way in which the bonds connect atoms to each other.

   Geometrical isomers: same bonds, differ in arrangement in space about rigid bond or ring.

3. Resonance: All atoms are in the same position. Only the position of π electrons is different.

   Isomerism: Atoms are in different locations in space.

   Isomers are distinctly different substances. Resonance is the use of more than one Lewis structure to describe the bonding in a single compound. Resonance structures are not isomers.

4. substitution: An atom or group is replaced by another atom or group.

   e.g. H on benzene is replaced by Cl. $C_6H_6 + Cl_2 \xrightarrow{\text{catalyst}} C_6H_5Cl + HCl$

   addition: Atoms or groups are added to a molecule.

   e.g. Cl_2 adds to ethene. $CH_2{=}CH_2 + Cl_2 \rightarrow CH_2Cl{-}CH_2Cl$

5. a. addition polymer: Polymer formed by adding monomer units to a double bond.

   Teflon: $n\ CF_2{=}CF_2 \rightarrow (CF_2 - CF_2)_n$

   b. condensation polymer: Polymer that forms when two monomers combine by eliminating a small molecule. Nylon and dacron are examples of condensation polymers.

   c. copolymer: Polymer formed from more than one kind of monomer. Nylon and dacron are also copolymers.

6. a. Free radical polymerization is a polymerization reaction that occurs by a free radical mechanism. These reactions require an initiator, such as a peroxide to generate free radicals. An example is polyethylene.

   b. Addition polymerization involves monomers adding to double bonds such as the formation of polyethylene. Most addition polymers occur by a free radical mechanism.

   c. Condensation polymerization is a polymerization in which two molecules condense to form a larger molecule by eliminating a small molecule such as water. Polyamides (nylon) and polyesters (dacron) are condensation polymers.

7. A thermoplastic polymer can be remelted; a thermoset polymer cannot be softened once it is formed.

8. The physical properties depend on the strengths of the interparticle forces. These forces are affected by chain length and extent of branching.

   longer chains = stronger forces

   branched chains = fewer crystalline regions (weaker forces)

9. Plasticizers make a polymer more flexible. Crosslinking makes a polymer more rigid.

10. The regular arrangement of the methyl groups in the isotactic chains allows the chains to pack together very closely. This leads to stronger intermolecular forces between chains.

11. a. Replacement of hydrogen with halogen results in fewer H and OH radicals in the flame, reducing flame temperature.

    b. With aromatic groups present it is more difficult to "chip off" pieces of the polymer in the pyrolysis zone.

    c. It is more difficult to "chip off" fragments of the polymer in the pyrolysis zone. The crosslinked polymer chars instead of burns.

12. A small amount of crosslinking can increase the elasticity of an elastomer by making the forces for snapping back greater. As crosslinking increases further, the elasticity will decrease as the polymer becomes more rigid.

13. For a given chain length, there are more hydrogen bonding sites in Nylon-46 than Nylon-6.

14. The stronger interparticle forces would be found in polyvinylchloride, since there are also dipole-dipole forces in PVC that are not present in polyethylene.

EXERCISES

Hydrocarbons

15. $CH_3—CH_2—CH_2—CH_2—CH_2—CH_3$ hexane or n-hexane (highest b.p., least branched)

$$\begin{array}{c} \quad CH_3 \\ \quad | \\ CH_3—CH—CH_2—CH_2—CH_3 \end{array}$$ 2-methylpentane

$$\begin{array}{c} \qquad\quad CH_3 \\ \qquad\quad | \\ CH_3—CH_2—CH—CH_2—CH_3 \end{array}$$ 3-methylpentane

$CH_3-C(CH_3)_2-CH_2-CH_3$ 2,2-dimethylbutane

$CH_3-CH(CH_3)-CH(CH_3)-CH_3$ 2,3-dimethylbutane

n-Hexane would have the highest boiling point. It is the least branched and, therefore, hexane will have the strongest dispersion forces between molecules.

16. a. 2,3,3-trimethylhexane b. 8-ethyl-2,5,5-trimethyldecane

c. 3-methylhexane

17.

a. $CH_3-CH(CH_3)-CH_2-CH_2CH_3$

b. $CH_3-C(CH_3)_2-CH_2-CH(CH_3)-CH_3$

c. $CH_3-CH(C(CH_3)_3)-CH_2CH_2CH_3$

d. The longest chain is 6 carbons long.

$\overset{1}{C}H_3-\overset{2}{C}(CH_3)_2-\overset{3}{C}H-\overset{4}{C}H_2-\overset{5}{C}H_2-\overset{6}{C}H_3$ (methyl on C-3)

2,2,3-trimethylhexane

18.

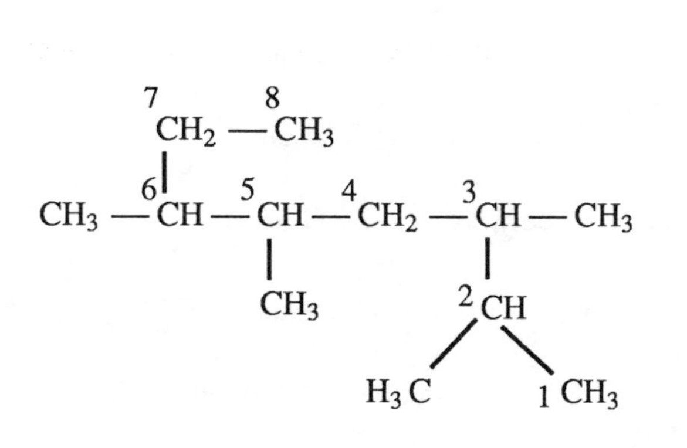

2,3,5,6-tetramethyloctane

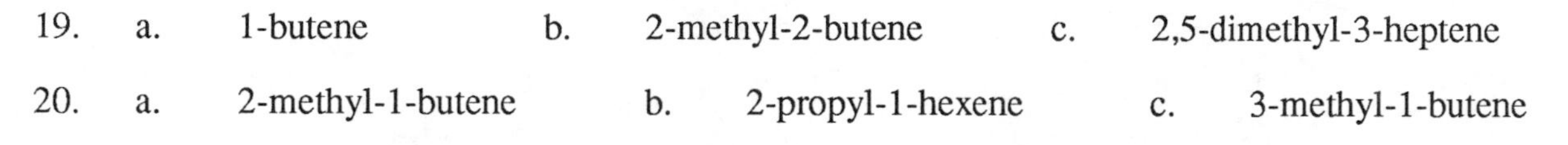

19. a. 1-butene b. 2-methyl-2-butene c. 2,5-dimethyl-3-heptene

20. a. 2-methyl-1-butene b. 2-propyl-1-hexene c. 3-methyl-1-butene

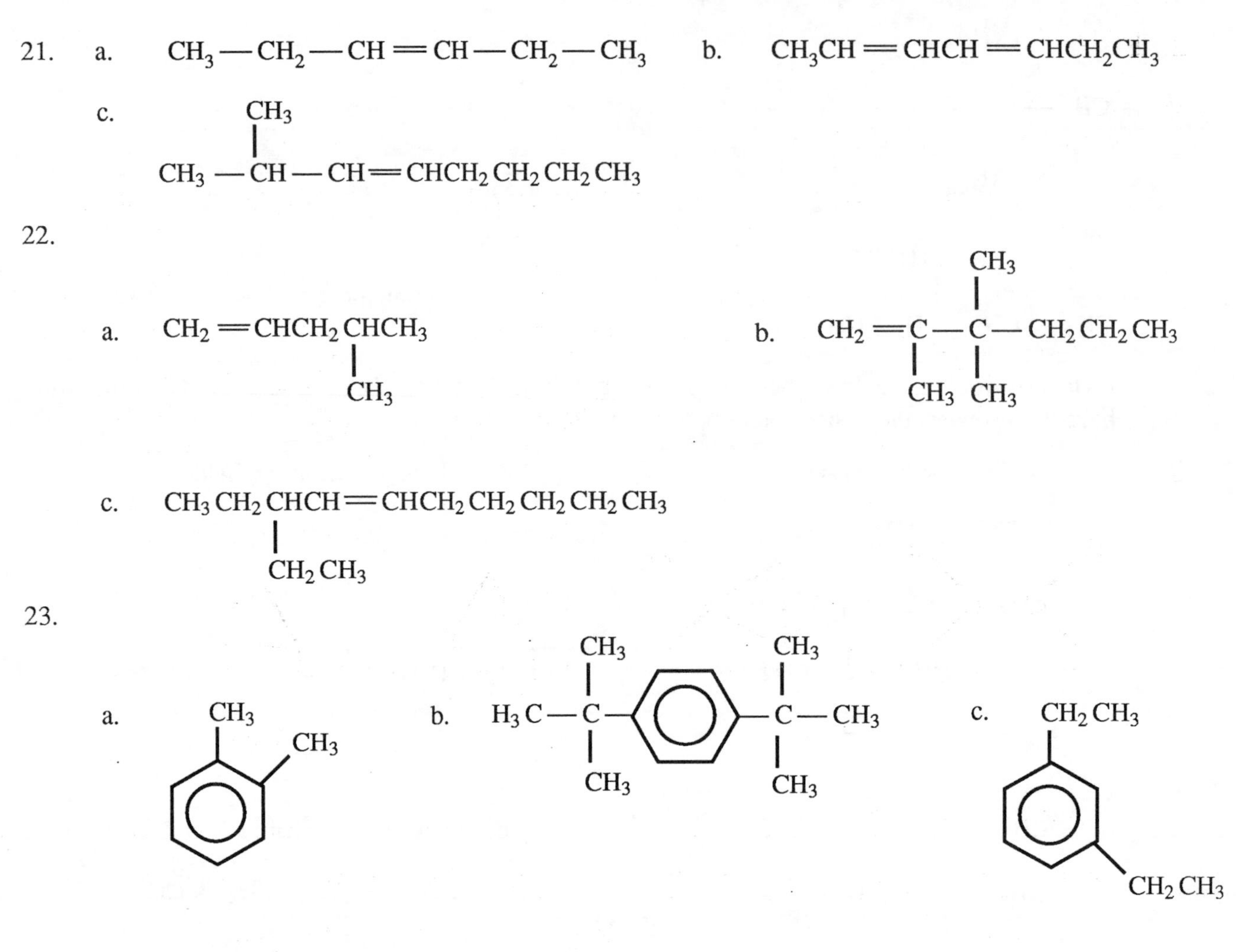

24. isopropylbenzene or 2-phenylpropane

25. a. 1,3-dichlorobutane
b. 1,1,1-trichlorobutane
c. 2,3-dichloro-2,4-dimethylhexane
d. 1,2-difluoroethane
e. chlorobenzene
f. chlorocyclohexane
g. 3-chlorocyclohexene (double bond assumed to be between C_1 and C_2).

26. a. methylcyclopropane
b. t-butylcyclohexane
c. 3,4-dimethylcyclopentene
d. chloroethene (vinyl chloride)
e. 1,2-dimethylcyclopentene
f. 1,1-dichlorocyclohexane

Isomerism

27.

28. HCBrCl — CH = CH_2

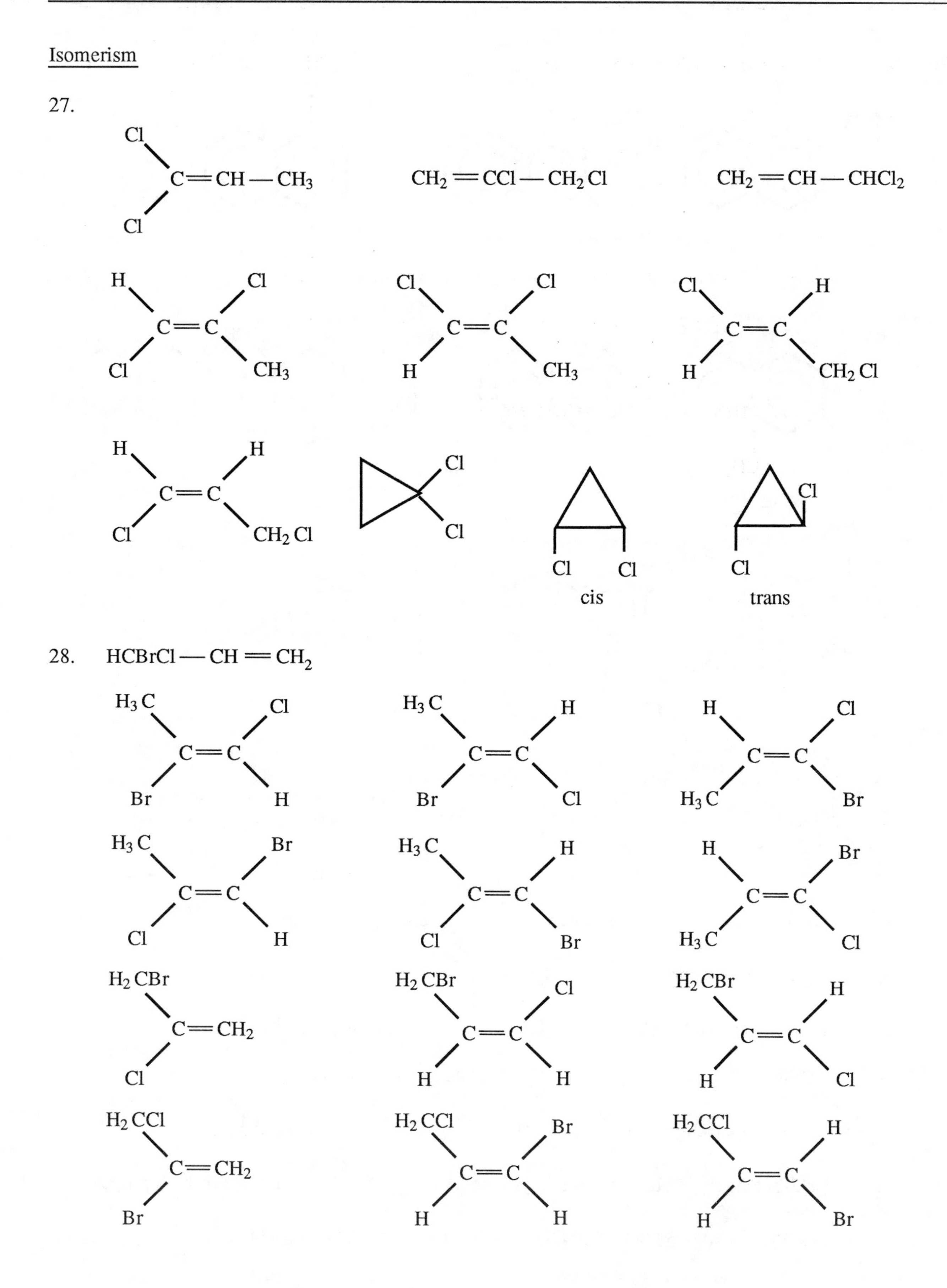

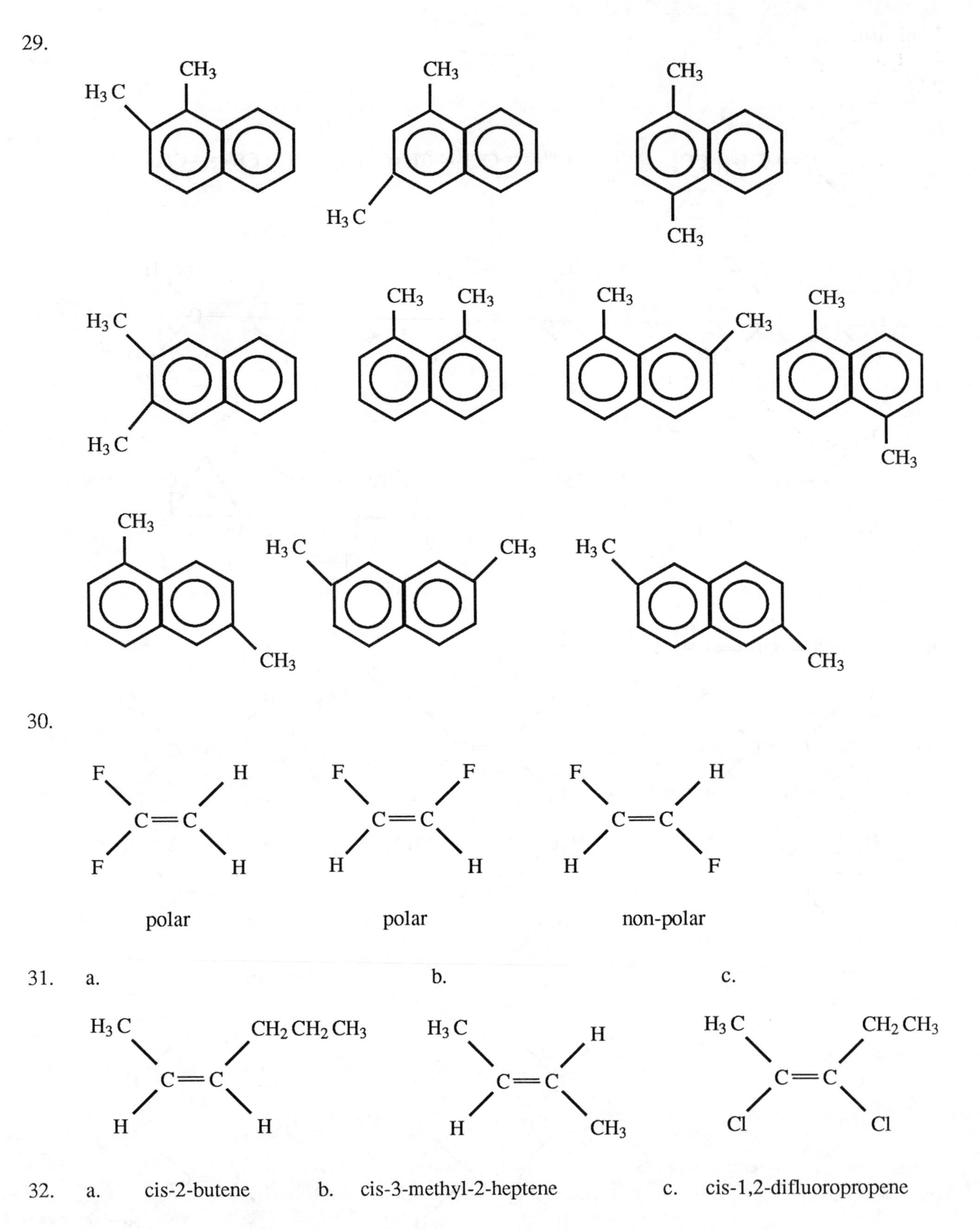

32. a. cis-2-butene b. cis-3-methyl-2-heptene c. cis-1,2-difluoropropene

Note: In general, cis/trans refers to the relative positions of the largest groups.

Functional Groups

33. a. ketone b. aldehyde c. ketone d. amine

34. a. b.

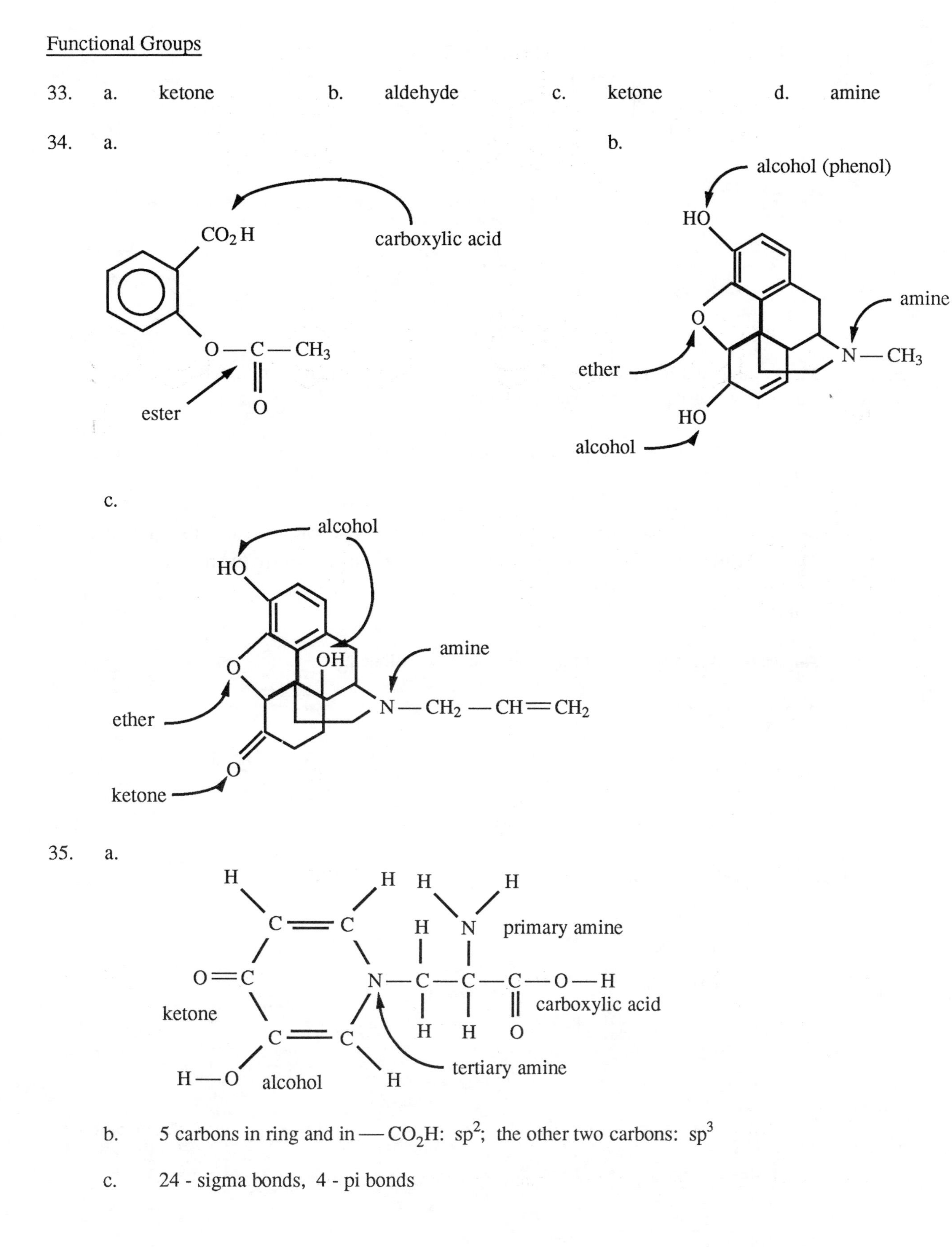

b. 5 carbons in ring and in —CO_2H: sp^2; the other two carbons: sp^3

c. 24 - sigma bonds, 4 - pi bonds

36.

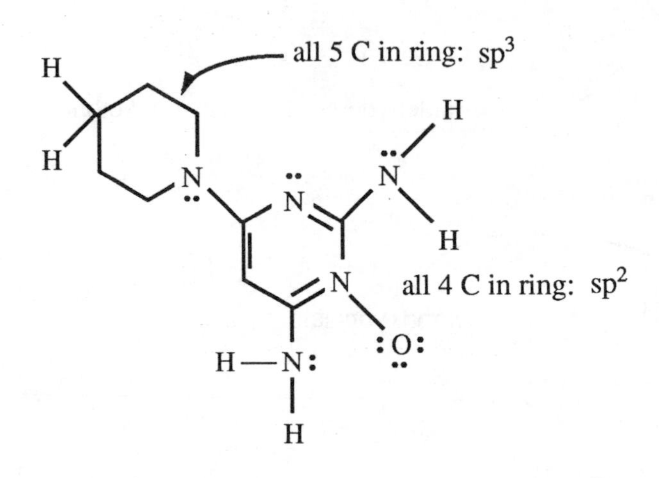

a. Minoxidil would be more soluble in an acidic solution. The four nitrogens with lone pairs can be protonated, forming a water soluble ion.

b. The two nitrogens in the ring with double bonds are sp^2 hybridized. The other three N's are sp^3 hybridized.

c. See structure: The five carbon atoms in the ring with one nitrogen are all sp^3 hybridized. The four carbon atoms in the other ring with double bonds are all sp^2 hybridized.

d. Angle a ≈ 109.5°

Angles b and e: slightly less than 109° (because of lone pair).

Angles c, d, and f ≈ 120°

e. thirty-one sigma bonds

f. three pi bonds

37.

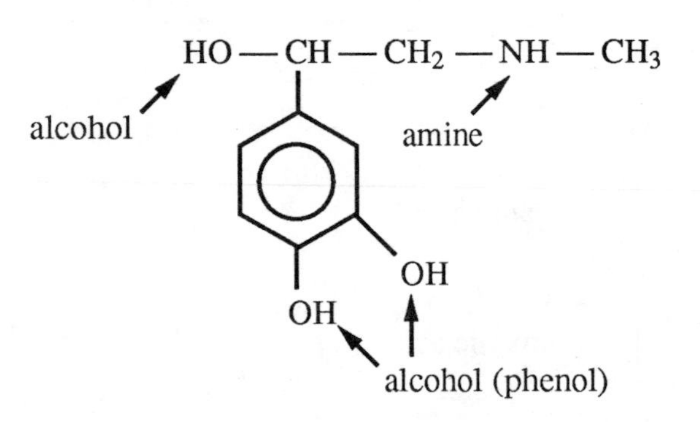

38. menthol: 2-isopropyl-5-methylcyclohexanol

The C with the — OH is number 1.

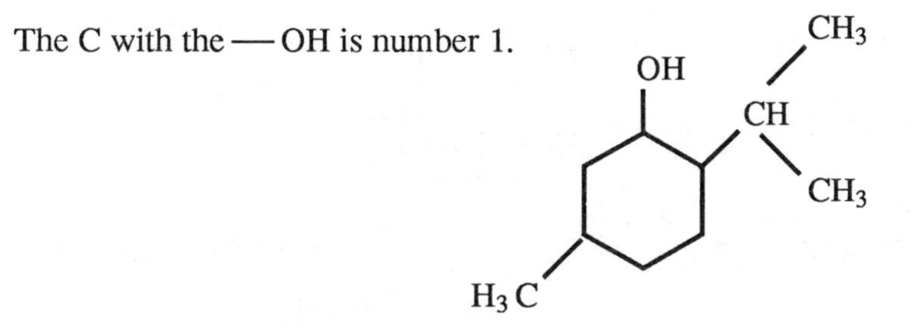

Reactions of Organic Compounds

39. a. $CH_2{=}CH_2 + Br_2 \rightarrow CH_2Br{-}CH_2Br$

b. $C_6H_6 + Br_2 \xrightarrow{Fe} C_6H_5Br + HBr$

c. $CH_3CO_2H + CH_3OH \rightarrow CH_3CO_2CH_3 + H_2O$

40. a. $CH_3CH{=}CH_2 + HCl \rightarrow CH_3CHClCH_3$

b. $CH_3CH{=}CH_2 + H_2O \rightarrow CH_3CH(OH)CH_3$

c.

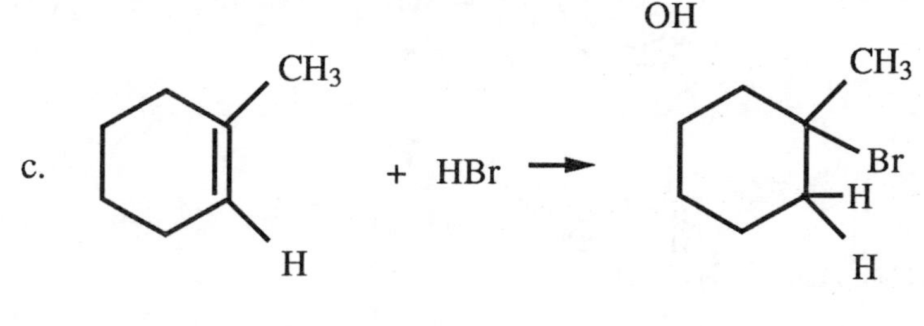

d. $H_2C{=}C(CH_3)_2 + H_2O \rightarrow H_3C{-}C(CH_3)_2{-}O{-}H$

e.

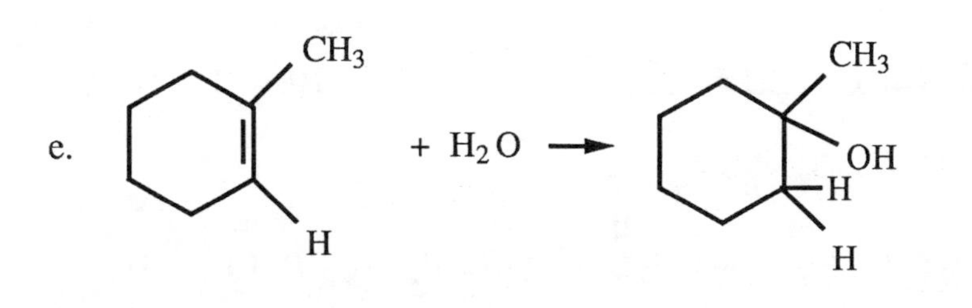

f. $CH_3C{\equiv}CH \xrightarrow{HCl} CH_3{-}C(Cl){=}CH_2 \xrightarrow{HCl} CH_3CCl_2CH_3$

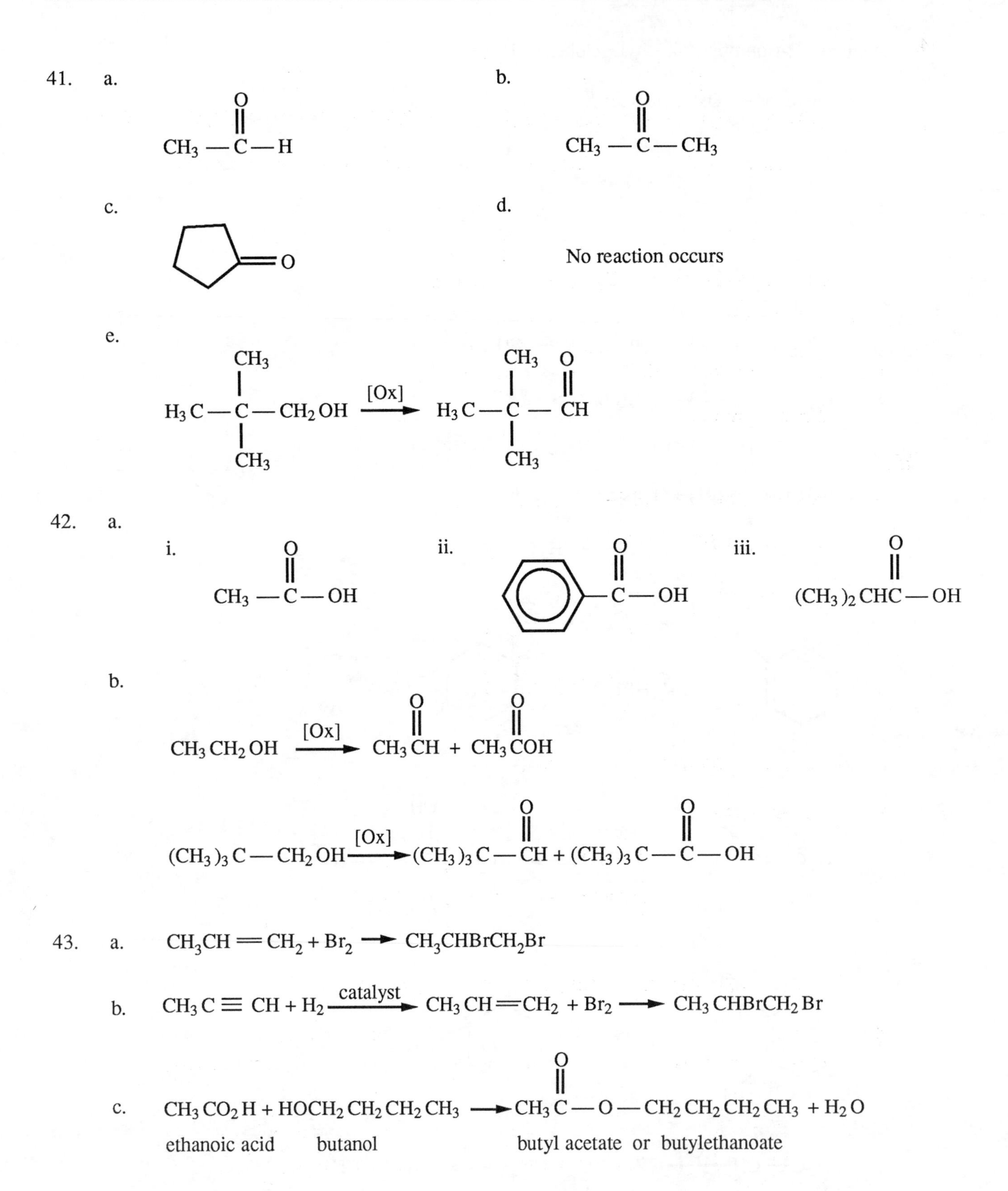
41. a.
CH3 — C — H
O
b.
CH3 — C — CH3
O
c.
O
d.
No reaction occurs
e.
CH3
H3C — C — CH2OH
CH3
[Ox]
CH3 O
H3C — C — CH
CH3
42. a.
i.
CH3 — C — OH
O
ii.
C — OH
O
iii.
(CH3)2CHC — OH
O
b.
CH3CH2OH
[Ox]
CH3CH + CH3COH
O O
(CH3)3C — CH2OH
[Ox]
(CH3)3C — CH + (CH3)3C — C — OH
O O
43. a.
CH3CH = CH2 + Br2 → CH3CHBrCH2Br
b.
CH3C ≡ CH + H2
catalyst
CH3CH = CH2 + Br2 → CH3CHBrCH2Br
c.
CH3CO2H + HOCH2CH2CH2CH3 → CH3C — O — CH2CH2CH2CH3 + H2O
O
ethanoic acid
butanol
butyl acetate or butylethanoate

44. a.

$$CH_3CH_2CH_2\overset{O}{\overset{\|}{C}}-OH + HO-CH_2CH_3 \longrightarrow CH_3CH_2CH_2\overset{O}{\overset{\|}{C}}-OCH_2CH_3 + H_2O$$

butanoic acid ethanol ethyl butyrate or ethylbutanoate

b.

$$CH_3-\underset{}{\overset{OH}{\overset{|}{CH}}}-CH_3 \xrightarrow{[Ox]} CH_3-\overset{O}{\overset{\|}{C}}-CH_3$$

2-propanol

c. $CH_2{=}C(CH_3)_2 + H_2O \xrightarrow{H^+} (CH_3)_3COH$

2-methylpropene

Polymers

45. a. repeating unit: $\left(CHF-CH_2\right)_n$

monomer: $CHF{=}CH_2$

b. repeating unit:

$$\left(OCH_2CH_2\overset{O}{\overset{\|}{C}}\right)_n$$

monomer: $HO-CH_2CH_2-CO_2H$

c. repeating unit:

$$\left(OCH_2CH_2-O-\overset{O}{\overset{\|}{C}}-CH_2CH_2-\overset{O}{\overset{\|}{C}}\right)_n$$

copolymer of: $HOCH_2CH_2OH$ and $HO_2CCH_2CH_2CO_2H$

d. monomer:

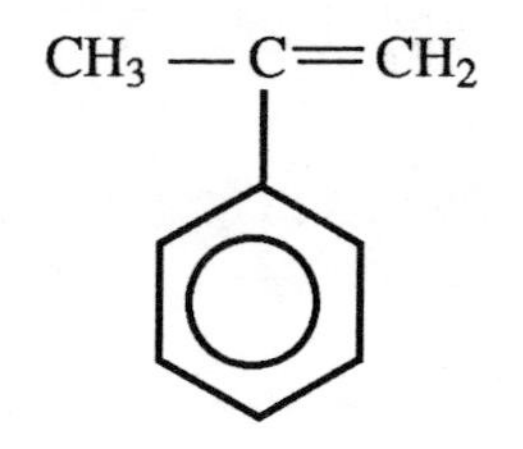

e. monomer:

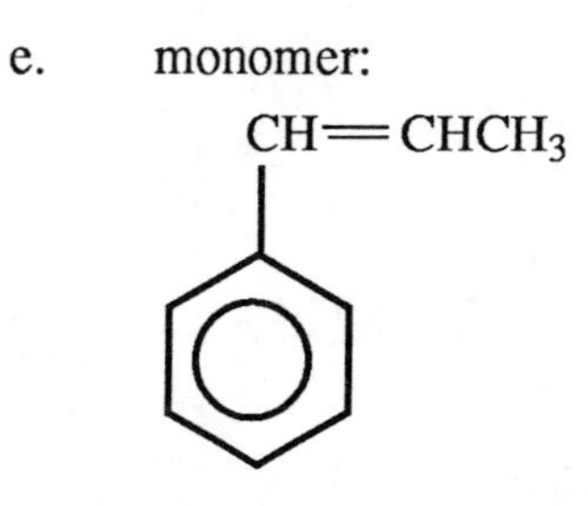

Addition polymers: a, d, and e

Condensation polymers: b anc c Copolymer: c

46. a. $CFCl=CF_2$

b. More reactive, C—Cl bonds are not as strong as C—F bonds.

47.

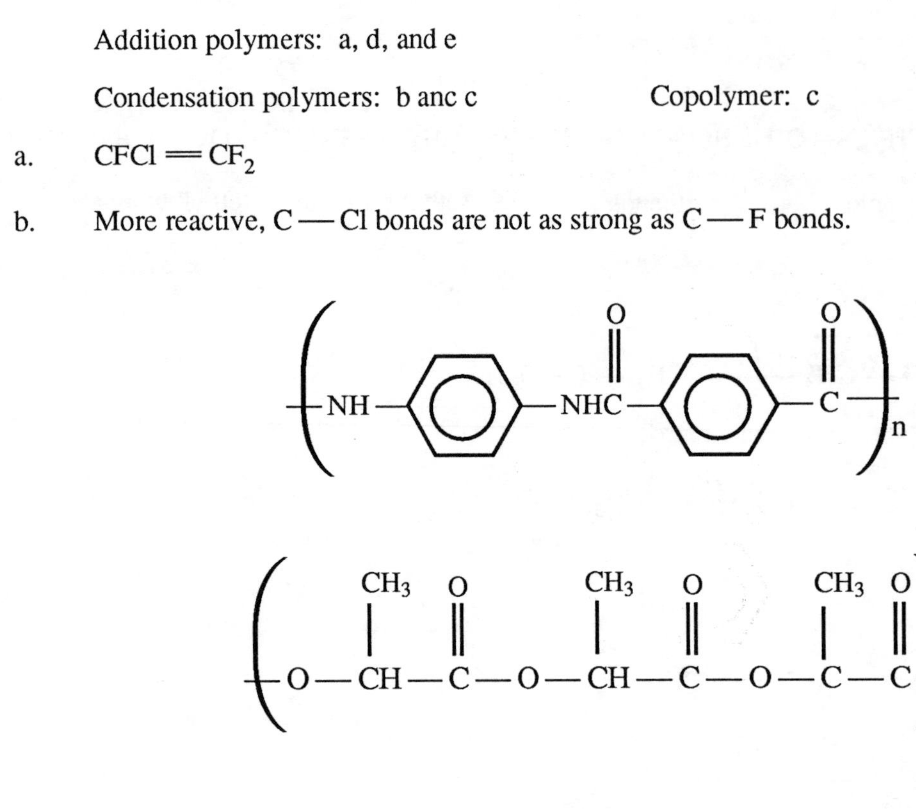

48.

49.

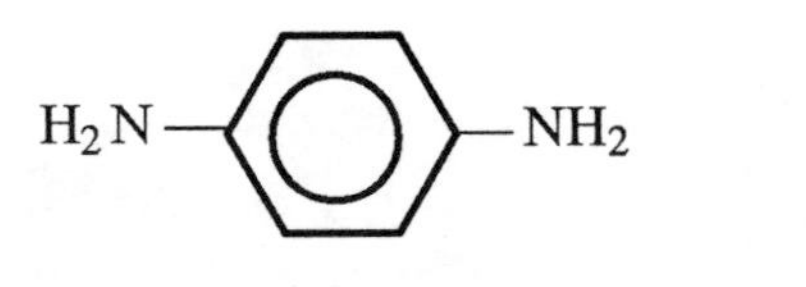

and

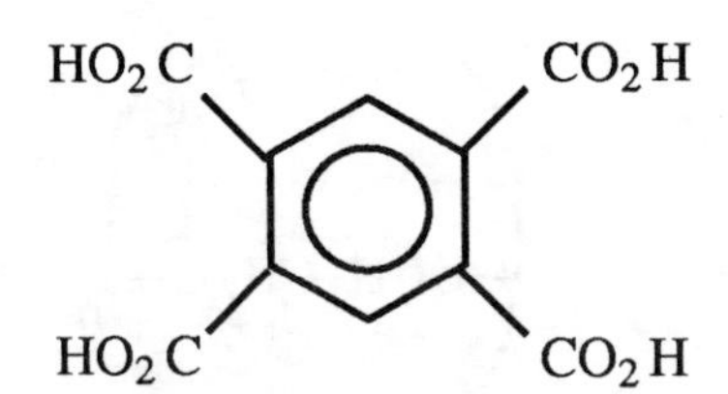

50.

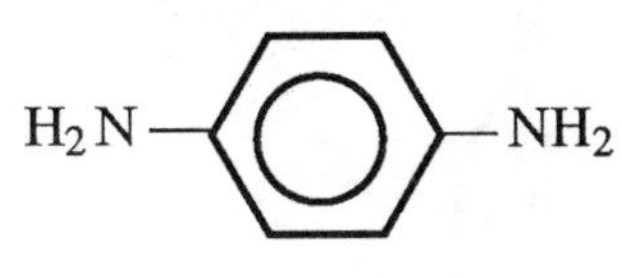

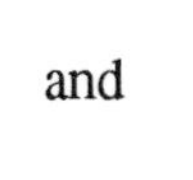

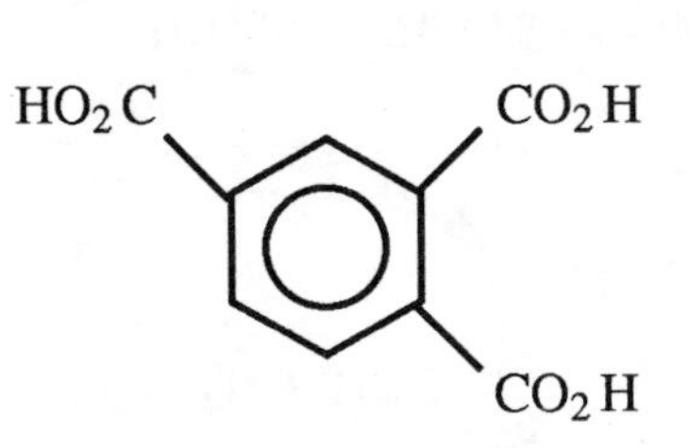

51. Divinylbenzene crosslinks different chains to each other. The chains cannot move past each other because of the crosslinks; thus, the polymer is more rigid.

52. a.

$$\left(-\mathrm{OCH_2CH_2O\overset{O}{\overset{\|}{C}}CH{=}CH\overset{O}{\overset{\|}{C}}}- \right)_n$$

b.

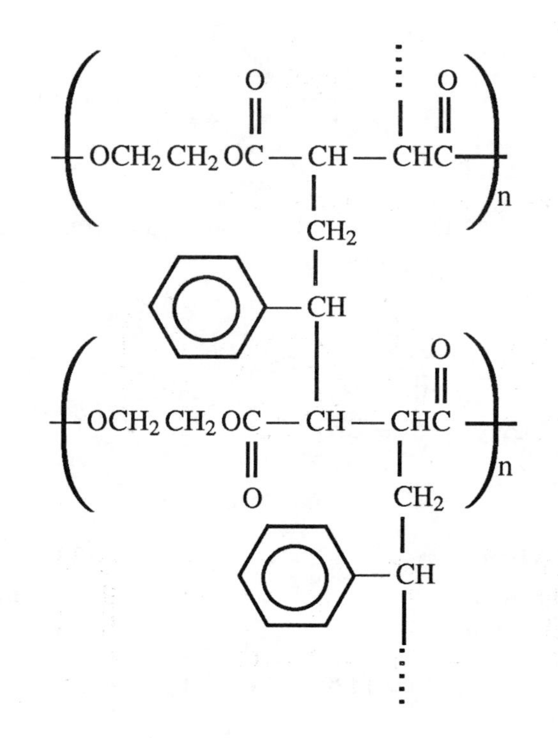

ADDITIONAL EXERCISES

53. 2-methyl-1,3-butadiene

54. a. b. c.

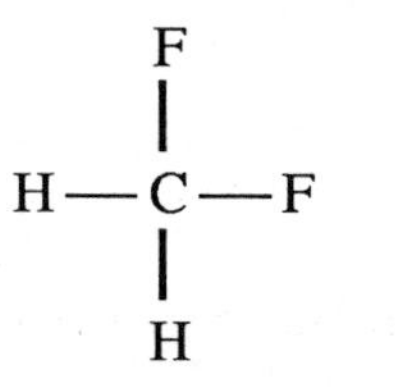

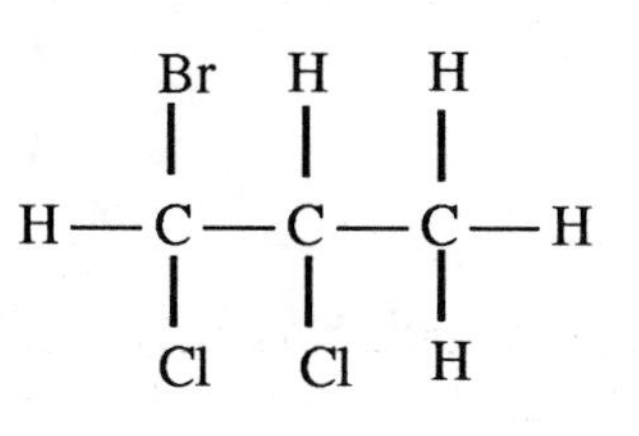

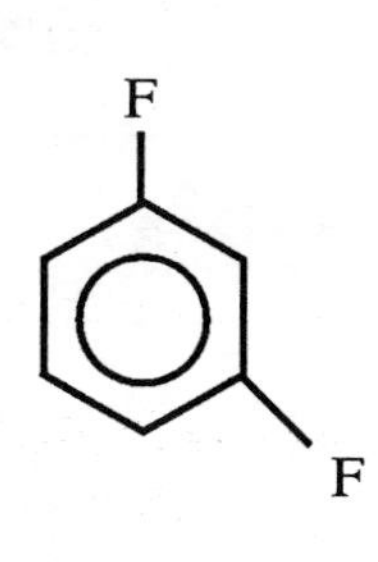

55.

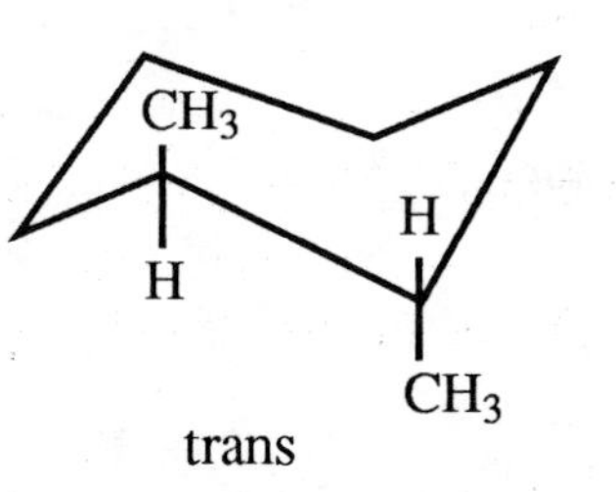

trans

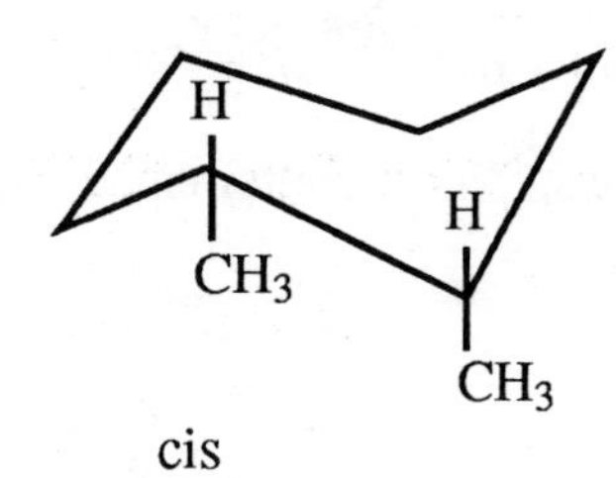

cis

56. $CH_2Cl—CH_2Cl$ 1,2-dichloroethane

There is free rotation about the C—C single bond.

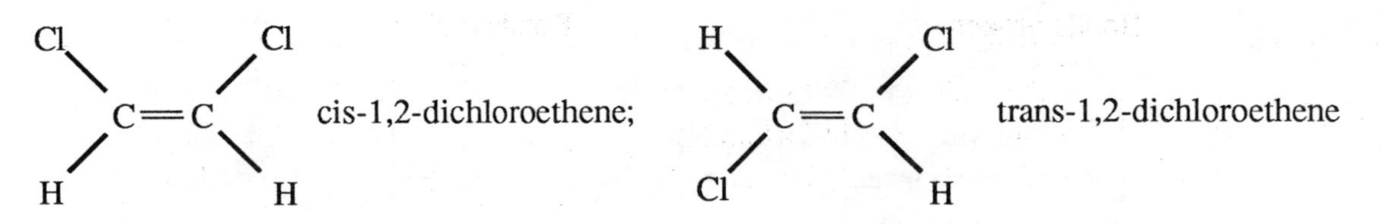

There is no free rotation about the C=C double bond. The cis and trans forms are different compounds.

57. a.

Bonds broken:		Bonds made:	
O—H	467 kJ/mol	C—H	413 kJ/mol
C—O	358 kJ/mol	C=O	799 kJ/mol
C=C	614 kJ/mol	C—C	347 kJ/mol

$\Delta H = 467 + 358 + 614 - (413 + 799 + 347) = 1439 - 1559 = -120.$ kJ

Ketone is more stable; it contains the stronger bonds.

b.

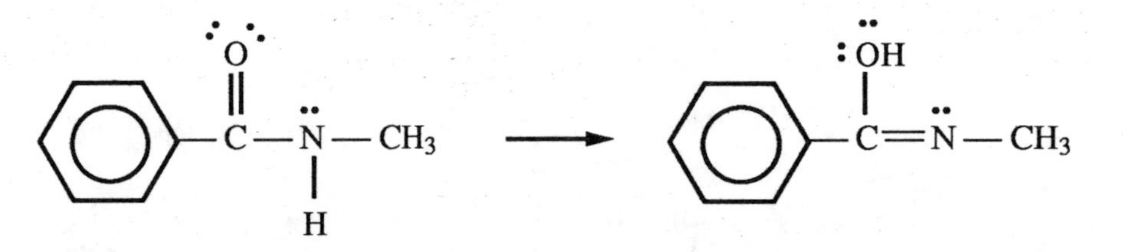

Bonds broken:		Bonds made:	
C=O	799 kJ/mol	C—O	358 kJ/mol
C—N	305 kJ/mol	C=N	615 kJ/mol
N—H	391 kJ/mol	O—H	467 kJ/mol

$\Delta H = 799 + 305 + 391 - (358 + 615 + 467) = 55$ kJ

Amide with C=O is more stable; it contains the stronger bonds.

c. For the reaction:

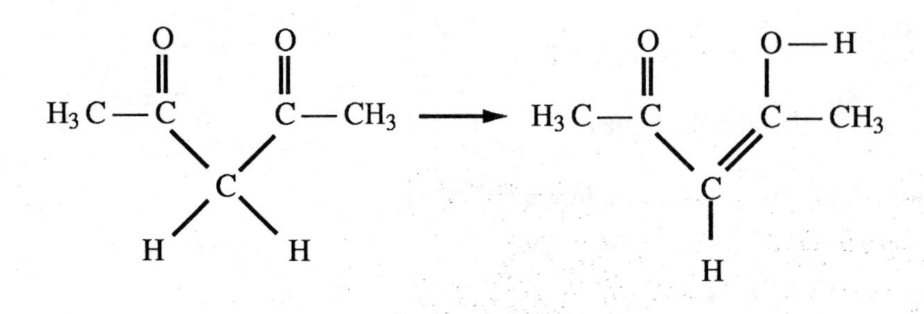

Bonds broken:		Bonds made:	
C═O	799 kJ/mol	O—H	467 kJ/mol
C—C	347 kJ/mol	C—O	358 kJ/mol
C—H	413 kJ/mol	C═H	614 kJ/mol

$\Delta H = (799 + 347 + 413) - (467 + 358 + 614) = 1559 - 1439 = 120.$ kJ

The form with two carbonyl groups is more stable because it contains the stronger bonds.

58. a. acetone:

$$CH_3-\overset{\overset{\displaystyle O}{\|}}{C}-CH_3$$

aldehyde that is an isomer of acetone:

$$CH_3-CH_2-\overset{\overset{\displaystyle O}{\|}}{C}-H$$ propanal

b. 2-propanol:

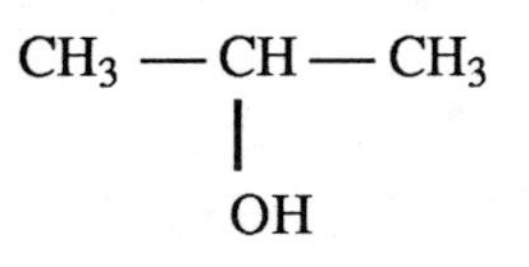

ether: $CH_3—O—CH_2—CH_3$ ethylmethyl ether

c. cis-2-butene:

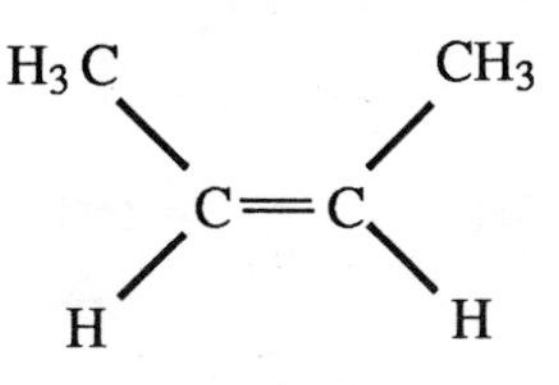

geometrical isomer:

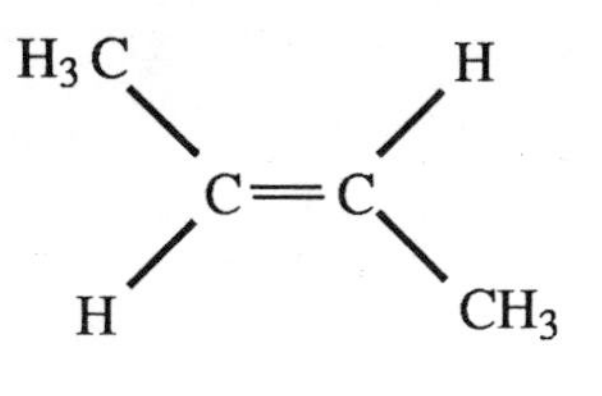

trans-2-butene

d. trimethylamine:

$$H_3C-\underset{\underset{\displaystyle CH_3}{|}}{N}-CH_3$$

primary amine: $CH_3CH_2CH_2NH_2$ propylamine or 1-aminopropane

e. ethylmethylamine (secondary amine):

$$CH_3-\underset{\displaystyle H}{\underset{|}{N}}-CH_2CH_3$$

f. 1-propanol (primary alcohol): $CH_3CH_2CH_2OH$

59. The amine group is protonated, making it a positive one charged cation.

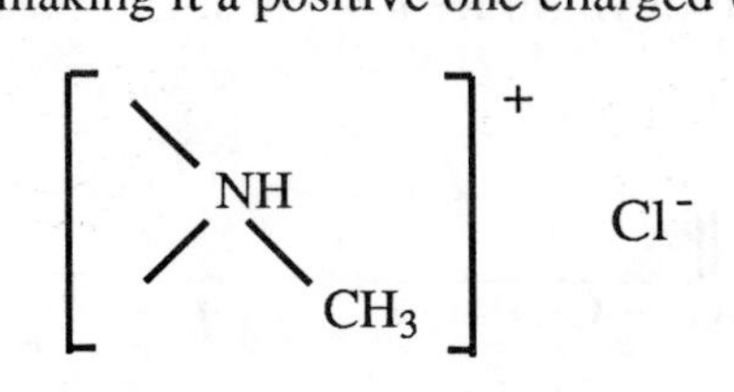

Morphine hydrochloride is ionic and, hence, more soluble in water.

60. a. Two monochloro products are formed:

$CH_2ClCH_2CH_3$ and $CH_3CHClCH_3$

b. $CHCl_2CH_2CH_3$, $CH_3CCl_2CH_3$, $CH_2ClCHClCH_3$ and $CH_2ClCH_2CH_2Cl$

Four dichloro products are formed.

61. a. The bond angles in the ring are about 60°. VSEPR predicts bond angles close to 109°. The bonding electrons are closer together than they want to be, resulting in stronger electron-electron repulsions. This makes ethylene oxide unstable (reactive).

b. $$\left(O-CH_2CH_2-O-CH_2CH_2-O-CH_2CH_2 \right)_n$$

c.

$$\underset{\diagdown\ \ O\ \ \diagup}{CH_2-CH_2} + H-O-H \longrightarrow HO-CH_2-CH_2-OH$$

Bonds broken:	Bonds made:
C—O	C—O
H—O	O—H

Since bonds broken = bonds formed, $\Delta H \approx 0$.

$$CH_2{=}CH_2 + H-O-H \longrightarrow CH_3CH_2OH$$

Bonds broken:		Bonds made:	
C═C	614 kJ/mol	C—H	413 kJ/mol
H—O	467 kJ/mol	C—C	347 kJ/mol
		C—O	358 kJ/mol

$\Delta H = 614 + 467 - [413 + 347 + 358] = 1081\text{ kJ} - 1118\text{ kJ} = -37\text{ kJ}$

d. $\frac{20{,}000\text{ g}}{\text{mol}} \times \frac{1\text{ ethylene oxide}}{44.05\text{ g}} = 454 \sim 500$ monomers

No. This is just an average. A sample of the polymer will consist of chains with a distribution of lengths.

62. a.

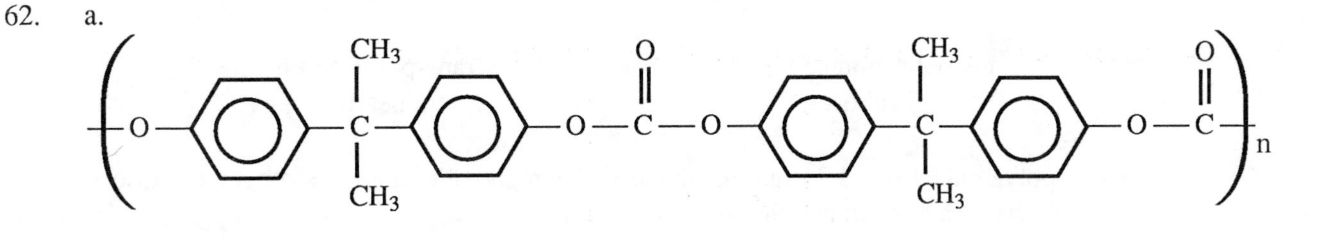

b. condensation

63.

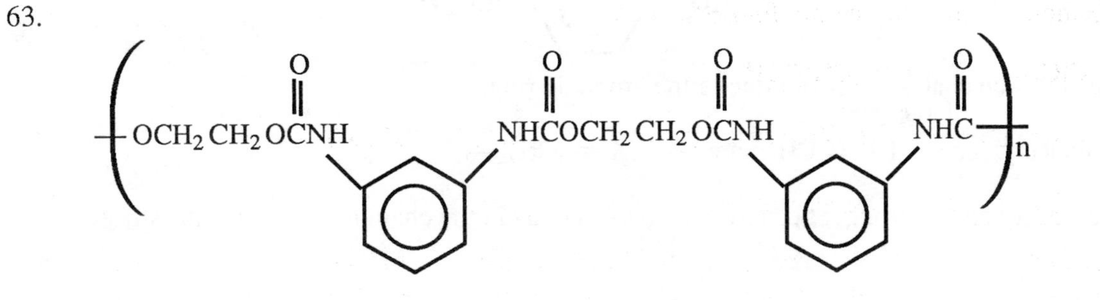

64. a.

polychloroprene:

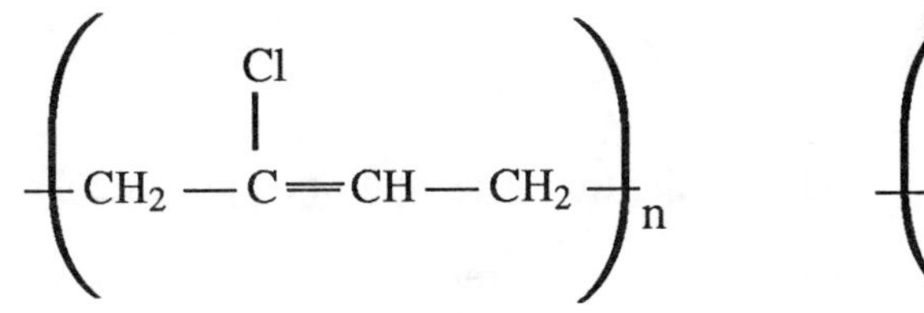

polyisoprene:

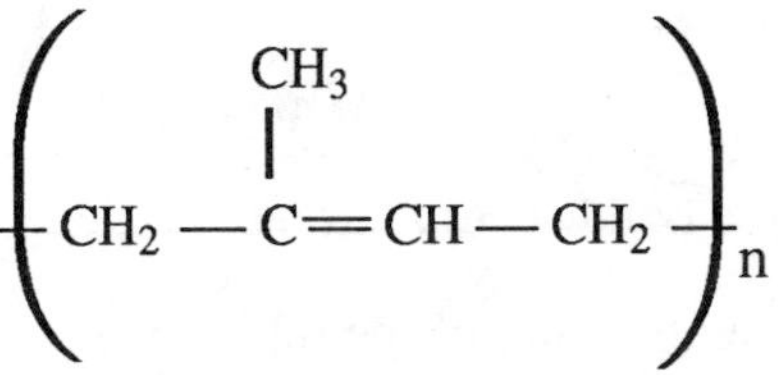

polynitrile:

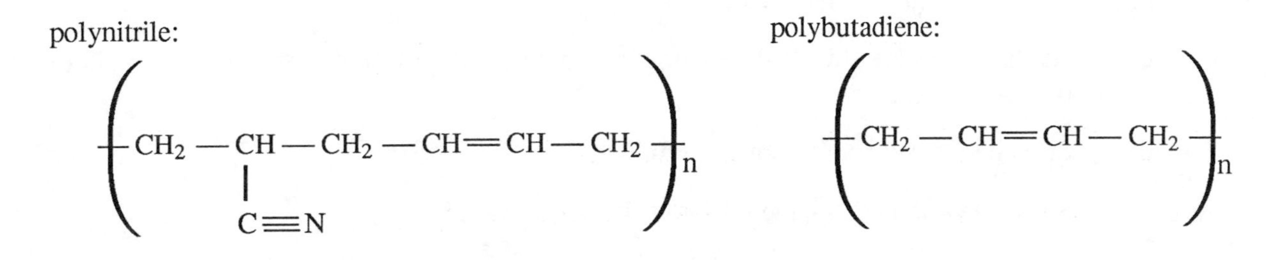

polybutadiene:

$$\left(\text{CH}_2 - \text{CH}{=}\text{CH} - \text{CH}_2 \right)_n$$

b.

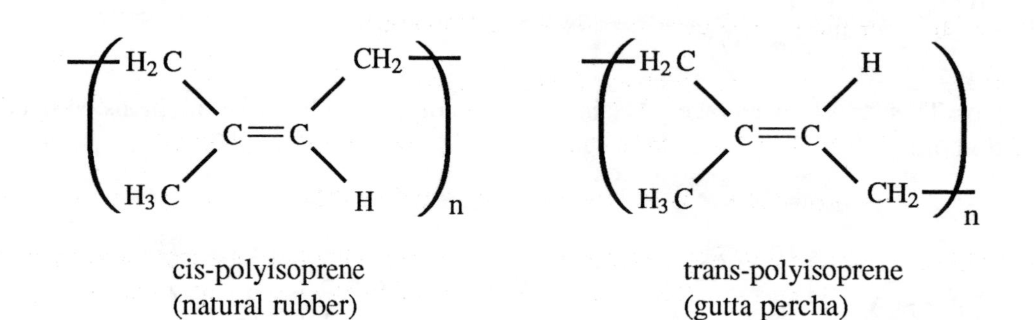

cis-polyisoprene
(natural rubber)

trans-polyisoprene
(gutta percha)

65. a. The polyamide from 1,2-diaminoethane and terephthalic acid is stronger because of the possibilily of hydrogen bonding between chains.

b. The polymer of

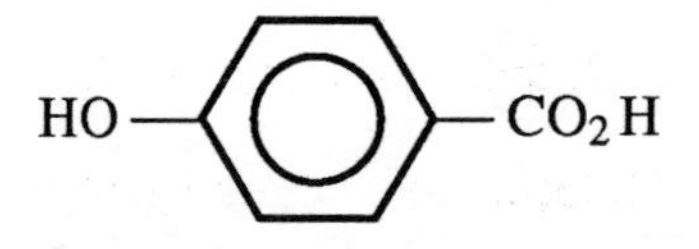

because the chains are stiffer due to the aromatic ring.

c. polyacetylene is n $HC \equiv CH \longrightarrow —(CH=CH)_n—$

Polyacetylene is stronger because the double bonds in the chain make the chains stiffer.

66. Polyacrylonitrile:

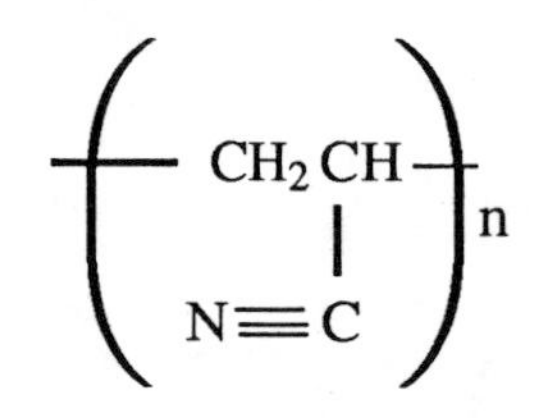

The CN triple bond is very strong. A likely product is the toxic gas hydrogen cyanide, HCN(g).

67. a. The temperature of the rubber band increases when it is stretched.

b. Exothermic (heat is released).

c. As the chains are stretched they line up more closely resulting in stronger dispersion forces between the chains.

d. Stretching is not spontaneous. ΔG (+)

Since $\Delta G = \Delta H - T\Delta S$, then ΔS must be negative. ΔS (-)
(+) (-)

e.

unstretched

stretched

The structure of the stretched polymer is more ordered (lower S).

68. As the car ages the plasticizers escape from the seat covers:

i) The waxy coating is the escaped plasticizer.

ii) The new car smell is the smell of the plasticizers (di-octylphthlate).

iii) Loss of plasticizer causes the vinyl to become brittle.

69. Compounds are numbered in *The Merck Index.* The number and page for each compound is given for the 10th edition. Uses are listed, look up structure and functional groups present.

a. Dopa, 3427, p. 497; anticholinergic, antiparkinsonian

b. Amphetamine, 606, p. 84; central stimulant

c. Ephedrine, 3558, p. 520; bronchodilator

d. Acetaminophen, 39, p. 7; analgesic, antipyretic

e. Phenacetin, 7064, p. 1035; analgesic, antipyretic

70. a. Ecotrin: aspirin

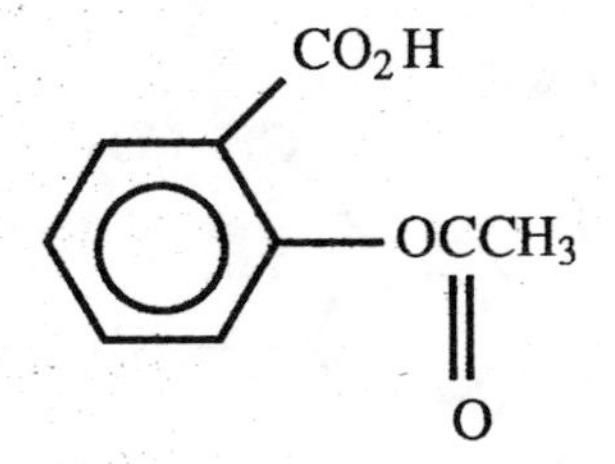

analgesic, anti-inflammatory

b. Tylenol: analgesic

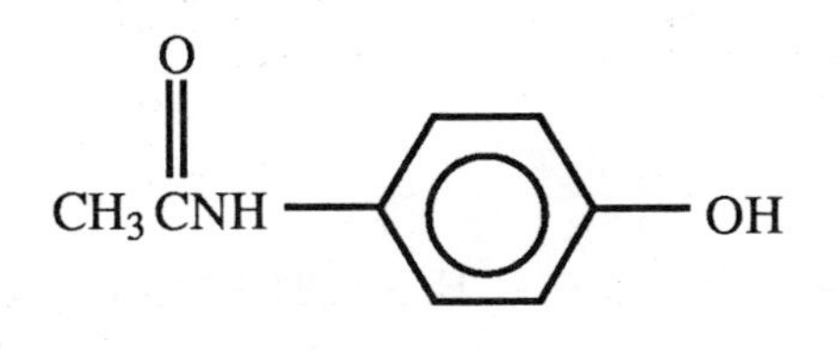

c. Lasix: diuretic, antihypertensive

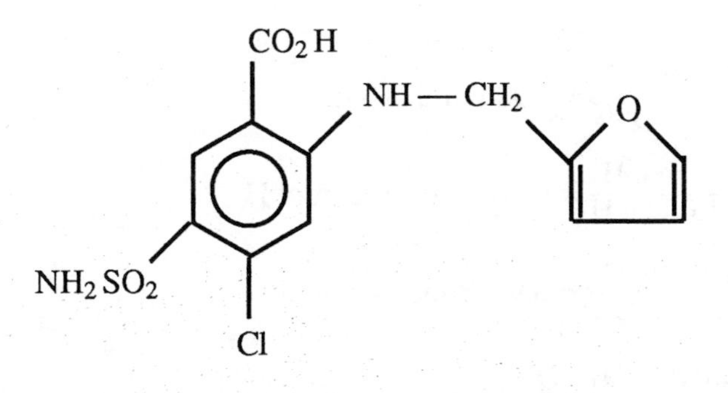

d. Keflex: antibacterial

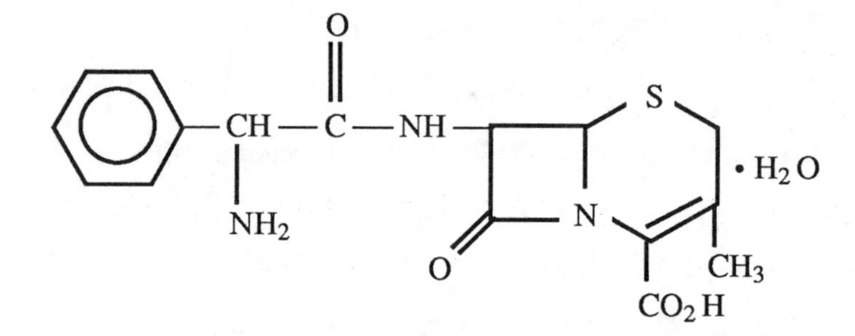

e. Nalfon: anti-inflammatory, analgesic

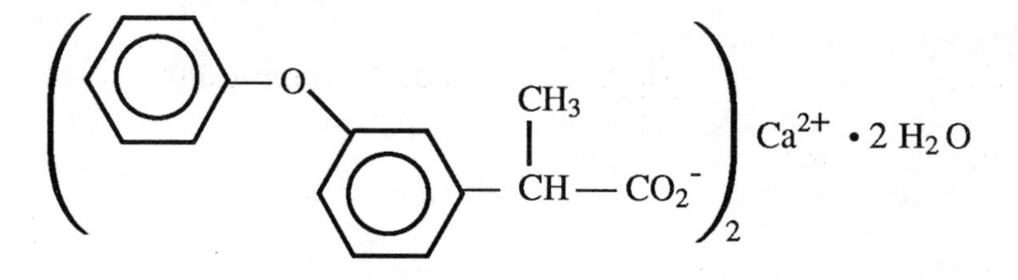

f. Voltaren: anti-inflammatory

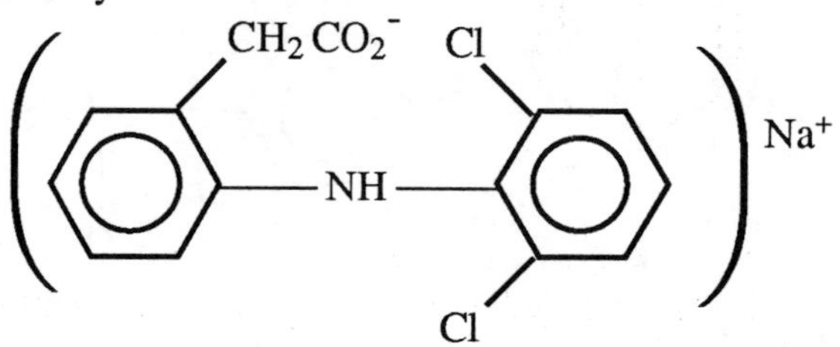

CHALLENGE PROBLEMS

71.

$$\underset{13}{HC}\equiv\underset{12}{C}-\underset{11}{C}\equiv\underset{10}{C}-\underset{9}{CH}=\underset{8}{C}=\underset{7}{CH}-\underset{6}{CH}=\underset{5}{CH}-\underset{4}{CH}=\underset{3}{CH}-\underset{2}{CH_2}-\underset{1}{\overset{\overset{\displaystyle O}{\|}}{C}}-OH$$

72. Out of 100 g compound:

$$71.89\ \text{g C} \times \frac{1\ \text{mol C}}{12.01\ \text{g C}} = 5.986\ \text{mol} \approx 6\ \text{mol C}$$

$$12.13\ \text{g H} \times \frac{1\ \text{mol H}}{1.008\ \text{g H}} = 12.03\ \text{mol} \approx 12\ \text{mol H}$$

$$15.98\ \text{g O} \times \frac{1\ \text{mol O}}{16.00\ \text{g O}} = 0.9988\ \text{mol} \approx 1\ \text{mol O}$$

empirical formula: $C_6H_{12}O$

$$R_1-\overset{\overset{\displaystyle O}{\|}}{C}-O-R_2 + H_2O \longrightarrow R_1\overset{\overset{\displaystyle O}{\|}}{C}-OH + HOCH_2CH_3$$

R_2 must be CH_3CH_2 since CH_3CH_2OH is one of the products.

Mass of — CO_2H is ≈ 45 g/mol.

Mass of R_1 is 172 - 45 ≈ 127

If R_1 is CH_3 — $(CH_2)_n$ —, then 15 + n(14) = 127 and $n = \frac{112}{14} = 8$

Ethyl caprate is:

$$CH_3(CH_2)_8\overset{\displaystyle O}{\overset{\|}{C}}—OCH_2CH_3$$

The molecular formula of $C_{12}H_{24}O_2$ agrees with the empirical formula.

73. For the first structure below,

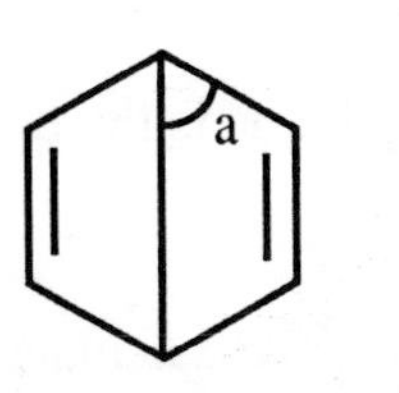

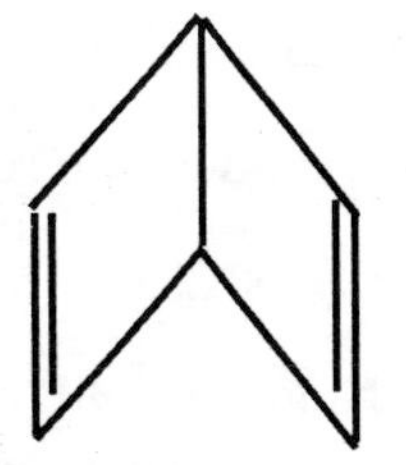

we would predict that the angle (a) should be close to a tetrahedral angle since there are four bonding pairs of electrons around the carbon atom. This would pull some of the C atoms out of the plane. Dewar's benzene has been synthesized. It is a distinct compound with the second structure. Because of the four membered rings, the angle (a) is closer to 90°.

74. For the reaction:

$3\ CH_2{=}CH_2 + 3\ H—H \rightarrow 3\ CH_3—CH_3$

Bonds broken:		Bonds made:	
3 C=C	614 kJ/mol	3 C—C	347 kJ/mol
3 H—H	432 kJ/mol	6 C—H	413 kJ/mol

ΔH = 3(614) + 3(432) - [3(347) + 6(413)]

ΔH = -381 kJ

Using enthalpies of formation:

$\Delta H° = 3\Delta H_f^\circ(C_2H_6) - 3\ \Delta H_f^\circ(C_2H_4)$

ΔH° = 3 mol(-84.7 kJ/mol) - 3 mol(52 kJ/mol)

ΔH° = -410. kJ/mol

The two values agree fairly well.

For $C_6H_6(g) + 3\ H_2(g) \longrightarrow C_6H_{12}(g)$

we would calculate the same ΔH from bond energies as the first reaction, since the same number and types of bonds are broken and formed. $\Delta H = -381$ kJ.

Using enthalpies of formation:

$\Delta H° = -90.3 \text{ kJ} - (82.9 \text{ kJ}) = -173.2 \text{ kJ}$

There is about a 208 kJ discrepancy. Benzene is more stable by about 208 kJ/mol (lower in energy) than we expect. This extra stability is evidence for resonance stablilization.

75. a. overall yield = $0.83 \times 0.52 = 0.43$ or 43%

b. $$\text{theoretical yield} = 1000. \times 10^3 \text{ g styrene} \times \frac{1 \text{ mol styrene}}{104 \text{ g}} \times \frac{1 \text{ mol anthraquinone}}{2 \text{ mol styrene}}$$

$$\times \frac{208 \text{ anthraquinone}}{\text{mol}} = 1.00 \times 10^6 \text{ g or } 1.00 \times 10^3 \text{ kg}$$

Actual yield = $0.43 \times (1.00 \times 10^3 \text{ kg}) = 430$ kg

c. i) Quantities are set up so that 208 lb anthraquinone is the theoretical yield.

$0.9 \times 208 = 187$ lb actual yield.

Note: We will carry extra significant figures and calculate all costs to the nearest cent.

$$\text{Cost} = 148 \text{ lb Phth. Anh.} \times \frac{\$0.35}{\text{lb}} + 293 \text{ lb AlCl}_3 \times \frac{\$0.67}{\text{lb}}$$

$$+ 78 \text{ lb C}_6\text{H}_6 \times \frac{\$0.61}{\text{lb}} = \$295.69$$

$$\frac{\text{cost}}{\text{lb}} = \frac{\$295.69}{187 \text{ lb}} = \$1.58/\text{lb anthraquinone produced}$$

ii) $$\frac{208 \text{ lb styrene} \times \frac{\$0.27}{\text{lb}}}{0.43 \times 208 \text{ lb}} = \frac{\$0.63}{\text{lb}}$$

d. The disposal of $AlCl_3$ and H_2SO_4 in i) would present a major problem.

e. iii) To produce $0.8 \times 208 = 166$ lb of anthraquinone would require:

$2 \times 78 = 156$ lb benzene.

The cost of raw materials per anthraquinone produced is (ignoring the cost of CO):

$$\frac{156 \text{ lb} \times \frac{\$0.61}{\text{lb}}}{166 \text{ lb}} = \frac{\$0.57}{\text{lb}} \quad \text{using benzene}$$

Or to produce 166 lb of anthraquinone would require 182 lb of benzophenone.

$$\text{Cost} = \frac{182\ \text{lb} \times \dfrac{\$2.40}{\text{lb}}}{166\ \text{lb}} = \frac{\$2.63}{\text{lb}} \qquad \text{using benzophenone}$$

iv) To produce 0.9 × 208 = 187 lb of anthraquinone would require 128 lb naphthalene and 54 lb 1,3-butadiene.

$$\text{Cost} = \frac{128 \times \dfrac{\$0.24}{\text{lb}} + 54\ \text{lb} \times \dfrac{\$0.12}{\text{lb}}}{187\ \text{lb}} = \frac{\$0.20}{\text{lb}}$$

f. Process (iv) would be best in terms of raw material cost.

76. a. For (i), we need 1 mol of furfural to produce one mole of diamine.

furfural: $C_5H_4O_2$ MW = 96.08 g/mol

$$1\ \text{mol}\ C_5H_4O_2 \times \frac{96.08\ \text{g}}{\text{mol}} \times \frac{1\ \text{lb}}{454\ \text{g}} \times \frac{\$0.79}{\text{lb}} = \$0.17 = 17¢$$

Furfural cost is 17¢ per mol of diamine produced.

For (ii), we need 1 mol C_4H_6 to produce 1 mol of diamine.

$$1\ \text{mol}\ C_4H_6 \times \frac{54.09\ \text{g}}{\text{mol}} \times \frac{1\ \text{lb}}{454\ \text{g}} \times \frac{\$0.12}{\text{lb}} = \$0.014 = 1.4¢$$

Butadiene cost is 1.4¢ per mol of diamine produced.

The butadiene process is economically advantageous.

b. Oil prices would have to increase by a factor of 17/1.4 or 12 times for the costs to equalize.

CHAPTER TWENTY-THREE: BIOCHEMISTRY

QUESTIONS

1. Primary: amino acid sequence: covalent bonds

   Secondary: features such as α-helix, pleated sheets: H-bonding.

   Tertiary: three dimensional shape: hydrophobic and hydrophillic interactions, salt linkages, hydrogen bonds, disulfide linkages, dispersion forces.

2. Secondary and tertiary

3. Both denaturation and inhibition reduce the catalytic activity of an enzyme. Denaturation changes the structure of an enzyme. Inhibition involves the attachment of an incorrect molecule at the active site, preventing the substrate from interacting with the enzyme.

4. An amino acid contains both acidic and basic functional groups. Thus, an amino acid can act as both a weak acid and a weak base; this is the requirement for a buffer.

5. Hydrogen bonding between the —OH groups of the starch and water molecules.

6. structural isomers: same formula, different functional groups or chain lengths (different bonds)

   geometrical isomers: same functional groups (same bonds), but different arrangement of some groups in space

   optical isomers: compounds that are mirror images of each other

7. Nitrogen atoms with lone pairs of electrons.

8. DNA: Deoxyribose; double stranded; A, T, G, C major bases

   RBA: ribose; single stranded; A, G, C, U are the bases

9. A deletion may change the entire code for a protein. A substitution will change only one single amino acid in a protein.

10. Lipids are non-polar and soluble in organic solvents. Carbohydrates contain several —OH groups capable of hydrogen bonding and are soluble in water.

11. A polyunsaturated fat contains 2 or more carbon-carbon double bonds.

12. Triglycerides (fats) are hydrolyzed to glycerol and fatty acids (soap) in the presence of base. This is how soap is produced. The products of the reaction (soap) feel slippery.

EXERCISES

Proteins and Amino Acids

13. a.

$$H_2NCH_2CO_2H \rightleftharpoons H^+ + H_2NCH_2CO_2^- \qquad K_a = 4.3 \times 10^{-3}$$

$$H_2NCH_2CO_2^- + H^+ \rightleftharpoons {}^+H_3NCH_2CO_2^- \qquad K_2 = 1/K_a(\text{amino}) = K_b/K_w = 6.0 \times 10^{-3}/10^{-14} = 6.0 \times 10^{11}$$

$$H_2NCH_2CO_2H \rightleftharpoons {}^+H_3NCH_2CO_2^- \qquad K = K_aK_2 = 2.6 \times 10^9$$

Equilibrium lies far to the right.

b. $^+H_3NCH_2CO_2H$ (1.0 *M* H^+)

$H_2NCH_2CO_2^-$ (1.0 *M* OH^-)

14. Crystalline amino acids exist as zwitterions:

$$^+H_3N—CH(R)—CO_2^-$$

The interparticle forces are strong. Before the temperature gets high enough to break the ionic bonds, the amino acid decomposes.

15. a. Aspartic acid and phenylalanine

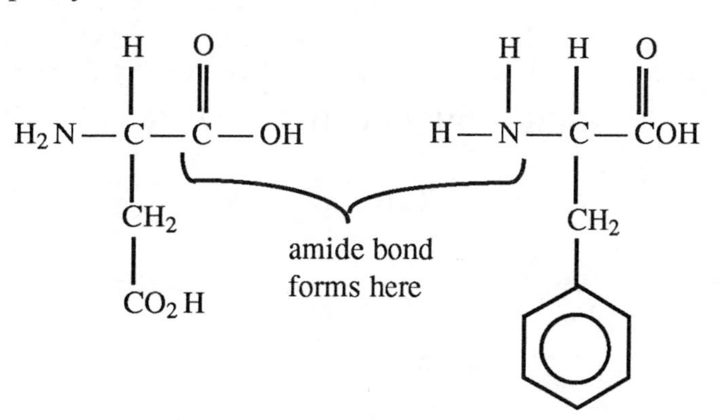

b. Aspartame contains the methyl ester of phenylalanine. This ester can hydrolyze to form methanol.

$$R—CO_2CH_3 + H_2O \rightleftharpoons RCO_2H + CH_3OH$$

16. They are both hydrophilic amino acids because both contain highly polar R groups.

17. a. ionic: Need —NH_2 (amine group) on side chain of one amino acid and - CO_2H on side chain of the other.

—NH_2 on side chain	—CO_2H on side chain
His	Asp
Lys	Glu
Arg	

b. Hydrogen bonding (X = O or N):

—X — H ·········· O ═ C (carbonyl group from peptide bond)

Ser	Asn	Any amino acid
Glu	Thr	
Tyr	Asp	
His	Gln	
Arg	Lys	

c. Covalent: cys . . . cys

d. London dispersion: All non-polar amino acids

e. dipole-dipole: Tyr, Thr and Ser

18. phenylalanine - isoleucine: London dispersion forces

aspartic acid - lysine: ionic

19.

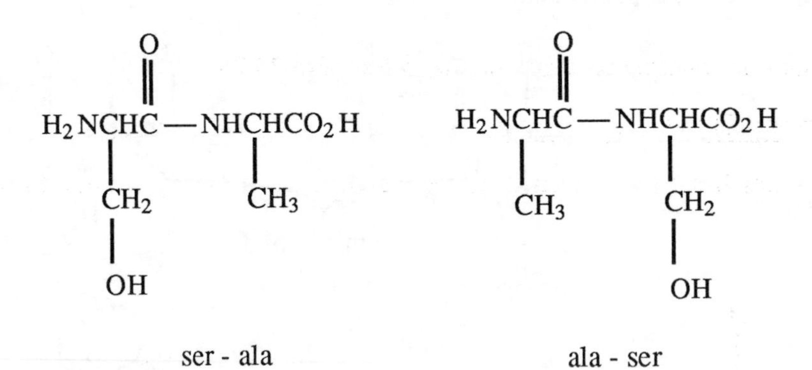

ser - ala ala - ser

20. Writing peptides from amino to carboxyl ends:

ala-ala-gly, ala-gly-ala, gly-ala-ala

So three peptides are possible.

21. From —NH_2 to —CO_2H end:

phe-phe-gly-gly, gly-gly-phe-phe, gly-phe-phe-gly

phe-gly-gly-phe, phe-gly-phe-gly, gly-phe-gly-phe

Six tetrapeptides are possible.

22. gly-phe-ala, gly-ala-phe, phe-gly-ala

phe-ala-gly, ala-phe-gly, ala-gly-phe

Six tripeptides are possible.

23. Glutamic acid: $R = —CH_2CH_2CO_2H$

Valine: $R = —CH(CH_3)_2$

A polar side chain of the amino acid is replaced by a non-polar group. This could affect the tertiary structure and the ability to bind oxygen.

24. The initial increase in rate is a result of the effect of temperature on the rate constant. At higher temperatures the enzyme begins to denature, losing its activity and the rate decreases.

Carbohydrates

25.

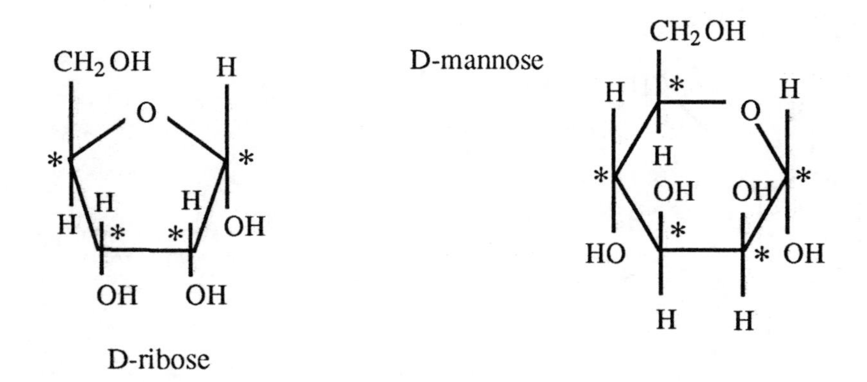

26. Chiral carbons are marked with an asterisk in Exercise 23.25.

Optical Isomerism and Chiral Carbon Atoms

27. A chiral carbon has four different groups attached to it. A compound with a chiral carbon is optically active.

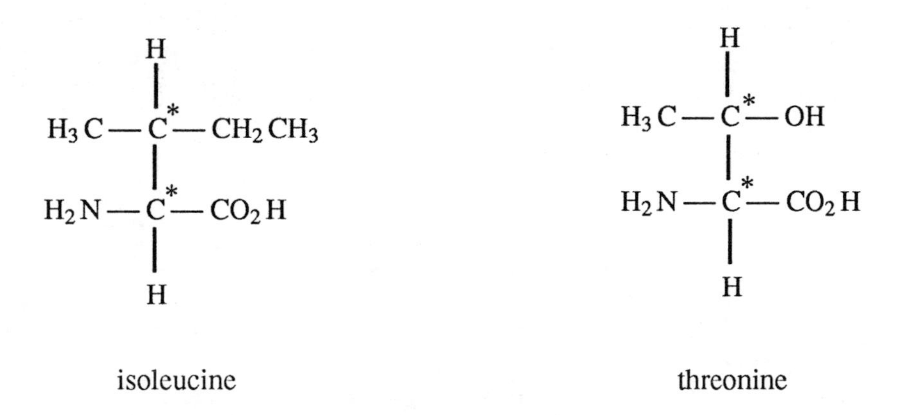

28. There is no chiral carbon atom in glycine.

29.

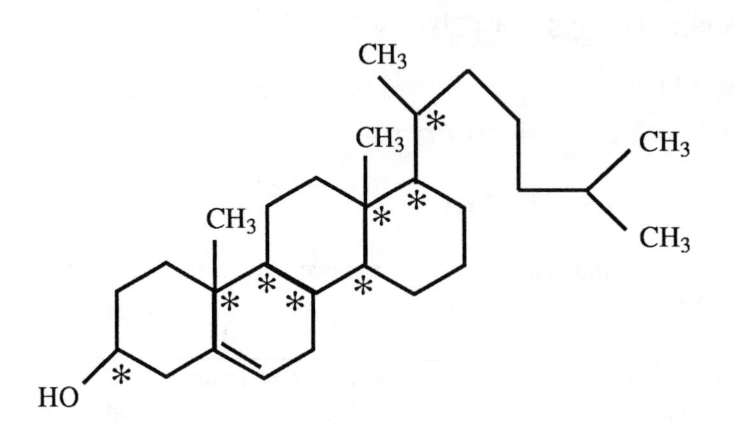

Eight chiral carbon atoms (marked with *).
Note: Some hydrogen atoms are omitted from the structure.

30.

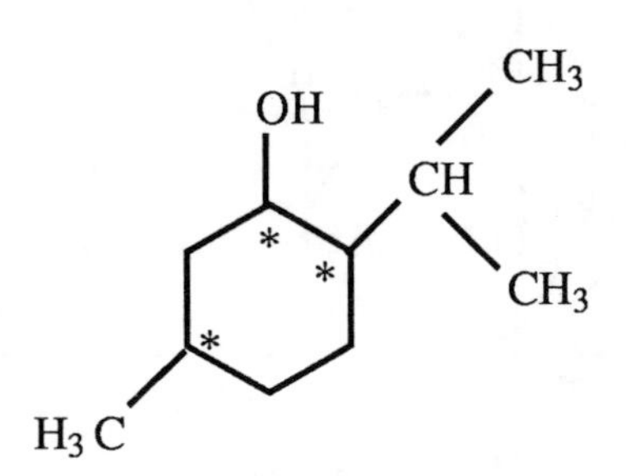

31.

```
       Cl
       |
 H —— C* —— Br
       |
      CH══CH2
```

is optically active. The chiral carbon is marked with *.

32. A chiral carbon has four different groups attached to it. The two chiral carbon atoms in α-pinene are marked with *. Since it has chiral carbons, α-pinene is optically active.

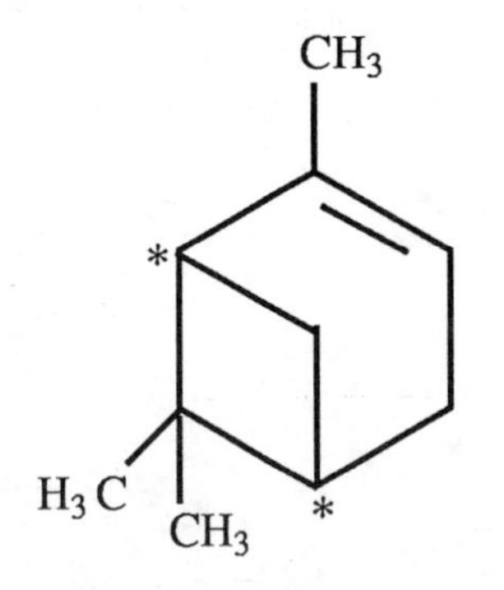

Nucleic Acids

33. T-A-C-G-C-C-G-T-A

34. For each letter, there are 4 choices; A, T, G, or C. Hence, the total number of words is: $4 \times 4 \times 4 = 64$.

35. Uracil: will H-bond to adenine.

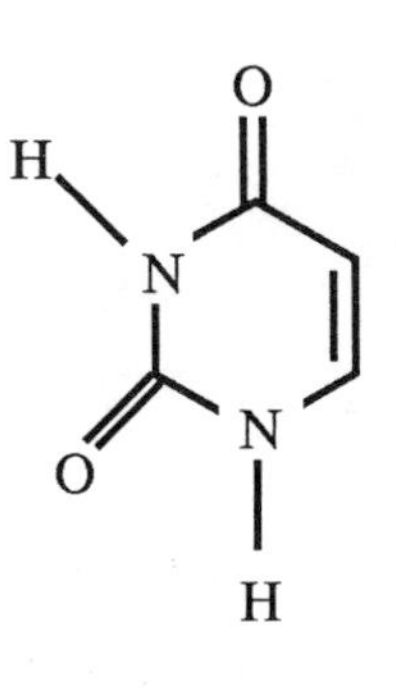

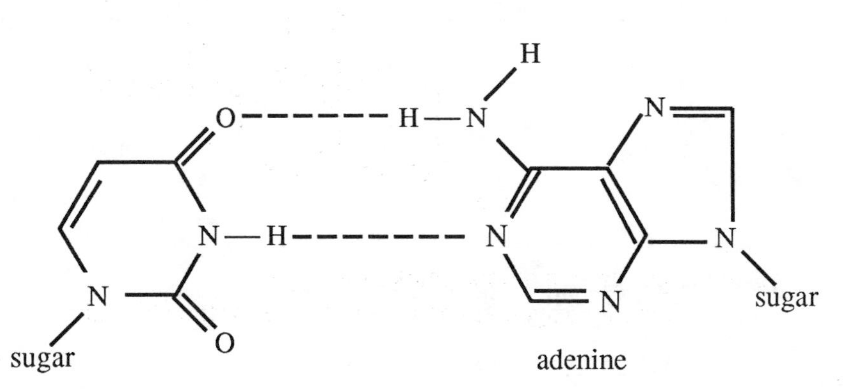

36. a.

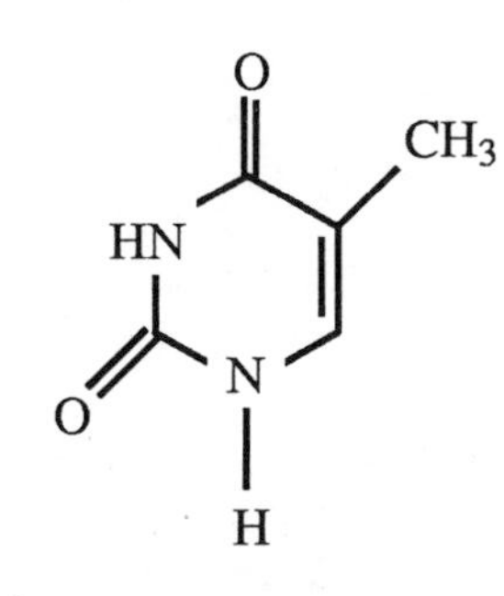

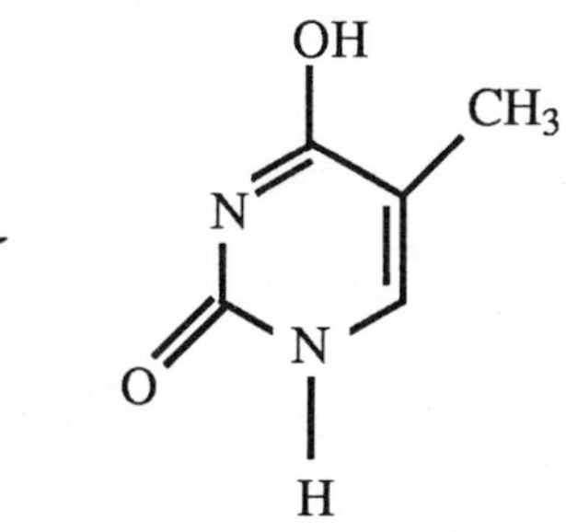

Bonds broken:		Bonds made:	
C═O	799 kJ/mol	C—O	358 kJ/mol
N—C	305 kJ/mol	N═C	615 kJ/mol
N—H	391 kJ/mol	O—H	467 kJ/mol

$\Delta H = 799 + 305 + 391 - (358 + 615 + 467)$

$\Delta H = 55$ kJ

The structure with two C ═ O bonds is more stable (has stronger bonds).

b. The tautomer could hydrogen bond to guanine, forming G-T base pair instead of A-T.

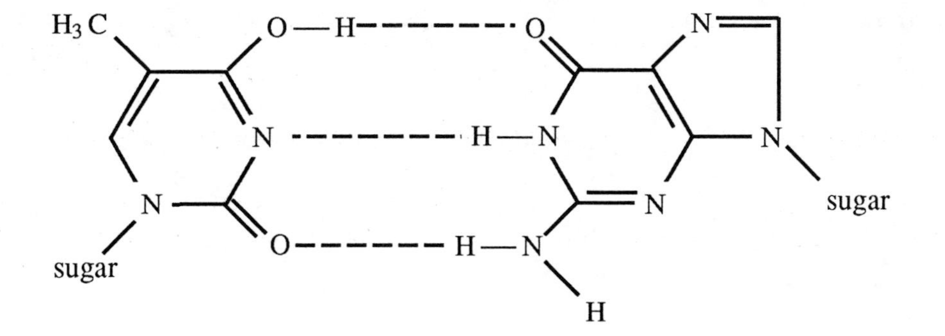

37. Base pair:

RNA	DNA
A	T
G	C
C	G
U	A

a.

Glu:	CTT, CTC
Val:	CAA, CAG, CAT, CAC
Met:	TAC
Trp:	ACC
Phe:	AAA, AAG
Asp:	CTA, CTG

b. DNA sequence for Met - Met - Phe - Asp - Trp:

TAC - TAC - AAA - CTA - ACC
or or
AAG CTG

c. four

d.

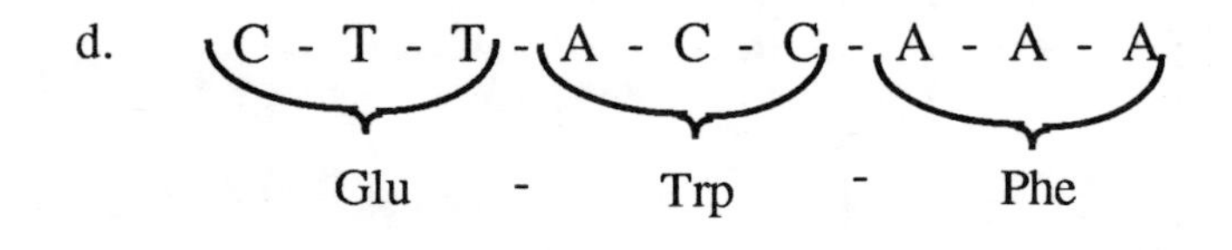

e. C - T - C - A - C - C - A - A - A

C - T - T - A - C - C - A - A - G

C - T - C - A - C - C - A - A - G

38. In sickle cell anemia, glutamic acid is replaced by valine.

DNA sequence: Glu: CTT, CTC; Val: CAA, CAG, CAT, CAC

Replacing a T with an A in the code for Glu will code for Val.

GTC	→	CAC	or	CTT	→	CAT
Glu		Val		Glu		Val

Lipids and Steroids

39.

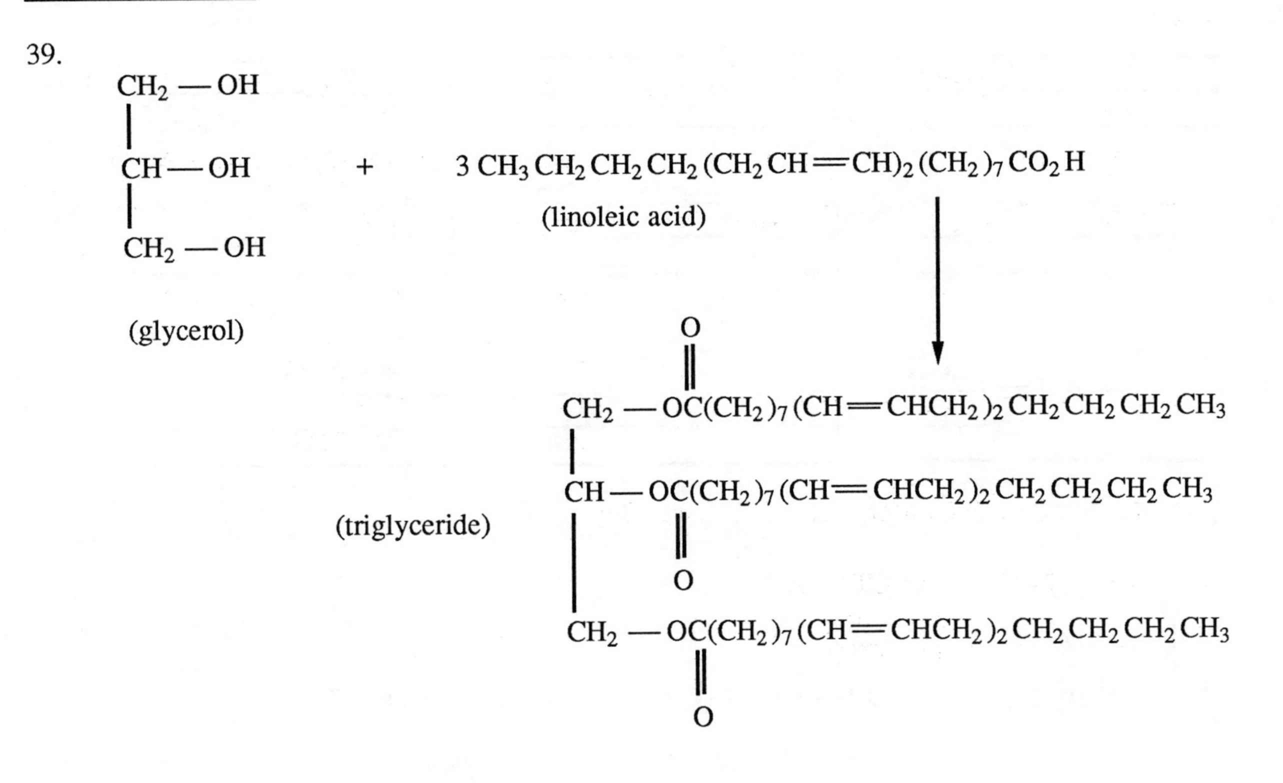

40.

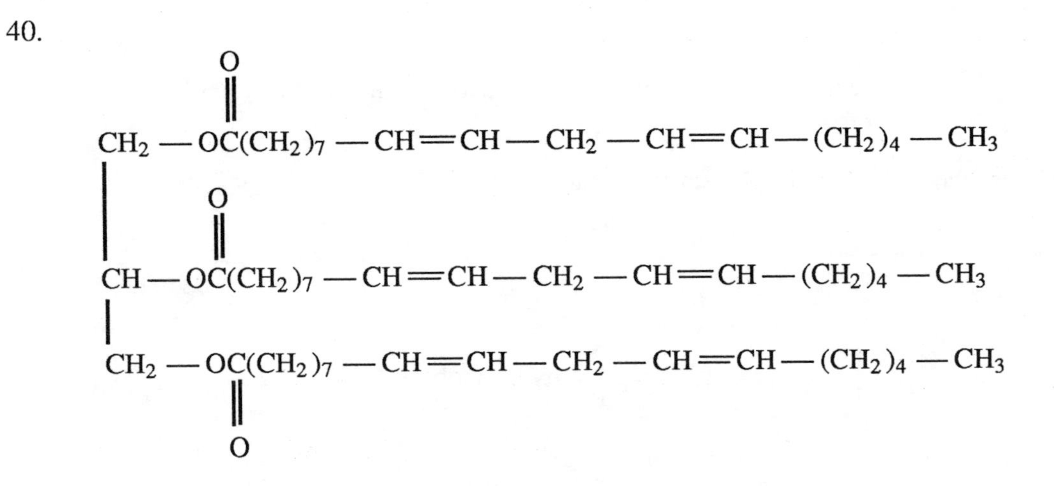

It will take 6 mol of H_2 to completely hydrogenate this triglyceride.

Nine products with 2 double bonds are possible. Shorthand notation for the different products are:

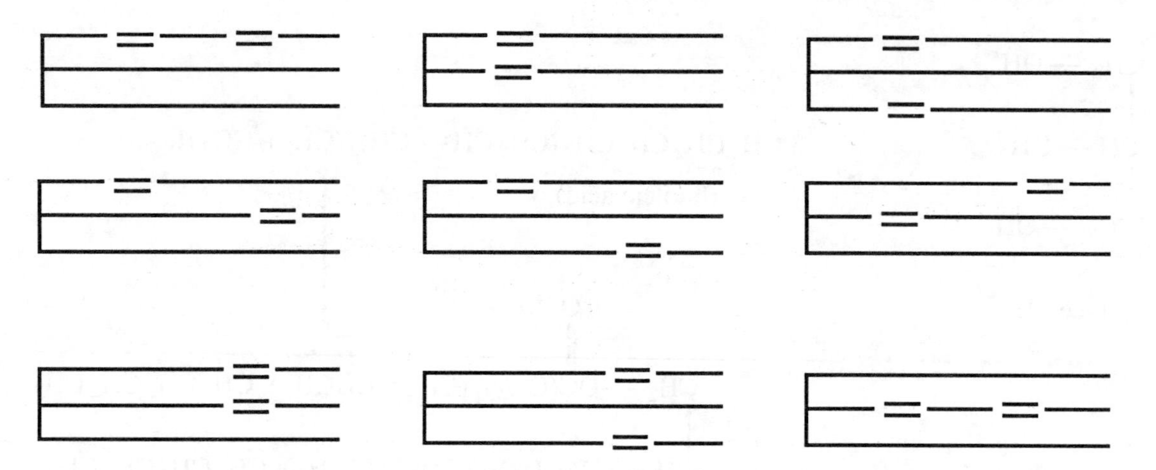

41.

$$HOC(=O)—(CH_2)_{11}—CH{=}CH—CH_2—CH_3$$ 16 carbon omega-3 fatty acid

$$HOC(=O)—(CH_2)_{13}—CH{=}CH—CH_2—CH_3$$ 18 carbon omega-3 fatty acid

42. R groups are hydrophobic, alkyl chains from fatty acids.

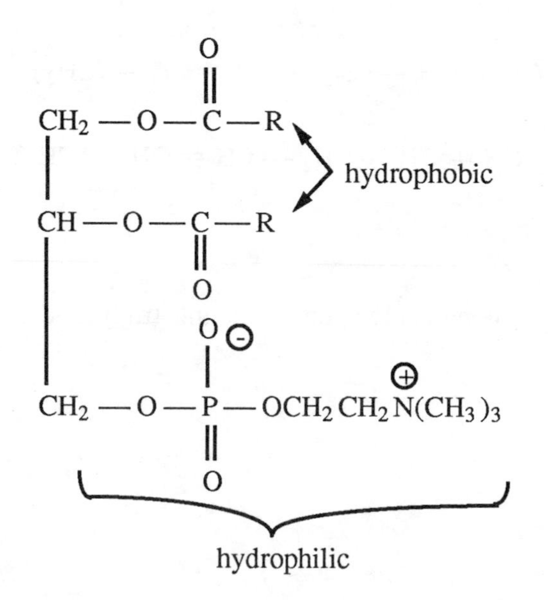

Lecithin: See #5271 p.779; *The Merck Index, 10th Ed.*

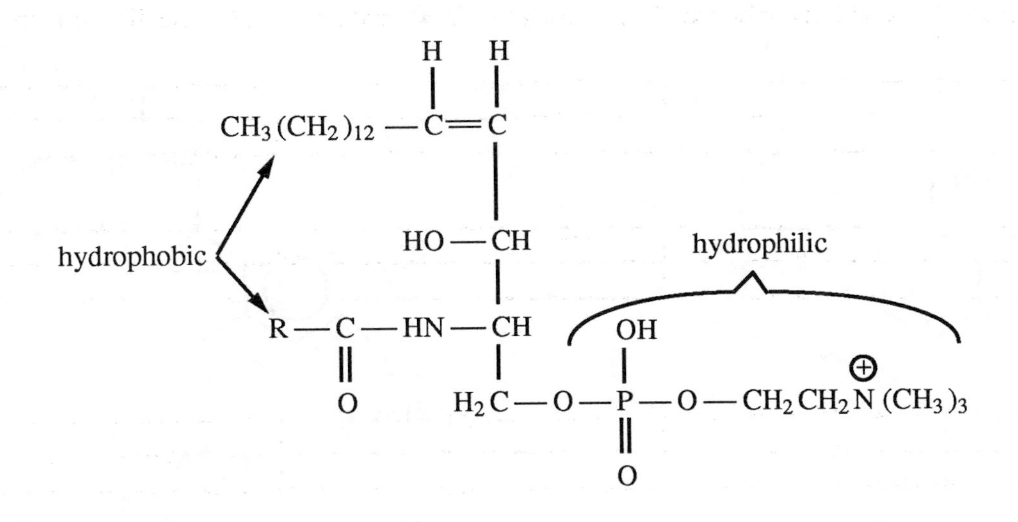

Sphingomyelin #8590; p. 1252; *The Merck Index, 10th Ed.*

43. See #1986; p. 282; *The Merck Index, 10th Ed.*

44.

a)

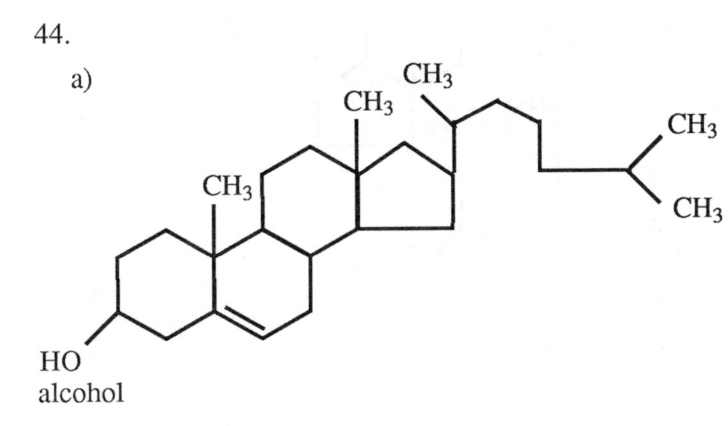

b)

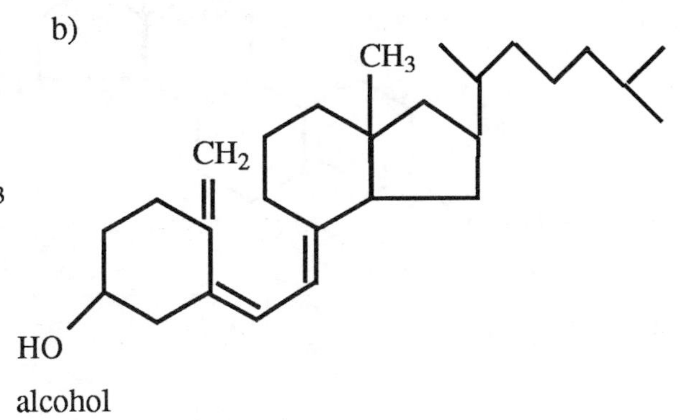

c)

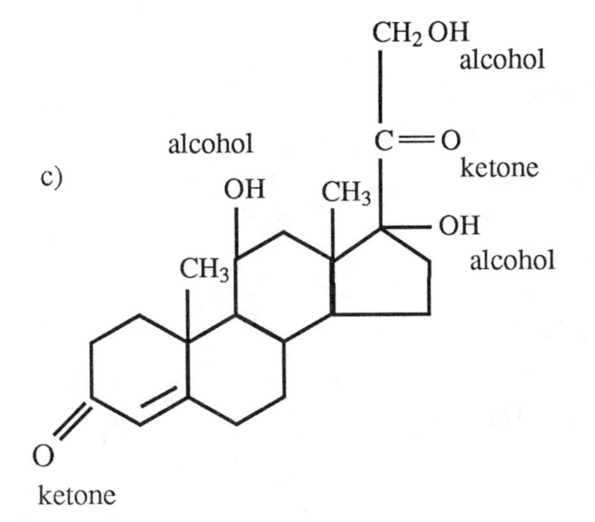

d)

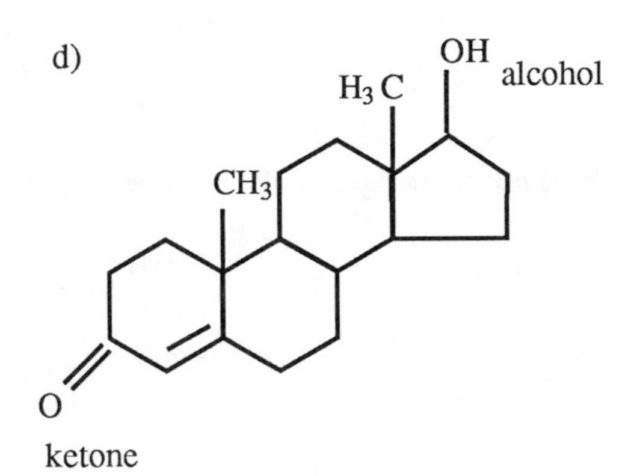

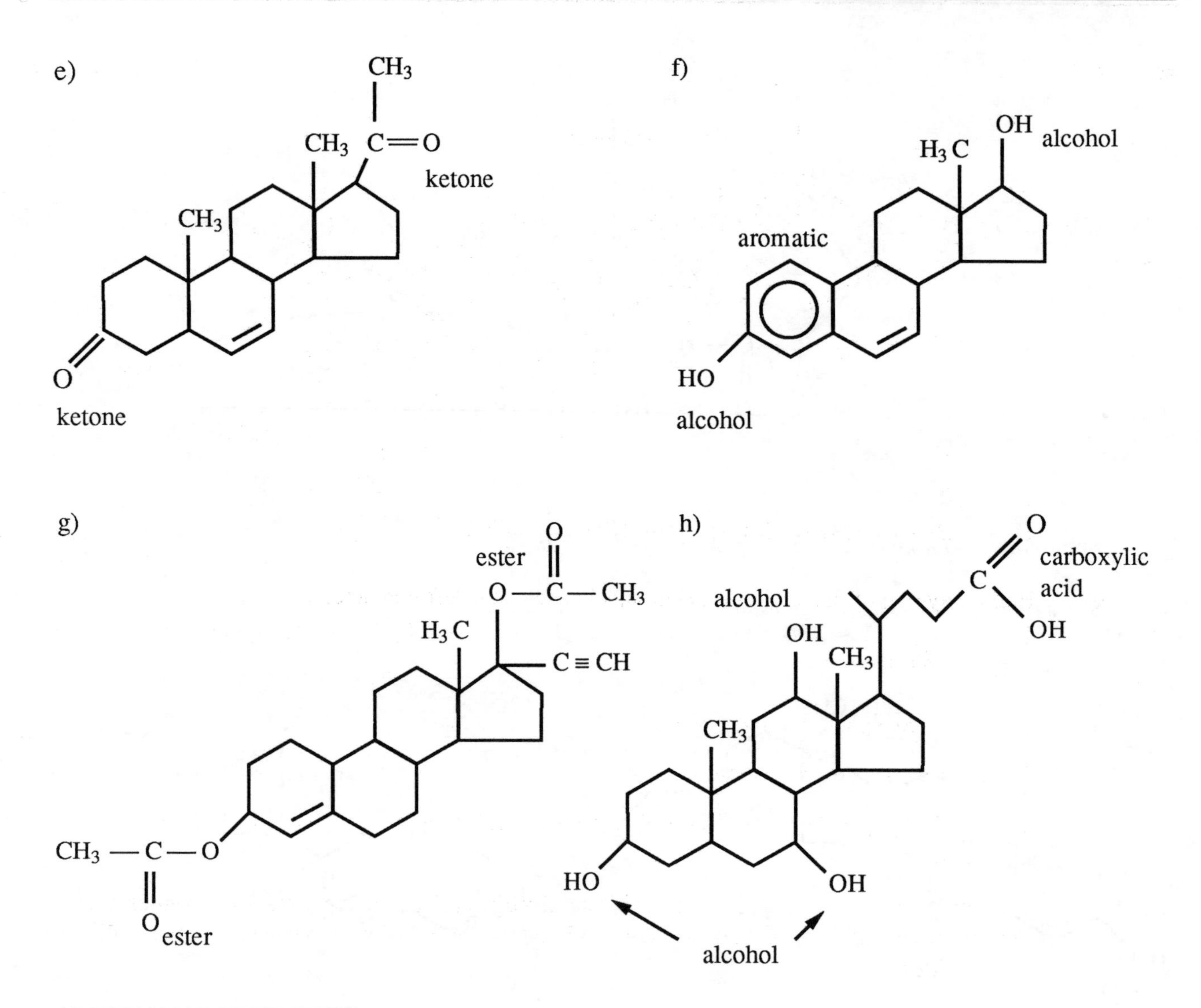

ADDITIONAL EXERCISES

45. Glutamic acid:

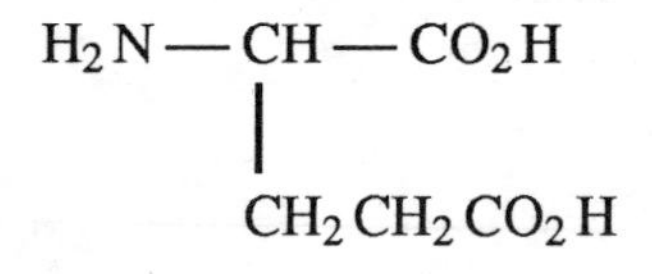

Monosodium glutamate: one of the acidic protons is lost.

$$H_2N-CH-CO_2H$$
$$|$$
$$CH_2CH_2CO_2^-Na^+$$

46. a.

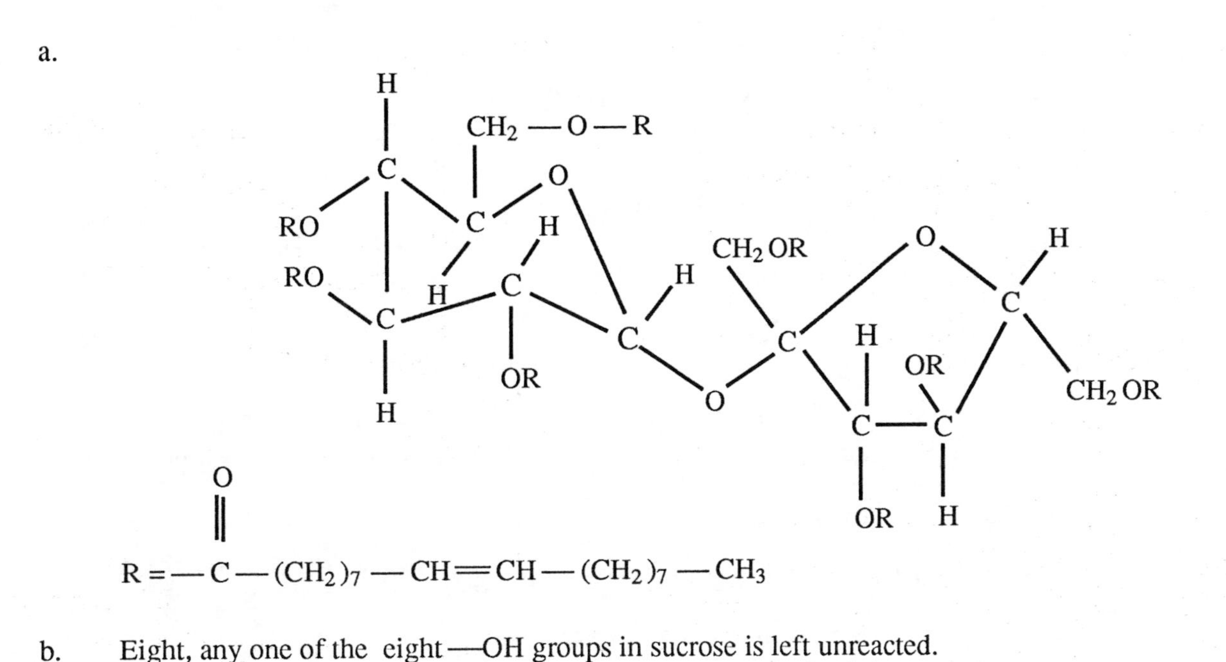

b. Eight, any one of the eight —OH groups in sucrose is left unreacted.

c. $\frac{8 \times 7}{2} = 28$ isomers

1st unreacted —OH = 8 possibilities

2nd unreacted —OH = 7 possibilities

We divide by two because it doesn't matter in which order we leave the two hydroxyls unreacted.

d. The long chains prevent the sucrose from fitting into the active sites of enzymes that metabolize carbohydrates and the fatty acid chains don't fit into enzymes that metabolize triglycerides.

47. $5 \times 10^9 \text{ pairs} \times \frac{340 \times 10^{-12} \text{ m}}{\text{pair}} = 1.7 \text{ m}$ (5'7") (Ignoring significant figures.)

48. The Cl^- ions are lost upon binding to DNA. The dimension is just right to bond to two adjacent bases in one strand of the helix.

49. Structures can be found on pp. 1434-1439 of *The Merck Index, 10th Ed.*

a. A - fat soluble

b. E - fat soluble

c. K_5 - water soluble

d. K_6 - water soluble

50. a.

$$H_2N—CH_2—CO_2H + H_2N—CH_2—CO_2H \rightleftharpoons H_2N—CH_2—\overset{\overset{O}{\|}}{C}—\underset{\underset{H}{|}}{N}—CH_2—CO_2H + H—O—H$$

Bonds broken:

C—O	358 kJ/mol
H—N	391 kJ/mol

Bond made:

C—N	305 kJ/mol
H—O	467 kJ/mol

$\Delta H = 358 + 391 - (305 + 467)$

$\Delta H = -23$ kJ for the reaction gly + gly $\rightleftharpoons$ gly - gly.

b. ΔS for this process is negative (unfavorable).

c. Overall ΔG is positive because of the unfavorable entropy change. The reaction is not spontaneous.

51. $\Delta G = \Delta H - T\Delta S$

For the reaction, we break a P—O and O—H bond and make a P—O and O—H bond. Thus, $\Delta H \approx 0$. $\Delta S < 0$, since 2 molecules going to one molecule. Thus, $\Delta G > 0$, not spontaneous.

52. Both proteins and nucleic acids must form for life to exist. From the simple analysis, it looks as if life can't exist, an obviously incorrect assumption. A cell is not a closed system. There is an external source of energy to drive the reaction. A photosynthetic plant uses sunlight and animals use the carbohydrates produced by plants as sources of energy. For a cell, $\Delta S_{sys} < 0$, but $\Delta S_{surr} > 0$ and $\Delta S_{surr} > |\Delta S_{sys}|$. Therefore, ΔS_{univ} increases (Second Law).

CHALLENGE PROBLEMS:

53. For a dipepetide there are $5 \times 5 = 25$ different cases.

For a tripeptide: $5 \times 5 \times 5 = 125$ different cases

So for 25 amino acids in the polypeptide there are $5^{25} = 2.98 \times 10^{17}$ different polypeptides.

54. a. $H_3N^+—CH_2—CO_2H + H_2O \rightleftharpoons H_2NCH_2CO_2H + H_3O^+$

$$K_{eq} = K_a(-NH_3^+) = \frac{K_w}{K_b(NH_2)} = \frac{1.0 \times 10^{-14}}{6.0 \times 10^{-3}} = 1.7 \times 10^{-12}$$

b. $H_2NCH_2CO_2^- + H_2O \rightleftharpoons H_2NCH_2CO_2H + OH^-$

$$K_{eq} = K_b(-CO_2^-) = \frac{K_w}{K_a(CO_2H)} = \frac{1.0 \times 10^{-14}}{4.3 \times 10^{-3}} = 2.3 \times 10^{-12}$$

c. $^{+}H_3NCH_2CO_2H \rightleftharpoons 2\,H^+ + H_2NCH_2CO_2^-$

$$K_{eq} = K_a(\text{-}CO_2H)\,K_a(\text{-}NH_3^+) = (4.3 \times 10^{-3})\,(1.7 \times 10^{-12}) = 7.3 \times 10^{-15}$$

55. We can use the reaction:

$$^{+}H_3NCH_2CO_2H \rightleftharpoons 2\,H^+ + H_2NCH_2CO_2^- \qquad K_{eq} = 7.3 \times 10^{-15}$$

$$7.3 \times 10^{-15} = \frac{[H^+]^2[H_2NCH_2CO_2^-]}{[^+H_3NCH_2CO_2H]} = [H^+]^2$$

$[H^+] = 8.5 \times 10^{-8}$ $\qquad$ pH = 7.07 = isoelectric point

56. a. No, since there is a plane of symmetry in this molecule, it is not optically active.

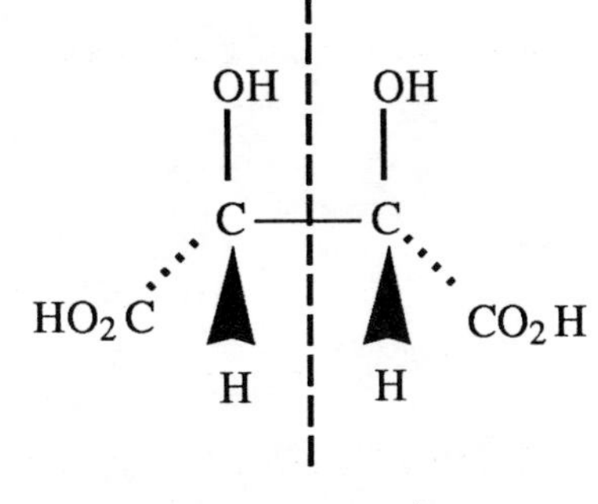

symmetry plane

b. The optically active forms of tartaric acid have no plane of symmetry. The structures are:

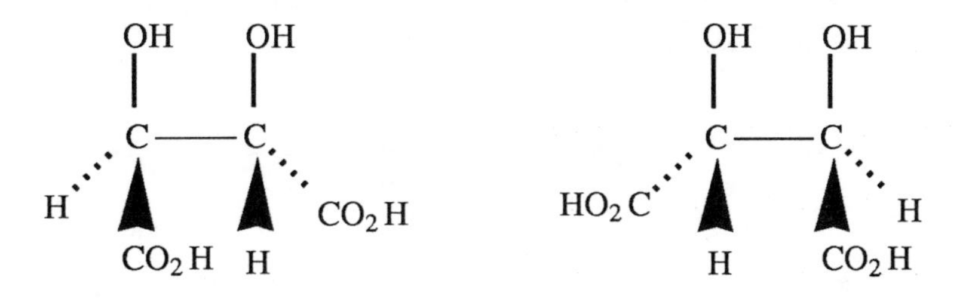